CONTENTS

PREFACE

There are several available textbooks devoted to 'The Invertebrates' and hence the production of yet another one requires some words of explanation and justification. Books already on the market tend to fall into one or other of two categories: they are either systematic treatments covering each group of animals phylum by phylum (e.g. R.D. Barnes's *Invertebrate Zoology*, Saunders, 1987), or are functional approaches reviewing the various invertebrate anatomical and physiological 'systems' (respiration, movement, co-ordination, etc.) mainly with reference to the better known groups (e.g. E.J.W. Barrington's *Invertebrate Structure and Function*, Nelson, 1979). Invertebrate courses therefore require one of each category as associated texts.

In general, however, the last 25 years have seen a great reduction in the teaching time devoted specifically to the various groups, in part to make room in courses of fixed length for new and expanded subject areas and in part because systematic reviews of the range and diversity of organisms have declined in popularity since the days of classical zoology. The end result has been that any pair of existing texts, and indeed many individual works, contain far more information than is required by shortened courses and, deluged by detail, students fail to see the wood for the trees.

We have therefore endeavoured to provide within one pair of covers the basic information on both the range and diversity of invertebrates and on their different functional systems that we feel most university courses actually require. Our main problem has thus been what to leave out rather than what to include, and here we have attempted critically to assess the essential features of each group or system and to bias our accounts towards these. Further, we believe strongly that the process of evolution is central to an understanding of all aspects of biology, and that too few existing texts present animals as other than static, mechanistic entities. Wherever possible, we have therefore adopted an evolutionary approach which aims to portray invertebrate diversity and function against a background of selective pressures and selective advantages now and in the past. This has also influenced our selection and treatment of material. The book is not therefore a summary of existing texts but is, we trust, a new, critical look at the essential features of invertebrate biology.

Since, as pointed out above, zoology courses contain less and less coverage of individual types of animals with each passing decade, it is perhaps not inappropriate here briefly to defend the place for a broad knowledge of the invertebrates in zoological education. Much of our present understanding of biological processes in general has been derived from research on invertebrates; we need only mention the fruit fly in respect of genetics and the squid with regard to neurophysiology to make the point. As yet, however, the number of animal phyla, let alone species, which have been studied in detail is extremely small, and certainly is no true reflection of the diversity of pattern and process that is the invertebrates. We believe that many future generalizations will emerge from studies of these so-far neglected groups, and that without an appreciation of the diversity, as well as the unity, of life, it is impossible to obtain a valid perspective both of biology in general and of the extent to which today's knowledge is based on a minute and biased sample.

With the exception of Chapter 16, this book reflects

a collaborative effort of its three authors. Although in practice first drafts of the various chapters, or parts of chapters, were prepared each by a single author, all were then recast in the light of communal discussion and criticism: all three of us therefore accept joint responsibility for Chapters 1–15 inclusive and Chapter 17. No book is a product solely of the authors, however, and we are most grateful to the many people who have helped or tolerated us during its preparation. In particular, we would like to record our gratitude to David Golding for taking on the task of preparing Chapter 16, and our appreciation of the labours of Helen Creighton, together with Bob Foster-Smith and Peter Kingston, who prepared all the final text figures. Several of our colleagues were so kind as to read drafts of material: Henry Bennet-Clark read the whole work, and Brian Bayne, Jack Cohen, Simon Conway Morris, Peter Croghan, Mustafa Djamgoz, David George, Peter Gibbs, Roger Hughes, Peter Miller, Todd Newberry, David Nichols, John Ryland, Ray Seed, Seth Tylor and Pat Willmer read various parts. Many others provided individual pieces of information and opinion. Their efforts have saved us from factual errors and textual infelicities. It is too much to hope that no unorthodoxies and inexactitudes remain, not least because we were sometimes intransigent in the face of just criticism. Robert Campbell and Simon Rallison of Blackwell Scientific Publications provided copious aid, advice, cajolery and administrative assistance: we owe a great deal to their iron fists and velvet gloves. The debt which we owe our families can only be appreciated by those who have also devoted most of their available 'spare time' to works of this kind.

Most of our illustrations are based on ones that have already appeared in the scientific literature, although all such have been redrawn. Citations to the original sources are given in the relevant figure legends and a list of these sources, other than those listed in the further reading sections, is provided on pp. 569–72.

R.S.K.B.
P.C.
P.J.W.O.

I
EVOLUTIONARY INTRODUCTION

The main undercurrent which permeates our survey of invertebrate diversity (Section II) and functional biology (Section III) is that of the evolutionary pressures and advantages which have influenced these animals in the past and which continue to mould invertebrate biology today. In this introductory section, we describe briefly this all-pervasive evolutionary ethos.

The word 'evolution' simply means 'change', and change can be analysed by two different approaches, which generally are related to each other, as is cause to effect or mechanism to manifestation: (a) there are the processes *ultimately responsible for such changes as are observed; and (b) there is the overall* pattern *or sequence of the changes which have occurred through time. In fact, although Charles Darwin is popularly credited with having demonstrated the fact of evolution, what he did was to propose a viable mechanism—natural selection—which could account for the evolutionary changes which others, before him, had already suggested had taken place. As indicated above, an evolutionary (or 'phylogenetic') tree of the invertebrate phyla and the process of natural selection are related to each other, but in practice a wide gulf and a large measure of controversy exist between, on the one hand, population geneticists studying processes of selection in living organisms, and, on the other, taxonomists classifying phylogenetic patterns and seeking to account for the origin of new taxa above the level of species.*

Here we treat these two subject areas largely separately, in that Chapter 1, besides serving as an introduction to the book as a whole, considers selection as a mechanism of change (this aspect is sometimes termed the 'Special Theory of Evolution'), whilst Chapter 2 deals with the phylogenetic interrelationships of the invertebrate groups (or the 'General Theory of Evolution') and the patterns of diversity and diversification through time. Nevertheless, within each of these chapters, we have found it appropriate to introduce elements of the subject matter of the other; for example to comment on such controversial matters as the manner of origin of invertebrate classes and phyla.

1

INTRODUCTION: BASIC APPROACH AND PRINCIPLES

1.1 Why invertebrates?

This book is about the invertebrates—animals *without* backbones. A definition such as this, based on the *absence* rather than the *presence* of a specific characteristic, is unusual and implies a deviation from a standard type that has the characteristic. If there were no standard or norm then such a definition would hardly make sense.

Thus, when Aristotle classified animals into sanguineous and non-sanguineous, the implication was that the presence of blood was a norm for animals. What he believed was that in its evolution, life had been directed towards a perfect animal form that involved having blood. He incorporated this idea into a hierarchical classification of living things called the 'scale of life' (*Scala naturae*) in which there was progression from the non-sanguineous state to the sanguineous *goal* (Table 1.1).

Similarly, when Lamarck (of acquired characters fame) first separated invertebrate from vertebrate animals (in his *Système des Animaux sans Vertèbres*, Paris 1801) there was an implication that the latter were the norm. And again this probably followed from Lamarck's peculiar theory of evolution in which he supposed that acquired characters were incorporated for hereditary transmission according to

Table 1.1 Aristotle's Scale of Life or *Scala naturae*

		Sanguineous
Viviparous		1 Man
		2 Hairy quadrupeds (land mammals)
		3 Cetacea (sea mammals)
Oviparous	With perfect egg	4 Birds
		5 Scaly quadrupeds and apoda (reptiles and amphibia)
	With imperfect egg	6 Fishes
		Non-sanguineous
		7 Malacia (cephalopods)
		8 Malacostraca (crustacea)
Vermiparous		9 Insects
Produced by generative slime, budding or spontaneous generation		10 Ostracoderma (molluscs other than cephalopods)
Produced by spontaneous generation		11 Zoophytes

principles that not only involved survivorship criteria but progress towards some higher form, of which the vertebrates and Man were closest representatives.

Modern zoology has forsaken concepts of goal-directed evolution (teleology) and yet the distinction between vertebrates and invertebrates has persisted and has influenced numerous generations of students. This is remarkable since the distinction is hardly natural or even very sharp; that is, it separates a group containing many phyla (the invertebrates) from one containing part of a phylum (some members of the phylum Chordata do not have true vertebral columns!).

However, there are two other main reasons for a continuing distinction being made between invertebrate and vertebrate zoology. First, a historical one—Lamarck created a precedent that, once established as a method of approaching zoology, was difficult to escape. Second, and probably more influential, there is still a feeling that because we ourselves have a backbone, vertebrate animals deserve more attention than their taxonomic status might merit.

By concentrating on the biology of invertebrate animals, we here perpetuate this distinction, but not because we have any philosophical commitment to goal-directed evolution or to the view that there are fundamental biological distinctions between invertebrates and vertebrates. Rather our position is pragmatic. We wish to demonstrate that:

1 All living things share in common a number of basic features of structure and function.

2 Major variations occur in these themes, and groups of taxa sharing these in common are referred to as phyla.

3 These variations have evolved and hence should be related by common descent.

4 Within the constraints of each major theme, animals have become adapted to the ecological circumstances in which they occur by natural selection. (The extent to which these microevolutionary processes can account for the macroevolutionary changes noted in points **2** and **3** is a matter of some debate and we shall return to this later.)

In examining these issues it is expedient to circumscribe the material in some way. We do this on the basis of historical precedent. Moreover, the invertebrates provide us with maximum diversity for examining the points raised under points **2–4**. Before doing this, however, we need to have an appreciation of the basic features common to all living things (point **1**), how they differ from those of non-living things and how they originated. This is the aim of the next few sections.

1.2 Properties of living things

1.2.1 Introduction

At a basic chemical level, less than a third of all naturally occurring elements are found in living things (Table 1.2). Around 75% of most animals is

Table 1.2 The elements found in living things

Element	Symbol	Approximate atomic weight (daltons)	Atomic number
Hydrogen*	H	1	1
Carbon*	C	12	6
Nitrogen*	N	14	7
Oxygen*	O	16	8
Fluorine	F	19	9
Sodium†	Na	23	11
Magnesium†	Mg	24	12
Silicon	Si	28	14
Phosphorus†	P	31	15
Sulphur†	S	32	16
Chlorine†	Cl	35	17
Potassium†	K	39	19
Calcium†	Ca	40	20
Vanadium	V	51	23
Chromium	Cr	52	24
Manganese	Mn	55	25
Iron†	Fe	56	26
Cobalt	Co	59	27
Copper	Cu	64	29
Zinc	Zn	65	30
Selenium	Se	79	34
Tin	Sn	119	50
Iodine	I	127	53

*The four most abundant elements in living things.
†The next most abundant elements in living things.

water and 50% of the remaining dry weight is carbon with little if any silicon. The earth's crust, in contrast, consists of more than 20% silicon and less than 0.1% carbon.

Despite the restricted number of chemical elements found in living organisms, the molecules they contain are structurally and functionally very diverse. This is because carbon is outstanding amongst all elements, with only silicon coming a close second, in being able to form diverse molecular strands and rings. These carbon-based, molecular building blocks of organisms are: sugars, amino acids, fatty acids and nucleotides and these associate as macromolecules to form, in turn, polysaccharides, proteins, lipids and nucleic acids. Free organic chemicals, of this complexity, form only rarely in non-living systems (p. 7).

An even more profound distinction, though, between the living and the non-living is the way that the organic chemicals are organized. In living systems the macromolecules make up membranes that bound further, non-random collections of macromolecules that react together in ordered metabolism. These packages are the cells. Many cells make a multicellular organism, and in this context are collected together as a highly ordered and organized, structural and functional unit. The very existence and persistence of this order, organization and complexity were considered for a long time to be *special,* even *mysterious,* features of organisms, created and maintained by mysterious *vital forces*; for the rule of the non-living world, summarized in physics by the Second Law of Thermodynamics, is that order and organization are unstable. Entropy, or disorder, should progressively increase in *all* reactions and processes.

However, we now recognize that the order and organization of biological systems arise from two non-mysterious features common to them all. *These are crucial in understanding the basic principles of biology and we shall continually allude to them in what follows:*

1 Organisms are *programmed.*

2 These *programmes* specify working systems and subsystems that are *open* to the input of material and energy.

1.2.2 The programme

At the most fundamental level, the genetic programme controls the properties of proteins by specifying the types and sequences of amino acids from which they are constructed. Only 20 different amino acids occur frequently in animals but with a chain of only 100 (which is short for a protein) there could in principle be 20^{100} possible configurations! This explains the great diversity of proteins. Some are enzymes that control all metabolic processes within organisms and others comprise the fabric of the cell and organism.

The programme itself is coded as nucleotide sequences in DNA. There are four different kinds of nucleotides (adenine, thymine, guanine and cytosine) but 20 amino acids so that there could not be one-to-one specification of the latter by the former. Only combinations of three (or more) nucleotides could give sufficient alternative combinations ($4^3 = 64$) and this triplet code has been found to be universal, with the excess number of alternatives (*c.* 40) being accounted for by redundancy (more than one triplet coding for a particular amino acid) and punctuation. However, there is not a direct translation of DNA to proteins. Instead the information is first translated to RNA (like DNA but uracil replaces thymine) which acts as a messenger taking the information to the cytoplasm where, on RNA particles—the ribosomes—it acts as a template for the assembly of specific amino acids. These are transported and 'plugged into' appropriate sites by another population of RNAs, the transfer RNAs. This complex system is depicted schematically in Fig. 1.1.

1.2.3 Openness

The ordered systems of organisms, right down to their macromolecules, are continually subjected to 'entropic insults', but disordered systems can be replaced according to specifications embodied in the genetic programmes. However, this is only possible if there is a continuous import of ordered raw materials (in a famous book called *What is Life?* (1944) Erwin Schrödinger referred to this as eating *negen-*

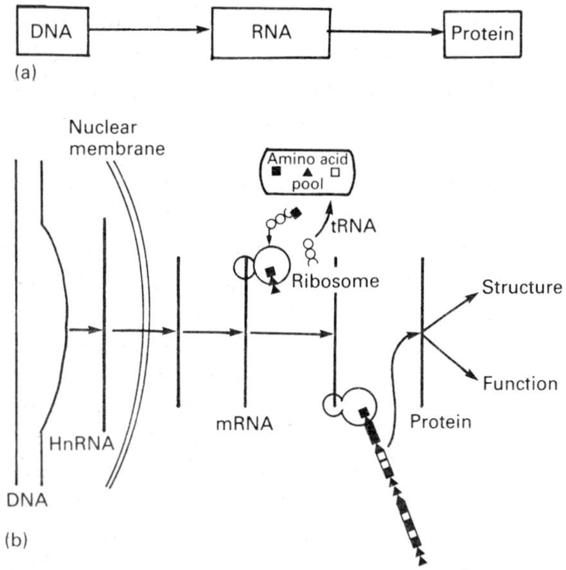

Fig. 1.1. The molecular basis of protein synthesis (after Calow, 1976).

tropy!) and an export of disordered material (excreta) and energy (largely heat). Hence, organisms and the cells that they contain have to be open systems and it is to be anticipated that even in non-growing animals there should be a continuous turnover of cells and/or molecules—what Schoenheimer described as the *dynamic steady-state* of the body in his book, *The Dynamic State of Body Constituents* (1946).

1.2.4 Evolution by natural selection is an inevitable consequence of systems that persist by replication

Whole organisms can also be replaced—*reproduced*—by replication of the genetic programme. This involves separation of multicellular or unicellular propagules that carry a whole or part of the genetic programme (genome) of their parents. Multicellular propagules invariably contain more or less complete replicas of the parent genome and processes of reproduction involving them are referred to as asexual or vegetative. Unicellular propagules (gametes) most often contain a part-replica of the parent genome and have to fuse with other gametes to reinstate the complete genome (fertilization) before development can proceed. This is sexual reproduction.

We have, then, systems that *replicate* a genome to *reproduce* an organism. However, the genome is not always replicated faithfully; even in asexual reproduction mutations introduce variation, and additionally in sexual reproduction the 'shuffling' processes associated with meiosis and the mixing associated with fertilization introduce considerable differences between parents and offspring. Variations in the genetic programme lead to variation in the form and function of the phenotype and this, in turn, influences the way that it interacts with its environment and hence its chances of survival and its rate of reproduction. It follows that those programmes that best promote survivorship and fecundity in the environment in which they occur, i.e. are best *adapted* to it, will become most common. Moreover, given that the world is finite and that the resources needed for the life processes are limiting, these programmes will tend to replace other less well-adjusted ones. This summarizes, in simple form, the process of evolution by *natural selection* that was first made explicit by Charles Darwin in his *Origin of Species* (1859). Borrowing from Herbert Spencer, he used the catchphrase 'survival of the fittest' to describe this process. However, the above description makes it clear that fitness is the *ability of one gene-determined trait to spread in a population in comparison with others* which involves both *survivorship* and *fecundity*.

1.3 Origins of life

The most fundamental feature of living systems is that they are able to persist in ordered and organized state by a process of programmed replication and reproduction. *Evolution by natural selection follows as an automatic consequence of this organization.* But how did such a system originate? Discovering how the organic molecules that make up organisms (p. 5) themselves originated is only a partial answer to this question. We need to imagine plausible ways whereby they become *organized* into self-replicating systems.

The carbon-based molecules that comprise living organisms were once thought to be so unique and special that they could only be synthesized by living things. Hence a distinction was made between organic (= from life) and inorganic chemicals. The first breach in this demarcation was made when Wöhler synthesized a very simple organic molecule (urea) from ammonium cyanate (an inorganic molecule) simply by applying heat (in 1832). And this initiated the rational and scientific treatment of the chemistry of life, that formed the foundation of modern biochemistry and molecular biology. Yet the controlled synthesis of urea is a far cry from the spontaneous origin of polysaccharides, lipids, proteins and nucleic acids that would be needed for the origin of living systems.

1.3.1 Prebiotic synthesis of organic polymers

Little is known with certainty about the early atmosphere of the earth, but it was probably formed by 'degassing' of the planet so that it is likely to have borne a close resemblance to the gas mixtures that escape from volcanoes. On this basis it was almost certainly devoid of oxygen (see Chapter 11). Experiments have now shown that with these conditions almost any energy source, lightning, shock waves, ultra-violet radiation (because there was no oxygen there was no ozone which filters these wavelengths out of the insolation) or hot volcanic ash, would have led to the prebiotic synthesis of a variety of 'organic' monomers: sugars, amino acids, fatty acids and even nucleotides. Given appropriate conditions, for example high concentrations of inorganic polyphosphates, it is possible for these to join in long chains to form, for example, polynucleotides and polypeptides. All these substances probably concentrated in the early ocean to form the famous 'primordial soup'.

There was a kind of selection going on in this prebiotic world since molecules that could polymerize fastest, and/or were most stable would be most common. But the tempo of these changes was not very rapid and could not be very 'adventurous', since the formation of each polymer was an independent event; there was no building on a genetic memory. The next step towards this was not a difficult one, because once certain polymers have formed these are able to influence the formation of other polymers. Polynucleotides in particular have the ability to specify the sequence of nucleotides by acting as a template for polymerization. If one polynucleotide acts as a template for its complement, which then acts as a template for the original, we have lineages linked by a kind of genetic memory. Those polynucleotides that do this most effectively increase in abundance relative to others, i.e. are selectively favoured.

The template systems would have been error prone and new polynucleotides would have been formed by 'mutation' and would compete with others for possibly limited building blocks. Since deoxyribonucleosides (precursors of nucleotides) are more difficult to synthesize than ribonucleosides and since RNA plays a central part in modern protein synthesis, a widely (but not universally) favoured hypothesis is that these early, self-replicating polymers were small RNAs.

1.3.2 Origin and evolution of cells

The next steps towards the system summarized in Fig. 1.1 are more difficult to imagine. At normal temperatures the spontaneous replication described above would have been slow and the error rate high. Association with a replicase, a protein capable of catalysing replication, would have speeded up the process. Just how this might have originated is not clear, but once it was present it would have been favoured. Moreover, there would have been some advantage in enclosing the template and replicase because then the advantages derived from this liaison could not be of benefit to other slightly different but competing templates. The cell was thus born and we begin to see a distinction between genotype and phenotype. Selection would operate on these primitive cells such that those in which the co-operation between genotype and phenotype enhanced replication rate and fidelity would spread more rapidly than others. Though difficult to specify precisely, it was in this context of 'co-operation' and selection that the complex system involving DNA, as well as various forms of RNA, originated and was refined.

The original cells were small and had a simple internal structure, something like modern bacteria, the so-called prokaryotes. In some cells a further membrane originated to enclose the genetic information and this is likely to have been favoured because it probably gave more protection against genetic damage. These cells, the so-called protoeukaryotes, also, probably later, acquired cytoplasmic organelles and prominent amongst these are the mitochondria. The latter show many similarities with free-living prokaryotes—resembling them in size and shape, containing their own DNA, and reproducing by dividing in two—and are now thought to have arisen through symbiotic associations between small prokaryotes similar to the surviving *Paracoccus* and larger nucleate protoeukaryotes. By breaking up eukaryotic cells it is possible to show that all the machinery for aerobic metabolism is located within mitochondria, so that this supposed symbiosis evolved in association with the accumulation of oxygen in the earth's atmosphere from the photosynthetic activity of early cyanobacteria.

1.3.3 Why doesn't spontaneous generation happen all the time?

If large biomolecules and indeed cells have originated once it is reasonable to consider why this does not happen continuously. The answer is probably that the living things themselves, once originated, created conditions which were unsuitable for this. For example, oxygen, a product of life, once formed would destroy the organic building blocks that form living things. In the presence of oxygen, solitary organic polymers are broken down into simple inorganic constituents by oxidation. Hence, once free oxygen became plentiful, the 'primordial soup' of organic molecules could no longer be sustained. Moreover, the complex organic molecules in the 'soup' were probably an excellent nutritional source for the early organisms, and were probably eaten or degraded more rapidly than they formed.

1.4 Levels of organismic organization

It is unlikely that the totality of physiological processes, when crammed together within a single cell (as they are in protists; Section 3.1), could be as effective as when they are divided between many cells in multicellular organisms. The multicellular condition provides more space for the reactions, and the division of labour between cells, by compartmentalization of function, means that at least within compartments, physiological conflicts can be minimized. Hence, there would have been considerable selection pressure for the evolution of multicellularity, e.g. for the origin of the Animalia.

The next chapter summarizes the major features of the invertebrate phyla and speculates upon their relatedness and evolution. It illustrates various possible trends in organization that have progressively opened up physiological potentialities of multicellular animals. For example:

1 The evolution of cellular differentiation.

2 The spatial localization of cells of the same type in tissues and then the organization of these into organs (collection of cells contributing to a common function).

3 The evolution of a through-gut which allowed more specialization of different regions.

4 The development of fluid-filled body cavities that allowed the gut and other organs (e.g. heart) to operate independently from body-wall musculature, facilitated diffusional distribution of nutrients and, by providing a hydrostatic skeleton (Section 10.5), more effective locomotion.

5 The evolution of specific systems for the distribution of nutrients and respiratory gases between tissues. This allowed escape from size constraints imposed by the diffusional distribution of these products (see Chapter 11).

6 The evolution of limbs, which opened up considerable potential for locomotion, particularly on land and in the air.

It is quite easy to see how natural selection might have improved function within levels of organization. But was it responsible for the major shifts between levels? Did these shifts occur *gradually* by a continuous sequence of small changes that improved physiological function and enhanced fitness, or by 'quantal' leaps between levels that were more to do with chance and developmental opportunities than natural selection? These alternatives are referred to in turn as the *Gradualist* and *Punctuated Equilibrium* hypotheses. There is certainly some evidence for a punctuated process in the evolution of the invertebrates (Chapter 2). But since the punctuations are usually measured in geological time, i.e. represent several million years, they could still alter under the influence of natural selection. It is likely that selection pressures change considerably from time to time and cause significant differences in the tempo of evolution. Hence, a punctuated pattern of evolution does not exclude a Darwinian mechanism. This has been a hotly debated area of evolutionary ecology (Calow, 1983) and we shall return to it later in this text.

1.5 Prospects

This chapter has briefly outlined what we consider to be the basic features of living systems:

They are organized systems that depend upon programmes, replication and openness for persistence.

The features of animal systems that follow, almost logically, from them are:

They acquire resources as food from their environment and use them in ways that promote survivorship and fecundity.

Various levels of organization of animal life have evolved and natural selection has operated within these phyletic constraints to cause adaptation in the acquisition and utilization processes. After a preliminary outline of these levels of organization in the next chapter, we describe them in more detail in Section II. This sets the scene for a more in-depth consideration of the behaviour patterns and physiologies of invertebrates in Section III, where we concentrate on individual aspects of their functioning. Hence, Section II adopts a phylum-by-phylum approach and Section III a cross-phylum, functional approach to the invertebrates. The reader, might, therefore, choose to concentrate on the phyletic approach in Section II and use Section III as a source of further information or alternatively to concentrate on the functional biology of the invertebrates in Section III and to use Section II as an 'index' to the taxa that are referred to within it. The two Sections are, nevertheless, integrated and aim to give a complete, holistic appreciation of invertebrate organisms.

1.6 Further reading

Calow, P. 1976. *Biological Machines*. Edward Arnold, London.

Calow, P. 1983. *Evolutionary Principles*. Blackie, Glasgow.

Maynard Smith, J. 1986. *The Problems of Biology*. Oxford University Press, Oxford.

Smith, D.C. & Douglas, A.E. 1987. *The Biology of Symbiosis*. Edward Arnold, London.

2

THE EVOLUTIONARY HISTORY AND PHYLOGENY OF THE INVERTEBRATES

Living animals are the products of their evolutionary pasts, and it is not possible fully to understand modern biology unless we have some appreciation of that past and of the constraints which it has placed upon their life styles and potentialities. In this chapter, we describe the major features of the evolutionary history of the animal kingdom, including those of its origin.

All too often, phylogenetic reconstruction has appeared a rather sterile pursuit—an ingenious but largely irrelevant series of armchair attempts to juggle with various body plans in different combinations in order to find an intellectually satisfying picture, a sort of jigsaw-puzzle game with most of the pieces lost. Entirely hypothetical organisms have been designed so as to bridge gaps in these schemes, often with little regard to the necessity of postulating animals which could conceivably have survived, captured food, escaped predation and reproduced. Part of the sterility of several such exercises has resulted from exclusively anatomical approaches, relying heavily on a few 'key features' (often embryological in nature), features which research into a wider range of animals has shown to be less clear cut and less diagnostic than was once thought.

Here we have endeavoured to set a modern phylogenetic account against a background of the selective pressures which must have operated at various times in the past, and of the likely responses to those pressures of organisms then alive. We have also stressed what the fossil record has to tell us about the manner in which diversification and extinction have occurred, highlighting how our knowledge of the past has influenced our understanding of the evolutionary process. It is in this field that palaeontologists have strongly re-entered evolutionary debates with their much-publicized hypotheses of 'macroevolution' and 'punctuated equilibria'.

Readers of this chapter should note two points. First, unavoidably it draws on some of the anatomical features which are described in the chapters of Section II. We feel that it is more appropriate to present an overview before detailed consideration of the individual groups, even though this may mean that some of the structures and concepts are not covered until later in the book. Description of the different phyla has therefore been kept to the minimum necessary. Secondly, as stressed in the chapter, phylogenetic relationships are contentious matters, but this book is not the right medium for discussion of all the various conflicting opinions. We have simply presented those views of animal interrelationships which seem to us to be most plausible without accompanying detailed justification of each and every scheme adopted. Again, the chapters of Section II will be found to contain relevant information (e.g. on the fundamental differences between the arthropod phyla), but those seeking such justification should consult the lists of 'Further reading' recommended at the end of this and the other chapters.

2.1 Introduction

It is certain that the multicellular animals, like the two other multicellular kingdoms, the Fungi and Plantae, are the descendants of the unicellular (or acellular) eukaryote protists. But there certainty ceases. Most of the animal phyla that are represented in the fossil record first appear, 'fully formed', in the Cambrian, some 550 million years ago. These include such advanced, anatomically complex types as trilobites, echinoderms, brachiopods and molluscs. Precambrian fossil animals are not numerous, but it is

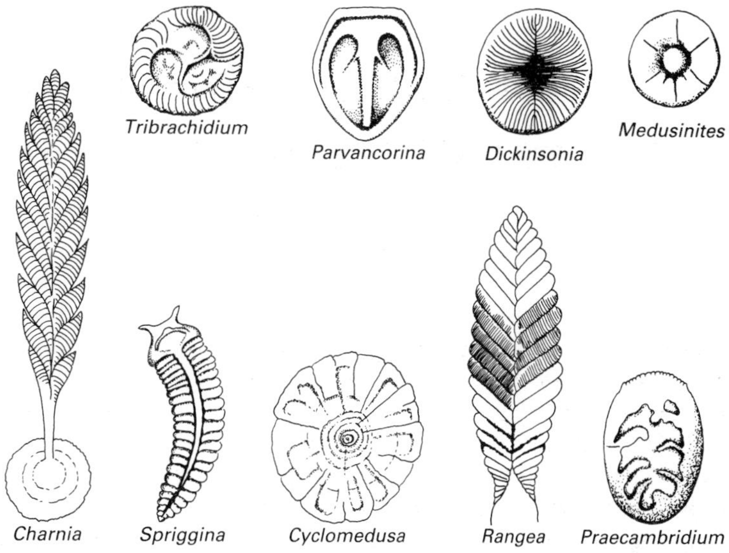

Fig. 2.1. Precambrian (Ediacaran) animals (after Glaessner & Wade, 1966).

generally accepted that cnidarians and segmented worms date back to that period (Fig. 2.1). The fossil record is therefore of no help with respect to the origin and early diversification of the various animal phyla, except in so far as to indicate that these must have occurred in the Precambrian, probably between 1200 and 900 million years before the present day. Life itself dates back 3500 million years.

Much early zoological debate centred on which animal group was the most primitive and from which group of protists it might have been derived, and that debate still continues. Excluding crank hypotheses, three animal phyla were—and are—the main contenders for the title of the ancestral animal group: the sponges (**Porifera**), the coelenterates (particularly the **Cnidaria**), and the flatworms (**Platyhelminthes**). Early proponents of the primitiveness of any one of these three then sought to derive the other two from their preferred candidate, together, ultimately, with all the other known phyla. Relatively early, the sponges came to be regarded as a rather peculiar, isolated group, possibly unrelated to other animals, and they were placed in a separate subkingdom, the Parazoa, distinct from all other phyla except the extinct **Archaeocyatha.** The question thereafter reduced to: were the coelenterates derived from the flatworms, or vice versa?

The first multicellular animals would presumably have been small, composed of relatively few cells, and without any hard parts. Since the fossil record is overwhelmingly one of organisms with hard tests, shells, plates or skeletons, it would therefore be unrealistic to expect that this record will ever be able to contribute to unravelling the ancestry of the animals; such ancestral forms are most unlikely ever to have been preserved. Zoologists have therefore been forced to argue solely on the basis of comparison of the structure and function of living members of the different animal and protist phyla, and to erect hypotheses of what existing characteristics of living organisms should be regarded as primitive features retained through to the present. Here it should be remembered that the living representatives of *all* groups of organisms are separated by at least 600 million years, and probably much more, from the origin of their group, with all the possible changes in biochemistry, physiology, embryology, anatomy, etc. that may have occurred during this interval.

To a modern zoologist, much of the early debate was characterized by adherence to now discredited 'laws', such as 'ontogeny recapitulates phylogeny' (Chapter 17), by the wholesale invention of purely hypothetical intermediate stages which owed more to the necessities imposed by the hypothesis being advocated than to any requirements of successful survival, and by a succession of ingenious armchair speculations which, because of the nature of the quest, were essentially untestable and unfalsifiable. As a reaction against this early rash of speculation and counter-speculation, several modern biologists regard any attempt to reconstruct phylogenies as being futile and unscientific; we will never know if a given answer is correct, so it is pointless attempting the question! Although this attitude might strictly be correct, it is unlikely that the human spirit of enquiry would permit itself to be shackled by this constraint. If we adhered too strictly to a definition of biological science which excluded any area not amenable to experimental refutation of hypotheses, much current biological research, including all that directed towards events in the past, would become inadmissable. Evidence from comparative study of living species and from the fossil record is available to render some suggested phylogenies more plausible than others, and it is upon such evidence that the following general account is based. The reader should note, however, that alternative interpretations are not only possible, but available in the phylogenetic literature.

One of the hallmarks of the relatively recent approach to unravelling phylogenetic history is a decreased reliance on a few, supposedly critical 'key features', such as type of larval stage or segmentation. In this connection, it may come as a surprise to learn in the following pages that pharyngeal clefts, for example, are not an exclusively chordate and hemichordate solution to the problem of the disposal of an ingested water current but that they are also found in the unrelated gastrotrichs; that the 'peculiarly nematode' form of locomotion involving a longitudinally incompressible skeletal element, longitudinal muscles, and consequent S-shaped wriggles is also displayed by the cephalochordates (both also have muscles which send processes to the nerve cord rather than vice versa); and that kinorhynchs are just as segmented as many arthropods. Even nematocyst-like structures are not an exclusively cnidarian feature in that similar organelles are present in a number of protists.

Animals faced with similar selection pressures have evolved similar solutions to common problems. There has therefore been an increasing realization that various levels of bodily organization ('grades') have been achieved in parallel, as a result of convergence, by different phylogenetic lines ('clades'). Segmentation is a specific case in point. Many different clades show serial repetition of muscular and associated organ systems along an essentially linear body and have, via this route, achieved different degrees of metameric segmentation, usually in association with their means of locomotion. When such has been coupled with the development of an external body covering used as an exoskeleton, evolutionary constraints have led to externally similar animals: the exoskeleton must be segmented and the legs, if present, must be jointed. There is therefore no logical or evolutionary necessity for all segmented animals to be related to each other, for example, nor any for all animals with jointed legs to be so either. Other criteria must be used to determine if this is so.

2.2 The first animals

There are theoretically three ways in which a multicellular organism could evolve from a protist. First, different types of protist could together symbiotically form a composite organism, rather after the fashion suggested by some for the origin of the eukaryote cell from different prokaryotes, and of lichens by algal and fungal partners (Fig. 2.2a). Second, the division products of a single individual protist could remain together after division, and multicellularity arise via a colonial stage (Fig. 2.2b) (on this basis, each protist individual would be equivalent to one cell of a multicellular organism and the protists could truly be regarded as being unicellular). Third, a multinucleate protist could evolve internal membrane partitions around each nucleus, confining the sphere of operation of each nucleus to certain regions of its body, and thereby become internally compartmented

(Fig. 2.2c) (here the multicellular organisms would be better regarded as being 'cellular' and the ancestral protist as being acellular, rather than unicellular).

The first of these three potential mechanisms presents serious genetic problems. How do the genetically distinct founding protists integrate into a single, reproducing, multicellular organism? Even the two or three distinct symbionts forming the composite lichens have to reproduce separately and then re-associate to form new colonies. In respect of the third potential mechanism, there are no indications of internal compartmentalization amongst living protists, and so there is no comparative evidence to suggest that it might have occurred in the past. It must be said, however, that were a living multinucleate protist to have evolved internal subdivision, biologists would probably regard it as a multicellular organism and not as a protist. Many protists, however, are known to form colonies, as indeed do many prokaryotic bacteria and cyanobacteria (blue-green algae) (Fig. 2.3), and in some colonies there is differentiation into distinct cell types. The second mechanism is, therefore, not surprisingly that favoured by most biologists and there is a wealth of indications of how unicellular protists might have formed multicellular organisms via coloniality.

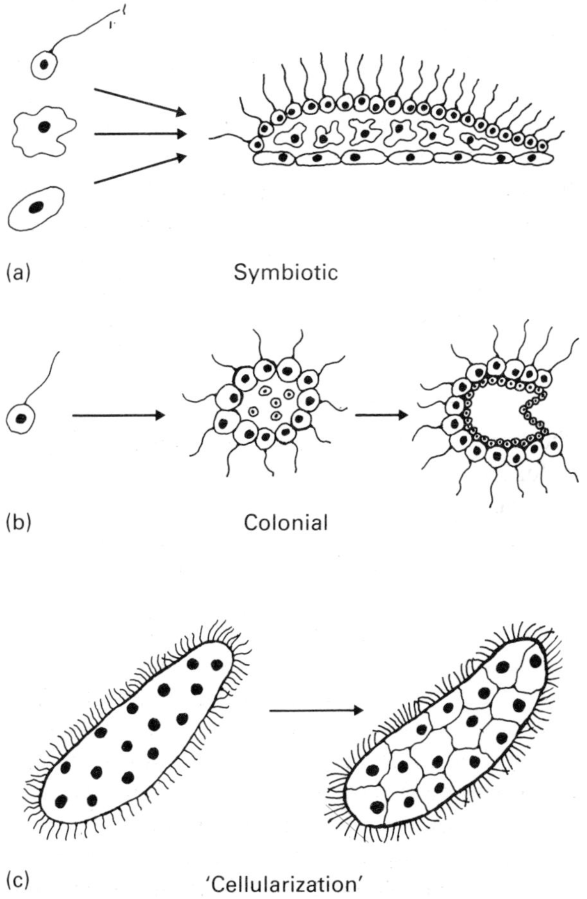

Fig. 2.2. Possible routes for the evolution of animal multicellularity from within the Protista.

It is, in fact, difficult to distinguish between a protist colony and a multicellular organism. Not all organisms classically regarded as being multicellular exhibit much co-ordination of their cells and, as we have seen, cellular differentiation is not confined to the multicellular. Often it is simply a matter of tradition and convenience. Of the 27 phyla of protists recognized in one recent classification, 16 include colony-forming species and in three the level of organization of some, or all, species is regarded as being truly multicellular. Excluding the animals, more than six living groups of multicellular organisms would appear to have arisen independently from within the Protista, and there is no necessary reason to assume that the multicellular animal condition could only have evolved once, even though all animals do share a number of cytological and biochemical features in common. (Neither, however, is there any necessity to attempt to derive each animal phylum separately from the protists, as has been tried.)

2.2.1 Sponges

Sponges come closest of all animals to being regarded as a colony of protists rather than multicellular. Had it not been traditional for many years to include them within the Animalia, they might now be placed in the Protista, in parallel with the retention of the multicellular algal phyla in that kingdom. The individual cells of a sponge show little (and no nervous) co-ordination; there is no system of symmetry of the individual/colony; and different types of individual

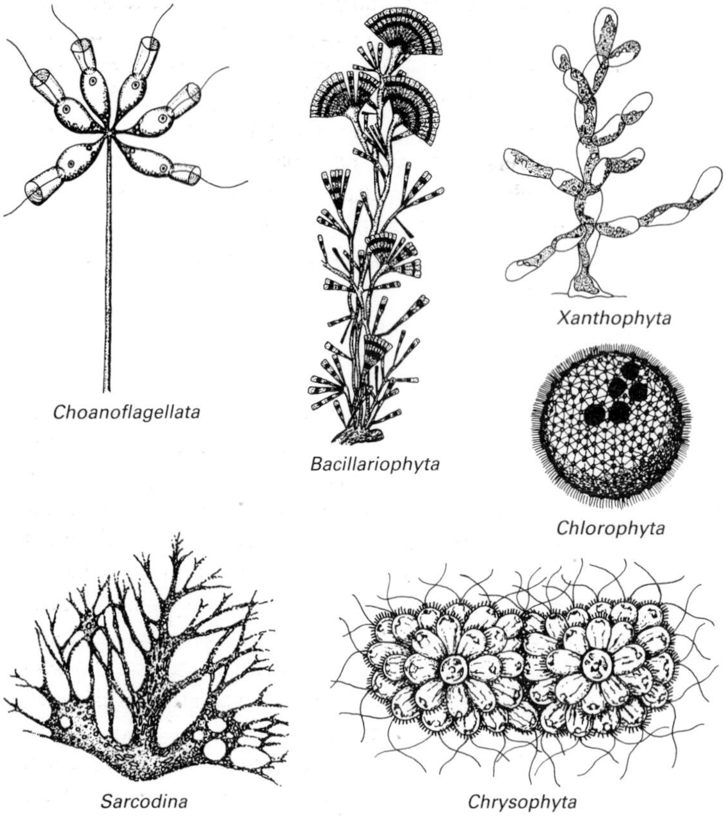

Fig. 2.3. Colonial protists from a variety of phyla (after various sources).

cells (there are few different types) do not together form well-defined tissues or organ systems, basement membranes being lacking. They are described as animals of the cellular grade of organization, in contrast to the tissue grade displayed by the coelenterates and the organ-system grade of the flatworms. That the individual sponge cell is the only essential subunit is shown by the ability of disaggregated sponges to reconstruct themselves, and sponges can also regenerate from very small fragments.

There is one phylum of protists, the **Choanoflagellata,** which contains solitary or colonial unicells effectively identical to the most characteristic of sponge cell types, the choanocytes (Fig. 2.4). Some choanoflagellates live attached to the substratum, and many secrete a membraneous or siliceous sheath or lorica. They feed, in so far as is known, in exactly the same manner as choanocytes. Most zoologists therefore believe sponges to be derived from choanoflagellate colonies. The fact that somewhat choanocyte-like cells are found in other animals might even indicate a choanoflagellate origin of non-sponge groups too.

The structure of sponges is so individual that it is impossible to derive any of the other living animal phyla from their body plan. This has led some people to regard sponges as an early, unsuccessful 'attempt' at multicellularity. Nothing could be further from the truth. Sponges are an extremely successful marine group, with more living species than the echinoderms and almost as many as those of the marine annelids; they have also been a prominent part of the marine fauna since the Cambrian. Their apparent simplicity need not be viewed as the result of some inability to

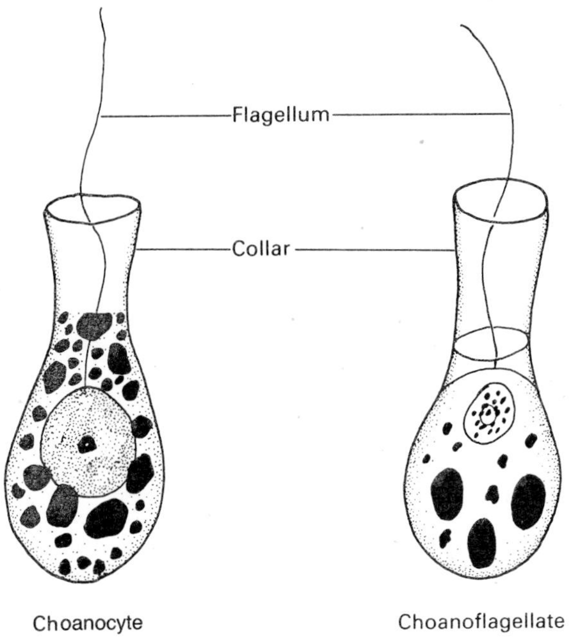

Fig. 2.4. The morphological similarity between a sponge choanocyte and a choanoflagellate protist.

evolve the organ systems found in other groups of animals, however, but can be related directly to their un-animal-like life style.

They are attached, sessile and completely immobile suspension feeders. Indeed, the function of their skeletal apparatus is the very antithesis of that in all other types of animal: in effect, it serves to prevent movement and to provide a rigid support for the body. Environmental water is induced to flow through the channels and chambers of the sponge mass by the (unorganized) beating of the choanocyte flagella; were the tubes not rigid and immovable, local reductions in water pressure could cause the tubing to constrict rather than serve to draw more water through. Granted that the body is incapable of movement, a nervous system, for example, would be functionless. Protection from predators is not brought about by detection and escape, but by distasteful chemicals or by the spicular or fibrous nature of the skeleton. Seen in the same light, it is difficult to imagine how any of the other organ systems possessed by more organized animals could in any way increase sponge efficiency or survival. Neither is it necessarily the case that sponges were a comparatively early multicellular form. The appearance of the first sponge spicules in the fossil record post-dates the occurrence of, for example, complex segmented animals and it is entirely possible that they evolved after several other animal groups.

The sponge is therefore an alternative animal, not a reject.

2.2.2 Coelenterates

Like the sponges, the coelenterates (i.e. the **Cnidaria** and **Ctenophora**) are generally regarded as a highly individual and dead-end group. The supposition that they did not give rise to any other phyla is—again similarly to the sponges—only another way of stating that their general body plan is so successful that no basic change to it would be likely to lead to greater success. The peculiarities of the coelenterate body plan are their radial symmetry, their generally tissue-grade of organization with co-ordination of the cells being achieved by a diffuse network of naked nerve-cell fibres, and the occurrence of only two layers of cells, one on either side of a gelatinous and often non-cellular mesoglea. Equally characteristic are the intracellular organelles, the nematocysts (in the Cnidaria) or colloblasts (in the Ctenophora), responsible for offence and defence. Organelles of this general type are otherwise known only in various protists (e.g. in the **Myxozoa, Microspora** and **Dinophyta**) (Fig. 2.5).

The body shape of a coelenterate is in essence that of a cup, sometimes flattened or elongated, enclosing a simple cavity communicating with the external environment through a single aperture (mouth/anus). This and the two-cell-layered state were seized on by early phylogeneticists, who pointed out that this 'diploblastic' condition is the same as that seen in the gastrula stage of animal embryology. Since, in the embryological development of most animals, the diploblastic gastrula becomes triploblastic through the formation of a third germ layer, the mesoderm, the coelenterates were viewed as the relict, perma-

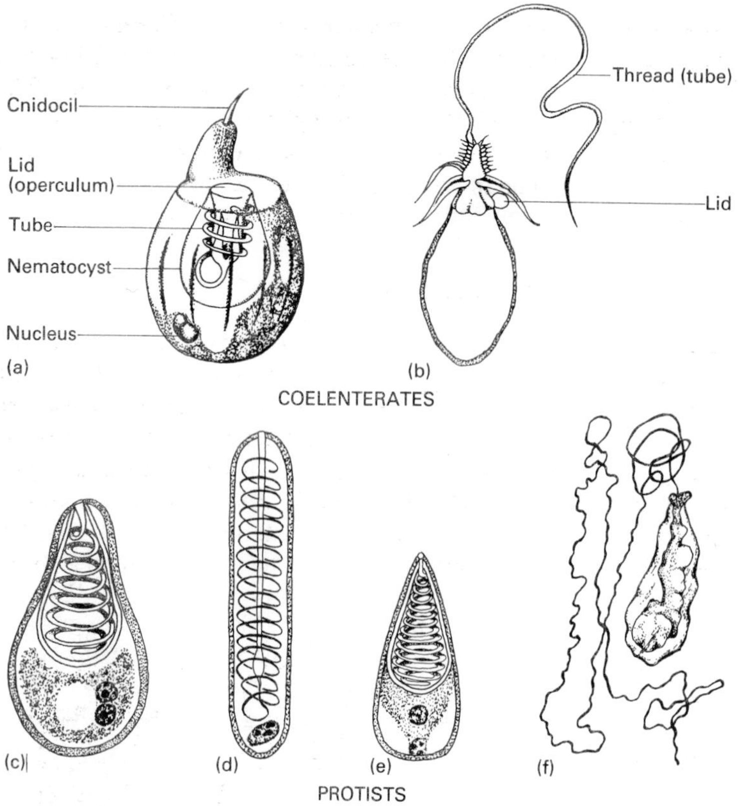

Fig. 2.5. Nematocyst-like organelles in cnidarian coelenterates and in myxozoan and microsporan protists. Cnidarian nematocysts: (a) cnidocyte containing an undischarged nematocyst; (b) discharged nematocyst. (c) Spore of a myxozoan protist with an undischarged organelle. (d, e) Spores of microsporan protists with undischarged and (f) discharged organelles. (After Hyman, 1940; Calkins, 1926; Wenyon, 1926 and others.)

nently diploblastic, ancestral forms which gave rise to all the other—triploblastic—animal phyla. The three stages of evolution of the animals were envisaged as: (a) a hollow ball of cells (= the blastula), a form in which several types of colonial flagellate are known to occur; (b) these, for some reason, became a double-walled cup (= the gastrula = the coelenterates); and (c) when mesoderm evolved, the triploblastic animals were formed. This argument by embryological analogy may have been ingenious, but there is no evidence which supports the notion that embryological germ layers can in any way be equated with the cells of superficially similar adult organisms. Embryological terms such as diplo- and triploblastic should not be applied to adult morphology. As mentioned above, it is unlikely that the coelenterates were ancestral to any other group (although there have been attempts to derive the flatworms from the 'planula' dispersal stage of several cnidarians—a free-swimming, non-feeding, solid blastula—which whilst being radially symmetrical is nevertheless elongated along a longitudinal axis).

By the same token, the basic body plan of the coelenterates is such as to make it unlikely that they were derived from any other known animal group. The ancestry of these animals is, therefore, most realistically to be sought amongst the Protista, although it is not possible to conclude which, if any, surviving type of protist is nearest to their origin. Because the cells of cnidarians are typically flagellated, rather than ciliated like the majority of animals, and because food particles are ingested by amoeboid phagocytosis, it has been suggested that the protists concerned are most likely to have been a group of heterotrophic flagellates. Although questions such as the precise bodily form of the colonial protist or protocoelenterate which could have been ancestral to the group, and whether the polyp or medusoid phase is the more primitive, have been much debated, all is here conjectural and little point would be served by attempting to argue the merits of any particular scheme.

In most cnidarians, the polyp phase can multiply by asexual budding and, in many, tissue contact is retained between the budded daughter polyps. The polyps are then all genetically identical and interconnected, and this has rendered possible polymorphism of polyp type: some can be specialized for feeding or defence, others for sexual reproduction, and so on. In a sense, specialization of individual polyps and retained medusae within a colony as a whole is the cnidarian equivalent of the organ systems of most other animals. Colonies of polyps, as a result of budding, can also achieve large size, as seen in many corals for example, as indeed can individual polyps and medusae, because each of the two cell layers is in contact with sea water (the outer layer with that in the external environment and the inner layer with that in the gastrovascular cavity) and hence diffusion of respiratory gases, excretory products, etc. is not affected by overall body mass. The same is also true of the sponge body plan, but not of that of the flatworms (see below).

One final point concerning the coelenterates deserves mention. All but a few known species are carnivorous and none can digest algae. They prey largely on planktonic animals which accidentally touch their nematocyst- or colloblast-containing tentacles. The ancestral coelenterates probably were consumers of bacteria and protists (see Chapter 9), and some living species also have such a diet, but the rise to prominence of the coelenterates in late Precambrian seas was probably consequent on an ability to switch to larger prey items, as various animal groups, particularly the arthropods, invaded the water column and formed the zooplankton. A further point of similarity with the sponges is therefore the possibility that the origin, and especially the adaptive radiation, of the coelenterates post-dated the appearance of other groups of animals.

(Note that although we follow the traditional view here that the cnidarians and ctenophores are related groups, there is an increasingly held opinion that their resemblances are a result of convergence. The 'mesoglea' of the ctenophores may be more akin to the flatworm mesenchyme than to the system of that name in the Cnidaria, and ctenophore cells are also multiciliate rather than monoflagellate.)

2.2.3 Flatworms

The modern flatworms are a large and diverse group, dominated today by parasitic species. The free-living turbellarians, from which it is generally considered that the parasitic flukes and tape worms evolved, are bilaterally symmetrical, with flat, solid bodies propelled by cilia. Organ systems occur, including a system of sheathed nerve fibres organized in all but some acoels into distinct, longitudinal cords.

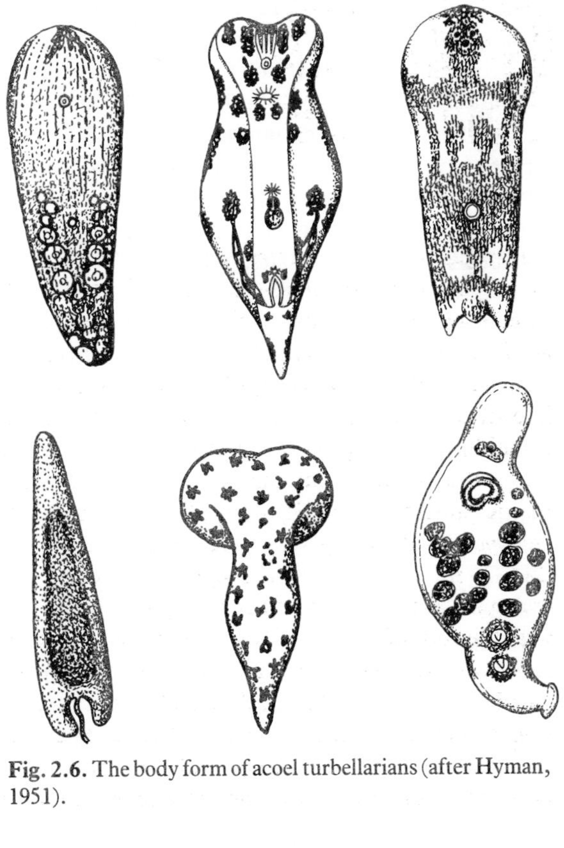

Fig. 2.6. The body form of acoel turbellarians (after Hyman, 1951).

Amongst the living turbellarians, the small acoels (Fig. 2.6) and some related groups are nearest to the presumed ancestral body plan, although they themselves are not necessarily ancestral. (One of the difficulties mentioned in Section 2.1 of arguing solely from modern forms is that all surviving groups may have diverged considerably from the now extinct ancestral forms, even those that appear to display many primitive features. It is therefore more appropriate to regard living animals as being analogous to the ancestral types. Acoels, for example, have specializations which debar them from being directly ancestral, and it is therefore their general level of bodily organization which appears primitive, not the surviving acoels themselves.)

The acoels lack a permanent gut and possess simple nervous and reproductive systems (the reproductive systems of other flatworms are usually extremely complex). In general, an acoel is as near to an integrated, colonial, ciliated protist as it is possible to imagine, and it is widely believed that the flatworms arose from ciliated protists of the general type that today are represented by the **Ciliophora** and, especially, the **Stephanopogonomorpha** (Fig. 2.7). An alternative view is that the ancestral protist was a flagellate which, when in the multicellular state, underwent multiplication of the number of flagella in each cell to form the ciliated condition (some surviving flatworm-like animals possess cells each with only a single cilium, see Section 2.3.1). The diets of modern acoels are not known in any detail, although most are thought to feed on bacteria and protists. Such was also presumably the ancestral diet, just as it is of many modern flagellates and ciliates.

As a group, however, the flatworms—like the coelenterates—are almost exclusively carnivorous. This may appear most paradoxical: the two groups suggested to be the most primitive of animals are themselves typically consumers of other animals or, more generally, of non-photosynthetic organisms. In fact, the school-biology notion that most animals should be herbivores, deriving from the classic terrestrial food chain of sun, carbon dioxide and water→plant→herbivore→carnivore, is at best only a half-truth. Although photosynthesis is clearly the ultimate source of most fixed energy, few terrestrial invertebrates and even fewer vertebrates can actually digest plant material. Most feed on it only after it has been processed by bacteria, heterotrophic protists or fungi, either via the decomposer food chain or by virtue of the possession of a culture of microbial organisms in their gut (see Chapter 9).

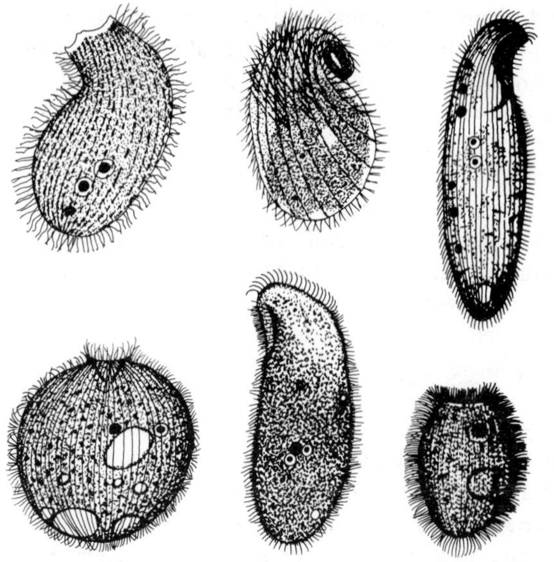

Fig. 2.7. Representative ciliate and stephanopogonomorphan protists (after Corliss, 1979).

In contrast, the marine photosynthesizers are almost entirely unicellular protists and they are relatively easily digestible by animals. The problem with herbivory here is a spatial one. Marine photosynthesis is confined to the comparatively shallow, sunlit surface waters, but most animal groups—with the possible exception of the coelenterates, should the medusa have been the ancestral body form—are likely to have evolved in association with the bottom sediments and most have remained benthic in habitat. They would have been dependent on a rain of dead algal material from above, a rain already colonized by decomposer organisms (principally bacteria) together with bacteria-consuming protists. Of necessity, the ancestral diet would have been largely the non-photosynthetic organisms of the decomposer food chain. As animal life continued to evolve, so progressively larger, and energetically more valuable, organisms were available for consumption, but, with the exception of the shallow and physically hostile margins of the sea, it would not have been until they were able to colonize the water column itself that photosynthetic protists were available for consumption. Hence direct consumption of living photosynthetic organisms probably appeared, as a specialist diet, relatively late in the phylogeny of animals.

2.2.4 Placozoans and mesozoans

Only two other groups of animals are serious candidates for independent evolution from the protists, the **Placozoa** and **Mesozoa**. Both could be argued to be nearer to colonial protists than to multicellular animals, lacking any nervous, muscular, digestive or skeletal cells, tissues or organs, although detailed knowledge of either group is sparse.

The placozoans look and behave like large, flat, flagellated and cellular amoebae. Their bodies are without any symmetry and are composed of a few thousand cells disposed in two layers, 'dorsal' and 'ventral', together with fibre cells in a matrix between the two. Their level of organization is equivalent to that of a sponge, and some have suggested that they are permanently larval sponges. Equally likely, however, is that they are a group of flagellate protists which have evolved multicellularity, perhaps even relatively recently. They are known to feed on protists by external digestion and subsequent absorption of the digestion products.

The mesozoans, which also lack any tissues or organs, nevertheless have a definite body shape and organization. The two rather dissimilar and perhaps unrelated groups included are minute, bilaterally symmetrical, ciliated endoparasites of marine invertebrates. Their body may comprise only some 20–30 ciliated cells surrounding one or more reproductive cells, although, in some, development in the host involves losing the external ciliated cells and forming a plasmodium. Their bilateral symmetry and generally worm-like body shape have suggested to some that one or both groups of mesozoans originated from the flatworms, and that their simplicity is a consequence of their parasitic nature; but the peculiarity of their cellular arrangement and life cycles, and the highly individual pattern of their reproductive processes, have suggested to others that they are not flatworm descendants. If they are not, then their origin must presumably be sought within the ciliated protists, but no relevant information is as yet available on this small and enigmatic phylum.

2.3 Flatworm derivatives

Multicellular animals may have arisen from within the Protista from three to five or six times, to consider only those groups which have survived through to the present, but it is almost certainly from only one of these lines—the bilaterally symmetrical flatworms—that all the other animal phyla have derived (see, e.g., Sleigh, 1979). Many of these flatworm derivatives have evolved new organ systems, such as blood systems and hydrostatic or hard skeletons, and have evolved more specialized versions of systems inherited from their ancestors; all features largely consequent on the inadequacy of diffusion to transport materials through large, solid-bodied animals, and on the limitations of the flatworm locomotory system. These notwithstanding, the essentially flatworm-like nature of their body plan remains obvious. Directly from a flatworm-like organism probably sprang the molluscs and a host of other types of worm, whilst from various of these secondary worm groups derived the arthropods, the lophophorates, and the deuterostomes (Fig. 2.8).

2.3.1 The vermiform host

A worm may be defined as any bilaterally symmetrical, legless and soft-bodied animal with a length greater than two or three times its breadth. On such a basis, there are some sixteen animal phyla composed solely of worms (in addition to the flatworms) and several other groups that contain various vermiform species. Although evidence of the existence of worms, such as burrows, tubes and crawling traces, are abundant in the fossil record, including in the

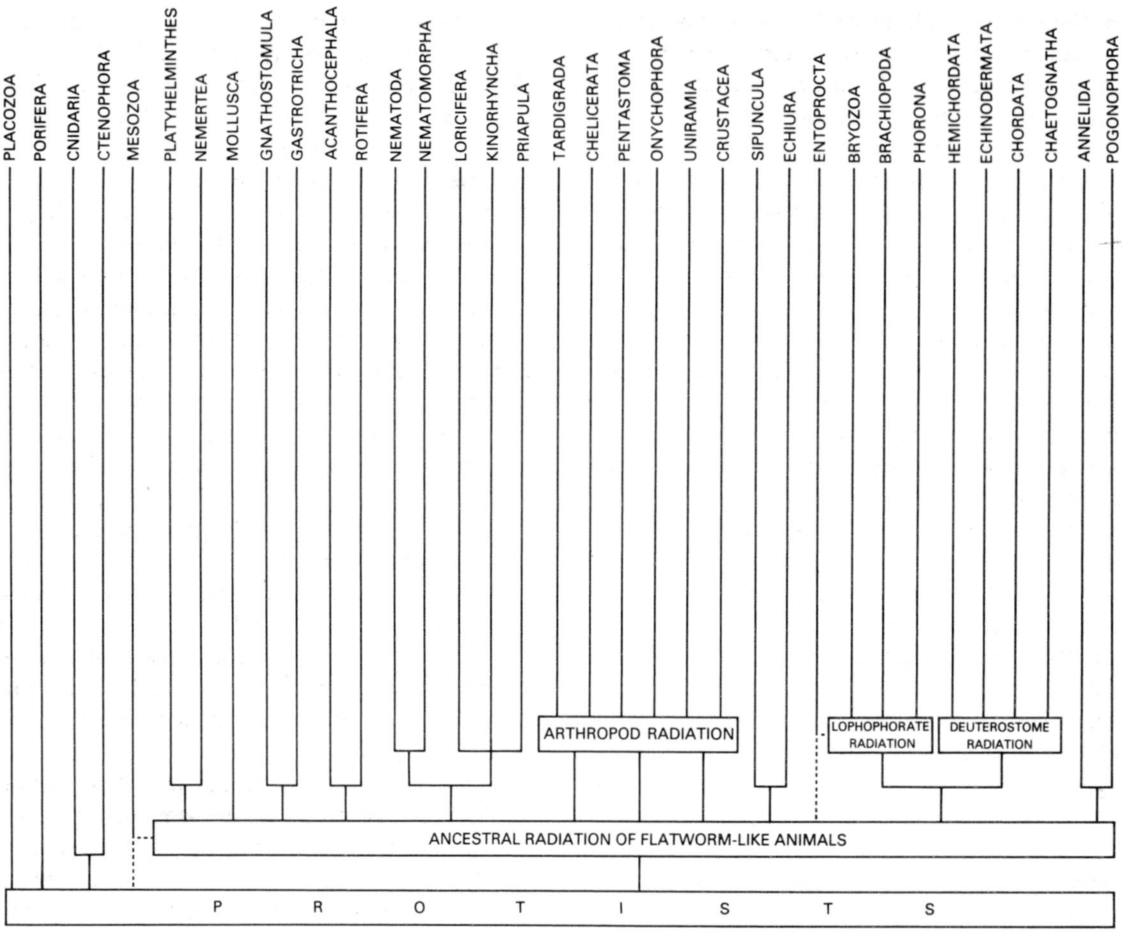

Fig. 2.8. The radiations which produced the living animal phyla (see also Fig. 2.20).

Precambrian, soft-bodied animals are not common as fossils. Priapulans and segmented worms are known to have been abundant in the Cambrian, but assignment of other fossil material to specific living phyla is fraught with contention. Be that as it may, representatives or evidence of several other phyla of worms have been claimed for the Cambrian and Precambrian. The fossil record is nevertheless of little assistance in disentangling the interrelationships of modern worms, and there is little general agreement on affinities within the group. Indeed, the worms undoubtedly do not form a natural group, except in the sense that they all have their origin in one way or another in free-living, flatworm-like animals. Nevertheless, seven broad groupings of worms can be distinguished, largely on the basis of life style and ecology, but often also with some degree of potential phylogenetic affinity. (These seven exclude the chaetognaths, which will be treated in a later section along with their deuterostome allies, and the secondarily legless pentastomes, which are probably related to the arthropods—see below.)

1 The larger, free-living turbellarian descendants of the early flatworms and a related group, the nemertines (**Nemertea**). These are flat, solid-bodied (i.e. 'acoelomate') carnivores which creep slowly over surfaces by means of ventral cilia or, in the larger types, by waves of muscular contraction passing along the ventral surface. The constraint of maintaining at least one bodily dimension small, for the diffusion of respiratory gases, etc., has necessitated increase in size only in the lateral or anterior/posterior planes. Large species are therefore either leaf- or ribbon-shaped; some can attain great length—one, only half unravelled, nemertine measured 30 m. Although basically marine, there are many freshwater and several terrestrial members of this group, and a variety of life styles occur, including free-swimming, commensal and parasitic forms. The larger turbellarians and nemertines are the largest and most active animals to have retained the ancestral solid-bodied form. They cannot burrow rapidly, but in spite of this they are successful carnivores preying on a wide variety of other animals.

2 **Gnathostomula** and **Gastrotricha**. These have also retained the general body structure of their flatworm ancestry, although both phyla have only a single cilium per epidermal cell. The small gnathostomulans and gastrotrichs (length generally less than 2 mm) have specialized in inhabiting the interstitial spaces within sandy sediments, moving through the interstices by means of their epidermal cilia without displacing the individual sand grains. Some inhabit relatively deep, anoxic layers and a few appear to be obligate anaerobes. Both groups feed on bacteria and protists occurring either free in the interstitial water or in association with particles of organic detritus.

3–7 All the remaining groups of worms share one common feature which is lacking in the flatworms and in the first two categories of their derivatives considered above: they possess liquid-filled body cavities serving a hydrostatic skeletal function. These body cavities are of many different forms and embryological origins, and they have probably been evolved independently by many separate lineages of vermiform animals. Broadly, however, three general developmental types can be identified (Fig. 2.9): 'pseudocoels', often formed from a persistent blastocoel (the cavity within the blastula which is usually obliterated during gastrulation); 'schizocoels', formed within blocks of mesodermal cells by cavitation; and 'enterocoels', formed as outpocketings of the embryonic archenteron or gut. A fourth category of body cavity, not seen in living worms but present in some non-worms and presumably present in some worm groups now extinct (see Section 2.4.1), is a 'haemocoel', a system of blood sinuses.

In most worms, this hydrostatic skeleton permits the achievement of forms of locomotion largely denied the solid-bodied flatworms, including active burrowing; these were probably originally of selective advantage in respect of the avoidance of surface-dwelling predators such as today are represented by the turbellarians and nemertines. Certainly, all the following groups of worms have invaded microhabitat types or have life styles not available to surface creepers. Usually, locomotion, including burrowing, is effected by peristaltic contractions of the body-wall musculature, which contains both circular and longitudinal elements, although the nematodes lack circular muscles and, accordingly, display a highly individual form of non-peristaltic locomotion (paralleling that of the early chordates).

3 **Nematoda** and **Nematomorpha**. Most members of this third group are of the same order of size as the gastrotrichs and gnathostomulans, and, like them, feed interstitially on bacteria and protists, although their body plan has proved as successful for movement through the tissues and fluids of other organisms as for movement through interstitial spaces in sediments, and many have adopted a parasitic existence for all or part of their lives. The parasitic species may be very much larger: up to 8 m in the case of some nematodes. Their distinctiveness consists of a locomotory system based on a large, high-pressure pseudocoel contained within a collagenous cuticle-covered body wall possessing only longitudinal muscles. (One nematode has recently been discovered to possess circular muscles and it may be assumed that the ancestral forms of this group were similarly endowed.) The collagen fibres are organized into a high-angle geodesic lattice which resists shortening of the length of the body. Contraction of the longitudinal muscles against the resistance of the

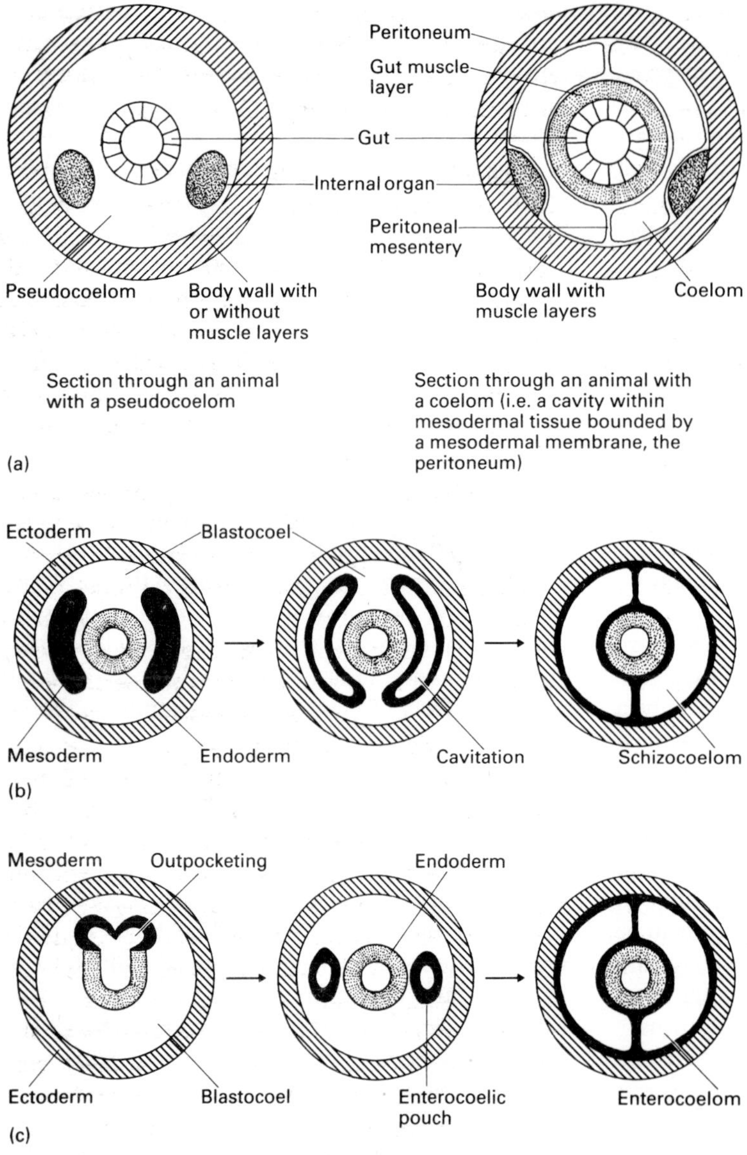

Fig. 2.9. Animal body cavities: (a) differences between a pseudocoelom and a coelom; (b) developmental formation of a schizocoelic coelom; (c) developmental formation of an enterocoelic coelom.

hydrostatic pseudocoel confined by this lattice causes a series of S-shaped thrashing movements in the dorso-ventral plane, and these worms can wriggle quite rapidly through narrow liquid-containing channels.

4 The fourth group (the **Priapula, Kinorhyncha** and probably also the **Loricifera**) also have a large pseudocoelomic* hydrostatic skeleton and a body covered by a cuticle, which is moulted, but in this case the cuticle is chitinous and circular muscles are present in the body wall. In the small kinorhynchs,

*The precise nature of the priapulan body cavity is contentious; it is certainly of a very individual type.

the cuticle is segmented and there is some segmentation of internal organs, i.e. in respect of the muscular and nervous systems. The priapulans and kinorhynchs actively burrow through sediments in search of prey, displacing the sediment particles ahead of them by means of an anterior end differentiated to form an eversible and retractable burrowing organ. Increase in body cavity pressure everts the burrowing organ, the spines on which grasp the surrounding sediment, and contraction of the longitudinal muscles then drags the rest of the body after and over the 'head' region. Little is yet known of the latest phylum to be discovered, the Loricifera (only described in 1983), although they appear to belong to the group under consideration on the basis of their general anatomy. A similar eversible, spiny anterior end is present, but the adult loriciferans may well be commensal or ectoparasitic on other interstitial animals.

From the flatworm ancestors of this group may have derived both the gutless and secondarily simplified parasitic **Acanthocephala** and, by a process of permanent larvalization, the free-living **Rotifera**. In any event, these two phyla are regarded by many as being related. Rotifers are mostly freshwater in habitat, moving and/or feeding on small particles by means of an anterior crown of cilia. Like several other animals with pseudocoels, rotifers have syncitial organs with a constant number of nuclei ('eutely'); and, as in rotifers, there is a strong possibility, based largely on their specialized forms of reproduction, that all the small-bodied pseudocoelomate groups have evolved from much larger worms by paedomorphosis.

5 **Sipuncula** and **Echiura**. These have adopted a sedentary, deposit-feeding life style with the large and somewhat sac-like body in a permanent burrow within the sediment, in a crevice or in an equivalent cavity. Similarly to the priapulans, etc., members of this group possess a voluminous body cavity although it is not pseudocoelomic but a schizocoel. Sipunculans and echiurans both feed using a specialized anterior region of the body: in the Sipuncula, this is the elongated, eversible and retractable anterior end, which terminally bears detritus-collecting lobes or tentacles; whilst in the Echiura, it is an extensible and contractile proboscis with a ventral gutter and a flared, ciliated, food-collecting tip. To some extent, both these phyla resemble giant, single-segmented annelids.

6 The third to fifth groups of worms above, inclusive, all possess one single, large body cavity (i.e. they are 'monomeric'), regardless of the method of formation of that cavity. The sixth group (the **Phorona**), however, has a tripartite body plan, each region with its own separate body cavity ('oligomeric') (Fig. 2.10). These sedentary filter-feeders will be considered in more detail below (Section 2.4.2) in connection with the origin of the lophophorate phyla.

7 Finally, the **Annelida** and **Pogonophora** have bodies largely comprising a chain of segments ('metameric') (Fig. 2.10), each segment with separate schizocoelic body cavities, permitting a much greater localization of changes in body shape. Individual segments are budded off in linear sequence from a single proliferation region just in front of the posterior end. This is in some ways equivalent to the transverse asexual budding displayed by several flatworms. Annelids move actively through or across sediments or through water in search of food, but many have also adopted a sedentary life with consequent reduction of several of the internal septa separating the individual segments. Although most phyla of worms are restricted to the sea, the annelids have in addition colonized both fresh waters and the land, and some have secondarily recolonized the interstitial habitat.

Most of the body of the pogonophorans conforms to the tripartite plan of the phoronans, but when complete specimens of these gutless deep-sea worms were finally obtained, it was found that a small fourth body region, probably serving an anchorage function, was present. This is divided into a number of segments (up to 30), each of which contains separate body cavities, and, after a somewhat chequered history at the hands of systematists, these worms are now regarded as being related in some way to the annelids.

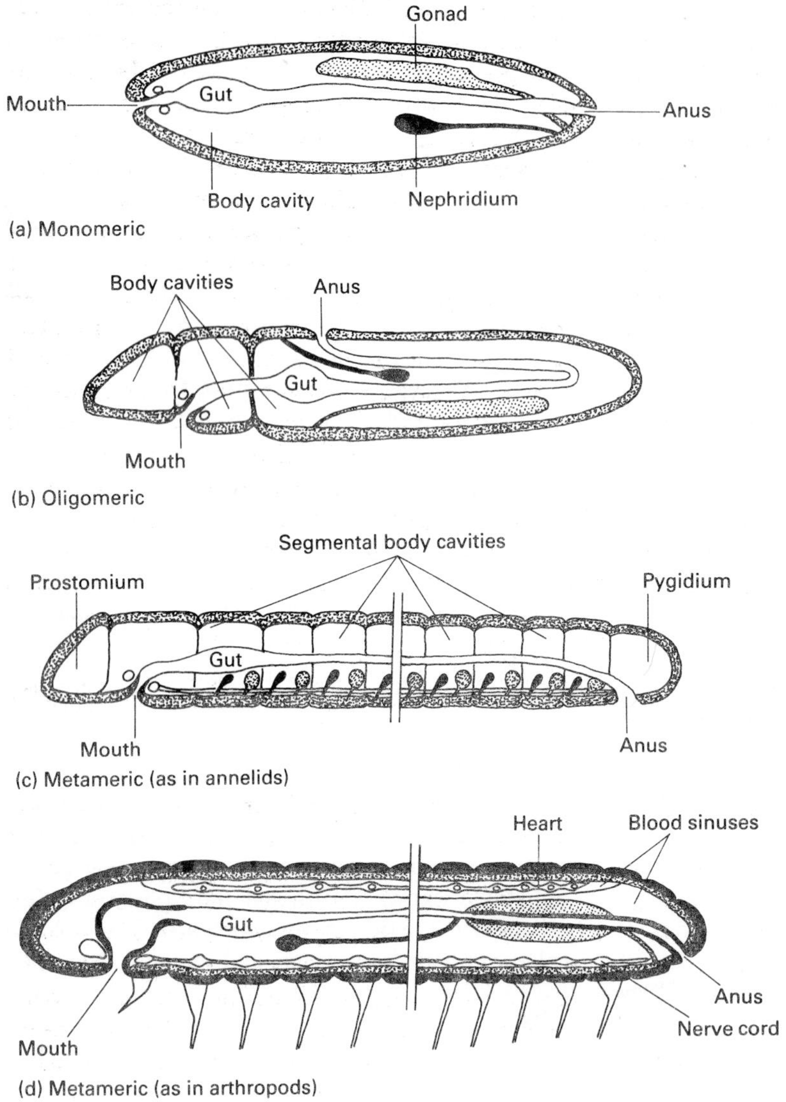

Fig. 2.10. Fundamentally different vermiform body plans in diagrammatic longitudinal section.

2.3.2 Molluscs

Molluscs have neither a coelomic hydrostatic skeleton, nor signs of metameric segmentation, yet because they share a type of larva—the trochophore—in common with the coelomic and segmented annelids, and because larval form was once held to be of overriding importance in establishing phylogenetic relationships, they were for many years assumed once to have had a coelomic body cavity and to have been segmented, only to have lost both in later evolution. It is certainly the case that some molluscs have multiple pairs of some organs (the monoplacophorans, for example, have eight pairs of foot-retractor muscles, six pairs of excretory organs, and five or six pairs of gills), but these multiple pairs of organs are not in any way associated with particular regions of the body which could be considered to represent segments. Several unsegmented animals have multiple pairs of organs in linear series without being regarded as having once been metamerically segmented: they are sometimes termed 'pseudo-metameric'. Molluscs do also have a coelomic cavity surrounding the heart, but this is more plausibly regarded as a space evolved by that group within which the heart can beat, than as a remnant of a once more extensive cavity which served as a hydrostatic skeleton.

There is, therefore, no compelling evidence for any such original presence and later great reduction or loss, and molluscs are no longer usually regarded as being closely related to the annelids (see, e.g., Trueman & Clarke, 1985). Indeed, the early molluscs are most likely to have been little more than flatworms with some form of dorsal protective covering (a protein coat with a series of calcareous scales or plates): like the surviving flatworms, they are acoelomate and non-segmented, and like the larger flatworms and nemertines, except in specialized species, movement is achieved by means of ciliary action or muscular contractions passing along the flat ventral surface.

Besides the dorsal protective shell, the other main feature likely to have differentiated the ancestral molluscs from the flatworms was the occurrence of a toothed, tongue-like organ, the radula, in the buccal cavity. This organ would have permitted algal films and colonies to be rasped from hard surfaces, and it is widely held that the early molluscs were shallow-water grazers of algal protists. Because of the development of a hard covering over the dorsal surface (the only surface exposed to the general environment and therefore to water flow as well as predators and wave action), the exchange of respiratory gases across the body surface—as is the case in flatworms—would have become greatly curtailed. In parallel with the evolution of the shell, therefore, must have occurred the development of specific respiratory gas-exchange organs (gills) and a circulatory system to transport the gases through the body. The paired gills of molluscs are of an individual and characteristic type, termed ctenidia, and these were housed, primitively, in a posterior mantle cavity concealed beneath the shell. Two other extinct groups, the **Hyolitha** and **Wiwaxiida**, appear to have reacted to the selection pressures favouring a protective body covering in an essentially similar fashion to the molluscs.

From such a shelled, flatworm-like animal, molluscs adaptively radiated to form the wide variety of types described later in this book (Chapter 5), including some which are the most complex of all the immediate derivatives of the flatworms and indeed of all invertebrates. Several, however, have retained the ancient browsing and grazing life style. This original mode of life could also relatively easily be adapted to terrestrial existence, and the molluscs are one of the very few groups of animals to possess the enzymes capable of breaking down the refractory structural carbohydrates which characterize the larger seaweeds and the land plants.

2.4 Animals of more remote flatworm ancestry

2.4.1 Arthropods

The essential anatomical differences between a worm with serially repeated organs along its length and an arthropod are that the latter group has legs and is covered by a hard, jointed exoskeleton. These differences are not as major as might appear at first sight. Of the two, it is likely that legs evolved before the exoskeleton. Some living animals, e.g. the onychophorans and tardigrades, are soft-bodied, worm-like organisms with ventral or near-ventral, unjointed legs and a hydrostatic skeleton. The onychophorans are covered by a flexible chitinous cuticle, whilst the tardigrades have external cuticular plates. It is therefore relatively easy to visualize how a cuticular protective covering could evolve to cover the body completely, with articulatory joints between the various sets of serially repeated plates (as seen, for example, in the unrelated kinorhynchs) and between different articles of the limbs, and then be utilized as a skeleton in place of the original hydrostatic one. In one group of arthropods, the chelicerates, extension of some joints of the legs is still achieved hydraulically and only leg flexure is by means of the exoskeletal/ muscular system.

This 'arthropodization' of the ancestral worms seems to have occurred, probably in the Precambrian, in the form of a major radiation (Fig. 2.11), in that numerous, very different types of segmented arthropod-like animals are known from the Cambrian (Fig. 2.12), many of them extremely bizarre. If all these arthropodized groups had survived through to today, they would probably be placed in as many, if not more, phyla than accommodate the vermiform host. The ancestral group of worms need not have been, and almost certainly was not, one which if alive today would be regarded as an annelid, the main surviving phylum of segmented worms. It is likely, for example, that the hydrostatic body cavity of the ancestral worms would have been in the form of a haemocoel, not the segmentally compartmented schizocoel characterizing the annelids, and that their bodies were not segmented in the annelid sense at all, but were monomeric with multiple pairs of locomotory organs in a linear series. They can perhaps be envisaged as elongate, superficially kinorhynch-like worms with their blood system (needed for the same reason as in the molluscs, see

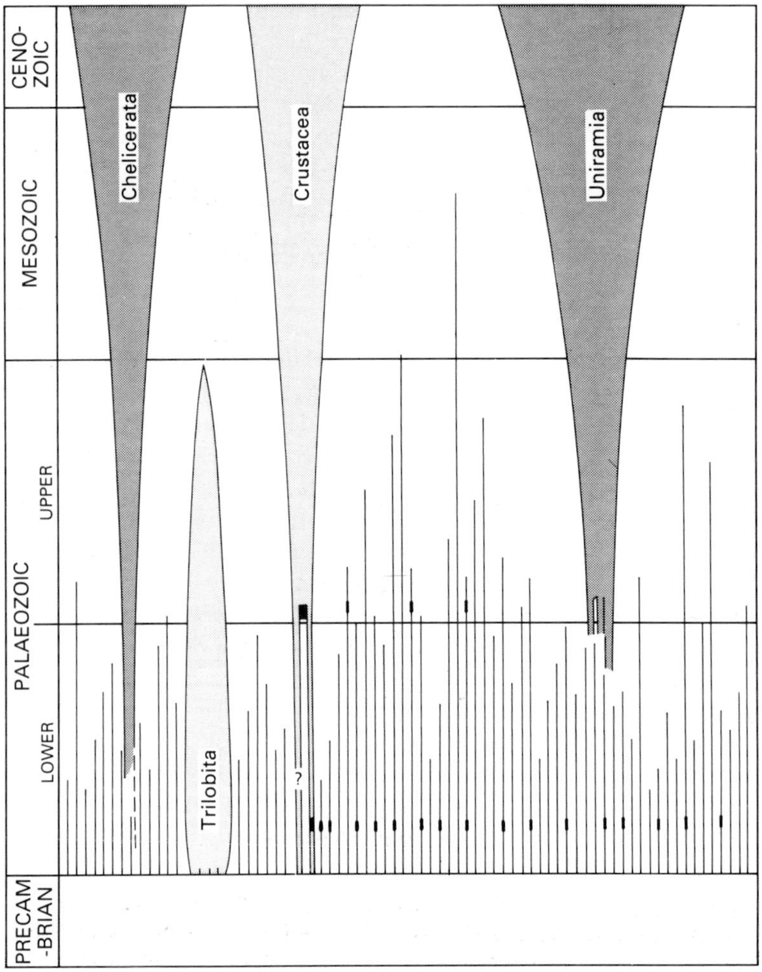

Fig. 2.11. The radiation of arthropod-like animals in the Precambrian/ Cambrian (after Whittington, 1979).

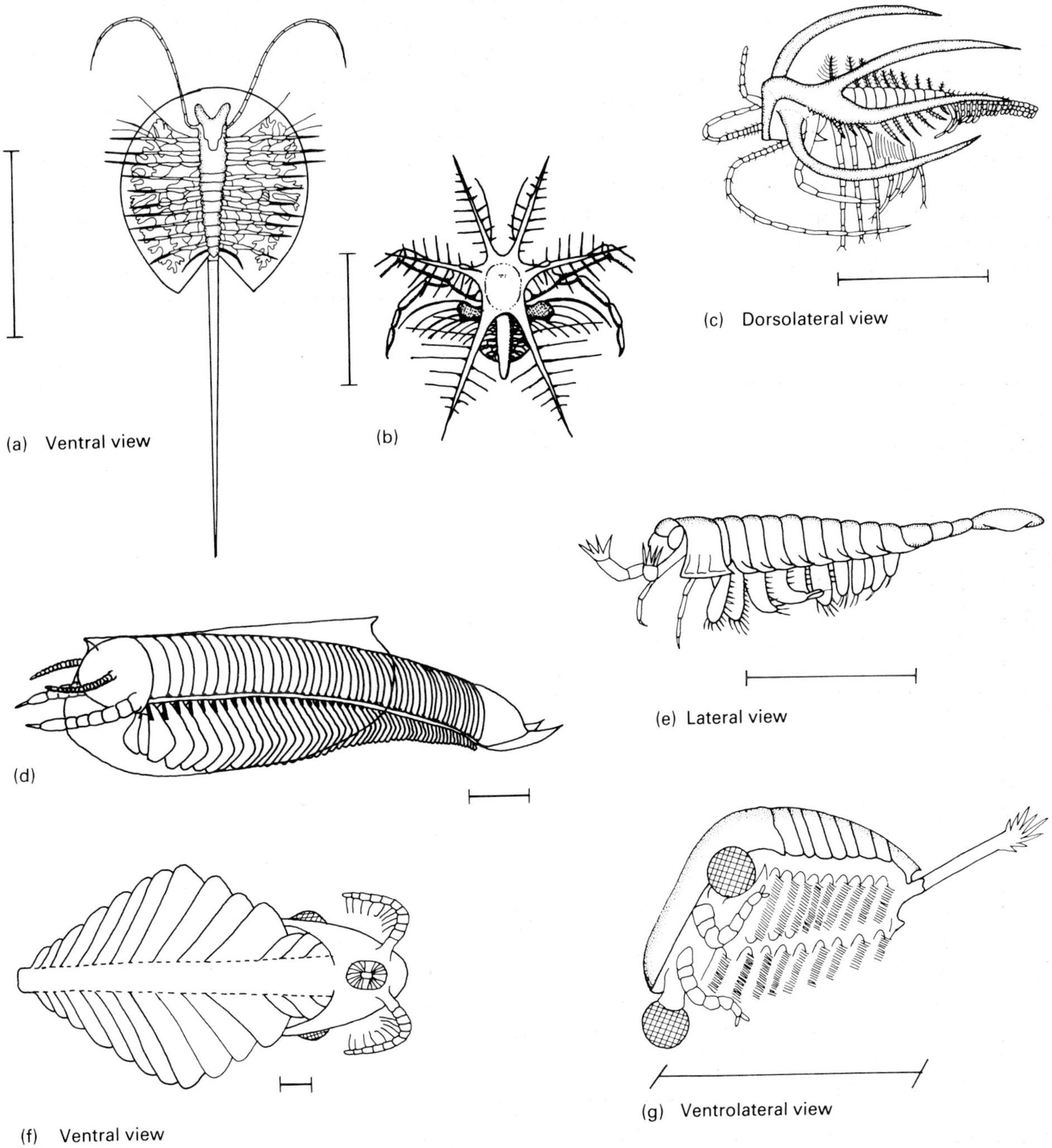

Fig. 2.12. Arthropod-like animals belonging to phylogenetic lines other than those of the Trilobita, Chelicerata, Crustacea or Uniramia: (a) *Burgessia* (Cambrian); (b) *Mimetaster* (Devonian); (c) *Marrella* (Cambrian); (d) *Branchiocaris* (Cambrian); (e) *Yohoia* (Cambrian); (f) *Anomalocaris* (Cambrian); (g) *Sarotrocercus* (Cambrian). (After Manton & Anderson, 1979 and Whittington, 1985.) Scale lines: 1 cm.

Section 2.3.2) greatly expanded to serve a hydrostatic function. There are also fundamental embryological differences between the annelids and the various arthropod groups, not that the latter show one single developmental plan (see, e.g. Manton & Anderson, 1979).

Few of the many arthropod groups survived beyond the Cambrian, and of those that did only four outlasted the Devonian. One of these, the trilobites, having survived for 300 million years at least, became extinct at the end of the Permian, leaving only three groups alive today. These three (the **Chelicerata, Crustacea** and **Uniramia**) differ particularly in the structure and arrangement of their legs and other appendages, as well as in their embryology and internal morphology, to such an extent that independent derivation from a vermiform ancestry is indicated. It is even likely that at least the crustaceans and the uniramians were derived from completely different groups of worms (Abele, 1982). (A case is sometimes made for regarding the pycnogonans as a fourth group of equivalent status; otherwise they are placed in the Chelicerata).

The ancestral arthropodized forms of all these surviving lines, however, were probably elongate, bottom-dwelling animals with many segments, each segment with one pair of appendages, those of the chelicerates and crustaceans being marine in habitat and that of the uniramians possibly being terrestrial. Uniramians are not known before the Silurian. Many members of all three groups have retained this ancestral, elongate body shape, with various degrees of differentiation of regions of the body, together with their appendages, to form a head and trunk, prosoma and opisthosoma, or head, thorax and abdomen, etc. Other arthropod lines, however, have greatly reduced the numbers of segments, by loss or fusion, and, to an even greater extent, the numbers of leg pairs. The ancestry of many of these lines, e.g. those of the marine and freshwater planktonic crustaceans, and of the uniramian insects, has been suggested to lie in the process of paedomorphosis. The copepods, for instance, may have arisen from the planktonic larvae of large, elongate, bottom-dwelling crustaceans. As we have already seen, there would be a selective advantage in remaining in the water column, where food is relatively abundant being near to the site of photosynthesis, rather than returning to the bottom sediments which are dependent on the rain of such unused organic material as remains.

Three living groups are regarded as being related in some way to the arthropods, perhaps as survivors of the original wave of arthropodization. Two of them (the **Onychophora** and **Tardigrada**) were mentioned above; the third is the **Pentastoma**. All three differ from the arthropods in their lack of an exoskeleton and jointed legs (the basic arthropod features). Neither are their bodies segmented, although pairs of fleshy legs, ending in claws, are repeated along part or the whole length of the body, each pair having its own ganglionic swelling on the nerve cord. The tardigrades possess four pairs of such legs, the onychophorans 14–43 pairs, and the pentastomes what can be interpreted as two highly reduced pairs (three pairs in their early larvae). In the onychophorans, other organs are also serially repeated, one pair to each leg 'segment'.

The pentastomes are worm-like endoparasites, and their presumed secondary reduction in bodily complexity renders phylogenetic assessment difficult—some recent research suggests an affinity with the crustaceans. Although the hydrostatic skeleton of all three groups is haemocoelic—the form suggested above for the ancestral worm groups—both the pentastomes and tardigrades have transient, paired enterocoelic pouches (three pairs in the former and five pairs in the latter), a feature otherwise confined to the lophophorates and deuterostomes (see below). These 'protoarthropod' phyla are therefore basically soft-bodied worms with some incipient arthropod features. All are now mainly terrestrial in habitat, although several tardigrades are marine and a marine onychophoran-like creature is known from the Cambrian. The Onychophora are regarded by many as being close to the origin of the Uniramia.

The nature of arthropod, protoarthropod, kinorhynch (and vertebrate) segmentation is thus essentially different from that of the annelids; the one being based on a monomeric non-compartmented body with multiple pairs of limbs (or, in vertebrates, of the muscle blocks powering swimming), and therefore with consequent serialization of the movement-associated skeletal, muscular, nervous and vascular elements, and the other on a linear chain of partially independent units, each with separate body cavities and sets of organs, including, primitively, gonads (Fig. 2.10). Further, the metamerism displayed by the various arthropod groups has probably been acquired independently from the ancestral group or groups of non-metameric worms. The end result, however, is superficially similar in the arthropod phyla and in the annelids, the body comprising a pre-segmental anterior portion ('acron' in arthropods, 'prostomium' in annelids), a series of metameric segments added during growth and/or development posteriorly, and a terminal post-segmental region ('telson' in arthropods, 'pygidium' in annelids). New segments are added during the development of kinorhynchs at a posterior proliferation region in an exactly comparable manner.

2.4.2 Lophophorates

All the flatworm descendants considered so far in this chapter, with one exception, tend to show certain common developmental features: (a) cleavage of the egg cells occurs in a spiral manner or in a modified version of this; (b) the blastopore forms the mouth, i.e. they are protostomes; (c) the developmental fate of the egg cells is fixed at a very early embryological stage (the 16-cell blastula), i.e. they are determinate; and (d) if the hydrostatic skeleton or cavities within or around specific organs are coelomic (are bounded by a mesodermal membrane), they form within mesodermal blocks of tissue by the process of schizocoely (Section 2.3.1, and see Sections 15.2–15.4)*. The exception is the group of worms mentioned briefly in Section 2.3.1 as having their bodies divided into three regions, each with its own body cavity—the **Phorona**.

Although protostome, the cleavage of phoronan eggs is radial and indeterminate, and some, at least, of their body cavities and mesodermal tissues are formed in a manner considered to be a modified form of enterocoely (Section 2.3.1, and see Sections 15.2–15.4). The living phoronans are small, sedentary worms which inhabit chitinous tubes. The first region of their tripartite body, the prosome, is very small and overhangs the mouth. The second part, the mesosome, is also small but it bears a large lophophore and contains a cavity, the mesocoel, which acts as the lophophoral hydraulic system; this lophophore comprises a series of hollow, ciliated, tentacle-like outgrowths of the body wall encircling the mouth which are primarily used for suspension feeding. The main region of the body is the appendage-less, non-segmented trunk or metasome containing a cavity, the metacoel, which forms the main hydrostatic skeleton. All three body cavities interconnect via pores. From such a worm as this are likely to have descended both the two other living lophophorate phyla and also possibly the deuterostomes. Both the lophophorates and deuterostomes show radial, indeterminate cleavage and tripartite, enterocoelic body cavities; and some of the lophophorates are possibly even deuterostomatous (the mouth being a secondary opening into the gut, not derived from the blastopore). Other common features include larvae with an upstream food-collecting system of tracts of monociliated cells around the mouth which, on metamorphosis, are transformed into the (monociliated) lophophoral arms, a mesosomal brain and mesoderm derived from the endoderm or vegetal-pole cells of the blastula, in contrast to the other bilaterally symmetrical animals in which it originates from the lips of the blastopore.

The **Brachiopoda**, which first appear in the Cambrian, have lost the small anterior body region (the prosome) and have greatly reduced the metasome, in comparison to the phoronans. Their mesosomal region, however, is large and supports a very well-developed lophophore. They are sessile, and instead of inhabiting tubes, the body is covered by a large calcareous or phosphatic shell like that of the bivalve molluscs except that the two valves are dorsal and ventral, not lateral. In some, extensions of the shell support the large and convoluted lophophore. It is

*Of the four, the first—spiral cleavage—is the least characteristic, occurring in only seven phyla; it is not displayed, for example, by gastrotrichs, nematodes or priapulans.

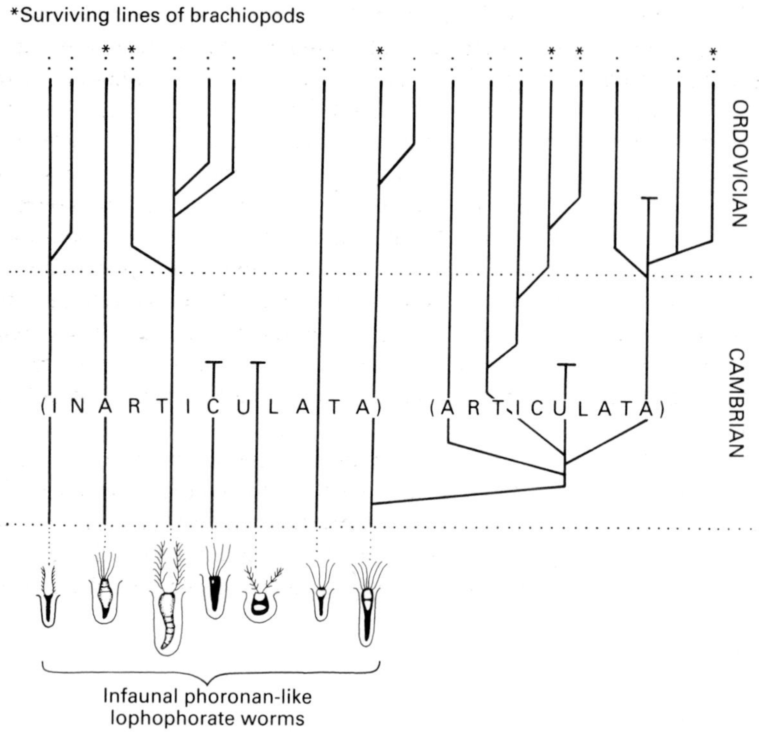

Fig. 2.13. Possible polyphyletic origin of the brachiopods as a wave of 'brachiopodization' of various phoronan-like worms in the Precambrian/Cambrian (after Wright, 1979).

possible that different groups of brachiopods are the descendants of different phoronan-like ancestors and that various phoronans underwent a phase of 'brachiopodization' in the Precambrian, in a similar radiation to the ancestors of the arthropods (Fig. 2.13).

Living phoronans can multiply asexually, one species by budding, and this is the norm in the third group of lophophorates, the **Bryozoa**. In the bryozoans, colonies, sometimes containing thousands of interconnecting individuals (zooids), are formed by repeated asexual budding of the originally founding individual and its descendants, in an exactly equivalent manner to that seen in several cnidarians (Section 2.2.2). Zooidal polymorphism is also a feature of several bryozoan colonies. Compared to the ancestral phoronan-like worm, individual bryozoan zooids are very small (increase in size being an attribute of the colony not the zooid), often lack the prosomal region, and secrete a shell, box or gelatinous matrix which encloses them apart from an aperture through which the lophophore can be extended for feeding and withdrawn for safety. The almost entirely sessile bryozoans can, therefore, be regarded as minute, colonial versions of a phoronan. Their appearance in the fossil record is relatively late: definite bryozoan material does not occur until the Ordovician.

Bryozoan relationships with the other lophophorate groups are still somewhat contentious, however. In part, this is because the method of formation of their body cavities is somewhat aberrant: in several species, the body cavities of the adults are formed *de novo* during metamorphosis. Some have suggested that their relationships lie with an obscure group, the **Entoprocta**, with which they share highly individual types of photoreceptor. These spirally cleaving, minute, sessile animals bear a tentacular ring like the lophophorates, although it does not conform to the structure of a true lophophore, and they possess a non-coelomic, mesenchyme-filled body cavity. Other suggested affinities of the Entoprocta include the rotifers, but there is little available evidence to support any specific relationships for this phylum.

2.4.3 Deuterostomes

Somewhat paradoxically, although the four deuterostome phyla (the **Hemichordata, Echinodermata, Chordata** and **Chaetognatha**) form a closely knit group, sharing many common features, almost nothing is known of their interrelationships. The ancestry of the group as a whole is rather more certain in that the differences between a phoronan-like lophophorate and the pterobranch hemichordates are not marked. The body of a pterobranch is divided into three regions, each containing enterocoelic cavities (a single one in the prosome, and paired cavities in the meso- and metasomes), and the mesosome supports a large hydraulically operated lophophoral apparatus. The only major differences are that in the pterobranchs the prosome is large and is used in movement, its enterocoelic cavity (protocoel) forming the required hydrostatic skeleton, and development is deuterostomatous. In view of this similarity, it is widely believed that the lophophorate pterobranchs are similar in essential body plan to the ancestral form of the deuterostomes; the other group of the hemichordates, the worm-like enteropneusts, are then regarded as having lost their lophophore in association with their return to a free-living, burrowing life style.

Although appearing markedly dissimilar, largely as a result of their early adoption of a spherical body with a near radial form of symmetry (in conjunction with a continued sessile, filter-feeding existence), the echinoderms retain in their body cavities clear indications of a basic tripartite bodily organization comparable with that of the lophophorates and hemichordates. One protocoelomic cavity forms the axial sinus; the left mesocoel and left protocoel together form the water vascular system, whilst the right mesocoel has been largely suppressed except in aberrant individuals; and the two metacoels form the main body-cavity compartments. The echinoderms are one of the many groups first to appear, clearly recognizable as to their phylum, in the Cambrian, although the early forms did not uniformly display the fivefold (pentaradial) symmetry that characterizes the surviving echinoderm classes. The ancestral body plan is purely a matter of conjecture. One of the more plausible hypotheses derives them from a sessile pterobranch-like animal in which the lophophore and its hydraulic mesocoel developed into the radial canal system and its hydrostatic water vascular skeleton—some modern pterobranchs even have a lophophore with five main trunks on each side.

Besides the common features shown by all deuterostomes, echinoderms share with the chordates a hard, mesodermal, protective, calcium-based system of plates in the dermis, and some extinct echinoderms may have had the equivalent of pharyngeal clefts, a

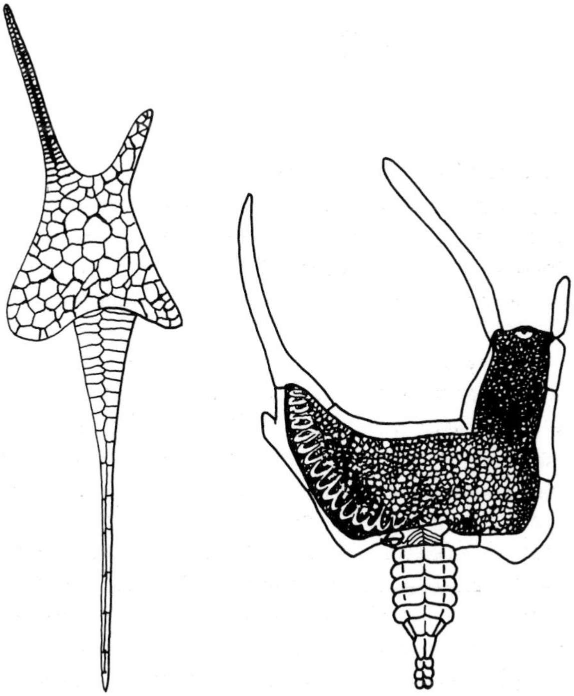

Fig. 2.14. Homalozoan echinoderms (after Clarkson, 1986): the ancestors of the chordates?

feature also found in the surviving hemichordates (and in some gastrotrichs). Not surprisingly, therefore, most attempts to derive the chordates from another animal group have implicated the echinoderms as being near to the point of origin. But, as with the hemichordate/echinoderm transition, all such suggested phylogenies come up against a lack of potential intermediate stages; no living species can serve as appropriate analogies and the fossil record is of no help unless the peculiar homalozoans of the Cambrian to Devonian (Fig. 2.14), which lack any form of symmetry, are ancestral chordates as has been claimed. Others seek to derive the chordates by processes of permanent larvalization of hemichordate/echinoderm larvae. Undoubted chordate material dates back to the Ordovician, although the case for chordate status of one Cambrian fossil and several late Cambrian phosphatic fragments has been argued, and, should it become generally accepted that the enigmatic conodonts are part of the rasping organs of agnathan vertebrates, as currently seems likely, then the Cambrian, and even Precambrian, status of the chordates is assured.

The fourth deuterostome phylum, the Chaetognatha, is as phylogenetically obscure as it is ecologically important; chaetognaths are perhaps the most important invertebrate predators of the marine plankton. The early embryology of these transparent, soft-bodied worms is typically deuterostome, and they possess a tripartite body plan of head, trunk and post-anal locomotory tail. Whether these three regions correspond to the hemichordate pro-, meso- and metasome, however, is more problematic, and some chaetognath features appear distinctly similar to those of the pseudocoelomate groups. On present knowledge, nothing can be surmised as to their precise relationships with the other deuterostomes. Soft-bodied members of the plankton are the most under-represented of all animals in the fossil record and it is quite possible—even highly likely—that there have been considerable numbers of these types of animals in the past (see, for example, Fig. 2.16d and e), the existence of which we are completely unaware (see Section 2.5); chaetognath ancestors probably belonged to one such group.

2.5 The origin, radiation and extinction of animal groups

Living animals can be placed in a total of 35 phyla; a phylum being defined pragmatically as a group of organisms that appear to be related one to each other, but whose relationships with other such groups are debatable or purely conjectural. In essence, therefore, phylum status is an admission of our ignorance, although the foregoing sections have endeavoured to trace likely patterns of affinity. These 35 living phyla, however, are not the end result of a long, slow process of evolutionary divergence which has culminated in the diversity that we see today. It seems probable that with very few possible exceptions, all the surviving animal phyla were in existence during the Cambrian, and that the diversification of the flatworm-like antecedents of the majority of groups occurred in the Precambrian. This presents something of a conundrum.

The fossil record documents that, over time, animals change slowly such that the forms present at the end and at the beginning of a given sequence may be sufficiently different as to be regarded as different species (termed different 'chronospecies'). The average time required for anatomical change of this magnitude across a wide range of invertebrates is of the order of 10 million years. If we now assume that different animal genera are, say, ten times as distinct from each other as are different species within any one genus, that different families are ten times as distinct from each other as are different genera within any one family, and so on, through orders, classes and phyla, then, on the basis of the average rate of change shown by chronospecies, it would require in excess of one million million years for two different species to diverge from each other to such an extent that they would be considered to belong to different phyla.

The first animals are likely to have evolved from their protist ancestors some 1000 million years ago, and so the above calculation would lead us to predict that their descendants will have diverged sufficiently to form a number of different phyla only some 1.1×10^{12} years into the future! By today, their descendants should only be as different from each other as are, for example, the various families of beetles. Even if we were to adopt an average time interval of only one million years between successive chronospecies, and to reduce the divergence factor between species and genus, genus and family, etc., from ten to five, it would still require over 3000 million years for the ancestral flatworms to diversify into a number of phyla. Yet the phyla that we know today were already present, as distinct from each other as they are now, only 400 million years after the time of likely origin of the first multicellular animals, and they probably diverged from each other in a much shorter interval of time. No wonder one recent author (Day in his book *Genesis on Planet Earth*, 1984) has commented: 'The abrupt appearance of the Cambrian fauna constituted a major biological problem . . . . The phyla that were present appeared quickly with no apparent origin.'

Something is clearly wrong with the evolutionary scenario presented in the above paragraphs, and that 'something' is that not all evolutionary change is as slow as that recorded in the fossil record within various species. Bursts of rapid change in morphology have occurred, and they have taken place so quickly that the intermediate stages have escaped fossilization. It may also have been that these events have occurred in relatively small populations (the rate of potential evolutionary change is inversely proportional to population size), further decreasing the likelihood of preservation in the fossil record. Permanent larvalization (see above), for example, could be accomplished between one generation and the next, producing in only a few years a change of such magnitude in the adult body form as to result in the permanently larval descendant being regarded as belonging to a different order or class from its parent stock. (Taxonomic categories are therefore often measures of morphological distinctiveness, as implied above, but are not necessarily measures of recency of common ancestry—this is a position that phylogeneticists of the 'cladistic school' seek to rectify; were they to achieve this, few if any of the phyla and classes in common use would remain unscathed—Fig. 2.15.)

These bursts of rapid change have often been associated with the generation of much diversity at the species, genus, family or even order level from one single founding type. Such 'adaptive radiations' also occur very rapidly in geological time, and the manner in which several new types of animals evolve in short spaces of time from a single stock has given rise to: (a) notions that the processes controlling change within a single population (microevolution) are qualitatively different from those resulting in the formation of taxonomic novelty and variety (macroevolution); and (b) theories that the rates at which micro- and macroevolutionary events occur are quantitatively different, macroevolutionary change being achieved more rapidly (see the 'gradualist' and 'punctuated equilibrium' models, Chapter 17).

Not all the phyla that were produced in the extensive Precambrian and Cambrian radiations are represented amongst living invertebrates. Some additional phyla are known only from the fossil record, e.g. those of the archaeocyathans and the trilobites, and especially, although they are not usually given formal

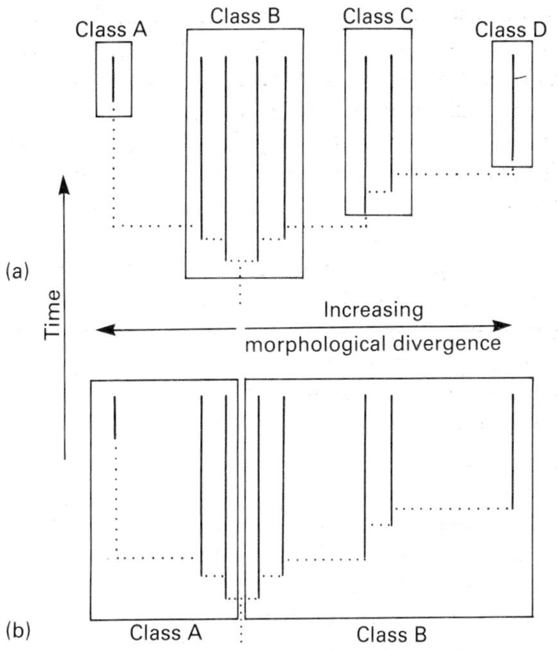

Fig. 2.15. The same hypothetical phylogenetic tree presented in two different forms: (a) classified along traditional lines biased towards morphological similarity and difference; (b) classified according to cladistic principles.

phylum status, the Cambrian arthropod host (Section 2.4.1). There are surprisingly few extinct phyla known, however, and most of them are characterized by the possession of hard parts—features which are likely to have appeared relatively late in any given lineage. The fossil record may, therefore, give a false impression of past animal diversity. This is dramatically emphasized by the abundance and variety of novel types of soft-bodied animals present in those very few deposits known in which this category of

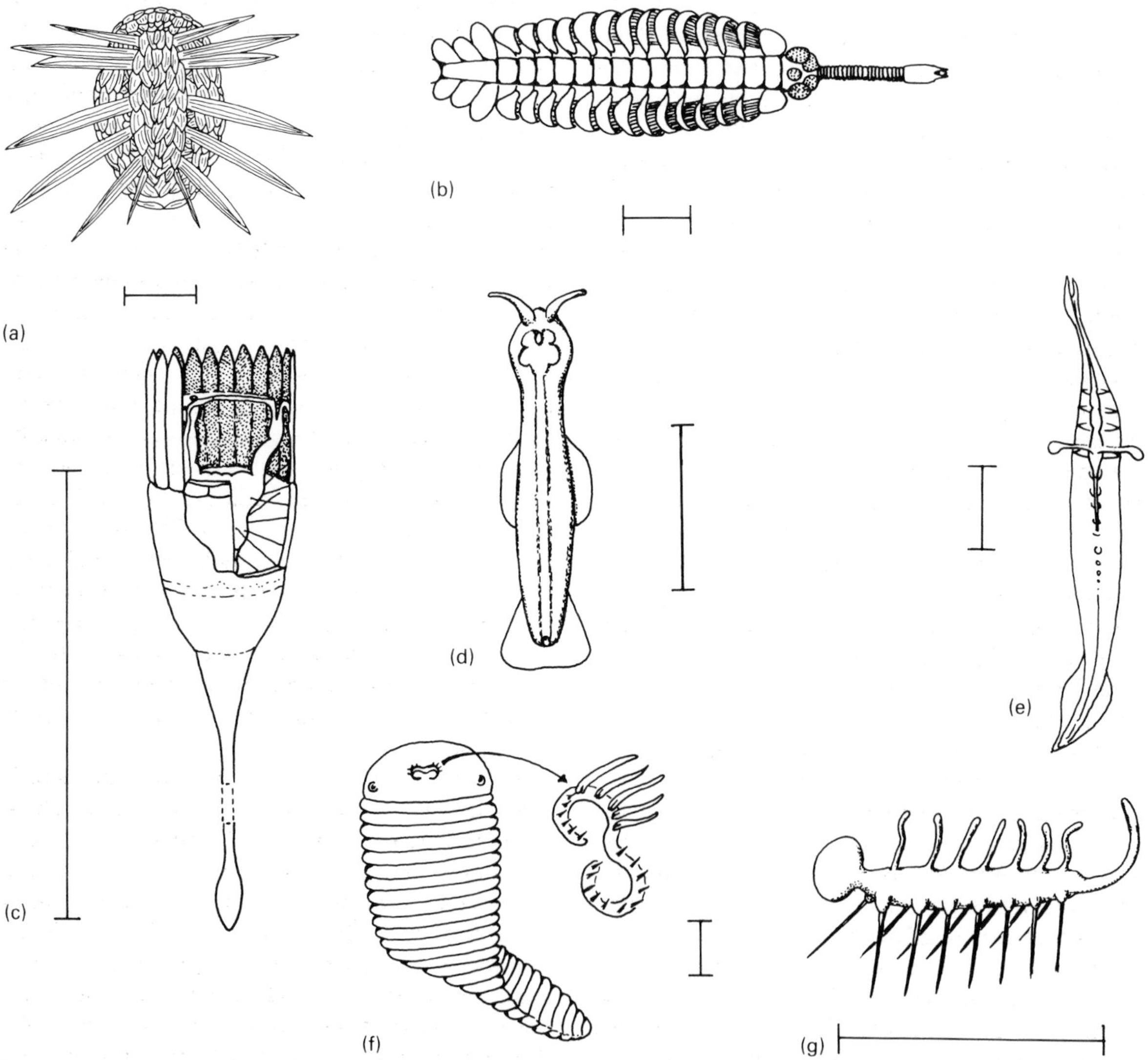

Fig. 2.16. Extinct and mostly soft-bodied animals which can be placed in no known phylum: (a) *Wiwaxia* (Cambrian); (b) *Opabinia* (Cambrian); (c) *Dinomischus* (Cambrian); (d) *Amiskwia* (Cambrian); (e) *Tullimonstrum* (Carboniferous); (f) *Odontogriphus* (Cambrian); (g) *Hallucigenia* (Cambrian). (After Clarkson, 1986; Conway Morris, 1979, 1985 and Whittington, 1985.) Scale lines: 1 cm.

organisms has been preserved (e.g. the Precambrian Ediacaran and Cambrian Burgess Shale beds) (Fig. 2.16).

Twenty living phyla, often termed the 'minor phyla', contain few species (less than 500 each) and, with one exception which proves the rule, being soft-bodied and worm-like have left little or no fossil record. The exception is the Brachiopoda with only some 350 living species but with over 25 000 fossil forms known from their hard, external shells. Attempts to account for the fact that these small phyla have managed to persist for so long whilst containing so few species have also suggested that this is statistically likely only if very many more similar (i.e. soft-bodied) phyla were once in existence but have since become extinct. (We have already noted this possibility in connection with the ancestors of the chaetognaths; Section 2.4.3.) The known extinct phyla may therefore represent a small and, by virtue of their possession of hard parts, atypical sample: the arthropod host is perhaps only the fossilized tip of the Precambrian and Cambrian iceberg of extinct groups.

A picture is therefore emerging of a large-scale radiation of vermiform animals in the Precambrian and Cambrian which produced maybe hundreds of groups equivalent to phyla. Most of these were probably small in terms of numbers of comprising species, and the majority of them became extinct without trace and without living descendants. The living vermiform host is but a small remnant of this one-time diversity. Some lineages evolved hard protective or skeletal parts, probably as a consequence of predation pressure, and of these we have considerable knowledge: it has been estimated in respect of fossil forms of these animals that some 12% of all species which have ever existed have now been found and described. A few groups radiated to give rise to large numbers of species and much intra-phylum diversity. Today, ten phyla are each represented by more than 10 000 living species, whilst, in addition, the brachiopods and echinoderms once achieved this level of diversity but have declined to less than this total now. These form the so-called 'major phyla'.

Even though some phyla have maintained high levels of species diversity over long periods of geological time, it has not usually been achieved by the same types of animals within any one phylum. Different component orders or classes have dominated different periods of time, and several once dominant classes are now extinct, having been replaced by others. Various subgroups have radiated in turn. This phenomenon is most familiar through the now extinct ammonites (Fig. 2.17) and almost extinct nautiloids within the molluscs, and even more so through the 'ages' of amphibians, reptiles and mammals amongst the vertebrates, but it is equally characteristic of other phyla, e.g. the cnidarians, brachiopods, echinoderms and non-cephalopod molluscs. In general, replacement of one subgroup by another does not seem to have occurred by competitive exclusion, such as might be expected to result from the evolution by the replacing subgroup of some morphological or other breakthrough that would confer a selective advantage. Instead, one type of animal has first become extinct, and only later, presumably under conditions of an ecological vacuum, has its replacement radiated to fill the ecological space created by the demise of the hitherto dominant group or subgroup. The fossil record indicates that such rapid radiations into numerous types (the punctuation phase of the punctuated-equilibrium model on a larger scale) is then followed by a much longer period of relatively little morphological change under the influence of stabilizing selection (equivalent to the equilibrium phase), until that group in turn suffers drastic reduction in numbers of species or complete extinction.

The extent to which different types of animals characterizing a wide range of niches and habitat types have become extinct at the same time is somewhat contentious, although a sizeable body of opinion favours the view that mass extinctions have occurred at various times in the past, especially towards the end of the Ordovician, Devonian, Permian, Triassic and Cretaceous periods (Fig. 2.18), but also to a greater or lesser degree every 26 million years or so. That which took place some 220 million years ago, at the end of the Permian, for example, involved the extinction of over half the then existing families of marine animals, and perhaps 80–95% of the then

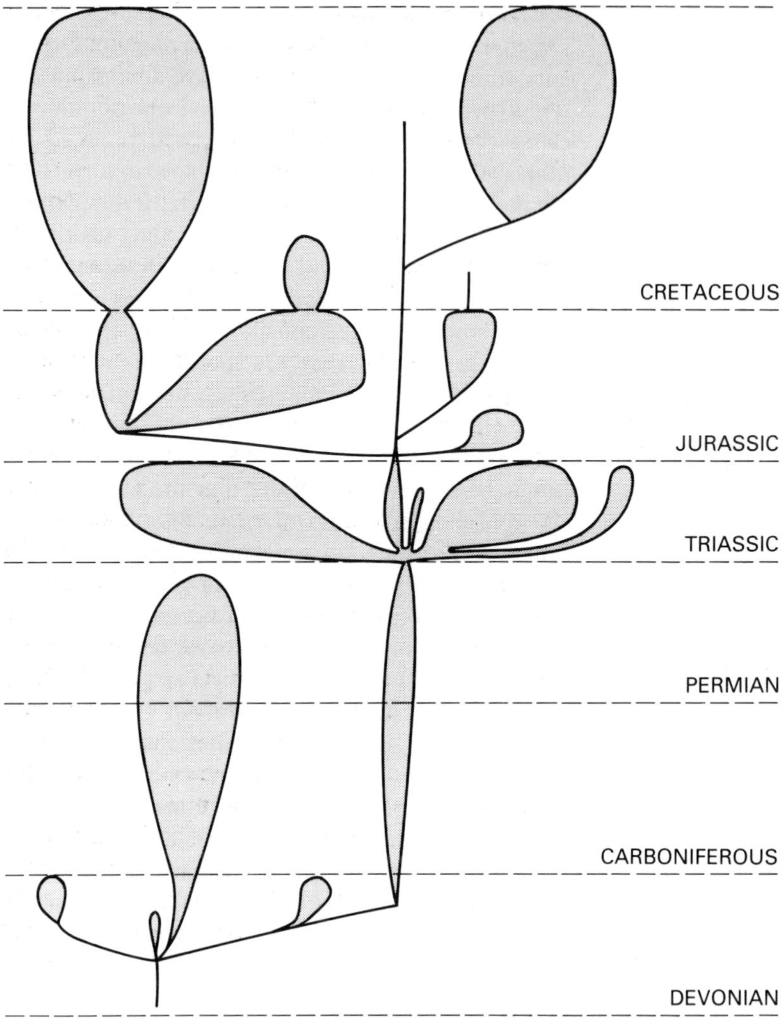

Fig. 2.17. Phylogenetic tree of the ammonites, showing radiation, extinction and replacement of one subgroup by another from the Devonian to the end of the Cretaceous, when the whole group became extinct (from data in Moore, 1957).

species. Hitherto dominant groups, including the eurypterid chelicerates and the trilobites, wholly disappeared, leaving at the beginning of the Mesozoic a fauna essentially modern in terms of the general types of animals dominating the seas. The causes of these extinctions are unknown, although they correlate well with major changes in sea level, which may have been brought about by glacial phases or continental movements, and/or with pulses of global cooling. There have, however, been numerous speculative attempts, both popular and scientific, to implicate various cosmic forces and events.

The history of animal life may therefore have been distinctly episodic. Each episode, including the one in which we are now living, being typified by a random mixture of expanding and declining groups of animals, with some which as a result of chance have managed to persist, albeit with very few species, and with others which although perhaps insignificant now will have an opportunity to increase and diversify

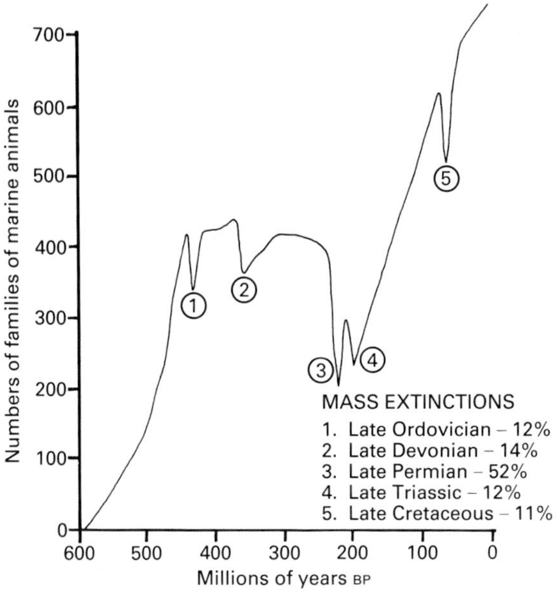

Fig. 2.18. The numbers of families of marine animals at different periods in the past, showing a series of mass extinctions (after Valentine & Moores, 1974 and others).

after the next phase of extinctions, human-induced or otherwise. With the passage of time, there has been an impoverishment in the number of animal phyla as some (and possibly many) have become extinct and few new ones have formed, although the number of animal species alive today is greater than it has ever been. The seas of the world are currently relatively fragmented, shallow semi-isolated shelf-seas are abundant both geographically and areally, and the ocean is oxygenated right down to its abyssal bed—circumstances which have not always prevailed; in the Cretaceous, for example, large areas of sea bed may have been anoxic. There is no reason to believe, therefore, that marine species are not as diverse now as at any time in the past, and indeed most estimates would make the late Caenozoic marine fauna the richest ever.

This notwithstanding, the enormous number of living species is almost entirely a reflection of the conquest of the land that began in the Silurian/Devonian and was consolidated in the Carboniferous. Somewhat surprisingly, the route which this colonization took appears largely to have been up the intertidal zone and thence on to the land, not the aquatic one of sea→estuary→river (→lake) →land, in that most terrestrial animals have high fluid concentrations and considerable dessication resistance, not vice versa. (The main groups to have used the less popular aquatic route were the vertebrates and some molluscs.) Invasion of the freshwater habitat was then largely accomplished by a suite of animals already adapted to terrestrial conditions. The small interstitial protists and animals (ciliates, nematodes, tardigrades, etc.) have never left the water films surrounding sediment or soil particles; they are terrestrial only in the sense that their microaquatic habitat also occurs on land. Which of the two routes they used is unknown and perhaps unknowable. There are many more barriers to dispersal on the land and this has resulted in wholesale speciation within the relatively few phyla which have become successfully terrestrial (for the same reason, there are far more benthic than pelagic marine species). Many terrestrial species are geographically replacing versions of one basic animal type. The land chelicerates and uniramians, especially the insects, are now the dominant invertebrates on the planet in terms of numbers of species (Fig. 2.19), so that in the last 350 million years the focus of organismal diversity at the generic and specific levels has moved away from the sea, in which the whole of the phylogenetic story told in this chapter has taken place, out onto the land.

2.6 Summary and evolutionary overview

Some 1×10^9 years ago, several lines of heterotrophic protists living on the sea bed probably crossed the extremely ill-defined boundary between coloniality and multicellularity, and thereby gave rise to 'animals'. During the next few hundred million years, descendants of one of those lines underwent a series of large-scale radiations, probably driven mainly by the existence of unexploited food sources, by predation pressure and by the ecological and evolutionary advantages attendant on larger body size (see Section 9.1). These radiations produced a wide variety of

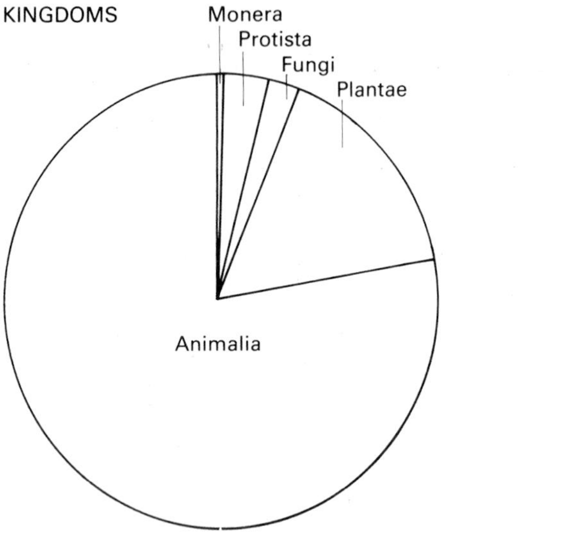

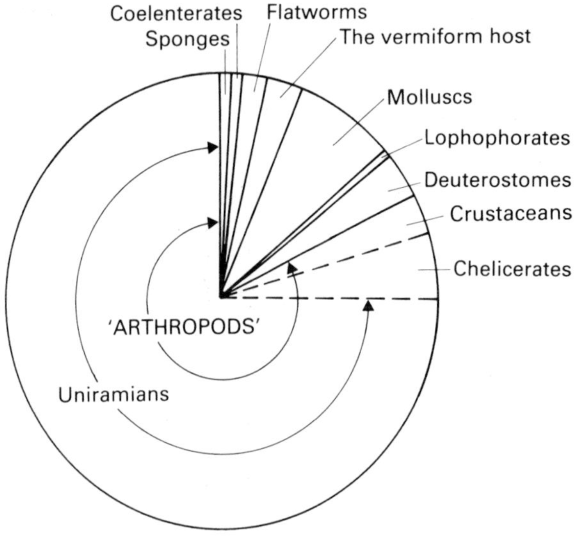

Fig. 2.19. The numbers of living species in the various kingdoms and in the animal phyla (from data in Barnes, 1984).

benthic worm- and mollusc-like animals, and from some of the worms evolved the arthropod, lophophorate and deuterostome groups (Fig. 2.20), so that by 6×10^8 years ago all the known phyla—and many unknown ones—were probably in existence. Since that time, however, different phyla and their component classes and orders have waxed and waned in ecological importance. Palaeoecologists, for example, have, on the basis of groups with hard fossilizable parts, identified three great marine faunas as having dominated the seas: a Cambrian/Ordovician one (of trilobites and other early arthropods, inarticulate brachiopods, and monoplacophoran molluscs); a late Palaeozoic (Ordovician–Permian) assemblage of shelled cephalopods, crinoid echinoderms, articulate brachiopods and stenolaematan bryozoans; and a Modern (Mesozoic and Caenozoic) fauna characterized by the importance of gastropod and bivalve molluscs, malacostracan crustaceans, echinoid echinoderms, gymnolaematan bryozoans and the vertebrates. Each fauna appears first to have become established in shallow-water environments and later to have been displaced (?) into deeper, outer shelf and Continental Slope zones.

Prominent in some of the later stages of these radiations of animal types was an invasion of the marine water column, permitting the phytoplanktonic production of the surface waters to be tapped at source. A major factor allowing animals to leave the sea bed was that of paedomorphosis (permanent larvalization—see Chapter 17), which carries with it the inherent reproductive advantages consequent on a reduced generation time. Changes to the relative timing of developmental processes (i.e. mutations of 'regulatory genes'), such as those accomplishing paedomorphosis, are a powerful means of achieving rapid major morphological change: within a few generations, adult anatomy could alter to such an extent that the ancestral form and its descendant would be so distinct as to be assigned to different classes or even phyla in classification schemes biased towards morphological distinctiveness. Some modern marine larvae are known in which, presumably in association with long-postponed settlement, gonad development has begun in advance of metamorphosis, indeed some larvae of one species of ctenophore have recently been observed spawning perfectly viable gametes, as have various bivalve mollusc spat; these perhaps indicate how such permanently juvenile groups might have

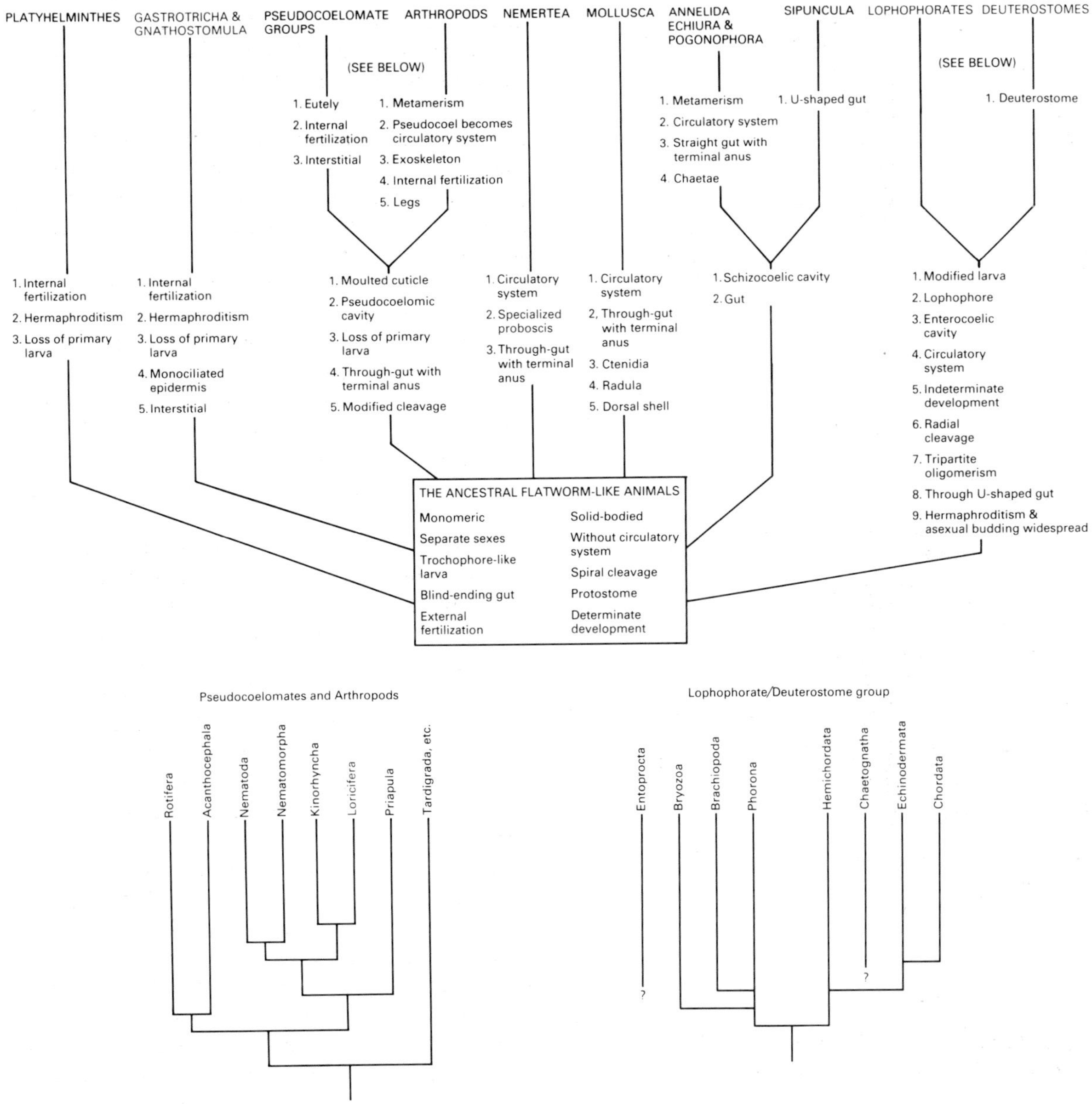

Fig. 2.20. Speculative general patterns of relationship within the descendants of the ancestral flatworm-like animals, showing ancestral features (retained in descendants unless stated otherwise) and the major secondary specializations characteristic of the various groups or clades (at least primitively). Detailed relationships within the pseudocoelomate/arthropod and lophophorate/deuterostome lines are shown separately.

originated. Thus might some major 'macroevolutionary events' occur.

The final chapter in animal diversification took place some $3\text{–}4 \times 10^8$ years ago when animals with legs effected their colonization of the land. Compared with the ancestral sea, the land is geographically a highly fragmented habitat with abundant barriers to animal dispersal. Land animals therefore speciated to an extent hitherto unknown, so that although very few major *groups* were able successfully to exploit this new environment, now the vast majority of animal *species* are terrestrial.

One rather surprising feature of the history of the major animal phyla is that although one component subgroup within each phylum has tended to succeed another in time, the replacing subgroup rarely seems to have displaced the preceding one by competitive exclusion. Instead, the earlier one has first become extinct, or nearly so, and only after its demise has the replacing subgroup radiated to fill the apparently vacant niches. Perhaps equivalently, there is little evidence from modern ecological studies that anatomically complex animals are ousting morphologically simpler ones from shared habitats: segmented arthropods and annelids, for example, still coexist in the same patches of sediment today with 'primitive' priapulans and sipunculans without any signs of competitive exclusion.

The temptation to regard flatworms and coelenterates, for instance, as in some way inferior to crustaceans or chordates with their much more complex structure and behaviour is therefore to be avoided. The variety of animal types produced by the great Precambrian and Cambrian radiations are all more or less equally ancient, alternative body plans; they do not form a ladder of increasing adaptedness or 'perfection'. To put it another way, all life probably evolved from early Precambrian prokaryotes and some of their descendants are now complex oak trees, elephants and men, but these have in no way 'superseded' the prokaryotes; bacteria still successfully dominate most habitat types!

The extinction of the phyla and of their subgroups referred to above appears to have had an important stochastic element, as indeed did the identity of the replacing types of organisms. As implied earlier, groups did not die out because they were in some way inferior body plans, but whole suites of unrelated and well-adapted animals seem to have succumbed to drastic changes of their environment—changes of an as yet unknown nature, but correlating well with major changes in sea level and global temperature. Many phyla with few component species probably failed to survive these phases of extinction, and the larger groups, purely by virtue of their more numerous and widespread members, and greater geographical and ecological ranges, managed to persist albeit with reduced diversity, and were therefore in a position to radiate again after the event. Thus, although there is no question that interspecific competition is an ecological reality, in general it is the more intense the more closely similar the competing species are; and hence it is perhaps not to be unexpected that the persistence through time of the higher taxonomic groupings, such as orders, classes and phyla, should be largely a matter of chance.

Members of the groups surviving these major extinctions would initially have experienced relatively

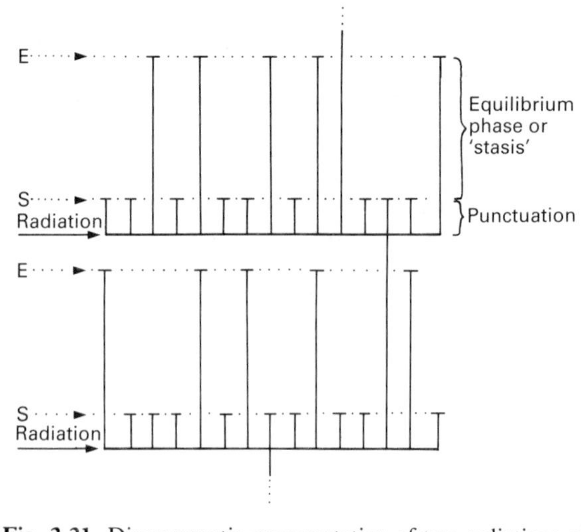

Fig. 2.21. Diagrammatic representation of two radiations of the descendants of one phylogenetic line to show the punctuation and equilibrium phases, and the declines in diversity of the group induced by environmental mass mortality (E) and by the reimposition of heavier selection pressures on saturation of the habitat (S)—see text.

empty habitats in which selection pressures are likely to have been, temporarily, greatly reduced. Under these conditions, subsequent radiation would have been largely unrestricted and much 'experimental' diversity generated could persist, only somewhat later to suffer a series of extinctions once saturation of the habitats was again achieved and heavier selection pressures were reimposed. Selection can therefore be envisaged as limiting the amount of diversity formed in each radiation which could outlive the phase of expansion itself (Fig. 2.21; see also Fig. 2.11). It was, then, these selected surviving lines within each group or subgroup which dominated the time interval until the next environmentally induced phase of mass extinction initiated the whole cycle again to a greater or lesser degree.

Chance, therefore, determines the play enacted, provides the initial potential cast and brings down the curtain; selection auditions for the filling of the vacant parts and oversees the continual adjustments made to the script and characters.

2.7 Further reading

Abele, L.G. (Ed.) 1982. *The Biology of Crustacea (Vol. 2). Embryology, Morphology and Genetics.* Academic Press, New York.

Barnes, R.S.K. (Ed.) 1984. *A Synoptic Classification of Living Organisms.* Blackwell Scientific Publications, Oxford.

Bryce, D. 1986. *Evolution and the New Phylogeny.* Llanerch, Dyfed.

Clarkson, E.N.K. 1986. *Invertebrate Palaeontology and Evolution,* 2nd edn. Allen & Unwin, London.

Cohen, J. & Massey, B.D. 1983. Larvae and the origins of major phyla. *Biol. J. Linn. Soc., Lond.,* **19,** 321–8.

Conway Morris, S., George, J.D., Gibson, R. & Platt, H.M. (Eds) 1985. *The Origins and Relationships of Lower Invertebrates.* Clarendon Press, Oxford.

Glaessner, M.F. 1984. *The Dawn of Animal Life.* Cambridge University Press, Cambridge.

Goldsmith, D. 1985. *Nemesis. The Death Star and Other Theories of Mass Extinction.* Walker, New York.

Hanson, E.D. 1977. *The Origin and Early Evolution of Animals.* Wesley University Press, Middleton, Connecticut.

House, M.R. (Ed.) 1979. *The Origin of Major Invertebrate Groups.* Academic Press, London.

Manton, S.M. & Anderson, D.T. 1979. Polyphyly and the evolution of arthropods. In: M.R. House (Ed.) *The Origin of Major Invertebrate Groups,* pp. 269–321. Academic Press, London.

McKerrow, W.S. (Ed.) 1978. *The Ecology of Fossils.* Duckworth, London.

Scientific American 1982. *The Fossil Record and Evolution.* Freeman, San Francisco.

Sleigh, M.A. 1979. Radiation of the eukaryote Protista. In: M.R. House (Ed.) *The Origin of Major Invertebrate Groups,* pp. 23–53. Academic Press, London.

Trueman, E.R. & Clarke, M.R. (Ed.) 1985. *The Mollusca (Vol. 10). Evolution.* Academic Press, Orlando.

Valentine, J.W. (Ed.) 1985. *Phanerozoic Diversity Patterns.* Princeton University Press, Princeton, New Jersey.

Whittington, H.B. 1985. *The Burgess Shale.* Yale University Press, New Haven.

For a phylogenetic view contrasting with the one presented here, especially in respect of the affinities of the Ctenophora, Bryozoa and Chaetognatha, and of the suggested place of origin of basal animal groups, see:

Nielsen, C. 1985. Animal phylogeny in the light of the trochaea theory. *Biol. J. Linn. Soc., Lond.,* **25,** 243–99. (Nielsen envisages that the ancestral forms of the major lines of animal evolution were radially symmetrical, planktonic organisms, and that bilateral symmetry arose separately in the lophophorate/deuterostome and protostome lines.)

II
THE INVERTEBRATE PHYLA

In the chapters of this section we illustrate and describe briefly all the known phyla of invertebrates with living representatives, together with their component classes.

You will find that our systematic survey is much less detailed than in many textbooks of invertebrate zoology, since, rather than describe all the various anatomical features of the different types of animals, we have endeavoured to distil only those essential characteristics of each group with which the student should be familiar, i.e. those responsible for the singularity of the phylum or class and for its evolutionary success or ecological significance. We have also provided lists of diagnostic features to permit comparison of the individual phyla.

Further, in this section our accent is on the diversity of invertebrate body plans, as reflected in the major animal taxa; we have therefore treated all groups equally by granting the same degree of coverage to all the classes of invertebrates, regardless of the number of species which they may contain. The systems-based chapters of Section III, however, will redress this bias by drawing their material largely from the bigger, better-known and, arguably, more important groups.

The phyla, classes, etc., recognized in this section are basically those of Barnes (1984) A Synoptic Classification of Living Organisms, *Blackwell Scientific Publications, Oxford. You may care to note that new types of animals are still being found–since 1980, one new phylum (the Loricifera) and two new classes in supposedly well-known phyla (Remipedia, phylum Crustacea, and Concentricycloidea, phylum Echinodermata) have been described. Whilst, therefore, our coverage of animal classes is complete as at 1987, there is no reason to believe that man has now discovered all the major groups of animals surviving on the planet; it is likely that as little-known habitats such as the deep sea and interstitial zones are investigated in greater detail, so will new, and sometimes radically different, types of animals be seen for the first time. The last word on the range and diversity of animals will not be said for a long time yet!*

3

PARALLEL APPROACHES TO ANIMAL MULTICELLULARITY

'The Protozoa'
Porifera
Placozoa
The Coelenterates (Cnidaria and Ctenophora)
Platyhelminthes
Mesozoa

The groups of animals considered in this chapter are mainly those which it was suggested in Chapter 2 could lay claim to have been derived directly from within the kingdom Protista, in that they possess fundamentally different forms of bodily organization. They can be viewed, therefore, as a series of parallel evolutionary lines, which may have achieved the multicellular animal condition independently of each other.

3.1 'The Protozoa'

Until the early 1970s it was customary to allocate all living organisms to one or other of two kingdoms. If an organism was photosynthetic, or grew in the ground, it was 'a plant'; if it did not photosynthesize and was freely mobile, it was 'an animal'. Whether an organism was prokaryotic or eukaryotic, or whether uni- or multicellular, were not considerations relevant to this fundamental division of life. Within this classification, the multicellular plants then formed one group, the Metaphyta, whilst the unicellular photosynthesizers were termed the Protophyta or Algae (a category which included various prokaryotes). Comparably, the multicellular animals comprised the Metazoa, and the single-celled animals the Protozoa.

Such a two-kingdom system clearly suffers from a number of drawbacks, of which two are major, First, it does not reflect any real phylogenetic dichotomy: there is no evidence at all to suggest that the cyanobacteria, diatoms, red seaweeds, toadstools and conifers, for example, all belong to one natural group, in contrast to the amoebae, gut flagellates, sponges and echinoderms which all belong to a second. Such artificiality need not be a significant problem, however, if the aim of a classification is purely pragmatic—the two kingdoms might still provide an unambiguous, convenient but arbitrary pair of mutually exclusive divisions of living organisms. But the second drawback is that the two categories of plant and animal are not really distinguishable. Within a group of apparently closely related organisms (the euglenoid flagellates, for example), some species within a given genus are apparently plants, others can be considered animals, and yet further species can be both at the same time. Some dinoflagellates also obtain some 5% of their requirements from photosynthesis and 95% from heterotrophic digestion of consumed materials. Moreover, a number of flagellates could even change the kingdom to which they were assigned by remaining in the dark for 24 hours! The two-kingdom system is thus neither a reflection of phylogenetic relationships nor is it a practical working scheme, in large measure because although many organisms are 'animal' and many are 'plant', many are neither.

During the last 15 years or so, an alternative basic classification, which seeks both to reflect phylogeny more accurately and to be unambiguous, has gradually been gaining adherents and is now that in most widespread use. This scheme also postulates a basic dichotomy of living organisms, although in this case between the prokaryote and the eukaryote superkingdoms, the eukaryotes being then subdivided into four kingdoms: the unicellular groups (the Protista), and the multicellular plant, fungal and animal kingdoms (Fig. 3.1). The Protista therefore includes the former algae and protozoans, neither of which any longer has any formal biological status except as a general term equivalent to 'shellfish' or 'worms'; in a nutshell, the 'Protozoa' has for some time ceased to be regarded as a valid or useful grouping of organisms. One of the latest general classifications of the Protista divides the kingdom into 27 phyla, of which more than half are obligately non-photosynthetic and of which 13 were once regarded as being groups of animals (the Parabasalia★, Kinetoplasta★, Opalinata★, Stephanopogonomorpha, Ciliophora, Apicomplexa★, Metamonada★, Sarcodina, Labyrinthomorpha★, Choanoflagellata, Microspora★, Ascetospora★, and Myxozoa★—Fig. 3.2—the asterisked phyla mainly or

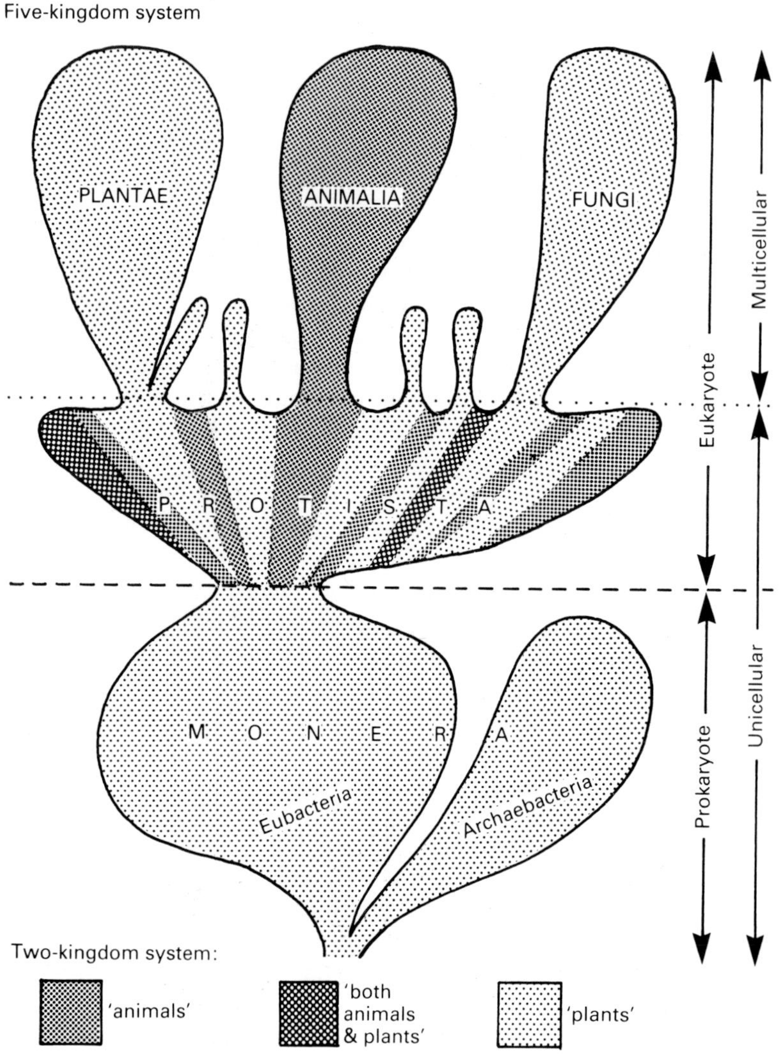

Fig. 3.1. The five-kingdom system of classification and its relations to the plant/animal dichotomy.

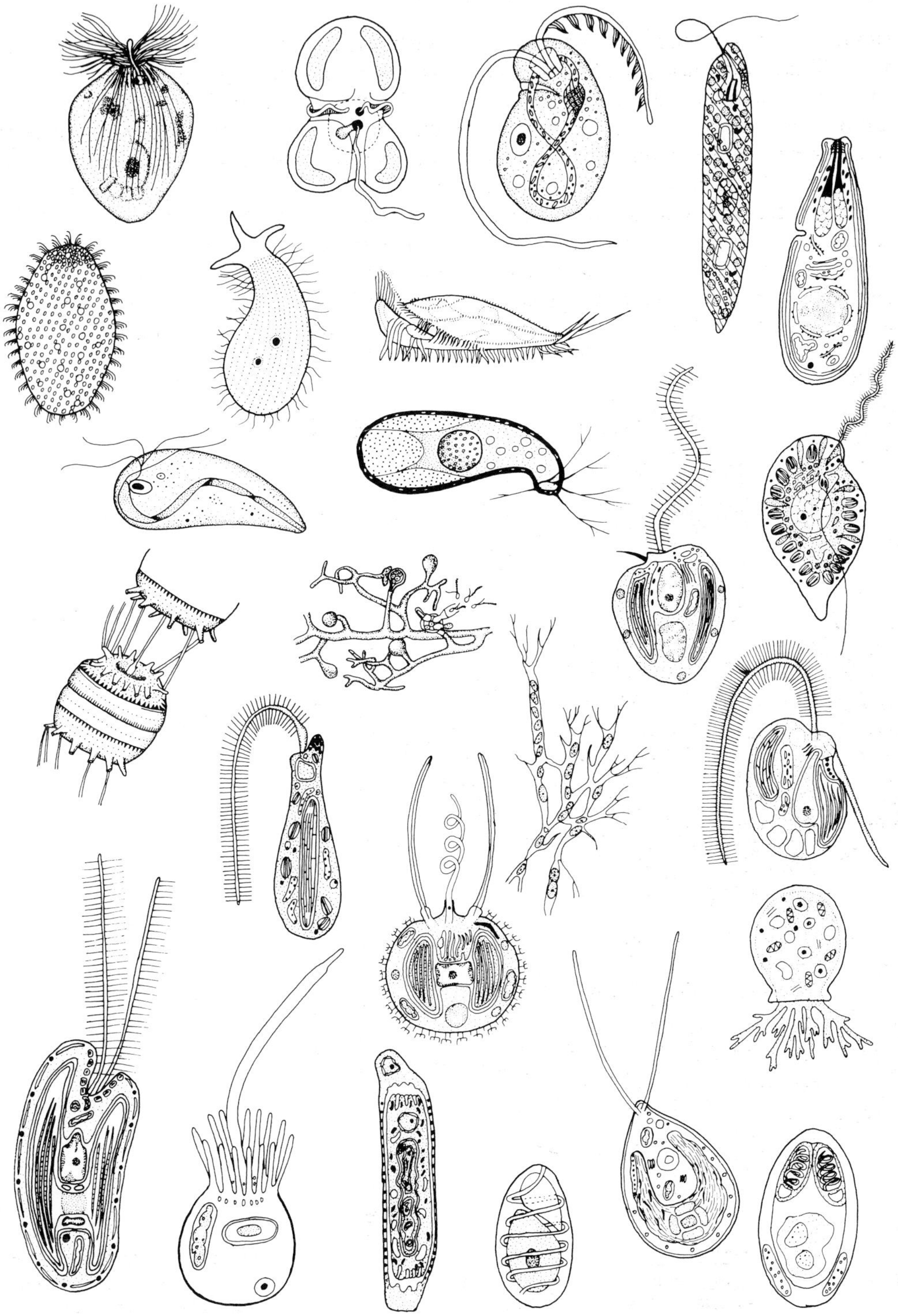

Fig. 3.2. The diversity of protistan body form (after Sleigh *et al.*, 1984).

entirely consisting of endosymbionts or endoparasites). Several other phyla (e.g. the Dinophyta, Euglenophyta and Cryptophyta) include some forms which were claimed by both zoologists and botanists.

The majority of these 'protozoan protists' are phylogenetically remote from the invertebrates and they share no characteristics uniquely with the multicellular animals. Shared negative features, such as lack of plastids, are also held in common with the 'fungus-like protists' and with the fungi themselves, and shared positive characters are solely those common to all the eukaryotes, including all the photosynthetic groups. By definition, of course, *the* difference between the protists, including the non-photosynthetic forms, and the other eukaryote kingdoms, including the animals, is that the latter are multicellular whilst the protists are solitary or colonial unicells. The individual protist cell is therefore capable of performing all the functions necessary for survival and multiplication—at one and the same time it may be the agent effecting locomotion, digestion, osmoregulation, reproduction, etc.—whilst in the cellularly differentiated animals, these, and the other functions, are carried out by different, specialized groups of cells usually arranged in discrete tissues or organs. This distinction, however, is purely phenotypic. All the many component cells of a multicellular animal are descended from a single, normally diploid cell and hence are genetically as totipotent as a protist individual. Some individual cells, e.g. the amoebocytes of sponges, may even remain undifferentiated and persist as a reservoir from which specialized cells can be differentiated as required during life.

Not surprisingly, the protist body is generally much smaller than that of an animal, although this is by no means always so. A considerable number of animals, composed of thousands of cells and with all the bodily complexity of lophophorates, arthropods or chordates, measure less than 1 mm in largest dimension; most members of the pseudocoelomate groups are smaller than 2 mm. On the other hand, some ciliated protists (phylum **Ciliophora**) attain 5 cm in cell length, and the pseudopodia of some xenophyophoreans (phylum **Sarcodina**) may span diameters of 25 cm. The only means by which a non-colonial protist can respond to any selective pressure favouring large size (see e.g. Section 9.1) is by increase in cell size itself, often accompanied by polyploidy or multiple nuclei, cytoplasmic streaming then forming an intracellular circulatory system. In a sense, large size is relatively easily achieved in a protist since, weight for weight, the basal metabolic requirements of a single-celled organism are less than those of a multicellular one (Fig. 3.3).

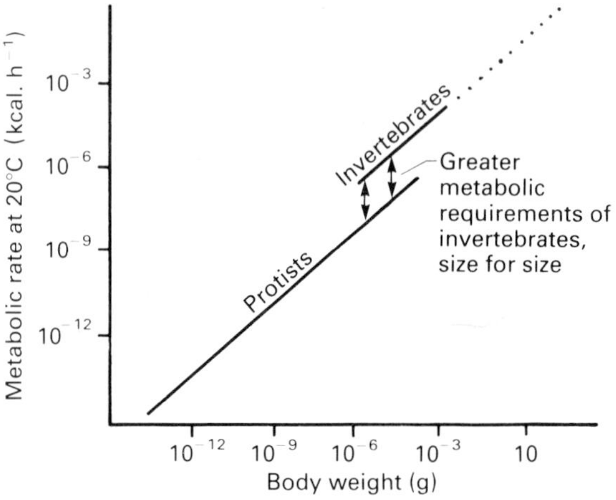

Fig. 3.3. The relationship between resting metabolic rate and body size in protists and invertebrates, showing the greater requirements of multicellular organisms, size for size (after Schmidt-Nielsen, 1984) (NB 1 kcal = 4.2 kJ).

The organisms formerly regarded as constituting the phylum or subkingdom Protozoa are therefore a diverse assortment of protists, each in some respects individually equivalent to one of the component cells of an animal and in other respects to a whole animal. The use of the word 'equivalent' in the above sentence is crucial, however; the surviving protists have had an even longer separate evolutionary history than have the various multicellular organisms and only the phylum **Choanoflagellata** has retained elements of structure that indicate that it might be considered directly related to any specific group of animals (Chapter 2).

3.2 Phylum PORIFERA (sponges)

3.2.1 Etymology

Latin: *porus*, pore; *ferre*, to bear.

3.2.2 Diagnostic and special features (Fig. 3.4)

1 No generally characteristic symmetry, even within species where form can sometimes depend upon local ecology.
2 Multicellular but with few cell types; tissue organization very restricted; no organs and little co-ordination between cells; no nervous system.
3 'Choanocytes' are characteristic and are concerned with water circulation and feeding.
4 A sometimes elaborate skeletal system of either calcareous or siliceous spicules or protein (collagen, sometimes called spongin) fibres or a combination of these.
5 Cells arranged around water chambers/canals of varying complexity but no true body cavity or gut.
6 Exclusively filter feeding; gas exchange by diffusion.
7 All sessile.
8 Reproduction: asexual or sexual, latter can be gonochoristic but usually hermaphrodite; cleavage complete; planktonic larvae.
9 Exclusively aquatic and mainly marine (*c.* 10000 species); *c.*50 freshwater species.

The functioning of sponges is dominated by one structural feature: the organization of their cells around a system of chambers and canals, through which water is circulated by the flagellar action of choanocytes. The environment is therefore moved through the organism rather than the other way round, so filter feeding is an inevitability, as is reproduction by propagules that can be released from the parent.

The simplest, asconoid, form is illustrated in Fig. 3.4: a single chamber, the atrium, surrounded by a body wall covered with pinacocytes, with incurrent pores, ostia, formed by porocytes, and a single excurrent pore, the osculum. Pinacocytes can contract, drawing the sponge in, and some secrete material for attaching it to the substratum. The mesohyl, lying below the pinacoderm, is a gelatinous protein matrix containing a supporting skeleton and amoebocytes, of which there are various types with functions ranging from digestion to the secretion of collagen and the formation of all other cell types, i.e. they are totipotent. To some extent the choanocytes and pinacocytes form tissues (the choanoderm and pinacoderm), but basement membranes are lacking.

This type of asconoid organization puts a limit on size, since the volume of water in the atrium increases as the cube of body length, whereas the flagellar surface increases as the square of body length. This limitation has been overcome in some sponges by folding the body wall and reducing the atrium, leading to syconoid and ultimately leuconoid forms (Fig. 3.5). Most leuconoids are either solid, with adjacent incurrent or excurrent canals running parallel to each other and with ostia and sometimes oscula scattered over the whole body, or hollow and tubular, in which excurrent canals never run to the surface direct but open into an inner cavity that leads to a distal osculum (Fig. 3.6). This classification of body forms transcends the systematic classification of the group (below). Throughout, the leuconoid form is dominant, providing some evidence of its efficiency. It evolved more than once in the sponges and in some instances might have involved an intermediate syconoid state. Leuconoid sponges often develop through asconoid- and syconoid-like forms before finally reaching the leuconoid condition.

Choanocytes are of central importance in the metabolic economy of sponges. They drive water currents through the sponge, particularly in quiet waters, by the beating of their flagella. They also trap the fine particles of food that are brought in with these currents, by passing water through collars that in fact consist of a microvillar palisade (Fig. 3.4). Larger food particles can be engulfed by amoebocytes. Both choanocytes and amoebocytes can transfer their engulfed particles to other cells, and amoebocytes rather than choanocytes seem to be the principal sites of digestion.

Sponges reproduce asexually by release of fragments or of an aggregate of essential cells, particularly

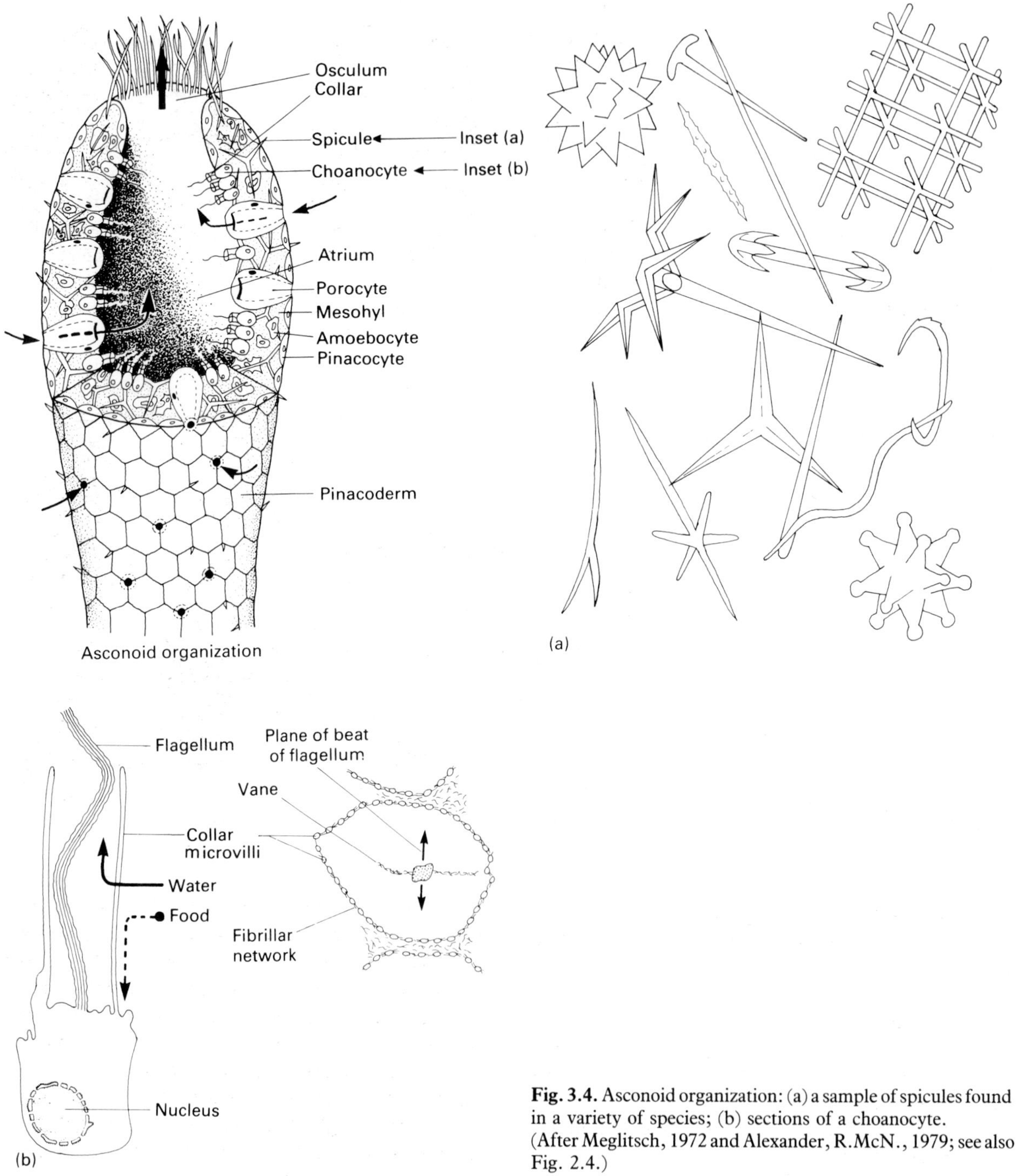

Fig. 3.4. Asconoid organization: (a) a sample of spicules found in a variety of species; (b) sections of a choanocyte. (After Meglitsch, 1972 and Alexander, R.McN., 1979; see also Fig. 2.4.)

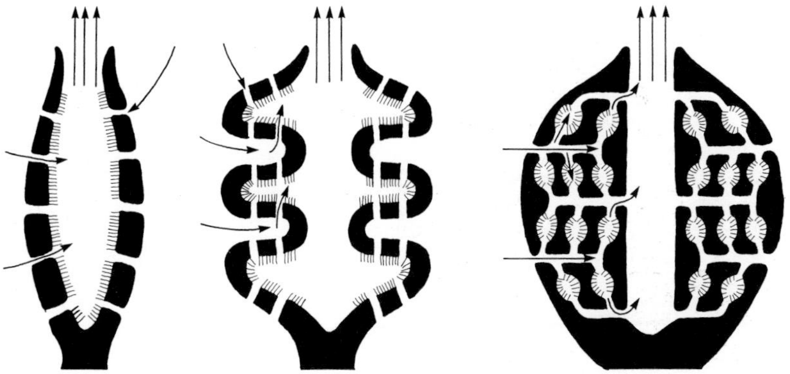

Fig. 3.5. Various types of folding of the body walls of sponges (after Buchsbaum, 1951).

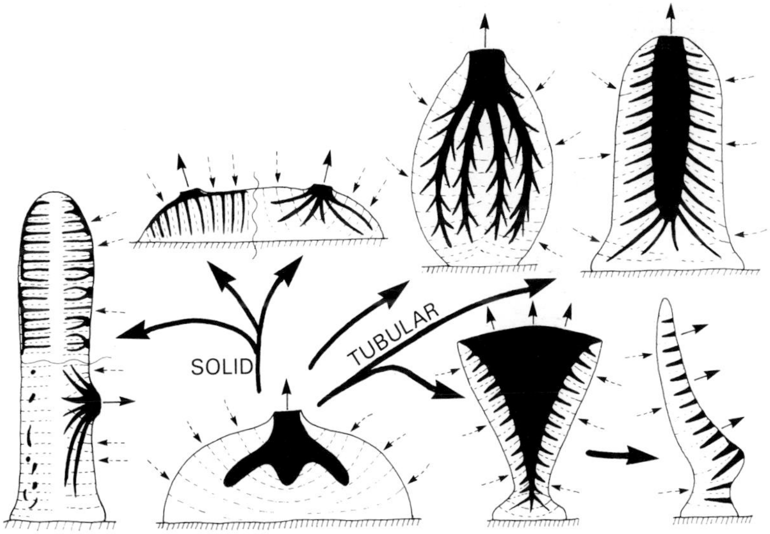

Fig. 3.6. Two types of architecture in sponges (after Barnes, 1980).

amoebocytes; for example, the gemmules (Fig. 3.7a) produced by freshwater sponges are packed with archeocytes, large specialized amoebocytes, and have a hard, resistant coat of spongin and spicules, which enables them to overwinter. Sexual reproduction, though commonly hermaphroditic (a feature correlated with the sessile habit), often involves cross-fertilization, the sperm leaving the osculum of one individual and entering the ostium of another is trapped by a choanocyte of the recipient and is then transported to an egg. Development to larval stages most often occurs within the parent. Larvae—usually a parenchymula larva with solid body and covered with flagella often except for one pole (Fig. 3.7b) but sometimes an amphiblastula larva with hollow body and one hemisphere consisting of small, flagellated cells and the other non-flagellated macromeres (Fig. 3.7c)—escape from parents and have brief planktonic existences before settling.

Not surprisingly, because of the autonomy of cellular units, sponges are capable of considerable regeneration, even from single cells.

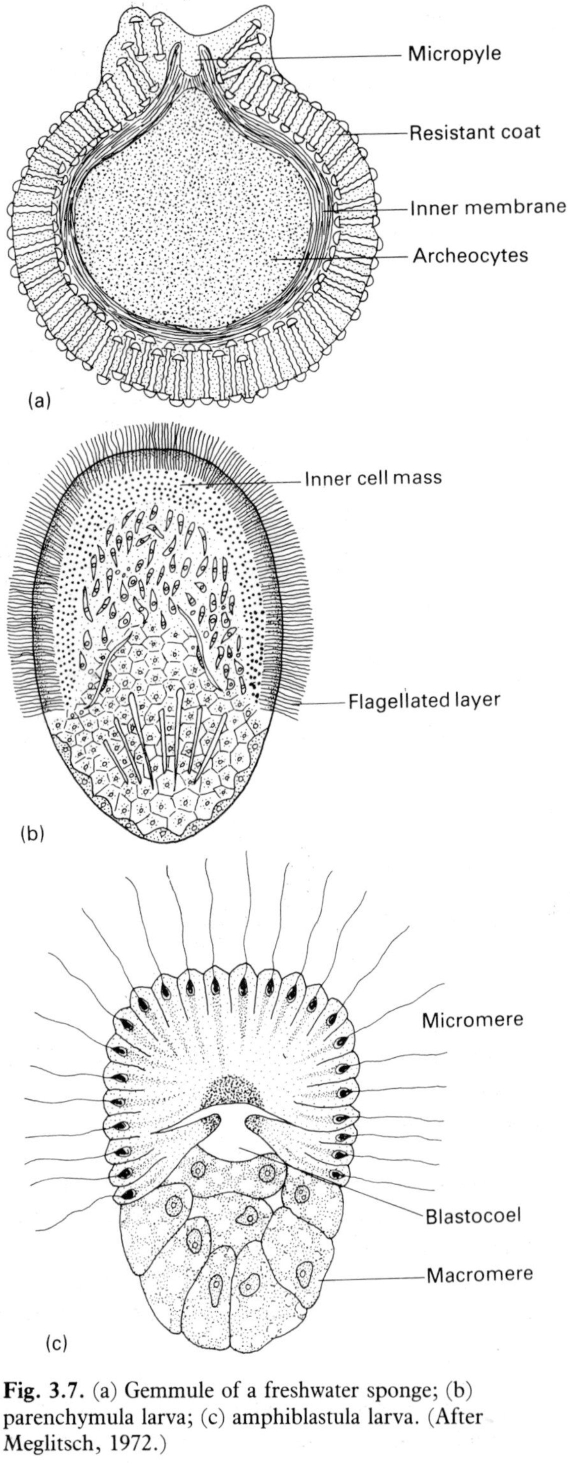

Fig. 3.7. (a) Gemmule of a freshwater sponge; (b) parenchymula larva; (c) amphiblastula larva. (After Meglitsch, 1972.)

3.2.3 Classification

The 10000 or so species that are currently described are distributed between four classes.

3.2.3.1 Class Hexactinellida

Members of this class are commonly known as glass sponges. They have a skeleton of siliceous spicules, mainly six-rayed (Fig. 3.8)—hence the name—and

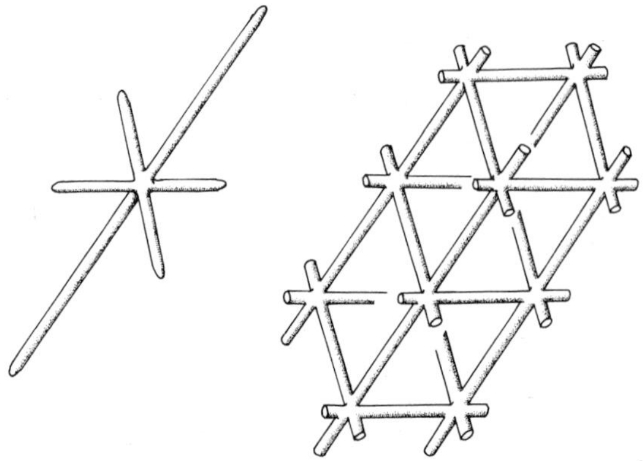

Fig. 3.8. Hexactinellid silica spicules (after Alexander, 1979).

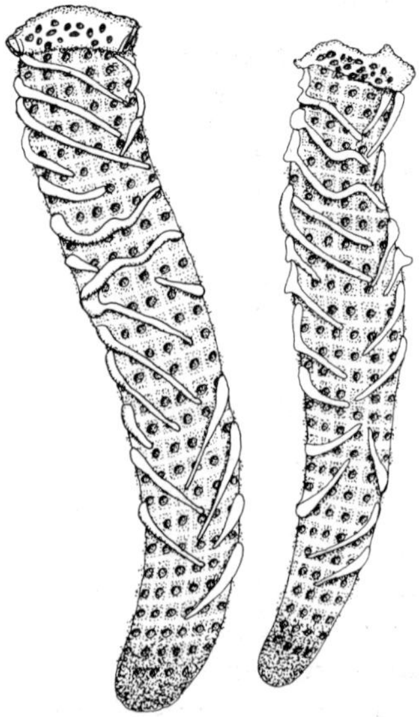

Fig. 3.9. Venus' flower basket (after Barnes, 1980).

are often fused into vase-shaped structures, averaging 10–30 cm in height (Fig. 3.9). The syconoid form dominates. There are no pinacocytes (present in all other classes); instead the epidermis consists of a net-like syncitium formed from interconnecting pseudopodia of amoebocytes. Species are entirely marine and chiefly occur in deep-water habitats. Note that these are separated off by some authorities as a distinct subphylum, or even phylum (the Symplasma).

3.2.3.2 Class Calcarea

Members of this class are unusual in having spicules of calcium carbonate, either calcite or aragonite. All three grades of structure—asconoid, syconoid, leuconoid—occur. Most are less than 10 cm in height. All are marine.

3.2.3.3 Class Demospongiae

This is the largest class, containing over 90% of all the species. The skeleton consists of spicules that are siliceous, of spongin or of a combination of the two; one family, the Spongiidae, contains the common bath sponges with skeletons of spongin only. All are leuconoid and some can achieve considerable size; the tropical loggerheads (*Spheciospongia*) form masses > 1 m in diameter and height. Coloration is frequently brilliant.

The class contains the boring sponges, able to bore into the calcareous structures of corals and molluscs. Representative also occur in fresh waters, particularly within the family Spongillidae.

3.2.3.4 Class Sclerospongiae

A class consisting of a very small number of specialist marine species found in grottoes and tunnels associated with coral reefs in various parts of the world. Representatives are very similar to Demospongiae but with the compound skeleton of siliceous spicules and spongin fibres restricted to a thin, superficial layer of living tissue supported on a massive basal layer of calcium carbonate (Fig. 3.10). All are leuconoid.

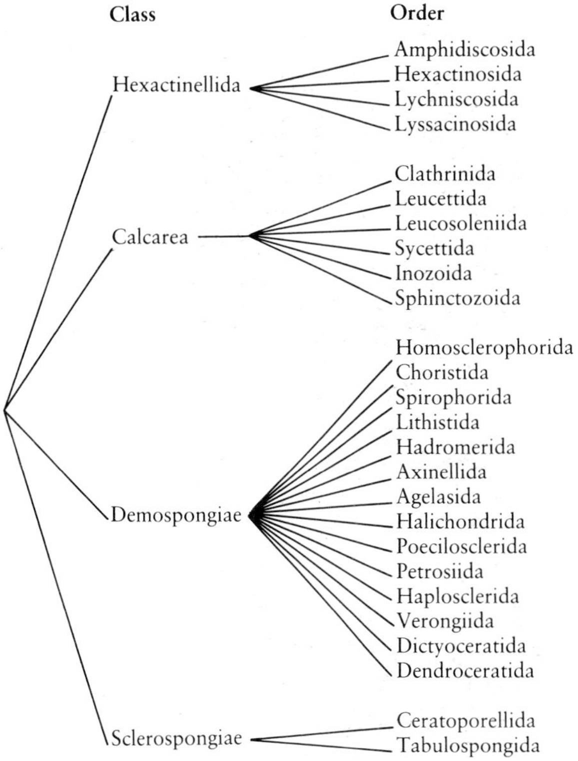

3.3 Phylum PLACOZOA

3.3.1 Etymology

Greek: *plakos,* flat; *zoon*, animal.

3.3.2 Diagnostic and special features (Fig. 3.11)

1 Without any system of symmetry, and capable of changing shape in an amoeboid manner.
2 Without distinct tissues or organs.
3 Without any body cavity or digestive cavity.
4 Without any system of nervous co-ordination.
5 Body in the form of a flat plate, which can move in any direction in the plane of the body.
6 A single outer layer of flagellated cells enclosing a fluid-filled mesohyl containing a network of stellate fibre cells.
7 Marine.

The upper surface of the flat bodily disc is formed of a thin layer of squamous cells, many of them bear-

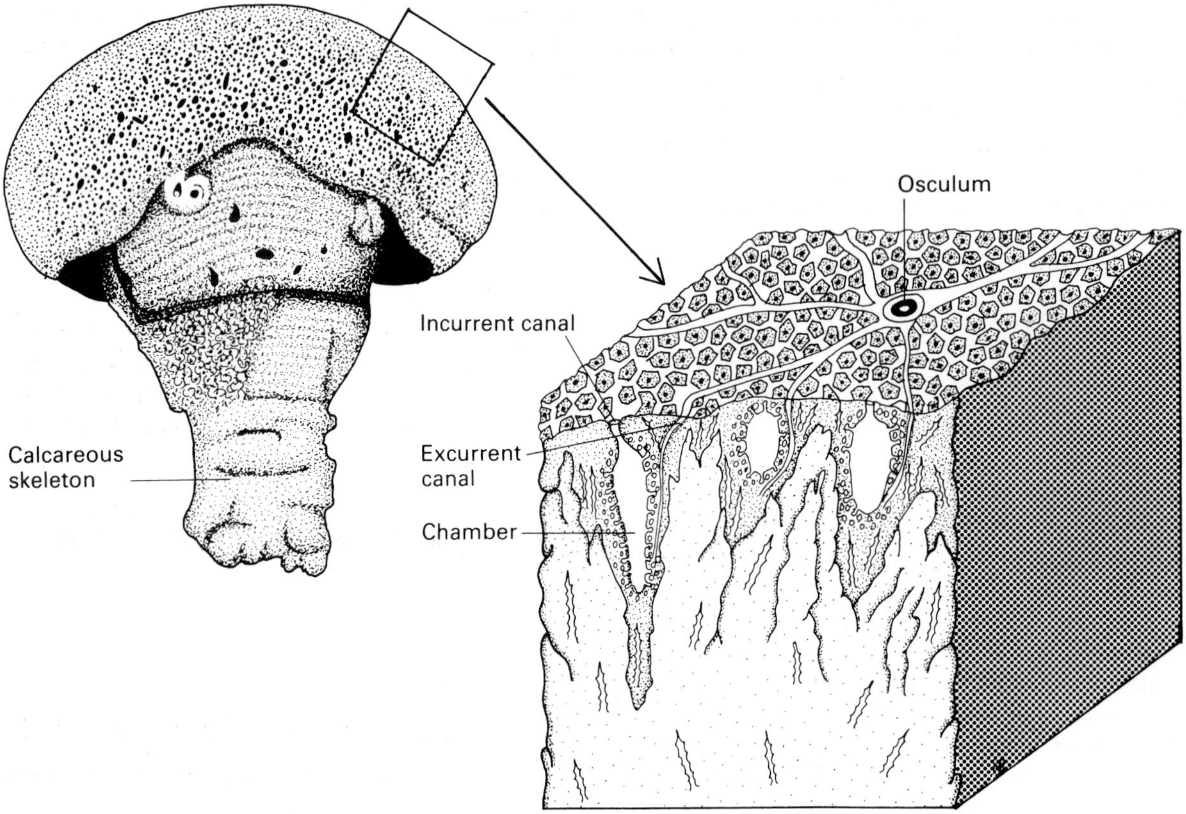

Fig. 3.10. A sclerosponge with details of the upper surface (after Bergquist, 1978).

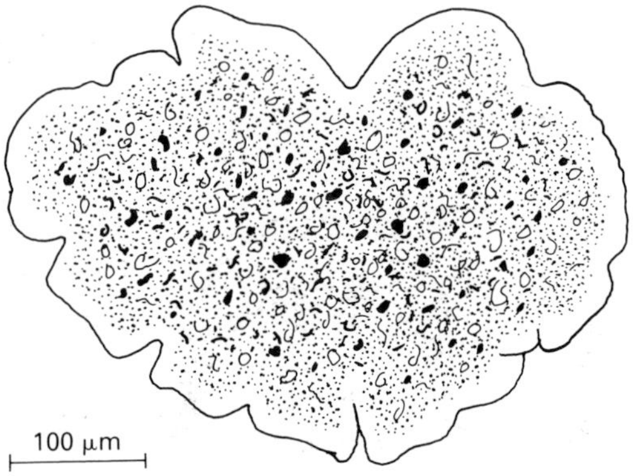

Fig. 3.11. The general appearance of a placozoan (after Barnes, 1980).

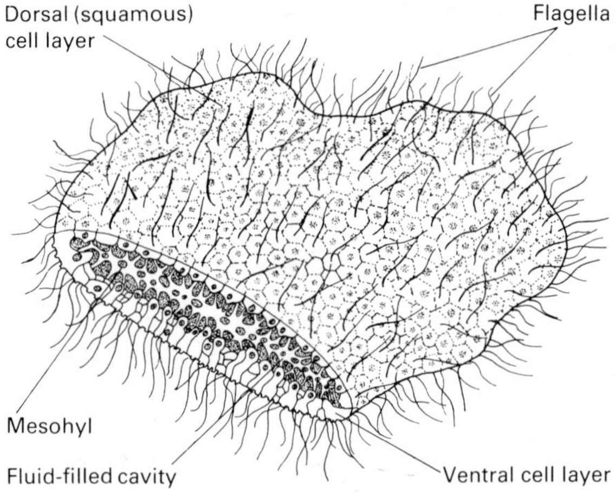

Fig. 3.12. A diagrammatic section through a placozoan (after Margulis & Schwartz, 1982).

ing a flagellum, whilst the thicker lower layer is composed of two cell types, flagellated columnar cells, interspersed with non-flagellated glandular cells (Fig. 3.12). These gland cells secrete enzymes on to their prey (various protists), which are therefore partly digested outside the body, and the digestion products are then absorbed by the same secretory cells. The ventral flagellated cells are responsible for a gliding form of locomotion, whereas the changes of shape which cause placozoans to resemble large amoebae are brought about by the co-ordinated contraction and relaxation of the fibre cells in the mesohyl.

Asexual multiplication, by both fission and budding, commonly occurs, and although sexual reproduction does take place, both it and the subsequent development of the embryo are very poorly known. Eggs, probably deriving from the lower cell layer, may be present in the mesohyl, and sperm have been reported from the water in which placozoans were being maintained, but sperm production and fertilization have not yet been observed.

The body contains a few thousand cells and can attain a diameter of 3 mm, although placozoans contain only four cell types and therefore are one of the simplest known multicellular organisms. Their component cells also contain less DNA than any other animal (of the same order of magnitude as in many bacteria) and their chromosomes are minute, less than 1 μm in length. For almost a century they were regarded as the planuloid larva of some type of sponge or cnidarian, until in the late 1960s it was discovered that they could achieve sexual maturity and a new phylum was created for them in 1971.

Placozoans have most often been recorded from marine aquaria and they are known in the wild from the intertidal zone. They are probably quite widespread in the sea, though easily overlooked.

3.3.3 Classification

Only one (or possibly two) species has been described. The phylum is sometimes placed in a separate subkingdom, the Phagocytellozoa (an unfortunate name since uptake of food is not usually by phagocytosis).

3.4 Phylum CNIDARIA (hydras, jellyfish, anemones, corals)

3.4.1 Etymology

Greek: *knide*, nettle.

3.4.2 Diagnostic and special features (Fig. 3.13)

1 Radially symmetrical animals with tentacles encircling a mouth at one end of the body.
2 An internal space for digestion, the 'gastrovascular cavity', with the mouth as the only opening.
3 Two different body forms may exist: a 'medusa' which is adapted for a pelagic existence and a 'polyp'

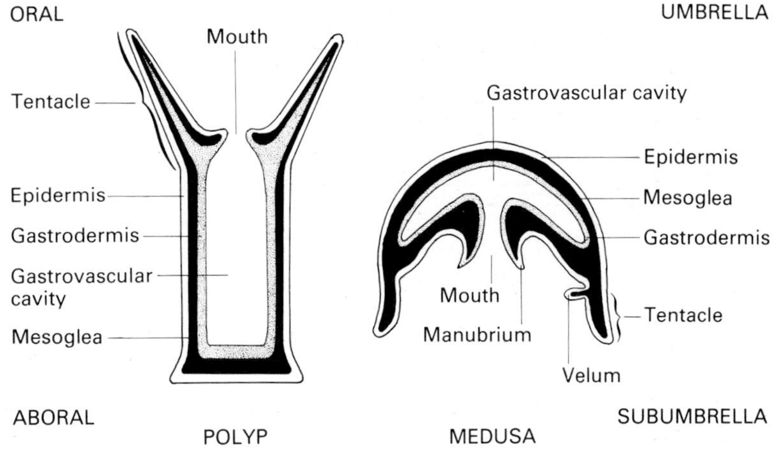

Fig. 3.13. The two body forms of cnidarians (after Fingerman, 1976).

which is adapted for benthic existence; polyps may be colonial and polymorphic within colonies.

4 Body walls consist of two tissues, 'epidermis' and 'gastrodermis', with several to many cell types sandwiching a less cellular 'mesoglea'; limited organ development.

5 'Cnidocytes', containing stinging organelles, particulary 'nematocysts', are unique amongst the metazoans and occur through all classes.

6 May be skeletal support of chitin and/or calcium carbonate.

7 Nerves arranged in nets, one or two nets per animal.

8 Most are carnivorous; gas exchange by diffusion.

9 Reproduction is asexual or sexual; after the latter, development often involves an almost bilaterally symmetrical, ciliated 'planula' larva.

10 All are aquatic; most are marine.

Two basically different forms can occur in cnidarian life cycles: a relatively mobile medusa that carries gonads, and a sedentary polyp that primitively probably did not (Fig. 3.13). Polyps are tubular with one end carrying the mouth and the other a basal disc that attaches the animal to the substratum. In medusae the mouth occurs at the end of a manubrium on the subumbrellar surface. Polyps bud off medusae asexually and medusae form polyps sexually, very often with an intermediate, ciliated and solid-bodied planula larva (see Box 14.5). This is a very generalized description and there are exceptions throughout the phylum: not all medusae are mobile; not all polyps are sedentary; some polyps carry gonads; some medusae bud other medusae; some life cycles do not involve medusae and others do not involve polyps; planula larvae are absent from some life cycles.

The cellular structures of both polyps and medusae are fundamentally similar but with differences in emphasis; in particular the mesoglea is a thin, mainly non-cellular layer in some polyps, but in medusae is thicker and often more fibrous, with or without a population of wandering amoebocyte cells.

A generalized diagram of the body wall, typical of hydrozoan polyps, is illustrated in Fig. 3.14.

The *epidermis* is composed of five main cell types:

1 Epitheliomuscular: columnar, 'cover' cells (occasionally squamous) but with basal extensions containing contractile myofibrils orientated parallel to the oral/aboral axis, i.e. longitudinal muscles.

2 Interstitial: with prominent nuclei and little cytoplasm, capable of forming other tissues.

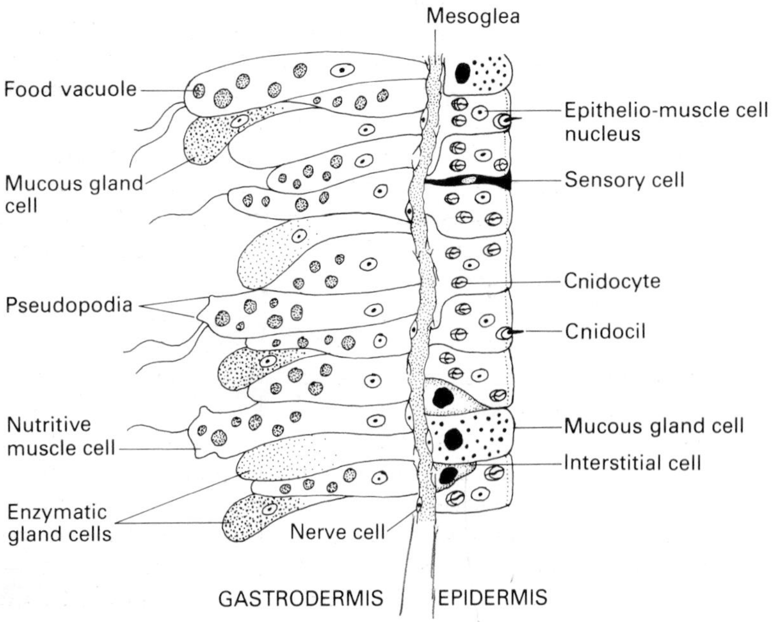

Fig. 3.14. Longitudinal section through the body wall of a hydrozoan (after Barnes, 1980).

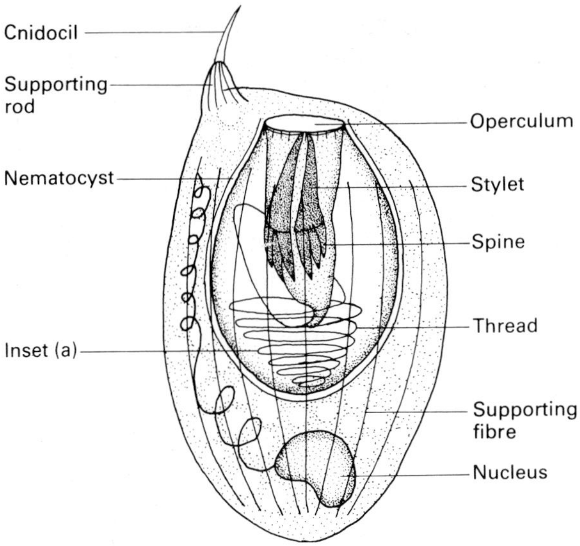

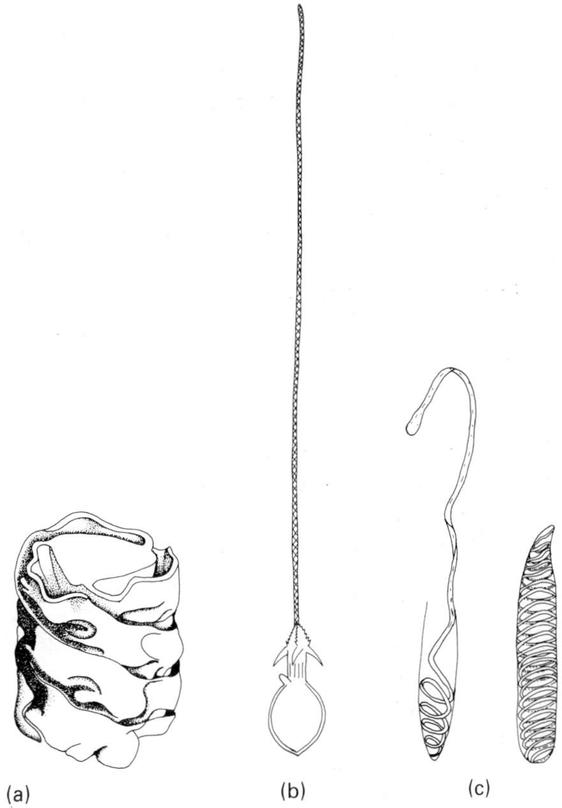

Fig. 3.15. Cnidocyte and nematocyst form; (a) the undischarged thread is pleated; (b) the discharged thread; (c) the spirocyst of an anemone (after various sources) (see also Fig. 2.5).

3 Cnidocytes: these contain everting structures, the nematocysts (Fig. 3.15 and see Fig. 2.5), more generally called cnidae. For hydras and jellyfishes, one end carries a stiff bristle, the cnidocil, which has the structure of a flagellum. This is not present in anemones but a ciliary cone complex, presumably of similar function, is sometimes present. The nematocyst contains a coiled, pleated and internally barbed inverted tube. Discharge is almost certainly effected by a rapid uptake of water into the cnidocyte, due to a change in the permeability of its membranes, possibly effected electrically by a signal from the cnidocil. The thread turns inside out and the pleats are straightened out as the cnidae are discharged and extended.

There are three main types of cnidae:

(a) *Nematocysts proper*—themselves varied in form but many with spines and capable of delivering a toxin; used in defence and the capture of prey.

(b) *Spirocyst*: an adhesive structure found in some anthozoans. The threads here are solubilized upon discharge to form a dense adhesive net that is used for attachment and prey capture.

(c) *Ptychocysts*: these only occur in ceriantharian anemones (p. 68); the threads are elaborately pleated and lack spines. The threads are also adhesive and assist in forming the tube in which ceriantharians are encased.

Cnidae can be used only once and are then jettisoned; cnidocytes can be regenerated from interstitial cells and the developing cnidocyte is referred to as a cnidoblast.

4 Mucous gland cells: gland cells that secrete mucus; for protection, prey capture and adhesion.

5 Sensory/nerve cells: elongated cells at right angles to the epidermal surface. They are more or less similar to multipolar neurons (Chapter 16). They synapse with other neurons to form an irregular net or plexus (see Section 16.4.1). In some cnidarians, though not hydras, there are two nerve nets in the epidermis; in others one net in the epidermis and another in the gastrodermis (below).

The *gastrodermis* consists of: (a) nutritive muscle cells that engulf food particles for intracellular digestion. They also have a muscular function, like epitheliomuscular cells, but with processes at right angles to the oral/aboral axis, to form a circular muscle layer. They are flagellated and are used to stir food in the gastrovascular cavity; (b) gland cells that secrete enzymes that effect some digestion in the gastrovascular cavity; (c) mucus-secreting cells; and (d) in some cnidarians (e.g. anemones, but not hydras) cnidocytes. The gastrodermis is complexly folded in medusae and in anemones.

Locomotion involves muscular contraction. Muscles contract against fluid in the gastrovascular cavity that acts as a hydrostatic skeleton (see Chapter 10), as may the mesoglea. Movement is usually limited in polyps, involving looping and shuffling, but is more extensive in medusoid forms. Even in the latter, though, active movement is in the oral/aboral plane, through well-developed circular muscles at the subumbrellar margin. This generally drives the animal vertically and horizontal movement is achieved passively in the ocean currents. However, some medusae, the *Cubomedusae* (see p. 63), are capable of rapid, active locomotion in the horizontal plane with their aboral surfaces anterior.

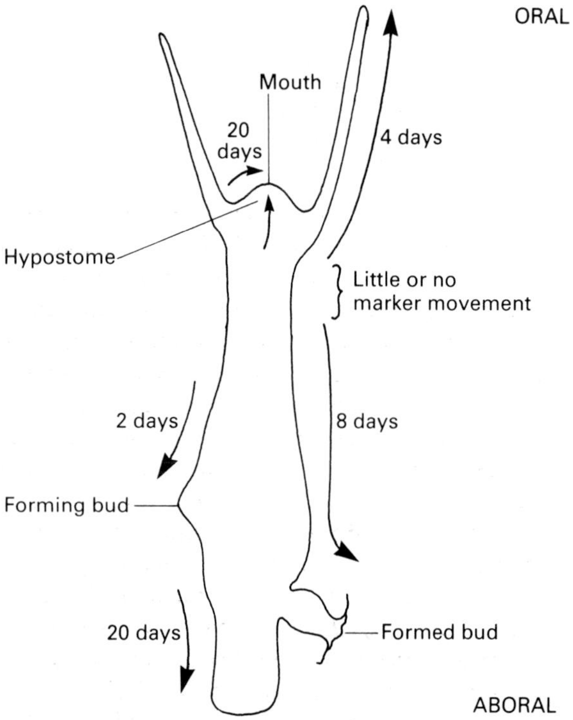

Fig. 3.16. Cell movement in hydra (after Campbell, 1967).

Notable, is the not unexpected capacity of cnidarians to regenerate after surgery. This has been exploited by developmental biologists to study the fundamental processes of cell division and differentiation—particularly in *Hydra* which is easy to keep and handle in laboratories (see Section 15.5). Here the oral/aboral polarity is retained after transection such that if a single individual is cut into several pieces along this axis, tentacles always form on the end that was closest to the oral end of the intact animal. Positional information appears to be associated with an oral/aboral gradient of some substance, to which cells can detect and respond appropriately (in terms of dividing or differentiating) dependent upon the concentration and hence the level in the gradient at which they occur.

Mitosis occurs throughout the stalk but there is a gradient from a high rate orally, just below the hypostome (a mound upon which is the mouth, Fig. 3.16), to a low rate aborally. There is therefore a continuous movement of cells in the oral to aboral direction and along the tentacles. These replace cells that die and that are sloughed off from these sites and they are also used to form new buds. As a result, there is a continuous turnover of cells—within a few weeks, all cells in the body of one individual are replaced—and this might allow these animals to be immortal (p. 398).

3.4.3 Classification

There are around 10 000 described living species. At the highest level of classification, the division of these into taxa depends upon whether polyps or medusae are a dominant feature of life cycles. In one taxon,

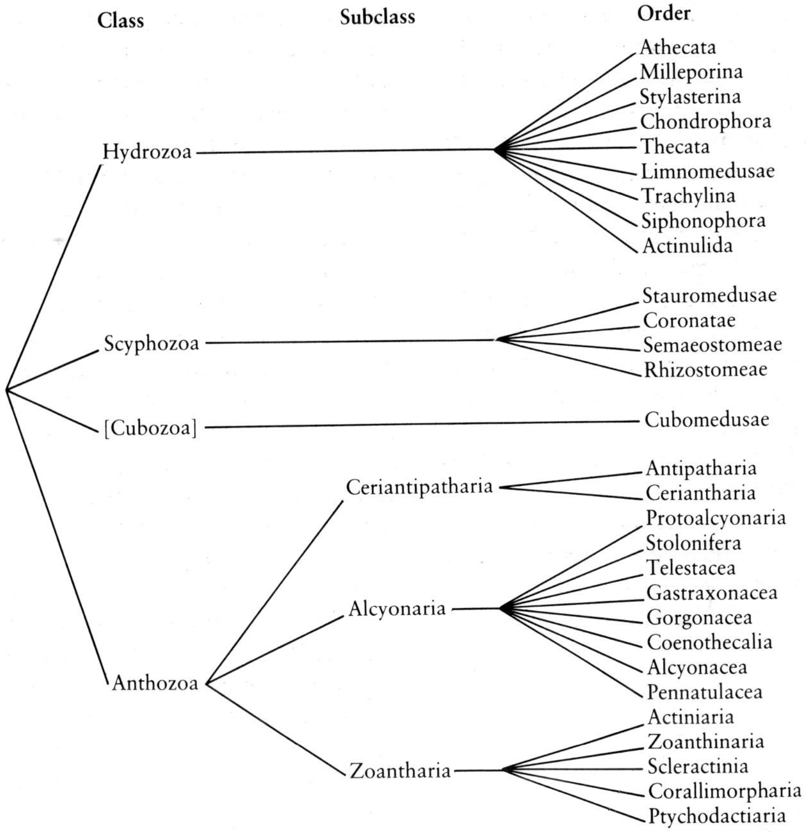

the Hydrozoa, polyps are generally more conspicuous than medusae. In another, Scyphozoa, the reverse is the case. In a third, Anthozoa, the form is polypoid and there are apparently no medusae. Yet the term 'polypoid' is used here advisedly since, though sessile, these forms possess many features in common with scyphozoan medusae: thick, cellular mesoglea; intucked manubrium (called a stomodaeum); endodermal gonads. Other sedentary medusoid forms are known (below). Hence anthozoans could be sedentary medusae without a true polyp phase!

Table 3.1 Classifications of Cnidaria

Classical	Alternative
Class Hydrozoa	Subphylum Medusozoa
Class Scyphozoa	Class Hydrozoa
Class Anthozoa	Class Scyphozoa
Subclass Alcyonaria	Class Cubozoa
[Subclass Zoantharia*]	Subphylum Anthozoa
	[Class Ceriantipatharia*]
	Class Alcyonaria
	Class Zoantharia

*Contains Ceriantharia.
[]Considered artificial by some (Schmidt, 1972).

Nevertheless, one classification considers the anthozoans to be polyps and takes the absence of a medusa so seriously that it separates Hydrozoa plus Scyphozoa from Anthozoa in different subphyla (Table 3.1). The same argument could be applied if anthozoans are considered to be medusae devoid of a polyp. This has the advantage of recognizing a number of classes within the Anthozoa which, as we shall see, *are* strikingly different. Yet lack of either polyp or medusa is not unusual in species in the other classes and so, it could be argued, should not be

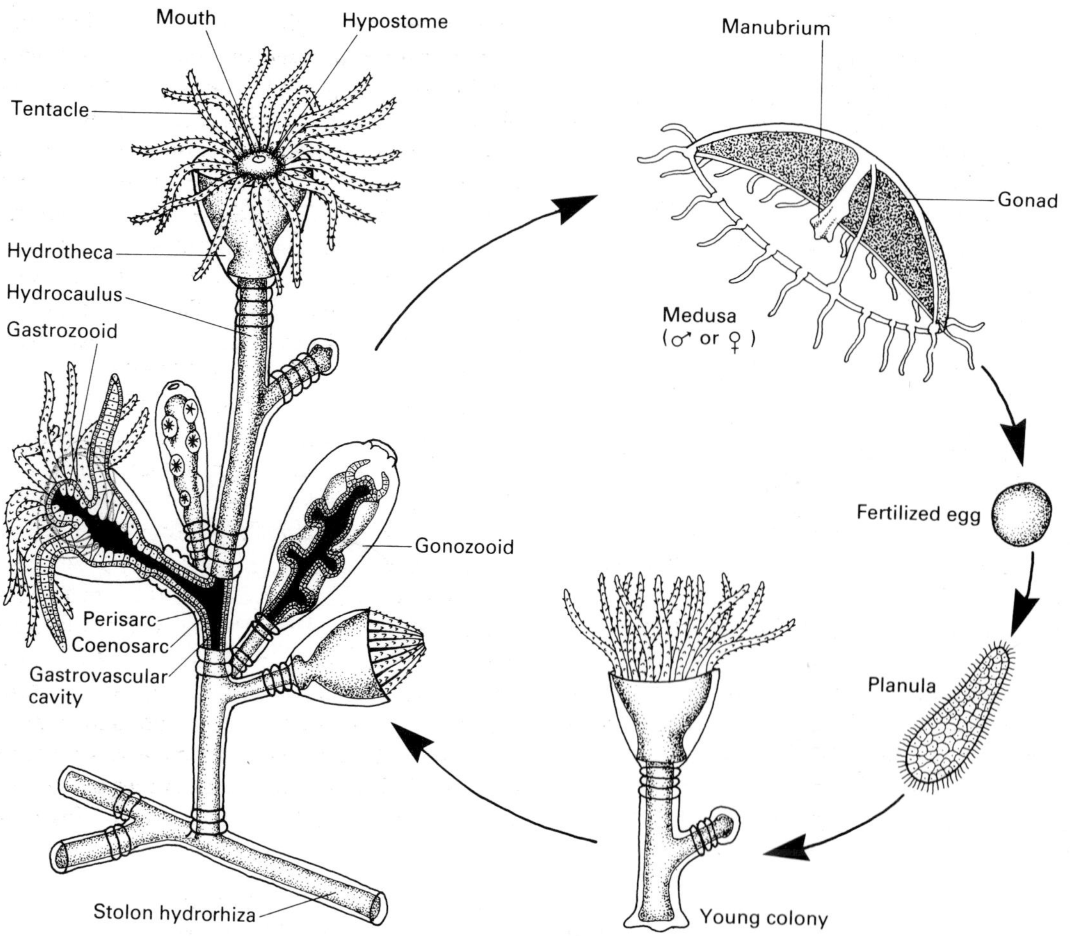

Fig. 3.17. Life cycle of *Obelia* (after Barnes, 1980).

taken too seriously in the Anthozoa. We have therefore adopted the more classical view of accepting three major cnidarian classes—Hydrozoa, Scyphozoa and Anthozoa—recognizing that this is not fully satisfactory or satisfying.

3.4.3.1 Class Hydrozoa

Hydra (Fig. 3.16), though familiar, is not a typical hydrozoan in that it is (a) solitary and (b) without a medusoid phase. More typical is *Obelia* (Fig. 3.17). There are three points to note about the polypoid phase of this animal: (a) it consists of a number of polyps (modules) connected to stolons (a root-like hydrorhiza and stalk-like hydrocaulus) so it is a colonial or modular organism; (b) not all the polyps are the same: some, with tentacles, have a feeding function (gastrozooid/trophozooid); others (gonozooids) a reproductive function, budding medusae; and, not present in *Obelia* but common in other hydrozoans are club-shaped polyps that have batteries of cnidocytes and are defensive (dactylozooids (p. 64); (c) there is a supporting skeleton of chitin (sometimes calcified) called the perisarc (the tissue it surrounds is called coenosarc and has a continuous enteron). There can be considerable quantitative and qualita-

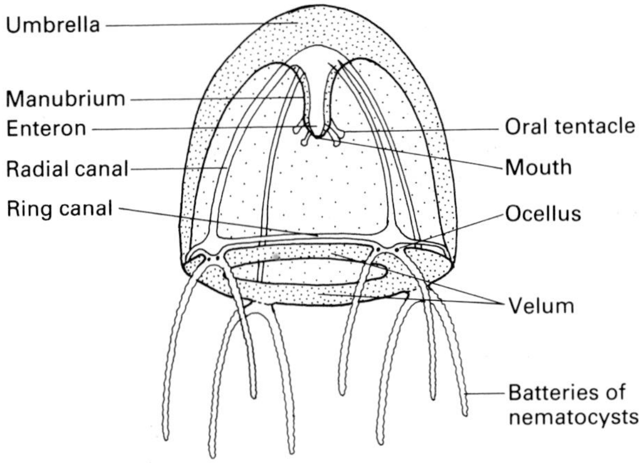

Fig. 3.18. Hydromedusa of *Bougainvillea* (after Bullough, 1958).

tive variation in all these features in different taxa.

The medusa of *Obelia* has many of the characteristics typical of hydromedusae. It is small (*c.* 0.5 cm) and also has epidermal gonads, different individuals carrying ovaries and testes. However, there is only a rudimentary velum on its subumbrellar surface which is better developed in many other species of hydromedusae (e.g. Fig. 3.18). The gastrodermis does not have muscles. Instead specialized muscles on the margin of the subumbrellar surface form circular sheets and give motive force; they contract forcing water down and the medusa up—a series of contractions appearing like pulsations. When present, the smaller opening produced by the velum produces a jet of water. The muscles work against the mesoglea, which is non-cellular, but contains protein fibres giving it elasticity. Downward movement is passive. The nervous system is also specialized with the plexus forming two nerve rings at the subumbrellar margin. They may even possess ganglia. They make connection with sense organs on the margin of the medusa: ocelli (light sensitive) and statocysts (acceleration sensitive) (Fig. 3.19).

In some species the medusae are not released, but remain attached to the polyp and it is from here that they release their gametes. The gonozooid is then referred to as a gonophore. This is the situation in *Tubularia*, which not only retains the medusae as gonophores on the gastrozooid but also retains the planula larvae, releasing instead actinula larvae—chubby, pelagic, polypoid forms (Fig. 3.20). In some species, the attached medusa regresses, ultimately leaving only gonads, as is the situation in *Hydra*. The retention and regression of the medusa have probably occurred independently several times in the Hydrozoa. On the other hand, in some species it is the

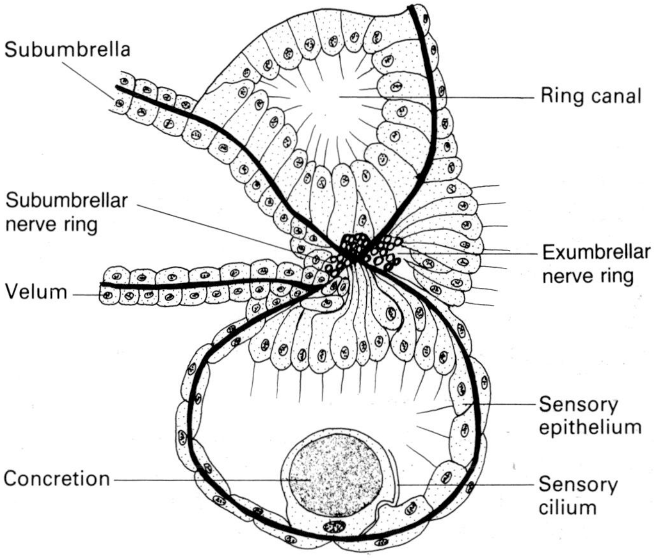

Fig. 3.19. Statocyst of a hydromedusa (after Barnes, 1980).

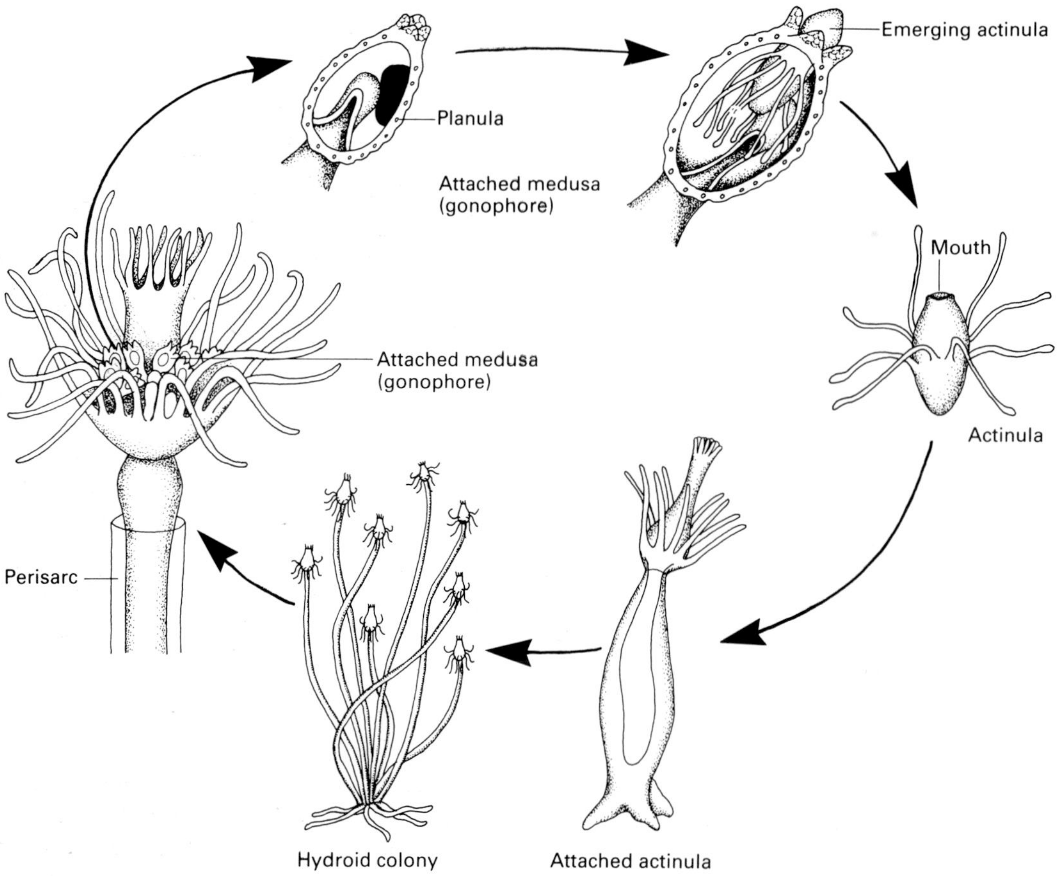

Fig. 3.20. Life cycle of *Tubularia* (after Barnes, 1980).

polyp, not the medusa, that is absent from the life cycle (Fig. 3.21).

Coloniality and polymorphism have reached their ultimate in certain, peculiar, pelagic hydrozoans: the famous 'Portuguese man-of-war' (*Physalia*) that consists of a gas float (modified medusa) and trailing tentacles arising from cormidia (i.e. clusters of variously modified polyps; Fig. 3.22a); the 'little sail' (*Velella*) with its saucer-shaped float (modified gastrozooid) carrying gonozooids and dactylozooids (Fig. 3.22b).

Finally, some colonial hydrozoans secrete an internal, epidermal skeleton that is calcareous. These are the so-called hydrocorals (orders Milleporina and Stylasterina) (Fig. 3.23). They have an upright and encrusting growth form and can attain considerable size. Dactylozooids are numerous and can inflict a serious sting. For this reason some members of the group are called fire or stinging corals.

3.4.3.2 Class Scyphozoa

Aurelia has a fairly typical scyphozoan life cycle (Fig. 3.24b). The medusa (Fig. 3.24a) is dominant with a diameter of around 10 cm; in some species it can be over 30 cm. Apart from its size, this scyphomedusa is very similar to the hydromedusae described above (p. 61), but with the following differences: (a) the manubrium is tentaculate (short in *Aurelia*, but very long in some species); (b) there is no velum

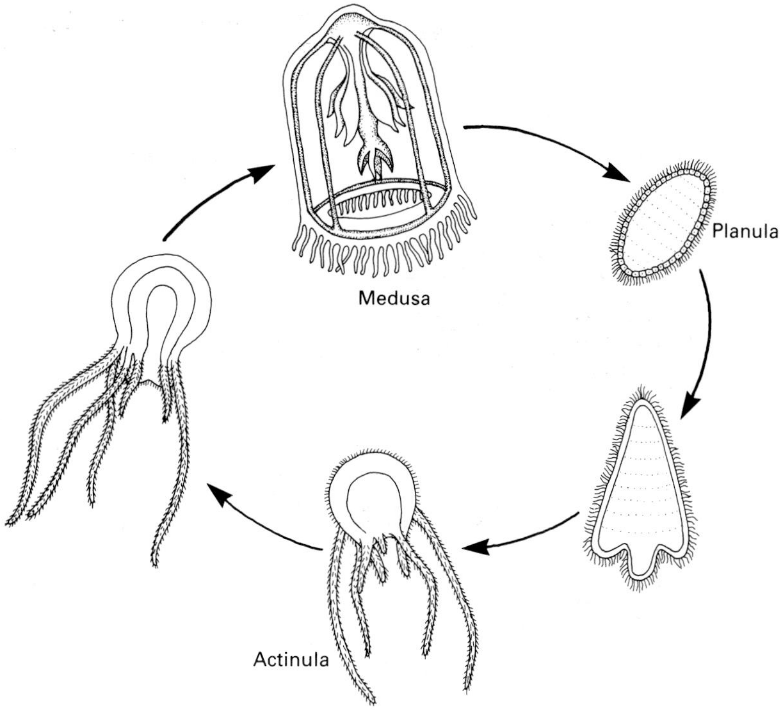

Fig. 3.21. Life cycle of a hydrozoan (*Aglaura*) that has no polyp (after Barnes, 1980).

(cubomedusae do have one; see below); (c) the mesoglea is thick and fibrous, but contains some wandering amoebocytes; (d) nerve rings are rare, instead the pulsation control is centred on marginal concentrations of neurons situated in small club-shaped structures, the rhopalia, that also carry statocysts and sometimes ocelli (Fig. 3.25); and (e) gonads are gastrodermal. Most scyphomedusae are carnivorous but some, including *Aurelia*, are filter feeders. Here the young filter water through the tentacular margin whereas older individuals use mucous strands, secreted on the subumbrellar surface, to trap plankton. The whole subumbrellar surface is flagellated, as are the arms of the manubrium, and these flagella pass food particles to the mouth.

The medusae are usually gonochoristic. The products of fertilization develop, through a planula, to a polypoid scyphistoma that buds by fission, to give juvenile medusae, the ephyrae larvae (Fig. 3.24). Both scyphistoma and ephyra feed.

There are a number of more aberrant scyphozoans. Some do not have a scyphistoma; others have a scyphistoma that develops gonads precociously but lacks a medusa. Stauromedusae are stalked and live attached to aquatic weeds (Fig. 3.26). Cubomedusae, already mentioned, are fast swimmers with a vicious sting. They have a velum and gastrodermal gonads, so they are often considered intermediate between hydromedusae and scyphomedusae and put in a separate class, the Cubozoa. *Cassiopeia*, common in the Caribbean, rests upside down in the quiet, shallow water of mangrove swamps. It possesses many small, secondary mouths that enter into the gastrovascular cavity directly. Small animals, trapped on the surface of frilly arms, are passed to the mouths in mucous strands.

3.4.3.3 Class Anthozoa

Members of this class are polypoid but (see also above, p. 59) larger and more complex than polyps in other classes (Fig. 3.27). They have a flattened

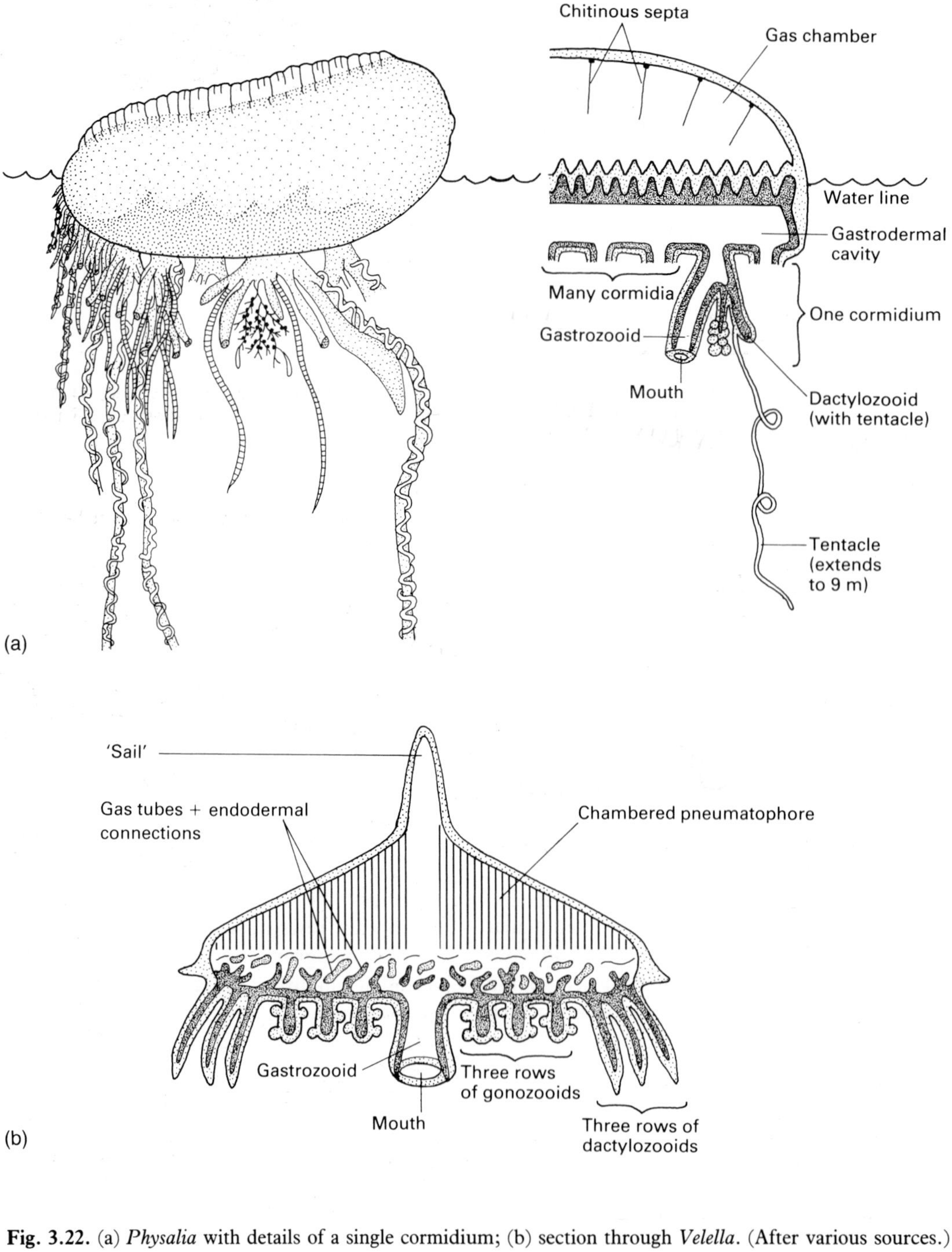

Fig. 3.22. (a) *Physalia* with details of a single cormidium; (b) section through *Velella*. (After various sources.)

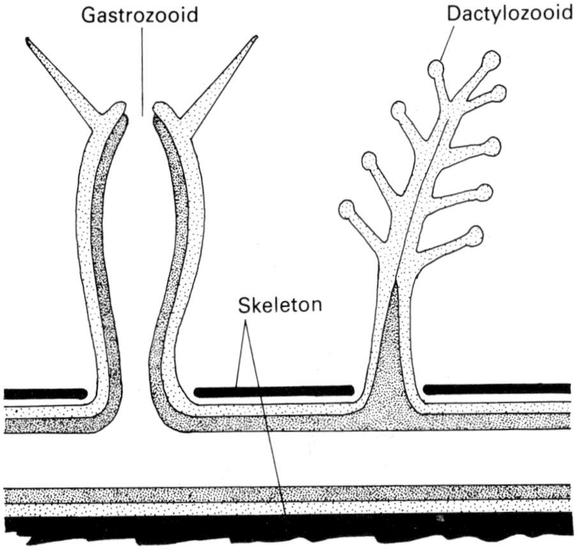

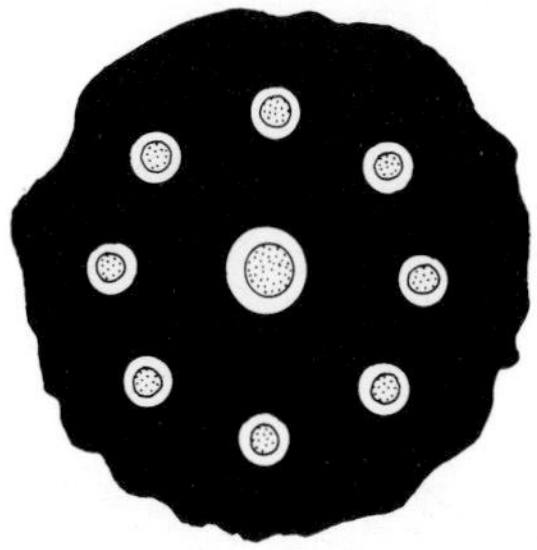

Fig. 3.23. Part of the colony of a hydrocoral, *Millepora*: (left) vertical section; (right) view from above (after Russell-Hunter, 1979).

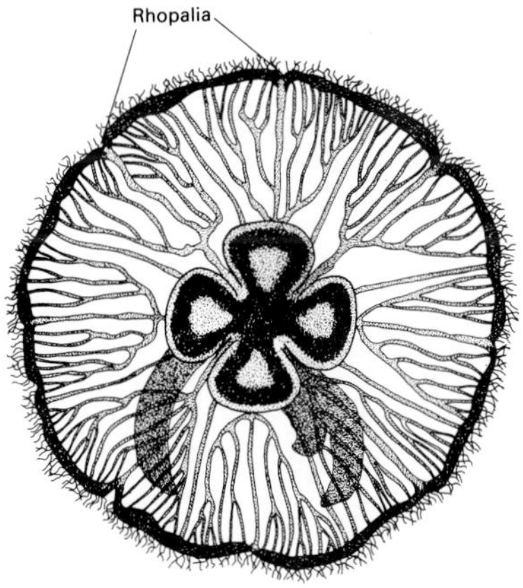

(a)

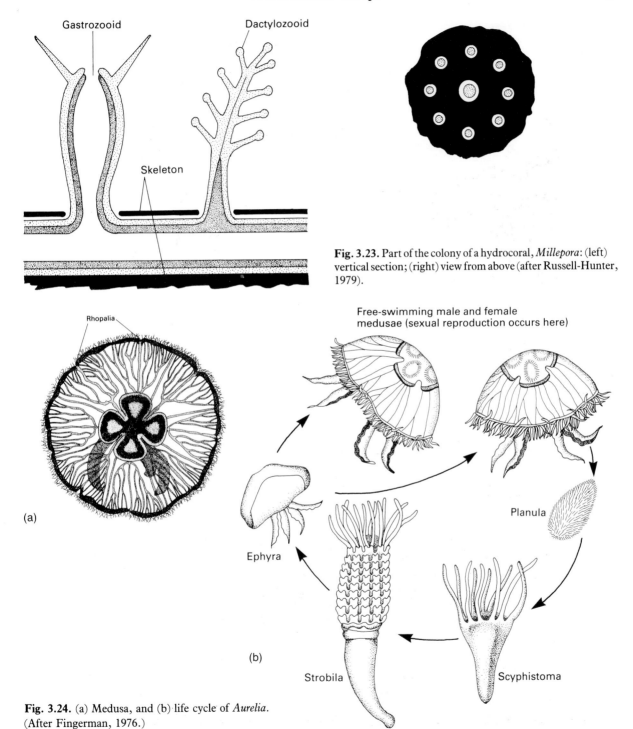

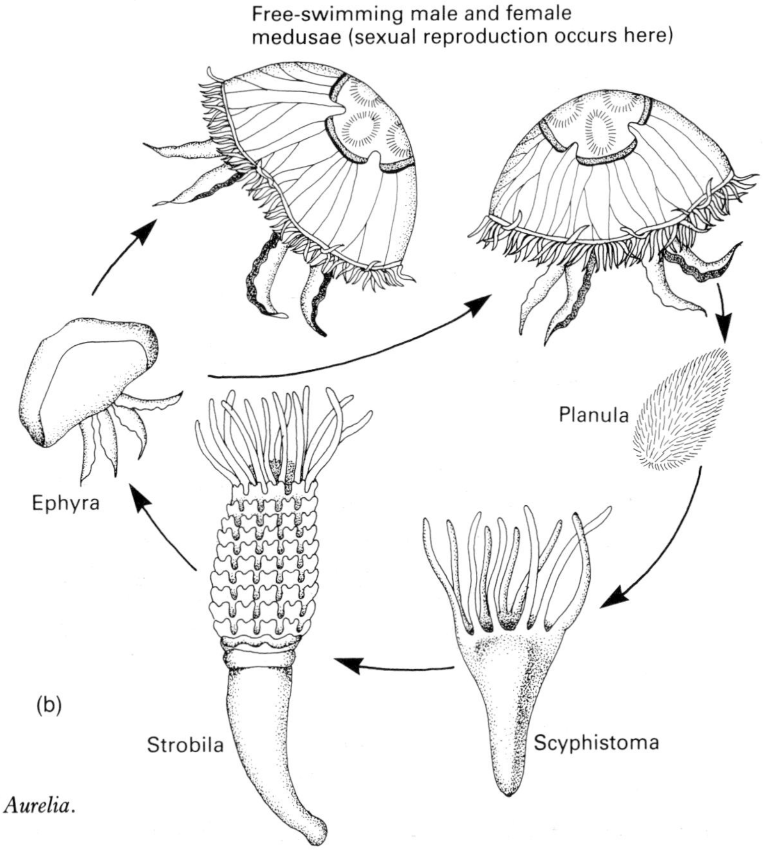

(b)

Fig. 3.24. (a) Medusa, and (b) life cycle of *Aurelia*. (After Fingerman, 1976.)

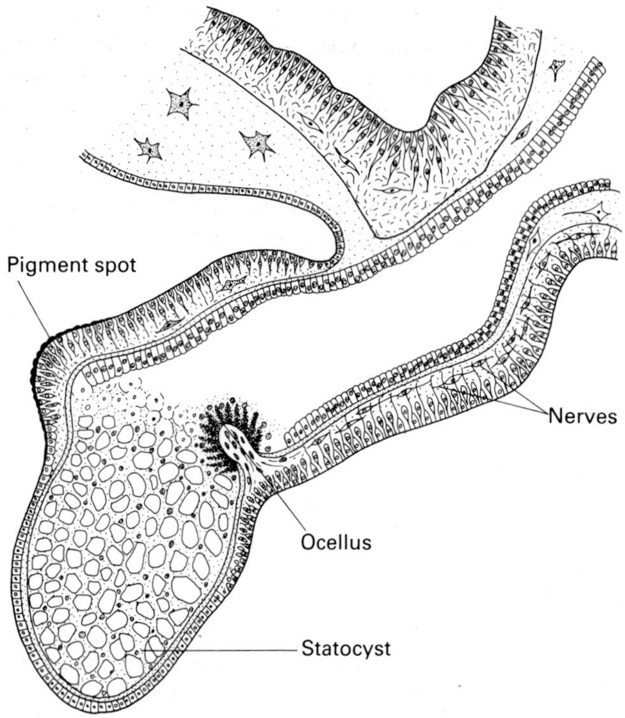

Fig. 3.25. Section through a rhopalium of *Aurelia* (after Hyman, 1940).

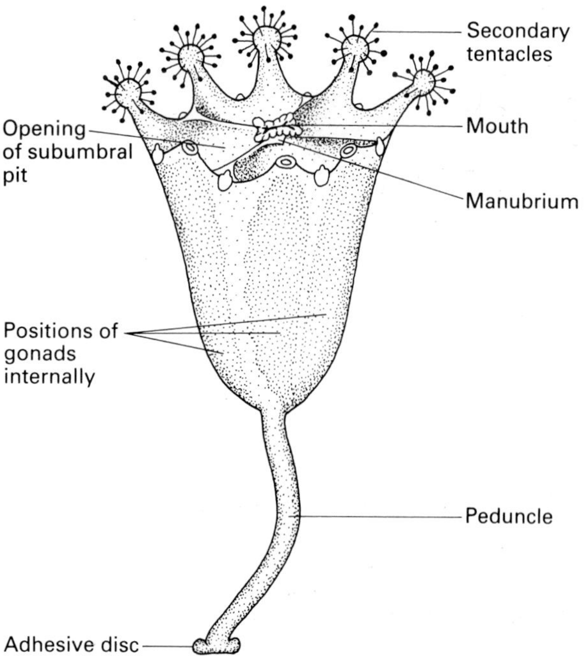

Fig. 3.26. *Haliclystus*, a stauromedusan (after Bullough, 1958).

oral disc bearing tentacles and a slit-like mouth leading to a laterally flattened stomodaeum. A ciliated groove, the siphonoglyph, can occur in one or both 'corners' of the mouth and extend into the stomodaeum. Siphonoglyphs probably direct water into the gastrovascular cavity, and maintain hydrostatic pressure. Together with the slit-like form of the mouth, these siphonoglyphs give a degree of biradiality to the symmetry, i.e. it is not possible to cut the body into two equivalent halves along any diameter. Mesenteries, intucked gastrodermis with mesoglea between, divide up the gastrovascular cavities. Some, so-called *completes*, attach to both the outer wall of the column and the stomodaeum, but have perforations in their walls, orally, to facilitate circulation of gastrovascular fluids; others, *incompletes*, are not attached to the stomodaeum at all (Fig. 3.27).

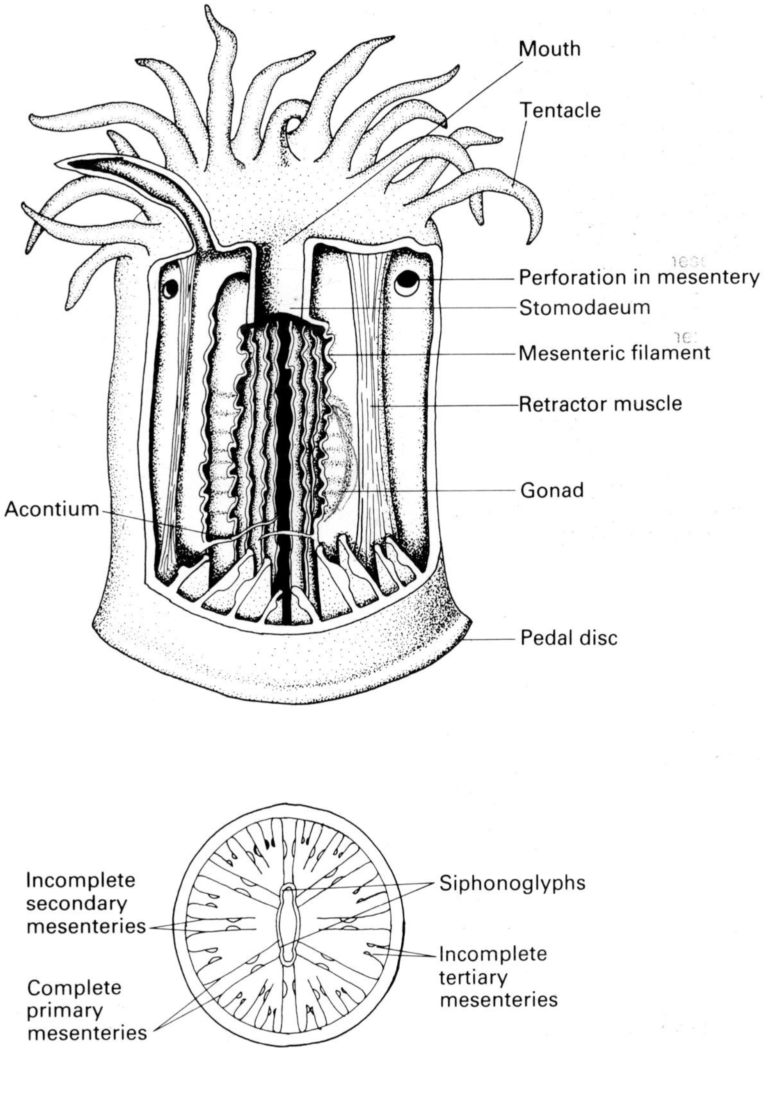

Fig. 3.27. Diagrams of a typical anemone. Lower figure is a transverse section through the stomodaeum. (After Alexander, 1979.)

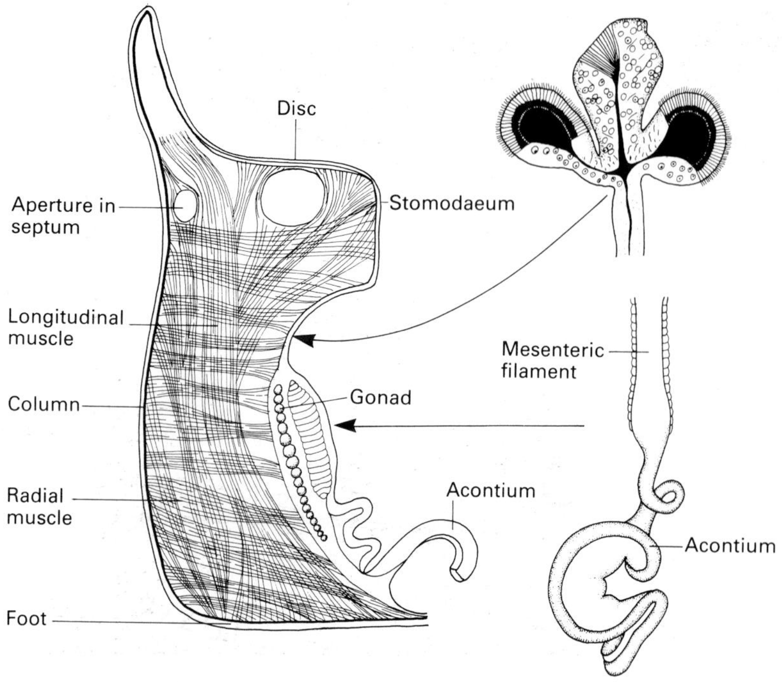

Fig. 3.28. Organization of a mesentery (after Fretter & Graham, 1976).

The edges of the mesenteries are trilobed filaments (Fig. 3.28). The outer filaments carry cilia and the central one carries cnidocytes and cells that secrete enzymes. The nematocysts, as is typical of Anthozoa, do not have an operculum (the filament often just ruptures out of the nematocyst through a three-flapped opening) and no cnidocil (see p. 57). The septa can also carry free threads, the acontia, which can wrap around food and help to kill it; they carry many cnidocytes. The mesoglea is thick, fibrous and cellular. Muscles are largely gastrodermal with longitudinal retractor blocks occurring in the mesenteries and circular muscles occurring in the columnar gastrodermis. Reproduction can be asexual with budding (sometimes to form colonies, see below), fission and laceration (Chapter 14), or sexual with gonads occurring in some or all mesenteries (of gastrodermal origin). Both gonochorists and hermaphrodites are known. Fertilization may be external or internal, and in the latter case some development may also occur internally (primitive viviparity). Planula larvae are typically produced and can either be planktotrophic or lecithotrophic (see Chapter 14) with a variable life-span. The mouth and mesenteries can begin to form in the larva but tentacles do not usually form until after attachment.

Sea anemones are familiar, solitary anthozoans and occur throughout the world in coastal waters. They commonly live attached to hard substrata in the littoral region. Some, such as *Metridium,* have well-developed muscles around the oral disc. These form a sphincter so that a collar can be drawn over the tentacles when the animal closes up between tides (Fig. 3.29) for protection and to economize on respiratory metabolism. A common North American and European species, *Stomphia coccinea*, can, under threat, release itself and swim by bending of the body. A few other species swim by lashing their tentacles. Members of the family Minyadidae are pelagic and float upside down by use of a gas bubble. Cerianthearians are also solitary anthozoans adapted for life in silt bottoms. They live in their mucous tubes formed from ptychocysts (above p. 57).

There are also various colonial, coral-forming

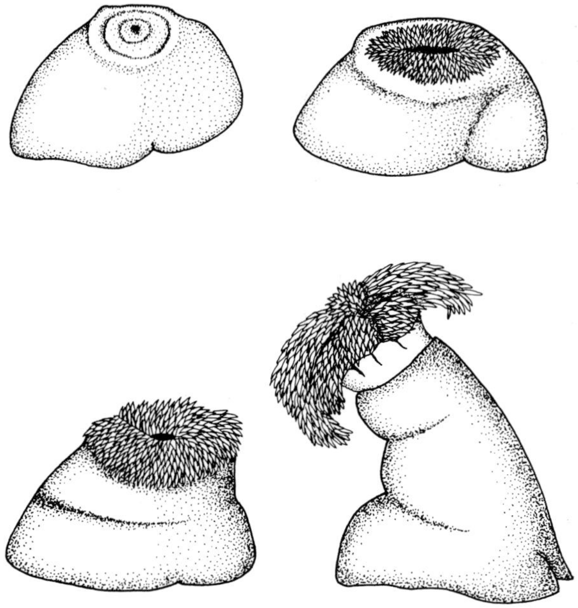

Fig. 3.29. *Metridium senile* in various states of contraction (modified after Robbins & Shick, 1980).

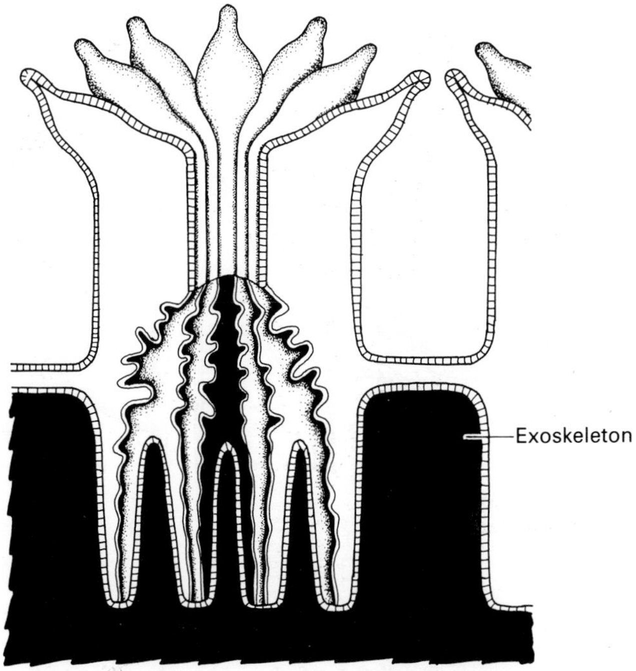

Fig. 3.30. Longitudinal section through a scleractinian coral (after Barnes, 1980).

anthozoans. The stony, scleractinian corals consist of polyps that are connected laterally and 'sit upon' a calcareous 'exoskeleton' (Fig. 3.30). This provides support and protection, for when the polyps contract down they project little above the skeletal plate. Octocorals (with only eight tentacles) also consist of colonies of polyps, attached laterally by a coenenchyme, but always in some branching form. The coenenchyme consists of thick mesoglea and connecting gastrodermal tubes (Fig. 3.31). The mesoglea contains amoebocytes that secrete a $CaCO_3$ skeleton, so here the skeleton is internal. In some, the coenenchyme contains a central supporting rod of either protein or calcareous spicules. The latter are extracted from the precious red coral, *Corallium*, to fashion the well-known coral jewellery.

Finally, alcyonacean anthozoans are soft corals. These occur in soft, leathery colonies that may be very large > 1 m. The coenenchyme does contain some calcium spicules but less than in other octocorals.

The form of all these corals can vary considerably and depends upon variations in modular growth patterns.

3.4.4 Symbiosis

A large number of cnidarians, through all classes, harbour symbiotic autotrophs: zoochlorellae (green algae) in the gastrodermis of hydrozoans; zoochlorellae in the mesoglea of the sedentary scyphomedusa, *Cassiopeia* (see above, p. 63); zooxanthellae (dinoflagellates of the genus *Symbiodinium*) in the gastrodermis of anthozoans, anemones and corals. Excess photosynthates, maltose in *Hydra* and mainly glycerol in anthozoans are used up by the host and can sustain it for long periods of starvation if not forming its only source of food. In anthozoan corals the symbiosis also facilitates the deposition of the calcareous skeleton, in a largely unknown manner. The importance of symbiosis in nutrition is discussed further in Chapter 9.

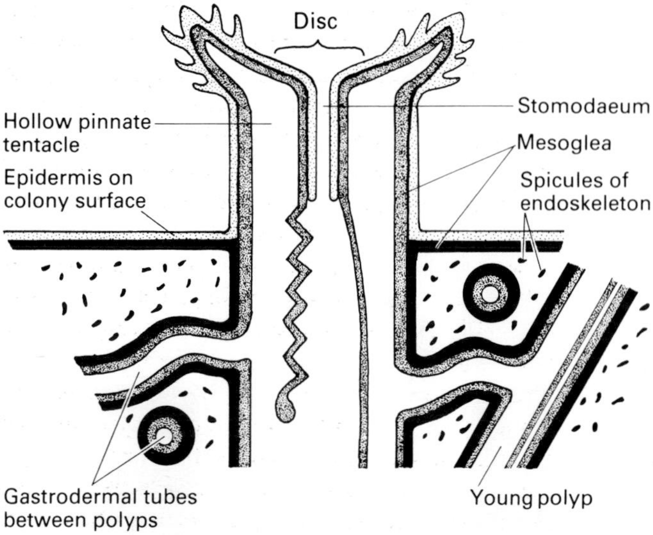

Fig. 3.31. Longitudinal section through an octocoral (after Russell-Hunter, 1979).

Symbioses, involving particularly anthozoans, are also important in providing defence for other organisms. Small fish of the genus *Amphiprion* live symbiotically amongst the tentacles of large anemones; something on the fish (possibly mucus derived from the anemone itself) inhibits nematocyst discharges. The anemone provides protection and in turn receives some protection from certain predators. Other animals that have a similar commensal relationship with anemones are cleaning shrimps, snapping shrimps, arrow crabs, brittle stars and several further species of fishes. Anemones also attach to other animals: scyphomedusae, ctenophores and the shells occupied by hermit crabs. They again offer protection and gain from being moved around into new feeding areas.

3.5 Phylum CTENOPHORA (comb jellies, sea gooseberries)

3.5.1 Etymology

Greek: *ktenos*, comb; *phoros*, bearing.

3.5.2 Diagnostic and special features (Fig. 3.32)

1 Medusoid; but radial symmetry rendered biradial by two 'tentacles'.

2 Gastrovascular cavity with mouth and anal pores.

3 Two-layered body wall, sandwiching thick 'mesoglea' that has amoebocytes and smooth muscle fibres.

4 Cnidocytes absent but tentacles carry 'lasso cells' (colloblasts).

5 Locomotion due to fused ciliary plates ('comb plates') in eight oral/aboral bands.

6 Nerves in subepidermal net, well-developed beneath comb rows.

7 Carnivorous; gas exchange by diffusion.

8 Reproduction: some asexual, but very limited; all are hermaphrodite with complete determinate cleavage leading ultimately to a 'cydippid' larva.

9 Exclusively marine; mainly pelagic.

Animals in this phylum so resemble cnidarian medusae that they are often thought to constitute an evolutionary offshoot of some ancestral medusoid cnidarian. Indeed for a long time the ctenophores were classed together with cnidarians in a single phylum: the Coelenterata (Greek: *koilos*, hollow; *enteron*, intestine). The two separate phyla, Cnidaria and Ctenophora, are still often united in a division of the animal kingdom known as Radiata (referring to their symmetry).

However, ctenophores have a number of important distinguishing features. First amongst these must be the comb rows, from which the phylum derives its name. Each comb row consists of short, transverse plates of long cilia, fused as combs, the comb plates (Fig. 3.33). The ciliary beat sweeps aborally, so the animal is driven, mouth forward, usually upward in the vertical plane, but they can also swim downward. Control is through the nerve net and an apical statocyst (Fig. 3.33). Tentacles emerge from deep, ciliated sheaths on the aboral surface. The tentacles bear colloblasts (Fig. 3.34)—pear-shaped structures with intracellular helical threads and carrying granules filled with mucous material—that are used for entangling prey. One species, *Euchlora rubra*, was thought, until recently, to have cnidocytes but new evidence suggests that it derives these from its cnidarian prey. The gut consists of canals like those of some medusae, but these can open to the outside world via anal pores. Digestion is both extra- and intracellular and indigestible residues are voided through both the mouth and the anal pores. All are hermaphrodite; the gonads comprise two bands, one of ovary and the other of testis, in the gastrodermis. The cydippid larvae closely resembles the ovoid adults and develops directly into them. In flattened species, though, there has to be extensive transformation to attain the adult state (see below).

A notable feature of ctenophores is their luminescence. Light production occurs in the gastrodermis.

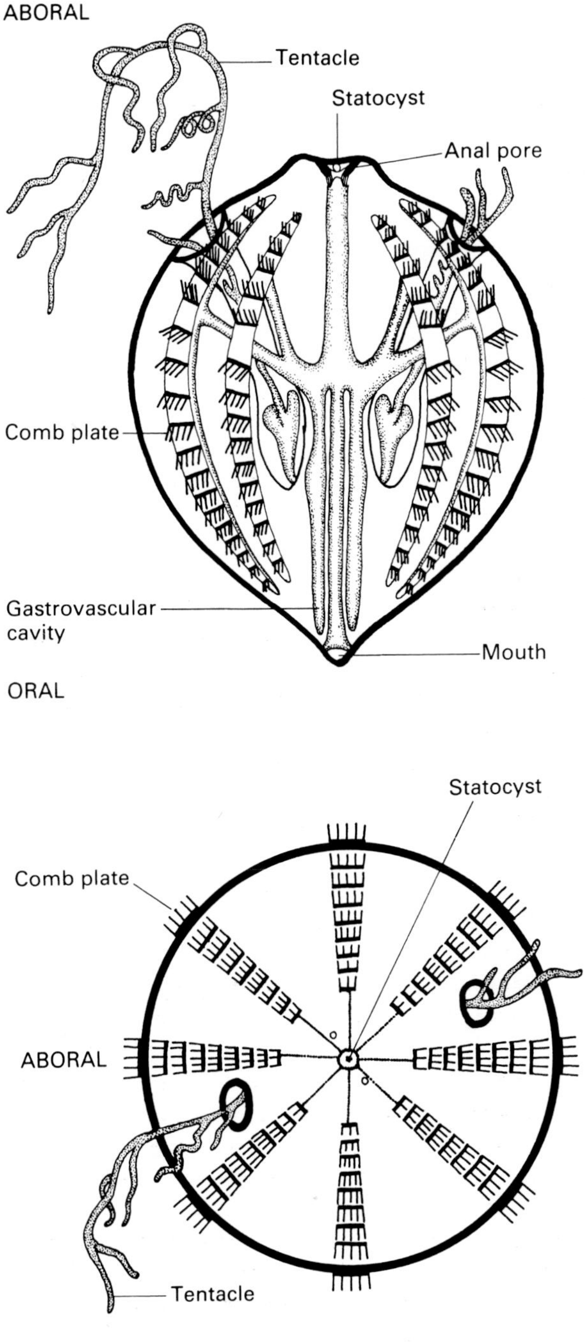

Fig. 3.32. A typical ctenophore, from the side and aboral surfaces (after Buchsbaum, 1951).

3.5.3 Classification

A small, 'minor' phylum, with only *c.* 100 known species. These are distributed between two classes as shown below.

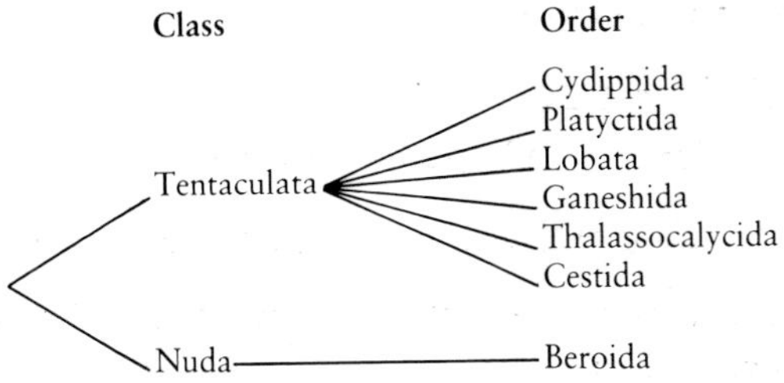

3.5.3.1 Class Tentaculata

Members of this class invariably possess two tentacles. It includes the familiar 'gooseberry-shaped' types (Fig. 3.32). But some are laterally flattened in the tentacular plane leaving the expanded outer portions as lobes (Fig. 3.35). Some are so expanded and flattened that they have the appearance of transparent ribbons. These animals not only use their combs but also body undulations for locomotion (Fig. 3.36). Individuals in another group are flattened along the oral/aboral axis, have greatly reduced or absent comb rows and have adopted a creeping existence (Fig. 3.37). All these flattened forms have reduced tentacles.

3.5.3.2 Class Nuda

These do not have tentacles at any stage in their life cycles. They are predators of other ctenophores, and have a large, flexible mouth and expanded stomodaeum to accommodate their ovoid prey. They are, therefore, somewhat conical in shape (Fig. 3.38).

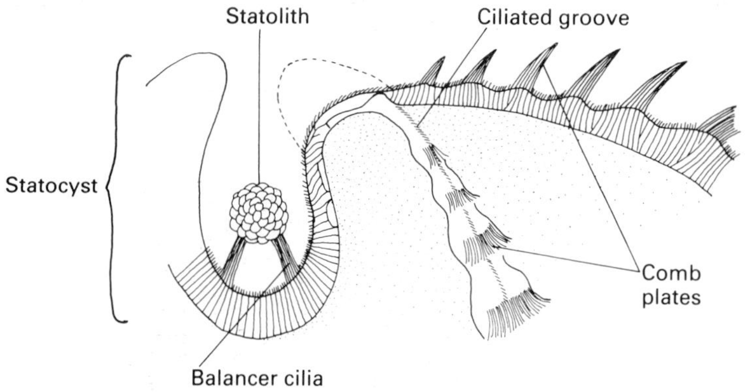

Fig. 3.33. Details of the aboral end of a ctenophore with statocyst and comb plates (after Barnes, 1980).

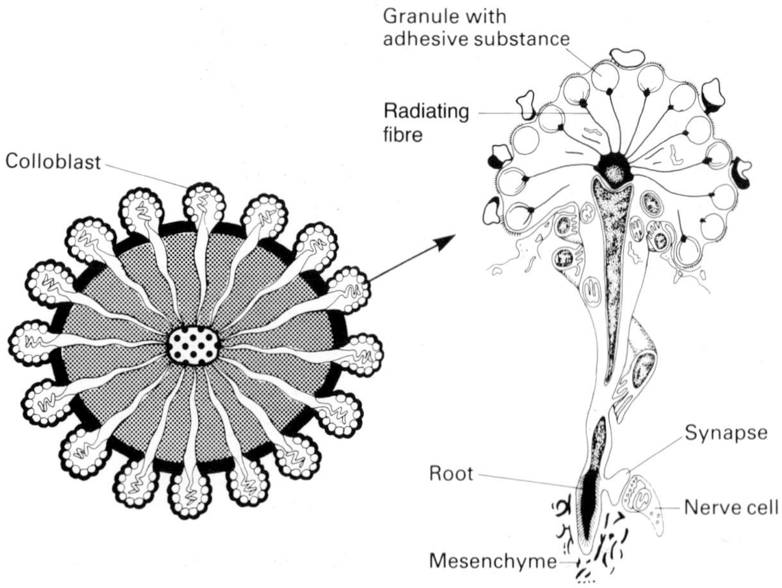

Fig. 3.34. Section through a tentacle of a ctenophore, with details of a colloblast (after various sources).

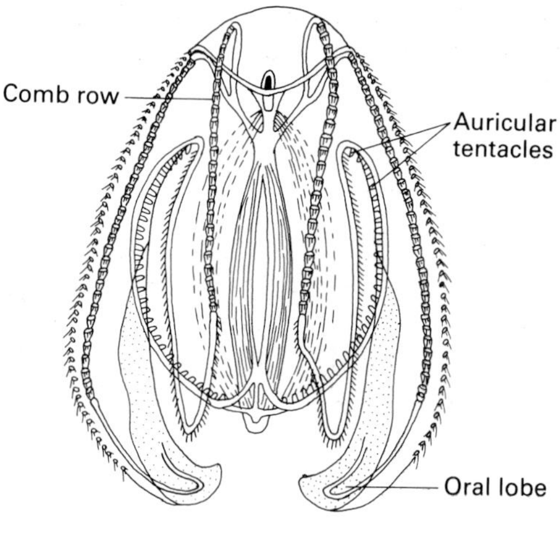

Fig. 3.35. Sagittal view of *Mnemiopsis*, a tentaculate ctenophore (after Meglitsch, 1972).

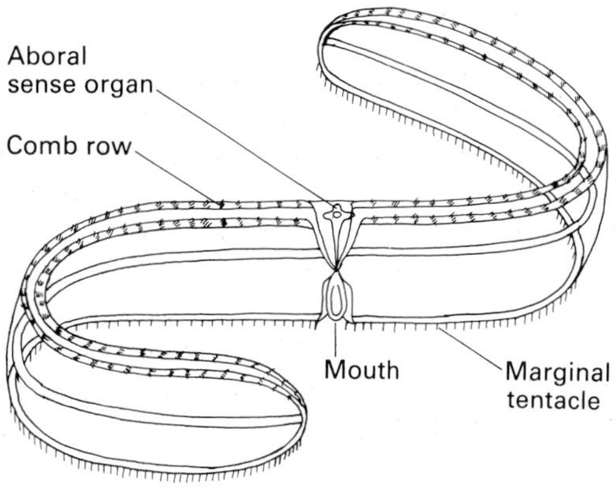

Fig. 3.36. *Velamen*, a very flattened tentaculate ctenophore (after Meglitsch, 1972).

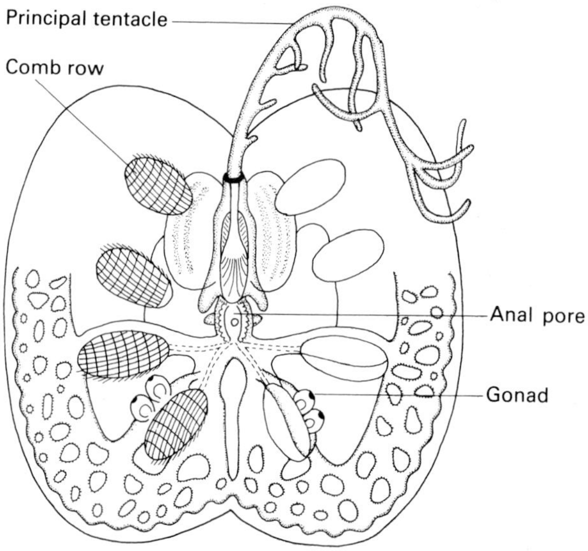

Fig. 3.37. *Ctenoplana*, flattened along the oral/aboral axis (after Meglitsch, 1972).

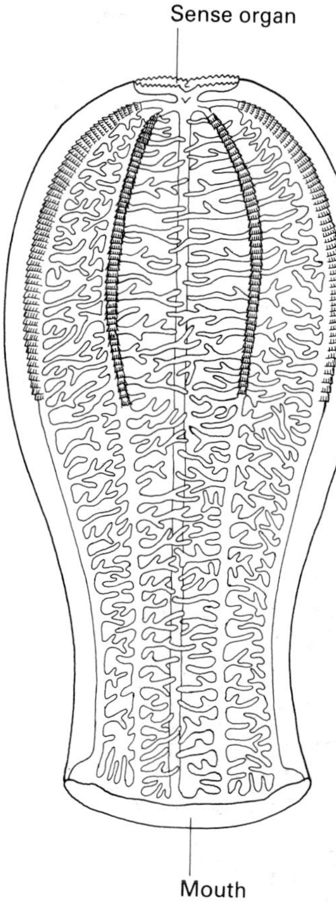

Fig. 3.38. *Beroë*, one of the Nuda (after Meglitsch, 1972).

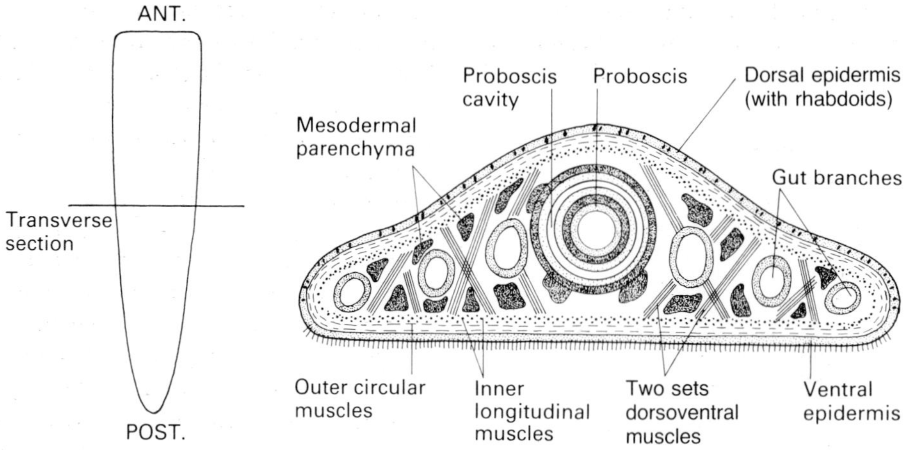

Fig. 3.39. A typical free-living platyhelminth (after Russell-Hunter, 1979).

3.6 Phylum PLATYHELMINTHES

3.6.1 Etymology

Greek: *platy,* flat; *helminthes,* worms.

3.6.2 Diagnostic and special features (Fig. 3.39)

1 Bilaterally symmetrical.
2 Body consisting of three tissue layers; organ systems occur in middle, parenchymatous tissue layer.
3 Without body cavity other than a blind-ending gut.
4 Feed almost exclusively from other animals.
5 Dorsoventrally flattened; vermiform, in free-living species with a head bearing sense organs.
6 Mainly soft-bodied, but with protective mucous secretions and a few species with internal spicules.
7 Gaseous exchange by diffusion; no blood system.
8 Protonephridial 'excretory' organs.
9 Nervous system can consist of a nerve net but usually concentrated in longitudinal fibres.
10 Generally hermaphrodite.
11 Eggs cleave spirally, but this may be very modified.
12 Development generally direct in free-living forms (in some there is a free-swimming Müller's or Götte's larva), but very elaborate (involving several larval stages) in some parasites.
13 Occur in all major habitats, aquatic and terrestrial, including the tissues of other animals.

Three major features of this phylum are: the occurrence of true bilateral symmetry; the evolution of an invariably cellular, third layer (parenchyma) between the epidermis and the lining of the gut (gastrodermis); and the dorsoventral flattening of the body. Bilateral symmetry is characteristic of all other phyla not so far mentioned—the so-called Bilateria (though echinoderms revert, secondarily, to a form of radial symmetry, see Section 7.3)—and this has evolved in response to the requirements of locomotion on solid surfaces (see also dorsoventrally flattened ctenophores, Section 3.5 and p. 72). The flattening is an inevitable consequence of the absence of a blood system and the dependence of these animals on diffusion for the distribution of nutrients and gases to and from tissues (Chapter 11).

Another system that appears in this phylum (all groups except acoels) is the protonephridial 'excretory' system. The quotes are used because its function is probably largely osmoregulatory, not excretory (Chapter 12), true excretory products still leaving the body mainly from the surface by diffusion. The protonephridia consist of blind-ending tubes that terminate in 'flame cells' (Fig. 3.40), which in fact consist of two cellular units, a tubule cell and a cap cell

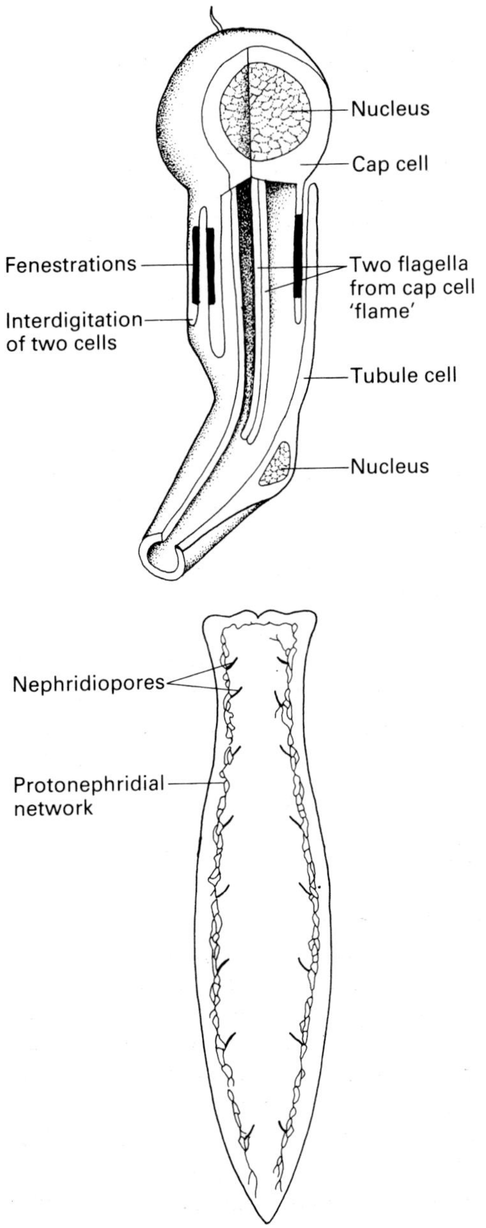

Fig. 3.40. Protonephridial system with details of a flame cell (shown above). (After Barnes, 1980.)

bearing two or more flagella. Fenestrations occur at the interdigitations of the two cells. The flagella beat, like a flame, and fluids are drawn in through the fenestrations (for details see p. 377). They are directed through tubules to an excretory pore and thence out of the body.

In general, excretory, feeding, reproductive and nervous and sensory *systems*, involving integrated sets of tissues and organs, become a more obvious and more elaborate feature of platyhelminths than they were of the poriferans and coelenterates.

With a blind-ending gut and simple 'mouth' (p. 74) it is not surprising that the majority of the members of this phylum feed from the tissues of other animals. A few species feed on algae, especially diatoms. In the carnivores there has been a general tendency for the association between food and feeder to become more and more intimate, even in the predominantly free-living turbellarians (Section 3.6.3.1) where trophic associations range from predator–prey (or scavenger–carrion) to forms of symbiosis, including ecto- and endocommensalism (symbiont benefits but host is not affected) and even ecto- and endoparasitism (symbiont benefits at the expense of its host). The majority of classes and of species of Platyhelminthes are exclusively parasitic. With parasitism have evolved: various specialized organs for attachment to the host; modification of the external surface; reduction or, in certain gut parasites that are surrounded by partially digested food, complete absence of the gut; expansion of reproductive systems.

3.6.3 Classification

Some 25 000 living species are distributed between four classes.

3.6.3.1 Class Turbellaria

Members of this class are mainly free-living and aquatic; but there are some terrestrial species, confined to humid areas, and mention has already been made of the commensalism within the group. One order, the Temnocephalida, is exclusively ectocommensal

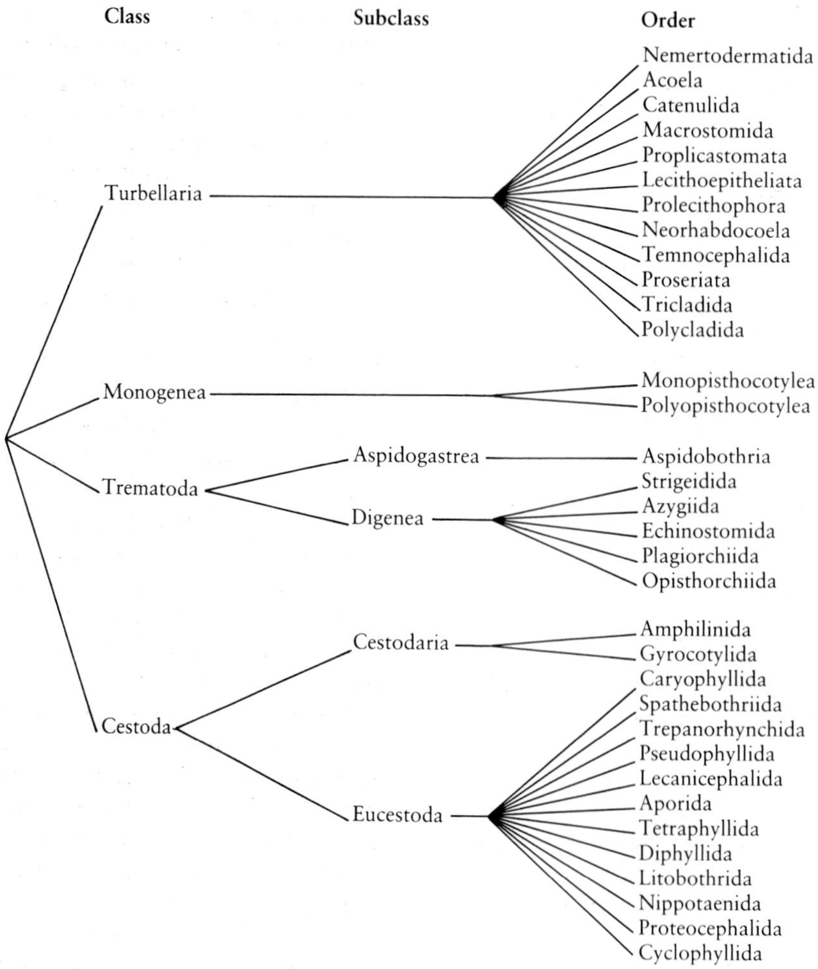

(possibly ectoparasitic) on the body surfaces of crustaceans, snails or turtles. These have a posterior adhesive disc and anterior tentacles (Fig. 3.41), and have no cilia on their body surface. So peculiar are they, that this order is sometimes raised to the rank of class.

Turbellarians range in size from a few millimetres—and many of these are found in the interstitial habitats of freshwater and marine systems—to giant land turbellarians, more than 0.5 m long. Cilia, lining the epidermis, provide the propulsive force in the small turbellarians; very small forms 'swim' through the interstitial fluids and the flattened body form of larger, creeping forms maximizes the ciliary surface area in contact with the substratum. Muscular undulations are important in the largest aquatic species and in all terrestrial species. Mucus, from gland cells, lubricates and protects the body surface but also forms trails that provide points of leverage for the cilia. It can also be used in the capture of food, to entangle prey. Hence mucus is very important in the biology and physiology of these animals and more than 50% of the energy obtained from the food can be invested in its production.

A transverse section through a fairly typical turbel-

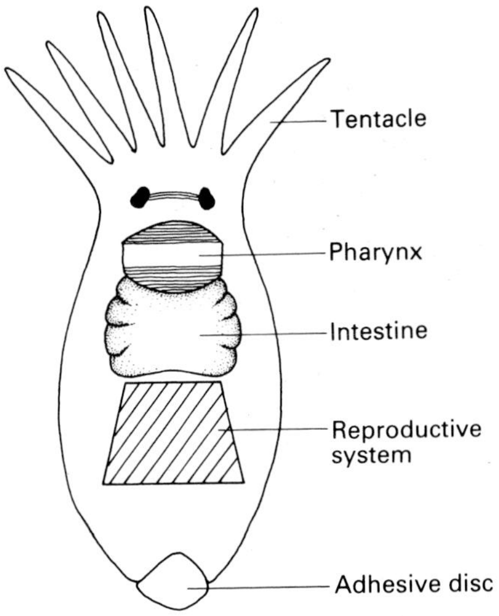

Fig. 3.41. A temnocephalid.

larian body is given in Fig. 3.42 (see also Fig. 3.39) to show cilia (may be absent from dorsal surface in some larger forms) and several muscle layers—consisting of circular, longitudinal, diagonal and transverse fibres. The muscle fibres are unstriated. In view of what has been said about mucus, it is not surprising that gland cells are prominent. They are often located in the parenchyma with ducts to the external surface. Also noticeable are various rod-like structures, the rhabdoids, some of which arise from epidermal cells, others (called rhabdites) from gland cells in the parenchyma. Their function is uncertain, but they are ejected when animals are disturbed and are used in defence (p. 389) and/or disintegrate to form a strong mucoid cover around the animal. They occur in almost all turbellarians. Parenchyma, an aggregation of irregularly shaped cells, fills the space between the internal organs. Amongst the larger cells may be small ones with a large nucleus and little cytoplasm, similar in structure and function to cnidarian interstitial cells and called neoblasts. They are important in replacing spent tissue and in regeneration after asexual reproduction or artificial transection (see Section 15.5.2).

The mobility of the turbellarians has led to some cephalization, i.e. modification of that anterior part of the body which moves first into new environments (Fig. 3.43). Eyes are commonly borne on the head—two to many—and are of the pigment-cup type (Section 16.7.3). Ciliated patches, sunken into grooves, are also prominent on the head and are probably chemoreceptors, important in the location of food. An anterior concentration of nerve tissue—'the brain'—has evolved to service these receptors. Emanating from this are pairs of longitudinal nerve cords, reduced to just two ventral cords in some groups (Fig. 3.44). There may or may not be an epidermal nerve net; though some acoels have only this.

The gut (Fig. 3.45) varies from a solid syncitial mass of cells (acoels) through a simple sac (the 'rhabdocoel' condition) to a series of lateral diverticulae (three in triclads and many in polyclads). This variation provided the basis for the classical system of turbellarian classification. The connection between mouth and gut shows correlated variation (Fig. 3.45).

The gonads are distinct (except in acoels) and connect to the outside via tubules that can have several dilations with various functions (Fig. 3.46). The male system invariably consists of many testes, and the sperm are usually biflagellate. The female system can either consist of many ovaries that produce entolecithal eggs or a single ovary with many separate vitellaria. The latter produce yolk cells, so the egg is ectolecithal (Section 14.4).

Those turbellarians that do not produce vitellaria are referred to as archoophoran and those with vitellaria as neoophoran.

Other notable features of turbellarians are their capacity for asexual reproduction, mainly by transverse fission and, correlated with this, their capacity for extensive regeneration after artificial transection. Like cnidarians they show polarity (this time anterior/posterior), possibly based upon an axial physiological/developmental gradient. Finally, with starvation, triclads are capable of extensive shrinkage (degrowth)—sometimes from adult size to size at hatching—and can regrow upon being refed. It has been speculated that this process may rejuvenate the tissues of triclads, and provide them with a capacity for immortality.

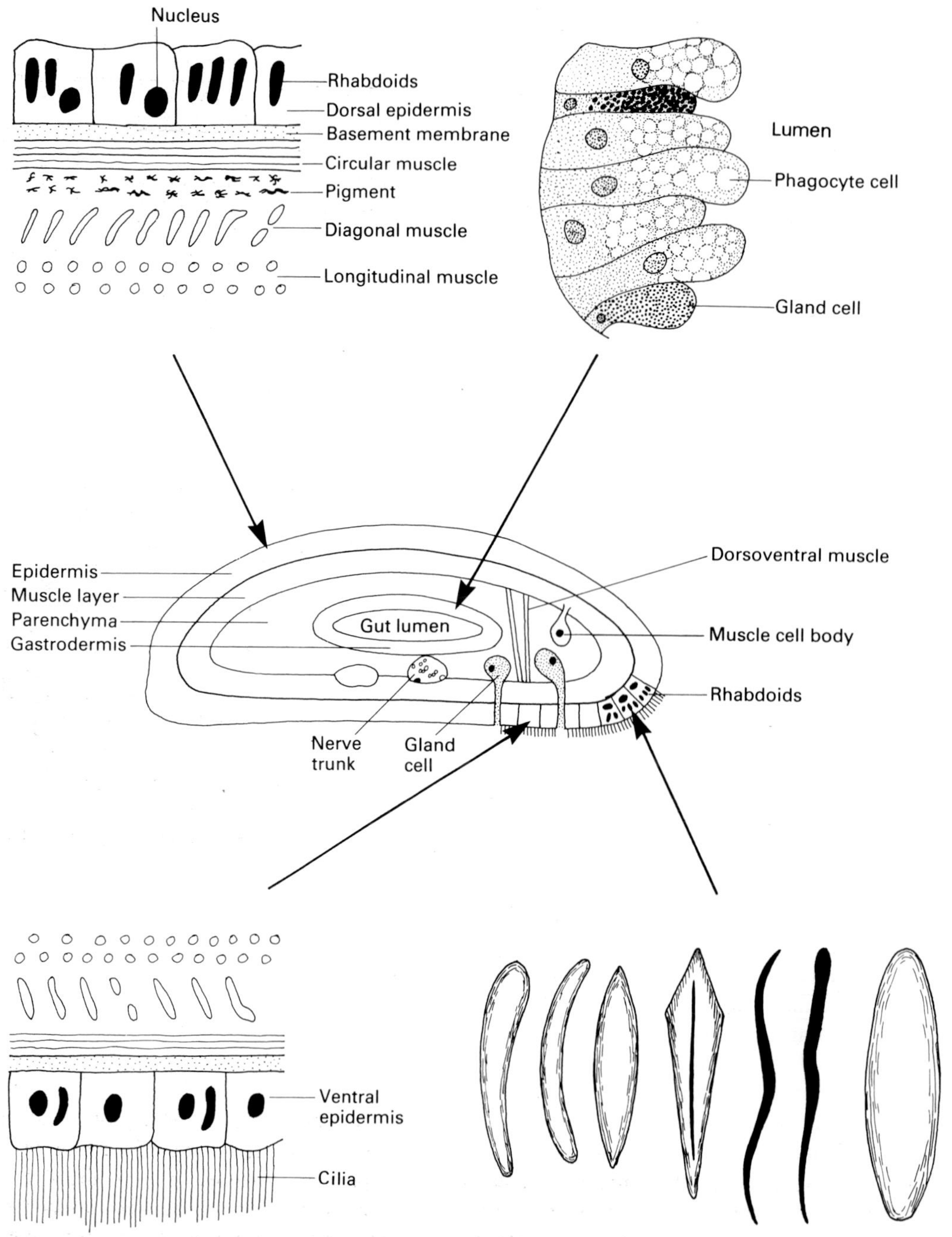

Fig. 3.42. Transverse section of a turbellarian with details of the rhabdoids, the dorsal and ventral surfaces and the lining of the gut (after various sources).

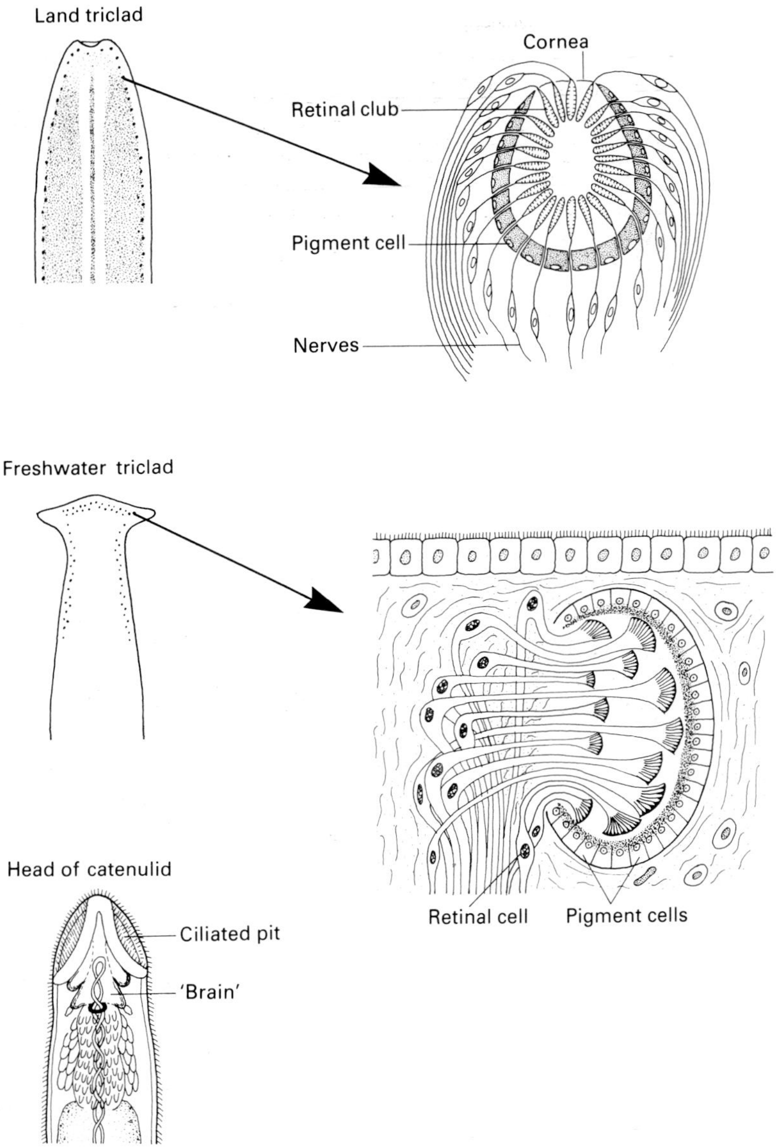

Fig. 3.43. Cephalization in Turbellaria with details of 'eye spots' (after various sources).

3.6.3.2 Flukes

The next two parasitic classes have many features in common, and were formerly treated as a single class (as they are still in some textbooks). They differ in that members of one class, the Monogenea, are largely ectoparasitic and have a simple life cycle, involving only one host (usually fishes), whereas members of the other class, the Trematoda, are almost exclusively endoparasitic. Moreover, members of the main trematode subclass, the Digenea, have complex life cycles always involving two or more hosts. Superficially, the basic body forms of both groups of flukes are similar to turbellarians, but important differences

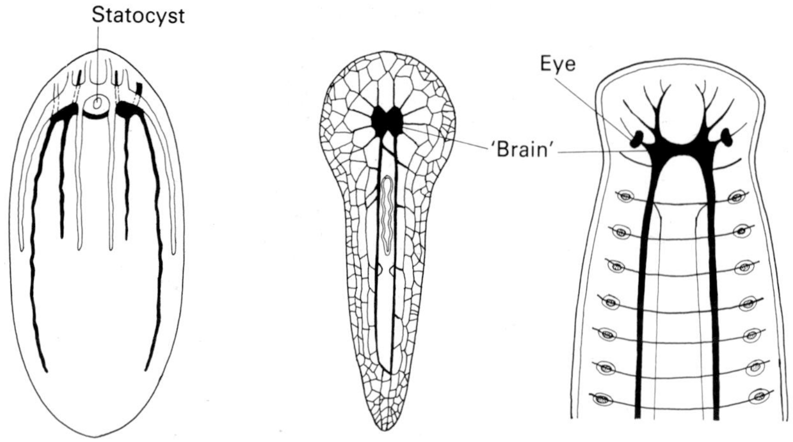

Fig. 3.44. Turbellarian nervous systems in Acoela (left), Polycladida (middle) and Tricladida (right). (After Barnes, 1980.)

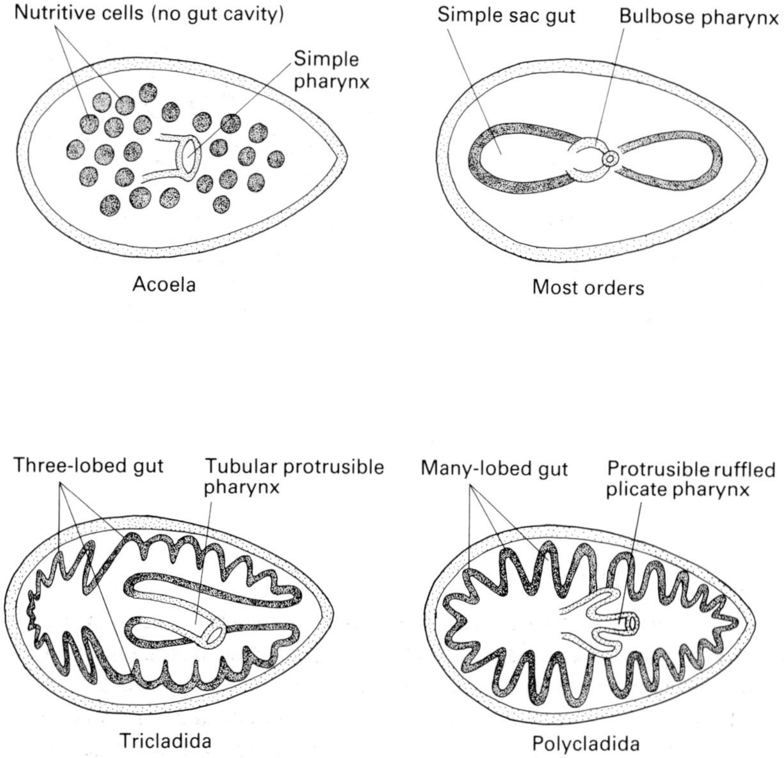

Fig. 3.45. Guts and mouths in Turbellaria (after Russell-Hunter, 1979).

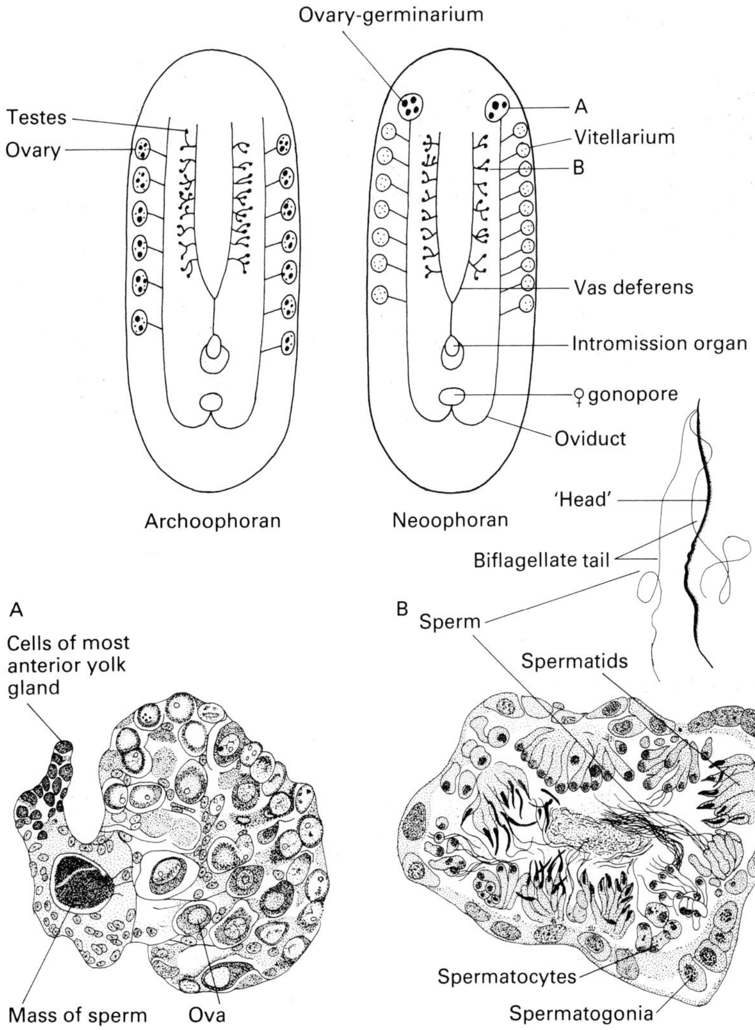

Fig. 3.46. Reproductive systems in Turbellaria.

include: a non-ciliated syncitium, the tegument, replaces the ciliated and cellular epidermis typical of most turbellarians (Fig. 3.47); a muscular pharynx, by which food is actively removed from the tissues of hosts, opens into a two-branched gut (Fig. 3.48); and there are no special sense organs in the adults.

Class Monogenea

As a rule, members of this group are *ecto*parasitic on *fish* hosts. As usual, though, there are many exceptions: some are found *inside* body chambers with openings to the exterior—mouths, gills, the urinogenital system—and a few even occur in the coelom;

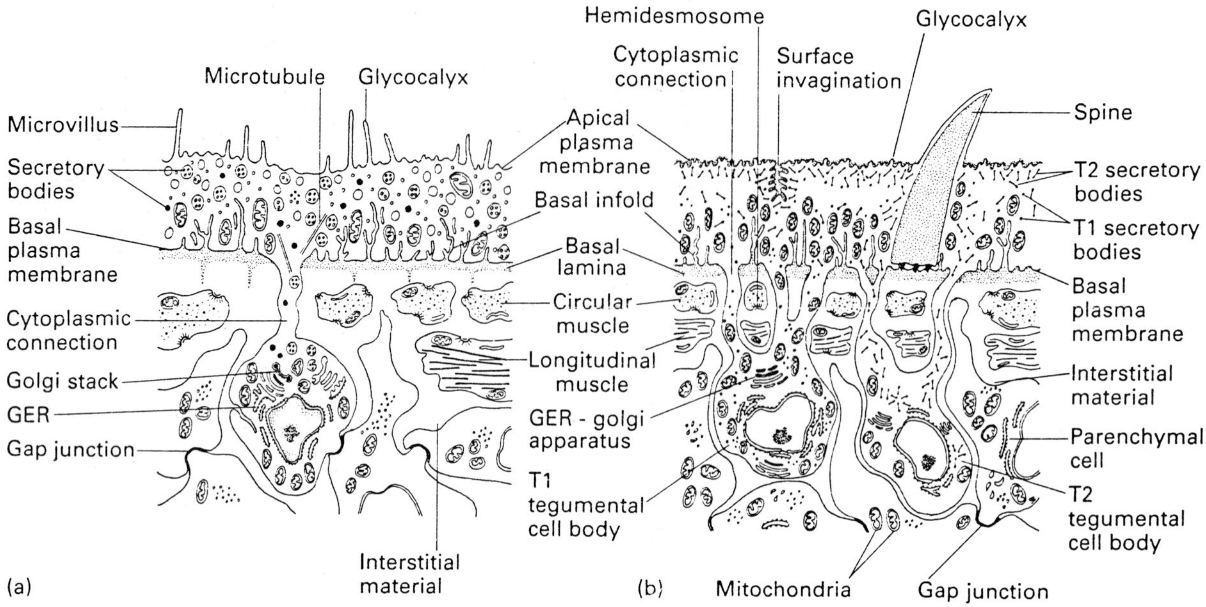

Fig. 3.47. Tegument of (a) monogenean and (b) trematode flukes.

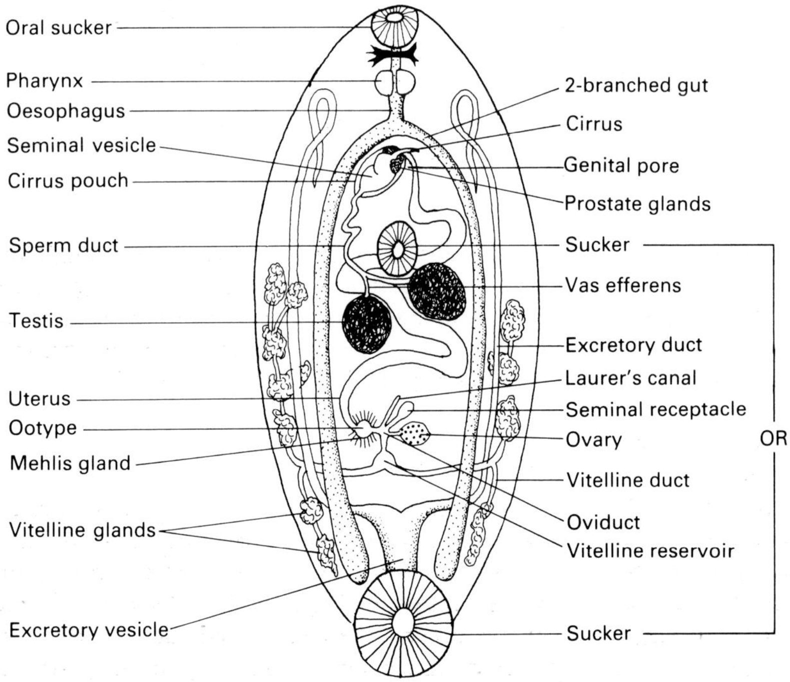

Fig. 3.48. Structure of a generalized fluke.

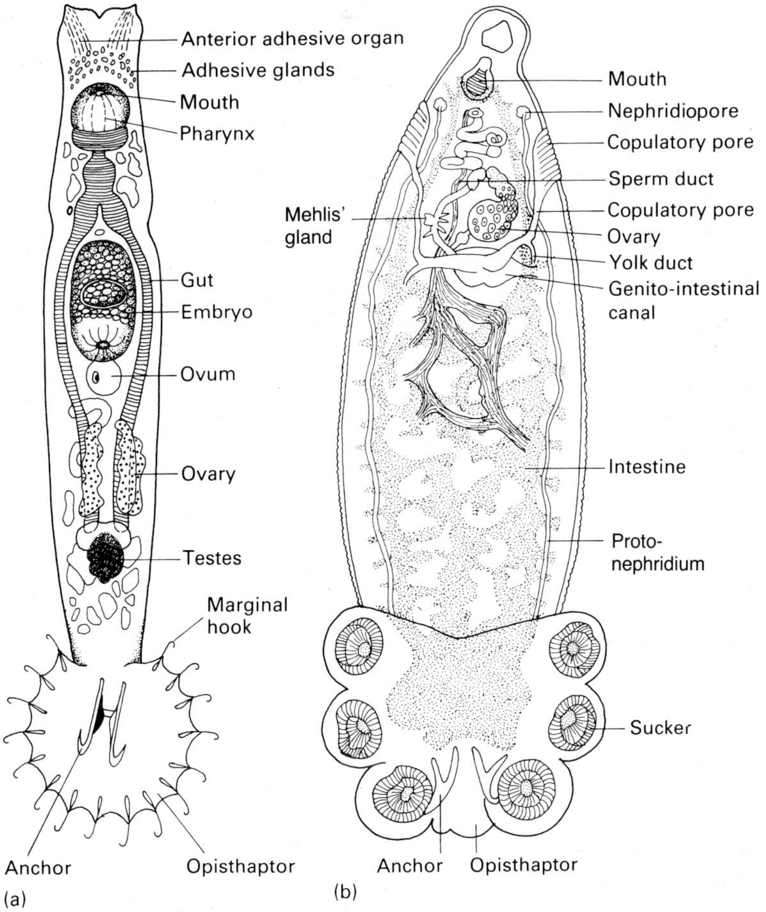

Fig. 3.49. Monogeneans: (a) *Gyrodactylus*, a monopisthocotylean; (b) *Polystoma*, a polyopisthocotylean. (After Meglitsch, 1972.)

amphibians, reptiles and cephalopod molluscs serve as hosts for a few species.

As might be expected for ectoparasites, where there is a need to attach to the external surface often on a mobile host, organs of attachment, including suckers, hooks and clamps, are an obvious feature at both anterior and posterior ends of the body (Fig. 3.49). Large posterior organs, called opisthaptors, are particularly prominent and are used as a basis for distinguishing between the two orders of this class: *Mon*opisthocotylea, with a *single* opisthaptor, but paired organs of attachment anteriorly; *Poly*opisthocotylea, with a divided opisthaptor but with a single oral sucker or a pair of eversible buccal suckers.

Monogenean life cycles are relatively straightforward. The eggs are shelled and on hatching release a free-swimming ciliated larva called an onchomiracidium. These larvae can disperse and allow exploitation of new hosts (Fig. 3.50).

Class Trematoda

Members of this class are endoparasitic, many causing diseases in man and domesticated animals. Like the monogeneans they have attachment organs, but in members of the large subclass Digenea these are situated around the mouth (oral sucker) and midventrally (acetabulum). The tegument plays an

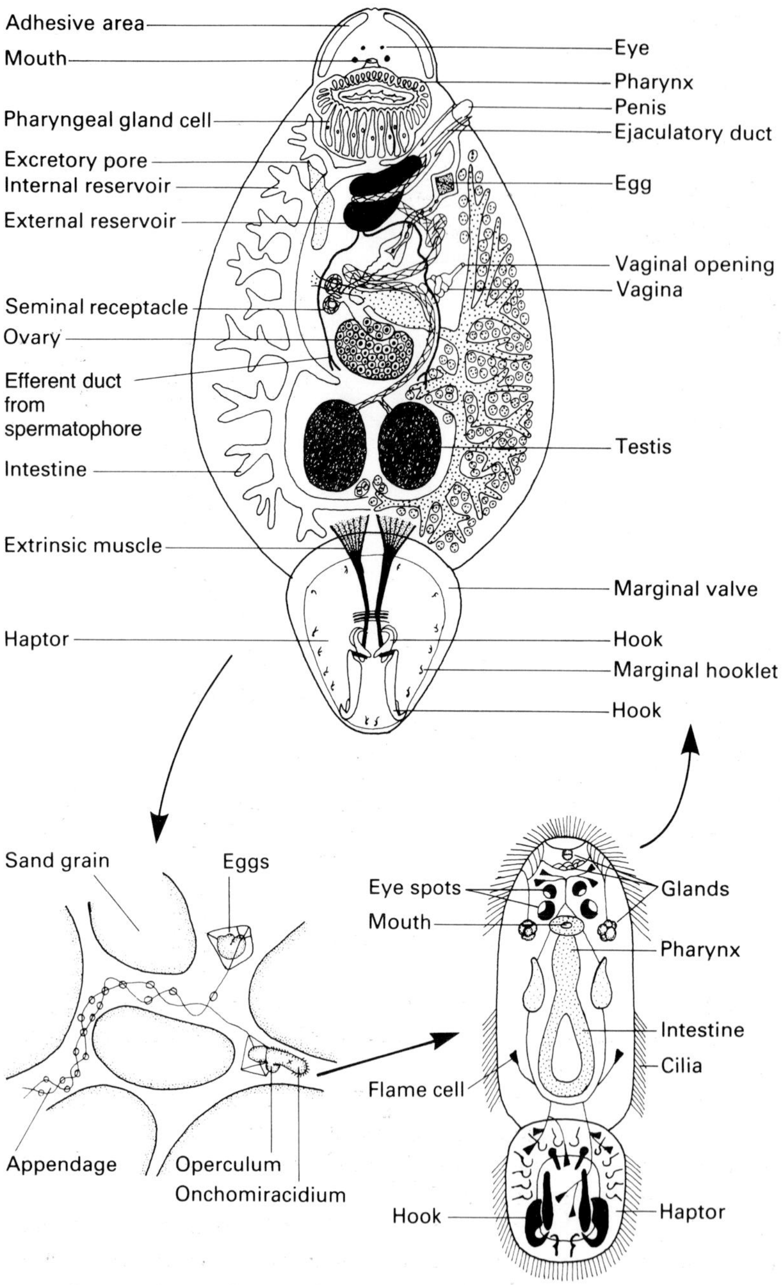

Fig. 3.50. A typical monogenean life cycle: the fish parasite *Entobdella soleae* (after several sources).

important part in the physiology of these animals. It provides protection, particularly against enzymes in gut-inhabiting species but also against host antibodies (p. 396). Some nutrients, in particular amino acids, can be taken up through it, directly from the host.

The digenean life cycle is complex: the host for the adult (gamete-producer) is the primary host and is a vertebrate; other hosts, containing components

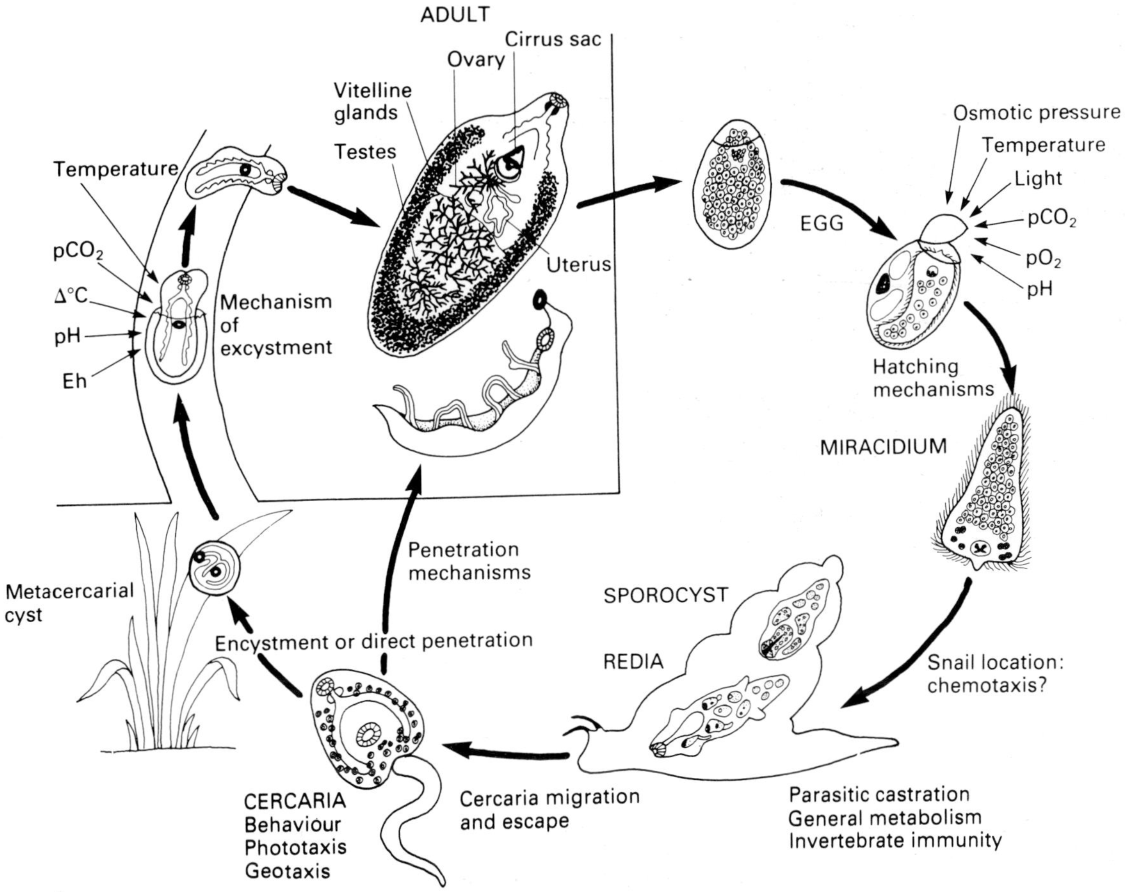

Fig. 3.51. Typical trematode life cycle (after Smyth & Halton, 1983).

of the life cycle capable of asexual reproduction, are known as secondary hosts and may number between one and three, of which one is usually a snail.

A summary of a typical life cycle is as follows (Fig. 3.51):

1 A shelled egg, with a lid, finds its way to the host gut and is passed out in the faeces (the eggs of some digeneans leave by way of the bladder and urine).

2 A free-living ciliated larva, called a miracidum, emerges.

3 This is eaten by a snail or penetrates its epidermis; the parasite then loses its cilia, and passes to the digestive gland of the snail.

4 There it develops into a sporocyst.

5 The sporocyst contains germinal masses that give rise, by mitosis, to a number of daughter sporocysts and/or redia.

6 These contain germinal cells that similarly develop to larvae called cercariae.

7 These leave the host by penetrating its tissues and are free-living.

8 They may penetrate, or be eaten by, a primary host direct, to complete the life cycle; or enter a secondary host and encyst to form a metacercaria.

9 If the secondary host, containing the metacercaria, is eaten by a primary host, the cycle is completed.

Parasitic adults produce vast quantities of eggs (e.g. the sheep liver fluke can release 20 000 per day). Moreover, each egg becomes multiplied to many propagules by a series of asexual stages. Such high levels of reproductive productivity (possible because endoparasites are, by definition, surrounded by superabundant food: the tissues of the host) can compensate for the massive losses of propagules in the hazardous transmission from one host to another.

The Schistosomatidae is a family of digeneans, notable on two counts. First it contains species producing the human disease schistosomiasis. This is seriously debilitating and can kill. In endemic areas, schistosomiasis can infect large fractions of the population and represents a major parasitic disease of human beings. Second, schistosomatids are gonochoristic, which differs markedly from most other platyhelminths. Why gonochorism should turn up in parasites, when the chances of the two sexes meeting must sometimes be slim, remains a mystery.

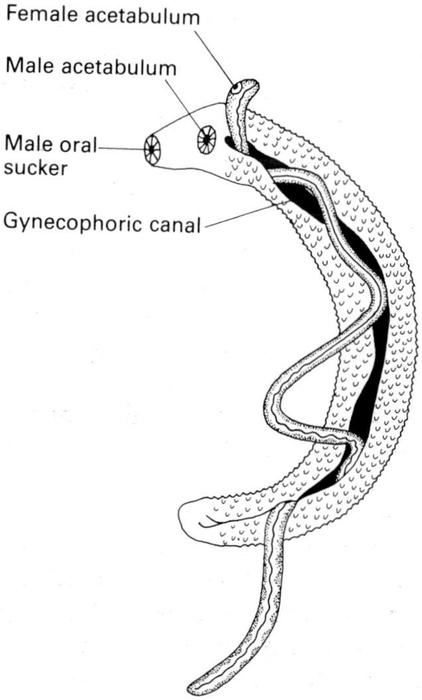

Fig. 3.52. *Schistosoma mansoni (after Meglitsch, 1972).*

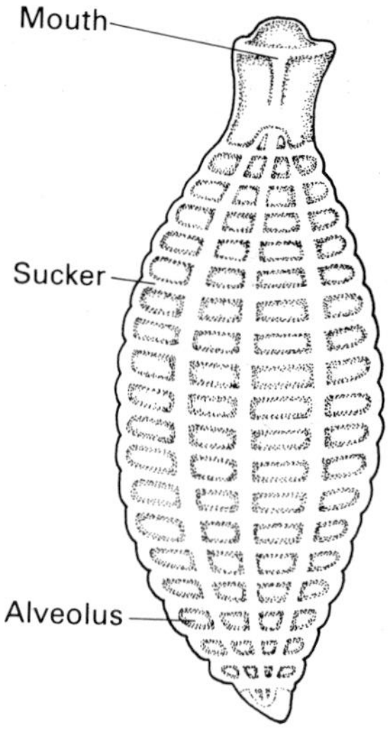

Fig. 3.53. An aspidogastrean (after Meglitsch, 1972).

However, male and females, when they do associate, can do so intimately, for the male has a ventral groove along the length of its body into which fits the more slender female, and mature males and females are usually found like this (Fig. 3.52).

Members of the second subclass, Aspidogastrea, show some similarities to digeneans as well as monogeneans (Fig. 3.53). They do not have an opisthaptor but they do have either a large sucker covering the whole ventral surface or a ventral row of suckers. They are endoparasites, occurring in the guts of fishes and reptiles and in the pericardial and renal cavities of molluscs, but their life cycles are simple, involving two and sometimes even just one host.

3.6.3.3 Class Cestoda

Members of this class, commonly known as tapeworms, have become adapted for parasitizing the vertebrate gut and as such are the most specialized group of Platyhelminthes. Surrounded by a digested slurry of hosts' food, they have no mouth or gut. Instead the tegument (Fig. 3.54), like that of the flukes (but more so), is capable of absorbing food

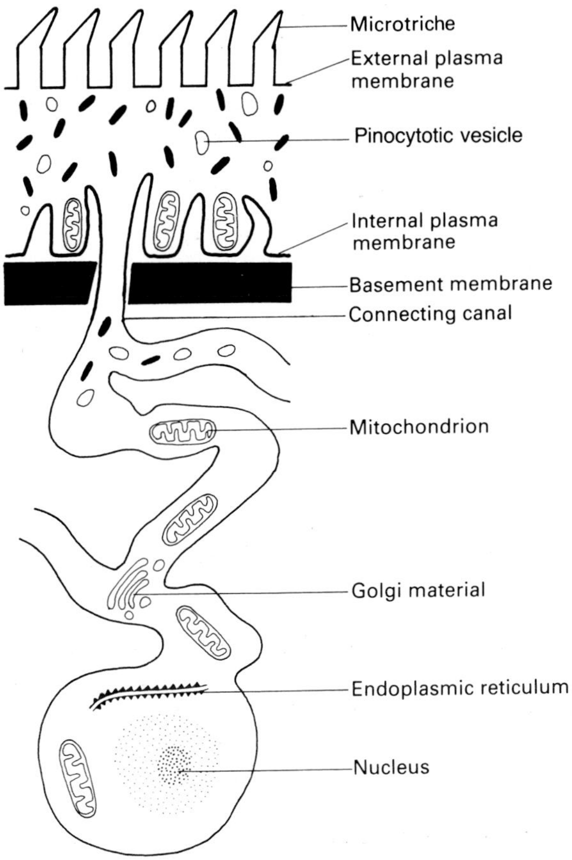

Fig. 3.54. Diagrammatic section of a cestode tegument (after Meglitsch, 1972).

directly. The surface is thrown into folds, the microtriches, to increase surface area, and the tegument is richly endowed with mitochondria, indicating the importance of active transport. The tegument may also facilitate digestion of food by the host enzymes, possibly by surface effects.

An elaborate organ of attachment, the scolex, with suckers and spines (Fig. 3.55), anchors the worm against a countercurrent of food produced by intestinal peristalsis. Life in this long tube, in a mobile medium of food, has favoured the evolution of an iterated body form (strobila) of segments (proglottids) budded at the scolex end (strobilization) in immature state and lost at the opposite end as bags of eggs (known as gravid proglottids). Between, the reproductive system develops (Fig. 3.55). It is very similar to that of digeneans and uses absorbed food to make gametes. Cross-fertilization occurs where possible but self-fertilization between two different proglottids in the same strobila or even within the same proglottid is known. (Note that segment formation in the development of annelids and arthropods is in the opposite direction to tapeworms, i.e. posterior, forward, not anterior, backward, p. 29). Fertilized eggs fill the uterus, and the gravid proglottid consists of little more than an expanded uterus full of eggs. This 'production-line' system is capable of astounding fecundity; for example, the human tapeworm, *Taenia saginata,* when fully grown develops 700–1000 proglottids and sheds 3–10 of these daily, each containing 100 000 eggs.

Eggs from gravid proglottids either hatch in the free environment or when gravid proglottids are eaten by hosts. They give rise to oncosphere larvae, sometimes surrounded by a ciliated membrane (embryophore), and these are referred to as a coracidia. These, in secondary hosts, can give rise to several larval types of which the main four are:

1 Procercoid: a small spindle-shaped larva with solid body and posterior hooks.

2 Plerocercoid: a solid larva possessing an adult scolex but lacking the embryonic hooks of **1** from which it develops.

3 Cysticercoid: a larva consisting of an anterior vesicle containing a scolex that is not evaginated, and a tail-like posterior region containing the larval hooks.

4 Cysticercus: a bladder enclosing a single scolex retracted and invaginated within itself.

Other larval types are variations of these four. Sometimes there can be asexual reproduction of scolexes in the secondary host by hydatid larvae. When a secondary host is eaten by a primary host, the larva infects the gut and develops into a new tapeworm. Figure 3.56a–c shows a number of typical life cycles each involving some of these stages.

Just because they are securely attached by the scolex, it should not be imagined that tapeworms cannot and do not move in the intestine of their host. They have muscles, consisting of subtegumental cir-

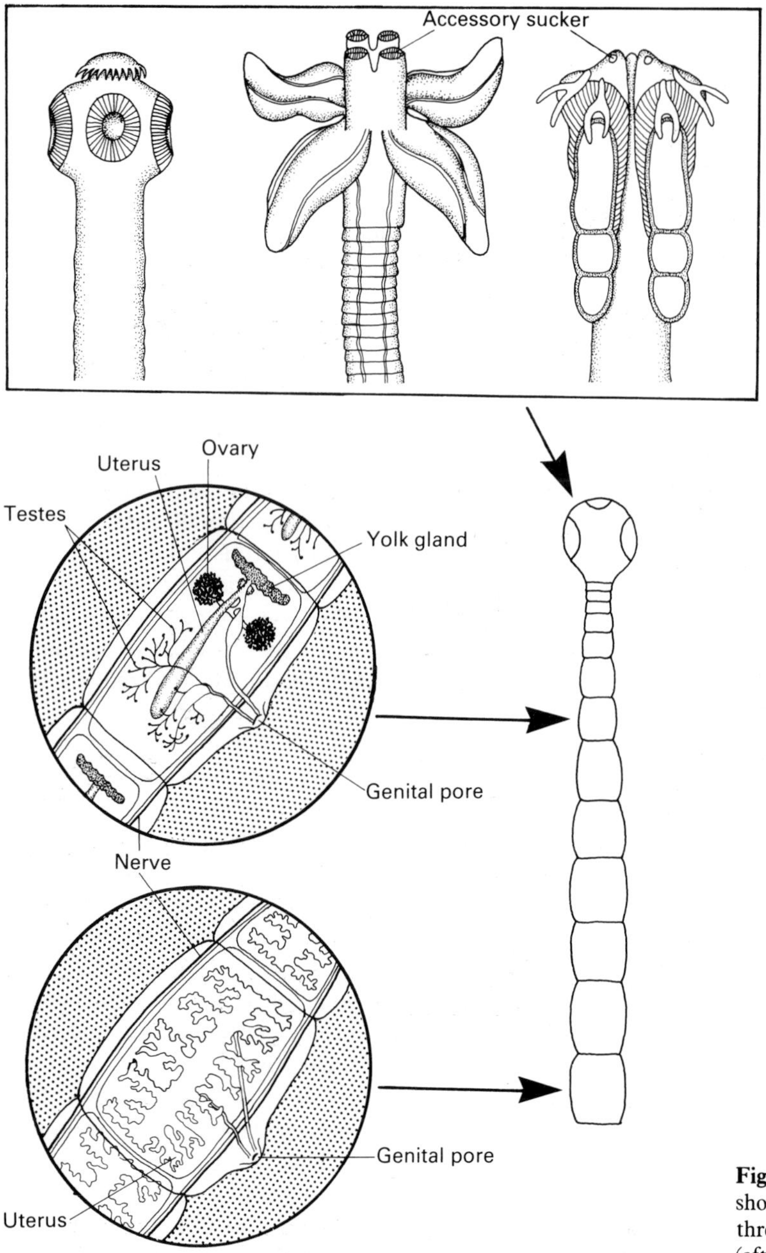

Fig. 3.55. Body form of tapeworms showing details of various regions—three examples of scolexes are given (after various sources).

cular and longitudinal fibres and a parenchymal musculature of longitudinal, transverse and dorsoventral fibres. Moreover, an anterior nerve mass occurs in the scolex and two lateral, longitudinal cords extend backwards through the strobila. Dorsal, ventral and also accessory lateral cords might occur, and ring commissures connect longitudinal cords in each commissure. The rat tapeworm, *Hymenolepis*, can

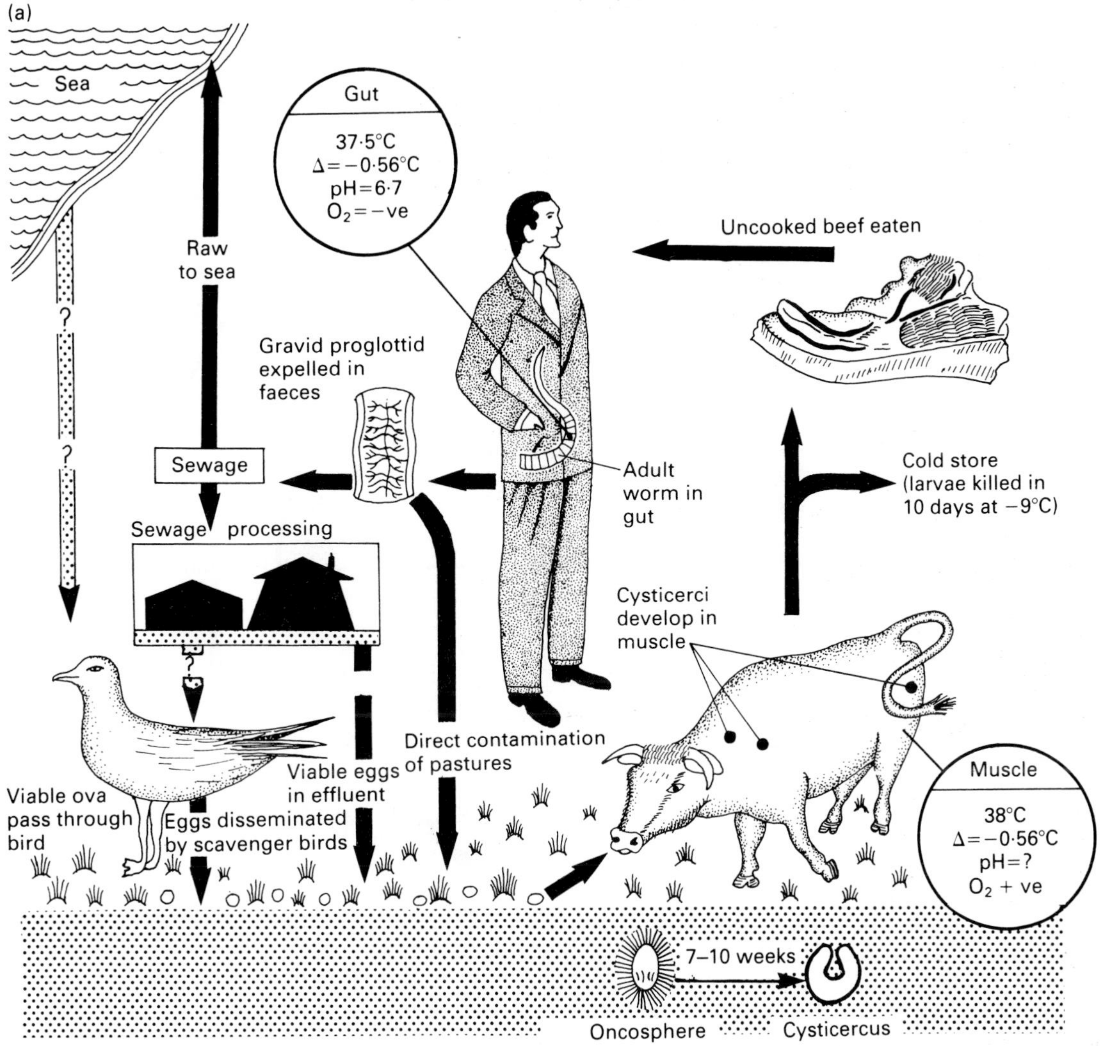

Fig. 3.56. Life cycles of: (a) *Taenia saginata*; (b) *Diphyllobothrium dendriticum* (p. 90); (c) *Hymenolepsis nana* (p.91). (After Smyth, 1962.)

undergo extensive migration in the gut of its host, presumably controlled and effected by this elaborate neuromuscular system.

Currently, the class Cestoda is divided into two subclasses, the Eucestoda and the Cestodaria. There are twelve recognized orders of the Eucestoda, tapeworms proper, varying from each other quite considerably in body form and in the hosts they parasitize. They exploit the whole gamut of 'warm- and cold-blooded vertebrates'. Deserving of special mention are members of the order Caryophyllida (Fig. 3.57). These parasitize freshwater fishes (primary hosts) and crustaceans (secondary host). The adults have a poorly developed scolex and undivided

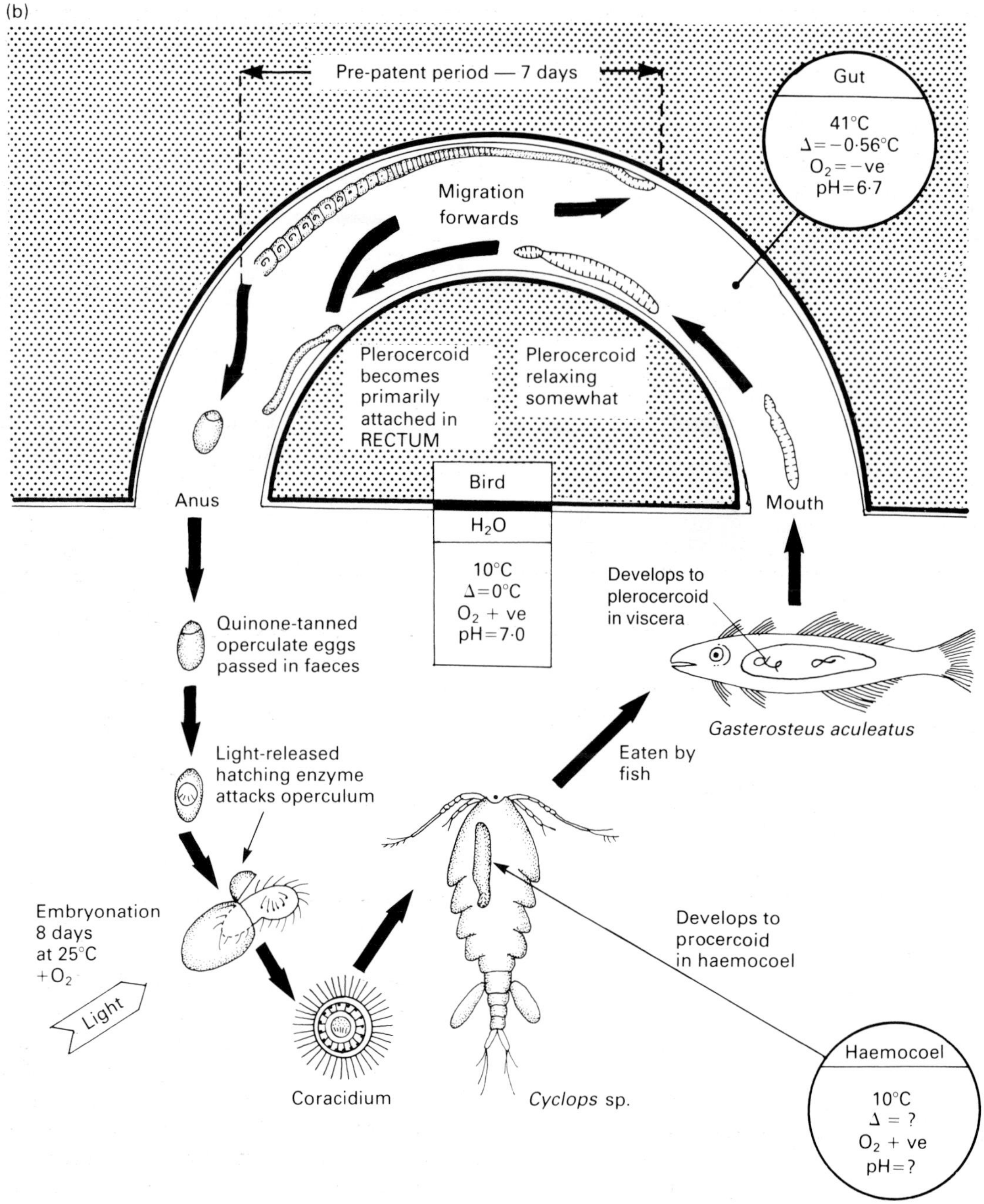

Fig. 3.56 continued.

(c)

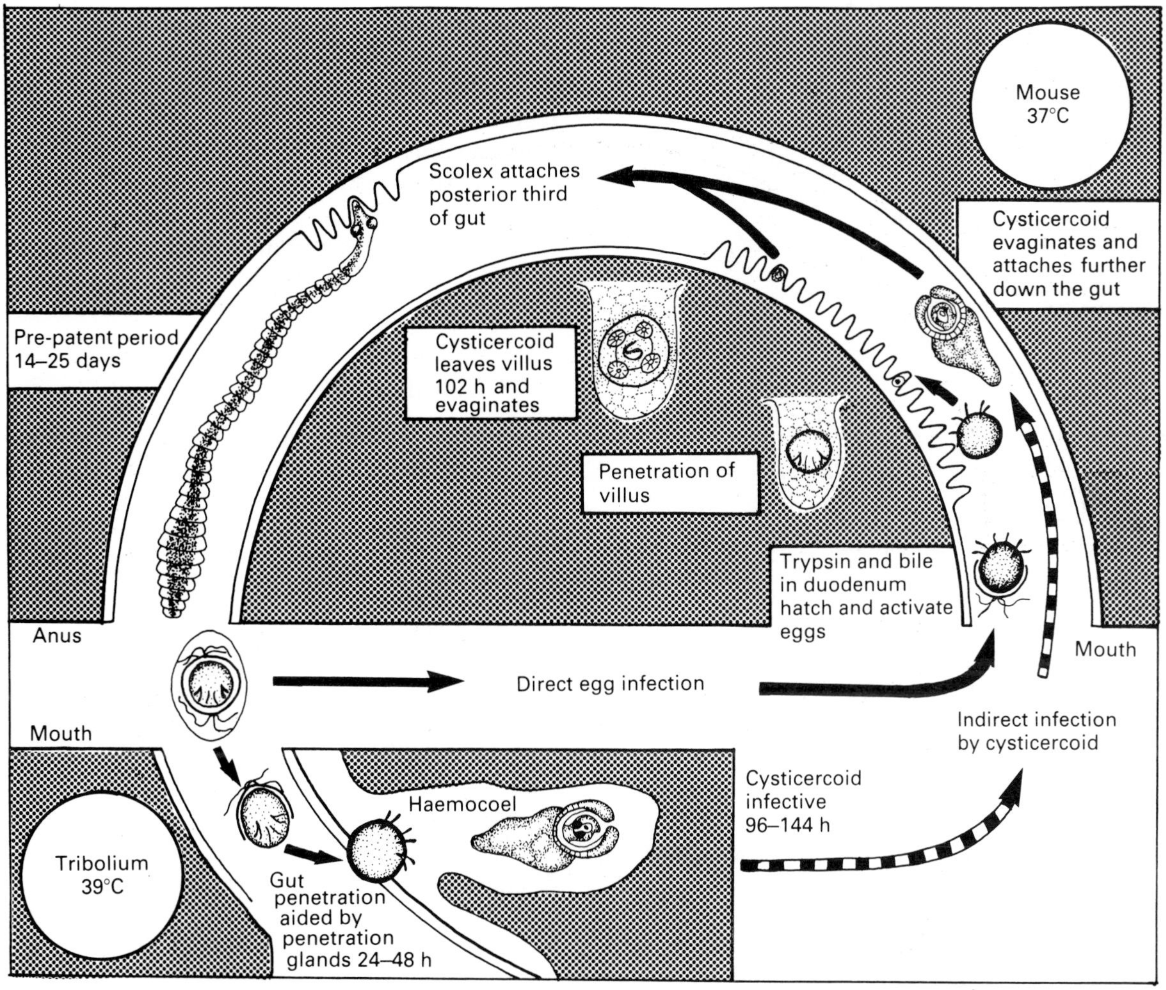

Fig. 3.56 continued.

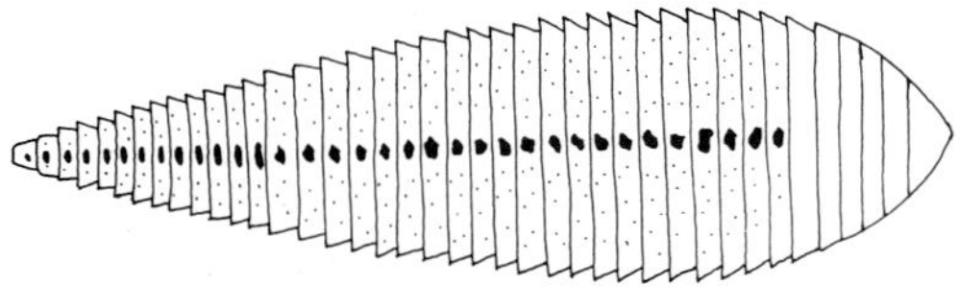

Fig. 3.57. *Schistocephalus solidus*—a caryophyllid that occurs in the body cavities of sticklebacks (after Smyth, 1962).

strobila with a single set of male and female reproductive organs (one ovary, many testes but one copulatory organ). They occur in the body cavity, not the gut, of their host. They are thought to be paedomorphic larvae (see Chapter 17) of an unknown group of tapeworms. Members of the Cestodaria show features in common with both cestodes and flukes (Fig. 3.58). They do not have a mouth and gut, but they lack strobilization and a scolex, instead having monogenean-like suckers. The larva is, nevertheless, like an oncosphere, but with ten hooks rather than

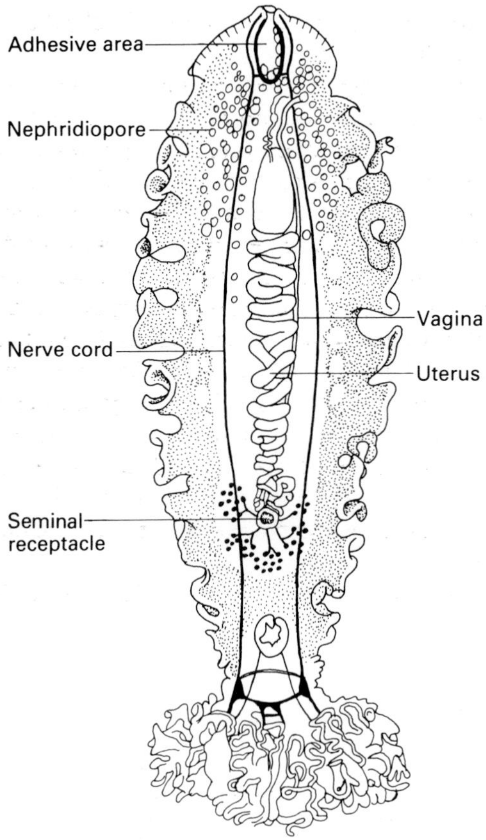

Fig. 3.58. *Gyrocotyle*, a cestodarian (after Barnes, 1980).

six. They parasitize the intestines and coelomic spaces of elasmobranchs and primitive teleost fishes.

3.7 Phylum MESOZOA

3.7.1 Etymology

Greek: *mesos,* middle; *zoon,* animal. (This name was bestowed because mesozoans were considered to be intermediate between the true multicellular animals (or Metazoa) and single-celled forms (the Protozoa).)

3.7.2 Diagnostic and special features

1 Bilaterally symmetrical.
2 Without any organs.
3 Without any body cavity or digestive cavity.
4 Very few cells disposed in no more than two 'layers', the outer layer ciliated, the inner cell or cells reproductive.
5 Some cells developing *within* other cells.
6 Endoparasitic in invertebrates during some stage of their life cycle.
7 Complex life histories involving both sexually and asexually reproducing phases.
8 Marine.

3.7.3 Classification

Two rather different types of animal, totalling 50 species, are included within this phylum, the rhombozoans and the orthonectidans. These may well prove to be unrelated to each other, and they are treated separately below.

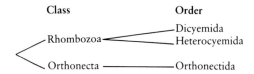

Class	Order
Rhombozoa	Dicyemida
	Heterocyemida
Orthonecta	Orthonectida

3.7.3.1 Class Rhombozoa

Rhombozoans, which are parasites of the kidney of cephalopod molluscs, have a 0.5–7.5 mm long vermiform body comprising between 20 and 30 ciliated somatic cells (each rhombozoan species having a constant cell number within this range) arranged spirally around a single elongate axial cell (Fig. 3.59). Within this axial cell are found from one to many (>100) axoblast cells, which give rise to future sexual or asexual generations.

The life cycle is complex and incompletely known, but it appears that the morphology described above applies to two completely different phases in the cycle: the asexually multiplying nematogen and the sexually reproducing rhombogen, distinguishable only by the developmental fate of their axoblast cells. The rhombogen is characteristic of the mature cephalopod host, and the axoblasts in its axial cell

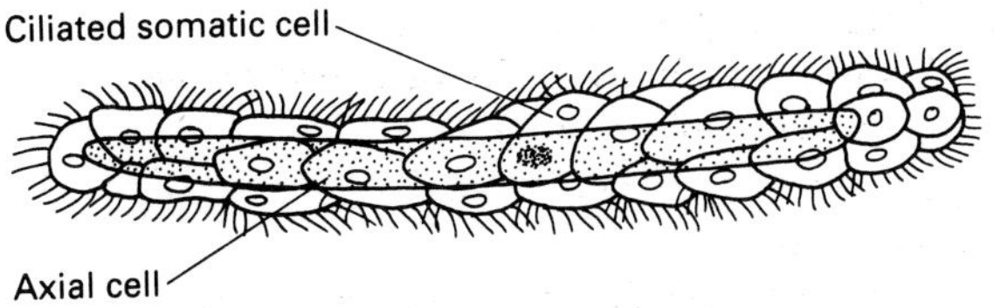

Fig. 3.59. The body plan of a rhombozoan (after Lapan & Morowitz, 1972).

produce 'infusorigens' which are equivalent to gonads and release egg and/or sperm cells. Fertilization occurs within the axial cell and gives rise to an infusoriform larva which soon possesses the full complement of cells so that development into later stages is by growth and differentiation of its cells, not by cell division. The future germinative cells are already intracellular at this larval stage. These infusoriform larvae leave the rhombogen and pass out of the cephalopod in its urine, and being negatively buoyant (as a result of high concentrations of magnesium inositol hexaphosphate in two special cells) they then sink to the sea bed.

In an unknown manner, what may be later growth stages of the infusoriform larvae, or another larval stage altogether, infect juvenile cephalopods and, having done so, migrate to the kidney, there developing into the nematogen phase. The axoblasts within the nematogen's axial cell develop, via vermiform larvae, into yet more nematogens, and many generations of these asexual nematogens may be produced. Finally, when the nematogen density in the kidney becomes high and/or when the host becomes sexually mature, the nematogens either change into or produce a new generation of rhombogens (Fig. 3.60). The nutrition of all the endoparasitic stages must presumably be by absorption by the somatic cells of substances present in the host's urine, but no details are known.

Only five genera, in two orders, have been described.

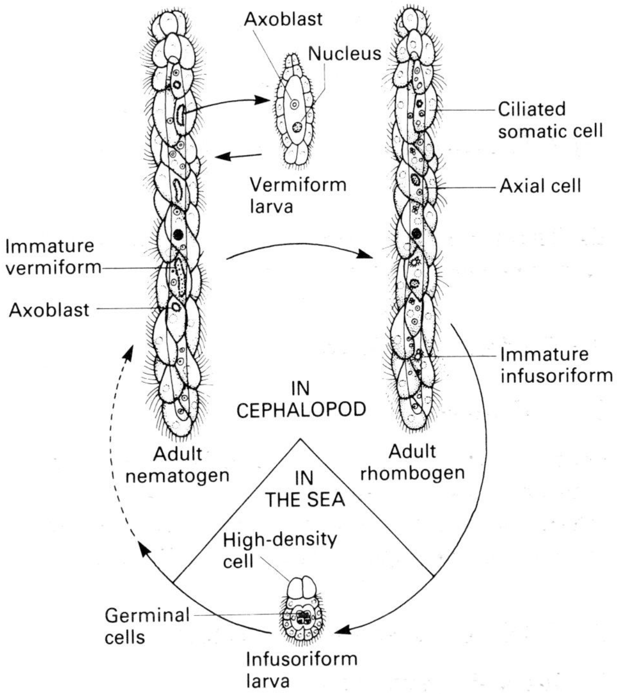

Fig. 3.60. The rhombozoan life cycle (after Margulis & Schwartz, 1982).

3.7.3.2 Class Orthonecta

The orthonectidans are endoparasitic in a wide variety of marine invertebrates: flatworms, nemertines, annelids, bivalve molluscs and echinoderms. The sexual phase is sexually dimorphic, free-living and minute (males, *c.* 0.1 mm; females, somewhat larger—Fig. 3.61). Both sexes have a basic structure not dissimilar to that of the adult rhombozoan: an annular outer layer of ciliated somatic cells, in this group surrounding numerous individual eggs or sperm, there being no axial cell. Males shed their sperm into the sea, and these sperm enter the bodies of the females there fertilizing the retained eggs. Infective ciliated larvae—a layer of ciliated cells surrounding a few germinal cells—then develop within the females.

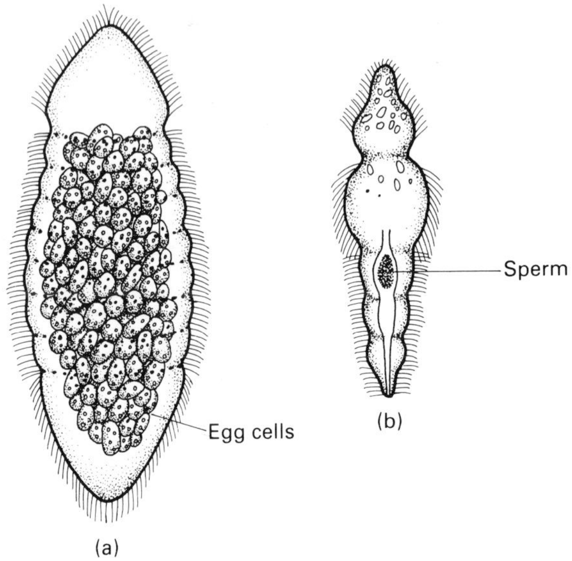

Fig. 3.61. Female (a) and male (b) orthonectidans (after Atkins, 1933).

After release and upon invasion of their host's tissues, the outer ciliated cells are lost and the germinal cells multiply to form a multinucleate plasmodium (Fig. 3.62), fragments detaching from this plasmodium and giving rise, asexually, to other plasmodia within the host. Some cells in the plasmodial phase then eventually produce both or only one sex of the sexual phase which leave the host to complete the cycle.

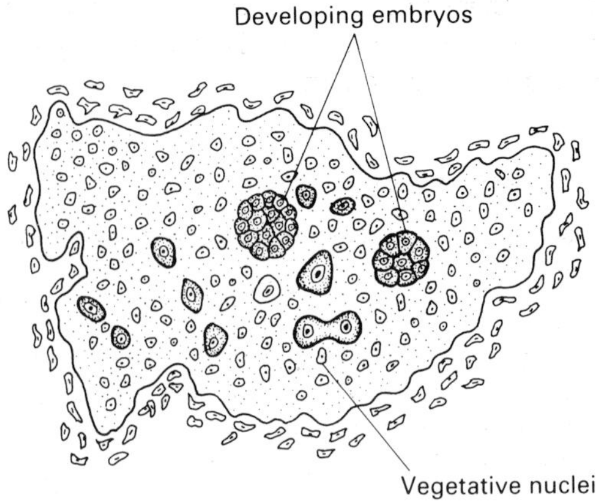

Fig. 3.62. Orthonectidan plasmodium (after Caullery & Mesnil, 1901).

Only three genera, in a single order, are known.

3.7.4 Relationships

Several endoparasites display a lack of various organ systems when compared with the free-living forms from which they are thought to have been derived, and the simplicity of the mesozoan anatomy has been interpreted in terms of such parasitic degeneration. But, if this is so, the mesozoans must form the most extreme example of the phenomenon known, since not only do they not possess any organs at all but they may have less than 30 cells in total. Further, some of their features, especially the intracellular development of both the sexual and asexual larval forms of the rhombozoans, are without parallel anywhere in the animal kingdom. There is no specific evidence to link the rhombozoans with any other known group, and they may well have been independently derived from the protists (see Chapter 2). If the latter has occurred, their origin might even have been much later than those of other multicellular groups since they are confined to a relatively advanced type of host.

The orthonectidans are somewhat less aberrant, and an origin within the flatworms is possible, although far from certain.

3.8 Further reading

Bergquist, P.R. 1978. *Sponges.* Hutchinson, London [Porifera].

Fry, W.G. (Ed.) 1970. *The Biology of Porifera.* Academic Press, New York [Porifera].

Grell, K.G. 1982. Placozoa. In: Parker, S.P. (Ed.) *Synopsis and Classification of Living Organisms,* (Vol. 1), p. 639. McGraw-Hill, New York [Placozoa].

Hyman, L.H. 1940. *The Invertebrates (Vol. 1). Protozoa through Ctenophora.* McGraw-Hill, New York [Porifera, Cnidaria, Ctenophora, Mesozoa].

Hyman, L.H. 1951. *The Invertebrates (Vol. 2). Platyhelminthes and Rhynchocoela.* McGraw-Hill, New York [Platyhelminthes].

Kaestner, A. 1967. *Invertebrate Zoology (Vol. 1).* Wiley, New York [Porifera, Cnidaria, Ctenophora, Platyhelminthes, Mesozoa].

Lapan, E.A. & Morowitz, H. 1972. The Mesozoa. *Scient. Am.*, **227,** 94–101 [Mesozoa].

Mackie, G.O. (Ed.) 1976. *Nutritional Ecology and Behavior.* Plenum, New York [Cnidaria].

Miller, R.L. 1971. *Trichoplax adhaerens* Schulze 1883: Return of an Enigma. *Biol. Bull. mar. Biol. Lab., Woods Hole,* **141,** 374 [Placozoa].

Morris, S.C., George, J.D., Gibson, R. & Platt, H.M. 1985. *The Origins and Relationships of Lower Invertebrates.* The Systematics Association, Special Volume No. 28. Clarendon Press, Oxford.

Reeve, M.R. & Walker, M.A. 1978. Nutritional ecology of ctenophores—a review of past research. *Adv. mar. Biol.* **15,** 246–87 [Ctenophora].

Schmidt, H. 1972. Die Nesselkapseln der Anthozoen und ihre Bedeutung für die phylogenetische Systematik. *Helgoländer Wiss. Meeresunters.*, **23,** 422–58 [Platyhelminthes].

Schockaert, E.R. & Ball, I.R. 1981. *The Biology of the Turbellaria.* Junk, The Hague [Platyhelminthes].

Smyth, T.D. 1977. *Introduction to Animal Parasitology,* 2nd edn. Wiley, New York [Platyhelminthes].

4
THE WORMS

Nemertea
Gnathostomula
Gastrotricha
Nematoda
Nematomorpha
Kinorhyncha
Loricifera
Priapula
Rotifera
Acanthocephala
Sipuncula
Echiura
Pogonophora
Annelida

The 14 phyla included in this chapter are all protostome derivatives of the flatworms although, that apart, they share no specific features and do not constitute a natural group of related animals (see Chapter 2). They are a group of 'convenience', distinguished more by the common absence of the various features which characterize other groups of phyla than by anything else, notwithstanding even their vermiform shape. That is to say that the worms covered here are animals which do *not* possess legs, are *not* covered by a protective shell, are *not* deuterostome, and do *not* bear a lophophore. The equally vermiform phoronans do have a lophophore and are therefore included in Chapter 6; the acorn-worms (enteropneust hemichordates) and arrow-worms (chaetognaths) are deuterostome and are, for that reason, included in Chapter 7; whilst the tongue-worms (pentastomans) are considered with their supposed arthropod allies in Chapter 8. In practice, therefore, these 14 groups of worms are simply animals which have retained the general vermiform shape of their ancestors more or less unchanged, having bodies some 2–3 to over 15 000 times longer than wide and flattened or rounded in section.

Although systems of relationship within these worms are not by any means universally accepted, it is nevertheless generally true that four clusters of phyla are broadly recognizable (see, for example, Fig. 2.20). The first and second groups (the Nemertea and Gnathostomula/Gastrotricha) are similar in essential body plan to the ancestral flatworms and show relatively little within-group diversity; both are characterized by highly individual specializations which prevent them from being accommodated within the Platyhelminthes.

The gastrotrichs may possibly provide a morphological link to the much larger and more diverse third cluster of phyla, variously known as the aschelminthes or pseudocoelomates. This group currently contains seven phyla distributed between two large subgroups. The Kinorhyncha, Loricifera and Nematomorpha all share, at some stage of their life history, a distinctive eversible introvert, bearing a stylet-surrounded mouth cone and whorls of 'scalids'; the Priapula possess a similar scalid-bearing introvert, although without oral stylets; and the Nematoda also appear linked to this subgroup in that one recently described species from Brazil shows an equivalent introvert with spine-like scalids, and the nematodes and nematomorphs share the same highly individual muscular/locomotory system. In contrast, the Rotifera and Acanthocephala lack any such introvert, and share the common presence of an intracellular cuticle, a locomotory and/or adhesive 'proboscis', and lemnisci (the latter two occurring only in the bdelloid rotifers). With the exception of their introvert and loricate larva, the priapulans are otherwise rather atypical aschelminthes; they could perhaps lay claim to be close to the ancestral body form and reproductive type. The other aschelminth phyla, however, are united by characteristics of their body wall construction, the nature of their pharynx and of their body cavity, besides the shared phenomena of eutely, internal fertilization and copulation, and a lack of the ability to regenerate body parts or undergo asexual multiplication. Many also share a protonephridial system, a moulted cuticle, thickening of the cuticle over the body to form a protective lorica, small size in the non-parasitic forms, segmentation of the cuticle (and associated body wall elements), flosculae, and duo-gland-cell adhesive organs.

The fourth assemblage also contains a diverse series of worm types. The Annelida, Pogonophora and Echiura, however, all share such features as schizocoelic body cavities, chitinous chaetae, closed blood systems and metanephridium-like organs, although the first two phyla are metameric whilst the echiurans are clearly monomeric. Also monomeric are the Sipuncula, of which the 'metanephridia' and schizocoel are often taken to indicate affinity with this fourth group of phyla, although they lack chaetae and any circulatory system. The sipunculans alone possess an introvert; but in contrast to the short, barrel-shaped, spiny organ of the aschelminthes, that of the sipunculans is a long, narrow tube, terminating in detritus-collecting lobes or tentacles.

Just as the position of the Priapula within the aschelminthes is aberrant—in respect of some of the peculiarly aschelminth features they are typical; in respect of others, they are radically unlike all the other phyla—so is that of the Sipuncula with regard to the other schizocoelomate worms. In both these large groups of worms, argument based on some anatomical features can be used to indicate the parallel acquisition of their common grade of organization, whilst other characters can be advanced in support of phylogenetic affinity. What is a gland and its duct to one author can be the pharynx and buccal canal to another (this example being cited from two views of the structure of the larval nematomorph). Even some individual phyla, for instance the Nematomorpha amongst the aschelminthes and the Pogonophora amongst the schizocoelomates, can convincingly be argued each to consist of two groups of worms with completely different origins: one subgroup of nematomorphs may have been derived from mermithoid nematodes, whilst the other is related to the kinorhynchs and loriciferans; and one class (or subphylum) of pogonophorans may be closely related to the annelids, the other class (or subphylum) may have arisen independently and acquired its schizocoel in parallel!

Clearly it would be premature to amalgamate any of these various groups of worms into a smaller number of larger phyla. A considerable amount of research will have to be carried out before any uncontentious phylogenetic conclusions can be drawn, and in this chapter we will treat all 14 groups as separate blocks of equal systematic rank.

4.1 Phylum NEMERTEA (ribbon- or proboscis-worms)

4.1.1 Etymology

Greek: *Nemertes,* a Mediterranean sea-nymph, the daughter of Nereus and Doris.

4.1.2 Diagnostic and special features (Fig. 4.1)

1 Bilaterally symmetrical, vermiform.
2 Body more than two cell layers thick, with tissues and organs.
3 A through gut and a terminal anus.
4 Without a body cavity, the skeleton being provided by parenchyma.
5 A ciliated outer, cuticle-less epidermis.
6 Body dorsoventrally flattened, often with serially repeated organs (e.g. gut pouches, protonephridial organs, gonads).

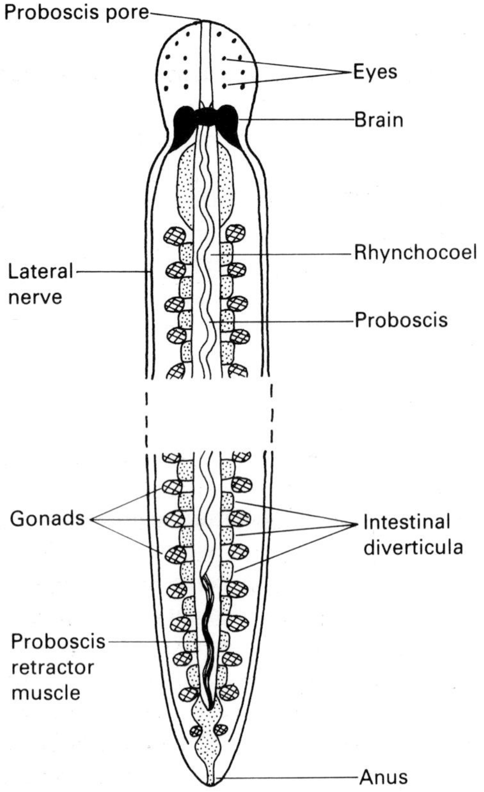

Fig. 4.1. Diagrammatic dorsal view of a nemertine, seen as if transparent (after several sources, especially Pennak, 1978).

7 Without a gaseous-exchange system, but with a circulatory system.
8 Nervous system with a brain and, usually, three longitudinal cords.
9 An eversible and retractable ectodermal proboscis housed in a longitudinal dorsal cavity, the rhynchocoel, which opens to the exterior near the mouth.
10 Eggs cleave spirally.

The < 0.5 mm to > 30 m long nemerteans or nemertines can be described as 'super-flatworms', the essential features of their bodily organization being basically those of the larger turbellarian platyhelminths (see Section 3.6.3.1). They differ from that group, however, in three important respects. First, they possess a more efficient through gut; second, they have a closed blood system, with blood flow (which is irregular) being driven both by contraction of the walls of the blood vessels and by bodily movements; and third, they exhibit a characteristic proboscis, which is a completely separate structure from the gut and not a region of it (Fig. 4.2).

This proboscis is housed in a tubular cavity extending almost the whole length of the body, with walls of the same construction as that of the body. Pressure generated within the rhynchocoel by contraction of the wall muscles everts the proboscis, and it is retracted by means of a longitudinal muscle running from its posterior end to the posterior margin of the

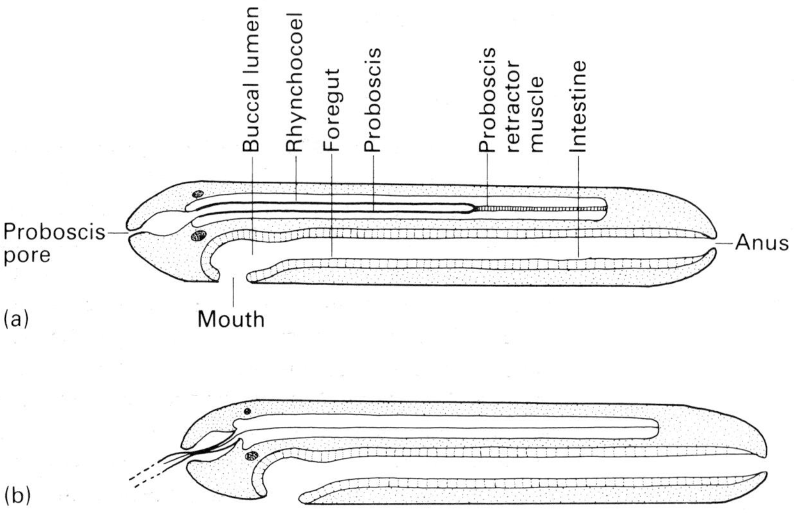

Fig. 4.2. Diagrammatic longitudinal section through a nemertine showing the rhynchocoel and the proboscis (a) retracted and (b) everted (after Gibson, 1982).

rhynchocoel. It is used mainly for prey capture, but also, in terrestrial species, for rapid locomotion: the proboscis being everted, attached to the ground ahead of the worm, and the animal then pulling itself forward over its own proboscis.

Nemertine locomotion is otherwise similar to that of the turbellarians, i.e. ciliary gliding, the cilia beating in secreted mucus, or waves of muscular contraction passing along the ventral surface in the larger forms. Some species can swim by means of dorsoventral undulations of the body.

Multiplication occurs both by asexual fragmentation and by sexual reproduction, the gonads being temporary structures formed from aggregations of differentiated mesenchyme cells which become enclosed within membranes and, when mature, connected to the exterior by temporary ducts. Most species are gonochoristic, although some freshwater and terrestrial forms are hermaphrodite, and capable of self-fertilization. Direct development is the norm, but members of one order possess a larval stage (Fig. 4.3); fertilization is usually external.

Almost all species are predatory, capturing organisms ranging in size from protists to molluscs, arthropods and fish. Nemertines are notable for including the longest of known animals: *Lineus longissimus* regularly attains 30 m and some individuals can probably achieve twice this length when fully extended.

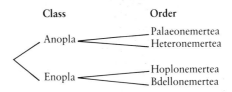

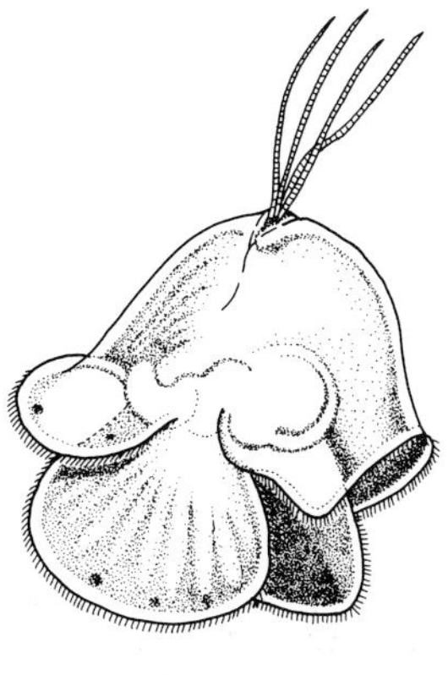

Fig. 4.3. Nemertine pilidium larva (after Kershaw, 1983).

4.1.3 Classification

The 900 known species are included in two classes.

Class	Order
Anopla	Palaeonemertea
	Heteronemertea
Enopla	Hoplonemertea
	Bdellonemertea

4.1.3.1 Class Anopla

In the anoplans, the central nervous system is located within the body wall, which often has three layers of muscle: either two layers of circular on either side of a single layer of longitudinal, as in the order Palaeonemertea, or two layers of longitudinal on either side of a single layer of circular, as in the order Heteronemertea. The anoplan proboscis is relatively simple, without regional differentiation or stylets, although large numbers of epithelial rhabdoids, of the type frequently found in the turbellarians, may be present; the proboscis pore opens completely separately from the mouth.

The majority of anoplans (Fig. 4.4) are marine and benthic, but three species occur in fresh waters and several inhabit brackish regions. Heteronemertines are the only group to possess a larval stage.

4.1.3.2 Class Enopla

Enoplans possess a central nervous system located internally to the body wall musculature, which is two-layered, and a proboscis differentiated into regions, the central one of which is a short muscular bulb bearing one or more stylets (except in the aberrant and commensal, filter-feeding genus *Malacobdella* which is placed in its own order, the Bdellonemertea). The enoplan gut is complex and opens to the exterior through an aperture in common with the proboscis: in the majority of species (the order Hoplonemertea), the intestine bears numerous pairs of lateral diverticula; whilst in the bdellonemertines, the large pharynx is the organ of filter feeding and it contains ciliated papillae, which trap food particles without the use of mucus.

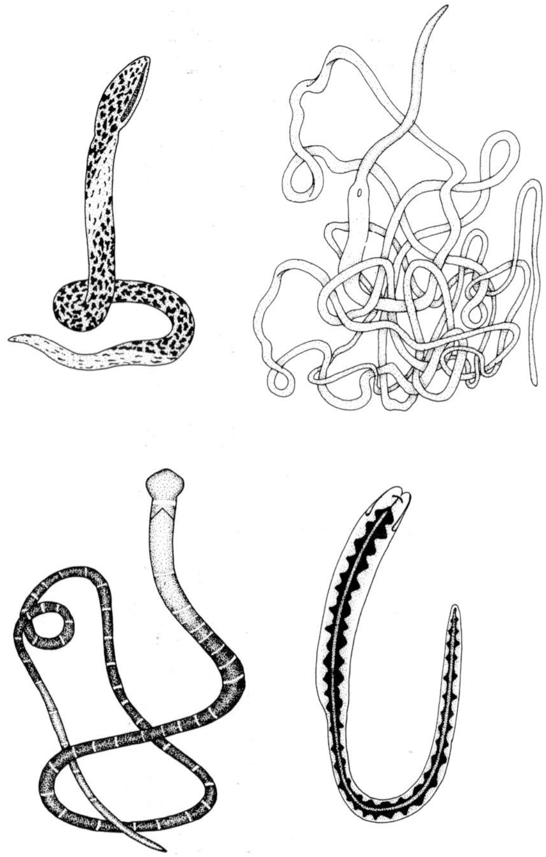

Fig. 4.4. Anoplan body forms (after Gibson, 1982).

Although most enoplans are marine and benthic, several are pelagic, including both swimming and floating forms (Fig. 4.5), some are freshwater and others terrestrial, and a few are commensal (in, for example, the atrial cavity of tunicates) or parasitic.

4.2 Phylum GNATHOSTOMULA

4.2.1 Etymology

Greek: *gnathos*, jaw; *stoma*, mouth.

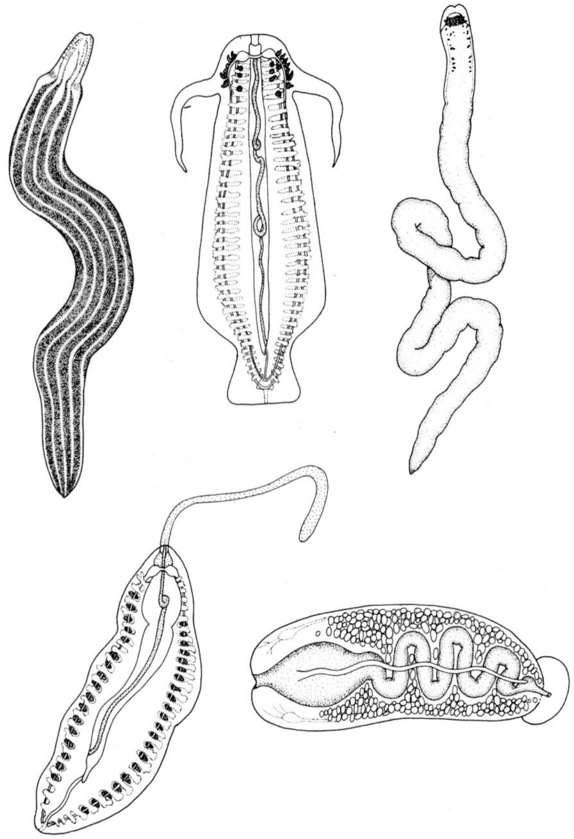

Fig. 4.5. Enoplan body forms (after Gibson, 1982).

4.2.2 Diagnostic and special features (Figs 4.6 and 4.7)

1 Bilaterally symmetrical and vermiform.
2 Body more than two cell layers thick, with tissues and organs.
3 With a blind-ending gut.
4 Without a body cavity, the skeleton being provided by poorly developed parenchyma.
5 Without circulatory or gaseous-exchange systems.
6 Excretory organs simple, two-celled protonephridia resembling epidermal cells in that the terminal cell bears only a single cilium.

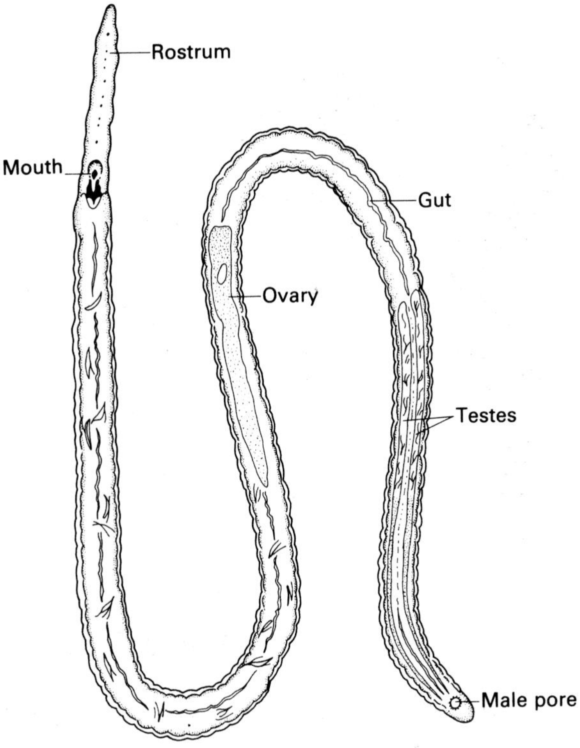

Fig. 4.6. The anatomy of a filospermoid gnathostomulan (after Sterrer, 1982).

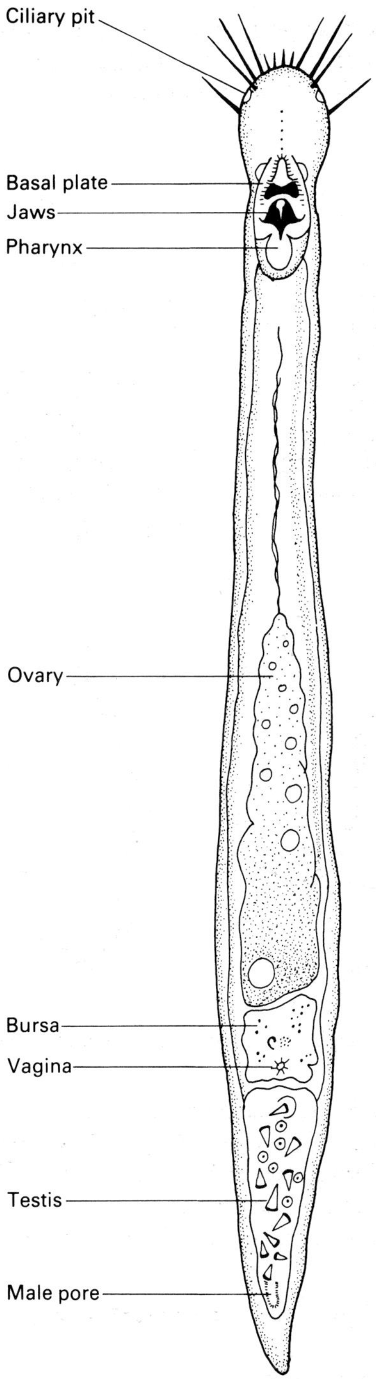

Fig. 4.7. The anatomy of a bursovaginoid gnathostomulan (after Sterrer, 1982).

7 With an outer monociliated, cuticle-less epidermis.
8 With a diffuse, epidermal nervous system.
9 With a complex feeding apparatus formed by paired jaws and a single basal plate.
10 Hermaphrodite (simultaneously so).
11 Eggs cleave spirally.
12 Development direct.
13 Marine, interstitial in often anoxic sands.

These minute (< 1 mm length), transparent worms possess a bodily organization essentially similar to those of the free-living turbellarian flatworms (Section 3.6.3.1) and of the gastrotrichs (Section 4.3). They differ from both these groups, however, particularly in respect of their highly specialized, muscular pharynx which bears a complex jaw system used

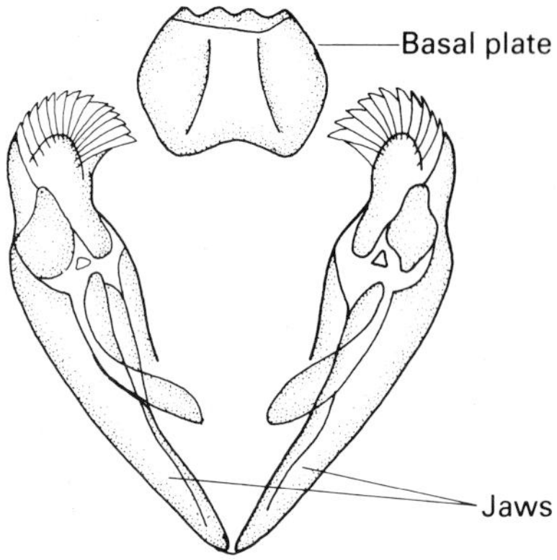

Fig. 4.8. The single basal plate and paired jaws of the Gnathostomula (after Sterrer, 1982).

in grazing bacteria, protists and fungi from the surfaces of sand grains (Fig. 4.8). In search of such prey, gnathostomulans move through sedimentary interstitial spaces by swimming or gliding using their ciliated epidermis, the cilia of which can beat either forwards or backwards, and by means of sudden contraction of the three or four pairs of longitudinal muscles which comprise the musculature of the body wall. The single ovary releases a single, large egg at a time, which is probably always fertilized internally. It is possible that during the life cycle, several gnathostomulans exhibit an alternation of forms between a non-sexual feeding stage and a sexual non-feeding one.

This phylum was not described until 1956 when the first gnathostomulan was discovered. This late discovery was not a result of rarity of the animals—they may achieve densities of 600 000 m^{-3}—but of previous lack of investigation of the anoxic layers of marine sediments, and of the extent to which gnathostomulans distort on attempted preservation.

4.2.3 Classification

The 100 or so described species are placed in two orders within a single class: the extremely elongate filospermoids (Fig. 4.6) with a long anterior rostrum but without anterior paired sense organs, and without a penis or a sperm-storage sac (the bursa); and the squatter bursovaginoids (Fig. 4.7) which lack an elongate rostrum but possess paired sensory bristles, cilia and pits anteriorly, and have a penis and a bursa. Many more species undoubtedly await description.

4.3 Phylum GASTROTRICHA

4.3.1 Etymology

Greek: *gaster*, stomach; *thrix*, hair.

4.3.2 Diagnostic and special features (Fig. 4.9)

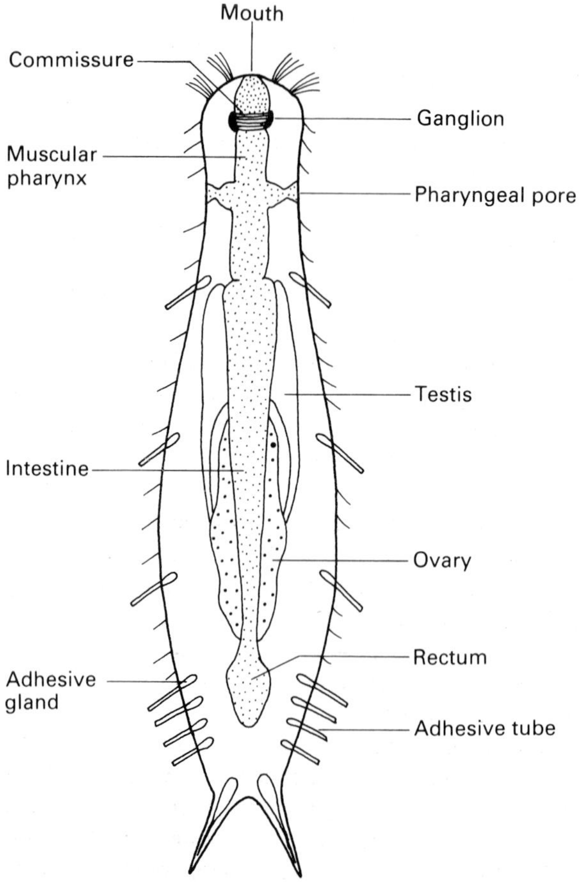

Fig. 4.9. Diagrammatic dorsal view of a gastrotrich, seen as if transparent (after several sources).

1 Bilaterally symmetrical, vermiform.
2 Body more than two cell layers thick, with tissues and organs.
3 A through gut and a subterminal anus.
4 Without a body cavity (notwithstanding the much quoted early view to the contrary).
5 Without circulatory or gaseous-exchange systems.
6 An outer non-chitinous cuticle and, usually, a monociliated dorsal and multiciliated ventral cellular epidermis; some layers of the cuticle extending as sheaths over each individual cilium.
7 Nervous system with a ganglion on each side of the anterior pharynx, connected dorsally by a commissure, and a pair of longitudinal cords.
8 Body covered by cuticular scales, spines or hooks, and bearing up to 250 adhesive tubes.
9 Hermaphrodite or parthenogenetic.
10 Eggs released by rupture of the body wall and cleave in a bilaterally radial manner, but development is determinate.
11 Development direct.
12 Aquatic; interstitial, surface dwelling or, rarely, planktonic.

Gastrotrichs possess a more or less transparent, dorsoventrally flattened body of up to 4 mm in length (although usually < 1 mm). Anteriorly, sense organs of the types seen in bursovaginoid gnathostomulans occur (paired bristles, cilia and sensory pits), together with, in some, eye-spots; posteriorly, the body may end in a stout fork or a thin tail. The adhesive tubes provide a means of temporary attachment to surfaces and, although the body wall contains both circular and longitudinal muscles, movement is by ciliary gliding. The cilia of the ventral surface are often non-uniformly disposed, e.g. in transverse or longitudinal bands, in patterns characteristic of the various genera (Fig. 4.10).

They feed on bacteria, protists and detritus swept into the mouth by the beating of buccal cilia or by pharyngeal pumping. In common with the nematodes, the pharynx is triradiate in section and is lined by a single layer of myoepithelial cells; and, as in rotifers and kinorhynchs, the freshwater species bear a pair of solenocytic protonephridia.

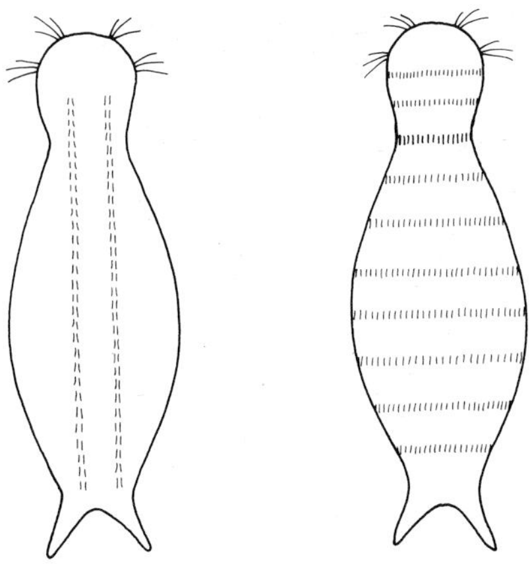

Fig. 4.10. Diagrammatic ventral views of two gastrotrichs showing different patterns of ciliation (after Hyman, 1951).

4.3.3 Classification

More than 400 species are known, apportioned between two orders within a single class: the strap-shaped marine macrodasyidans (Fig. 4.11) which

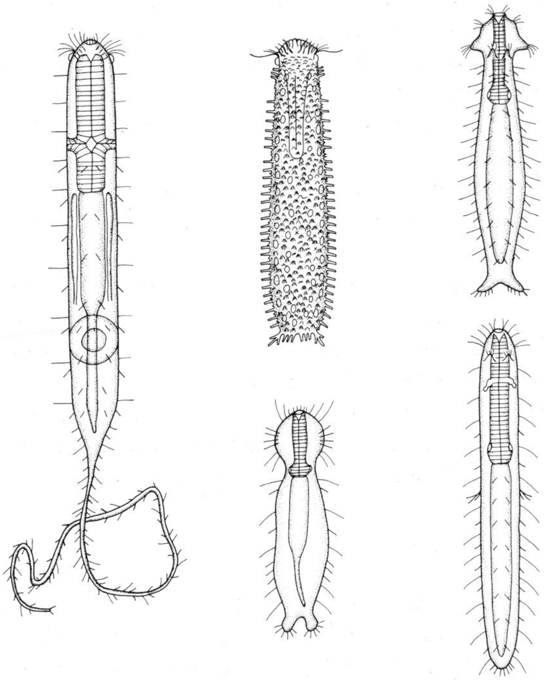

Fig. 4.11. Macrodasyidan body forms (after Grassé, 1965; Hummon, 1982).

possess a pharynx perforated through to the exterior by a pair of pores, probably serving as an exit for water taken in whilst feeding as in the pterobranch hemichordates (Section 7.2.3.2, and see Section 9.2.5); and the marine and freshwater chaetonotidans, (Fig. 4.12) which are normally fusiform in shape, lack pharyngeal pores, have adhesive tubes, if at all, only posteriorly and may be parthenogenetic.

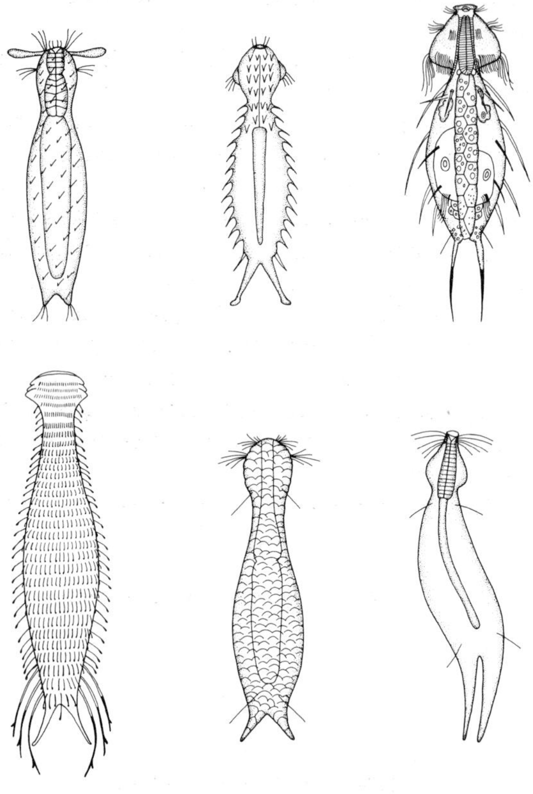

Fig. 4.12. Chaetonotidan body forms (after Grassé, 1965; Hummon, 1982).

4.4 Phylum NEMATODA

4.4.1 Etymology

Greek: *nema*, thread; *eidos*, form.

4.4.2 Diagnostic and special features

1 Bilaterally symmetrical, vermiform, but with a tendency to radial symmetry around the longitudinal axis.

2 Body more than two cell layers thick, with tissues and organs.

3 A complex cuticle.

4 Body wall without circular muscles.

5 Body cavity a pseudocoel, usually derived from the blastocoel.

6 Muscular gut leading from an anterior mouth via a muscular pharynx to the subterminal anus.

7 Longitudinal muscles arranged in four zones; ventral, lateral and two dorsal epidermal chords.

8 Nervous system with longitudinal nerves in the mid-ventral and mid-dorsal epidermal chords and with direct contacts with the muscle cells in the contralateral muscle fields.

9 Cross-sectional area always circular; body fluid always maintained under high pressure.

10 Without a circulatory system.

11 Excretory system without flame cells or nephridia. Excretory tubules in one, or a small number of, renette cells.

12 Embryonic cleavage pattern neither spiral nor radial, but highly determinate with T-shaped arrangement of cells at the four cell stage.

13 Development always direct, but in parasitic forms larvae may be infective stages.

The phylum Nematoda is one of the great success stories of the animal kingdom. More than 15 000 species have been described, of an estimated 1 million living species. In contrast with the other large phyla, this diversity of species is not based upon great diversity of structure. All nematodes are constructed along the same fundamental plan. The many species represent minor modifications of a successful formula. As this includes the ability to withstand potentially harmful environments, many nematodes are parasitic.

A key feature of their structural organization is the complex cuticle (Fig. 4.13). As many as nine cuticular layers can be detected but conspicuous among these are three layers of crossed fibres forming a spiral

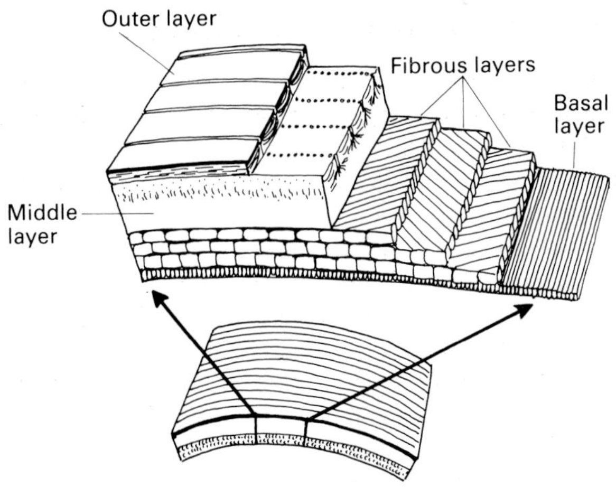

Fig. 4.13. Structure of the nematode cuticle. The cuticle consists of an outer striated layer, an inner homogeneous layer and a complex of fibrous layers (after Clark, 1964).

network about the body. This network and the high-pressure body cavity are fundamental to the nematode locomotory system (see Section 10.4).

A nematode is a 'tube-within-a-tube', pointed at both ends and round in cross-section (Fig. 4.14). The terminal mouth opens into a muscular pharynx: an important feature of nematode organization. Since the body fluids are maintained at high pressure, internal tubules tend to collapse, and in order to feed there must be some mechanism to introduce food into the inner tube, the alimentary system. The anterior opening of the gut is termed the *stroma*. It leads into a muscular pharynx, at the other end of which is a valve. The pharynx pumps food against a pressure gradient, and the unidirectional flow towards the intestine depends on the relative dimensions of the fore and hind parts of the anterior food duct (Fig. 4.15).

Much of the body cavity is filled with the paired reproductive organs, ovaries or testes. These are serially arranged, and often coiled. The positions of the openings are shown in Fig. 4.14a and b.

The sexes are usually separate, and fertilization is internal. The 'eggs' laid by female nematodes are, in fact, fertilized zygotes or early stage embryos. Secondary oocytes pass along the oviducts to the receptaculum where fertilization takes place. They are then invested by a layer of protein secreted by the uterus, and the zygote itself secretes two inner coats on to the egg case. The 'eggs' can be extraordinarily resistant and are produced in vast numbers by some parasitic species.

The structure of the body wall is best seen in cross-

Fig. 4.14. The general structure of a nematode (based on the parasitic genus *Rhabditis*): (a) a female; (b) a male.

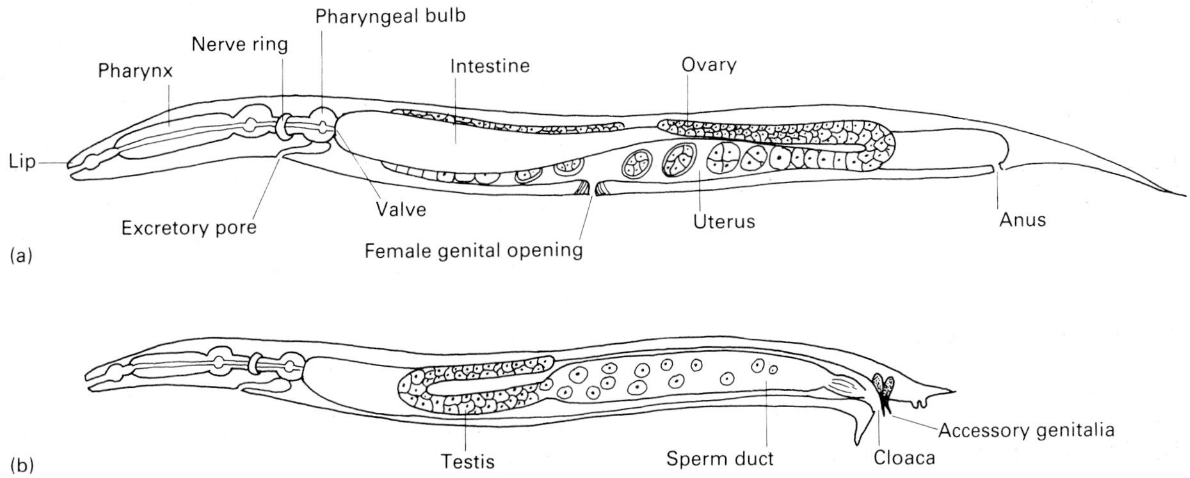

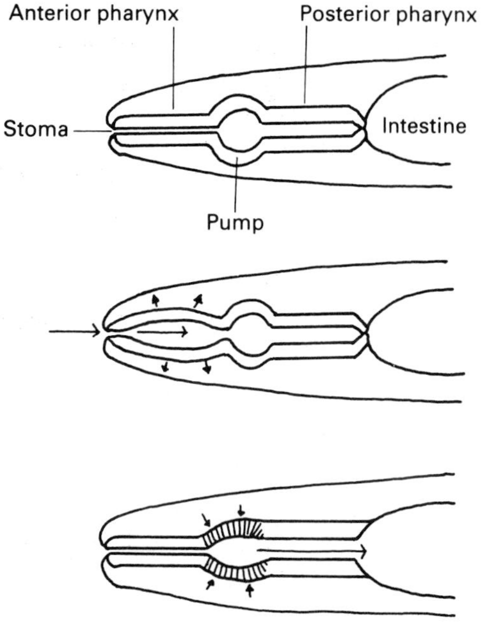

Fig. 4.15. The mechanism of the nematode pharyngeal pump. The radius of the anterior pharynx is much smaller than that of the posterior pharynx. The valve at the posterior is closed when dilation of the pharynx occurs. Consequently liquid flows into the anterior pharynx. When the pharynx contracts the posterior valve is open and liquid flows backward because of the greater radius of the posterior pharynx (after Croll & Matthews, 1977).

section and is unique to the phylum (Fig. 4.16). Beneath the complex cuticle lies the ectodermal epidermis containing the cells of the nervous and excretory systems, the principal elements of which are located in four longitudinal thickened strands or chords. The dorsal and ventral chords contain the prominent dorsal and ventral nerves. In the quadrants between the epidermal chords lie the cells of the longitudinal muscles; these are few and fixed in number. After the earliest stages of embryonic development, bodily growth involves an increase in cell size rather than cell number.

The muscle cells could be more correctly described as neuromotor units (Fig. 4.17). The contractile fibres of each cell are situated distally and rest on the epidermis. An elongated innervation process leads directly to nerve fibres in the dorsal or ventral chords, according to the position of the muscle cell. As it approaches the nerve fibres, each muscle cell is drawn out into a number of processes, the tips of which form synapses with other muscle fibres and the nerve fibres. This may provide a system which permits simultaneous contraction of all the muscle cells. This arrangement of muscles sending processes to nerves is unusual, since in most other animals it is usual for nerve cells to send processes to contact the motor cells (but see Section 16.4.2).

Fig. 4.16. The structure of a nematode as seen in transverse section: (a) stereoscopic view of the head showing the triangular arrangement of the mouth, lips and the associated sense organs; (b) cross-section in the pharyngeal region; (c) cross-section through the mid-body region.

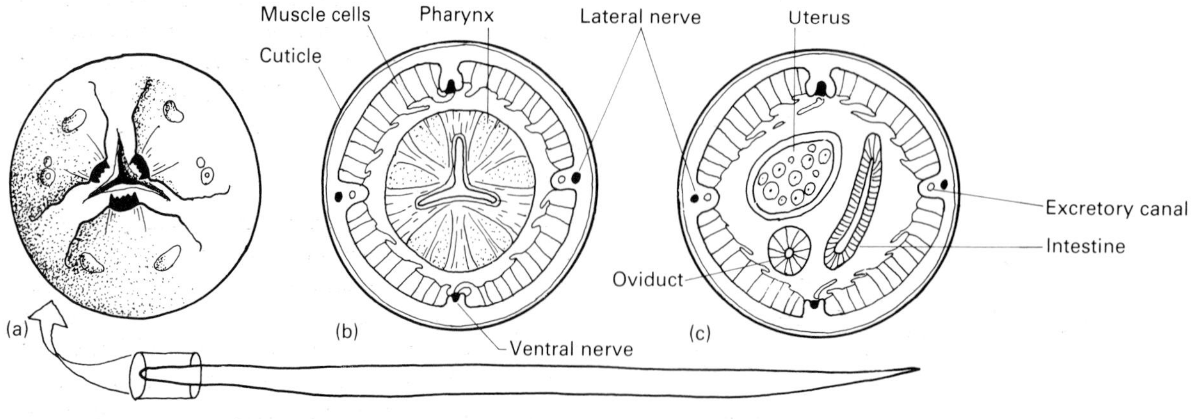

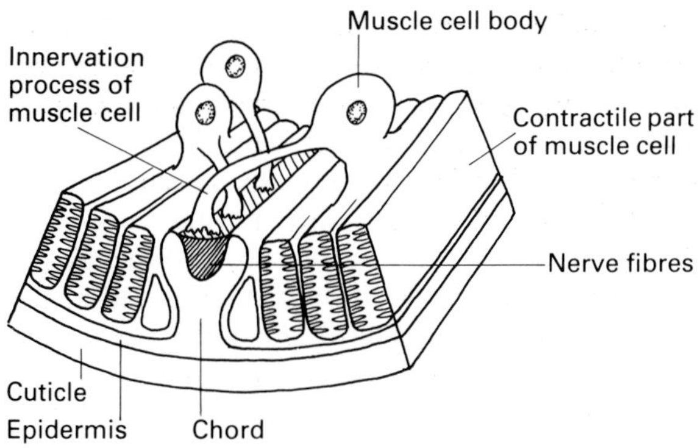

Fig. 4.17. Stereoscopic view of the nematode neuromuscular complex. Note the contractile filament of the muscle cells, the cell bodies and innervation process which make direct synaptic contact with the nerve fibres in the lateral, dorsal or ventral cords.

Nematodes are equipped with various sensory devices. These include: *cephalic papillae; amphids* (chemosensory pits often associated with anterior amphid glands); *cephalic setae; somatic setae; caudal papillae* and *spicules; ocelli* and *phasmids* (Fig. 4.18). Phasmids are caudally located sensory pits, not unlike amphids, and sometimes these two are associated with glands. The Nematoda is subdivided into two classes by presence or absence of phasmids.

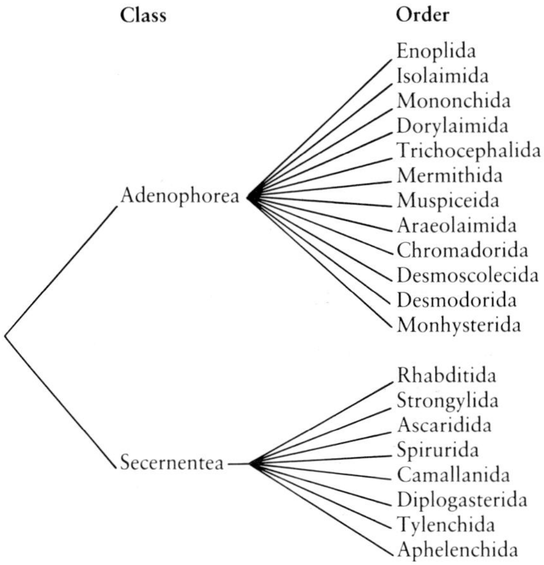

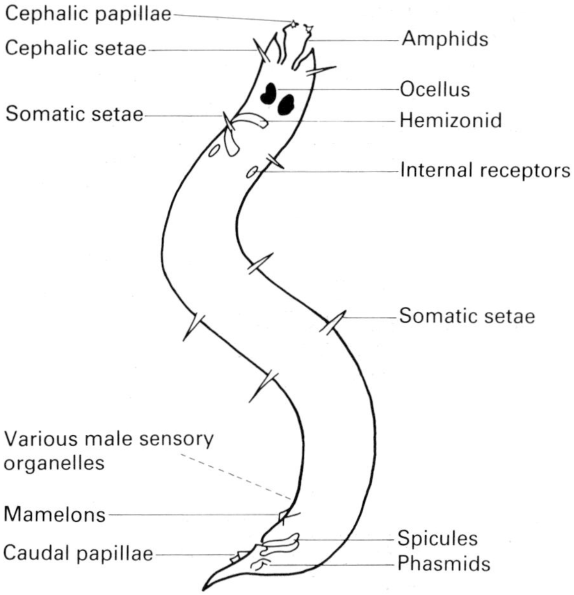

Fig. 4.18. A generalized scheme showing different types of sense organs found in nematodes. No single species is known to have all of these (after Croll & Matthews, 1977).

The excretory system exemplifies the extreme economy of cell numbers characteristic of the phylum. The primitive condition involves one or two specialized *renette* cells. In more advanced arrangements a system of tubes is developed within the glandular renette cells, which in some forms is lost (see Section 12.3.1).

Nematodes exhibit a uniform fundamental design which is reflected in their general appearance but which can be adapted to an enormous diversity of circumstances. This is best illustrated by their feeding mechanisms and by the diversification of their life cycles associated with parasitism. Many free-living nematodes feed on bacteria ingested as a suspension or on detritus. Some, however, are predators, especially of other nematodes and micro-organisms, and they have developed a variety of tooth or jaw-like plates at the anterior surfaces of the stroma. These plates pierce the prey, the contents of which are then sucked out by the pumping action of the pharynx.

A number of free-living species are phytophagous.

Some ingest fungi and yeast cells whole, but most pierce the cells of the plants and suck out the contents. To feed in this way, many phytophagous nematodes possess a special structure in the buccal cavity: spear-like to open up the cells, or hollow to use as a stylet under suction from the pharyngeal pump.

The feeding mechanisms exhibited by free-living nematodes can all be exploited by parasitic forms. The gut cavities of vertebrates, for instance, are filled with a rich mixture of bacteria and cell debris which forms a substrate for nematodes feeding by bacterial ingestion. Similarly, plate-like jaws can be used to tear the linings of the gut releasing the blood corpuscles on which the nematodes feed.

Debilitation caused by nematode infections of the alimentary system of vertebrates depends in part on this feeding biology. The human 'pinworm', *Enterobius vermicularis,* is relatively harmless while feeding on bacteria in the gut and is only pathogenic if the population increases sufficiently to cause blockage of the alimentary system. The human hookworm, *Ankylostoma duodenale,* is a more serious pathogen, causing intestinal bleeding and then ingesting the blood corpuscles. A human patient with 100 such worms could lose 50 ml of blood a day.

All phytoparasitic nematodes have a stroma with a stylet or spear. Many form permanent cysts in which the sedentary females feed on giant cells which they cause the host plant to produce, perhaps by interfering with the production of natural plant growth substances or their inhibitors. The most important genera are *Heteroda* and *Meloidogyne.*

Parasitic habits have arisen in nematode phylogeny on numerous occasions. This reflects a number of features fundamental to the structural organization of the phylum.

1 The complex impermeable and resistant cuticle.
2 Internal fertilization and the capacity to produce highly resistant eggs.
3 Microphagous feeding habits.
4 Small body size.
5 Chemical diversity and the mechanisms necessary to avoid the defence systems of the hosts.

Parasitism in many invertebrates involves extreme modification of adult form and the development of specialized life cycles with asexual multiplication phases, as seen in the many parasitic platyhelminths (Sections 3.6.3.2 and 3.6.3.3). This modification has not, however, occurred with the nematodes. The adults are 'preadapted' to the parasitic mode of life and their pattern of development has remained unchanged. The basic life history is illustrated in Fig. 4.19a. It involves cross-fertilization of separate males and females, and the release of highly protected 'eggs'. Four larval stages occur.

From this basic life cycle it is possible to recognize a series of adaptations to parasitism. In no way, however, should the various stages be conceived as representing steps in an evolutionary sequence.

In the least modified life cycle the larvae are free-living in the soil. Occasionally adults of a species may be either free-living or parasitic. The life cycle of *Strongyloides,* for instance, can be entirely free-living, but some third-stage larvae enter a mammalian host through the skin and move via the trachea to the gut, where females live and reproduce parthenogenetically (see Chapter 14). The eggs either enter the free-living life cycle or give rise to further infective third stage larvae (Fig. 4.19b).

The hookworm *Ankylostoma* has a similar life cycle but with no free-living adults. Both males and females occur in the enterine population (Fig. 4.19c). The first- and second-stage larvae are typical rhabditiform microphagous larvae feeding in the soil. The third-stage, infective larvae are encysted and do not feed.

In the unusually large, human gut nematode *Ascaris,* the free-living larval stage is suppressed. The eggs are voided with the faeces but the first- and second-stage larvae remain encysted within the eggs, which are ingested with contaminated food. The third-stage larvae hatch in the gut lumen but do not remain there. Instead they pass via the circulation to the lungs and thence back to the intestine (Fig. 4.19d). This complex life cycle may reflect an evolutionary history in which there was a secondary host, now lost.

Similar life cycles occur in the human gut parasites *Trichurus* and *Enterobius* (Fig 4.19e) but with no signs of the dual circulation prior to establishment in the

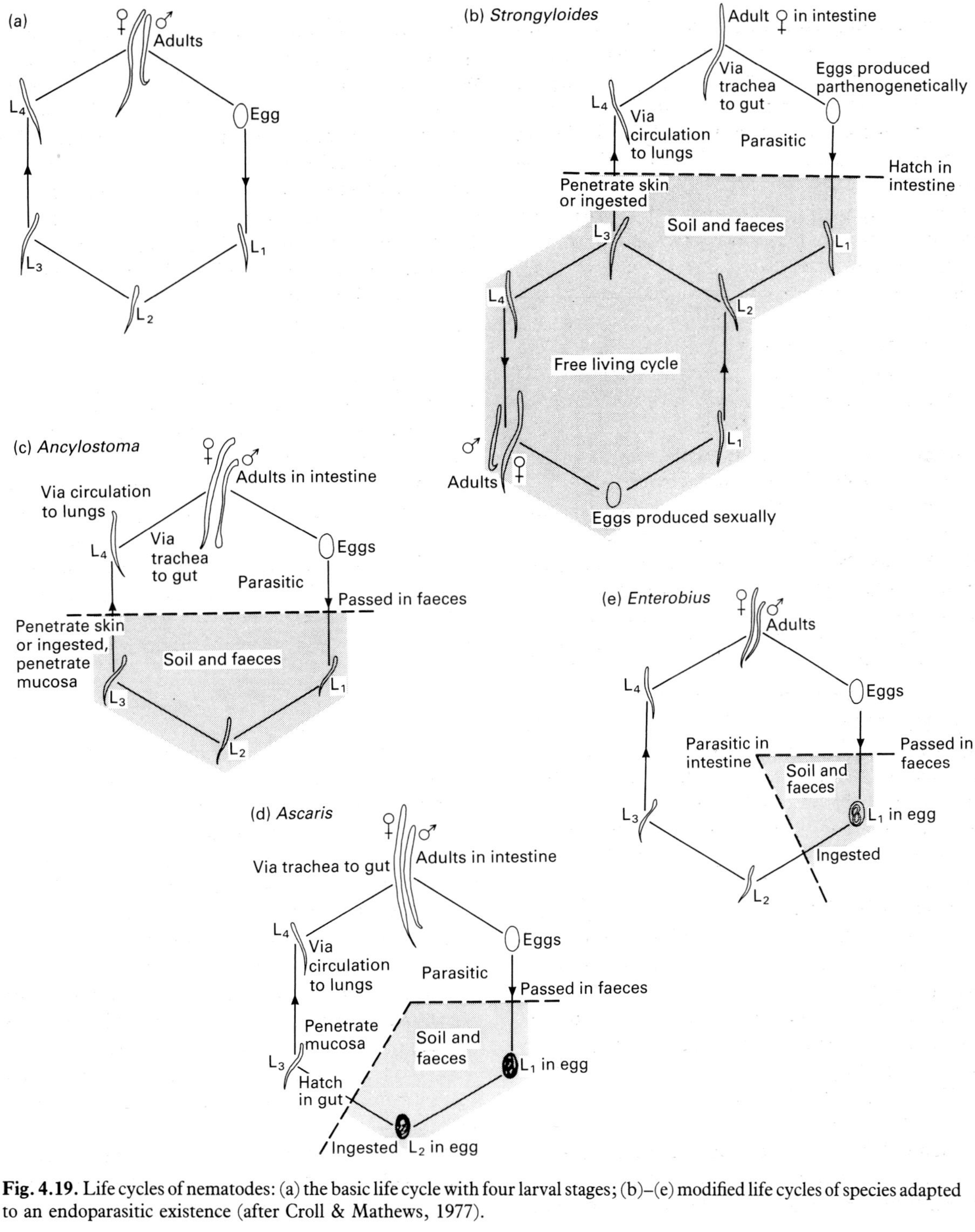

Fig. 4.19. Life cycles of nematodes: (a) the basic life cycle with four larval stages; (b)–(e) modified life cycles of species adapted to an endoparasitic existence (after Croll & Mathews, 1977).

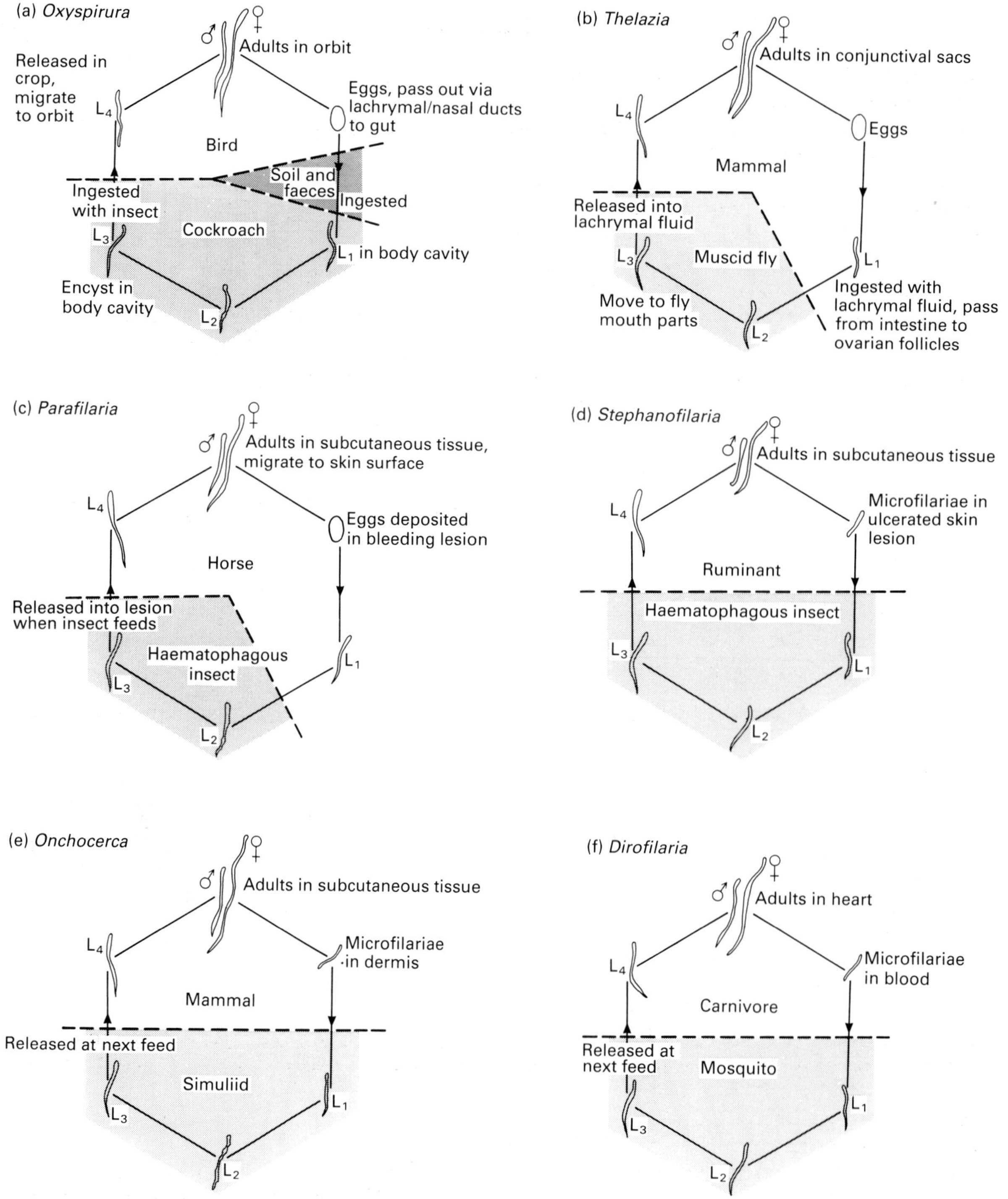

Fig. 4.20. Life cycles of endoparasitic nematodes involving an alternation of primary and secondary hosts: (a) *Oxyspirura*; (b) *Thelazia*; (c) *Parafilaria*; (d) *Stephanofilaria*; (e) *Onchocerca*; (f) *Dirofilaria* (after Croll & Matthews, 1977).

gut. The females of *Enterobius* can produce severe itching around the anus. This causes scratching and leads to direct reinfection and transfer to other potential hosts via the host's fingers.

Some nematode parasites of vertebrates do have secondary hosts and may exploit food chain relationships (Fig. 4.20). Other nematode life cycles include those of the mermithids; the larvae of these are parasitic in insects but the adults are free-living (cf. the phylum Nematomorpha, Section 4.5).

4.5 Phylum NEMATOMORPHA (horsehair worms)

4.5.1 Etymology

Greek: *nematos*, thread; *morphe*, form.

4.5.2 Diagnostic and special features (Fig. 4.21)

1 Bilaterally symmetrical, thin, elongate worms (< 3 mm diameter, 10 cm to > 1 m length).
2 Body more than two cell layers thick, with tissues and organs.
3 With a through, straight gut which is often degenerate and probably always non-functional.
4 Body monomeric, with a pseudocoelomic cavity, often occluded by mesenchyme.
5 Body wall with flexible, collagenous cuticle, epidermis, and layer of longitudinal muscle; without circular muscle layer.
6 Without circulatory, excretory or gaseous-exchange organs.
7 Nervous system intraepidermal, with an anterior nerve ring and one or two non-ganglionated longitudinal cords.
8 Gonochoristic, with elongate single or pair of gonads; fertilization internal via spermatophores.
9 Adult free-living but short-lived.
10 Larval stage infects arthropods (or, rarely, leeches), developing in their haemocoels; juveniles with three oral stylets and three whorls of hooked scalids (Fig. 4.22).
11 Freshwater or in moist soil; one genus marine.

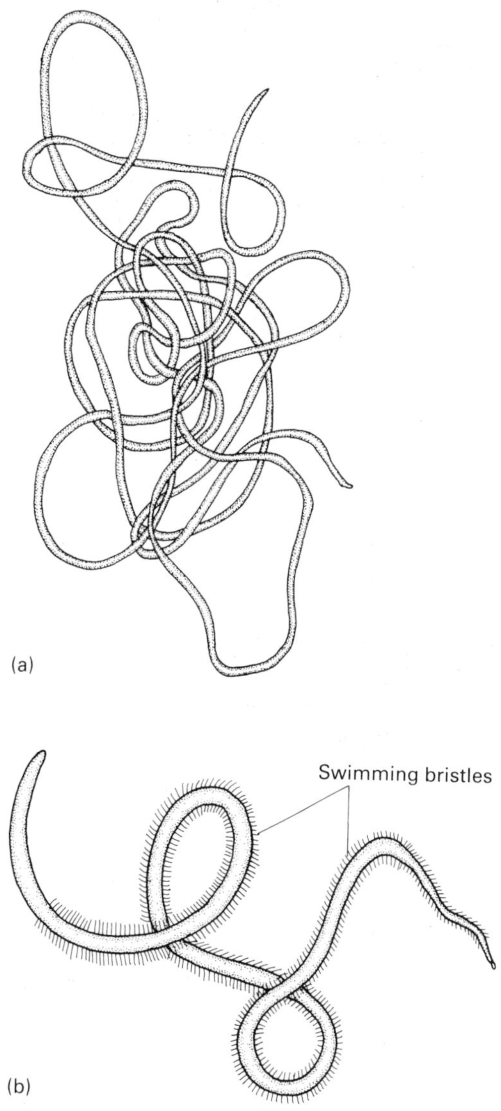

Fig. 4.21. Body form of (a) adult gordioid and (b) nectonematid nematomorphs (after Margulis & Schwartz, 1982 and Fewkes, 1883).

The bodily organization of adult nematomorphs is essentially similar to that of the nematodes (Section 4.4) and, like them, they move by means of undulatory waves passing along the body in the dorsoventral plane. The marine species and the male sex in the

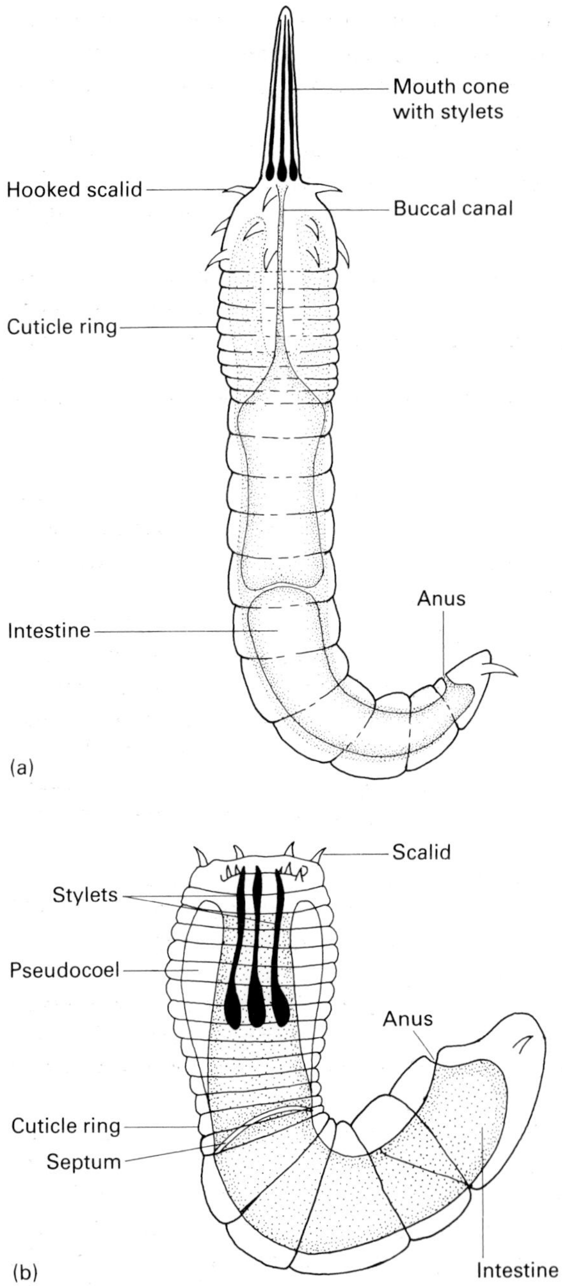

Fig. 4.22. The larva of a gordioid nematomorph: (a) with introvert everted; (b) with it retracted (after Hyman, 1951 and Pennak, 1978).

freshwater/terrestrial group are most mobile; the larger freshwater females usually lead a relatively sedentary life, curled or coiled on the substratum. The most obvious difference between the adults of this group and the nematodes is the degenerate nature of the gut and/or of its orifices, which correlates with the reduction of the adult stage to a brief, dispersive and reproductive phase in the life history.

The feeding phase is the parasitic juvenile that, although it possesses a gut, nevertheless probably obtains most or all of its food requirements by absorption from the host's haemocoel across its body surface. The larvae and juveniles are most unlike nematodes, instead resembling adult kinorhynchs (Section 4.6) and loriciferans (Section 4.7) in their oral stylets, scalids and anterior gut anatomy (Fig. 4.22). As the larval introvert is everted (Fig. 4.22a), aided by a transverse septum isolating the anterior pseudocoel, these stylets and scalids presumably aid penetration of the host's tissues into the haemocoel, but whether this is via the integument or the gut is as yet undecided.

4.5.3 Classification

The 250 known species are placed in two orders within a single class. By far the more numerous group, the freshwater gordioids (Fig. 4.21a), parasitize uniramians. They are distinguished by a single (ventral) nerve cord, paired gonads, and, at least when pre-reproductive, a body cavity almost totally filled with mesenchyme cells. The few known marine nectonematids (Fig. 4.21b), on the other hand, which parasitize decapod crustaceans, possess a second (dorsal) nerve cord, a single discrete (males) or diffuse (females) gonad, and a pseudocoel unoccluded by mesenchyme. They also bear a double row of topographically lateral bristles along most of the body which increase the undulatory surface during swimming. Some zoologists consider these two orders to be only distantly related, regarding the nectonematids as being derived from the nematodes, and the gordioids as having affinity with the kinorhynchs and loriciferans.

4.6 Phylum KINORHYNCHA

4.6.1 Etymology

Greek: *kinema*, motion; *rynchos*, snout.

4.6.2 Diagnostic and special features (Fig. 4.23)

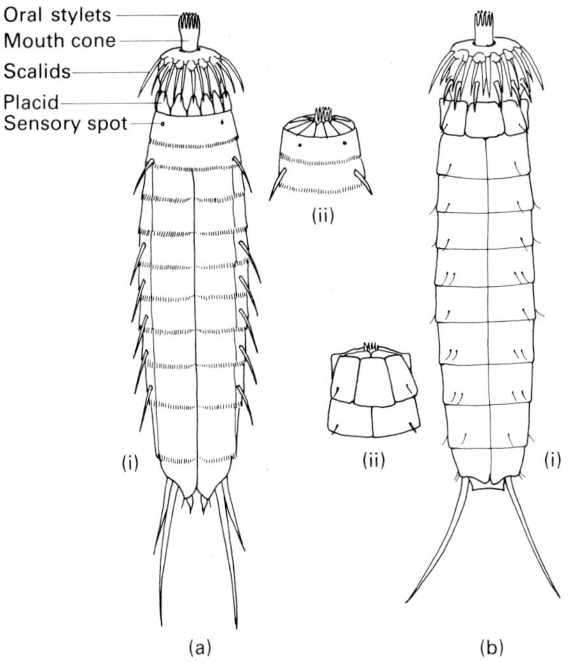

Fig. 4.23. The external morphology of Kinorhyncha: (a) Cyclorhagida: (i) ventral view; (ii) anterior view with introvert withdrawn. (b) Homalorhagida: (i) ventral view; (ii) anterior view with introvert withdrawn and protected by plates of the third zonite (after Higgins, 1983).

1 Bilaterally symmetrical; vermiform but rather short.
2 Body more than two cell layers thick, with tissues and organs.
3 Alimentary system tubular with a posterior anus, and a muscular pharynx and protrusible mouth cone containing the buccal cavity.
4 Body divided externally into a fixed number (13 or 14) of segments or 'zonites'.
5 Epidermis with a spiny cuticle comprising single dorsal and paired ventral plates on each zonite.
6 Body cavity a pseudocoel, derived from the persistent blastocoel.
7 Body wall with circular and diagonal muscles.
8 Excretory system of protonephridia in the 11th zonite; the solenocytes with paired cilia and multiple nuclei.
9 Nervous system with anterior circum-enteric ring and clusters of ganglion cells in a ventral nerve cord, reflecting the segmental appearance of epidermis and musculature.
10 Invariably small; almost always members of the marine meiofauna.
11 Without external ciliation.

The Kinorhyncha are one of the least-known groups of animals. They have strong affinities with other aschelminth phyla, especially the Loricifera and Nematomorpha.

The most characteristic feature is the spiny cuticle which is clearly subdivided into a fixed number of transverse zonites. The segmentation of the cuticle affects the epidermis, musculature and nervous tissues, giving them a segmented structure. This segmentation is not homologous, however, with that of the segmented coelomate worms, the Annelida (Section 4.14).

The Kinorhyncha are all minute, living as meiofauna in muddy marine substrata. About 100 species have been described, all similar in basic functional design.

The mouth cone with its stylets forms a protractable introvert which can be withdrawn into the second segment or neck. This introvert is armed with a circlet of recurved scalids.

The body region consists of ten unmodified trunk zonites, each with a single dorsal plate, the *tergite*, and paired ventral plates, the *sternites*. The body wall musculature flattens the trunk as dorsal and ventral plates are pulled together and pressure in the pseudocoel everts the introvert. The animals move by thrusting the introvert forwards, the recurved spiny scalids serving as anchors, while the body segments are pulled forward by retractor muscles.

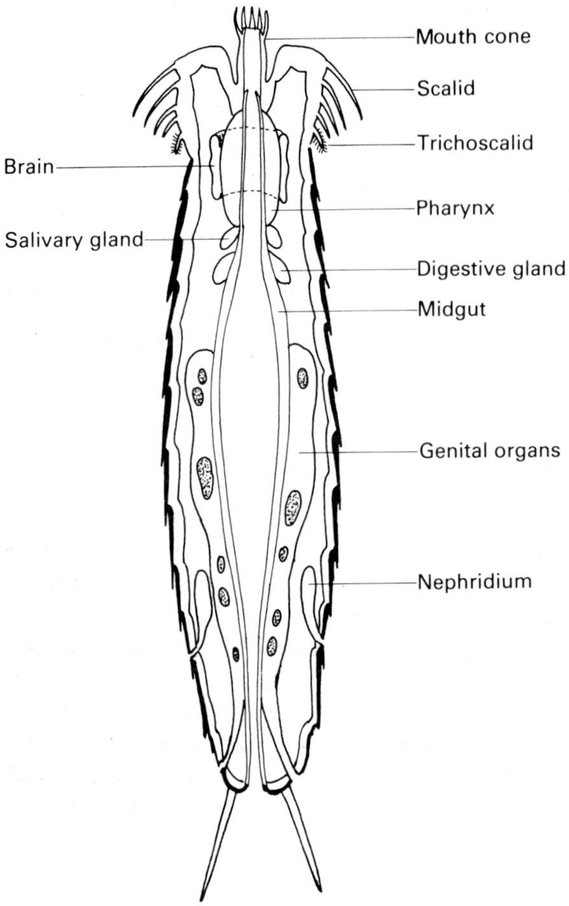

Fig. 4.24. The internal anatomy of a generalized kinorhynch with the introvert extended.

Internally (Fig. 4.24) they share some features with the other aschelminth phyla. The epidermis is somewhat syncitial in structure, forms longitudinal thickenings mid-dorsally and laterally (as in the Nematoda) and enters the major body spines. It also forms cushions projecting into the pseudocoel. There is a muscular pharynx, which, with its triangular cross-section, resembles that of many aschelminths.

4.6.3 Classification

The arrangement whereby the introvert is retracted is used to distinguish the two orders. In the Cyclorhagida, only the first zonite can be withdrawn and it is then protected by large plates on the second zonite (see Fig. 4.23a and b). Both the first and second zonites of the Homalorhagida can be retracted, and they are usually protected by ventral plates on the third zonite (Fig. 4.23c and d).

4.7 Phylum LORICIFERA

4.7.1 Etymology

Latin: *lorica*, a breastplate or corselet; *ferre*, to bear.

4.7.2 Diagnostic and special features (Fig. 4.25)

1 Bilaterally symmetrical.
2 Body more than two cell layers thick, with tissues and organs.
3 A straight through gut and a terminal anus.
4 A body cavity.
5 Body with three regions: an anterior, eversible head or introvert bearing nine whorls of backwardly directed spine- or tooth-like scalids and a protrusible mouth cone surrounded by eight to nine rigid oral stylets; a short, externally segmented neck region, bearing rows of plates; and a trunk covered by a cuticularized lorica composed of six longitudinal plates and bearing anteriorly hollow, forwardly directed spines. In adults, the introvert and neck can be withdrawn into the loricate trunk, the neck plates probably serving as a protective covering after withdrawal.
6 Gut with a large, muscular pharyngeal bulb and a telescopically extrusible buccal canal; whole gut possibly lined with cuticle.
7 One pair of gonads; gonochoristic.
8 Well-developed brain and ganglionated ventral nerve cord.
9 A larval stage similar to a small version of the adult (Fig. 4.26), except for the lack of oral stylets and the presence of two sets of locomotory organs: (a) two to three pairs of spines on the anteroventral region of the lorica, using which it can crawl; and (b) a pair of movable, foot-like caudal appendages with which it swims. A number of larval instars occur, separated by moults.

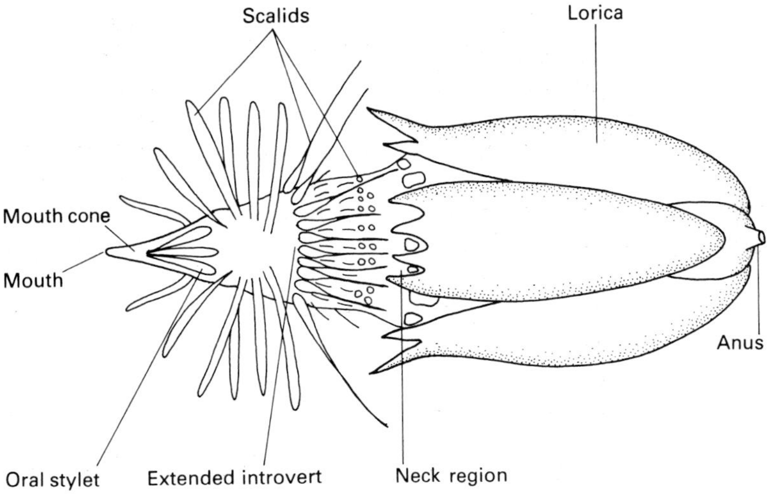

Fig. 4.25. Dorsal view of a loriciferan (simplified, after Kristensen, 1983).

10 Interstitial in marine shell-gravel.

The structure of these minute animals (< 0.3 mm) is as yet known only in outline fashion, and no information is available on their embryology, life style or habits; they were first described only in 1983. One reason for their late discovery would appear to be that they cling tightly to sediment particles, or possibly to other organisms, and are not susceptible to the standard extraction techniques used to collect interstitial marine species; another may be the superficial resemblance of the withdrawn animal to a rotifer or to a priapulan larva.

The most obvious feature of their body plan is its division into an anterior introvert and neck, and a posterior trunk encased by the lorica. In the adult, the whole of the anterior region can be retracted into the loricate trunk, although the larva can only withdraw the introvert into the neck. This system is

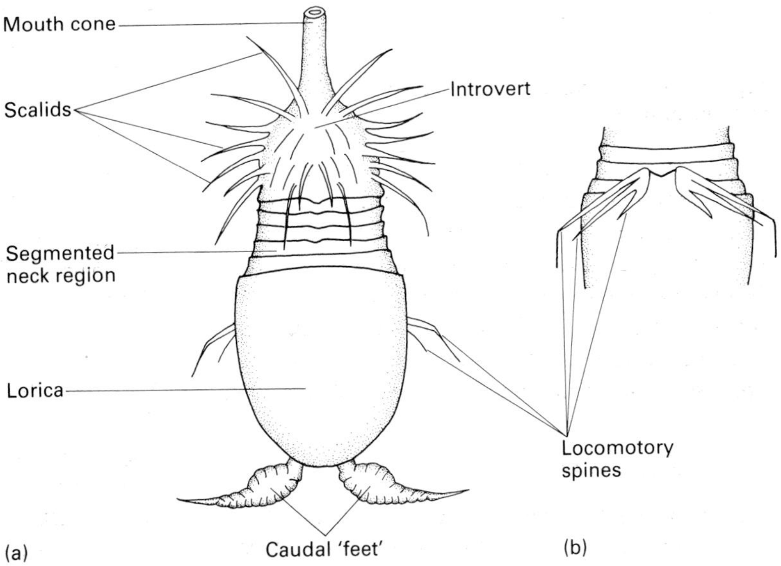

Fig. 4.26. (a) Dorsal view of the last-instar larva, and (b) a ventral view of its locomotory spines (simplified, after Kristensen, 1983).

closely similar to that in the priapulans, larval nematomorphs, the nematode *Kinochulus*, and, especially, the kinorhynchs (Section 4.6). In both kinorhynchs and loriciferans, the 'head' is everted by body cavity pressure and withdrawn by specific retractor muscles, whilst the mouth cone with its oral stylets is operated independently, being protracted and retracted muscularly. The function of extension and withdrawal of the anterior end of the body in loriciferans is not known; in other groups it accomplishes locomotion either through sediments or through animal tissues, as well as prey capture and consumption, and the loriciferans may be ectoparasitic.

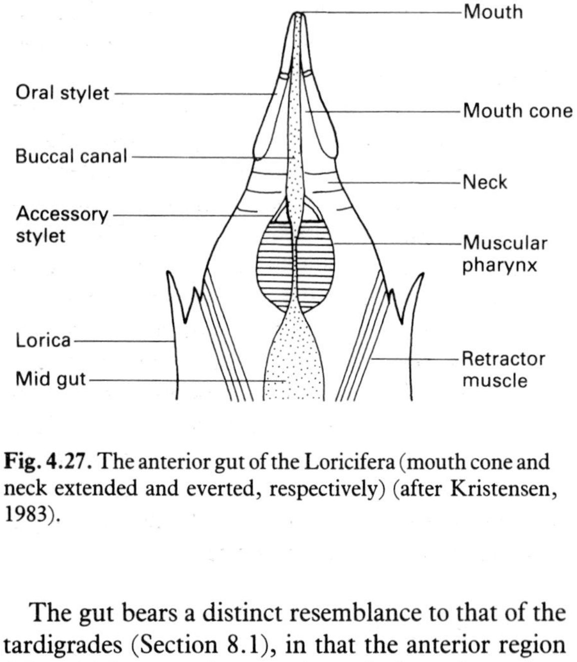

Fig. 4.27. The anterior gut of the Loricifera (mouth cone and neck extended and everted, respectively) (after Kristensen, 1983).

The gut bears a distinct resemblance to that of the tardigrades (Section 8.1), in that the anterior region (Fig. 4.27) comprises a telescopic buccal canal, a muscular, placoid-bearing pharyngeal bulb, a pair of accessory stylets, and two large salivary glands (the only glands to issue from the gut). Whether this resemblance is due to convergent evolution is not yet known, although it presumably indicates that the loriciferans feed by piercing with the oral stylets and then sucking fluids by means of a pharyngeal pump.

4.7.3 Classification

Although only one species has been formally described, several others are known and await description.

4.8 Phylum PRIAPULA

4.8.1 Etymology

Greek: *Priapos*, a phallic deity personifying male generative power.

4.8.2 Diagnostic and special features

1 Bilaterally symmetrical.
2 Body more than two cell layers thick, with tissues and organs.
3 A large retractable introvert or *prosoma* and a spiny or scaly, undivided *trunk*; sometimes a branching caudal appendage.
4 Alimentary system with anterior mouth surrounded by spine- or hook-like scalids, and a posterior anus sometimes surrounded by a crown of spines.
5 Excretory and genital systems closely associated in a urogenital organ with multiple solenocytes.
6 Nervous system with circum-oral ring and ganglionated ventral cord.
7 Body cavity spacious, possibly a true coelom.
8 Separate sexes, with external fertilization.
9 Larva without cilia and loricate (i.e. enclosed within plates).
10 Without a circulatory system but with corpuscles containing hemerythrin in the body cavity.

The Priapula contains a small but characteristic group of marine worms whose affinities are far from clear. They have some resemblance to the aschelminth pseudocoelomate phyla, with which they have often been classed, but functionally they are more similar to the coelomate worms of Sections 4.11–4.14. Some authors have viewed their body cavity as a coelom in that it has a mesodermal lining, but electron microscopical studies show this lining to be unique and unlike that of the coelomates. More detailed knowledge of the origins of the body cavity is required. What little is known of their embryology suggests that the priapulans show radial not spiral cleavage.

The range of form is shown in Fig. 4.28. The body of most species is composed of a large bulbous prosoma or introvert and a trunk. The introvert

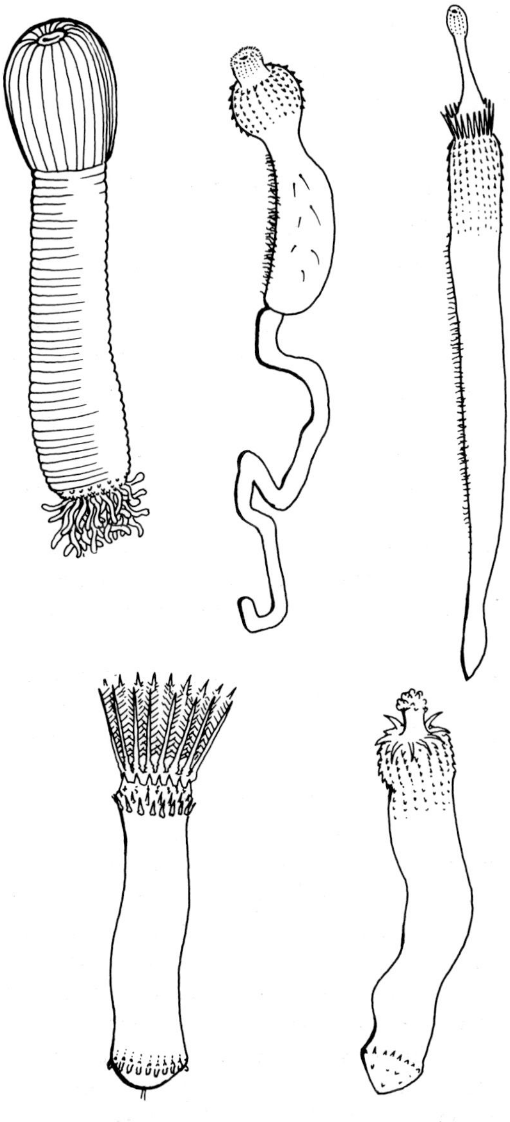

Fig. 4.28. The range of form in the Priapula.

terminates in a mouth surrounded by a ring of five circum-oral spine-like scalids. This large barrel-shaped structure with its longitudinal rows of scalids is usually everted, but can be retracted into the trunk, from which it is usually separated by a distinct collar. The often annulate trunk is usually scaly or spiny and bears a cuticle that is moulted periodically. At the posterior end of *Priapulus* is a curious branching structure, the caudal appendage, which may have a respiratory role (Fig. 4.29a). The cuticle overlies an epidermis in which there are fluid-filled spaces between the cells; it differs in structure from those of both the coelomare and aschelminth worms. The body wall has both circular and longitudinal muscles which exert pressure on the fluid-filled body cavity to evert the prosoma. The gut is a simple, straight tube with a horny, toothed buccal region and a muscular pharynx (Fig. 4.29b); there is no mouth cone, nor any oral stylets.

A unique feature is the conjunction of the genital system and the solenocytes of the protonephridia to form the paired urinogenital organs (Fig. 4.29b).

The loricate larva may be pelagic but is enclosed within cuticular plates, hence its name (Fig. 4.30). The larva already has the retractable scaly prosoma and is able to burrow effectively, whereas the adults find it difficult to re-enter the substratum once displaced. The larva may be the only stage in the life cycle normally to enter the substratum. The superficial resemblances between the spiny introvert of the Priapula and that of the Kinorhyncha, and between the loricate larva and rotifers, are not generally considered sufficient grounds for considering these phyla as being particularly closely related.

4.8.3 Classification

The 16 species of priapulans can be divided between two orders, the Priapulida containing the larger, more common predatory forms (Fig. 4.29a), and the Seticoronaria which have thin scalids in the form of a crown of stiff 'tentacles' (see Fig. 4.28, bottom left) and possess a circum-anal ring of hooks. Priapulans have a long history, several being known from the Cambrian.

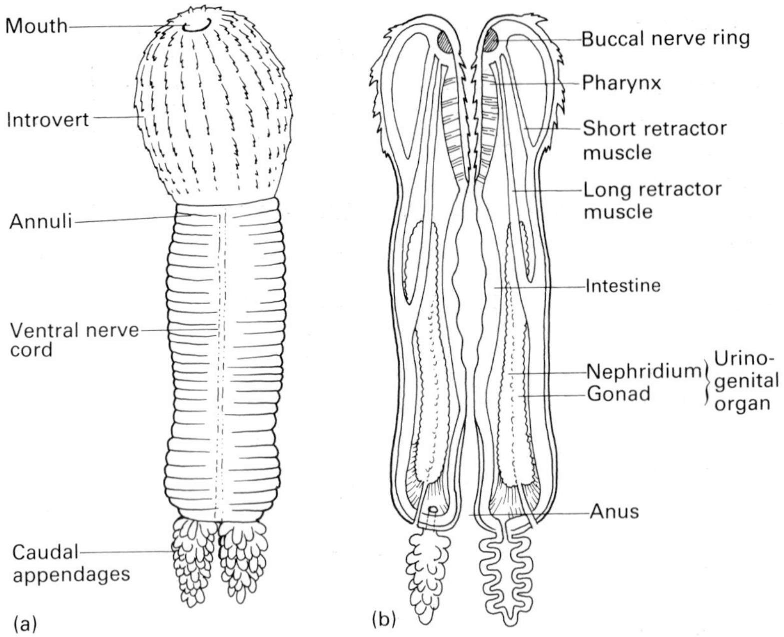

Fig. 4.29. The structure and anatomy of *Priapulus*: (a) external morphology; (b) schematic internal anatomy.

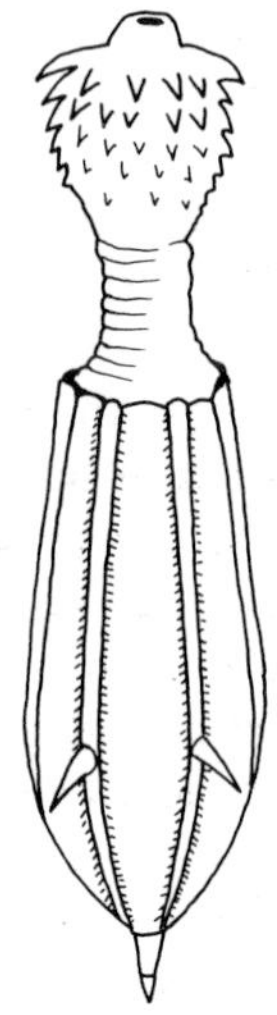

Fig. 4.30. A loricate larva of a priapulan.

4.9 Phylum ROTIFERA

4.9.1 Etymology

Latin: *rota,* a wheel; *ferre,* to carry.

4.9.2 Diagnostic and special features

1 Bilaterally symmetrical.
2 Body more than two cell layers thick, with tissues and organs.
3 With a crown of cilia in the anterior part of the body in the form of pre-oral and post-oral bands. Often organized into two wheel-like ciliary organs from which the name of the group is derived.
4 Alimentary system with anterior mouth, complex jaw apparatus, muscular pharynx, and posterior anus opening in a common cloaca with urinogenital system.
5 Epidermis with a fixed small number of nuclei; it includes an intracellular cuticle often thickened to form an encasement or lorica.
6 Body not segmented.
7 Excretory system, protonephridia.
8 No circulatory system or respiratory organs.
9 Body cavity pseudocoelomic.
10 Sexes separate, but males often rare or absent and when present almost always dwarf.
11 Minute, rarely reaching 3 mm in length.
12 Development direct, following modified spiral

cleavage. No further nuclear divisions occur after embryonic stages.

The Rotifera are among the smallest animals; their dimensions when adult are similar to the ciliated protists and larvae of many other phyla. Their bodies are composed of a small and strictly determined number of cells, or, strictly speaking, nuclei, since many tissues are syncitial (a feature shared with several other aschelminth phyla).

The Rotifera as adults characteristically pursue a mode of life similar to that of the ciliated larvae of larger invertebrates; indeed they are superficially like the trochophore larvae of marine Annelida, Sipuncula and Mollusca. The characteristic feature of a trochophore is the equatorial ring of cilia, the prototroch, which has two bands of cilia—pre-oral and post-oral—beating in contrary motion. A similar arrangement occurs in most planktonic Rotifera where the cilia of the corona form two bands—pre-oral (the trochus) and post-oral (the cingulum) (Fig. 4.31).

These structural similarities, however, do not imply any close phylogenetic relationship and other features suggest closer affinities with the aschelminth phyla, especially with the Acanthocephala. Some authors regard the Rotifera as being reproductively active, permanent larvae, but of what line of descent is unknown. No other aschelminth phylum has ciliated larvae, from which rotiferans could have derived.

Externally the body is often covered by a sculptured cuticle forming a cup-like lorica, the open end of which bears the ciliated corona and the mouth. The corona may be highly specialized and in sessile forms it can be much reduced. In these (see Fig. 4.37), the cilia are often modified as stiff sensory hairs. The corona can usually be withdrawn into the lorica. Posteriorly the body narrows to form the movable narrow 'foot'; this is often ringed or annulate to give the appearance of pseudo-segments. Like the corona the foot can usually be withdrawn, the pseudo-segments sliding telescopically one within another. The foot ends in a pair of toes which serve to anchor the organisms permanently or temporarily to the substratum; it may be absent or reduced in permanently

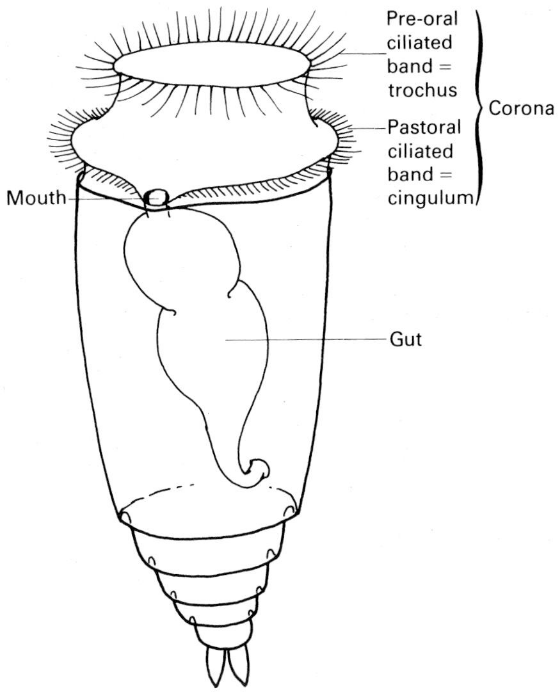

Fig. 4.31. A diagrammatic representation of the rotiferan body plan showing the two ciliated girdles.

planktonic forms. Internally the animals are relatively simple (Fig. 4.32), one of the consequences of small size. The mouth parts, however, are complex and vary according to the diet. The basic plan of the *trophi*, the hard parts of the *mastax* or feeding apparatus, is illustrated in Fig. 4.33. The fulcrum, which lies below the middle line of the trophi, supports two rami which branch symmetrically from it. Upon these are hinged the paired *unci* and *manubria*. Different trophi are illustrated in Fig. 4.33. They can be adapted for piercing and sucking cell contents, seizing prey in a pincer-like manner, grinding, crushing or tearing it.

Most rotifers have two protonephridia, each composed of a syncitium with a small number of cell nuclei. Inhabiting fresh waters, there is a high rate of fluid flow through the nephridia and the fluid is hypo-osmotic.

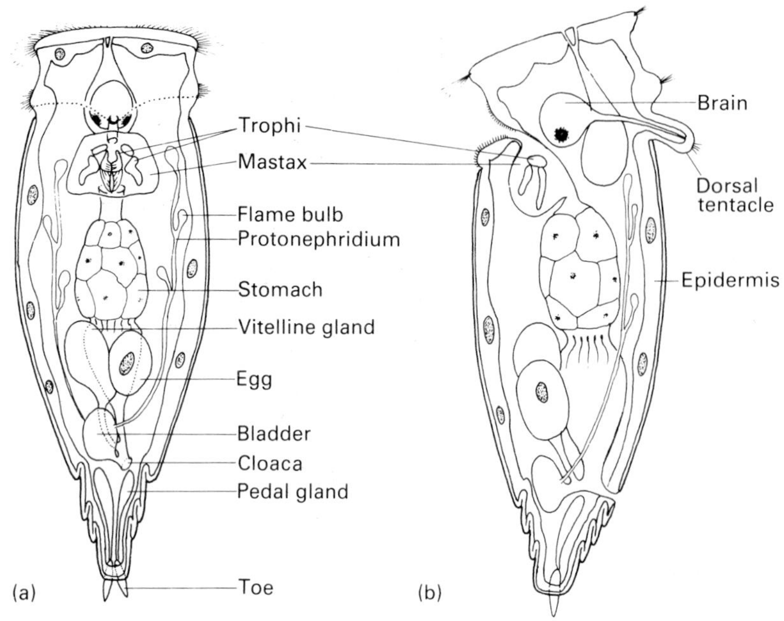

Fig. 4.32. General features of rotiferan anatomy (class Bdelloidea): (a) ventral view; (b) side view.

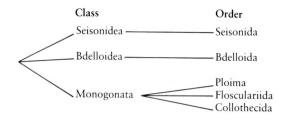

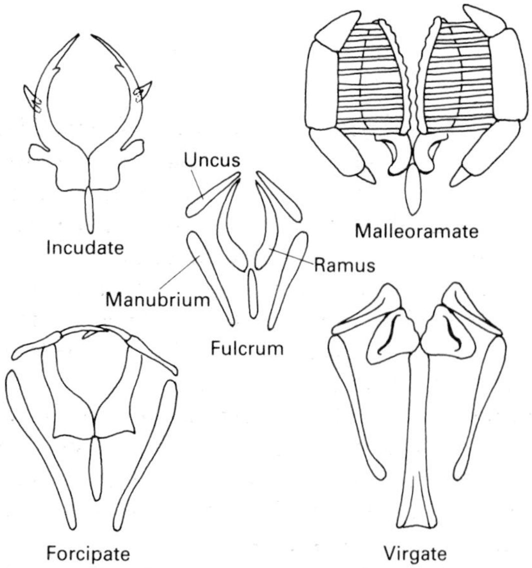

Fig. 4.33. The range of form of the mouth parts of Rotifera (after Donner, 1966).

4.9.3 Classification

The phylum Rotifera is subdivided into three component classes.

Class	Order
Seisonidea	Seisonida
Bdelloidea	Bdelloida
Monogonata	Ploima
	Flosculariida
	Collothecida

4.9.3.1 Class Bdelloidea

The class Bdelloidea for the most part have a characteristic 'two-wheeled' corona (Fig. 4.34) and the body is not enclosed within a lorica. Most species creep in a looping manner but also swim well by means of the unfolded corona. The bdelloids are adapted to impermanent and cryptic habitats; they are common

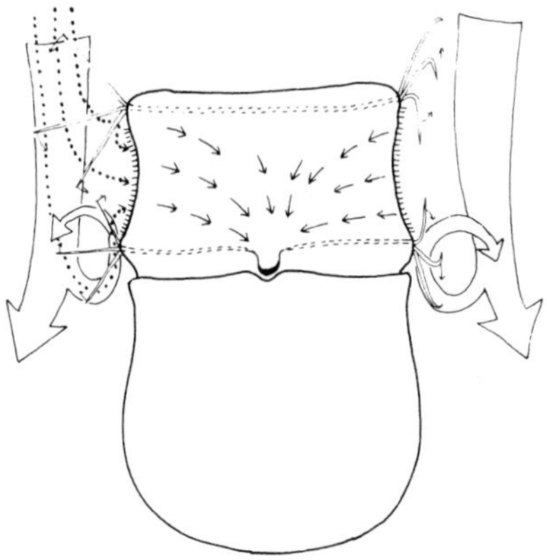

Fig. 4.34. The feeding mechanism of a pelagic rotiferan. The large cilia of the anterior girdle (trochus) draw a current of water in an anterior/posterior direction. This provides support and carries particles. The cilia of the more posterior girdle (cingulum) beat in a counter direction. This creates a vortex which traps food particles which are then transported to the mouth.

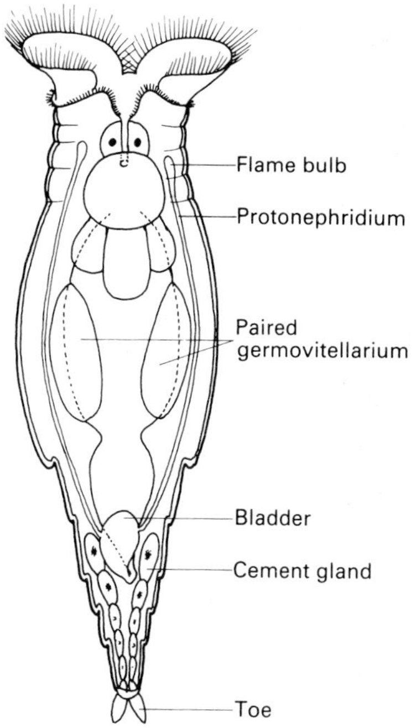

Fig. 4.35. The internal anatomy of a rotiferan of the class Bdelloidea, family Philodinidae. Note the paired ovaries. These animals are exclusively amictic; males have never been observed.

in the interstices of the damp sand and soil of lake and river beaches, and in association with mosses in periodically damp terrestrial habitats. They creep rather than swim, although the ciliary discs are well developed, and have massive, powerful, pedal cement glands (Fig. 4.35).

Their reproductive biology is most unusual. The entire group appears to be amictic, i.e. is without meiosis or crossing over in any form. Males have never been found and the whole population comprises obligately parthenogenetic females (see Section 14.2.1). They are also characterized by an ability to pass into a state of anabiosis when they become dried out, and can survive in this desiccated state for many years, then withstanding extremes of temperature, +40 to −200°C.

4.9.3.2 Class Monogonata

The majority of rotifers belong to this class which is distinguished by the single (not paired) ovaries. Several orders can be recognized.

The largest order, the Ploima, contains a wide variety of sessile and free-swimming forms (Fig. 4.36a,b,g). The foot, when present, has two toes, and the mouth usually lies within, not anterior to, the cingulum. These rotifers often have a stiff lorica.

The order Flosculariida (Fig. 4.36c,e) is a second group of either free-swimming or sessile rotifers in which the ciliary bands of the trochus are clearly separated into trochal and cingular bands. The foot, if present, does not have paired toes.

In the Collothecida the trochus is a large funnel often with the cilia forming stiff sensory hairs (see Figs 4.36d and 4.37). The females are always sessile, attached by cement glands, and the body is often encased in a gelatinous mass.

Unlike the bdelloids, monogont rotifers do exhibit sexual reproduction; nevertheless populations are dominated by females for most of the year; the rela-

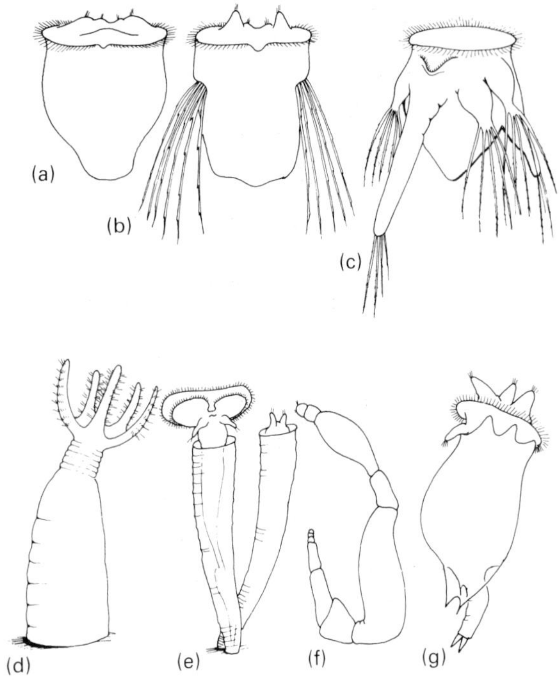

Fig. 4.36. The range of form among rotiferans of the classes Monogonta and Seisonidea (after Donner, 1966).

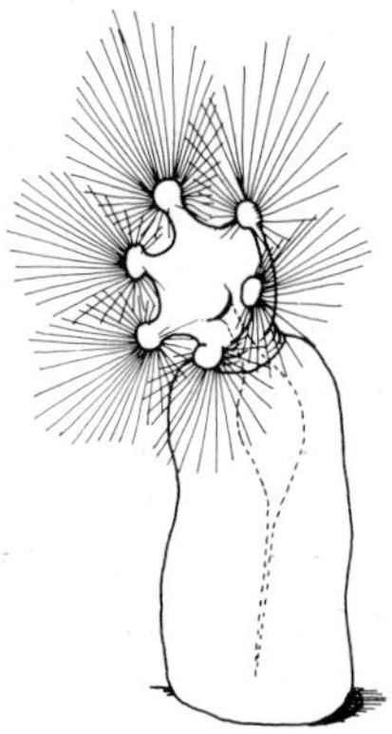

Fig. 4.37. A sessile collothecid rotiferan. Note the modified cilia of the corona.

tively simple, often dwarf but free-swimming, males occur only briefly and cyclically at certain times.

During most of the year diploid amictic females produce eggs that develop without fertilization into young females. On occasion, however, perhaps as a response to environmental conditions, morphologically distinct mictic females appear which lay haploid eggs. These either develop rapidly into haploid males, or if fertilized, become zygotes that develop into amictic females. The mictic females therefore display facultative parthenogenesis (see Section 14.2.1).

4.9.3.3 Class Seisonidea

This small class of marine rotifers live on the gills of crustaceans—*Nebalia* and some isopods. There is only one genus, *Seison* (Fig. 4.36f). Individuals are relatively large (a few millimetres) and have a much reduced corona and prominent mastax. The ovaries, like those of the Bdelloidea are paired, but unlike that group there is a normal pattern of gonochoristic sexuality in which fully developed males and females are equally common in the population.

4.10 Phylum ACANTHOCEPHALA

4.10.1 Etymology

Greek: *akantha*, prickle; *kephale*, head.

4.10.2 Diagnostic and special features

1 Bilaterally symmetrical, vermiform.
2 Body more than two cell layers thick, with tissues and organs.
3 Alimentary system lacking.
4 Body without segmentation, but superficial transverse annulation sometimes present.
5 A prominent hooked proboscis.
6 Body cavity a pseudocoel.
7 Epidermis syncitial with a small, fixed number of relatively large nuclei.
8 Nervous system a single ventral/anterior ganglion with paired or single nerves to the organs.
9 Nephridia occasionally present.

10 Respiratory and circulatory systems absent.
11 Separate sexes, with internal fertilization and viviparous development.
12 Infective larvae occupy a secondary insect host.
13 Adults always parasitic in the alimentary tract of vertebrates.

The general body form is illustrated in Fig. 4.38. Most species are small, 1 mm to a few centimetres in length, but a few attain lengths up to 1 m. Their most striking feature is the proboscis with its recurved spines. This, together with the neck region, can be inverted but it normally forms a permanent attachment to the host tissues. As in other pseudocoelomates, certain tissues tend to be constructed from a strictly determined small number of cells.

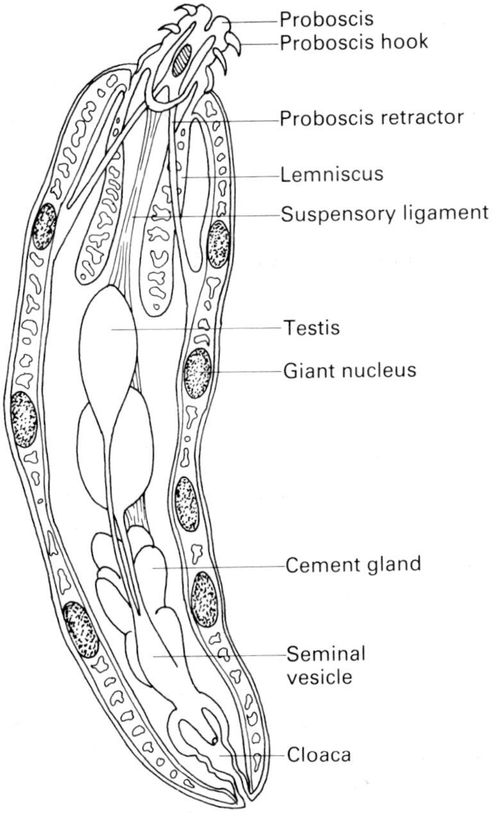

Fig. 4.38. The anatomy of a typical acanthocephalan. Based on *Neoechinorhyncus* (male).

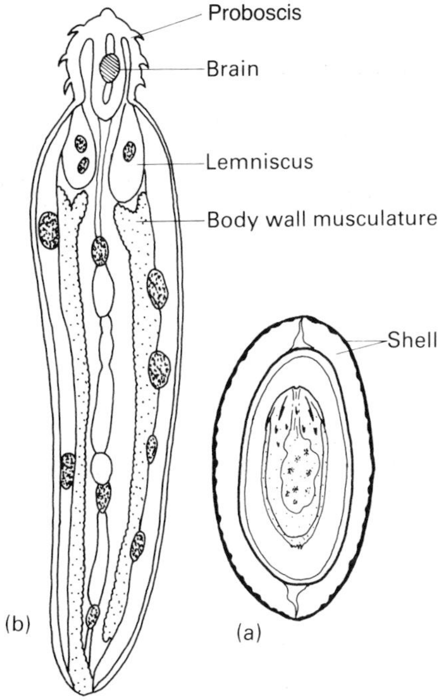

Fig. 4.39. Stages in the life history of the Acanthocephala: (a) the egg containing a larva; (b) the hatched and infective acanthella stage taken from a beetle larva.

The body wall is composed of a thin cuticle overlying a syncitial epidermis in which there is a further intracellular cuticle and a small number of precisely positioned and rather large nuclei. Indeed the position of the giant nuclei is a useful diagnostic feature. Beneath the epidermis is a thin layer of circular and longitudinal muscles overlying the pseudocoel. Two vertical flaps of tissue (the lemnisci) project into the pseudocoel and in these run fluid-filled channels from the epidermis. The endoderm is reduced to a ligament along which are suspended the reproductive tissues.

As in many parasites the adults show extreme simplification, with loss of many of the organ systems characteristic of free-living animals, but with hypertrophy of the reproductive organs. The Acanthocephala are gonochoristic, an unusual feature for endoparasites (but see Section 4.4); because of this they exhibit complex copulatory behaviour to ensure internal fertilization. Sperm are injected into the

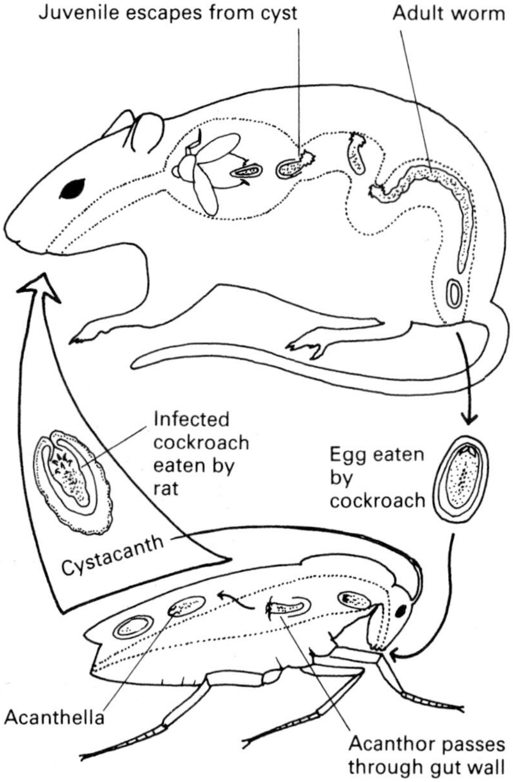

Fig. 4.40. The life cycle of *Moniliformis*, an acanthocephalan with a mammalian primary host (after Noble & Noble, 1976).

female genital tract which is plugged by cement glands after copulation. Fertilization takes place in the pseudocoelom.

The larvae of many specialized parasites betray the phylogenetic relationships of the adults, but this is not the case in this phylum. The larvae too are highly specialized and clearly share the structure of the adult. Following internal fertilization, the early larvae stages are encapsulated as 'eggs' (Fig. 4.39a, p. 123). These shelled larvae pass out with the faeces and must be eaten by an insect secondary host before further development can take place. In the insect the shelled larva hatches and makes its way to the haemocoel where it develops into a juvenile *acanthella* larva. (Fig. 4.39b). Such a larva may become encysted in the insect and this, when eaten by the vertebrate, re-establishes the adult in its definitive host. Figure 4.40 above illustrates the life cycle of *Moniliformis*, a parasite of mice, rats, cats and dogs.

4.10.3 Classification

The 1000 species are distributed between three orders in a single class. Two of the orders contain parasites of freshwater fish, the third those of terrestrial tetrapods.

4.11 Phylum SIPUNCULA

4.11.1 Etymology

Latin: *sipunculus*, little pipe.

4.11.2 Diagnostic and special features (Figs 4.41 and 4.42)

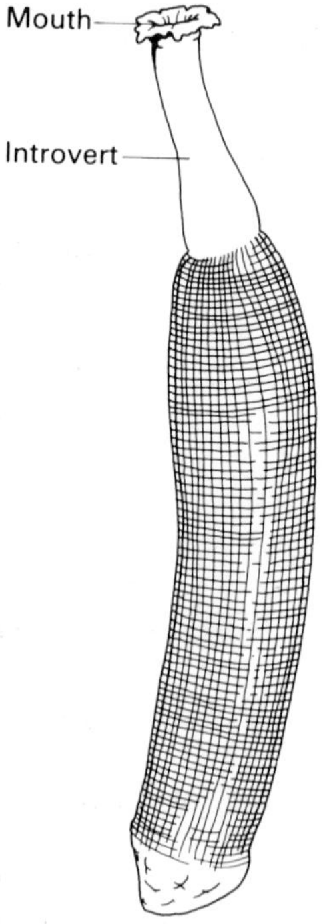

Fig. 4.41. The external appearance of a typical sipunculan.

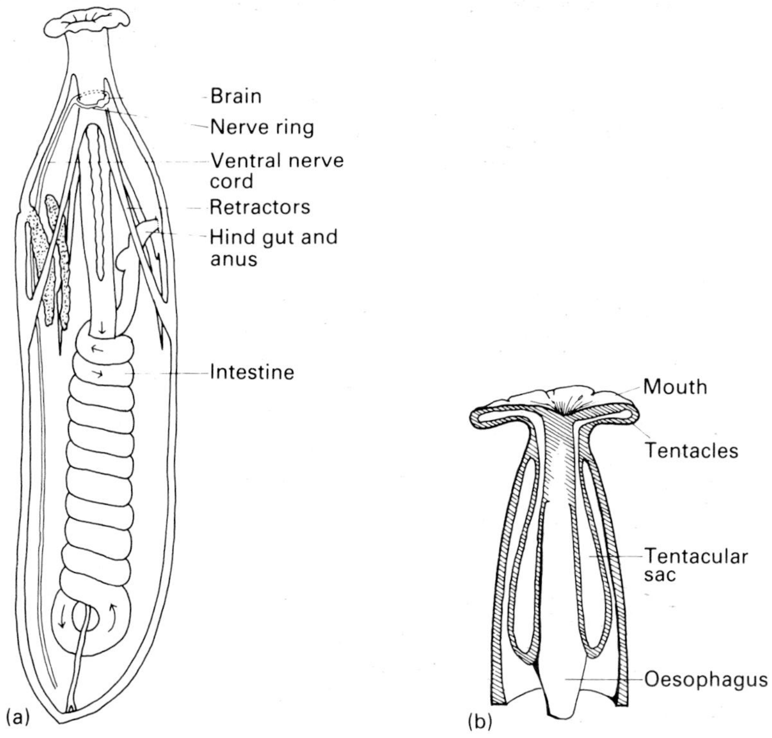

Fig. 4.42. (a) The internal anatomy of a sipunculan shown by dissection from the left side. (b) Detail showing the separate hydraulic system of the tentacles and introvert.

1 Bilaterally symmetrical, vermiform.
2 Body more than two cell layers thick, with tissues and organs.
3 A muscular U-shaped gut having both a mouth and an anus, the latter situated dorsally in the anterior part of the body.
4 Body divided into an anterior part, the introvert containing the mouth, and a stouter posterior trunk.
5 Body not divided into segments.
6 Body cavity a schizocoel, but without internal septa.
7 Outer epithelium covered by a cuticle, but without bristles or chaetae.
8 Without circulatory or differentiated respiratory systems.
9 Nervous system with an anterior brain, circum-oesophageal ring and a ventral nerve cord without ganglia.
10 A single nephridium or pair of nephridia.
11 Development with spiral cleavage, typically to a trochophore larva that in some species is superseded by an oceanic 'pelagosphaera' larva unique to the phylum.

The body is divided into two distinctive regions, of which the introvert can be retracted into the posterior trunk by powerful retractor muscles (see Fig. 4.42), and be everted by hydraulic pressure generated by the trunk musculature. Local relaxation of the circular muscles can cause dilations that wedge the worm in the substratum while permitting eversion of the introvert.

Sipunculans are generally non-selective deposit feeders, using the tentacles which surround the tip of the introvert to gather food. These tentacles are operated by a hydraulic system separate from that of the general body cavity. A pair of ducts lead from each tentacle into a ring-like canal from which extend one or two sacs running parallel to the axis of the introvert. The dorsal and ventral compensation sacs are compressible and serve as hydraulic reservoirs for the canal system (Fig. 4.42b).

Many species live in muddy substrata in a non-permanent burrow but some inhabit empty mollusc shells or worm tubes. Boring forms construct a tube in calcareous coral material. The range of form is illustrated in Fig. 4.43.

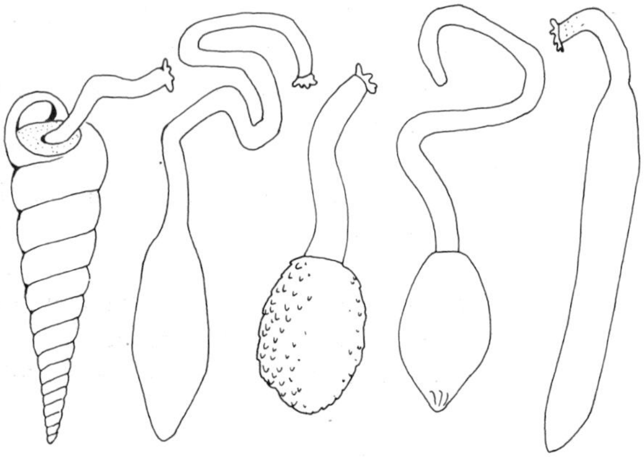

Fig. 4.43. The range of body form in the Sipuncula.

The nervous system of the sipunculans resembles that of the annelids and echiurans, with an anterior ganglion (the brain) in the introvert, a circum-oesophageal ring and a ventral nerve cord. There are, however, no signs of segmental ganglia.

The spacious hydraulic coelom also serves for the accumulation of gametocytes, shed into it from the simple gonads at an early stage of differentiation and stored there until large numbers of mature gametocytes have been accumulated. The gametocytes are then shed to the exterior via the nephridia. This pattern is modified in the minute form *Golfingia minuta*, a protandric hermaphrodite producing large eggs which develop directly. This is a consequence of its small size, the pattern of reproduction exhibited by most sipunculans being generally characteristic of large-bodied marine invertebrates.

The majority of Sipuncula hatch into a pelagic trochophore larva essentially similar to that of some annelids. In some sipunculans the trochophore larva metamorphoses directly to the juvenile adult, but usually it metamorphoses to a secondary larva, the pelagosphaera, which may be benthic but is usually pelagic. There are four developmental pathways in Sipuncula as illustrated in Fig. 4.44.

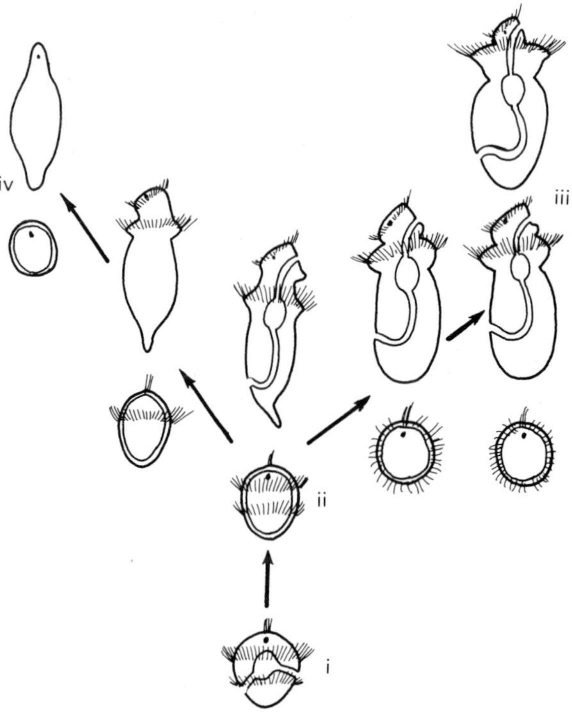

Fig. 4.44. Developmental pathways of the Sipuncula: (i) development via the trochophore larva; (ii) development via a non-feeding trochophore larva and a new phylum-characteristic pelagosphaera; (iii) development with large planktotrophic pelagosphaera larvae characteristic of the open oceans; (iv) development via yolky non-feeding larvae with eventual suppression of the pelagic larval stages. (Modified, after Rice, 1985.)

4.11.3 Classification

All 250 known species can be placed in a single class and order.

4.12 Phylum ECHIURA

4.12.1 Etymology

Greek: *echis*, viper; *ura*, tail.

4.12.2 Diagnostic and special features

1 Bilaterally symmetrical, vermiform.
2 Body more than two cell layers thick, with tissues and organs.
3 Muscular gut bearing both anterior and posterior openings.
4 Body with a large extensile and contractile anterior projection or 'proboscis' with lateral folds leading to the mouth.
5 A single undivided schizocoelomic body cavity present between muscular components of the body wall and the gut.
6 Body not divided into segments.
7 Chaetae present as a single pair in anterior ventral region and sometimes elsewhere.
8 Closed blood system with dorsal and ventral vessels; open vascular system in one group.
9 Nervous system with circum-oesophageal ring and ventral nerve cord, without definite ganglia.
10 Excretory system with up to 400 nephridia not arranged in a metameric manner.
11 Development via spiral cleavage to a trochophore-like larva.

The Echiura are unsegmented coelomate worms with affinities with the Annelida (Section 4.14) within which they were once classified. Their most distinctive feature is the proboscis, which contains the anterior lobe of the nervous system and which is probably homologous with the annelid prostomium (Fig. 4.45). The paired ventral chaetae or hooks in the anterior region of the body are also annelid-like features, as are the nephridia with a ciliated funnel, the arrangement of muscular layers in the body wall and the structure of the alimentary system. The Echiura differ from the annelids, however, in the complete absence of segmentation or metamerism. They can be regarded as being close to the annelid line of evolu-

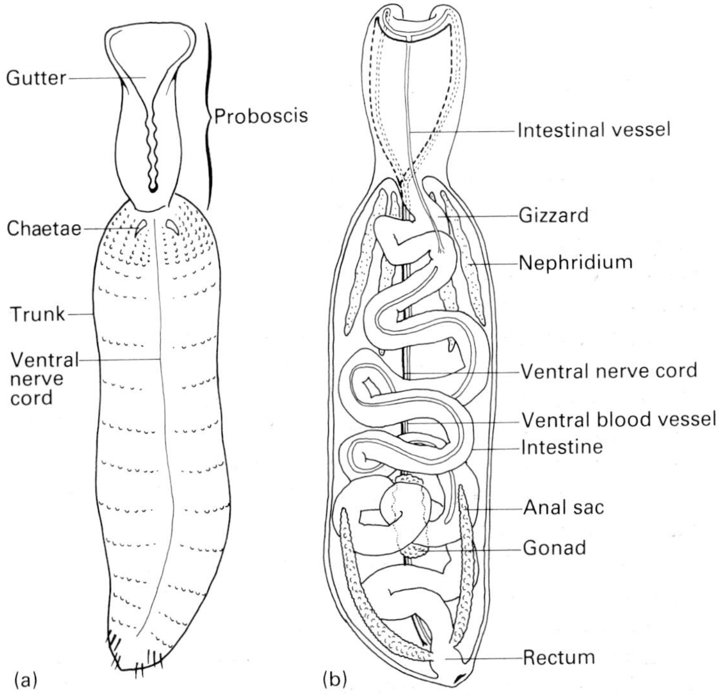

Fig. 4.45. The external morphology and anatomy of Echiura: (a) ventral view (*Echiurus*); (b) dissection from the dorsal side.

tion. Indeed there is evidence from their development that there may have been metamerism at an early stage in their evolution.

All are marine, living a sedentary existence in soft substrata. Most are detritus feeders living in permanent burrows, with the proboscis forming a non-selective food-collecting device directing material along its gutter, towards the mouth (see Fig. 9.6). The range of body form is illustrated in Fig. 4.46. The genus *Urechis* exhibits a different mode of feeding. It lives in a deep U-shaped burrow through which it drives water by waves of peristaltic muscular activity.

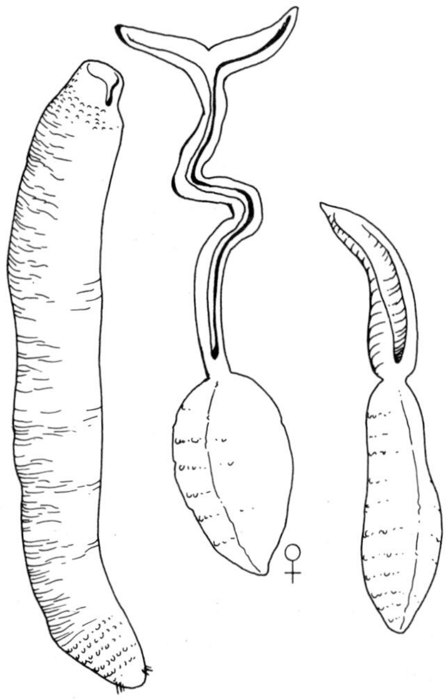

Fig. 4.46. Examples of the range of body form in the Echiura.

The reproductive biology of the Echiura is that characteristic of large-bodied coelomate worms in the marine environment. Sexes are separate and the coelom forms a cavity in which gametocytes can develop prior to mass discharge via the nephridia, in a spawning crisis.

One genus of Echiura exhibits a specialized mode of sexual reproduction discussed in Section 14.2.3.3. The males are dwarf individuals living on the proboscis of a female. Dwarf males were formerly thought to be a feature of only the bonellid echiurans but the phenomenon has been discovered recently in other echiurans, indicating perhaps that it has arisen more than once.

4.12.3 Classification

The 150 species are placed in three orders within a single class dependent on the disposition of the body wall muscle layers, the numbers of nephridia and the open or closed nature of the blood system. Members of one order, the Xenopneusta, have adapted the posterior part of the gut as a respiratory organ.

4.13 Phylum POGONOPHORA

4.13.1 Etymology

Greek: *pogon*, beard; *phoros*, bearer.

4.13.2 Diagnostic and special features (Fig. 4.47)

(Note that the interpretation of various anatomical features of the pogonophorans is still in somewhat of a state of flux, not least in respect of the homology, or otherwise, of the structures of the recently discovered vestimentiferans with those of the other members of this phylum. It is, for example, still far from agreed which is the dorsal and which the ventral surface.)

1 Bilaterally symmetrical, elongate, metamerically segmented worms (0.5 mm–3 cm diameter, 5 cm–3 m length) which permanently inhabit chitinous and protein tubes within which they can move.

2 Body more than two cell layers thick, with tissues and organs.

3 Without any mouth or gut at any stage of the life history.

4 With single and/or paired body cavities of uncertain nature (although usually assumed to be schizocoels, they lack a peritoneal lining).

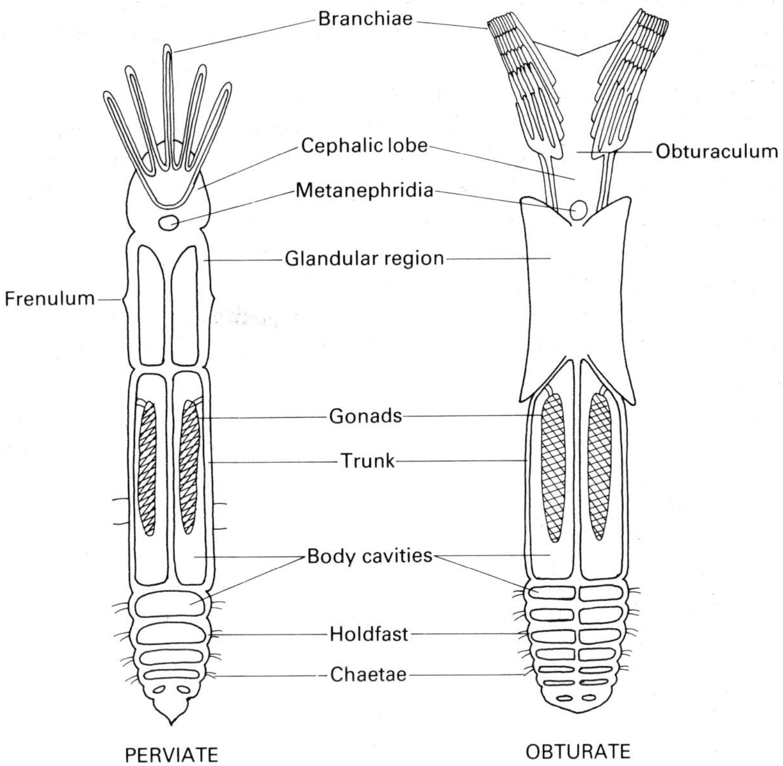

Fig. 4.47. Highly schematic dorsal views of generalized members of the two pogonophoran classes, showing the main regions of the body in diagrammatic form and their body cavities (after Southward, 1980 and Jones, 1985).

5 Body with four regions: an anterior 'cephalic lobe' bearing 1 to > 1000 tentaculate 'branchiae' supported hydrostatically; a short 'glandular region' with its paired cavities sometimes largely occluded by muscular tissue; an extremely elongate 'trunk' with a single pair of large hydrostatic compartments and, often, with various adhesive or secretory papillae; and a short 'holdfast' composed of up to 30 segments each with a single or paired body cavity and each budded off from a terminal proliferation zone.

6 Body wall with cuticle, epidermis, circular and longitudinal muscle layers; the holdfast and, in some, the trunk with chaetae; some regions with tracts of cilia.

7 With a closed blood system including a heart enclosed within a pericardial cavity in the cephalic lobe.

8 With a pair of metanephridium-like organs excretory organs in the cephalic lobe.

9 Gaseous-exchange organs assumed to be the branchiae, which may also function in nutrition.

10 With an epidermal nervous system comprising an anterior nerve ring or mass and usually a single longitudinal cord lacking ganglia (this is generally regarded as being ventral); multiple cords may occur in some regions.

11 Gonochoristic, with a pair of elongate gonads in the trunk, those of the male producing spermatophores.

12 Fertilization assumed to be external; development indirect, eggs and young larvae being brooded within the maternal tube.

13 Marine, usually in deep waters, i.e. 100–4000 m.

Although first discovered in 1900, little is known of these deep-sea worms. It was not until 1964, for example, that complete specimens were obtained (until then the metameric holdfast region was unknown), and up to that date they were generally thought to be related to the lophophorate and deuterostome groups, not least because of their apparently tripartite, oligomeric bodies. The discovery of the segmented and chaetae-bearing terminal section of the body, together with the description of the first vestimentiferan in 1969, however, led to a changed consensus that they had affinities with the annelids. Indeed, some authorities would place one or both pogonophoran classes in the Annelida; but, as indicated above, this and many other questions raised by these unusual animals await resolution.

Pogonophorans dwell within erect, closely fitting tubes, secreted by the glandular region and thickened by secretions of the trunk, within which they are supported by means of (a) the trunk and/or holdfast chaetae, (b) some of the trunk papillae, and, whilst the branchiae are extended into the surrounding water, (c) by ridges or wings of the glandular region. The branchiae vary greatly in their number and arrangement; often their bases are fused together or mounted on a tongue-like or ridged structure. In the largest class, they are possibly held so as to enclose a central tubular cavity, and, in those species with only a thick single 'tentacle', this is coiled achieving the same effect (see Fig. 4.49c). Into this cavity project paired rows of 'pinnules'—elongate extensions of the epidermal cells—flanked by long epidermal cilia (Fig. 4.48). The action of these cilia is most likely to be the creation of a water current down the intrabranchial cavity, but the function of the pinnules is uncertain.

Lacking a gut, the manner in which pogonophorans feed has excited much speculation. External predigestion of food seems unlikely since no appropriate enzyme-secreting cells have been detected. Absorption of dissolved organic compounds is a possibility, although whether sufficient concentrations of these occur has been questioned. It is also possible that small particles of organic matter, brought by the water current within the crown of branchiae, might be trapped by the pinnules and taken in by pinocytosis; certainly the epidermis of at least some species possesses microvilli similar to those of the epithelia of animal mid-guts. All that is definitely established to date is that those species which live near submarine hydrothermal fumaroles or 'cold seeps' derive most, if not all, their requirements from intracellular chemoautotrophic bacteria which metabolize the reduced sulphur compounds and methane issuing from such vents, and that equivalent bacteria are also present in such perviate species as have been investigated.

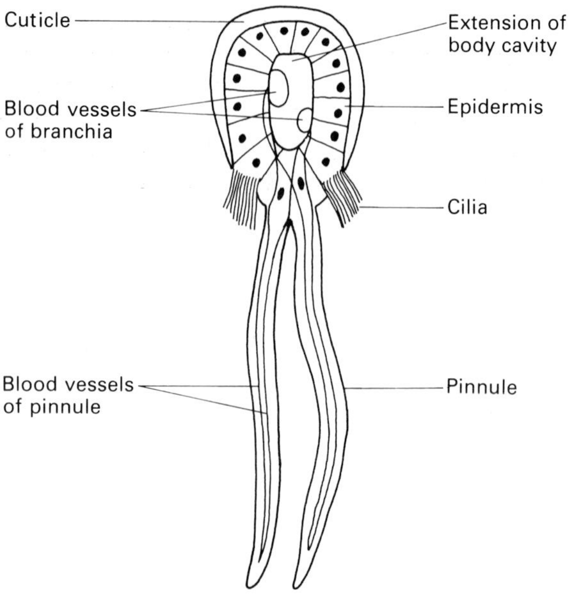

Fig. 4.48. Transverse section of a perviate branchia showing the two pinnules and tracts of cilia (after Ivanov, 1963).

4.13.3 Classification

The 100 species of pogonophorans are divided between two classes.

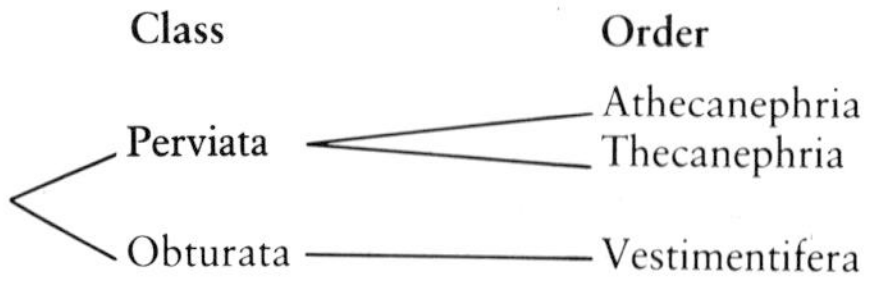

4.13.3.1 Class Perviata

These, the typical pogonophorans, are relatively small species (with diameters of <3 mm and lengths of <85 cm) living in tubes anchored in soft sediments. They possess few (1–250) elongate branchiae, and a raised ridge, the 'frenulum', running obliquely around the glandular region (Fig. 4.49)—this probably supports the anterior part of the body at the mouth of its tube. The 95 species are placed in two orders.

4.13.3.2 Class Obturata

The single order (Vestimentifera) (Fig. 4.50) comprising this class contains much bigger worms (1–3 cm diameter, and lengths of more than 2 m in some) with shorter, more numerous branchiae—in excess

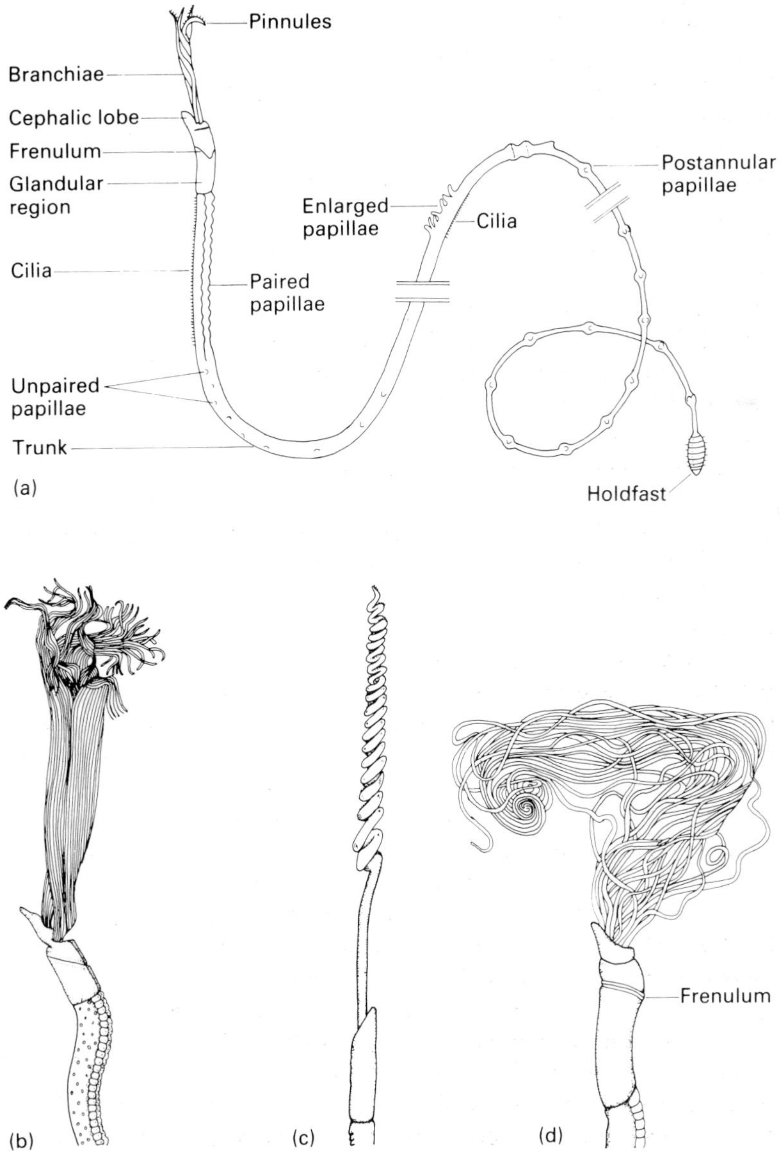

Fig. 4.49. Perviate pogonophorans: (a) a diagrammatic representation of the external appearance of a typical perviate (much shortened), showing its main features (after George & Southward, 1973) (note that in life the body is oriented straight and vertical); (b)–(d) the anterior ends of three perviates showing different numbers and arrangements of branchiae (after Ivanov, 1963).

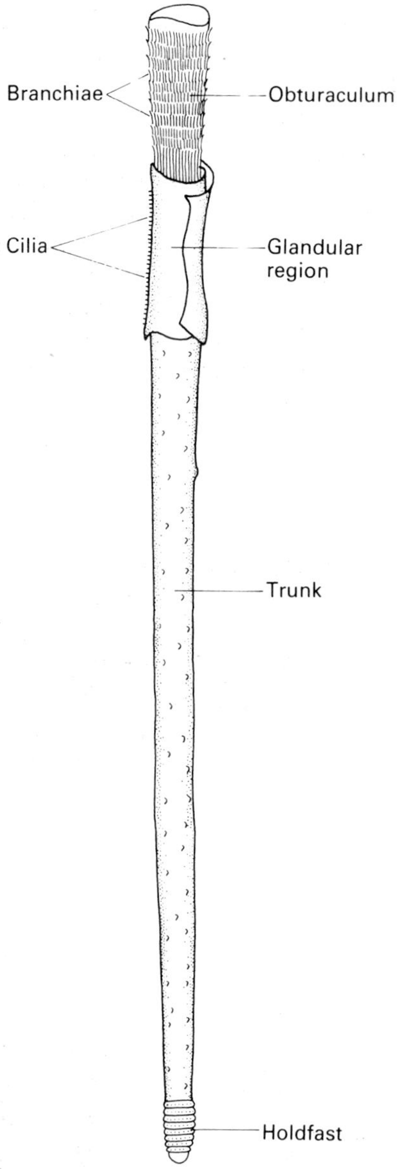

Fig. 4.50. Diagrammatic representation of the external appearance of a typical vestimentiferan (after Southward, 1982).

of 1000—mounted on an 'obturaculum', part of which closes the aperture of the tube after the animal has withdrawn. In contrast to the Perviata, vestimentiferans lack chetae on the trunk and do not possess a frenulum, the latter being replaced by two large wings which meet in the dorsal mid-line and extend anteriorly over the base of the obturaculum.

As more and more seeps and hydrothermal vents are being explored, so the number of known vestimentiferan species is showing an almost exponential increase. The first species was announced in 1969; the second in 1975; the third in 1981; six new ones in 1986; and perhaps a dozen further species are currently being described. Although a recent review of the group raises it to the level of phylum, and divides it into two classes and three orders, in view of the continual change in knowledge and opinion which will undoubtedly accompany the discovery of more material it seems advisable at the moment to retain the traditional system of a single class and order until the classification of the group stabilizes somewhat.

4.14 Phylum ANNELIDA

4.14.1 Etymology

Latin: *annellus* or *annelus*, a diminutive of *anulus*, a ring.

4.14.2 Diagnostic and special features

1 Bilaterally symmetrical, vermiform.
2 Body more than two cell layers thick, with tissues and organs.
3 A muscular gut with mouth and anus.
4 Body divided into segments (the segmentation may not be visible externally, but is always evident in the nervous system).
5 A presegmental prostomium containing a nervous ganglion, and a post-segmental pygidium.
6 Body cavity a series of schizocoels, obscured in specimens with anterior and posterior suckers.
7 Body cavity often subdivided by transverse septa,

but frequently suppressed or obscured in some or all segments.

8 Outer epithelium covered by a cuticle and with epidermal bristles or chaetae in bundles or singly, except in specimens with anterior and posterior suckers.

9 Body wall muscular, often with complete circular muscle layers and four blocks of longitudinal muscles.

10 A closed blood system.

11 Nervous system with presegmental supra-oesophageal ganglion, circum-oesophageal ring and a ventral nerve cord with segmental ganglia.

12 Segmental ducts of mesodermal and ectodermal origin, which may be combined, restricted to one or a few segments and/or partially suppressed.

13 A varying degree of cephalization.

14 Development with spiral cleavage, but with modification of this and epibolic gastrulation in forms with yolky eggs.

15 Planktonic development in marine forms sometimes via a free-living trochophore larva but this stage frequently encapsulated. Freshwater and terrestrial forms with encapsulated eggs.

The structural grade of organization of the Annelida combines the hydrostatic and functional properties of the coelomate grade of organization with a segmented body plan. At its most primitive level of expression, the segmentation affects the ectoderm and the mesoderm and includes segmental septation of the coelom. Many annelids, however, have suppressed or modified this pattern and are secondarily aseptate or acoelomate.

Segmentation (Fig. 4.51, see also Fig. 4.54), is

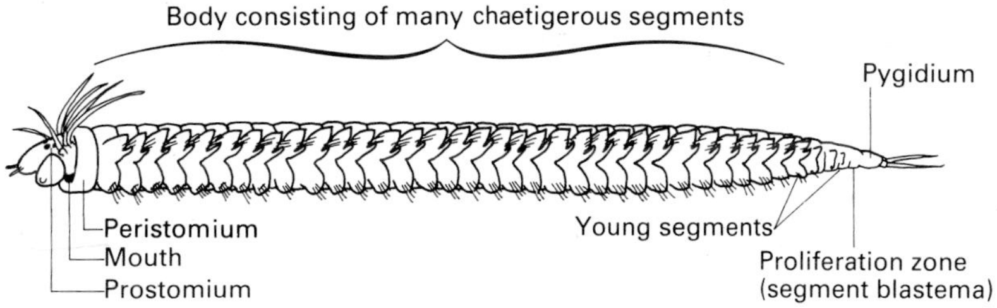

Fig. 4.51. The basic body regions of an annelid illustrated with the polychaete *Nereis*.

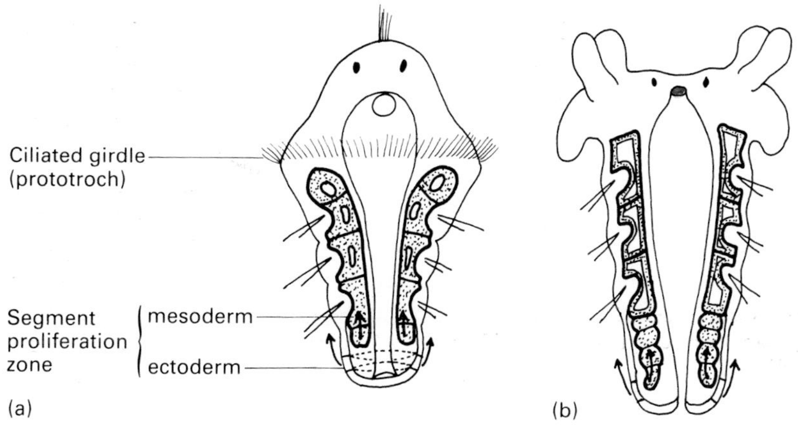

Fig. 4.52. The formation of segments in early embryonic development: (a) ciliated post-trochophore larva; (b) metamorphosed larva.

established during development from paired mesodermal growth zones and a corresponding ectodermal ring in front of the pygidium, the posterior region through which the endoderm opens to the exterior. New segments are formed at the anterior face of the pygidium, so that the last-formed segment is always the most posterior (Fig. 4.52a). In some Polychaeta the first three segments are formed precociously and simultaneously, and may be specialized for planktonic life, but subsequent segments are produced successively (Fig. 4.52b). Annelids may continue to add segments throughout adult life, though a definitive number is usually reached. In adult Polychaeta and Oligochaeta, renewed production of segments can be stimulated by transection, and the loss of posterior segments results in the formation of a new pygidium and prepygidial growth zone (Fig. 4.53) (see Section 15.5).

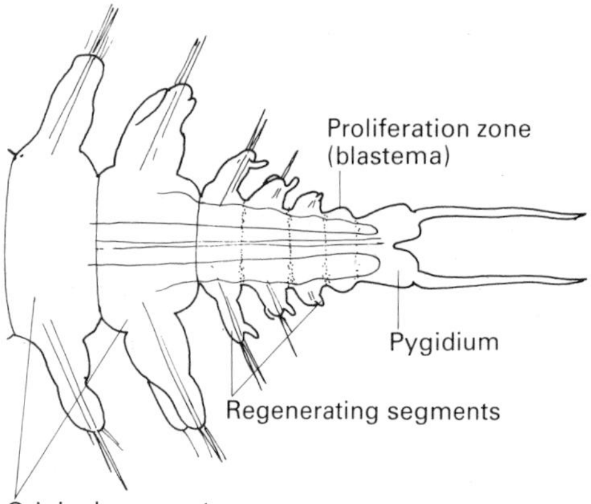

Fig. 4.53. Regeneration in *Nereis*. Loss of segments causes formation of a new segment proliferation zone and the production of a series of segments, the most posterior of which is the youngest.

At the most fundamental level, the annelid body can be thought of as comprising a presegmental prostomium containing the supra-oesophageal ganglion, a series of more or less identical structural units, the segments, and a posterior post-segmental region, the pygidium, in the anterior ventral region of which is the segment blastema or growth zone (Fig. 4.54). The fundamental features of this segmentation are as follows:

1 The ectoderm: segmental arrangement of the chaetae; the nervous system with a segmental ganglion and associated nerves; and the ectodermal nephridia.

2 The mesoderm: segmentation of the musculature associated with the chaetal sacs/parapodia, and of the corresponding vessels; septation of the coelomic space, creating isolated body cavities; segmental arrangement of the germinal epithelia and the associated paired coelomoducts.

The proto-annelids probably had complete septa and a complete series of ectodermal protonephridia and mesodermal coelomoducts. This arrangement has not persisted in any living annelid. Protonephridia, with a group of flame cells, are retained in a few families of polychaetes but the nephridia usually have an open ciliated funnel, a nephrostome. Nephridia and coelomoducts have different embryological origins (Section 12.3.1) and are separate in the oligochaetes. In the essentially marine Polychaeta the nephridium and the coelomoduct are usually combined to form a compound structure, the nephromixium, and in those families lacking complete septa and having open coelomic cavities the number of germinal epithelia and segmental ducts is greatly reduced.

4.14.3 Classification

There are at least 75 000 described species of annelid, which are readily subdivided into three major groups, the Polychaeta, Oligochaeta and Hirudinea, and two minor ones. The Hirudinea diverged from an early line of oligochaete evolution, and these two groups, together with the much smaller Branchiobdella, form the class Clitellata.

4.14.3.1 Class Polychaeta

The mainly marine polychaetes are remarkable for their morphological and anatomical diversity; they take their name from the numerous chaetae or

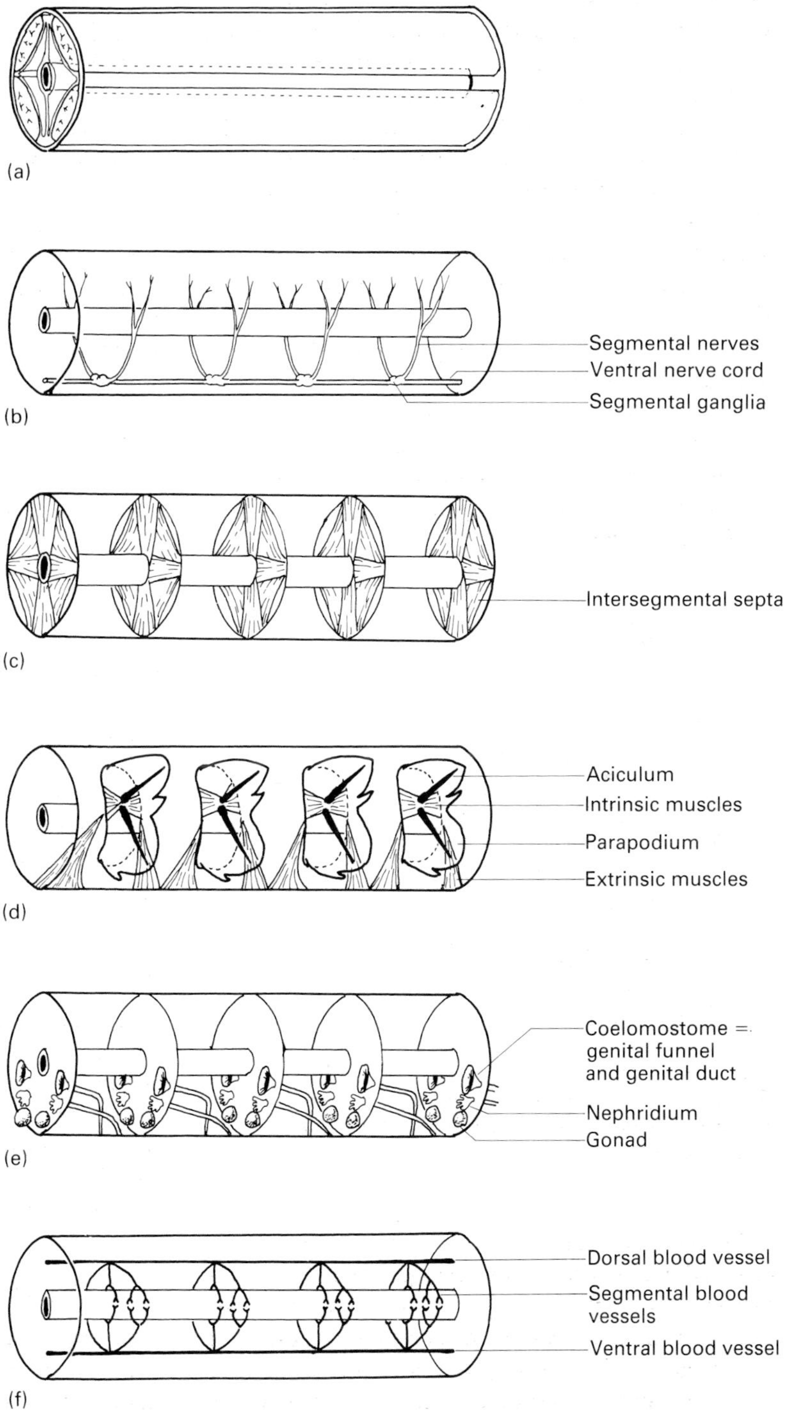

Fig. 4.54. Diagrammatic representation of the components of the segmented body plan of a hypothetical proto-annelid ancestor. Living annelids exhibit some, but not necessarily all, of these features. (a) The longitudinal muscle blocks and the gut are not segmented. The body cavity is a true coelom with peritoneal lining. (b) The segmentation of the ventral nervous system, with segmented ganglia and nerves. This is the most conserved element of segmentation in living annelids (see section on Hirudinea below). (c) Subdivision of the coelom by transverse septa. This feature is retained in most Oligochaeta and some Polychaeta, but can be much modified and is lost in Hirudinea. (d) Segmentation of the ectoderm and mesoderm due to the development of parapodia or segmentally arranged chaetae. This feature is prominent in Polychaeta, less prominent in Oligochaeta and suppressed in Hirudinea. (e) Segmentation of genital ducts, excretory ducts and germinal epithelia. Proto-annelids are supposed to have had a complete series of mesodermal coelomoducts (for the discharge of gametes) and ectodermal nephridia. They are also supposed to have had a pair of germinal epithelia in each coelomic compartment. These conditions are not found in any living annelids and are frequently much modified. (f) Segmentation of the blood vascular system.

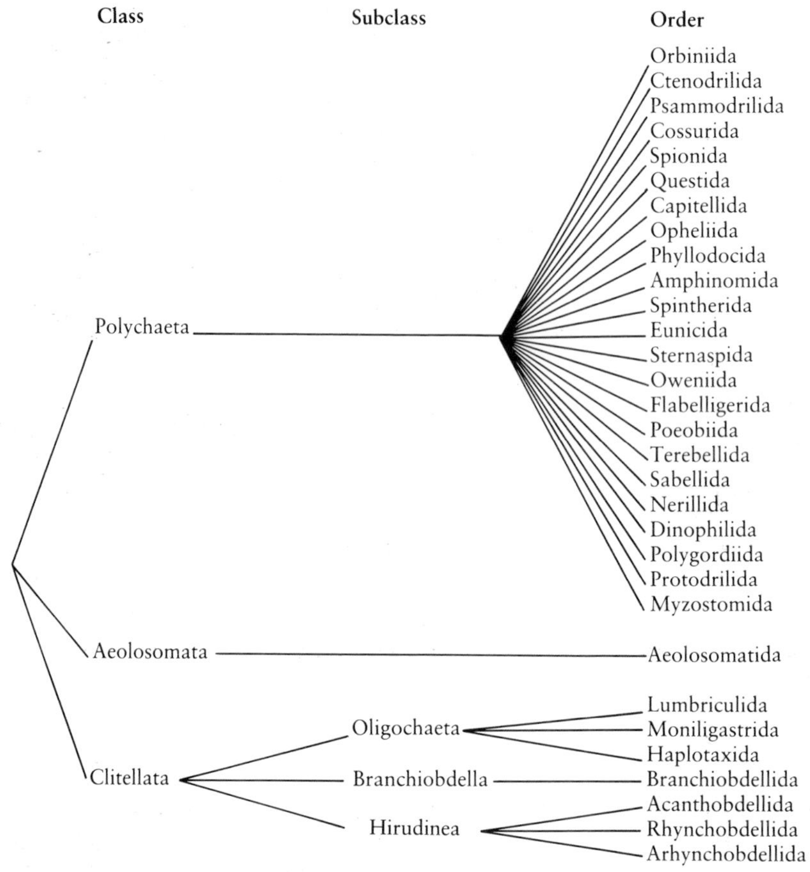

bristles, inserted in two groups rather than singly, in the biramous segmental parapodia (Fig. 4.55). Many exhibit indirect development (Fig. 4.56), though this may be suppressed.

About half the polychaete families and species can be conveniently grouped as 'errant' forms; these belong mainly to two large and easily defined orders, the Phyllodocida and the Eunicida. The Phyllodocida have an axial proboscis (the anterior part of the alimentary canal) everted hydraulically and withdrawn by retractor muscles. This is often armed with a small number of jaws of hardened proteins that may contain a high proportion of heavy metal ions, but which are never calcified (Fig. 4.57).

The Eunicida are superficially like the Phyllodocida, but the proboscis is not axial but an eversible buccal mass, armed with a complex array of calcified jaws, a pair of chisel-like mandibles and several pairs of maxillae (Fig. 4.58). The range of form among the errant polychaetes is illustrated in Fig. 4.59. Most have a well-developed head with a variety of sense organs and a complex brain. Many are carnivorous but some are detritivores, filter-feeders or omnivores. They have well-developed parapodia, with the dorsal and ventral lobes supported by a stiff proteinaceous rod—the aciculum.

Some errant polychaetes display *epitoky* or swarming behaviour associated with sexual reproduction.

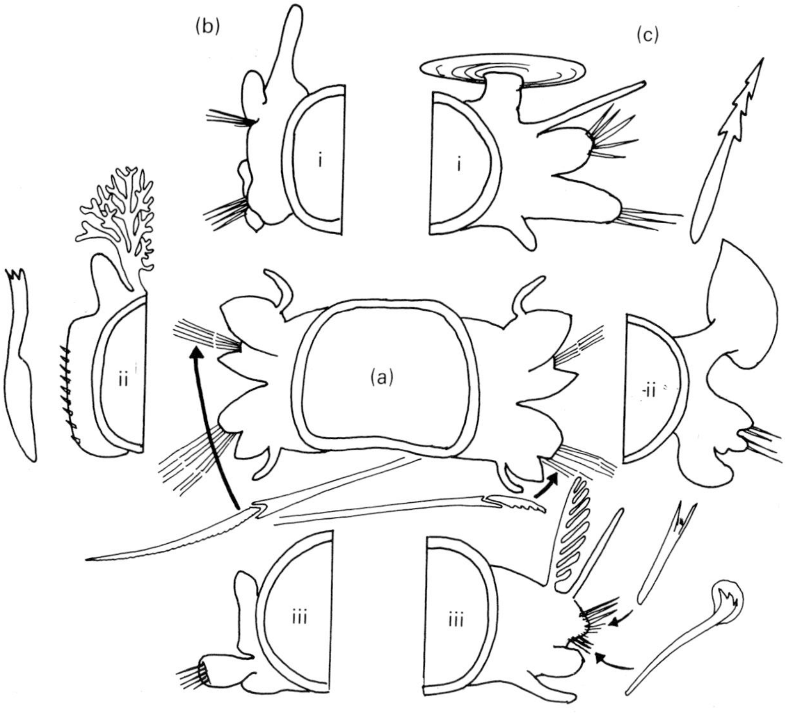

Fig. 4.55. Examples of the range of form among polychaete parapodia.
(a) The fundamental biramous type as found in *Nereis*. Details show the morphology of some chaetae. (b) Examples of parapodia among sedentary families: (i) Spionidae; (ii) Arenicolidae; (iii) Sabellidae.
(c) Examples of parapodia and chaetae among errant families: (i) Polynoidae; (ii) Phyllodocidae; (iii) Eunicidae.

This often involves a complex metamorphosis of the adult worms. Epitoky takes place in two fundamentally different ways (Figs 4.60 and 4.61). In epigamy the whole worm is transformed into the swarming epitoke whereas in schizogamy (stolonization) the posterior segments of sexually mature worms become detached as migratory gamete-bearing stolons.

The remaining polychaetes are frequently grouped as the sedentary forms and 'archiannelids'. Both are diverse and artificial assemblages, in which it is difficult to recognize relationships. Most biologists refer directly to the rather distinctive families rather than to orders.

In the 'sedentary polychaetes' there is usually regional variation in structure of the segments and distinct thoracic and abdominal regions (Fig. 4.62). They are all microphagous deposit or suspension feeders and do not have the wriggling or walking locomotory patterns exhibited by the 'errant' species (see Section 10.5). Their diversity arises from their widely different modes of life. Some live in burrows, others in tubes of secreted parchment-like material or calcium carbonate, or sand grains. Several families exhibit a functional convergence with Oligochaeta. They burrow in soft substrata and are either 'swallowers' or sand-grain 'lickers'. They have a small prostomium without prominent sense organs, simple chaetae in reduced parapodia, well-developed circular muscles and complete septa. Several families have soft prehensile tentacles which collect fine particles of organic sediment and convey them to the mouth (see Fig. 9.6). Such tentacles are often outgrowths of the prostomium. Some rather advanced families of tubiculous worms have a crown of stiff prostomial tentacles which act as a true filter-feeding device (Fig. 9.3). Other feeding specializations are discussed in Chapter 9 (see, e.g. Fig. 9.5).

The archiannelids (Fig. 4.63) are an unrelated

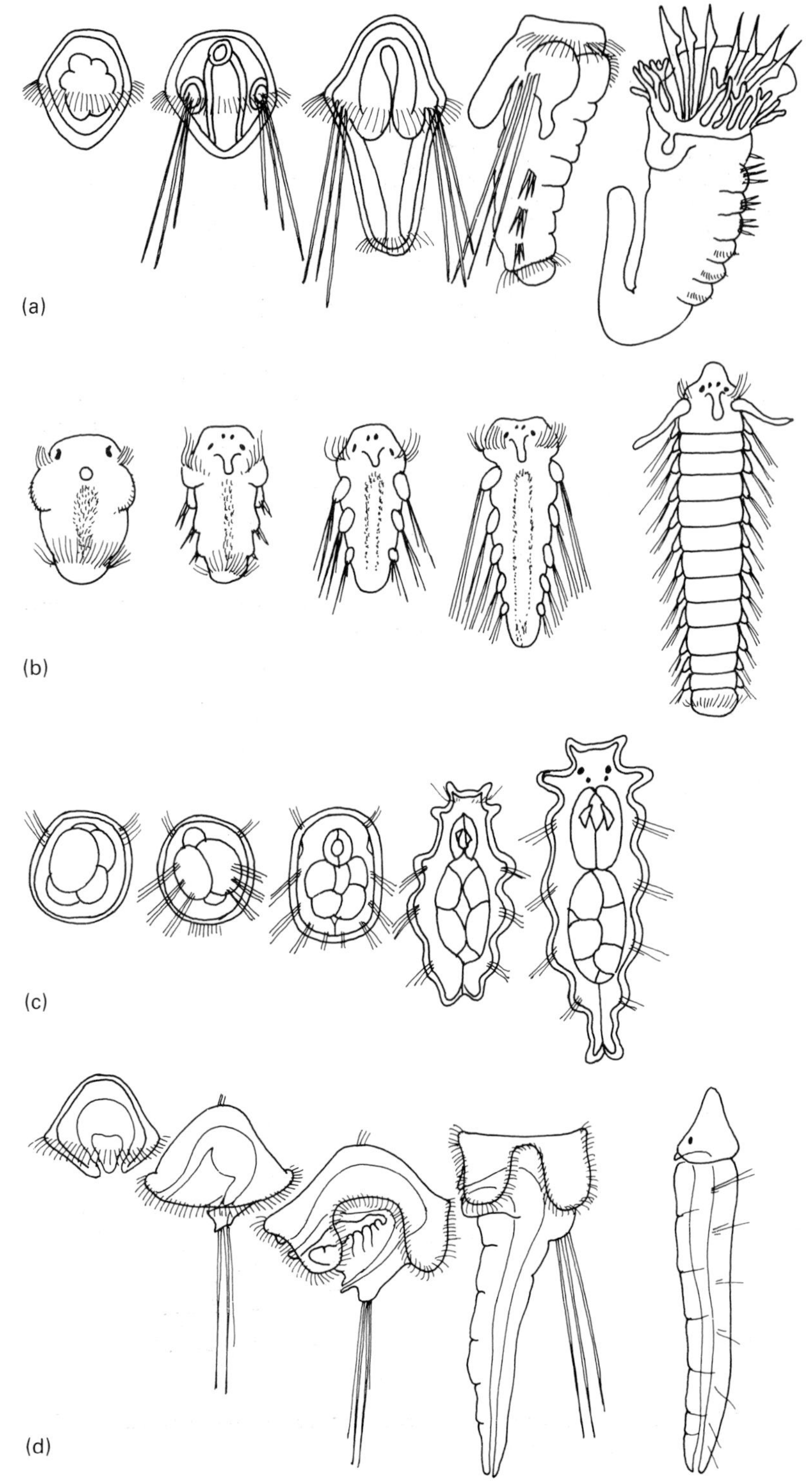

Fig. 4.56. Examples of planktonic development stages among the Polychaeta: (a) *Sabellaria*; (b) *Spio*; (c) *Nereis*; (d) *Owenia*.

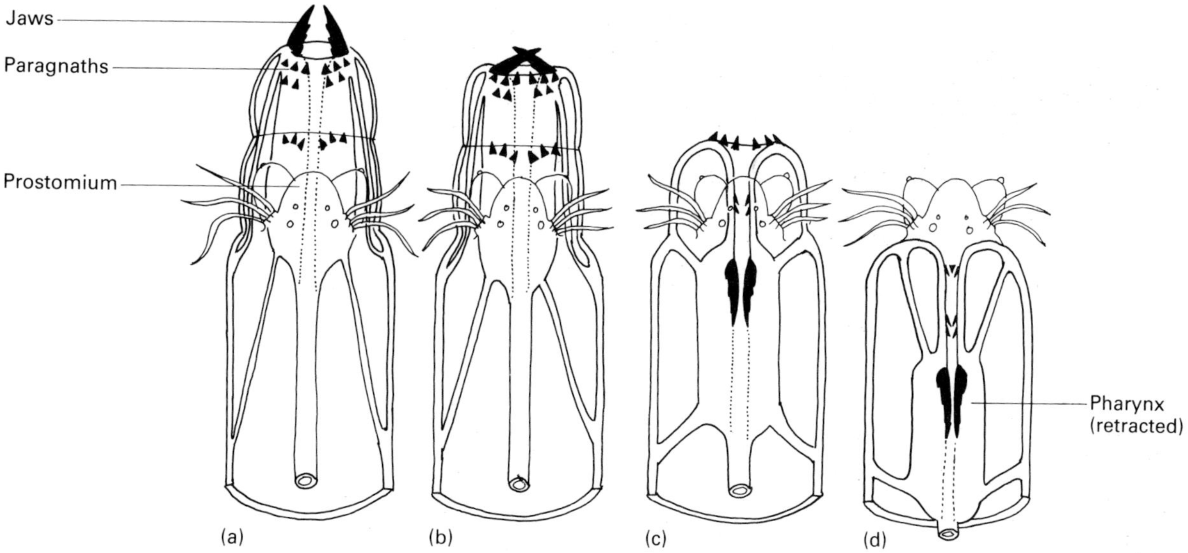

Fig. 4.57. The proboscis of the phyllodocid type. This is a muscular pharynx which is everted by coelomic pressure and withdrawn by the action of retractor muscles as shown in (a)–(d). It is often, as in *Nereis*, armed with proteinaceous jaws.

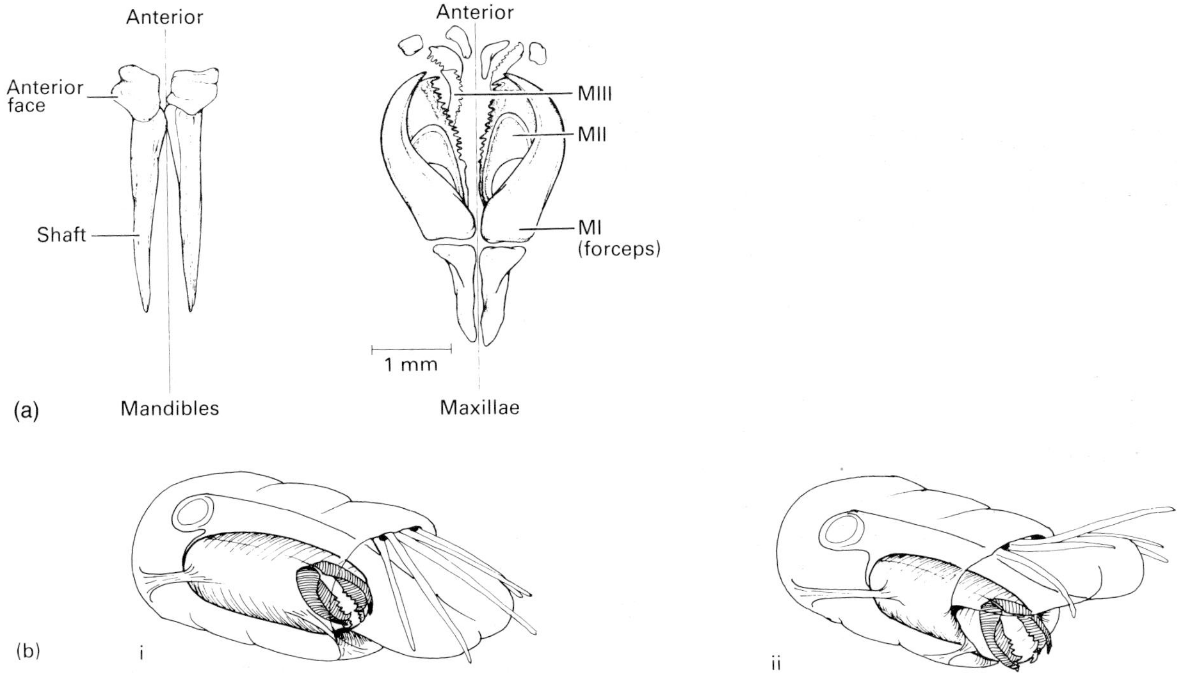

Fig. 4.58. The jaw apparatus of the eunicid type. (a) The isolated jaw complex. (b) (i) The jaws retracted in the muscular ventral tongue-like floor of the pharynx; (ii) the everted jaws. (After Olive, 1980.)

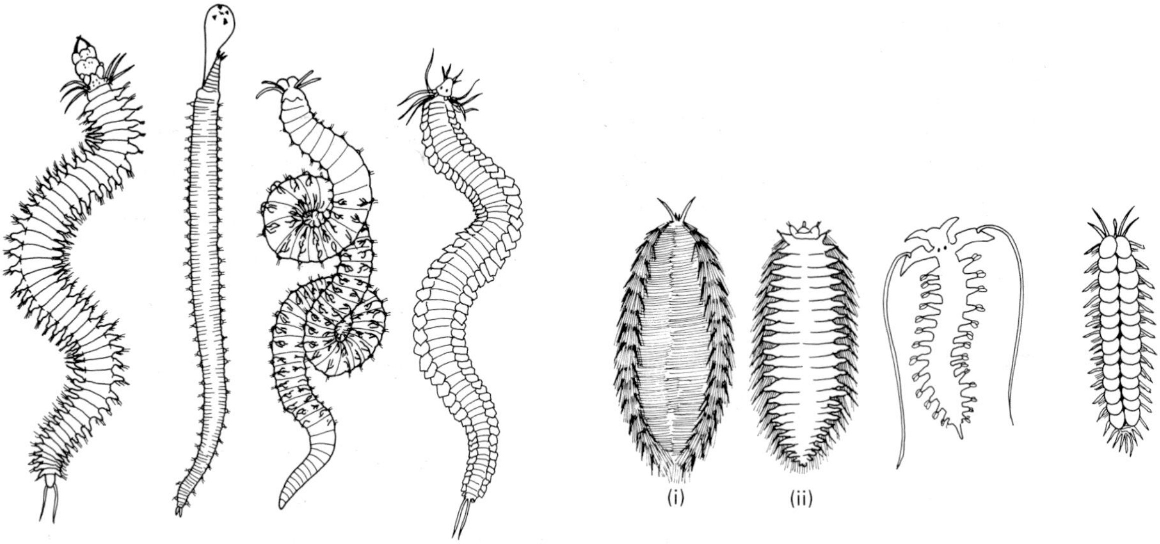

Fig. 4.59. The range of form among the 'errant' Polychaeta; (i) and (ii) are the dorsal and ventral views, respectively, of the animal.

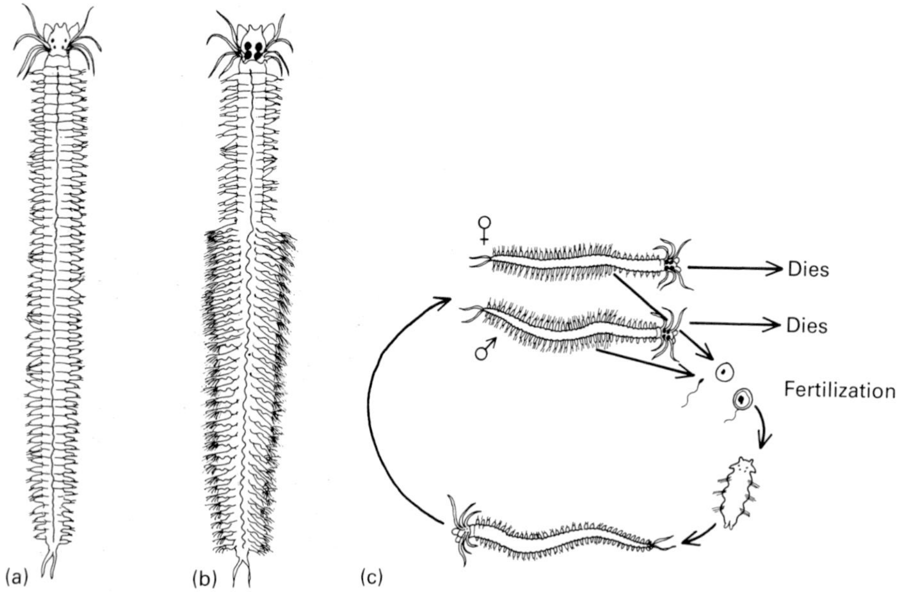

Fig. 4.60. Epitoky: epigamy. Mature specimens undergo metamorphosis, swarm and die: (a) the unmodified atokous form of *Nereis*; (b) the metamorphosed epitokous *Heteronereis* form of *Nereis*. (c) Diagrammatic representation of the life cycle.

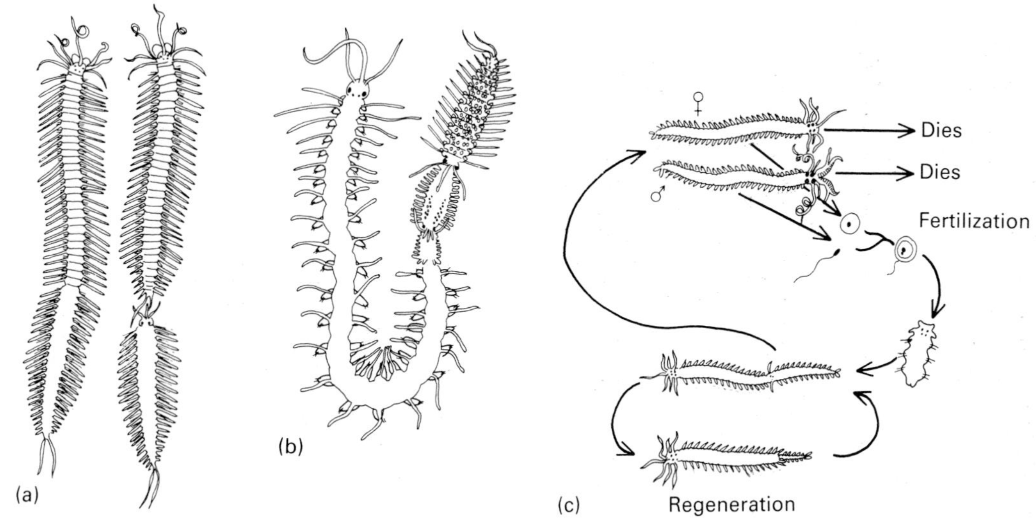

Fig. 4.61. Epitoky: schizogamy: (a) production of a single stolon by modification of posterior segments; (b) production of multiple stolons by terminal budding, the terminal stolon being the oldest; (c) diagrammatic representation of the life cycle.

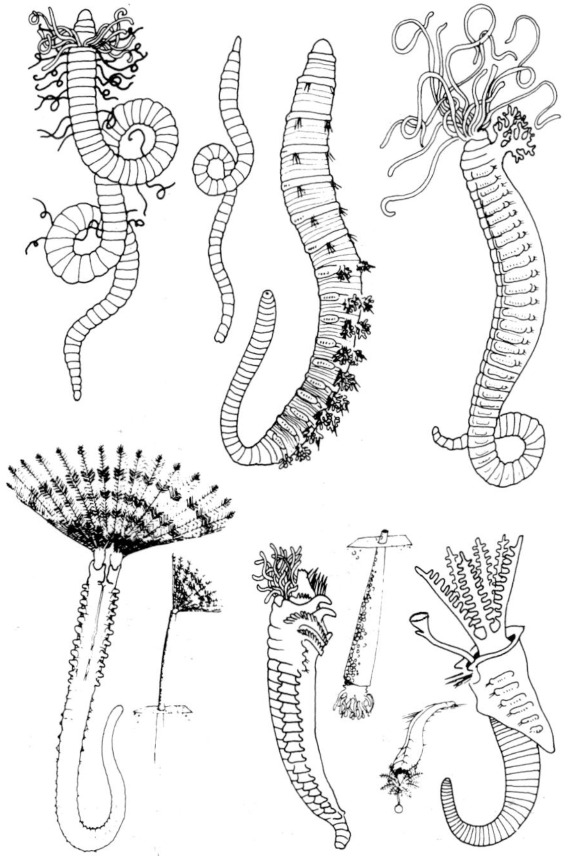

Fig. 4.62. The range of body form among the sedentary polychaetes. (Smaller figures show the characteristic mode of life. See also Fig. 9.6.)

assemblage of minute polychaetes independently adapted to the interstitial environment in marine sands.

4.14.3.2 Class Clitellata

Subclass Oligochaeta

The oligochaetes are predominantly terrestrial or freshwater annelids, with secondarily marine representatives especially in estuarine and interstitial environments.

In their locomotion, oligochaetes exploit to the full the hydrostatic properties of a completely septate and segmented grade of organization. They have therefore retained a body plan close to that attributed to the proto-annelids (see Fig. 4.53), but they are not primitive. The oligochaetes are specialized in being simultaneous hermaphrodites, in shedding small numbers of large, yolky eggs into a protective nutrient-filled envelope, the cocoon, secreted by the *clitellum*, and in having a greatly reduced number of gonads. The arrangement of the oligochaete genitalia is the basis for their classification into three orders.

Functionally and ecologically, there are essentially two types of oligochaetes: the mainly aquatic *microdrile* species and the terrestrial *megadriles* or earthworms.

The functional anatomy of the earthworms is remarkably constant and can be illustrated with reference to *Lumbricus terrestris* (Fig. 4.64a).

A cross-section (Fig. 4.64b) shows the following features: waterproof cuticle lubricated by secretory goblet cells in the epidermis; the epidermis; a nervous layer; a complete layer of circular muscles; blocks of longitudinal muscles; chaetae inserted singly, or in small groups as in *Lumbricus*; a coelom divided by complete septa and lined with peritoneum; muscles of the gut wall; and the endodermal lining of the alimentary system.

The worm is constructed on a segmental plan, but there is considerable specialization in the anterior region, especially in the structure of the alimentary and reproductive systems (Fig. 4.64a and b, and Fig. 4.65). Earthworms engage in complex pseudocopulatory behaviour during which paired worms become coated in mucus, and sperm are transferred externally to the spermathecae or sperm sacs in the region of the clitellum (Fig. 4.66). Fertilization takes place within the cocoon and once mated an individual earthworm can release many egg cocoons.

The microdrile oligochaetes are smaller than the

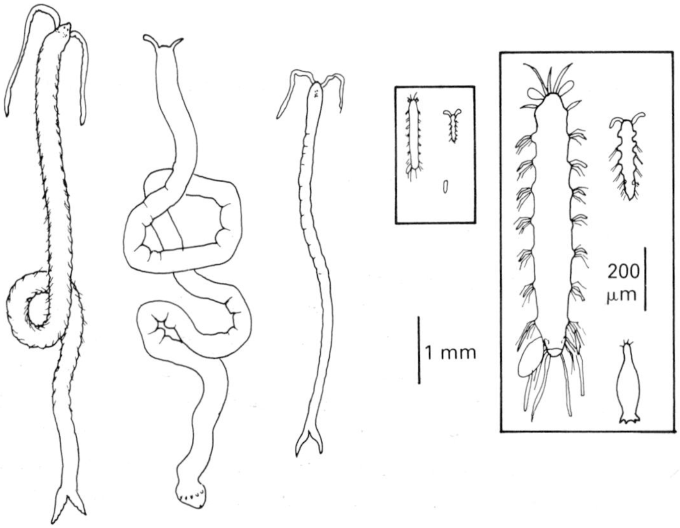

Fig. 4.63. The range of form and size among the polychaete families formerly grouped together as archiannelids. These are now thought to be secondarily simplified as an adaptation to small size.

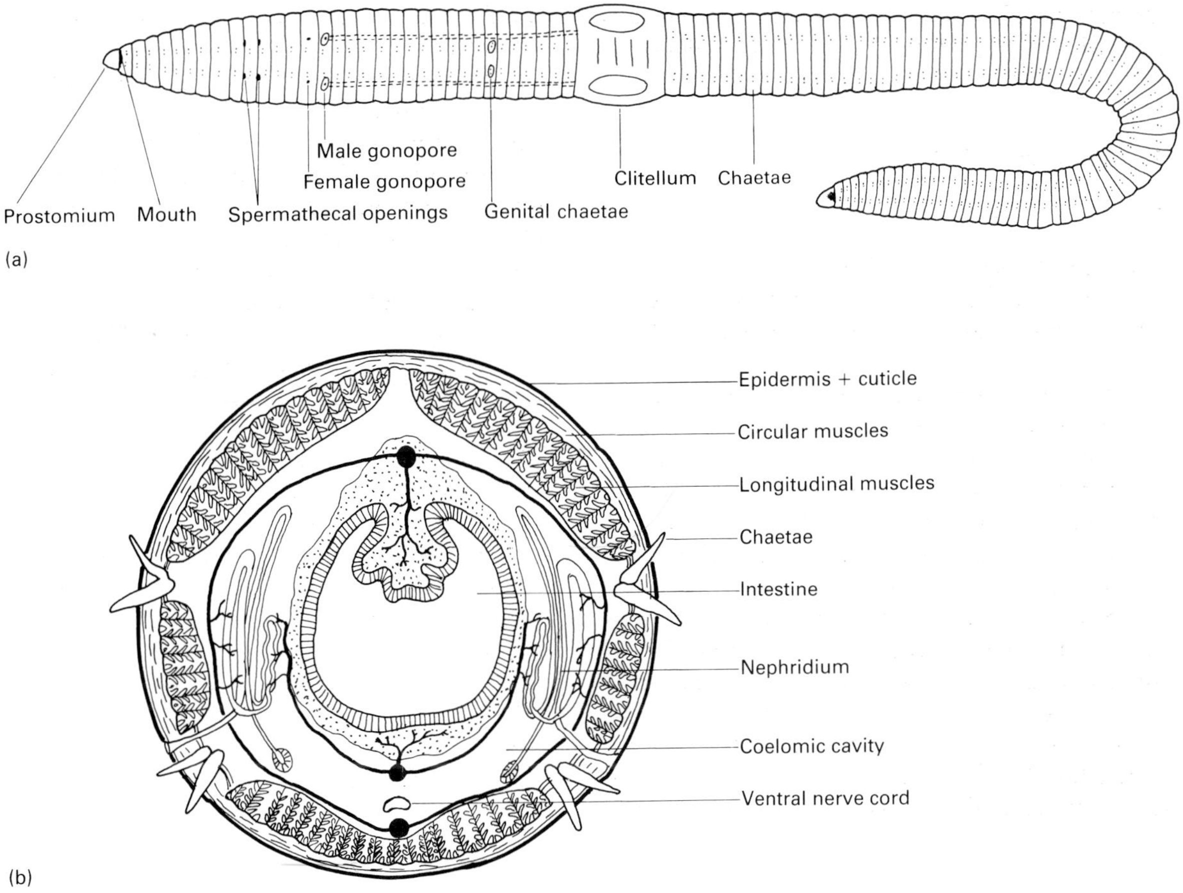

Fig. 4.64. (a) The external morphology of the earthworm *Lumbricus terrestris* as seen in ventral view. (b) Diagrammatic cross-section of the earthworm *Lumbricus terrestris*.

earthworms. They comprise several families living mainly in fresh water, although the Enchytraeidae are primarily terrestrial. Some even occur in the deep sea. They are more variable in body form than earthworms, and some have prominent hair-like chaetae which resemble those of some polychaetes (Fig. 4.67). Their reproductive biology is frequently highly specialized. Asexual reproduction by spontaneous fission (schizogenesis) is common and indeed sexual reproduction has never been observed in some forms.

Subclass Hirudinea (leeches)

The special features of leeches follow the development of suckers which enable them to form fixed attachments at the two ends. The body then acts as a single functional hydrostatic unit. Consequently the segmentation of the coelomic spaces, ectoderm and mesoderm has largely been lost. Functionally leeches operate in a similar manner to flatworms, but with the advantages of coelomic sinuses for the trans-

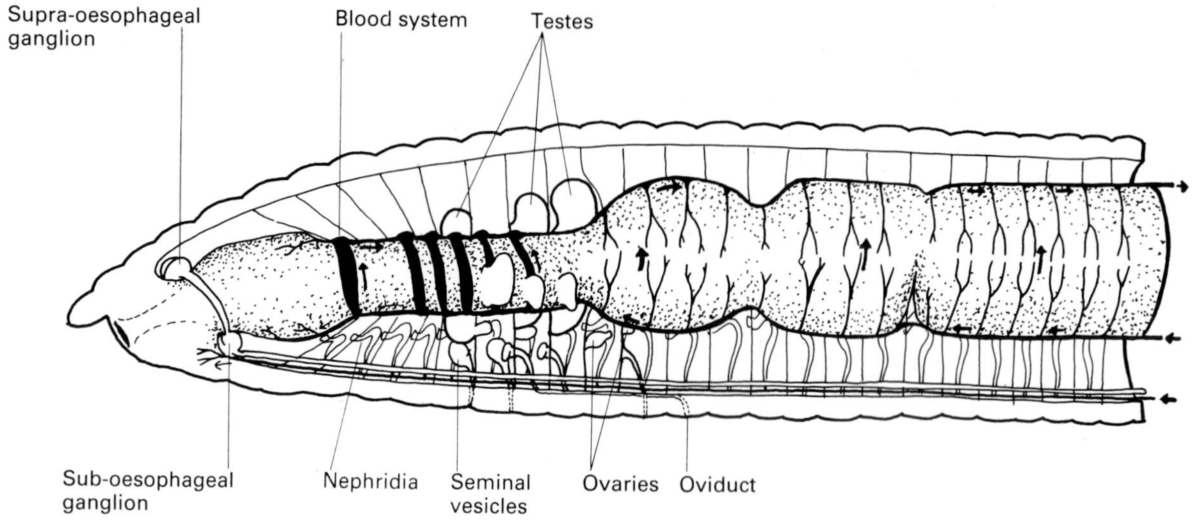

Fig. 4.65. The internal anatomy of the anterior segments of the earthworm *Lumbricus terrestris*.

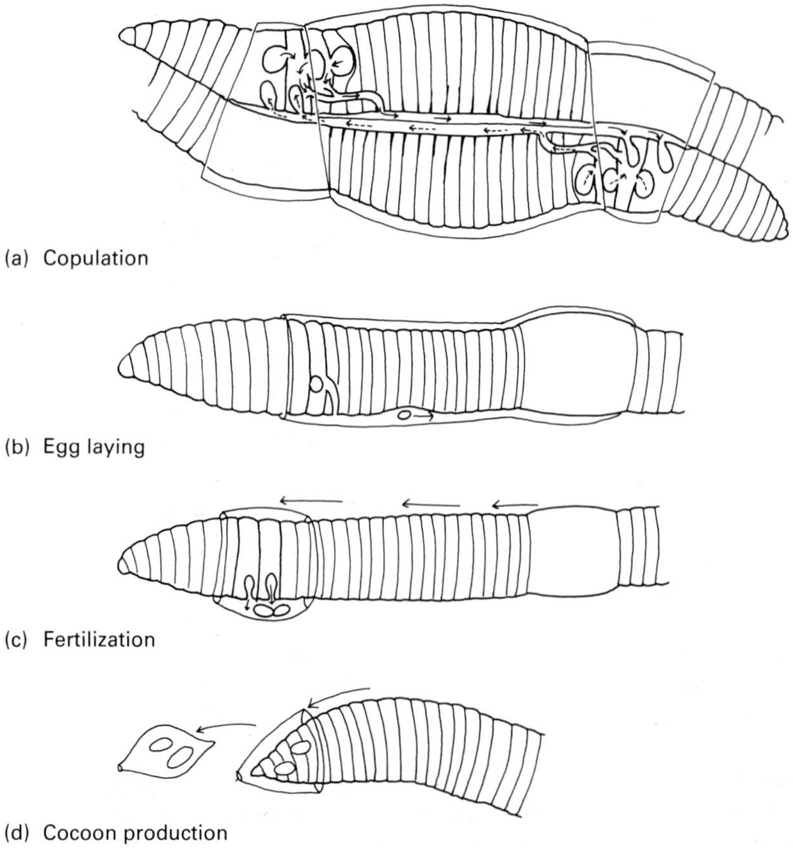

Fig. 4.66. The copulatory behaviour of the earthworm *Lumbricus terrestris*. (a) The animals leave their burrows and form a couple in a coating of mucus. As they couple, mutual exchange of spermatozoa occurs, the sperm being transferred from the seminal vesicles to the spermathecae as shown by the arrows. (b) A mature egg is released from the female genital duct and passes back to the clitellum. (c) The cocoon secreted by the clitellum around a small number of eggs passes forward and the eggs are fertilized as the cocoon passes the opening of the spermathecae. (d) The cocoon containing the developing embryos is deposited in the soil.

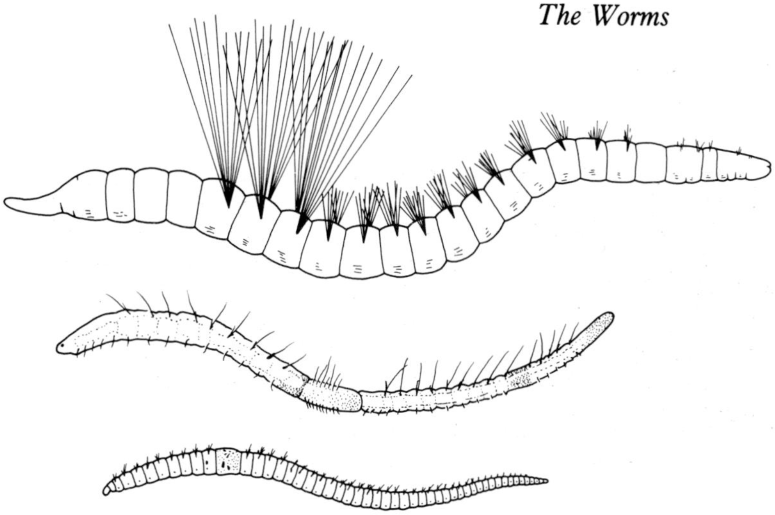

Fig. 4.67. The range of form among microdrile oligochaetes.

mission of forces, a nervous system with segmental ganglia for the co-ordination of complex behavioural and locomotory patterns and a closed vascular system with pigments for respiration.

All leeches have precisely 33 segmental ganglia. The head sucker is formed from segments 1–4 and the posterior sucker from segments 25–33 (Fig. 4.68). There are no chaetae and externally segmentation is obscure, although the body is divided transversely by a series of annuli which do not correspond to segmental boundaries. A cross-section shows that the space between the mesoderm and the muscular wall of the gut is filled with a loose 'parenchyma-like' botryoidal tissue (Fig. 4.69). Some internal organs, especially the germinal epithelia and the nephridia, do reflect the ancestral segmental condition (Fig. 4.70).

Leeches are hermaphrodite, with several pairs of testes, a pair of ovaries, and a single genital opening. Fertilization is internal and reproduction requires transfer of spermatozoa between partners, often with complex copulatory behaviour. In many species the structurally complex spermatozoa are injected hypodermically and make their way to the ovaries, between the cells of the recipient animal. After mating, adult leeches will lay numerous cocoons, each with one or a few eggs. Some of the large species survive for several years but the majority are annual, overwintering as young embryos in the cocoons.

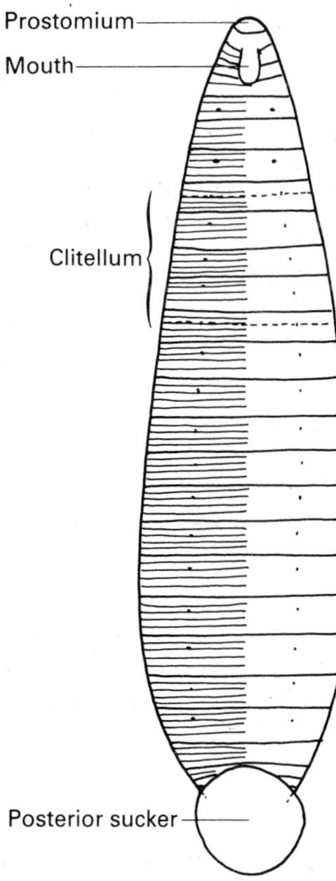

Fig. 4.68. The external anatomy of the leech *Hirudo*. The external annulation is shown on the left and the true segmentation on the right.

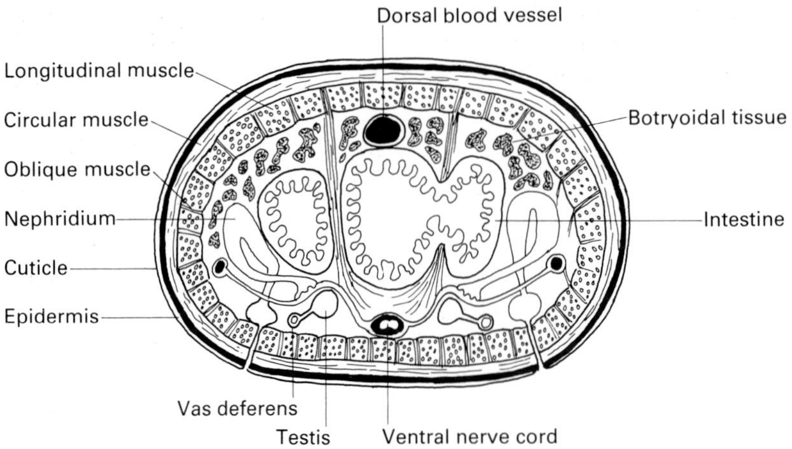

Fig. 4.69. A semi-diagrammatic transverse section of the leech *Hirudo*. The coelomic space is filled with a parenchyma-like tissue. The spaces between these cells may be organized into well-defined sinuses.

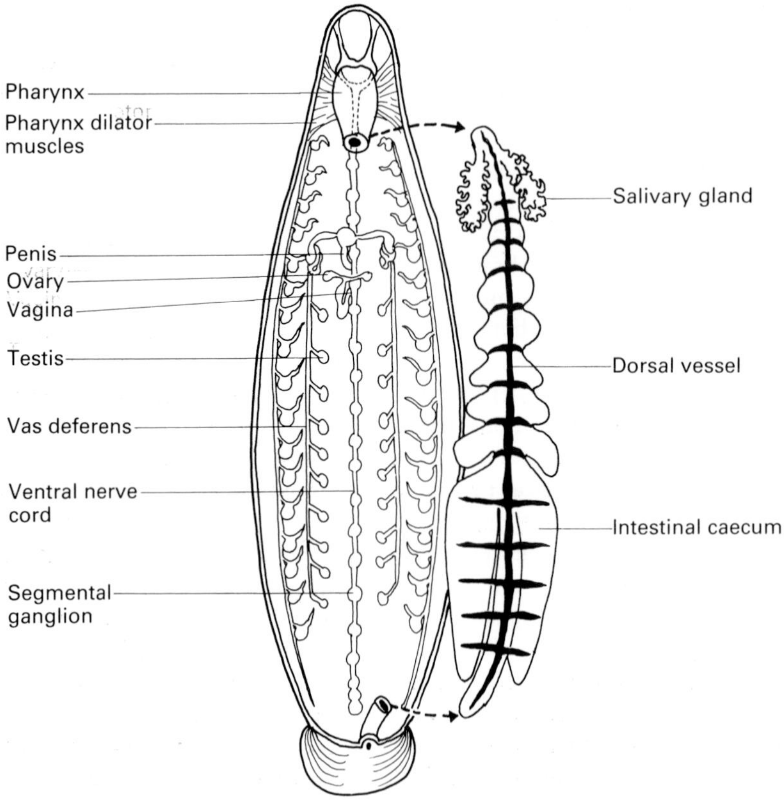

Fig. 4.70. The internal anatomy of the leech *Hirudo*, as revealed by dissection from the dorsal side. The gut is shown displaced.

The range of body form is illustrated in Fig. 4.71. All are carnivorous but not all are external parasites or blood-sucking forms. Classification is primarily based upon the structure of the mouthparts and the mode of feeding.

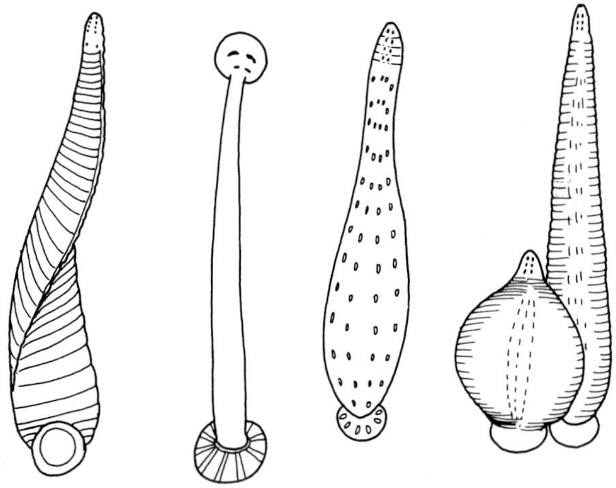

Fig. 4.71. The range of form in the leeches: the form on the extreme right shows the characteristic ability of leeches to show great changes in shape.

The Acanthobdellida is a peculiar primitive group of ectoparasites of salmonid fish. They lack the anterior sucker, have a few segments with chaetae and have a segmental coelom. They thus provide a link between the oligochaete and leech-like grades of organization.

Subclass Branchiobdella

These are minute ectoparasites of freshwater Crustacea. They have 15 or 16 segments, the first four fused to form a cylindrical head with a sucker. These little-known clitellates seem to form an independent line from an oligochaete-like ancestor, paralleling in many ways the evolution of the Hirudinea (Fig. 4.72).

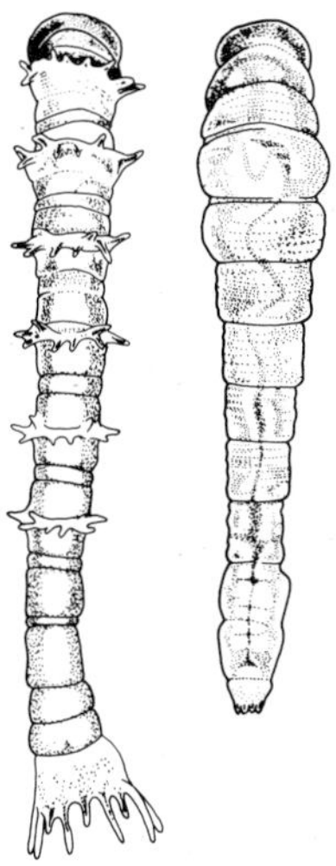

Fig. 4.72. Branchiobdella. The external morphology of typical species.

4.14.3.3 Class Aeolosomata

These minute annelids were formerly thought to be primitive oligochaetes but are now regarded as independent of clitellate evolution. All are small, and most are minute, living in the interstitial environments of fresh and brackish waters. They are hermaphrodite but have a single ovarian segment, and testes in segments both anterior and posterior to this segment. The so-called clitellum is composed of ventral glands and is not homologous to the dorsal clitellum of the clitellate annelids.

A typical species is illustrated in Fig. 4.73. It has a ciliated prostomium forming the main locomotory organ and rather long hair-like chaetae. The Aeolosomata, like the archiannelid families, exhibit a simplified structure due to their minute size, and their relationship to other annelids is unknown. There are only about 25 species in this class.

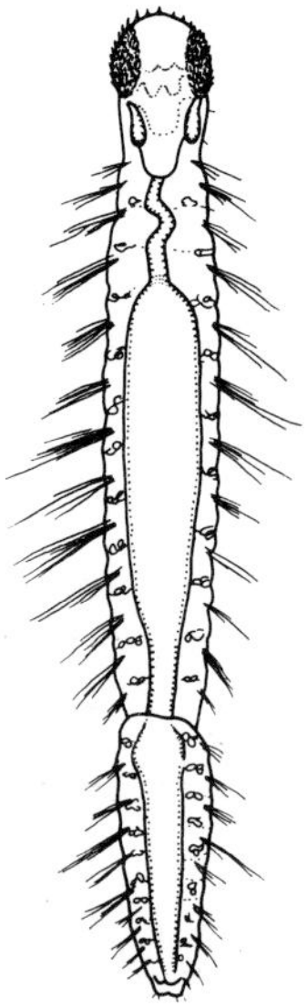

Fig. 4.73. Aeolosomata. A typical species.

4.15 Further reading

Bird, A.F. 1971. *The Structure of Nematodes*. Academic Press, New York [Nematoda].

Boaden, P.J.S. 1985. Why is a gastrotrich? In: Conway Morris, S. *et al.* (Ed.) *The Origins and Relationships of Lower Invertebrates*, pp. 248–60. Clarendon Press, Oxford [Gastrotricha].

Croll, N.A. 1976. *The Organisation of Nematodes*. Academic Press, London [Nematoda].

Croll, N.A. & Mathews, B.G. 1977. *Biology of Nematodes*. Blackie, London [Nematoda].

Dales, R.P. 1963. *Annelids*. Hutchinson, London [Annelida].

D'Hondt, J.-L. 1971. Gastrotricha. *Oceanogr. mar. Biol., Ann. Rev.*, **9**, 141–92 [Gastrotricha].

Donner, J. 1966. *Rotifers* (transl. Wright, H.G.S.). Warne, London [Rotifera].

Edwards, C.A. & Lofty, J.R. 1972. *The Biology of Earthworms*. Chapman & Hall, London [Annelida].

Gibson, R. 1972. *Nemerteans*. Hutchinson, London [Nemertea].

Hyman, L.H. 1951. *The Invertebrates (Vol. 2). Platyhelminthes and Rhynchocoela*. McGraw-Hill, New York [Nemertea].

Hyman, L.H. 1951. *The Invertebrates (Vol. 3). Acanthocephala, Aschelminthes and Entoprocta*. McGraw-Hill, New York [Gastrotricha, Nematoda, Nematomorpha, Kinorhyncha, Priapula, Rotifera, Acanthocephala].

Ivanov, A.V. 1963. *Pogonophora*. Academic Press, London.

Kaestner, A. 1967. *Invertebrate Zoology (Vol. 1)*. Wiley, New York [Nemertea, Gastrotricha, Nematoda, Nematomorpha, Kinorhyncha, Priapula, Rotifera, Acanthocephala, Sipuncula, Echiura, Annelida].

Kristensen, R.M. 1983. Loricifera, a new phylum with aschelminthes characters from the meiobenthos. *Z. zool. Syst. Evolutionsforsch.*, **21**, 163–80 [Loricifera].

Mill, P. 1978. *Physiology of the Annelids*. Academic Press, London [Annelida].

Nørrevang, A. (Ed.) 1975. *The Phylogeny and Systematic Position of Pogonophora*. Parey, Hamburg [Pogonophora].

Rice, M.E. & Todorovic, M. 1975. *Proceedings of the International Symposium on the Biology of Sipuncula and Echiura*. Smithsonian Inst., Washington [Sipuncula & Echiura].

Sterrer, W. 1972. Systematics and evolution within the Gnathostomulida. *Syst. Zool.*, **21**, 151–73 [Gnathostomula].

Sterrer, W., Mainitz, M. & Reiger, R.M. 1985. Gnathostomulida: enigmatic as ever. In: Conway Morris, S. *et al.* (Ed.). *The Origins and Relationships of Lower Invertebrates*, pp. 181–99. Clarendon Press, Oxford [Gnathostomula].

5

THE MOLLUSCS

The molluscs are a distinctive and individual phylum without any close resemblance or phylogenetic affinity to any other living group, except in so far as they have retained a number of features indicative of their flatworm ancestry: they were portrayed in Chapter 2 as being effectively chunky flatworms with dorsal protective shields. Nevertheless, the majority of species have diverged very far from flatworm body form and life style. Indeed, the members of this highly successful phylum have radiated into as much morphological diversity as have the various assemblages of phyla described in the other chapters in this section, and as such deserve a chapter to themselves.

5.1 Phylum MOLLUSCA

5.1.1 Etymology

Latin: *molluscus*, a soft nut or soft fungus.

5.1.2 Basic and special features (Fig. 5.1)

1 Bilaterally symmetrical.
2 Body more than two cell layers thick, with tissues and organs.
3 A through gut.
4 Without a body cavity other than that provided by blood sinuses.
5 Body monomeric and highly variable in form, but basically squat and often conical, frequently elongated in the dorsoventral plane to form a 'visceral hump'; essentially with an anterior head bearing eyes and sensory tentacles, a large flat ventral foot, and a posterior mantle cavity, but all of these subject to considerable modification.
6 A protective, external dorsal shell of protein (conchiolin) reinforced by calcareous spicules or from one to eight calcareous plates, secreted by the dorsal and lateral epidermis (the mantle); sometimes shell secondarily reduced, covered by tissue, or lost, and sometimes enlarged so as to cover whole body.

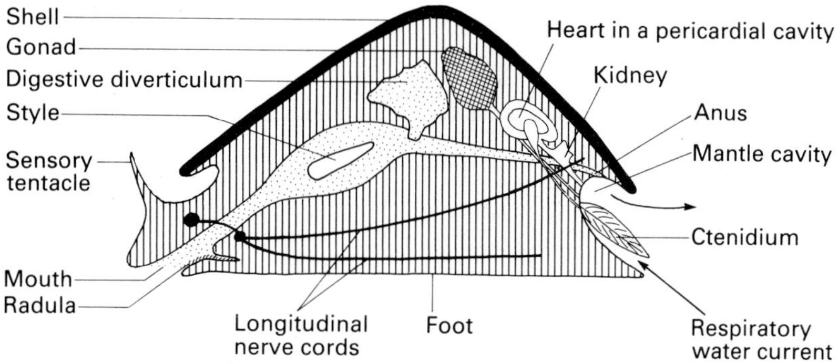

Fig. 5.1 Diagrammatic longitudinal section through a generalized basic mollusc.

7 A toothed, chitinous, tongue-like ribbon, the radula, can be protracted from the buccal cavity through the mouth; rasped particles are borne in a mucous cord which is winched into the stomach by a style housed in a style sac (see Fig. 9.17 and Section 9.2.5).
8 Gaseous exchange effected by one or more pair/s of ctenidial gills housed in the mantle cavity (sometimes lost).
9 An open blood system and a heart enclosed within a mesodermal cavity, the pericardium, through which the intestine also passes.
10 A pair of sac-like 'kidneys', opening proximally into the pericardium and discharging into the mantle cavity.
11 Nervous system with a circum-oesophageal ring and two pairs of ganglionated longitudinal cords, sometimes highly concentrated.
12 Typically with a single pair of gonads, discharging into the mantle cavity, primitively via the pericardium and kidneys.
13 Eggs cleave spirally.
14 Development indirect, via trochophore and veliger larval stages (Fig. 5.2), or secondarily direct.

With the possible exception of the Nematoda (of which thousands of species probably await discovery and description), the Mollusca, with almost 100 000 species, is the second largest animal phylum. Their success is probably not so much attributable to any particular special anatomical or ecological features of the group as to the extreme plasticity and adaptability of the basic molluscan body plan, the list of features given above having been extensively modified in a wide variety of fashions by the different component classes. This plasticity can be illustrated, on the one hand, by the variation displayed in the function of any single molluscan structure (e.g. the shell, besides being protective, may form a buoyancy device, a burrowing organ, or an endoskeletal plate), and, on the other, by the multiplicity of structures which have been adapted to serve the one function (food-catching organs include ciliated tentacles or palps, greatly enlarged gills, sucker-bearing arms, radular teeth, etc.).

In body shape, molluscs range from cylindrical, burrow-dwelling 'worms' lacking both a foot and a shell, and greatly elongated along their anteroposterior axes, to almost spherical, effectively headless clams encased within large bivalved shells; whilst in size they extend from 2 mm long planktonic and interstitial species up to the giant squids which measure, including their arms, more than 20 m in length, in their case as a result of elongation in the ancestral dorsoventral plane. Molluscs occupy all major habitats and include representatives of all known feeding types; they include the most and the least mobile of all free-living invertebrates; and within the

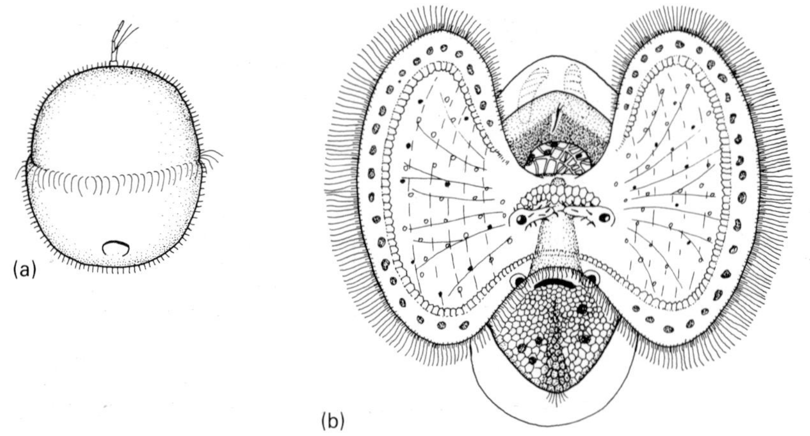

Fig. 5.2. Molluscan larvae: (a) trochophore; (b) veliger (after Hyman, 1967).

one phylum are placed animals with some of the least developed brains and sense organs, together with the most intelligent of invertebrates. ('If the Creator had indeed lavished his best design on the creature he shaped in his own image, creationists would surely have to conclude that God is really a squid.' (Diamond, 1985).)

As discussed in Chapter 2, the basic molluscan morphology is really that of a flatworm (Section 3.6) with two distinctive additional features and a few others consequent on these two): the radula and the dorsal shell. The radula (Fig. 5.3) is housed in a posteroventral diverticulum of the buccal cavity and it comprises a cartilaginous skeletal element, the odontophore, on which is mounted a movable linear radular ribbon bearing transverse rows of backwardly pointing teeth in a longitudinal series. Protractor muscles can cause both the odontophore to be everted through the mouth and a degree of movement of the ribbon over its basal support. This action also causes the individual teeth to be erected. These teeth then rasp any surface to which they are applied and detached particles are conveyed into the buccal cavity as the retractor muscles bring the odontophore and ribbon back through the mouth. Notwithstanding that they may be mineralized by incorporation of SiO_2 or Fe_3O_4, the chitinous teeth are subject to considerable wear, and therefore the whole ribbon is moved slowly forward over the odontophore, bringing unused tooth rows into play, whilst replacement ribbon is secreted on to the proximal end. The radula is thus essentially an organ of browsing, grazing or

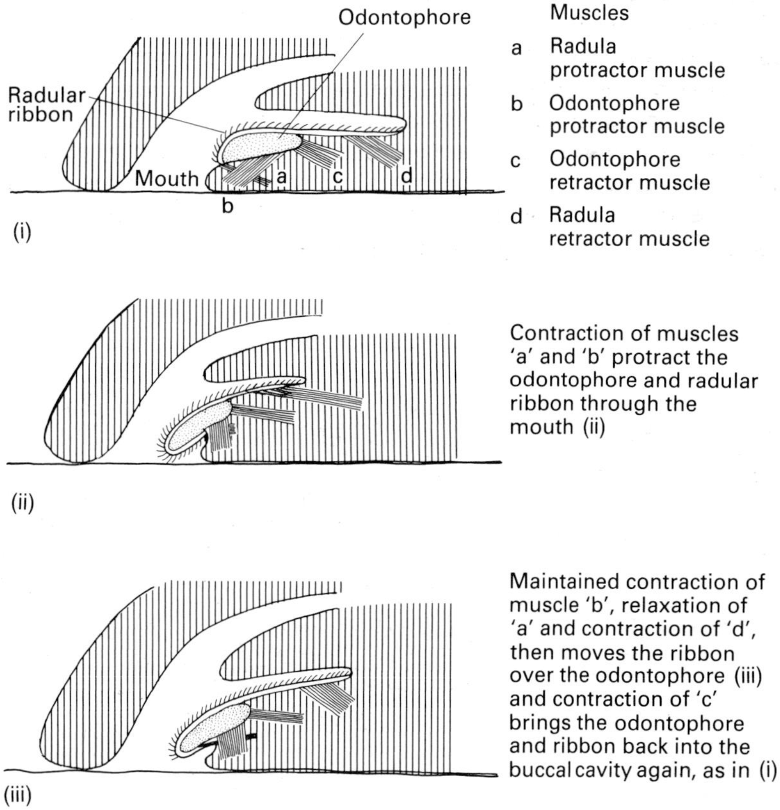

Fig. 5.3. Radular rasping in molluscs (after Russell-Hunter, 1979 and other sources).

boring, although it has been modified in some predatory species for prey capture by enlarging the individual teeth (and correspondingly reducing the number in each row) and, in some, connecting them to poison-secreting glands.

The dorsal protective covering of the ancestral dorsoventrally flattened mollusc was probably a proteinaceous and chitinous cuticle reinforced by calcareous ossicles or scales and secreted by the epidermis. In most descendant forms, however, the calcium carbonate content has been greatly increased and deposited as a large, individual plate or plates covered by a thin conchiolin periostracum. Typically, the calcite or aragonite is laid down in distinct layers, each in an organic matrix: an outer prismatic layer and a variable number of inner nacreous layers. Once a distinct shell had been evolved, the major axis of molluscan growth appears no longer to have been along the anteroposterior axis, as it is in most vermiform animals, but dorsoventral, so that the shell and the animal within it assumed—and in most surviving groups still assumes—a generally conical shape, sometimes elongately so and then coiled into a planar or helical spiral. The mantle which secretes the shell usually extends as a fold over the unshelled regions of the body, forming a covering 'skirt'; so therefore may the shell, and retractor muscles can pull the shell down over the unprotected head and locomotory ventral foot for safety. At the same time, the foot can cling tightly to the substratum by generating suction.

The presence of a dorsal covering has immediate consequences in respect of gaseous exchange across the body surface, and concomitantly with the development of the shell must have proceeded the elaboration of a specialized unshelled region of the body surface into a gill, together with the evolution of a transport system for circulating the respiratory gases through the body (if such was not already present). The overlapping of the molluscan body by the mantle and shell is most marked laterally and/or posteriorly, where a mantle cavity is enclosed (Fig. 5.4). In this cavity are located one or more pairs of distinctive ctenidial gills; each gill is composed of a central flattened longitudinal axis, from each side of which issue flat triangular filaments, each filament supported by a small chitinous rod along the frontal surface. Cilia on the gills drive water between the individual filaments, through which blood diffuses in a countercurrent flow (see Fig. 11.9). Water normally enters at the sides of the mantle cavity and flows out dorsally along the mid-line, the anus and kidneys discharging into the outgoing current.

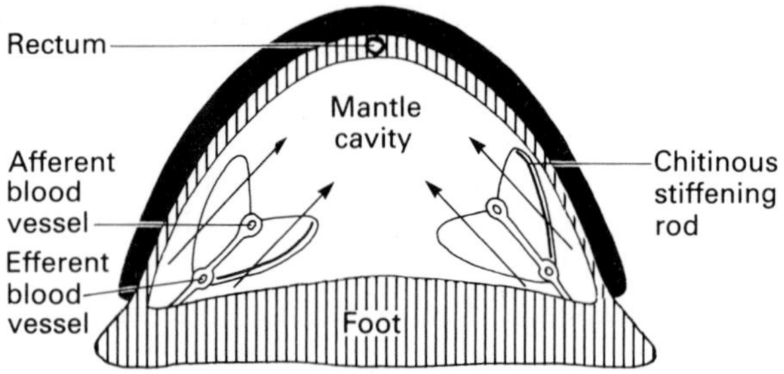

Fig. 5.4 Cross-section through the mantle cavity of a generalized basic mollusc showing the bipectinate ctenidial gills and the direction of the inhalent water currents resulting from ciliary action.

As stressed above, however, the nature and development of the radula, shell and ctenidia, together with most other anatomical features of the molluscs, vary considerably in the different component classes, and hence further characteristics of the phylum will be treated separately in the following sections. Because molluscs, more than any other invertebrate group, have been the favoured subject of the attention of neurobiologists, the reader should also consult Chapter 16.

5.1.3 Classification

The phylum Mollusca consists of eight classes, of which two—the gastropods and bivalves—contain between them over 98% of the known living species.

5.1.3.1 Class Chaetodermomorpha

The chaetodermomorphs are a peculiar group of shell-less and vermiform molluscs, much elongated

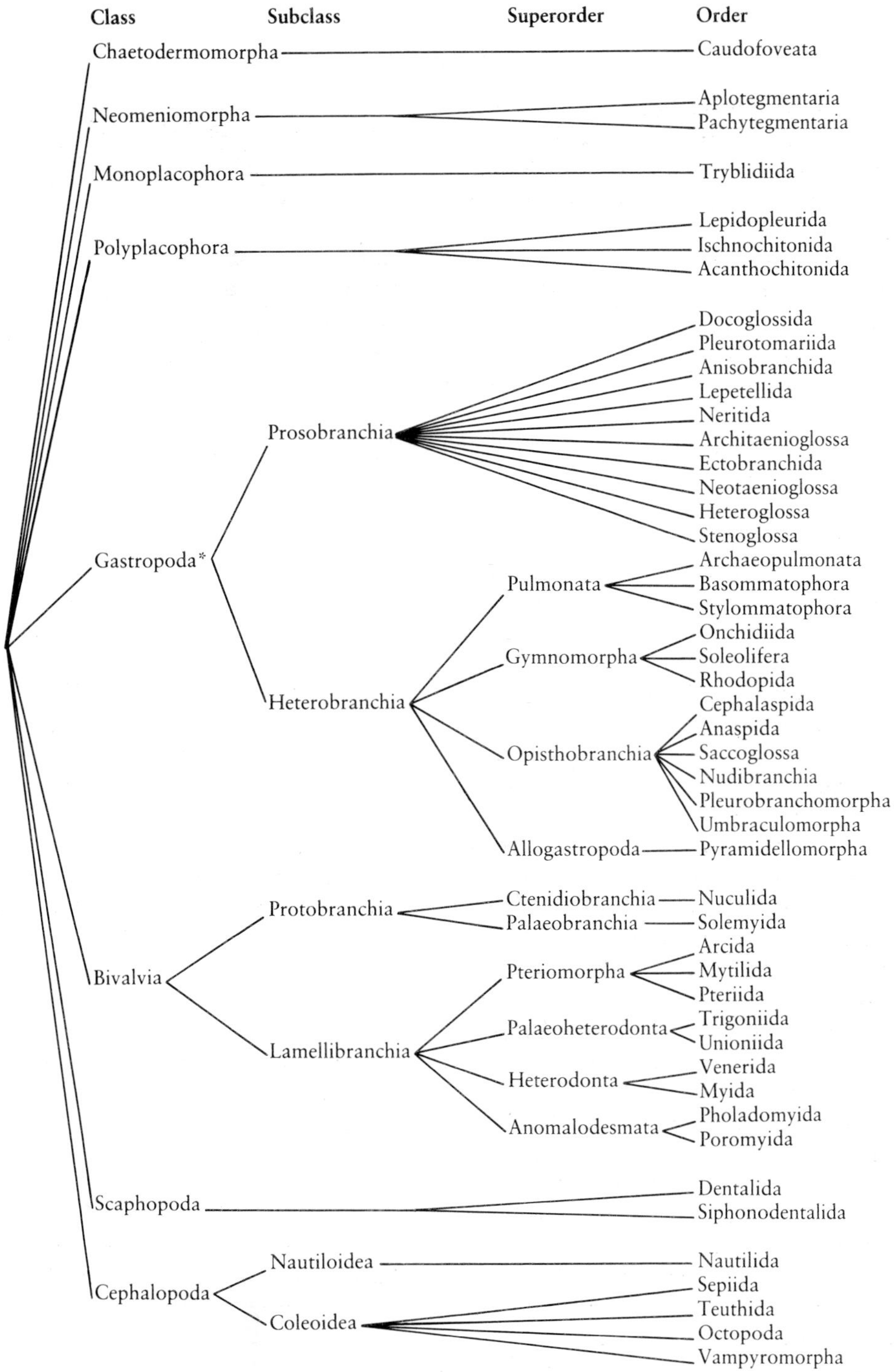

*The classification of the Gastropoda is currently in a state of flux as the old pre-1930 system, based upon different grades of organization, is slowly giving way to more phylogenetically based approaches. That given above should perhaps be regarded as illustrative of modern classifications, in that no individual scheme has yet had time to become generally accepted.

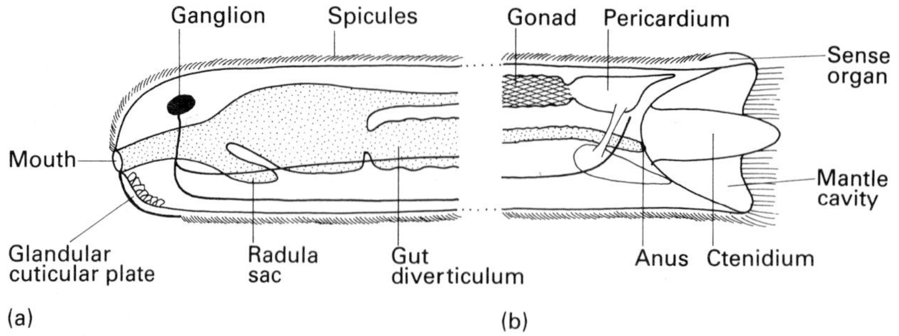

Fig. 5.5. Diagrammatic longitudinal section through the anterior (a) and posterior (b) ends of a chaetodermomorph mollusc (after Boss, 1982).

along the anteroposterior axis, which construct and inhabit vertical burrows in soft marine sediments. The cylindrical body, from 2 mm to 14 cm in length, is positioned head downwards in the burrow, the terminal mouth ingesting sediment. The mantle cavity, containing one pair of bipectinate ctenidia, is posteroterminal (Fig. 5.5) and therefore positioned in the mouth of the burrow.

A foot is lacking, the mantle covers the entire body; movement is therefore atypical and is effected by peristaltic contractions of the well-developed body wall musculature. Instead of a shell, the epidermis secretes a chitinous cuticle in which are embedded imbricating scales which all point posteriorly; these also cover the whole body surface. Several characteristically molluscan organ systems are absent: the very poorly marked head is without eyes or sensory tentacles; there are no excretory organs or gonoducts (the gonads discharging via the pericardial cavity); and some species lack a radula. The head does possess a terminal cuticular plate associated with various glands (Fig. 5.6), but as yet its function is uncertain.

The 70 species of these gonochoristic, deposit-feeding molluscs are grouped in a single order.

5.1.3.2 Class Neomeniomorpha

Members of this class are superficially similar to the chaetodermomorphs in being shell-less, effectively

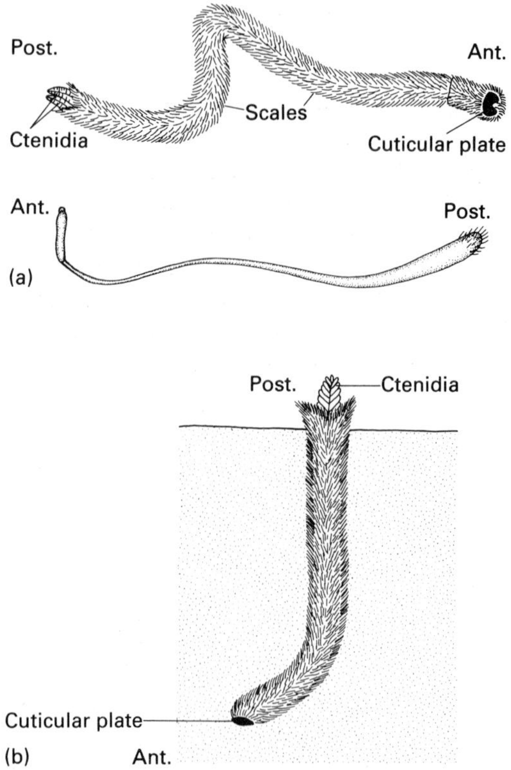

Fig. 5.6. Body form of chaetodermomorphs: (a) external appearance; (b) life style in marine sediment (after Hyman, 1967 and Jones & Baxter, 1987).

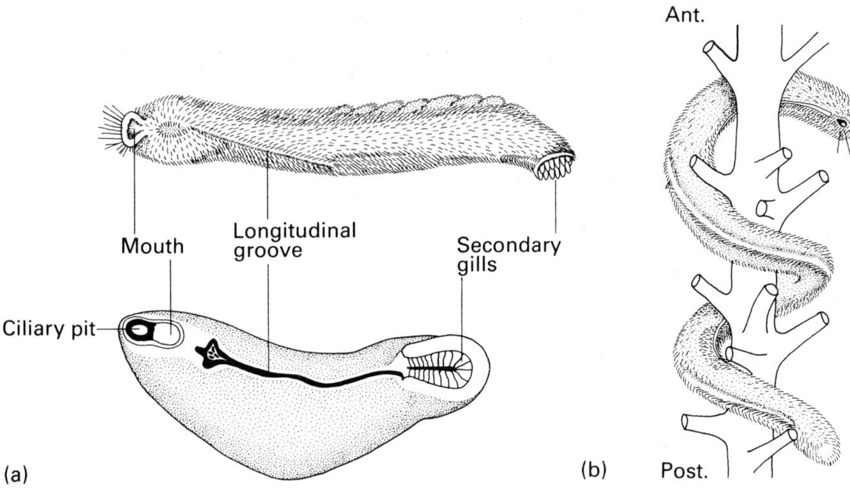

Fig. 5.7. Body form of neomeniomorphs: (a) external appearance; (b) life style, crawling over a colonial hydroid (after Jones & Baxter, 1987).

headless, and vermiformly elongated along their anteroposterior axis (Fig. 5.7), and, like them, they lack excretory organs, gonoducts, and, in some, the radula. They differ, however, in numerous other respects, and are generally regarded as being more closely related to the shell-bearing mollusc groups. The 1 mm to 30 cm long body is laterally compressed and possesses a longitudinal ventral groove in which are located one or more small ridges thought to represent a highly reduced foot. The mantle covers the body, apart from this groove, and has embedded in it one or more layers of separate calcareous scales or spicules beneath the cuticle. At the anterior end of the groove are located a ciliary pit and the anteroventral mouth; and at the posterior end is the posteroventral mantle cavity which lacks ctenidia, although secondary gills in the form of folds or papillae are often developed (Fig. 5.8).

Fig. 5.8. Diagrammatic longitudinal section through a neomeniomorph (after Boss, 1982).

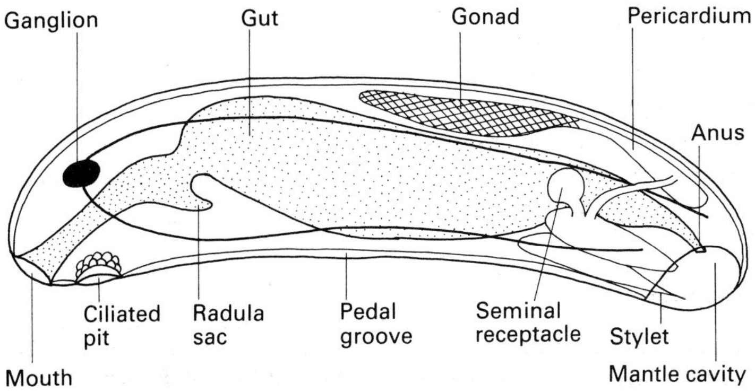

Neomeniomorphs are carnivores, feeding on the cnidarians on which they are usually to be found; their method of ingestion, however, is somewhat uncertain because of the frequent reduction or absence of the radula—in some species the fore-gut is protruded to engulf the food. Although the body wall musculature is well developed, movement is effected not by muscular means but by gliding, using the cilia on the eversible ridges within the longitudinal ventral groove. All species are hermaphrodite; copulation occurs with the assistance of stylets, in several species the sperm being stored by the recipient in seminal receptacles.

The 180 known species are all marine and are divided between two orders dependent on the number of layers of calcareous bodies in the mantle and on the presence or absence of epidermal papillae.

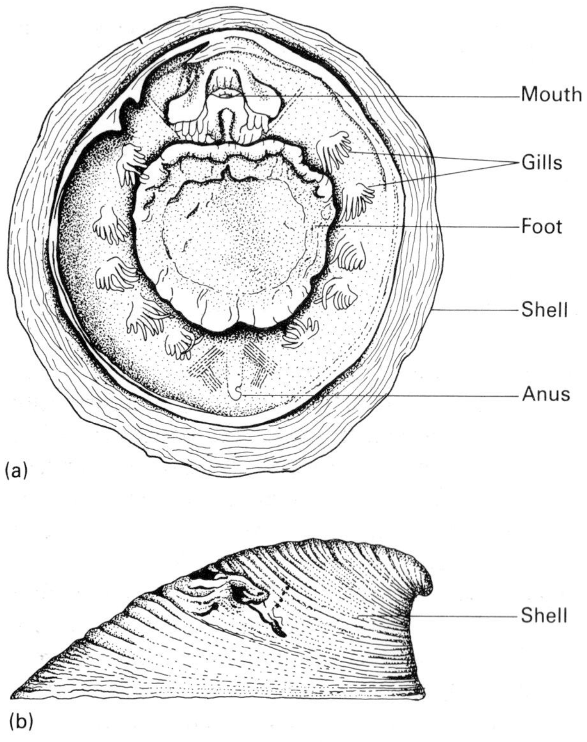

Fig. 5.9. A monoplacophoran mollusc in (a) ventral and (b) lateral view (after Lemche & Wingstrand, 1959).

5.1.3.3 Class Monoplacophora

The monoplacophorans were thought to have become extinct in the Devonian, until, in 1952, living specimens were obtained from an oceanic trench in the Pacific Ocean. Further material from other localities has since been discovered, bringing the number of known surviving species to eight, all referable to a single order.

All are small (3 mm–3 cm), marine, gonochoristic, deposit feeders with a body covered dorsally by a single conical or cap-shaped shell, beneath which is a weakly muscular, circular foot surrounded laterally and posteriorly by an extensive mantle cavity (Fig. 5.9). Within the mantle cavity, around the sides of the foot, are located five or six pairs of monopectinate ctenidia. Other organs also occur in multiple pairs: there are six pairs of lobular kidneys, discharging separately into the lateral regions of the mantle cavity; two pairs of gonads; and eight pairs of foot-retractor muscles (Fig. 5.10). The head is distinct, but poorly developed and without sensory tentacles (except around the mouth) and eyes; the radula, however, is well developed. The anus opens into the posterior section of the mantle cavity.

5.1.3.4 Class Polyplacophora (chitons)

The oval to somewhat elongate, dorsoventrally flattened polyplacophorans are distinguished by their possession of a linear chain of eight, serially overlapping, dorsal shell plates, and by the development of the surrounding mantle into a thick 'girdle', the cuticle of which often bears spines, scales or bristles. In several species, this girdle partially or even completely covers the shell plates (Fig. 5.11). The division of the dorsal protective covering into a number of separate elements enables chitons to roll up into a ball in the same fashion as several woodlice and millipedes.

The head is poorly developed and hidden beneath the anterior girdle; eyes and sensory tentacles are lacking, but the radula is large and bears many teeth in each transverse row. Most of the ventral surface is occupied by a large, muscular, elongate foot with

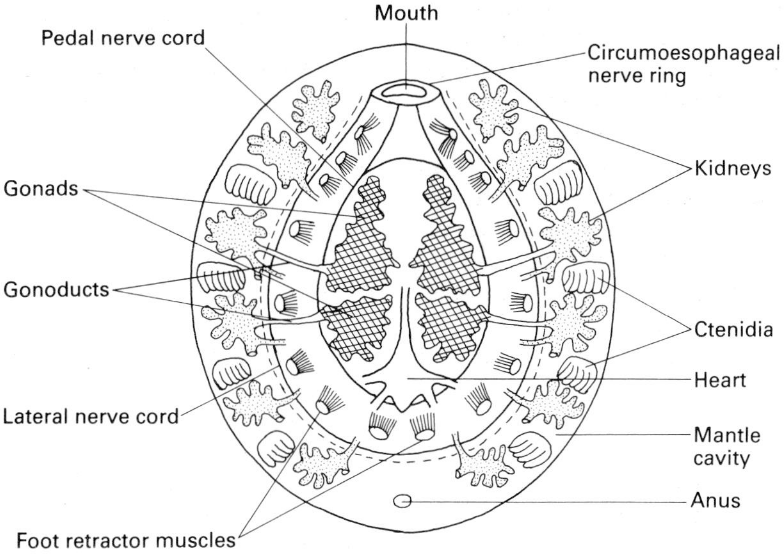

Fig. 5.10. Diagrammatic view of the anatomy of a monoplacophoran, seen from the ventral side with the foot and body wall removed, showing the serially repeated internal organs and ctenidia (after Lemche & Wingstrand, 1959).

which chitons slowly crawl over hard substrata. The foot and girdle can also be used to generate considerable suction, permitting the animals to adhere tightly to surfaces. The narrow mantle cavity surrounds the foot on all sides except immediately anteriorly and contains from six to 88 pairs of bipectinate ctenidia; posteriorly it receives the anal opening and the discharges of the single pair of kidneys. These are large and, together with the single (fused) gonad (which has paired gonoducts and gonopores), often extend the whole length of the body (Fig. 5.12). The nervous system is relatively simple; the cords, for example, are not ganglionated.

The 550 species of polyplacophorans are all marine and gonochoristic; most are grazers of algae. They range in length from 3 mm up to 40 cm. Three orders are differentiated on the location within the mantle cavity of the gills, on the presence or absence of attachment teeth on the shell plates, and on the extent to which the plates are covered by the girdle.

5.1.3.5 Class Gastropoda

The gastropods are a very large and diverse group sharing the common feature that during development the visceral hump is rotated through some 180° in an anticlockwise direction relative to the head and foot, so that the mantle cavity occupies a forwardly facing position and the anus and kidneys therefore discharge anteriorly: gastropods undergo 'torsion' (Fig. 5.13). This is brought about by the asymmetrical development of the two pedal retractor muscles and/or by differential growth (see Section 15.4.1), and may have had as its selective advantage the ability of the well-developed gastropod head to be accommodated in the mantle cavity on retraction of the animal into its shell, and of the chemoreceptory sense organs in the mantle cavity, the osphradia, to sense the water ahead of, rather than behind, the moving animal. An automatic consequence of this torsion, however, would have been that the exhalent water current from

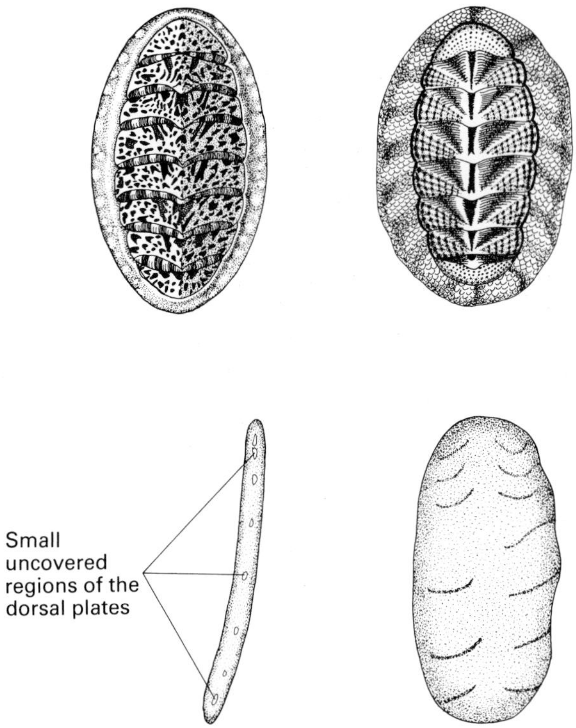

Fig. 5.11. Dorsal views of four polyplacophoran molluscs showing differing degrees of cover of the shell plates by the girdle (after sources in Hyman, 1967).

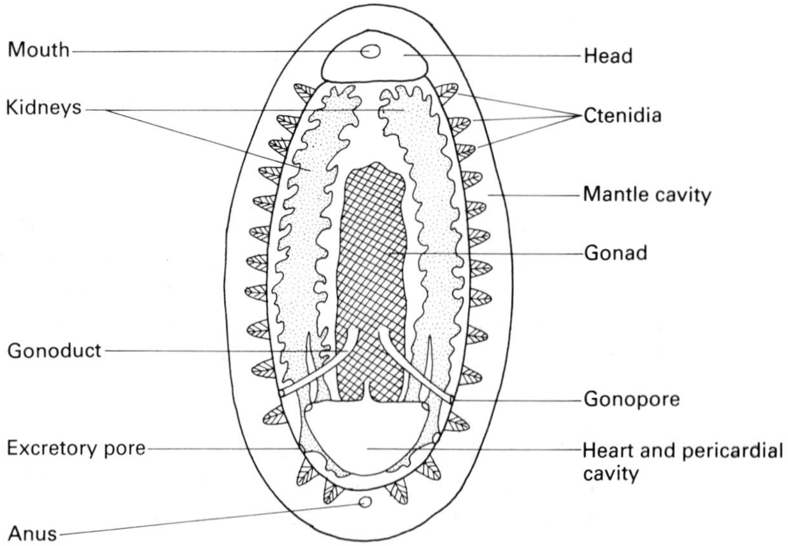

Fig. 5.12. Diagrammatic view of the anatomy of a generalized polyplacophoran, seen from the ventral side with the foot, body wall and gut removed (after Hescheler, 1900).

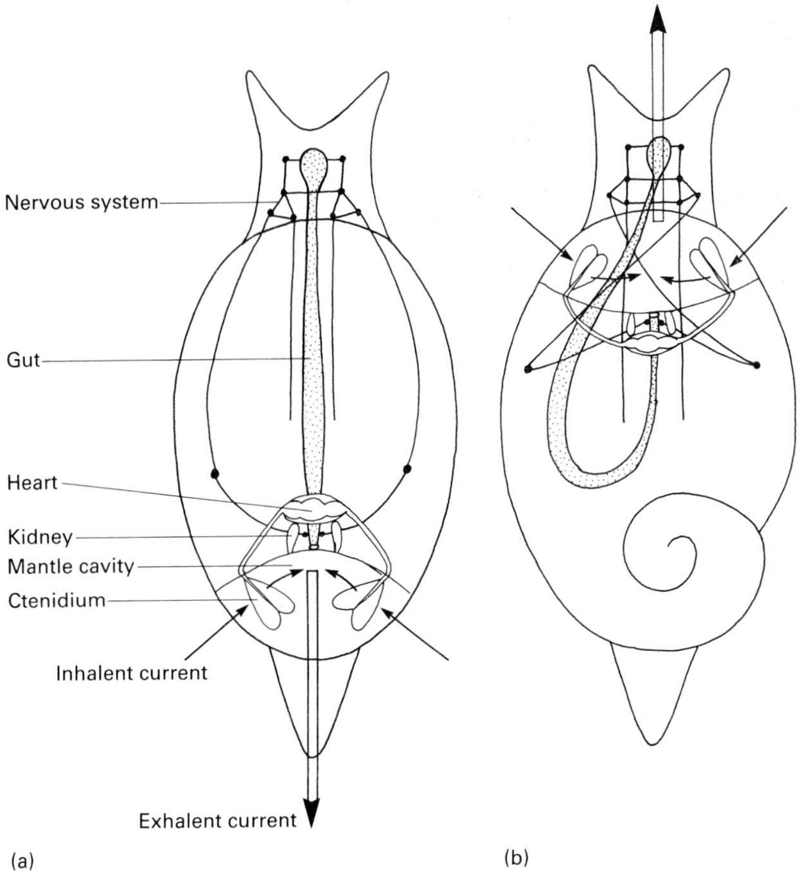

Fig. 5.13. Torsion in gastropod molluscs: (a) before torsion; (b) after torsion.

the mantle cavity, bearing the excretory and alimentary discharges, flowed out immediately over the head. Such fouling of the cephalic sense organs has been circumvented by modifying the path of the exhalent stream; for example, by evolving a slit or aperture(s) in the dorsal shell through which the exhalent current can leave, or by the loss of the ctenidium on the right-hand side of the body and the production of a unidirectional through current, passing into the mantle cavity to the left of the head and out to the right (Fig. 5.14). In association with the latter system, the (inhalent) left-hand side of the mantle cavity is often drawn out into a manoeuvrable siphon. Many phylogenetic lines of gastropods, however, have secondarily undergone detorsion, the visceral hump being detorted through some 90° so that the mantle cavity (if retained) lies on the right-hand side of the body, and the anus usually once again becomes posterior.

The basic body plan of the gastropods—from which many species have departed radically—is that of a squat mollusc, with a well-developed crawling foot and a well-defined head bearing a radula, a pair of jaws, a pair of eyes, and one or more pairs of sensory tentacles, both head and foot being completely retractable into a single, thick, helico-spirally

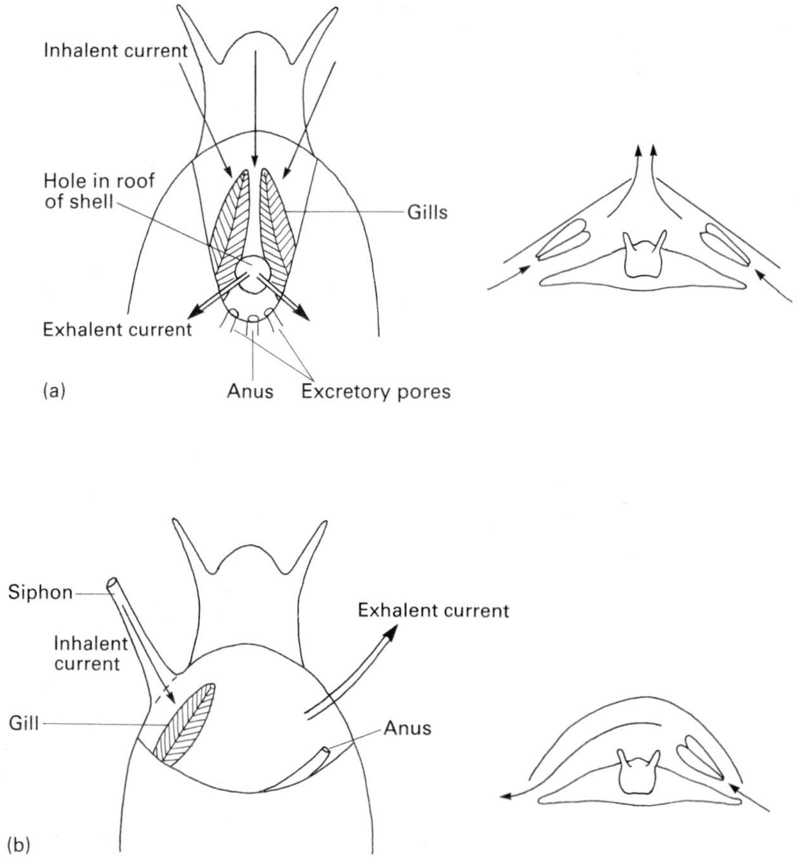

Fig. 5.14. Two modifications of the path of the respiratory water current in gastropods, necessitated by torsion: (a) the exhalent current leaving via a dorsal hole in the shell; (b) loss of the right ctenidium and production of a cross-directed current, the anus being displaced in the direction of the current.

coiled shell (Fig. 5.15), of which the aperture can be plugged, after retraction, by a calcareous or organic operculum borne on the posterodorsal part of the foot. Primitively, at least, the mantle cavity contains a single pair of bipectinate ctenidia, and these gonochoristic molluscs develop via trochophore and veliger larval stages.

The 55 000 species of the mainly marine subclass Prosobranchia have largely retained this basic body form, including the shell, operculum, and torsional state, although the ctenidia display a trend towards reduction from the ancestral bipectinate pair, through a single (left) bipectinate ctenidium, to a single (left) monopectinate gill. Correlated with this, the right kidney, right auricle of the heart, and right osphradium are also lost with the right ctenidium. Most prosobranchs remain gonochoristic, although a few are sequential hermaphrodites. The large majority are benthic snails which graze algae (limpets, winkles, topshells, etc.) and sessile animal colonies (e.g. cowries) or deposit feed (e.g. mudsnails); one group (order Stenoglossa) includes many predatory forms, which characteristically have a radula with only a few, large, fang-like teeth in each radular row;

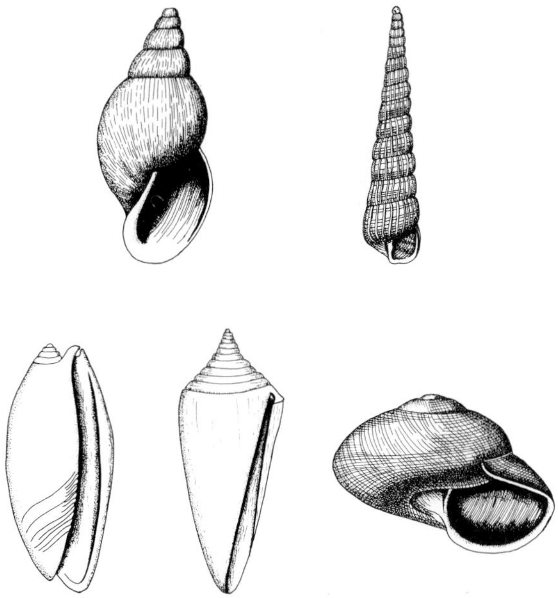

Fig. 5.15. Spiral gastropod shells (after sources in Hyman, 1967).

and a few species suspension feed using external mucous nets (see Fig. 9.5) or enlarged ctenidia (e.g. slipper limpets) paralleling the bivalves in this respect. Amongst prosobranchs and indeed all gastropods, the limpets (order Docoglossida) form a very isolated group with several peculiarities of the shell, radula and gut.

From a prosobranch origin evolved the second gastropod subclass, the Heterobranchia: several lines of gastropods which display marked trends towards (a) reduction, internalization or loss of the shell, (b) loss of the operculum, (c) detorsion, (d) loss of the ancestral ctenidium and its replacement by secondary gaseous-exchange surfaces, and (e) simultaneous hermaphroditism. Two major superorders of heterobranchs are the marine opisthobranchs (with some 1000 species) and the largely terrestrial and freshwater pulmonates (20 000 species). The opisthobranchs (the sea-slugs, sea-hares, pteropods, etc.) tend to replace the ctenidium by secondary gills or by gaseous exchange across the often papillate body surface. Most of these species are carnivores, including via ecto- and endoparasitism and planktonically as well as on or in the bottom, although several consume algae (often by suctorial feeding) and one planktonic group suspension feeds using secreted mucous nets (Fig. 9.5) and cilia on their broad, winglike feet. Up to four pairs of cephalic tentacles may be present.

The pulmonates are principally characterized by conversion of the mantle cavity into an air-breathing lung with a contractile opening, the pneumostome. The majority are grazers of land plants, and all have abandoned larval stages, directly developing into young slugs or snails. Many pulmonates have retained the spiral gastropod shell, although it is usually thin, but several members of the terrestrial order Stylommatophora have lost it and thereby form the land-slugs. A third superorder of heterobranchs, the Gymnomorpha, are also slugs, but these three probably unrelated groups of marine or terrestrial gastropods (totalling 200 species) each show a different mixture of opisthobranch and pulmonate features; in much the same way, the final superorder, the allogastropod snails (with 500 species), show an amalgam of prosobranch and opisthobranch characteristics. Clearly, the Heterobranchia are gastropods typified mainly by the parallel evolution of many 'progressive' features and it is difficult to disentangle their various interrelationships.

In total, therefore, the Gastropoda includes some 77 000 species of slug- or snail-like molluscs (Fig. 5.16) reaching up to 60 cm in height. They can be distributed between 23 orders.

5.1.3.6 Class Bivalvia

The bivalves are essentially laterally compressed molluscs completely enclosed within a pair of shell valves. Being so covered, they are relatively sedentary or even sessile animals, several being cemented or otherwise attached to the substratum, and, as in other animals 'hidden' within a thick, protective casing, the head is greatly reduced (Fig. 5.17). It is without radula, eyes and tentacles, although the missing cephalic sense organs may effectively be replaced by

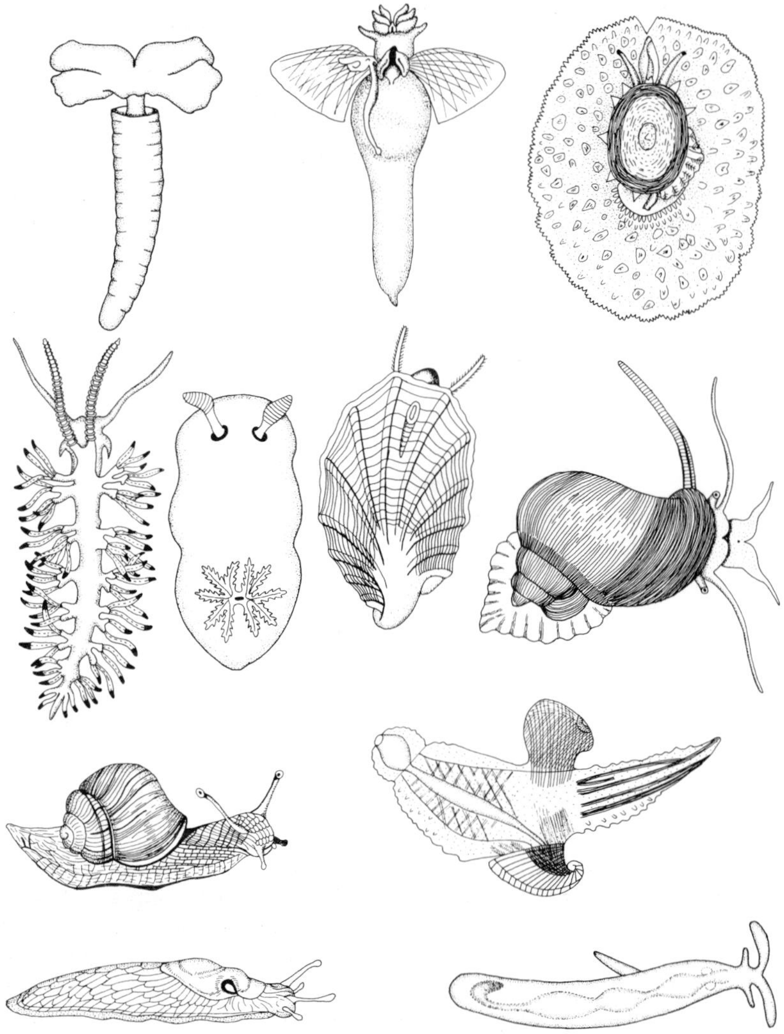

Fig. 5.16. Gastropod body forms (principally after sources in Hyman, 1967).

tentacles and, in a few, eyes positioned around the margins of the mantle. The two shell valves, which are lateral and articulate along the dorsal mid-line, open passively by virtue of an elastic dorsal ligament; they must, therefore, be kept closed actively. This is achieved by two adductor muscles (in some, reduced to one) which possess a catch-fibre mechanism so that they can remain contracted without the need for continual expenditure of energy and for rotational relaxation and contraction of the individual fibres.

The body, which in the largest species may exceed 1 m in length, is located dorsally within the shell, the large mantle cavity occupying the lateral and ventral remainder. When the adductor muscles relax, the shell valves gape slightly, permitting a water cur-

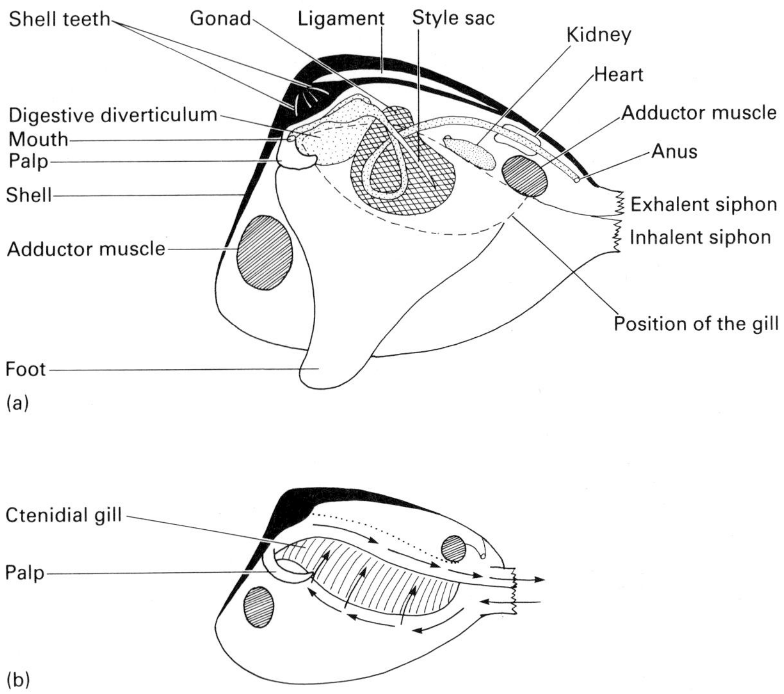

Fig. 5.17. Diagrammatic longitudinal section through a bivalve mollusc: (a) with gill omitted (for clarity); (b) with the gill *in situ*, showing the path of the feeding current (after several sources).

rent to be drawn into and through the cavity, and in forms which burrow into soft sediments allowing the laterally compressed foot to be protruded outside the confines of the shell. Frequently, the ventral and lateral margins of the mantle fuse together, leaving only a gap for the foot and small apertures through which water can flow—in these regions, the mantle is often drawn out into inhalent and exhalent siphons which may or may not be retractable into the shell.

Three broad categories of bivalves can be distinguished on the basis of the nature and function of the ctenidia. In the primitive forms (the subclass Protobranchia), the paired ctenidia serve mainly the ancestral gaseous-exchange function, and feeding, when not entirely dependent on symbiotic chemoautotrophs, is accomplished by means of labial palps, on either side of the mouth, which in one order (Nuculida) bear long tentacles which roam into or over the surrounding sediment and convey food particles to the palps for sorting before ingestion (see Fig. 9.6). These species are mucociliary deposit feeders. In the large majority of species (the superorders Pteriomorpha, Palaeoheterodonta and Heterodonta), however, the paired ctenidia are greatly enlarged and folded each into a W-shape in cross-section to form the organs of filter feeding (Fig. 5.18). Surface deposits are sucked into suspension or material already in suspension is drawn into the mantle cavity with the inhalent current, particles are filtered as the current passes through the gills and then passed to the mouth via the palps, all by (muco)ciliary means. Finally, in the 'septibranch' members of the third group (the poromyidan Anomalodesmata), the gills have been greatly reduced, forming muscular pumping septa by means of which, together with an enlarged and raptorial inhalent siphon, small animals are sucked into the mantle cavity, there to be captured by the muscular palps and ingested.

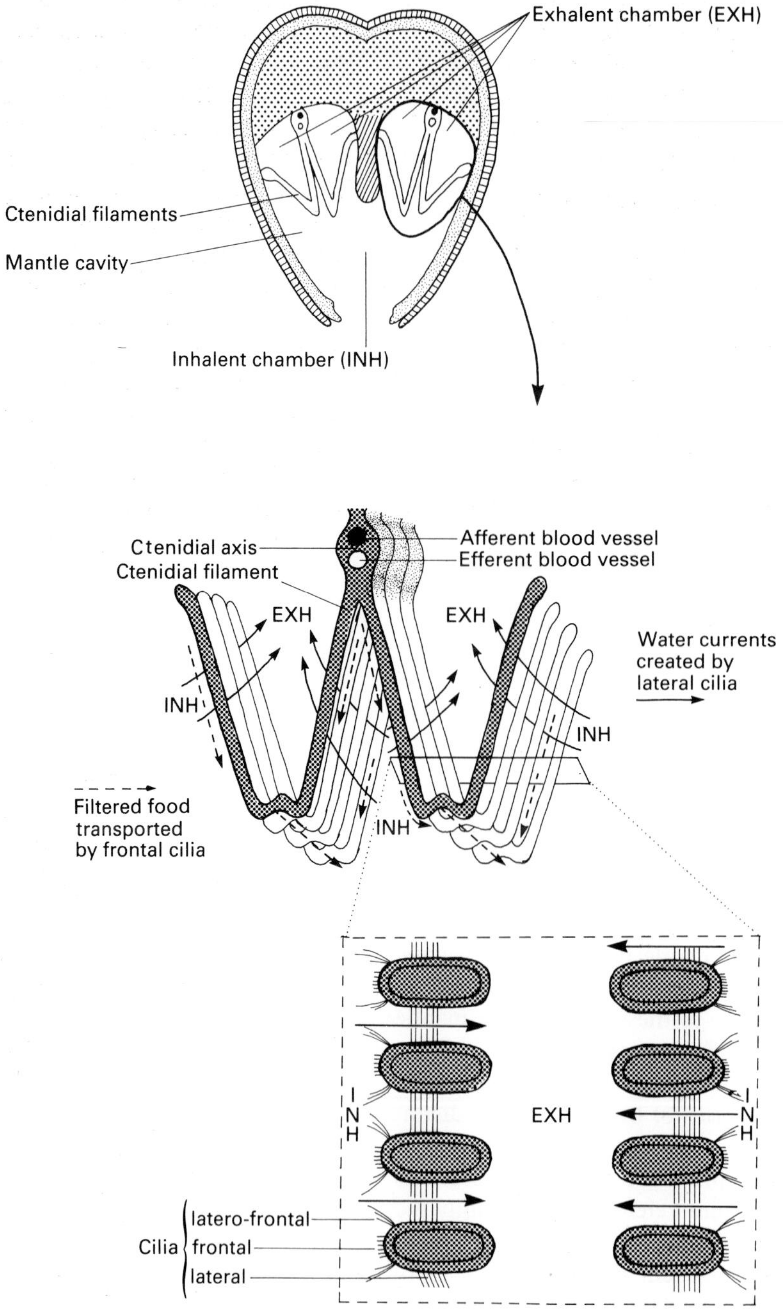

Fig. 5.18. Transverse section through a lamellibranch bivalve, illustrating filter-feeding by means of the ctenidial gills (after Russell-Hunter, 1979).

The 20000 bivalve species are common benthic animals in the sea and in fresh waters. Most are gonochoristic and undergo indirect development, although the larval stages of some freshwater species are aberrant in being adapted to parasitize fish. Eleven orders may be recognized.

5.1.3.7 Class Scaphopoda (tusk-shells)

The 2–150 mm long scaphopods are elongate, cylindrical molluscs almost completely enclosed by the mantle, which secretes a single, tubular, calcareous shell open at each end. They burrow in soft marine sediments, living with the somewhat narrower end of the shell-tube projecting slightly above the sediment surface (Fig. 5.19). From the larger ventral aperture of the shell project the conical or cylindrical, burrowing foot and the small proboscis-like head, which lacks eyes and sensory tentacles, but possesses a radula, a single median jaw, and paired clusters of narrow, clubbed, contractile filaments, the captacula. The numerous captacula are the organs of deposit feeding, smaller food particles being conveyed back to the mouth along the filaments by cilia, the larger ones adhering to the sticky clubbed tips and being brought directly to the mouth (see Fig. 9.6).

The plane of elongation of the body is difficult to determine. If the anus and mantle cavity are regarded as being posterior, then, like the gastropods and cephalopods, scaphopods are greatly elongated in the dorsoventral plane (Fig. 5.20); alternatively, the mantle cavity and anus are often considered to be ventral, in which case the scaphopods are lengthened along the anteroposterior axis. The mantle cavity extends the posterior height (or ventral length) of the shell and lacks ctenidia. Water is drawn into it through the dorsal (or posterior) aperture by mantle

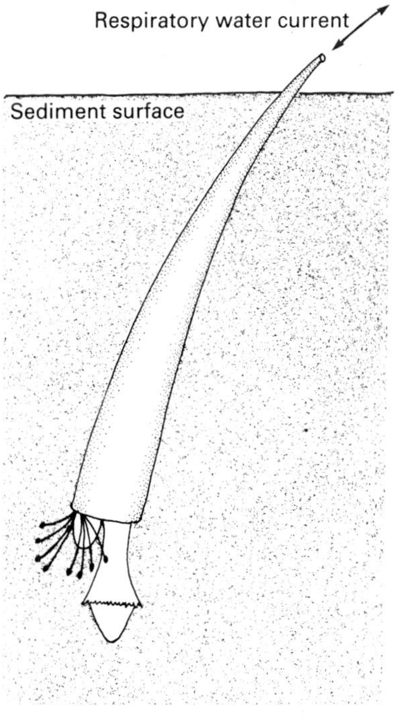

Fig. 5.19. Life style of scaphopod molluscs.

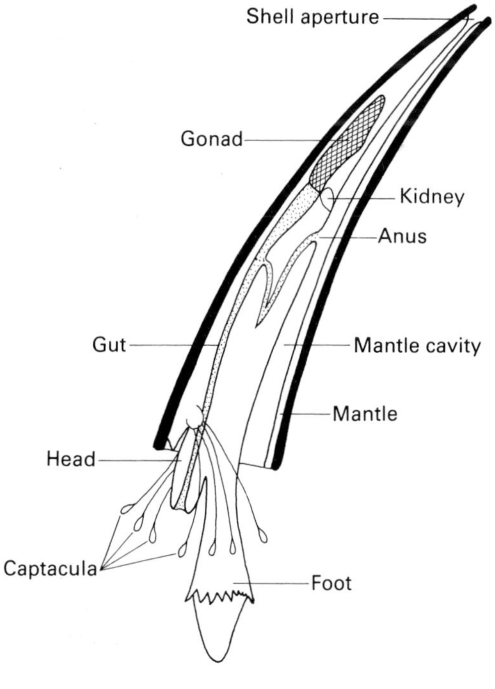

Fig. 5.20. Diagrammatic longitudinal section through a scaphopod (after various sources).

cilia, and is periodically pumped out through the same aperture by muscular contraction, gaseous exchange being effected across the mantle. All species are gonochoristic and possess a single gonad that discharges via the right kidney.

The 350 species are placed in two orders, dependent on the number and shape of the captacula and on the shape of the foot.

5.1.3.8 Class Cephalopoda

The wholly marine cephalopods are, anatomically and behaviourally, the most sophisticated of the molluscs and arguably of all invertebrates. They also include the largest species, giant squid achieving total lengths in excess of 20 m. The diagnostic features of cephalopods are the structures produced by what in other molluscs would be the foot, which as such is lacking. The anterior region of the embryonic foot develops into a series of prehensile arms or tentacles around the mouth, and the posterior portion forms a muscular funnel around part of the opening of the mantle cavity. Water can be taken into the mantle cavity around the margins of the head and forcibly expelled through the manoeuvrable funnel by muscular contraction. Cephalopods are essentially pelagic animals which swim by jet propulsion after mobile prey, these being caught by use of the arms. Like the gastropods, cephalopods are elongated in the dorsoventral plane, but in association with their swimming life style, the ancestrally ventral region has functionally become anterior, and the dorsal visceral hump effectively forms the posterior end of the animal (Fig. 5.21). The mantle cavity therefore opens anteriorly and thus during rapid swimming cephalopods move backwards; during slower locomotion, movement of the funnel can permit progression in a variety of directions. Prey having been caught with the arms, it is macerated by a pair of beak-like dorsoventral jaws and then by the radula. Some of the arms are also modified for copulation (spermatophore transfer) in the male, most species being gonochoristic, with a single gonad and direct development. Somewhat surprisingly considering their large size, all cephalopods, except the nautiloids, are relatively short-lived animals, which breed once and then die ('semelparous'—see Section 14.3).

Other features of the group differ quite markedly between the two surviving subclasses. The six living species of the Nautiloidea possess a single external planospiral shell, comprising many chambers in a linear series, only the last of which is inhabited, although a thin filament of living tissue, the siphuncle, extends through the other chambers (Fig. 5.22). As the animal grows, new material is laid down so as to add a roughly cylindrical section to the old shell; it then moves forwards, secretes a new partition behind it, and thereby adds another chamber to the existing

Fig. 5.21. Diagrammatic side view of a cephalopod mollusc showing the changed orientation of the body from that in the ancestral mollusc: the orientation of the living cephalopod is shown in parentheses by the ancestral states.

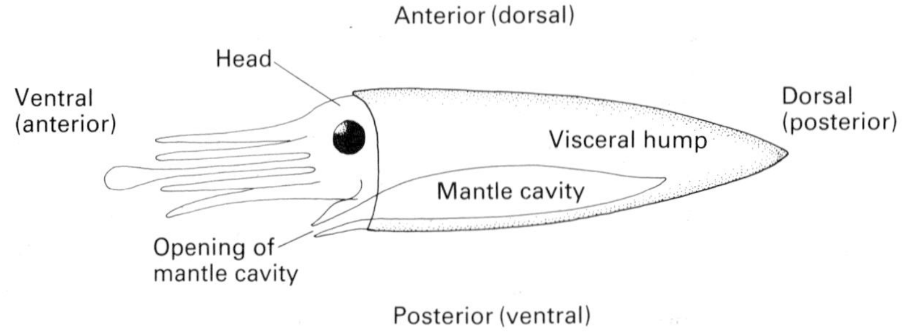

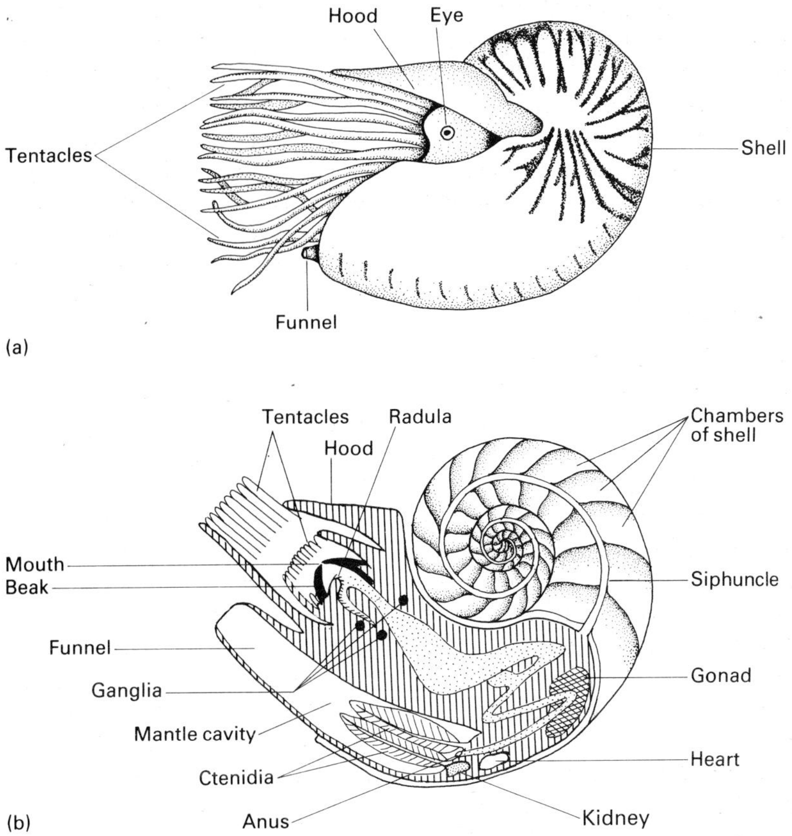

Fig. 5.22. Nautiloid cephalopods: (a) external appearance; (b) diagrammatic longitudinal section (after Boss, 1982 and others).

series. Nautiloids are poor swimmers, water being expelled from the mantle cavity only by contraction of the funnel muscles, and they rely on buoyancy to remain suspended in the water column. The pressure-insensitive buoyancy system is provided by the uninhabited shell chambers, the siphuncle absorbing the water originally contained in a chamber and secreting gas instead with the result of neutral buoyancy (see Section 12.2.2). Many (80–90) suckerless tentacles surround the mouth, of which four are modified for spermatophore transfer; there are two pairs of ctenidia and of kidneys; and the nervous system and eyes are relatively simple.

In the Coleoidea, however, there is a marked trend of shell reduction correlated with a more active swimming life. In this group, water is expelled from the mantle cavity more forcefully by the simultaneous contraction of powerful circular muscles in the mantle wall, co-ordinated by large stellate ganglia from which issue giant fibres, of increasing diameter the greater the distance between the innervated muscles and the ganglion. The cuttlefish (order Sepiida) have retained the calcareous shell as a buoyancy device (Section 12.2.2), but it is reduced in size and lies within the body; in the squids (orders Teuthida and Vampyromorpha), it is further reduced to a thin internal cartilaginous element with no buoyancy function; whilst in most octopuses (order Octopoda) it has been lost. Many pelagic coleoids are streamlined and torpedo shaped (Fig. 5.23), and have continually

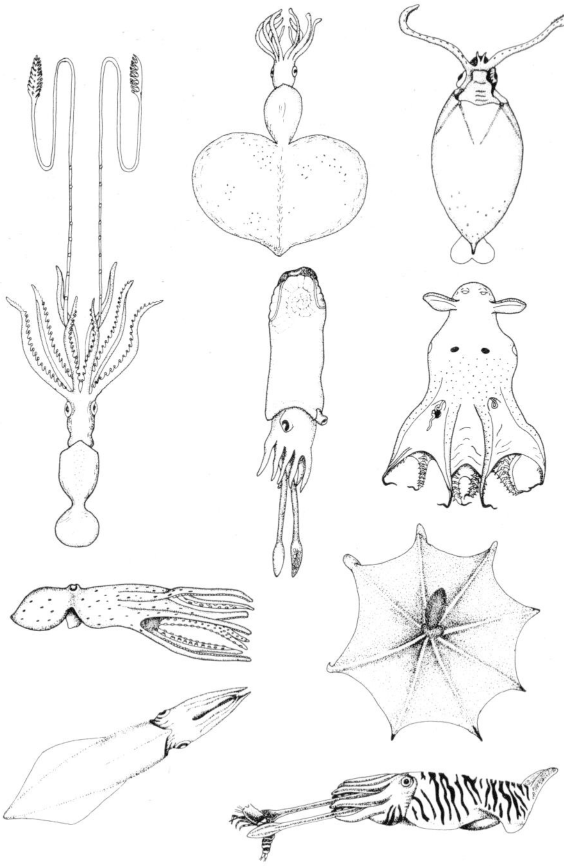

Fig. 5.23. Coleoid cephalopod body forms (after various sources).

to swim to remain within the water mass, undulations of lateral fins permitting a slow but efficient form of locomotion. Some, however, have evolved alternative buoyancy aids, such as the replacement of heavy divalent cations by ammonia in an expanded pericardial cavity, coupled with low-density gelatinous tissues. The octopuses have adopted a largely benthic existence, although some forms have a web of skin between the arms that can be used as a sail to drift with near-bottom currents. Such species have reduced or vestigial ctenidia. In contrast with the nautiloids, coleoids otherwise have a single pair of sturdy, folded ctenidia, a single pair of kidneys, and eight short sucker-bearing arms (two of them usually modified for spermatophore transfer) together, in the squids and cuttlefish, with two elongate, tentaculate, prey-catching arms bearing pads of suckers at their tips (Fig. 5.24).

Perhaps the most distinctive features of the 650 species of coleoids, however, are the closed nature of the blood system and the highly concentrated state of the nervous system which includes well-developed eyes. The ctenidia lack cilia (water being driven through the gills and mantle cavity by the muscular pumping which effects the jet-propelled locomotion), but contain capillaries—not arranged in a counter-current flow—through which blood is pumped by branchial hearts. The circum-oesophageal nerve ring is enlarged into a highly complex brain, to a greater degree than seen in any other invertebrate, and the

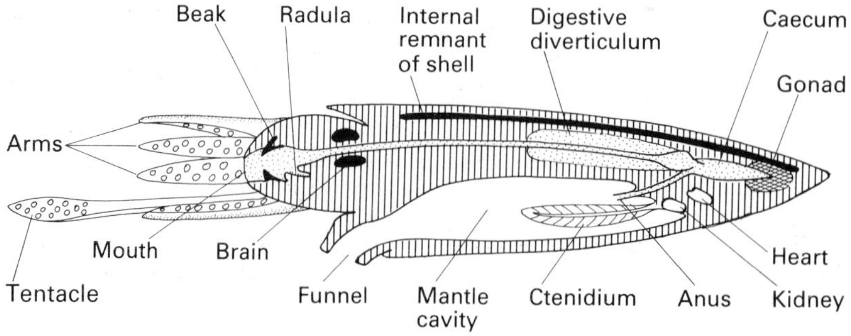

Fig. 5.24. Diagrammatic longitudinal section through a coleoid (after Boss, 1982 and others).

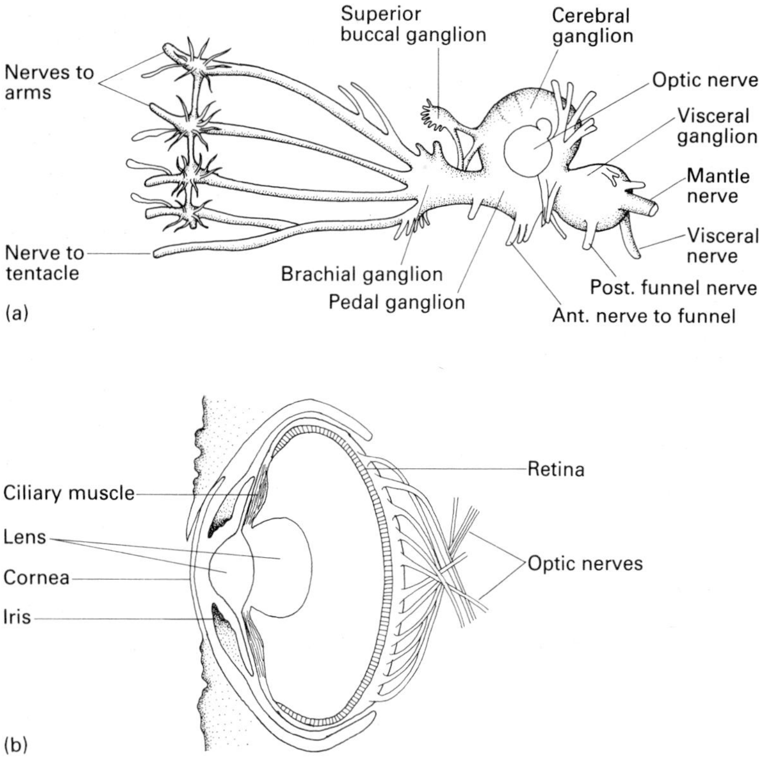

Fig. 5.25. (a) The brain and (b) eye of coleoids (after Wells, 1962 and Kaestner, 1967).

paired eyes innervated from the brain are of the same general form as possessed by the vertebrates, with cornea, iris diaphragm, lens and retina (Fig. 5.25). In contrast to the vertebrates, however, the photoreceptors of the coleoid eye are directed *towards* the incoming light. The nervous system also controls the numerous surface chromatophores, that unusually are operated by small muscles and hence can respond rapidly, giving rise to almost instantaneous changes in colour pattern.

5.2 Further reading

Fretter, V. & Peake, J. (Eds) 1975–78. *Pulmonates* (2 Vols). Academic Press, London.

Hughes, R.N. 1986. *A Functional Biology of Marine Gastropods*. Croom Helm, Beckenham.

Hyman, L.H. 1967. *The Invertebrates (Vol. 6): Mollusca 1*. McGraw-Hill, New York.

Kaestner, A. 1967. *Invertebrate Zoology* (Vol. 1). Wiley, New York.

Morton, J.E. 1979. *Molluscs*, 5th edn. Hutchinson, London.

Purchon, R.D. 1968. *The Biology of the Mollusca*. Pergamon Press, Oxford.

Runham, N.W. & Hunter, P.J. 1970. *Terrestrial Slugs*. Hutchinson, London.

Solem, A. 1974. *The Shell Makers*. Wiley, New York.

Wells, M.J. 1962. *Brain and Behaviour in Cephalopods*. Heinemann, London.

Wells, M. J. 1978. *Octopus*. Chapman & Hall, London.

Wilbur, K.M. (Ed.) 1983–87. *The Mollusca* (12 Vols.). Academic Press, New York.

6
THE LOPHOPHORATES

Phorona
Brachiopoda
Bryozoa
Entoprocta

The three phyla usually regarded as comprising the lophophorates, the Phorona, Brachiopoda and Bryozoa, share, as their name suggests, the common feature of a lophophore: a circular, horseshoe-shaped or complexly whorled or folded ring of ciliated hollow tentacles serving as a suspension-feeding apparatus. This lophophore, which encircles the mouth but not the anus, is supported by its own separate hydrostatic body cavity.

Basically, the lophophorate body is tripartite in form, with a minute pre-oral section (the prosome), a small second region (mesosome) bearing the mouth and lophophore, and a much larger third part (metasome), which comprises the major part of the body, contains the other organ systems and bears, near the base of the lophophore, the anus. A further, rather distinctive, common feature of the lophophorates is that their 'gonads' are not discrete organs, merely loose peritoneal aggregations of germ cells. In the majority of lophophorates, the prosome has been lost, and even when present it is insignificant; its significance for the relationships of the lophophorates, however, would appear far to outweigh its importance to the living animals.

Although they are protostomes in that the mouth forms from the blastopore, the lophophorates share a number of characteristics with the deuterostomes (Chapter 7) including the tripartite body plan (in some deuterostomes the prosome is well developed), in which each—or all three—body regions contains a separate body cavity which arises, in some at least, by enterocoely (Sections 15.2–15.4). Further shared features include radial and indeterminate cleavage and a basiepidermal nervous system; some deuterostomes even possess temporary peritoneal 'gonads' and a mesosomal lophophore. The lophophorates, therefore, provide a link between the protostome and deuterostome phyla (see also Section 2.4.3).

Lophophorates are sedentary or sessile, and their bodies are protected within a secreted external covering. This may take the form of a chitinous tube within which the animal is free to move, or a gelatinous, chitinous or calcareous shell or box to which the epidermis is permanently attached.

The lophophorates, therefore, appear a well-knit group and some authorities have argued that a single phylum Lophophorata should include them all. That this has not become widely accepted is in part due to continuing debate on the relationships of the bryozoans to the other groups, and in part to the affinities of the fourth phylum, the Entoprocta, which are still mysterious. The feeding tentacles of the entoprocts do not conform to the definition of the lophophore, and yet they do show similarities to the bryozoans in particular, although their cleavage pattern is spiral and their development is determinate—markedly non-lophophorate characteristics. Their inclusion in this chapter is therefore largely a matter of convenience.

6.1 Phylum PHORONA

6.1.1 Etymology

Derived from the generic name *Phoronis*, one of the epithets of the Egyptian goddess Isis.

6.1.2 Diagnostic and special features (Fig. 6.1)

1 Bilaterally symmetrical, vermiform.
2 Body more than two cell layers thick, with tissues and organs.

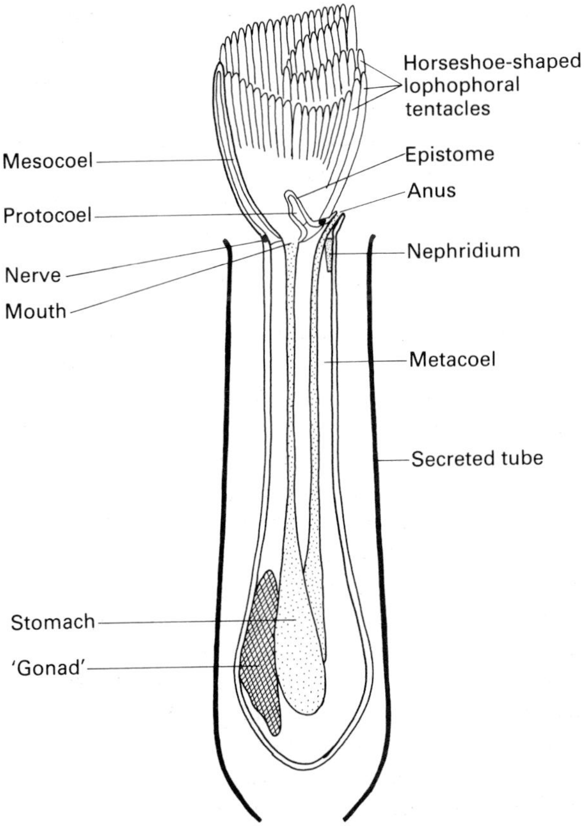

Fig. 6.1. Diagrammatic longitudinal section through a phoronan (after Emig, 1979).

3 A through, U-shaped gut.
4 An oligomeric, tripartite body, each region with a single body cavity.
5 Prosome very small; mesosome small but with a large lophophore supported by the mesocoel; metasome large and elongate.
6 Body wall without a cuticle, with muscle layers.
7 A closed blood system, with haemoglobin in corpuscles.
8 One pair of metanephridia-like organs (protonephridia in the larva).
9 Nervous system diffuse, basiepidermal.
10 Eggs cleave radially.
11 Development indirect, usually via an actinotroch larva.
12 Marine.

Phoronans are sedentary worms permanently inhibiting secreted chitinous tubes partially buried in soft sediments or, less commonly, attached to hard substrata. Although exceptionally up to 50 cm in length, most are less than 20 cm long of which the majority is formed by the metasome or trunk. The only other externally visible structure is the large and topographically terminal, suspension-feeding lophophore, which comprises up to 15 000 tentacles arising from the mesosome in the form of a horseshoe (Fig. 6.2); in species with many tentacles, the free arms of the horseshoe are curled round into spirals.

Most phoronans are hermaphrodite, although a few are gonochoristic. The sex cells originate from the peritoneum and are shed into the metacoel, from which they escape via the nephridia. Fertilization is external although the eggs and developing larvae (Fig. 6.3) may be brooded within the cavity enclosed by the lophophore, or within the tube, for between 40 and 75% of their developmental period. One species is capable of asexual multiplication by fission and budding, the other species by transverse fission only.

Phoronans share a number of features in common both with the other lophophorate phyla (Sections 6.2 and 6.3) and with the deuterostomes (Chapter 7), and are often regarded as being close to the origin of both these assemblages (Chapter 2). Some would even

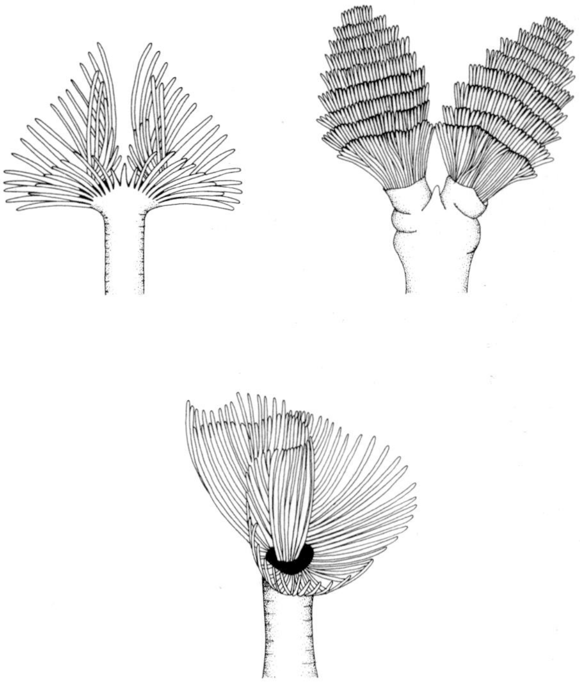

Fig. 6.2. Lophophore form in phoronans (after Emig, 1979).

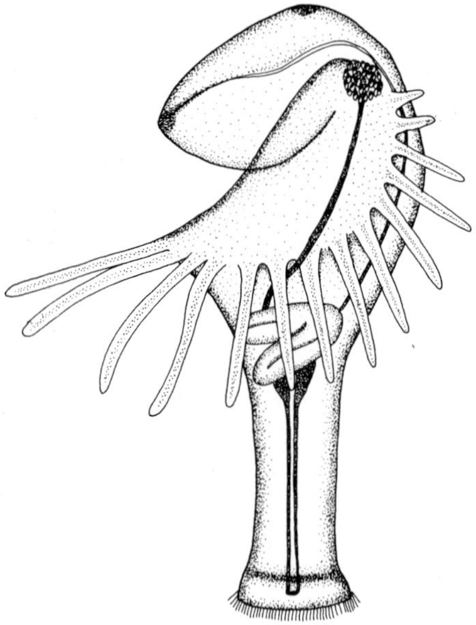

Fig. 6.3. Phoronan actinotroch larva (after Emig, 1979).

regard them as being more properly included within the deuterostome superphylum itself, notwithstanding that their mouth develops from the blastopore.

6.1.3 Classification

Only ten species are known, placed in two genera within one family.

6.2 Phylum BRACHIOPODA (lamp-shells)

6.2.1 Etymology

Greek: *brachion*, arm; *pous*, foot.

6.2.2 Diagnostic and special features (Fig. 6.4)

1 Bilaterally symmetrical.
2 Body more than two cell layers thick, with tissues and organs.
3 Either a through, U-shaped gut or one secondarily blind-ending.
4 An oligomeric, bipartite body, each region with a single, essentially enterocoelic body cavity.
5 Prosome absent; mesosome small but with a large complex lophophore supported by the mesocoel (see Fig. 9.3) and, in some by calcium carbonate extensions of the shell; metasome small.
6 Body completely enclosed, apart from the stalk, within a bivalved shell which may be cemented to the substratum, attached to rock or anchored in soft sediments by a stalk (pedicle), or be unattached; the small body is located posteriorly within the shell, most of the enclosed cavity being occupied by the lophophore.
7 An open blood system with a heart or hearts.
8 With one or two pairs of metanephridia-like organs.

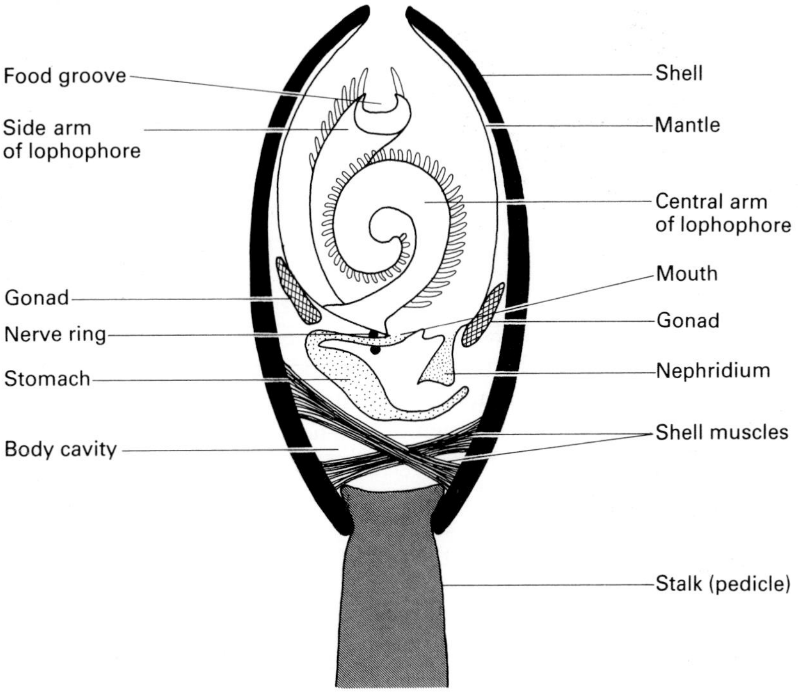

Fig. 6.4. Diagrammatic longitudinal section through a generalized brachiopod.

9 Nervous system in the form of a ganglionated circum-oesophageal ring, from which individual nerves issue.
10 Without asexual multiplication; mostly gonochoristic, without discrete gonads but with four aggregations of sex cells associated with the peritoneum, gametes being shed into the metacoel and leaving via the nephridia.
11 Eggs cleave radially.
12 Marine.

The surviving brachiopods, which are but a small remnant of this once important phylum (335 living species as opposed to 26000 fossil ones), are in their external appearance and general life style remarkably similar to the bivalve molluscs (Section 5.1.3.6). Both groups are sedentary or sessile, suspension-feeding animals enclosed within a bivalved shell, which is secreted by an epidermal mantle and covered by an organic periostracum. The suspension-feeding apparatus of the brachiopods, however, is a circular, looped or spiral lophophore (not ctenidial gills), and the two valves of their shell are dorsal and ventral (not lateral, as in bivalve molluscs). A further point of convergence is the reduction of the ancestral head region (in brachiopods, loss of the lophophorate prosome) consequent on the complete encasement of the body within a thick external shell.

All living brachiopods are relatively small (<10 cm shell length or width) and benthic in habitat. In evolutionary terms, they may be equivalent to shelled phoronans (Chapter 2) and indeed the brachiopod condition may have arisen polyphyletically from such worms (Fig. 2.13).

6.2.3 Classification

Two classes are recognized which display many divergent features.

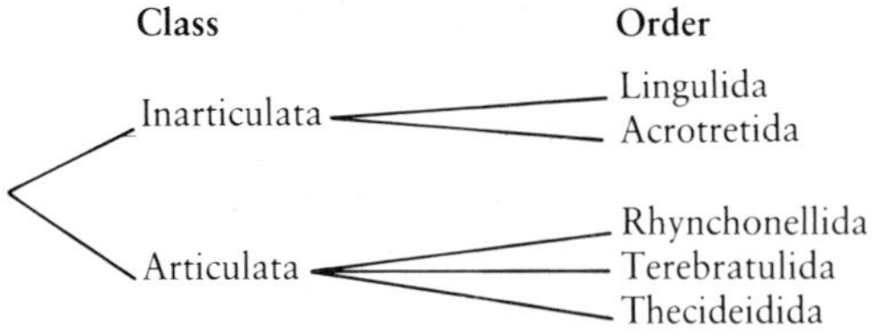

6.2.3.1 Class Inarticulata

Inarticulate brachiopods, as their name suggests, are characterized by having shell valves, which are usually composed of calcium phosphate and chitin, held together solely by muscles. Posteriorly, from between the two valves, emerges the pedicle, which is an integral part of the body formed from the larval mantle; it contains an extension of the metacoel and intrinsic muscles; in some groups it is absent, the ventral valve then being cemented to the substratum. In contrast to other brachiopods, an anus is present (Fig. 6.5) and the lophophore is relatively simple and unsupported by shelly material.

The larva is a small free-swimming version of the adult, complete with shell and lophophore, the cilia on which provide the means of propulsion through the water; correspondingly, there is no metamorphosis on settlement. During development, mesoderm develops from the larval archenteron in the enterocoelic manner, but the body cavities form schizocoelically within the enterocoelic mesodermal cell masses.

The 47 species are divisible into two orders: the burrowing Lingulida and the sessile and limpet-like Acrotretida (Fig. 6.6).

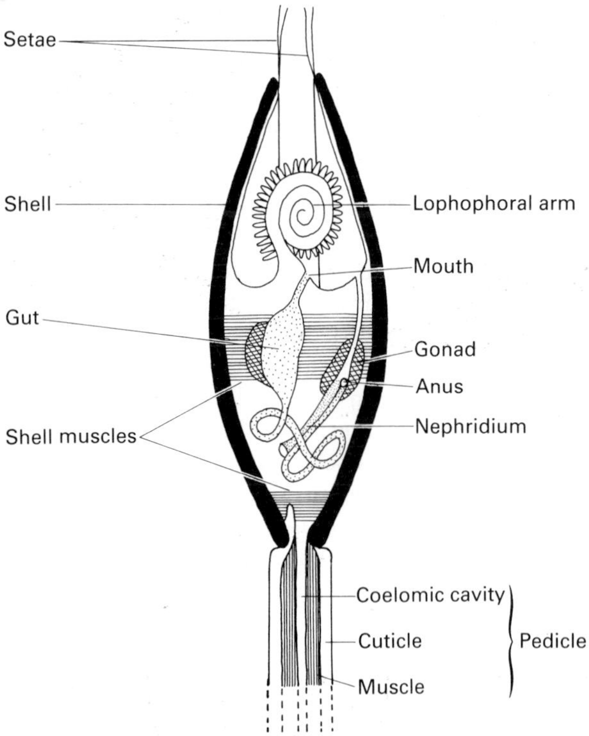

Fig. 6.5. Diagrammatic longitudinal section through an inarticulate brachiopod.

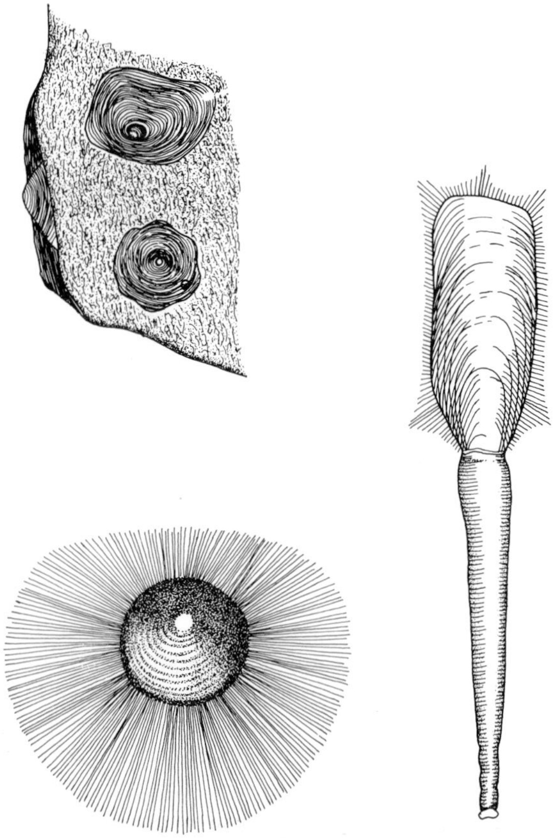

Fig. 6.6. Body forms of inarticulate brachiopods (after sources in Hyman, 1959).

6.2.3.2 Class Articulata

The articulates possess calcium carbonate shells, the valves of which articulate by means of teeth present on the ventral valve that insert into sockets on the dorsal. The ventral valve also bears a slit or notch through which the pedicle, if present, emerges; the dorsal valve often includes inwardly directed supports for the large and complexly looped or spiralled lophophore (Fig. 6.7). Neither intrinsic muscles nor an extension of the metacoel are present in the pedicle, which is derived not from the mantle but from one of the three body regions of the larva. This larval stage (Fig. 6.8) is markedly different in

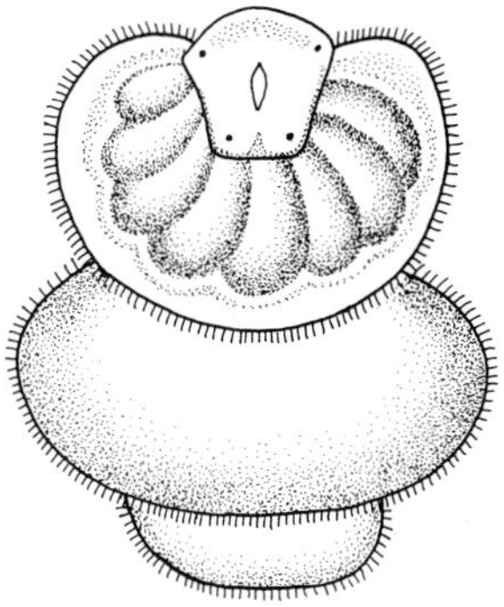

Fig. 6.8. The larva of an articulate brachiopod (after Lacaze-Duthiers, 1861).

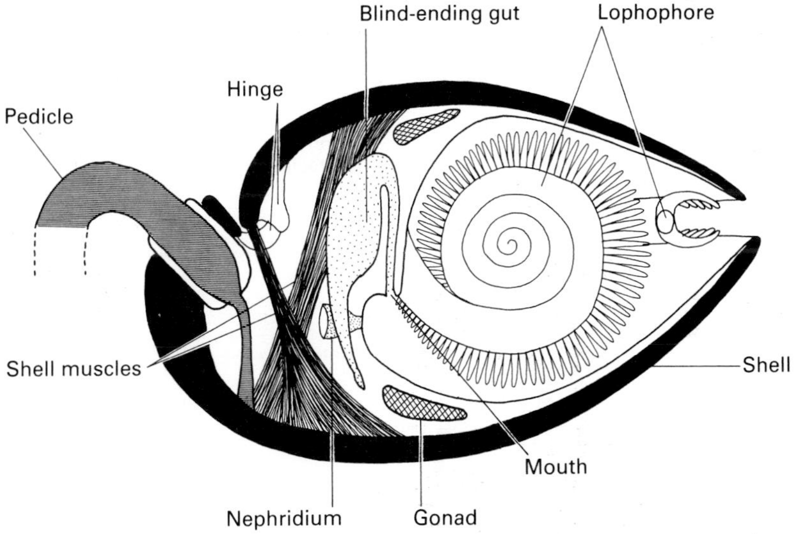

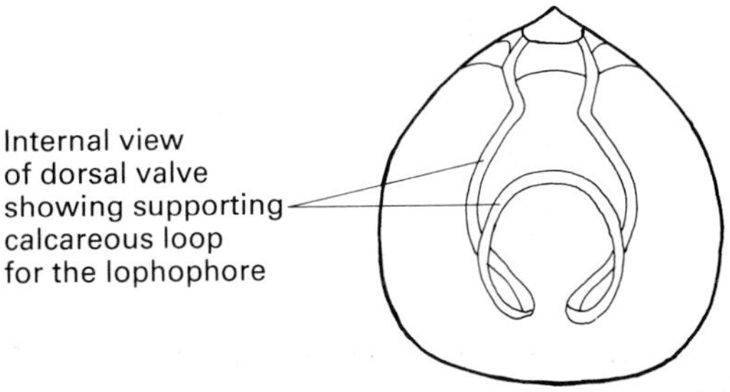

Fig. 6.7. Diagrammatic longitudinal section through an articulate brachiopod (after Moore, 1965).

appearance from the adult, and undergoes metamorphosis. In contrast to the inarticulates, the gut is blind-ending and both mesoderm and the contained body cavities develop enterocoelically.

Articulates appeared relatively late in brachiopod evolution (in the Ordovician) and nearly 300 species remain extant. These are placed in three orders on the basis of the structure of the lophophore and of its supporting calcareous system.

6.3 Phylum BRYOZOA

6.3.1 Etymology

Greek: *bryon*, moss; *zoon*, animal.

6.3.2 Diagnostic and special features (Fig. 6.9).

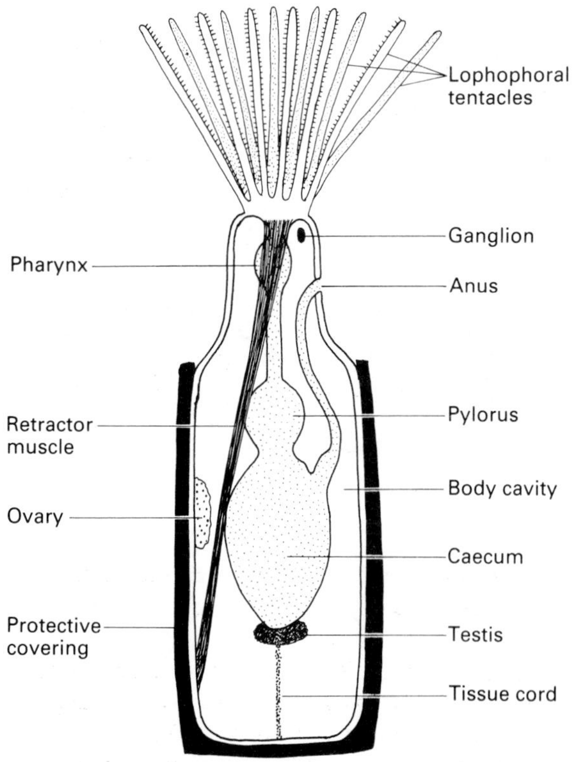

Fig. 6.9. Diagrammatic longitudinal section through a generalized bryozoan.

1 Modularly colonial animals; each colony with from a few to a million individuals (zooids) formed by asexual budding from a founding ancestrula; each zooid is in tissue contact with its immediate neighbours.
2 Zooids often polymorphic, with feeding, defensive, brooding types, etc.
3 Colonies are hermaphrodite, although zooids may be hermaphrodite or gonochoristic.
(Features below refer to the characteristic individual zooid.)
4 Bilaterally symmetrical.
5 Body more than two cell layers thick, with tissues and organs.
6 A through, U-shaped gut.
7 An oligomeric, bi- or tripartite body, each region with a single body cavity formed *de novo* during metamorphosis; some with an additional metasomal body cavity.
8 Prosome present only in one group, small; mesosome small, and with a relatively small, circular or horseshoe-shaped lophophore supported by the mesocoel; metasome large, sac-like.
9 Body enclosed or embedded in a chitinous, gelatinous or calcareous tube, box or communal matrix, except for an orifice through which the lophophore can be protruded (by hydrostatic pressure) and withdrawn (by retractor muscles).
10 Without circulatory or excretory systems.
11 Nervous system in the form of a ganglion between mouth and anus, from which issue a circumpharyngeal ring and individual nerves.
12 Gonads are 'peritoneal', gametes rupturing into the metacoel and leaving through 'coelomopores' associated with the lophophore.
13 Eggs cleave radially and are usually brooded.
14 Development normally indirect.

Bryozoans appear similar to minute (*c.* 0.5 mm) phoronans (Section 6.1) which form extensive stoloniferous, mat-like, arborescent or foliaceous colonies by asexual budding (Fig. 6.10). Similarly to the other lophophorate phyla (Sections 6.1 and 6.2), they exhibit, besides the suspension-feeding lophophore, an oligomeric body plan, radial cleavage, and an apparently similar gonadal system in that the

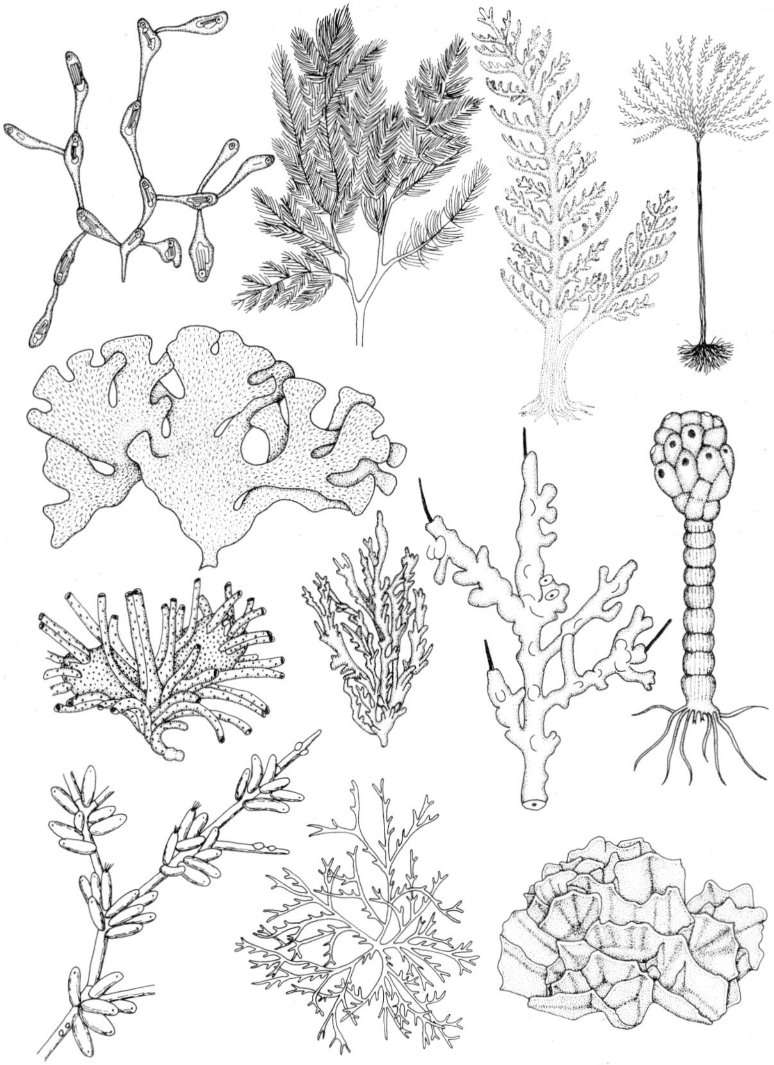

Fig. 6.10. Colony form in bryozoans (after several sources).

sex cells develop from cells in the membrane lining the body cavity. The nature of their body cavities is debatable, however, because they have no embryological precursors nor temporal continuity with the embryonic germ layers. During metamorphosis, the tissues of bryozoan larvae are histolysed and then completely reorganized into the adult body. Mesoderm and the body cavities of the other lophophorates are basically enterocoelic, deriving from the embryonic gut; but the larval gut of bryozoans, if present, is broken down at metamorphosis. There is, therefore, no obvious mesodermal origin of any of the adult cavities, and the coelomic status that they are usually given is based mainly on analogy with the phoronans and brachiopods. This is not uncontested: an alternative school of thought regarded them as being equivalent to the supposedly pseudocoelomic cavity of the entoprocts (Section 6.4).

As a group, bryozoans are a very successful line of suspension feeders, more so today than any other lophophorate. With a few exceptions, colonies are sessile, encrusting or attached to relatively firm substrata; one exception, the freshwater *Cristatella*, can creep slowly (c. 10 cm day^{-1}) by means of a muscular 'foot'; whilst another, the marine *Selenaria maculata* can walk quite quickly (1 m h^{-1}) on the long setae of its avicularia zooids (see Section 6.3.3.3). *S. maculata* lives in shallow, coral-reef habitats and it has been recorded as moving towards patches of sunlight. The colony is green and it is therefore possible that this behaviour is related to the presence of symbiotic zooxanthellae in its tissue, as seen in many other reef invertebrates (Section 9.1).

6.3.3 Classification

Most of the 4000 described species are marine, but one class and various individual species in the two other classes are freshwater.

Class	Order
Phylactolaemata	Plumatellida
Stenolaemata	Cyclostomata
Gymnolaemata	Ctenostomata
	Cheilostomata

6.3.3.1 Class Phylactolaemata

The phylactolaemes are relatively unspecialized bryozoans. The zooidal body has retained the presumed ancestral tripartite form, in that a prosomal flap of tissue, the epistome, containing a hydrostatic protocoel, overhangs the mouth, as in the phoronans. The body wall also bears well-developed layers of circular and longitudinal muscle; contraction of the circular layer generates the pressure required to protract the lophophore, which is large with up to 120 tentacles issuing from a horseshoe-shaped ridge (Fig. 6.11). Individual zooids are monomorphic, cylindrical and among the largest of the bryozoans; the body cavity compartments of adjacent individuals often interconnect.

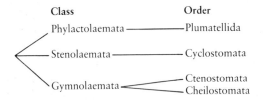

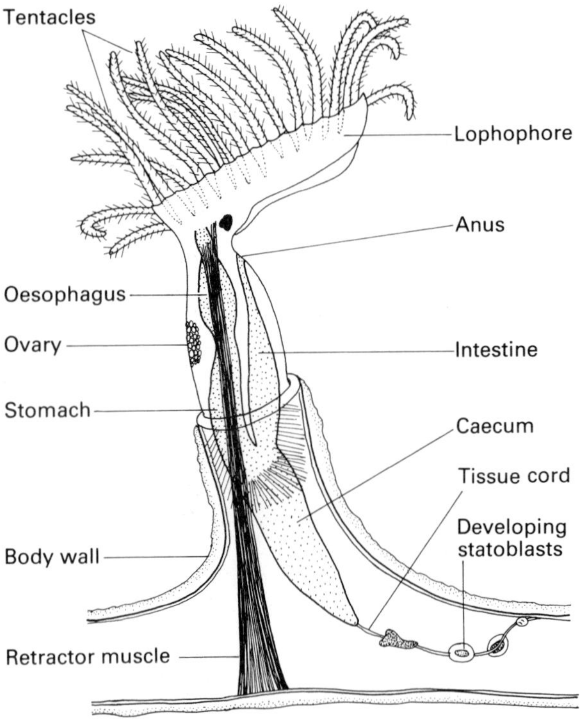

Fig. 6.11. An extended phylactolaeme zooid, seen as if transparent (after Pennak, 1978).

All species are freshwater and, calcium salts being less abundant in fresh than in salt water, the epidermis secretes a non-calcified, chitinous or gelatinous case, and, in common with other freshwater invertebrates, they have evolved a resting-stage mechanism for dispersal and/or overwintering. This is highly individual in nature: on a cord of tissue running from the gut to the body wall, aggregations of epidermal and 'peritoneal' cells develop into 'statoblasts' (Fig. 6.12)—yolk-rich cells enclosed within (usually) disc-shaped chitinous valves. These may remain attached to the body wall and serve to reinstate the colony *in situ* after die-back, or be liberated during life or after death of the zooid. The latter type often can float and may be dispersed over large distances. Statoblasts are highly resistant both

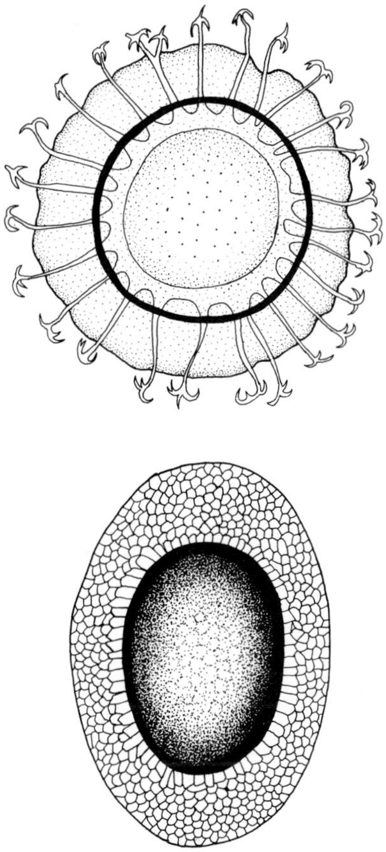

Fig. 6.12. Phylactolaeme statoblasts (after Hyman, 1959).

to freezing and to desiccation, and, in temperate latitudes, after winter they 'germinate' and reform into founding zooids.

A single order containing some 50 species is recognized.

6.3.3.2 Class Stenolaemata

The stenolaeme zooids are also cylindrical, but they lack a prosome and muscle layers in the body wall. Their cylindrical tubes, bearing a terminal circular orifice, are heavily calcified beneath an outer hyaline cuticle or cellular layer; the orifice is closed after retraction of the lophophore by a membrane and not an operculum. The pressure required to protract the lophophore, which is circular and comprises only some 30 tentacles at most, is generated by dilator muscles acting on the fluid both in the metacoel and in an exosaccal body cavity generally regarded as being pseudocoelomic (Fig. 6.13). Stenolaeme zooids display a limited degree of polymorphism.

The reproductive system of stenolaemes exhibits several peculiarities, the chief of which is a form of polyembryony. After fertilization, the zygote cleaves to produce a ball of blastomeres, but this ball then buds off a series of secondary embryos, which in turn may bud off tertiary embryos: more than 100 blastulas may be derived from a single zygote.

Most groups of stenolaemes are extinct; a single order containing some 900 marine species survives.

6.3.3.3 Class Gymnolaemata

Today, the gymnolaemes are the most abundant and successful of the bryozoans, with over 3000 species, mainly in the sea but also in brackish and fresh waters. Similarly to the stenolaemes, their zooids lack a prosome and muscle layers in the body wall, and possess a relatively small, circular lophophore; but the generally short and squat zooids generate their lophophore-protracting pressure by muscular deformation of the body wall, parietal muscles acting on a specific membraneous region (Fig. 6.14). The body wall, therefore, if calcified, is only partially so. After retraction of the lophophore, the orifice can, in many species, be closed by an operculum. Zooidal polymorphism reaches its greatest degree in this group, with feeding (autozooids), packing (kenozooids), grasping (avicularia), cleaning (vibracula), and brood-chamber (ooecia) individuals.

Two orders are distinguished: the Ctenostomata, which have non-calcified walls and lack opercula (and are often cylindrical with a terminal orifice) (Fig. 6.15); and the Cheilostomata, which possess flattened, box-shaped zooids with partially—often largely—calcified walls, an operculum, and a frontal orifice (Fig. 6.16).

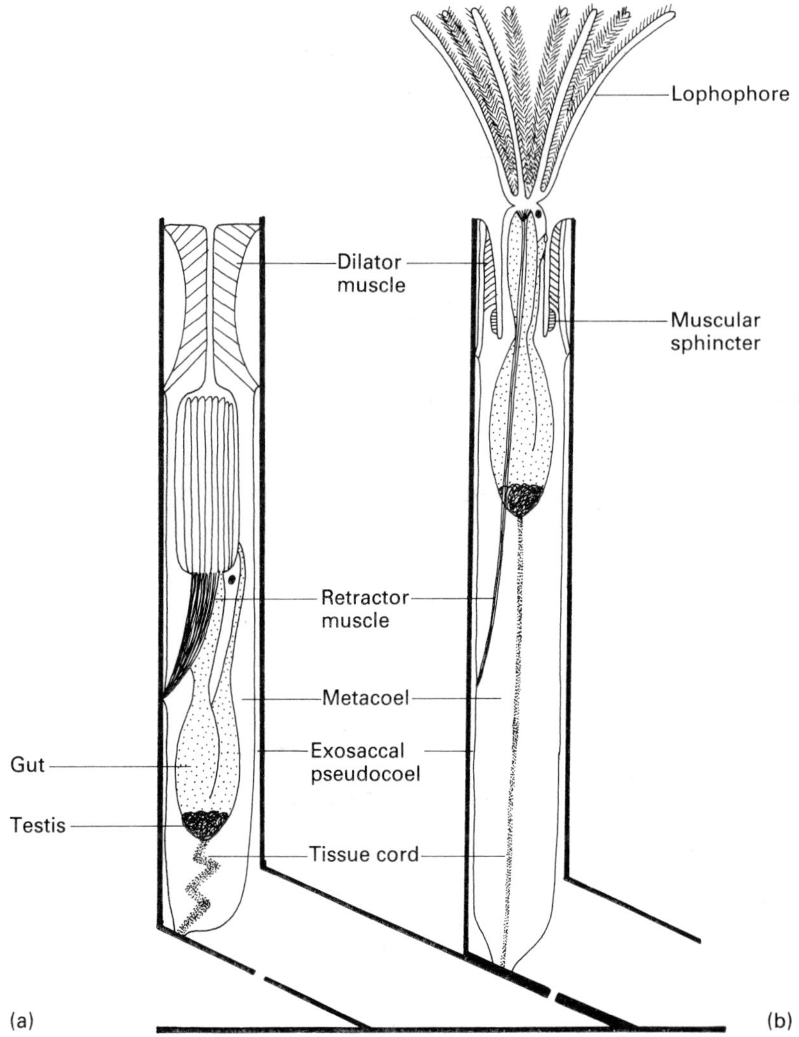

Fig. 6.13. Stenolaeme zooids: (a) retracted; (b) extended (after Ryland, 1970).

6.4 Phylum ENTOPROCTA

6.4.1 Etymology

Greek: *entos*, inside; *proktos*, anus.

6.4.2 Diagnostic and special features (Fig. 6.17)

1 Bilaterally symmetrical, goblet-shaped.
2 Body more than two cell layers thick, with tissues and organs.
3 A through, U-shaped gut.
4 Space between body wall and gut filled with a gelatinous 'mesenchyme'—interpreted by some as an occluded pseudocoelomic body cavity.
5 Body in the form of a hemispherical to ovoid calyx, bearing a ring of tentacles around both mouth

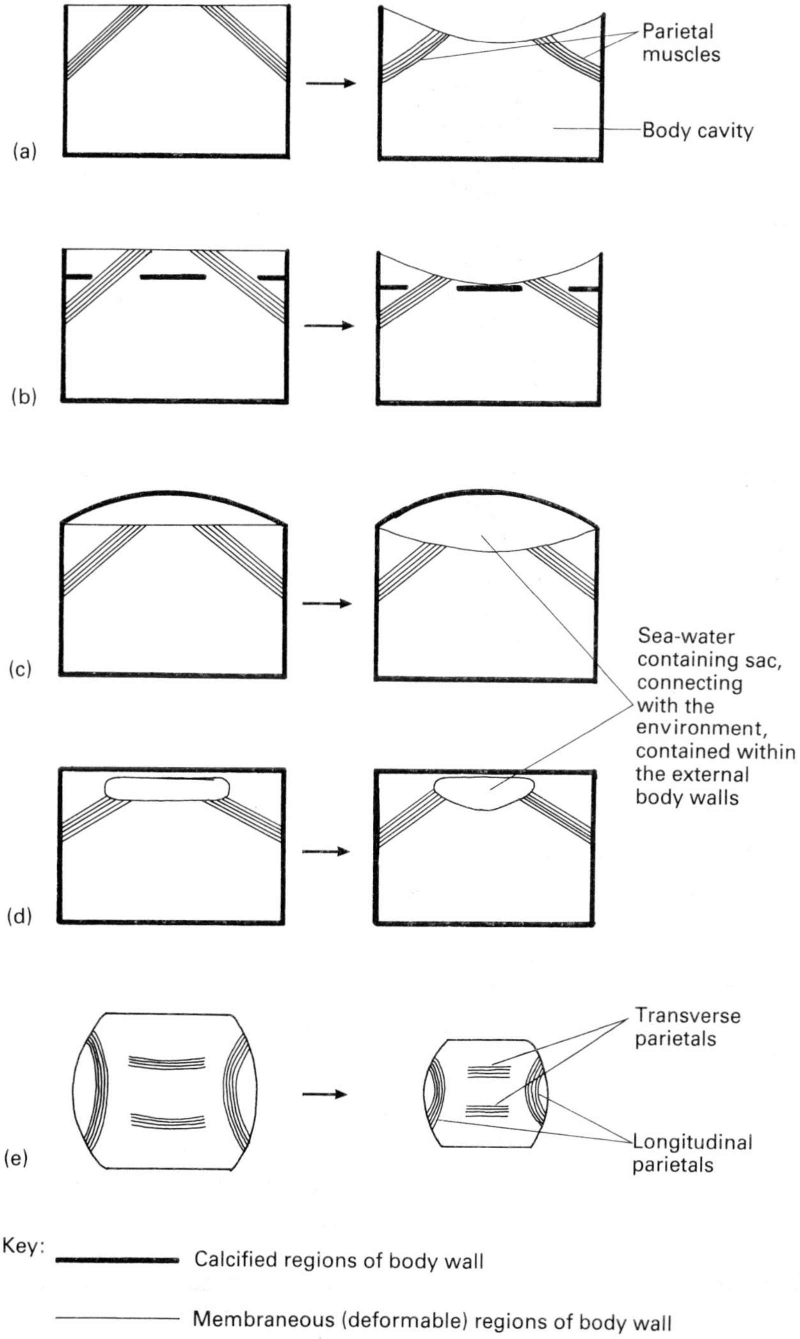

Fig. 6.14. Sections through the zooidal boxes of various gymnolaeme bryozoans, showing how contraction of parietal muscles causes deformation of the body walls and thus generates increased pressure in the body cavity: (a)–(d) cheilostomes; (e) ctenostome (after Ryland, 1970).

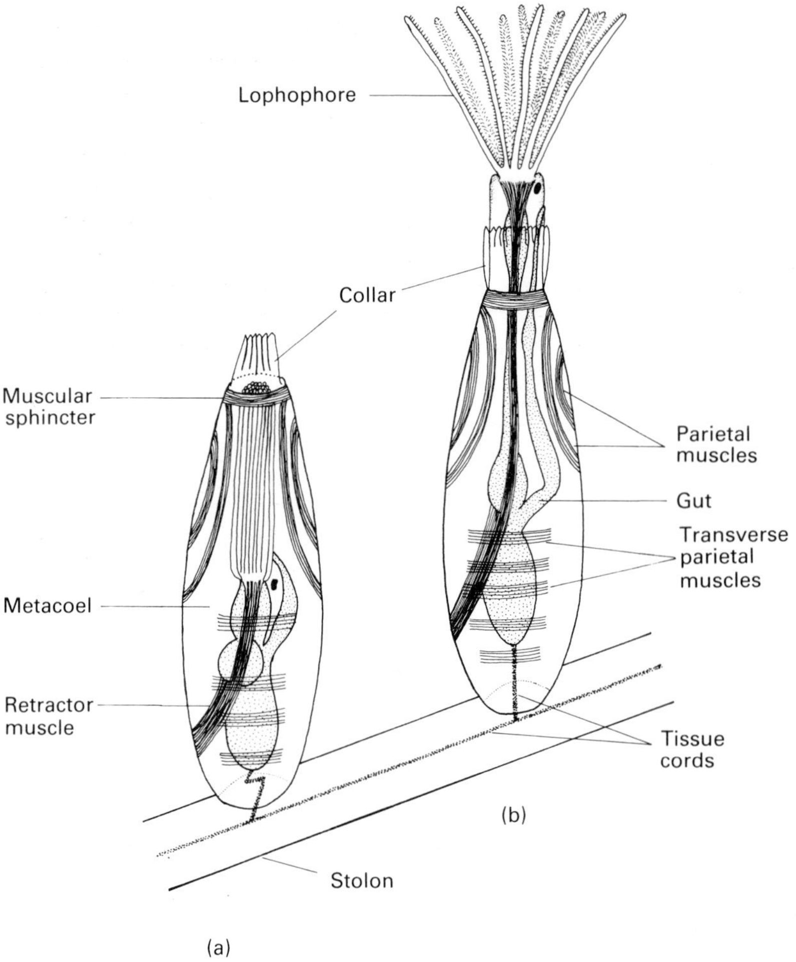

Fig. 6.15. Ctenostome zooids: (a) retracted; (b) extended (after Ryland, 1970).

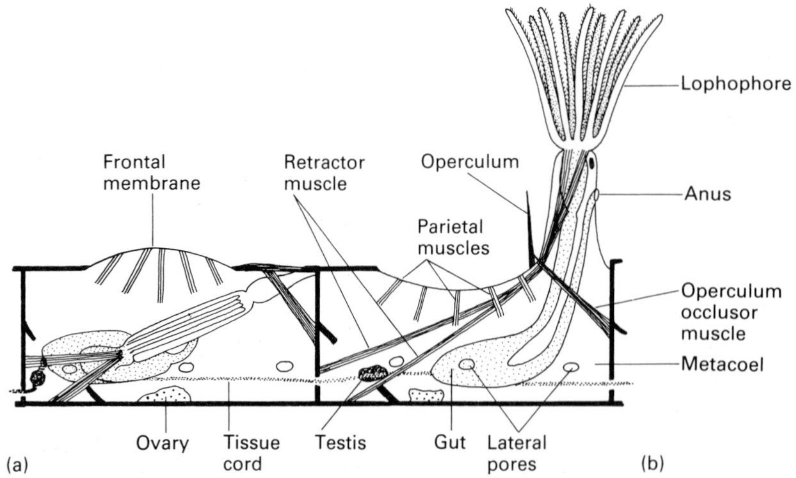

Fig. 6.16. Cheilostome zooids: (a) retracted; (b) extended (after Ryland, 1970).

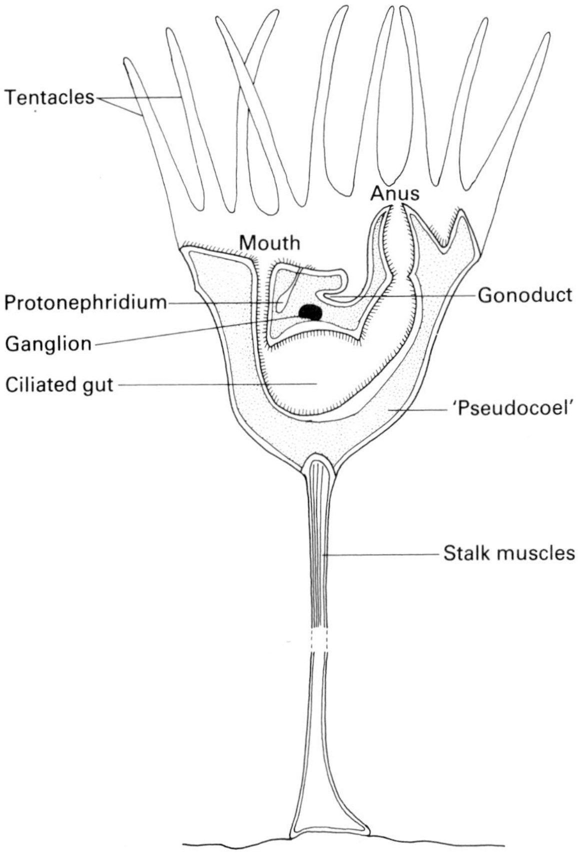

Fig. 6.17. Diagrammatic longitudinal section through an entoproct (after Becker, 1937).

and anus, and attached to a substratum by a contractile stalk.

6 Body wall with a cuticle but without muscle layers.

7 Without circulatory or gaseous-exchange systems.

8 One pair of protonephridia (or, in the freshwater genus, many protonephridia).

9 Nervous system in the form of a ganglion between mouth and anus, from which individual nerves issue.

10 Tentacles not retractile but they can contract and fold inwards to occlude the intratentacular cavity.

11 Eggs cleave spirally.

12 Development indirect.

Entoprocts are small (0.5–5 mm high), solitary or colonial animals living temporarily or, more commonly, permanently attached to a substratum, including that provided by other organisms. All are suspension feeders, the 6–36 extensions of the body wall which form the tentacles bearing cilia that collect food particles from the water and transport them, in mucus, to the mouth. In forms creating their own feeding current, water passes from outside the tentacular ring, between the tentacles and out via the intratentacular space (Fig. 6.18): the opposite system

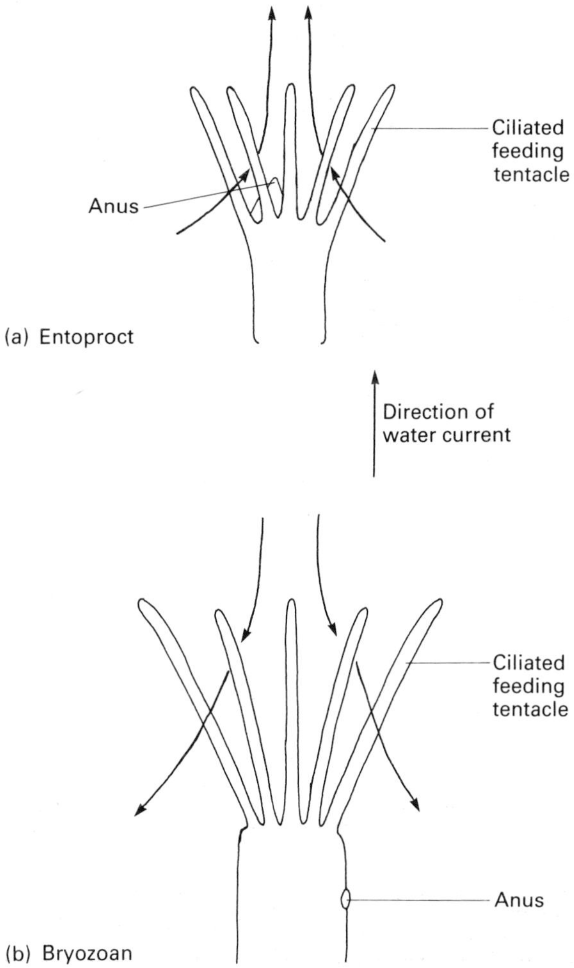

Fig. 6.18. Comparison of the feeding currents of entoprocts (a) and bryozoans (b) in relation to the position of the anus.

to that operated by the lophophorates. The anus, which is mounted on a cone, discharges into the central exhalent current.

Asexual multiplication by budding is widespread and can give rise to modular colonies; sexually, entoprocts are probably all hermaphrodite, apparently gonochoristic species being differently aged sequential hermaphrodites. Sperm are shed into the water, but fertilization is likely always to be internal. The larva is a planktotrophic or lecithotrophic trochophore (Fig. 6.19), which in most species metamorphoses into the adult; in some, however, the adult develops from a bud produced by the larval stage.

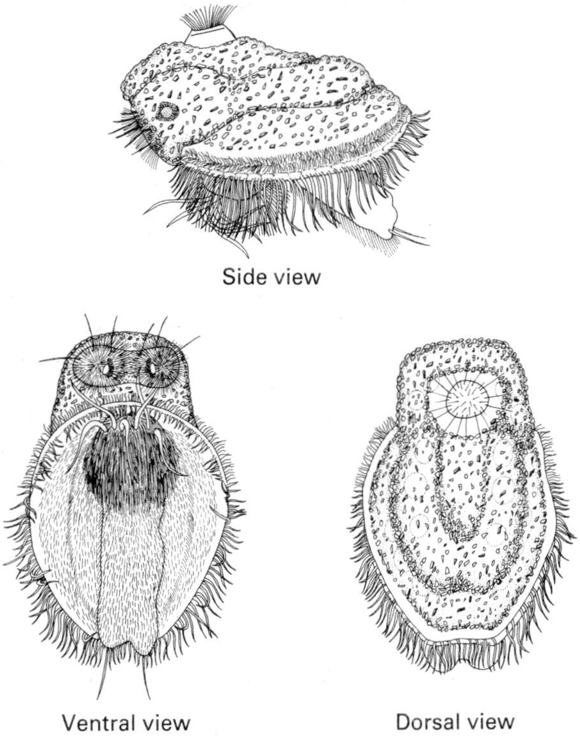

Fig. 6.19. An entoproct larva (after Nielsen, 1971).

The relationships of this small phylum are contentious. Some authorities interpret the extensive 'mesenchyme' tissue as a haemolymph-filled pseudocoelom and ally the entoprocts with the other pseudocoelomate phyla (Chapter 4); others regard them as being related to the bryozoans, because of some developmental similarities, the body cavity of the bryozoans also being of a contentious nature (Section 6.3.2).

6.4.3 Classification

The 150 described species are placed in a single class and order. All but one freshwater genus are marine.

6.5 Further reading

Emig, C.C. 1979. *British and Other Phoronids*. Academic Press, London [Phorona].

Hyman, L.H. 1951. *The Invertebrates (Vol. 3). Acanthocephala, Aschelminthes and Entoprocta*. McGraw-Hill, New York [Entoprocta].

Hyman, L.H. 1959. *The Invertebrates (Vol. 5). Smaller Coelomate Groups*. McGraw-Hill, New York [Phorona, Brachiopoda & Bryozoa].

Nielsen, C. 1971. Entoproct life-cycles and the entoproct/ectoproct relationship. *Ophelia* **9**, 209–341 [Entoprocta].

Rudwick, M.J.S. 1970. *Living and Fossil Brachiopods*. Hutchinson, London [Brachiopoda].

Ryland, J.S. 1970. *Bryozoans*. Hutchinson, London [Bryozoa].

Wright, A.D. 1979. Brachiopod radiation. In: House, M.R. (Ed.) *The Origin of Major Invertebrate Groups*, pp. 235–52. Academic Press, London [Brachiopoda].

7
THE DEUTEROSTOMES

That the deuterostomes are a natural group of related phyla is suggested by their common possession of a number of unusual developmental, structural and biochemical features, and of a series of trends unknown in other groups. During the early embryological development of most deuterostomes, for example, (a) the blastopore does not form the mouth, which is therefore a secondary opening into the gut (and hence the name 'deuterostome'), (b) cleavage of the cells of the blastula occurs in a radial pattern, (c) the developmental fate of the cells is not fixed until a relatively late stage of morphogenesis ('indeterminate development'), and (d) their body cavities are formed by outpocketings from the embryonic gut, creating a series of 'enterocoelic pouches' (see, for example, Fig. 2.9). These are typical of deuterostomes particularly in combination, since individually they (especially (b) and (d)) are also known in various other phyla. Further deuterostome characteristics include photoreceptors of the ciliary type (in contrast to the rhabdomeric photoreceptors of the protostomes), the prevalence of monociliated cells in the epidermis, creatine phosphate as the phosphate store (rather than the more usual arginine phosphate of the protostomes), and the virtual absence of chitin.

The ancestral deuterostome was probably not dissimilar to the modern pterobranch hemichordates (see Section 7.2.3.2), themselves not unlike phoronan lophophorates (see Section 6.1). That is they would have been short, sedentary, worm-like animals with tripartite oligomeric bodies, the second (mesosomal) region bearing a lophophore, and the third (metasomal) compartment the temporary aggregations of peritoneal cells which probably comprised the gonads; further, their nervous system would have largely been in the form of a diffuse subepidermal plexus concentrated into one or more longitudinal cord-like thickenings. In contrast to all known lophophorates, however, although similar to some gastrotrichs (Section 4.3.3), the water currents used to convey those food particles trapped by the lophophore into the anterior gut left the body via a number of lateral apertures extending from the pharynx right through to the body surface. The ancestral gut, therefore, was connected to the external environment by more orifices than just mouth and anus.

Most subsequent lines of deuterostome evolution lost or greatly modified the lophophore, lost the original tripartite body plan in association with the evolution of a sessile, attached existence (echinoderms and some chordates) or of a paedomorphic, free-swimming life style (most chordates), and two groups, at least, modified the pharyngeal perforations to serve alternative functions associated with pharyngeal filter feeding (early chordates) or with gaseous exchange (enteropneust hemichordates and many later chordates). Other trends within the deuterostomes not shown by protostome phyla include: enrolling the subepidermal nerve plexus to form a hollow, dorsal nerve tube (chordates and some hemichordates); developing a propulsive post-anal tail moved in S-shaped undulations by longitudinal muscles (chaetognaths and chordates); and the deposition of protective dermal calcareous plates, which in two separate lines subsequently became used as part of a hard internal skeletal system against which locomotory muscles could act (some echinoderms and some chordates).

7.1 Phylum CHAETOGNATHA (arrow-worms)

7.1.1 Etymology

Greek: *chaite*, long hair; *gnathos*, jaw.

7.1.2 Diagnostic and special features (Fig. 7.1)

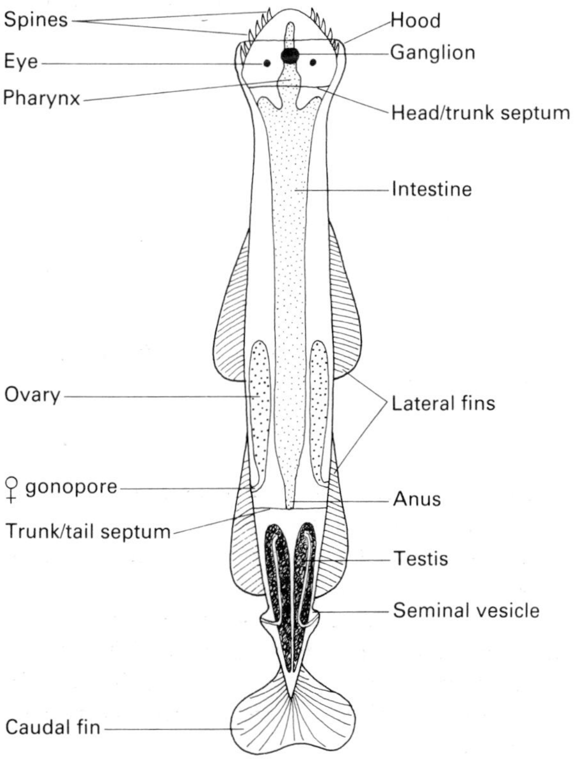

Fig. 7.1. Diagrammatic dorsal view of a chaetognath (after several sources).

1 Bilaterally symmetrical, vermiform.
2 Body more than two cell layers thick, with tissues and organs.
3 A straight, through gut and a non-terminal, ventral anus.
4 An oligomeric, tripartite body divided by septa into a head, trunk and post-anal tail; the trunk and tail bearing lateral and caudal non-muscular fins.
5 Each body region with one or two body cavities, which are enterocoelic in origin but in the post-juvenile stages lack a peritoneal lining.
6 Body wall with a non-chitinous cuticle and bands of longitudinal muscle.
7 Without circulatory, gaseous-exchange and excretory systems.
8 Nervous system with a circum-pharyngeal ring ganglionated dorsally and laterally, from which individual nerves issue.
9 Hermaphrodite, with large paired testes and ovaries, the latter with gonoducts.
10 Eggs cleave radially.
11 Development deuterostomatous and direct.
12 Marine.

Chaetognaths are torpedo-shaped carnivores which swim by means of rapid flicks of the post-anal tail in the dorsoventral plane, the lateral fins having a stabilizing function. Prey are grasped by series of movable, non-chitinous spines flanking and in front of the ventral chamber which leads into the mouth (Fig. 7.2), although, when not feeding, the head with

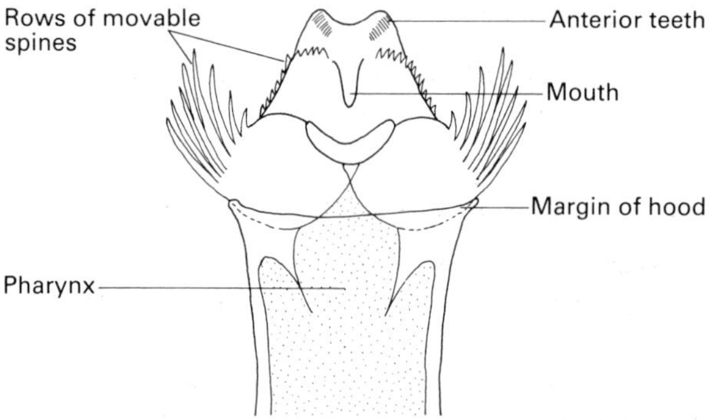

Fig. 7.2. Ventral view of the head of a chaetognath (after Ritter-Zahony, 1911).

its spines is covered by a fold of the body wall, the hood, which probably serves mainly to reduce drag as well as for protection.

A pair of eyes, formed by the fusion of individual ocelli, are present on the head. The photoreceptors in these eyes are of the ciliary type, as in the other groups of deuterostomes, but the chaetognaths are not typical deuterostomes. In particular, none of the adult body cavities possesses a peritoneum (otherwise a diagnostic feature of a coelom), and the tail enterocoels, of which there are one or two, are secondary derivations of the paired trunk cavities and therefore do not originate as separate pouches from the archenteron as in the other tripartite oligomeric groups. The head cavity is single.

Chaetognaths may attain lengths of up to 10 cm and form the dominant group of planktonic predators in the sea, preying on organisms ranging in size from protists to young fish as large as themselves.

7.1.3 Classification

The 70 known species are placed in a single class containing two orders on the basis of the presence (Phragmophora) or absence (Aphragmophora) of a transverse ventral musculature. The Phragmophora includes, *inter alia*, the only non-planktonic genus, the benthic *Spadella* (Fig. 7.3).

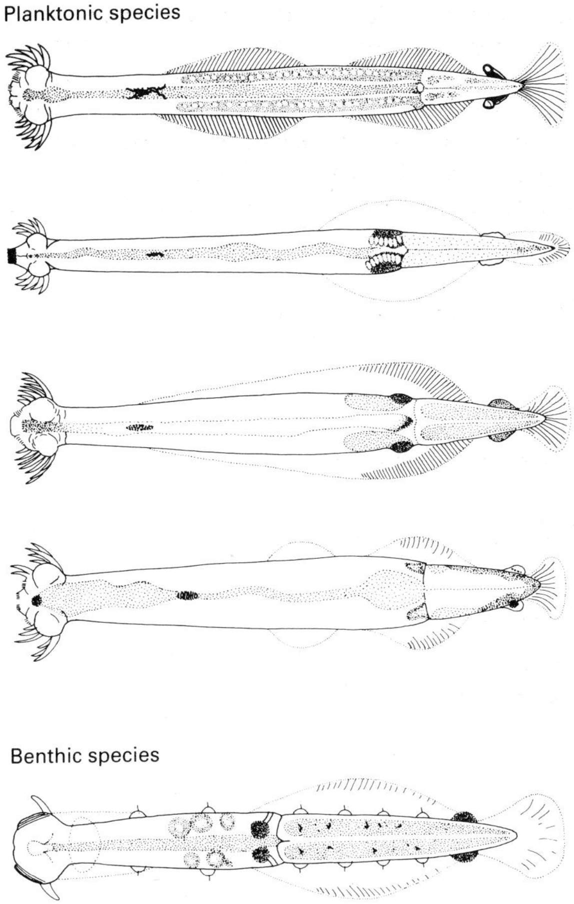

Fig. 7.3. The body form of chaetognaths (after Pierrot-Bults & Chidgey, 1987).

7.2 Phylum HEMICHORDATA

7.2.1 Etymology

Greek: *hemi*, half; Chordata, referring to the phylum of that name (Section 7.4).

7.2.2 Diagnostic and special features

1 Bilaterally symmetrical.
2 Body more than two cell layers thick, with tissues and organs.
3 A through, straight or U-shaped gut.
4 An oligomeric, tripartite body, comprising a large prosomal proboscis, a small mesosomal collar, and a large, elongate or sac-like metasomal trunk.
5 Each body region with one (proboscis) or two (collar and trunk) enterocoelic body cavities.
6 In some, a lophophoral organ on the collar is supported by the mesocoels.
7 In all but one group, the upper half of the pharyngeal wall is perforated by 1 to >100 pairs of slits through which a water current is discharged; only the lower half of the pharyngeal tube is then alimentary.
8 A ciliated, cuticle-less epidermis, in some secreting an external non-chitinous tube.

9 Circulatory system partially open.
10 Excretory organ a glomerulus formed by peritoneal evaginations into the protocoel.
11 Nervous system diffuse and basiepidermal, in some with a hollow dorsal neurocord in the collar; basiepidermal network concentrated mid-dorsally and mid-ventrally.
12 Gonochoristic.
13 Eggs cleave radially.
14 Development deuterostomatous and indirect.
15 Marine.

Hemichordates are of two main types, which differ considerably in their body forms and life styles: the vermiform, burrowing enteropneusts and the sessile, tubicolous and often colonial pterobranchs. Both, in their different ways, employ mucociliary feeding mechanisms, however. In the pterobranchs, a lophophoral organ is present dorsally on the collar, the arms of which are held so as to form a cone flaring away from the ventral mouth (in the lophophorate phyla and in all other groups feeding by means of a tentacular ring, the tentacles are held so as to surround and encircle the mouth), and water is induced to flow in through the circlet of tentacles and out through the intralophophoral cavity (as also seen in, for example, the sabellid polychaetes and the entoprocts). Particles then pass down the outside faces of the tentacular arms, by ciliary means, towards the mouth (Fig. 7.4). In addition, food particles may be collected by cilia over the general body surface. The latter is the sole method of ciliary feeding of the lophophore-less enteropneusts, the ciliary tracts being especially well developed on the proboscis (Fig. 7.5).

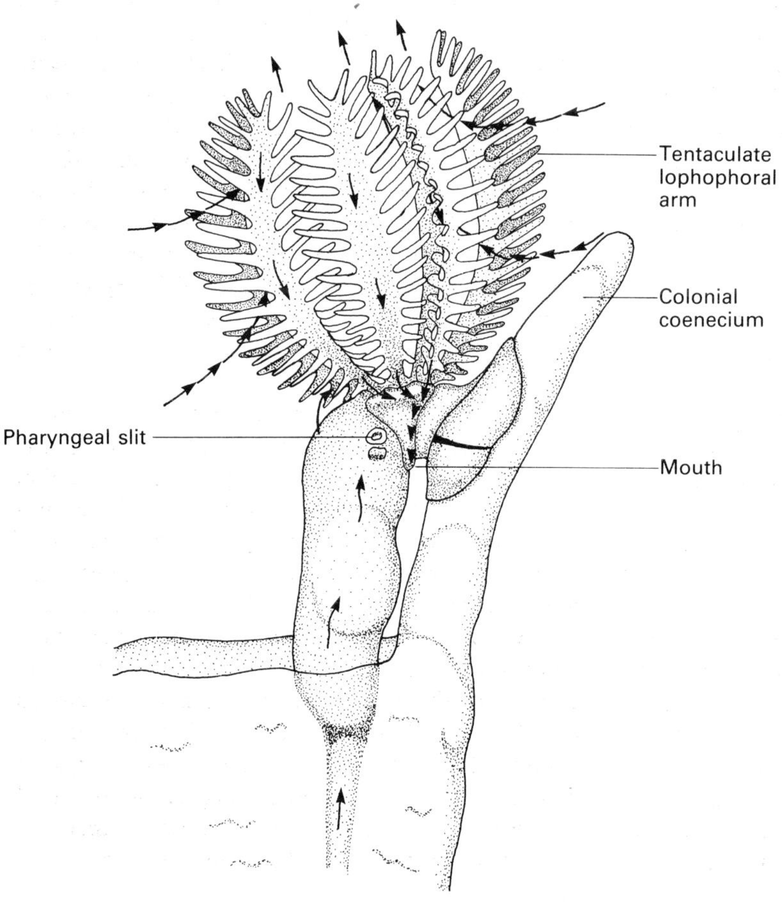

Fig. 7.4. Feeding currents in a pterobranch hemichordate (*Cephalodiscus*). Curved arrows indicate water moving towards lophophoral arms and into the rejection current; upwardly pointing arrows represent the rejection current in the centre of the lophophoral apparatus; arrows on the body show particle movements caused by epidermal cilia; and arrowheads show food moving towards mouth. (After Lester, 1985.)

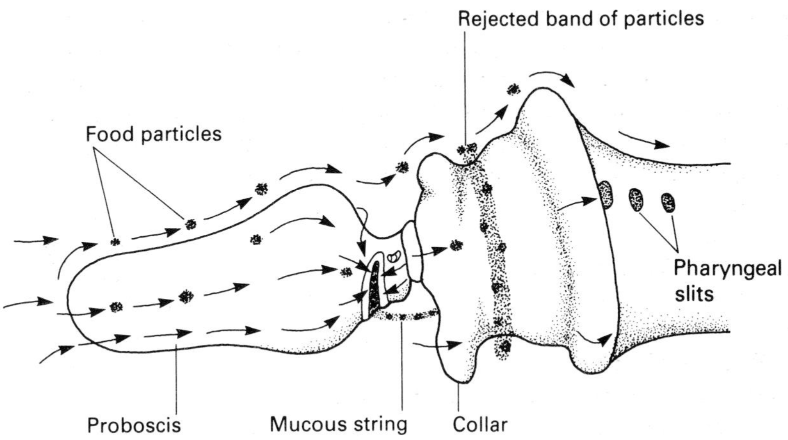

Fig. 7.5. Ciliary feeding currents on the proboscis and collar of an enteropneust hemichordate (after Barrington, 1965).

In both forms, ingestion of food particles is probably aided, as in other microphagous feeders, by a water current drawn into the buccal cavity, and, as in some gastrotrichs (Section 4.3.3), this water current is discharged through pharyngeal pores opening on the body surface. Such a unidirectional flow is a more efficient system than seen in the lophophorate phyla, for example, which periodically have to expel through the mouth the excess water taken in with the food particles. Although this was probably the ancestral function of the pharyngeal perforations (see Section 9.2.5) and remains so in the pterobranchs, in the enteropneusts (which can also ingest sediment particles and deposit feed), the water current has been developed to serve a gaseous-exchange function instead, and the number of pores has been greatly increased.

As their name suggests, hemichordates share a number of special features with the phylum Chordata. Chordates are generally considered to be characterized by four such special features: the presence of a notochord; a pharynx perforated to the exterior by slits or pores; a post-anal tail and a hollow dorsal nerve tube (Section 7.4). Of these, at least the enteropneusts share two (the hollow dorsal nerve tube and the perforated pharynx), and arguably a third in that the young of some species have been described as possessing a post-anal tail. Nevertheless, as in the chaetognaths, the chordate tail is essentially a locomotory structure, which is not the case with any post-anal hemichordate body region (of course, many invertebrates, particularly those with U-shaped guts, possess post-anal body regions, and the juvenile enteropneust 'tail' is probably homologous with the pterobranch stalk). A notochord is definitely lacking in the hemichordates, however, although a forwardly directed diverticulum of the gut, issuing from the buccal cavity, was for many years misinterpreted as a notochord. Some hemichordates do then possess half of the special chordate features, and would thus appear to be aptly named.

As will have been evident above, the pterobranchs in particular also share a number of features with the lophophorates, especially the phoronans (Section 6.1), including their mesosomal lophophore supported by mesocoelic pressure, a tripartite system of enterocoelic body cavities, a diffuse basiepidermal nerve plexus, etc. The only major difference between the two are the deuterostomatous development of the hemichordates and the paired nature of their meso- and metacoels. The two main hemichordate groups therefore provide an anatomical link between the protostome lophophorates and the other groups of deuterostomes. It could even be argued—and indeed has been—that the pterobranchs, rather than the phoronans, are ancestral to both phylogenetic lines.

7.2.3 Classification

The 100 known species of hemichordate are divided between three classes.

Class	Order
Enteropneusta	Helminthomorpha
Pterobranchia	Rhabdopleurida
	Cephalodiscida
Planctosphaeroidea	Planctosphaerida

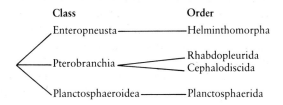

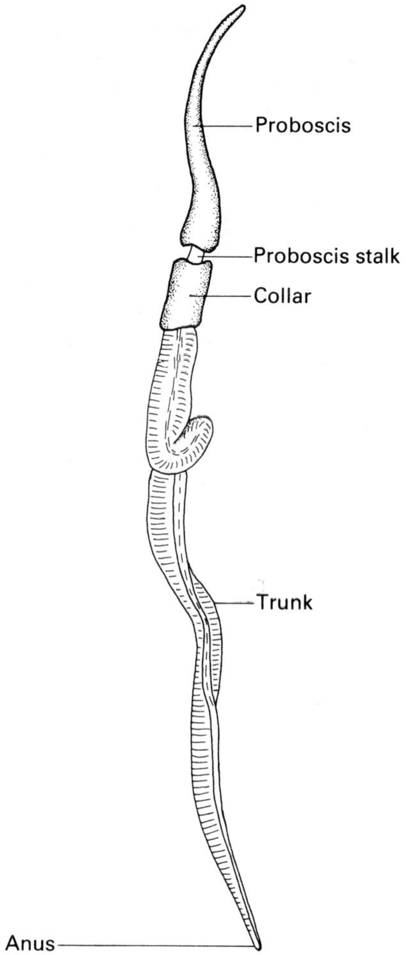

Fig. 7.6. The external appearance of an enteropneust (after Marion, 1886).

7.2.3.1 Class Enteropneusta (acorn-worms)

These solitary, mobile, elongate worms, of up to 2.5 m in length, occupy burrows in soft sediments, live under stones, etc., where they deposit- and suspension feed. Their bodies are in the form of (a) a long proboscis, the cilia on which provide both the main propulsive force and the transport system for food particles, (b) a short, lophophore-less collar, and (c) a very long trunk bearing a terminal anus and many pharyngeal pores (Fig. 7.6), the number increasing throughout life. In the pharyngeal wall, these perforations are U-shaped, with a supporting system of skeletal bars, but they open on the body surface as small, circular dorsal pores. The neurocord and glomerulus are well developed in enteropneusts, but the body cavities are largely occluded by muscle fibres and connective tissue formed from the peritoneum, which replace much of the body wall musculature.

Both asexual multiplication by fragmentation and sexual reproduction occur, many pairs of gonads, each discharging by means of a duct, being located in the anterior region of the trunk. Fertilization is external and proceeds through developmental stages closely similar to those of many echinoderms to, in most species, a tornaria larva (Fig. 7.7).

All 70 enteropneust species are placed in the one order.

7.2.3.2 Class Pterobranchia

The pterobranch hemichordates are sedentary, tube-dwelling animals, of up to 12 mm in length, with a short body comprising (a) a shield- or disc-shaped proboscis, which is responsible both for the (ciliary) locomotion within the tube and for secreting this collagenous structure, (b) a collar bearing ventrally the mouth and dorsally from one to nine pairs of lophophoral arms from each of which issues a double row of ciliated and mucus-secreting tentacles, and

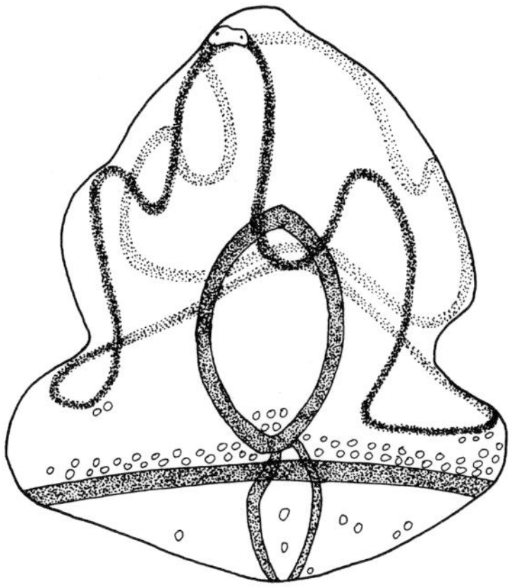

Fig. 7.7. The tornaria larva of an enteropneust (after Stiasny, 1914).

(c) a short sac-like trunk with, near the collar, an anal papilla, and, subterminally, a contractile stalk (Fig. 7.8). This stalk may end in an organ of temporary attachment, may pass into a communal stolon, or be wrapped around a support, like the prehensile tail of many an arboreal mammal. In contrast to the enteropneusts, at most only one pair of pharyngeal perforations are present, the neurocord is absent and the glomerulus is poorly developed, the coeloms are not occluded, and only one or a pair of gonads occur.

Asexual budding is widespread, leading in one group to colonies of individuals joined by their stolonic stalks and, in the other, to clones of separate individuals living within a common 'coenecium' (Fig. 7.9). Little is known of their sexual reproduction,

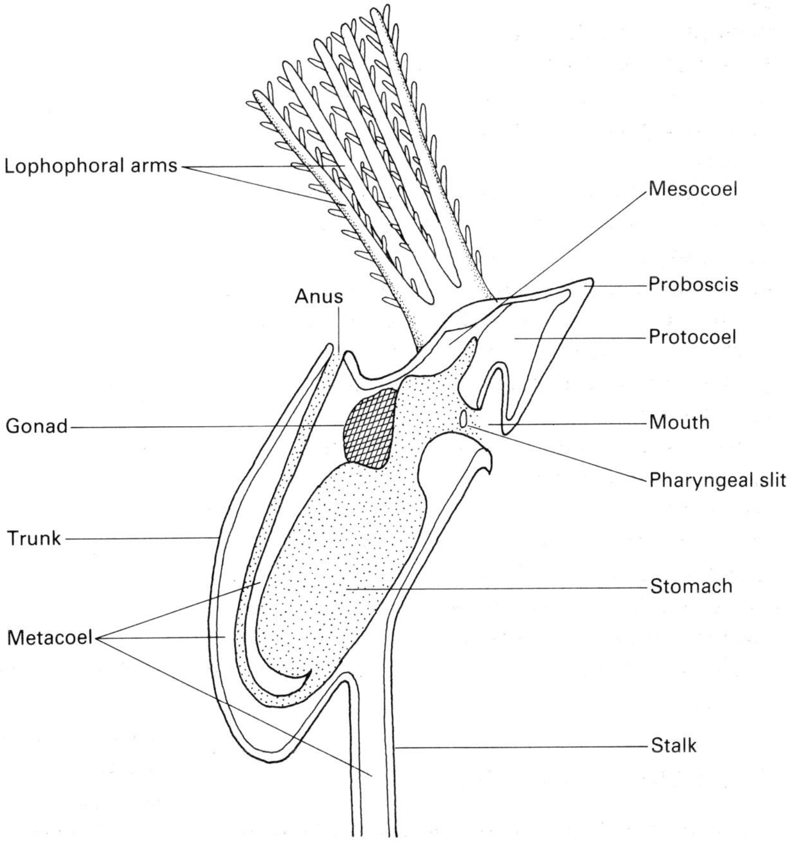

Fig. 7.8. Diagrammatic longitudinal section through a pterobranch (after McFarland *et al.*, 1979 and others).

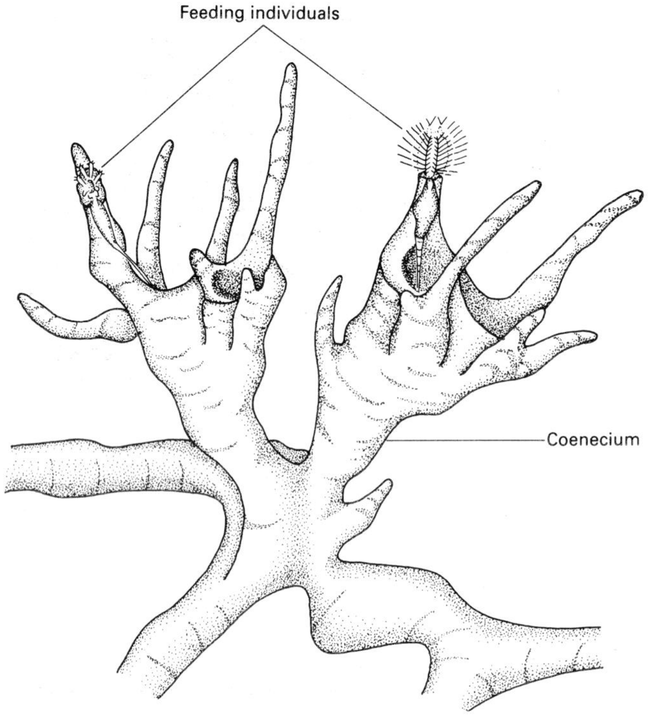

Fig. 7.9. Individual pterobranchs (*Cephalodiscus*) feeding whilst clinging to spines projecting from the common coenecium (after Lester, 1985).

but the larval stage (Fig. 7.10) does not resemble that of the enteropneusts.

The 10 species of pterobranchs are divided between two orders. The Cephalodiscida contains the non-colonial forms which either creep over the surfaces of hydroid colonies or live within a communal coenecium, climbing up to feed near its apertures (Fig. 7.9). They possess one pair of pharyngeal perforations, from four to nine pairs of lophophoral arms, and a pair of gonads. The Rhabdopleurida are the colonial pterobranchs (Fig. 7.11): they have a single gonad, a single pair of lophophoral arms, and no pharyngeal slits.

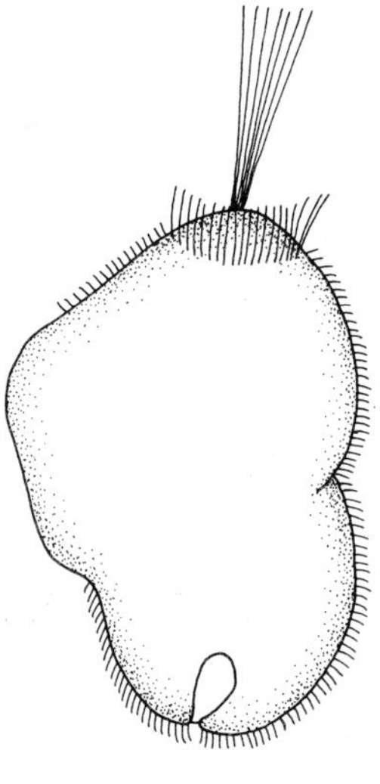

Fig. 7.10. The trochophore-like larva of the pterobranchs (after Schepotieff, 1909).

7.2.3.3 Class Planctosphaeroidea

This class was erected to receive, hopefully only temporarily, some giant larvae (up to 2.2 cm in diameter) resembling in their general form those of

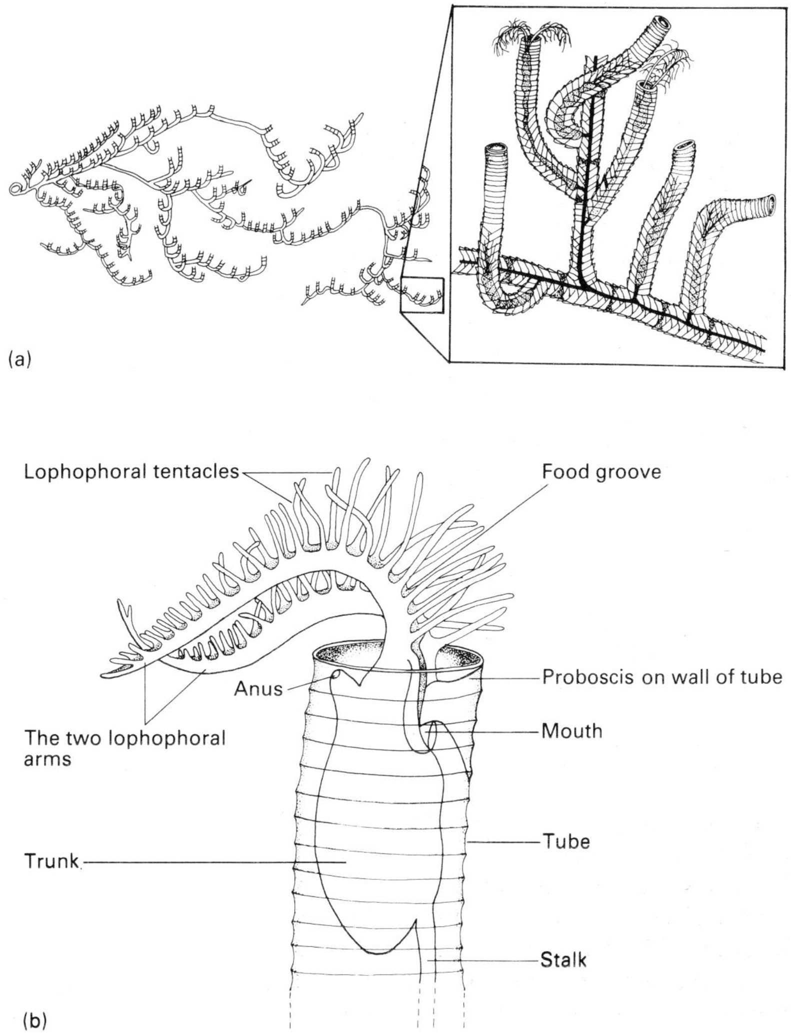

Fig. 7.11. Part of a rhabdopleuran colony (a), with enlarged detail of an individual zooid (b) (after Grassé, 1948 and others).

enteropneusts (Fig. 7.12), which have been taken by plankton tows in the Atlantic Ocean since the early 1930s. As yet the adult form is still unknown, and their status and affinities therefore remain uncertain.

7.3 Phylum ECHINODERMATA

7.3.1 Etymology

Greek: *echinos*, hedgehog; *derma*, skin.

7.3.2 Diagnostic and special features (Fig. 7.13)

1 Adult with five-rayed symmetry, effectively radial in most, bilateral in some.
2 Body more than two cell layers thick, with tissues and organs.
3 A through gut (secondarily blind-ending in some, lacking in one group).
4 Body shape highly variable: ancestral and sessile forms with spherical or cup-shaped body attached

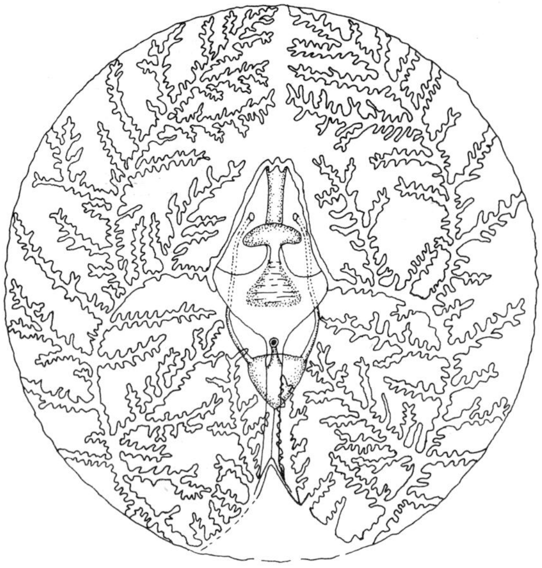

Fig. 7.12. Ventral view of *Planctosphaera* (after Spengel, 1932).

aborally to the substratum by a stalk and with the upwardly directed mouth surrounded by a circle of five arms; derived free-living forms, without stalk, with body oriented so that the mouth is on the lower surface (or, more rarely, is anterior), with or without discrete arms, sometimes bilaterally symmetrical.

5 Without any differentiated head.

6 Basically with oligomeric, tripartite paired enterocoelic body cavities, the metacoels ('somatocoels') forming the main body cavity; in a few forms with direct development, the body cavities are formed schizocoelically.

7 A water vascular system derived mainly from the left mesocoel (left 'hydrocoel') and partly from the left protocoel (left 'axocoel'), which contains and is in indirect communication with sea water and which operates hydraulically the locomotory tube feet and/or feeding tentacles ('podia'); part of the system forms a ring canal encircling the oesophagus, from which issue an ambulacral canal along each arm.

8 A mesodermal, subepidermal system of calcareous ossicles or plates, often bearing projecting tubercles or spines.

9 Without excretory organs.

10 Circulatory 'haemal' system poorly defined, its function largely being accomplished by coelomic fluids.

11 Nervous system subepidermal, in the form of a circum-oesophageal ring from which issue diffuse nerves along each ambulacrum.

12 Usually gonochoristic.

13 Eggs cleave radially.

14 Development deuterostomatous, characteristically indirect via ciliated, bilaterally symmetrical larvae (Fig. 7.14).

15 Marine.

The echinoderms are a highly individual group, with three very distinctive features: their symmetry, calcareous mesodermal skeleton, and water vascular and other coelomic systems. All three of these can be related directly to their sessile, suspension-feeding origins. Sessile suspension feeders of whatever ancestry tend towards radial symmetry of their food collection apparatus (Section 9.1), and in the echinoderms this takes the form of a circle of five arms (often dichotomously branching), which in the ancestral forms were equivalent to, and perhaps derived from, the lophophorate arms of animals such as the pterobranch hemichordates (Section 7.2.3.2). Like the latter, the hydraulic system of the arms and of the ciliated tentacles borne on them is provided (mainly) by mesocoelic pressure. A number of echinoderm lines, however, have secondarily become free-living, and in most of these the position of the body axis in life has been turned through 180° so that the erstwhile upper mouth- and arm-bearing surface is in contact with the substratum. The former feeding podia now function in locomotion, with relatively little anatomical change being developed into tube feet (one common adaptation to this changed function being the development of a hydraulic reservoir, the ampulla, for each tube foot, permitting them individually to dilate and contract). The arms have also often become much larger and broader (as in the asteroids) or have been incorporated into—wrapped around—the body (as in the echinoids and holothu-

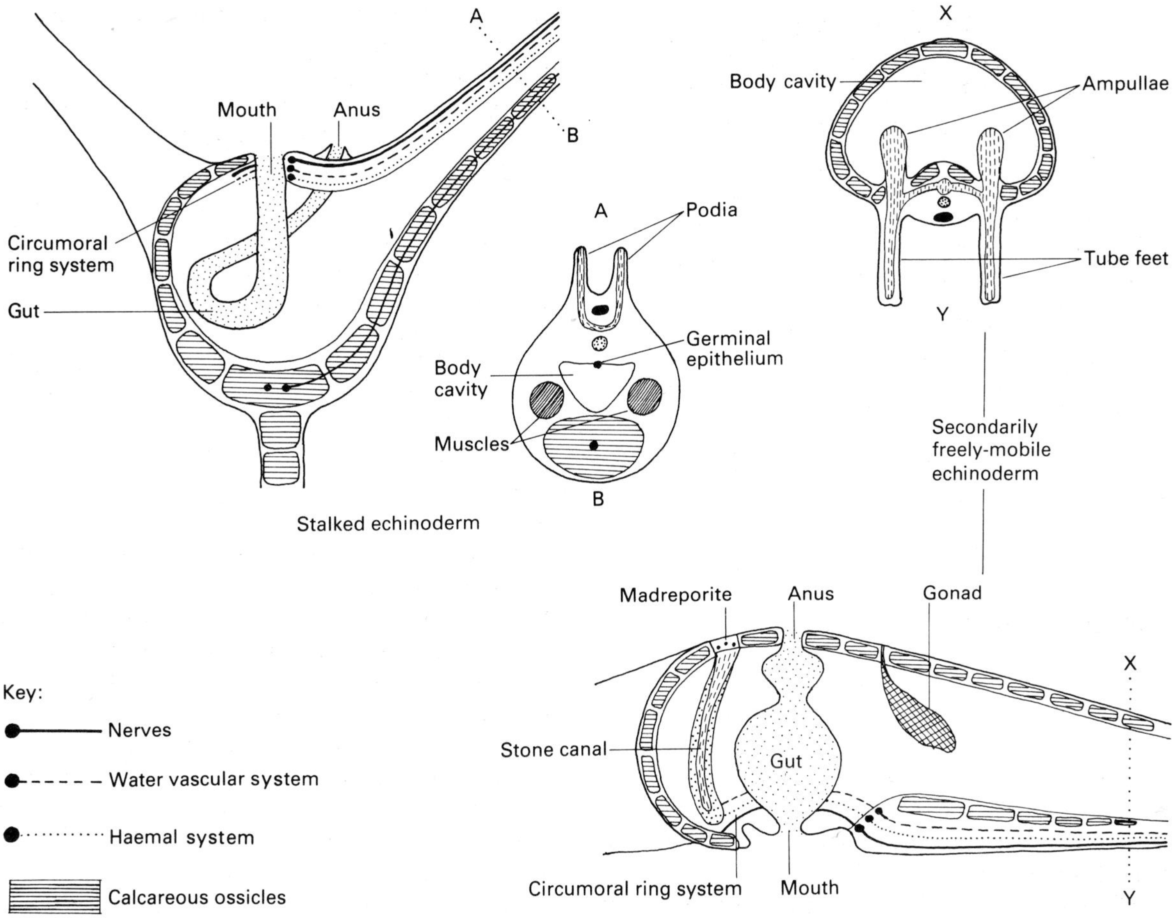

Fig. 7.13. Diagrammatic sections through two generalized echinoderms (after Nichols, 1962).

rians). In any event, the five arms still dominate the body plan, so that in the majority of free-living forms the ancestral, effectively radial symmetry has been retained and consequently the animals can move in any direction. (Not all the early types of echinoderm displayed the pentaradial symmetry characteristic of all surviving groups: some were helicoidally symmetrical, and others showed no basic pattern of symmetry at all—see Fig. 2.14.)

Sessile animals also require means of support and protection from predators and wave action. The echinoderms have achieved these by depositing a system of ossicles in their dermis, each ossicle being a porous lattice of calcium carbonate which behaves as if it was a single crystal of calcite. Individual ossicles are often increased in size by accretion to form plates which may interlock with other such plates, even to the extent of forming a rigid almost external box or test, and in common with other such enclosed animals (e.g. bivalve molluscs and brachiopods) any ancestral head has been lost. The overlying, generally ciliated epidermis may be missing and parts

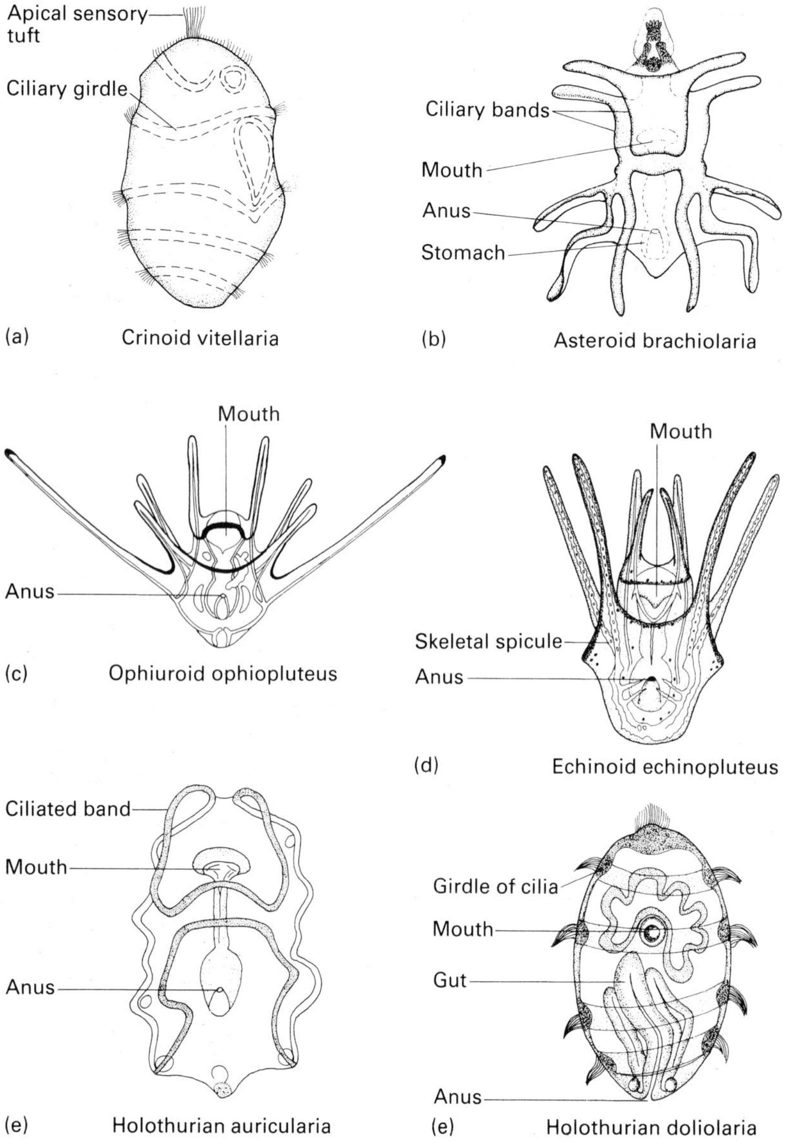

Fig. 7.14. Echinoderm larvae (after Barnes, 1980).

of the test thereby become the outer margin of the body. Further, the calcareous ossicles may secondarily assume a role in locomotion, replacing the tube feet in this respect, with muscles running between different ossicles effecting movement of the arms or of spines against the substratum (see Chapter 10).

Although the body of an echinoderm is monomeric, three pairs of enterocoelic pouches develop (sometimes in modified form), as in the oligomeric, tripartite groups to which echinoderms are generally regarded as being related (see Section 15.4.1). And, as in the lophophorates and hemichordates, the metacoels ('somatocoels') form the main body cavity, and (in the echinoderms, *one* of the) mesocoels ('hydrocoels') provide the hydraulic system of the lophophore-like arms; the right mesocoel normally atrophies to a small pulsatile sac. The right protocoel ('axocoel') forms the axial sinus, a space around part of the haemal system of debatable function; whilst the left one is incorporated into the mesocoelic water vascular system. The protocoels and mesocoels of the oligomeric enterocoelic animals communicate with the external environment each by means of a small pore, and this is classically regarded as having been greatly developed by the echinoderms as a means of varying the quantity of liquid in the water vascular system. The common axohydropore is in the form of a porous plate, the madreporite, through which it is suggested water is exchanged with the environment (effectively the water vascular system contains sea water with a few coelomocytes, rather than coelomic fluid), a 'stone canal' running from that plate to the circum-oesophageal ring. Actual passage of water across the madreporite has never been observed, however, and it may serve instead to equalize hydrostatic pressure inside and outside the animal.

The peculiarities of the echinoderms are therefore not so much in their anatomy *per se*, but result from the fact that they are the only successful free-living animals to have descended from an attached, sessile group probably capable of moving only their tentaculate arms and of bending their stalk.

7.3.3 Classification

Although some 6250 living species are known, the diversity of echinoderms is much lower now than it was in the Palaeozoic: only six of the 24 component classes survive.

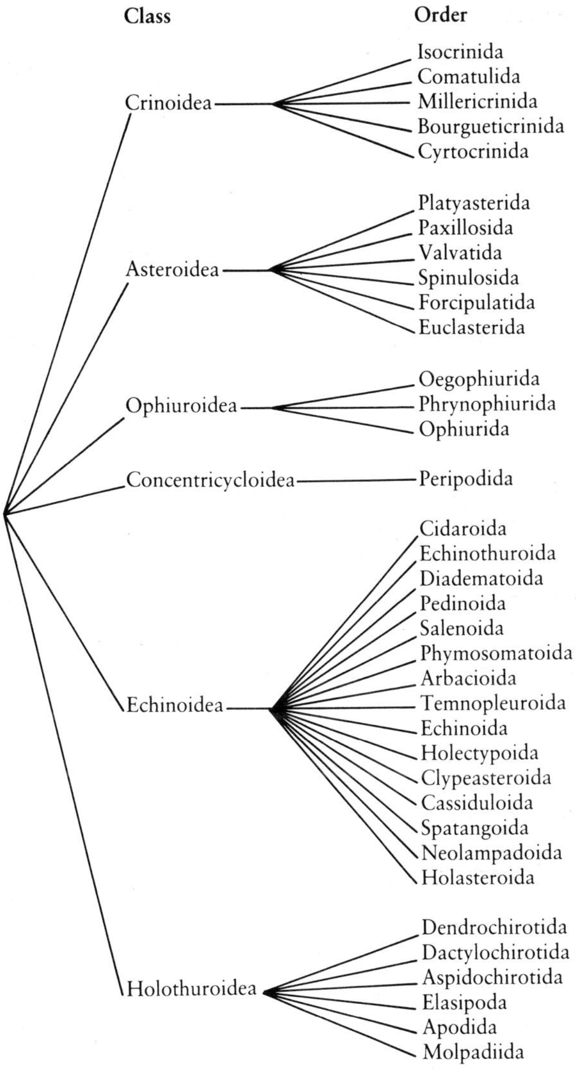

7.3.3.1 Class Crinoidea (feather-stars and sea-lilies)

Crinoids are the only surviving echinoderms to have retained the ancestral body posture of an upwardly directed mouth in the centre of a circle of mucociliary, suspension-feeding, podia-bearing arms. Although basically five arms are present, they may divide repeatedly to form from ten to more than 200 apparent arms, each with numerous side branches (pinnules) on to which the upwardly facing ambulacral grooves extend (Fig. 7.15). Particles are trapped by the ciliated and mucus-secreting podia and conveyed along ciliated ambulacral grooves to the mouth, and thence into the U-shaped gut (Fig. 7.16). The water vascular and other coelomic systems are simple, there being no ampullae to operate the podia and no madreporite, the often numerous stone canals

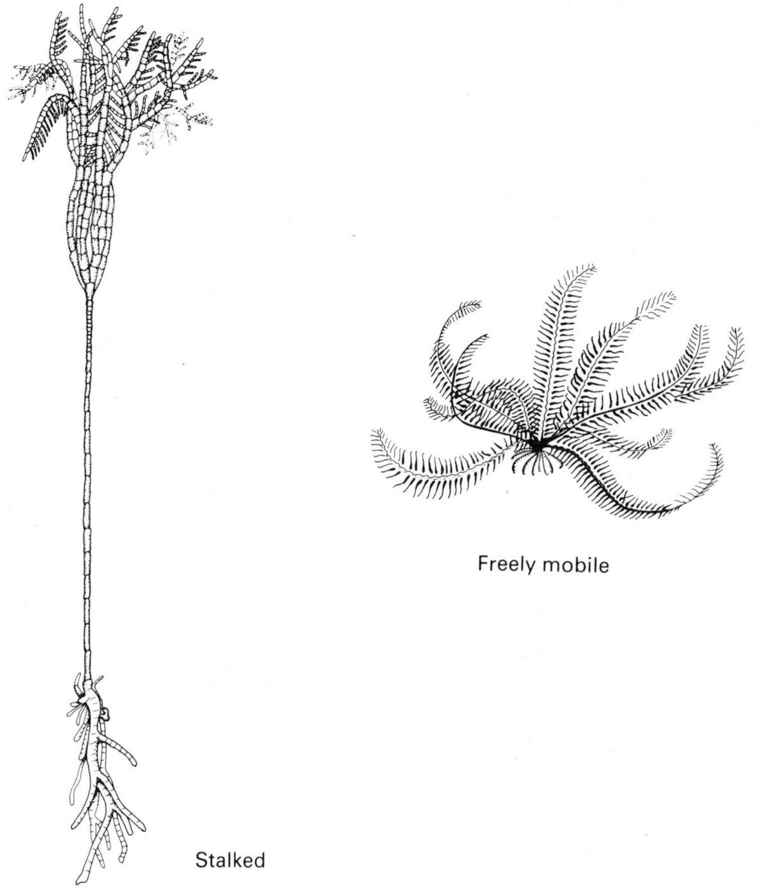

Fig. 7.15. Crinoid body forms (after Danielsson, 1892 and Carpenter, 1866).

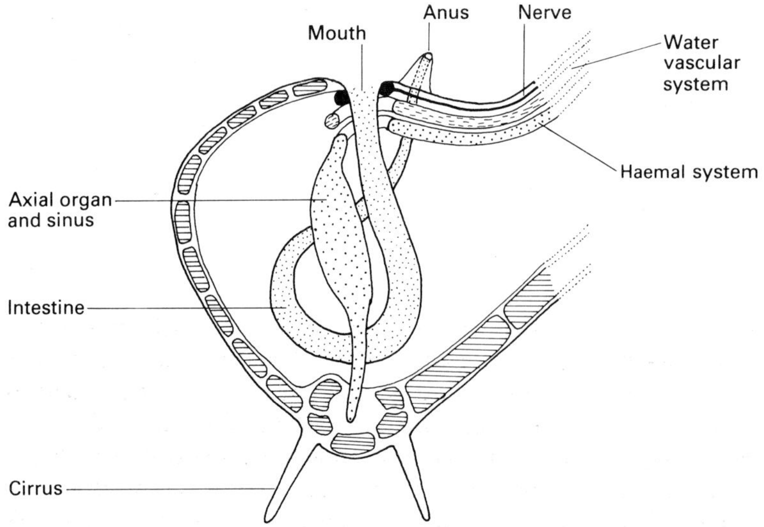

Fig. 7.16. Diagrammatic section through the body and part of an arm of a free-living crinoid (after Nichols, 1962).

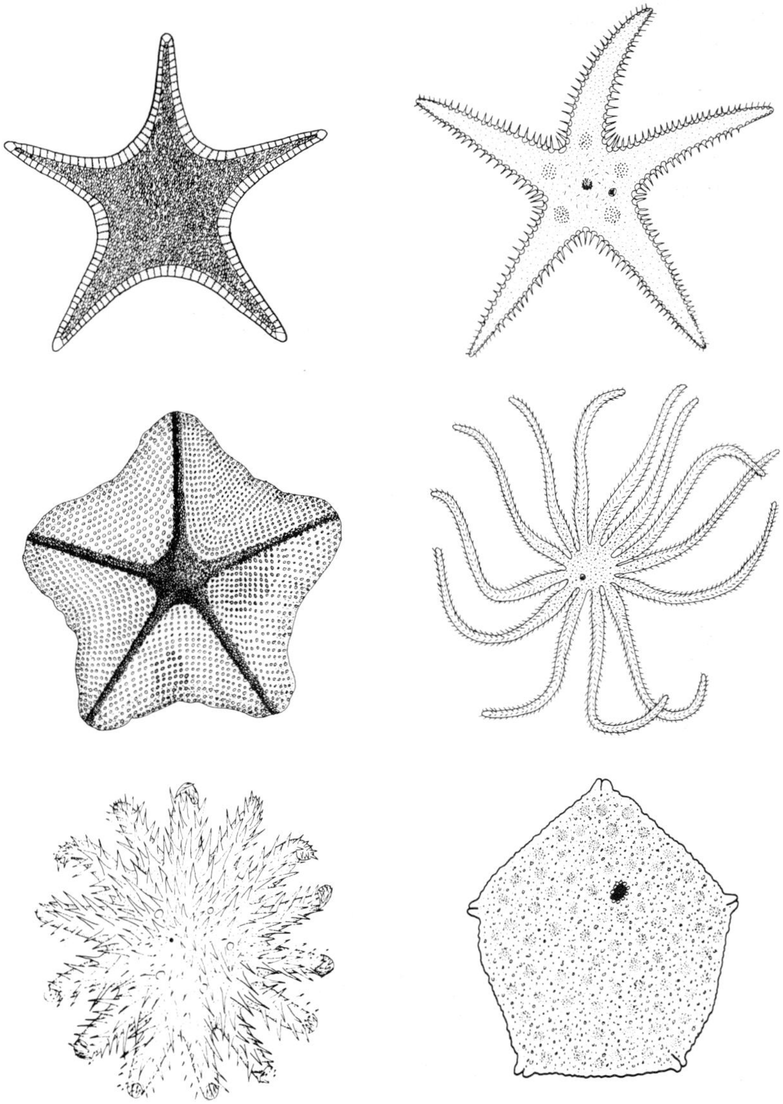

Fig. 7.17. Asteroid body form (after sources in Hyman, 1955).

opening into the somatocoels. (The podia are extended by contraction of the sphinctered water vascular canals.)

The globose body is permanently or temporarily attached to the substratum by a non-contractile aboral stalk (which can attain a length of 1 m), often bearing whorls of cirri and terminating in an attachment disc or in a system of cirri or rootlets which can grasp or anchor it in the sea bed. In many living species, after a brief period of attachment, the animal becomes free-living, swimming by use of its feather-like arms and temporarily attaching itself to the substratum by aboral cirri: the ancestral, mouth-upwards posture is nevertheless maintained. The arms, body and stalk possess heavily calcified plates, which take up most of the body volume so that the tissue space is small.

Gonadal tissue is diffuse; gametes form from areas of peritoneum within the arms, develop in small proximal 'gonads', and are released by rupture of the pinnule walls. A vitellaria larva (Fig. 7.14a) is characteristic; this metamorphoses into a miniature stalked version of the adult.

Five orders are distinguished, mainly on details of the construction of the stalk and of its system of attachment; they contain a total of 625 species.

7.3.3.2 Class Asteroidea (starfish)

The asteroids are generally characterized by a flattened body which grades imperceptibly into the five, or sometimes more (up to 40), arms. These may be short and broad, giving the animal a pentagonal outline, through to very long and slender (Fig. 7.17). The animals are unattached and freely mobile, the mouth and ambulacral grooves being positioned on the lower surface and the madreporite and anus (if present) aborally. The ambulacral grooves are well developed and protected along each margin by a double row of unfused ossicles, through notches in which pass ducts from the internal ampullae to the locomotory tube feet, which are often suckered. Otherwise, the calcareous skeleton is somewhat loosely organized, although it bears external tubercles and spines, sometimes in definite arrangements. Some of the spines may be modified into pedicellariae (Fig. 7.18): groups of usually three small ossicles which can interact in the manner of scissors or forceps, and thereby remove other organisms attempting to settle on its body surface.

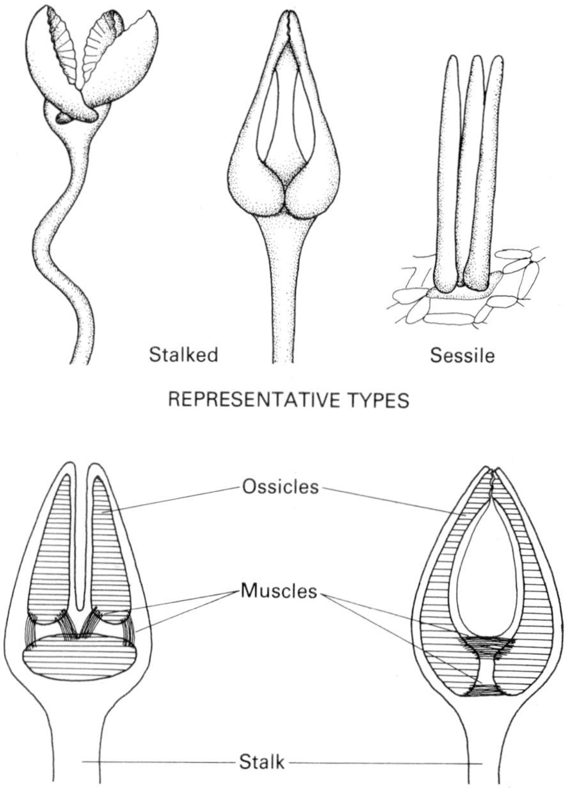

Fig. 7.18. Pedicellariae (after Mortensen, 1928–51 and Hyman, 1955).

The body cavities, including the water vascular system, are extensive, and include small extensions of the somatocoels, papulae, which project through the body wall aborally and serve a gaseous-exchange function (Fig. 7.19). Some asteroids are hermaphrodite and several forms of multiplication and development occur, including asexual fission, brooding and direct development (especially in high latitudes), and indirect development via bipinnaria and brachiolaria larvae (see Fig. 7.14b). Most species are scavengers or predators on sessile or sedentary prey, but deposit and suspension feeders are also represented. In macrophagous feeding, the prey may be ingested whole or the stomach, which is very large, everted through the mouth on to the prey tissues; from the stomach issue a pair of large caeca into each arm.

Six orders, containing 1500 species, are recognized on the basis of the type of pedicellariae present, the arrangement of the ambulacral ossicles, and the form of the tube feet. On its discovery in 1962, the genus *Platasterias* was hailed as a surviving somasteroid, an

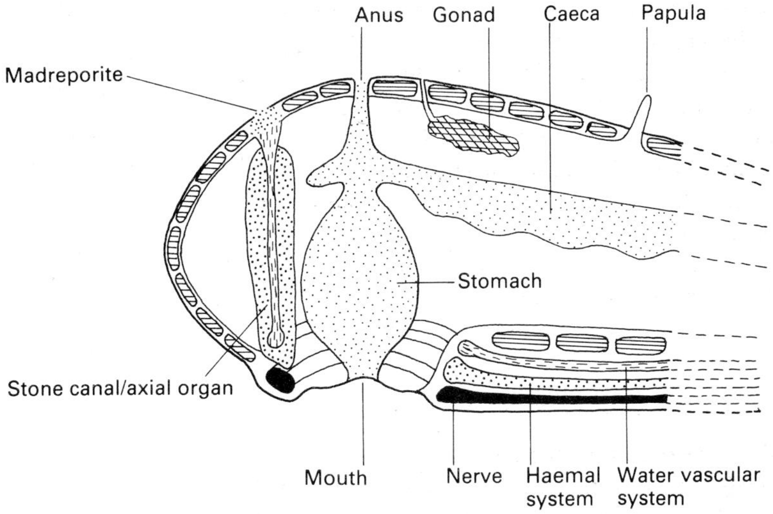

Fig. 7.19. Diagrammatic section through the body and part of an arm of an asteroid (after Nichols, 1962).

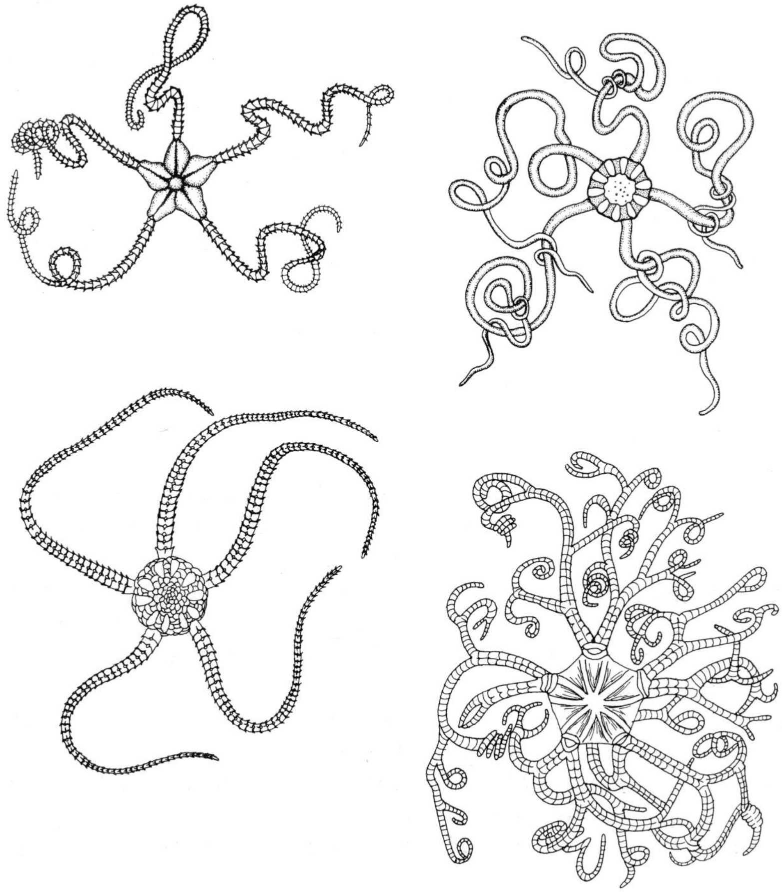

Fig. 7.20. Ophiuroid body forms (after sources in Hyman, 1955).

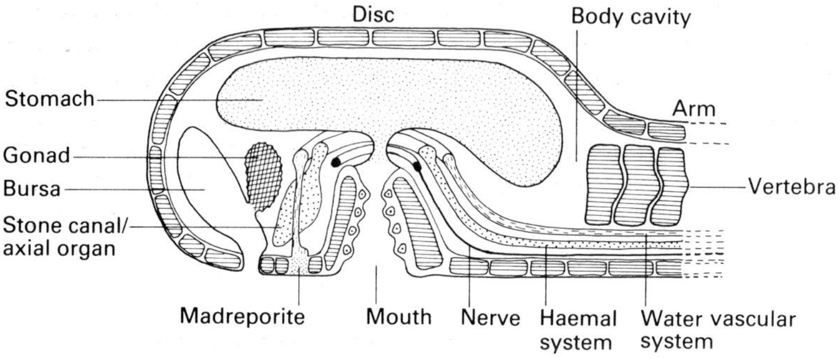

Fig. 7.21. Diagrammatic section through the disc and part of the arm of an ophiuroid (after Nichols, 1962).

echinoderm class otherwise long extinct. The consensus view today, however, is that it is not a somasteroid but a specialized member of the asteroid order Platyasterida.

7.3.3.3 Class Ophiuroidea (brittlestars and basket-stars)

Like the asteroids, to which they are related, the ophiuroids are flat, free-living echinoderms with a mouth located on the lower surface of the body; but in this group, the central body is small, disc-shaped and sharply demarcated from the long, narrow arms (Fig. 7.20). The calcareous skeleton occupies most of the arm volume (and is similarly well developed in the central disc), individual arm ossicles being fused together to form longitudinal series of 'vertebrae' which articulate with each other and which can be moved by intervertebral muscles. By these means, ophiuroids walk on their arms, the central disc often being held off the ground; accordingly, the tube feet have no locomotory function, except to help the arms gain purchase on the substratum, and are used mainly for feeding, the ambulacral grooves being enclosed within the skeletal plates. In contrast to the asteroids, the madreporite is on the oral surface.

Many ophiuroids are mucociliary suspension feeders, in several of them the five basic arms repeatedly bifurcating, but deposit feeding and omnivorous scavenging are also common; all possess a large stomach but lack an intestine and an anus. Between the stomach and the oral surface are positioned ten invaginations of that lower surface, the bursae (Fig. 7.21), which function as the specialist organs of gaseous exchange, water being drawn in through slits around the mouth by ciliary beating or

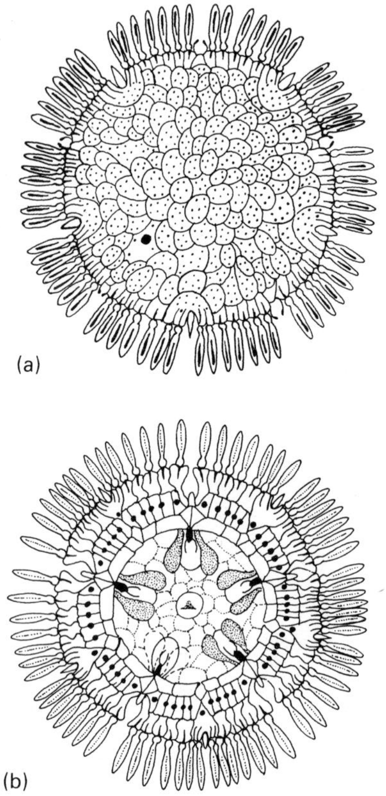

Fig. 7.22. Dorsal (a) and ventral (b) views of the concentricycloid *Xyloplax* (after Baker *et al.*, 1986).

by muscular pumping. Into these bursae discharge the gonads, and eggs are often brooded there; in some species, the embryo is actually attached to the bursal wall and develops viviparously. Development may be direct or via an ophiopluteus larva (see Fig. 7.14c).

Three orders are distinguished, with a total of 2000 species.

7.3.3.4 Class Concentricycloidea

In 1986, a new class of bizarre, medusa-like echinoderms was described from nine specimens of a single species found at some 1000 m depth off the coast of New Zealand on waterlogged wood. The 8 mm diameter body of *Xyloplax medusiformis* is in the shape of a flat disc surrounded by a circlet of spines (Fig. 7.22); it lacks arms and any indication of a gut. The upper surface of the disc is covered by scale-like plates, and the lower surface bears two concentric water-vascular ring canals from which issues a single ring of marginal tube feet. Five pairs of gonads are present, and the eggs appear to be brooded within bursae, in which they develop directly into young adults.

7.3.3.5 Class Echinoidea (sea-urchins, sand-dollars, etc.)

Echinoids are spherical or secondarily flattened, free-living echinoderms which lack arms (Fig. 7.23) (in a sense, the arms have been incorporated into the body). They are characterized by the development

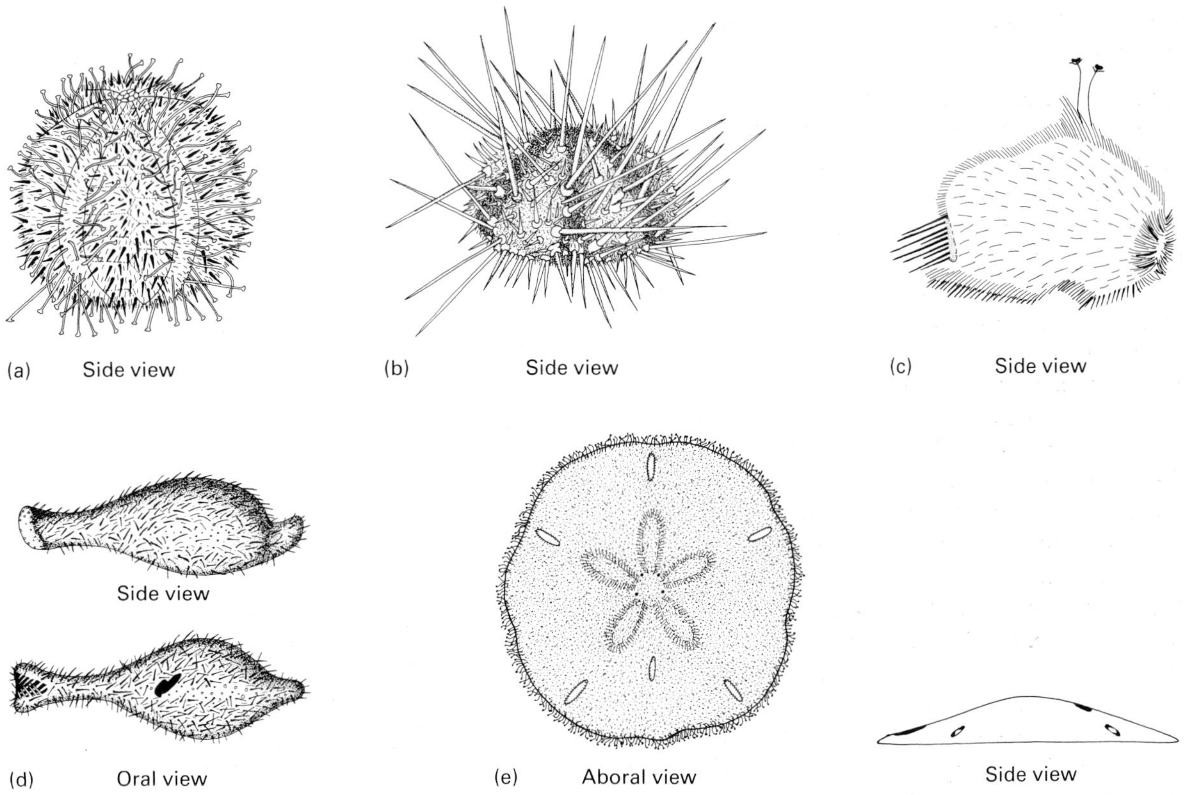

Fig. 7.23. Echinoid body forms: (a) and (b) are 'regular'; (c)–(e) 'irregular' (after several sources, principally in Hyman, 1955).

of their calcareous ossicles into a rigid test of attached plates, each ambulacral and interambulacral zone possessing essentially two vertical columns of such plates which curve around the body aborally from the mouth. This test bears pedicellariae (see Section 7.3.3.2) and movable spines on or with which some species can walk or burrow; otherwise locomotion is by means of tube feet that emerge through pores in the ambulacral plates.

The mouth, which is on the lower surface, bears a distinctive grazing apparatus of five large and several small plates, Aristotle's lantern (Fig. 7.24), with which algae and sedentary or sessile animal prey can be chewed, the lantern being protracted through the mouth for this purpose. In the epifaunal 'regular' species, the mouth is in the centre of the lower surface and the anus is mid-aboral. A number of species burrow into soft sediments and have become secondarily bilaterally symmetrical and often greatly flattened (these comprise the 'irregular' species); in these, the anus is displaced markedly towards the 'posterior' end and the mouth may move somewhat 'anteriorly'. Their diet is detrital, in association with which the lantern is often reduced or absent and feeding is effected by specialized, mucociliary tube feet.

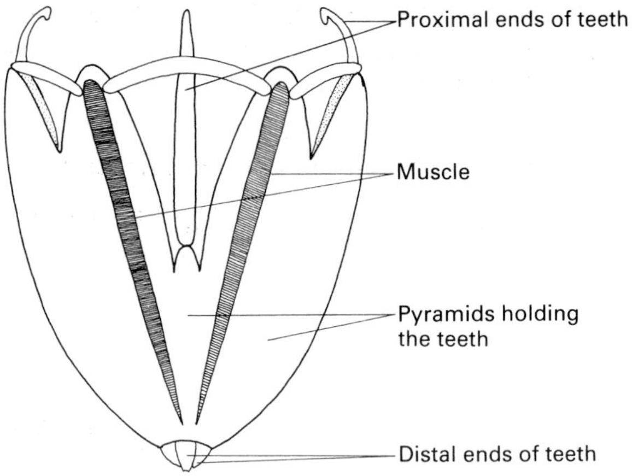

Fig. 7.24. Aristotle's lantern (after Hyman, 1955).

The body cavity is large (Fig. 7.25) and bears extensions in the form of yet further modified tube feet (in irregular species) or peristomial gills (in regular forms) for gaseous exchange. The four (irregular) or five (regular) large gonads also project into the body cavity, discharging via aboral gonopores. Brooding occurs in some; in most, however, development is indirect, via an echinopluteus larva (Fig. 7.14d).

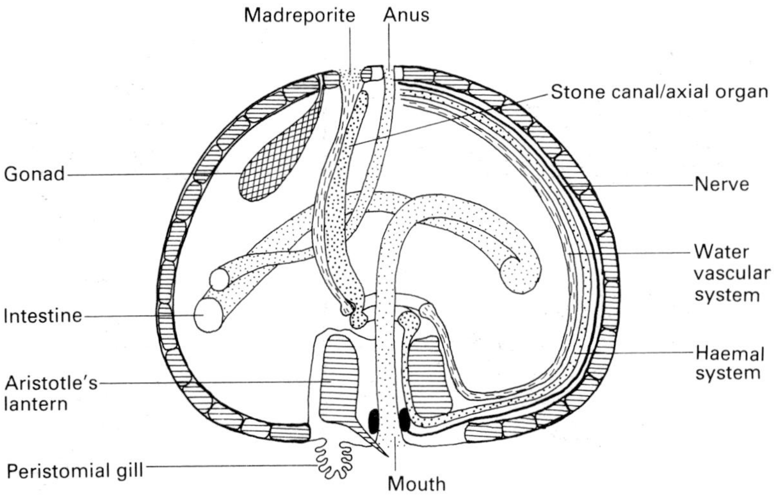

Fig. 7.25. Diagrammatic section through the test of an echinoid (after Nichols, 1962).

The ordinal classification of echinoids is based largely on fossil forms; 15 surviving orders contain a total of 950 living species.

7.3.3.6 Class Holothuroidea (sea-cucumbers)

In comparison with other free-living echinoderms, holothurians lack arms (equivalently to the echinoids, the arms have been incorporated into the body) and possess a bilaterally symmetrical body greatly elongated along the oral/aboral axis, so that they lie on their sides with three ambulacra 'ventrally' and the remaining two in a 'dorsal' position (Fig. 7.26). The body wall is also unusual in being leathery, with

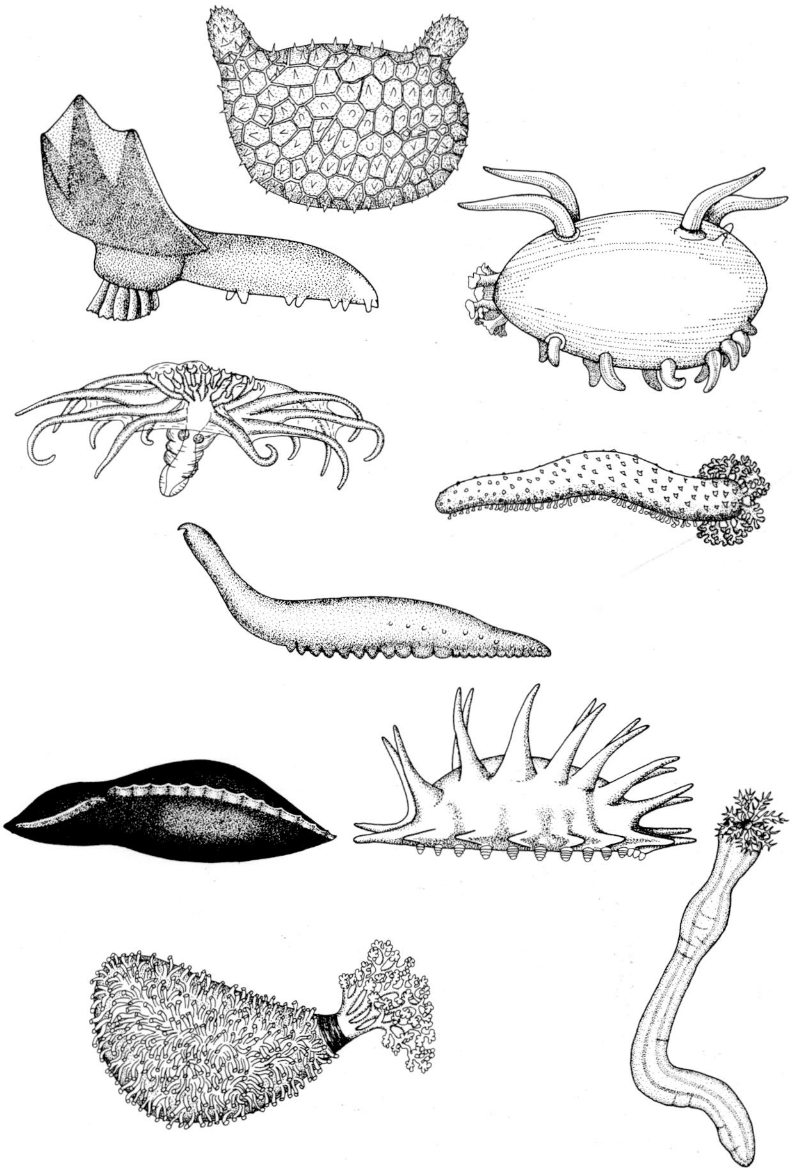

Fig. 7.26. Holothurian body forms (after several sources, principally in Hyman, 1955).

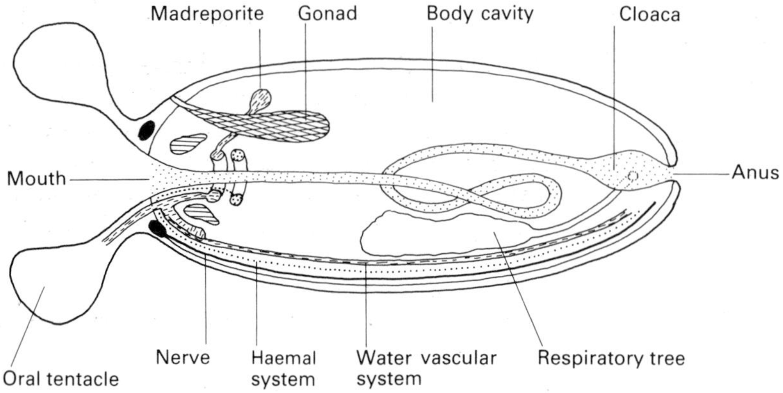

Fig. 7.27. Diagrammatic section through a holothurian (after Nichols, 1962).

well-developed circular and five longitudinal muscles, the calcareous skeleton being reduced to separate microscopic ossicles.

Around the 'anterior' mouth, the water vascular system supports from eight to thirty finger-like, branched or shield-shaped oral tentacles which are used in deposit, or more rarely, suspension feeding (Figs 9.3 and 9.6). The through gut, which is often long, terminates in a 'posterior' cloaca or rectum and anus, the cloaca receiving in many species a pair of large diverticula lying in the body cavity, the respiratory trees, into which water is pumped for gaseous-exchange purposes (Fig. 7.27). Some species eviscerate the posterior gut, including the respiratory trees, through the anus on provocation, whilst a few possess specific sticky or toxin-containing Cuverian tubules associated with the respiratory trees, with the same function. The madreporite lies free in the body cavity.

Locomotion is very slow and is achieved in most species by tube feet, which are often scattered over the body surface, rather than being confined to the ambulacra; some groups, however, completely lack tube feet, and in them movement is effected by peristaltic muscle contractions, aided by small, pointed ossicles projecting through the body wall and anchoring dilated regions of the body against the surrounding sediment. Several are sedentary and use the tube feet more for attachment than for locomotion, and some deep-sea forms have greatly enlarged and elongated tube feet on which they walk like stilts (Fig. 7.28). Most species are epifaunal, although several burrow in soft sediments and one group includes planktonic forms.

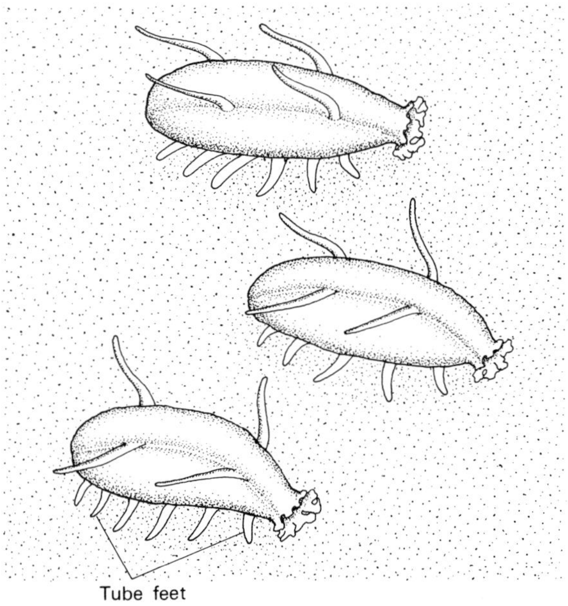

Fig. 7.28. Holothurians walking across the ocean floor by means of large, elongate tube feet.

A single gonad is normally present, discharging through a gonopore near the oral tentacles. Brooding of the embryos is common, in some actually within the body cavity and in one species within the ovary itself. A few are protandrous hermaphrodites. In non-brooding species, development is indirect, proceeding via vitellaria or auricularia and doliolaria stages (see Fig. 7.14e).

Six orders are differentiated on the basis of the form of the oral tentacles, the nature of the tube feet, and the shape of the body. There are some 1150 living species.

7.4 Phylum CHORDATA

7.4.1 Etymology

Latin: *chorda*, a cord (of cat gut).

7.4.2 Diagnostic and special features (Fig. 7.29)

The phylum Chordata comprises three subphyla, of which only two are invertebrate and therefore fall within the compass of this book. These two differ quite markedly, as might be expected from their subphylum status, and hence will be treated separately below. Nevertheless, they share the following characteristics:

1 Bilaterally symmetrical.

2 Body more than two cell layers thick, with tissues and organs.

3 A through gut and a non-terminal anus.

4 Body essentially monomeric, without a distinct head and without appendages or jaws.

5 A large pharynx, the wall of which is perforated through to the exterior by from a few to very many pharyngeal slits.

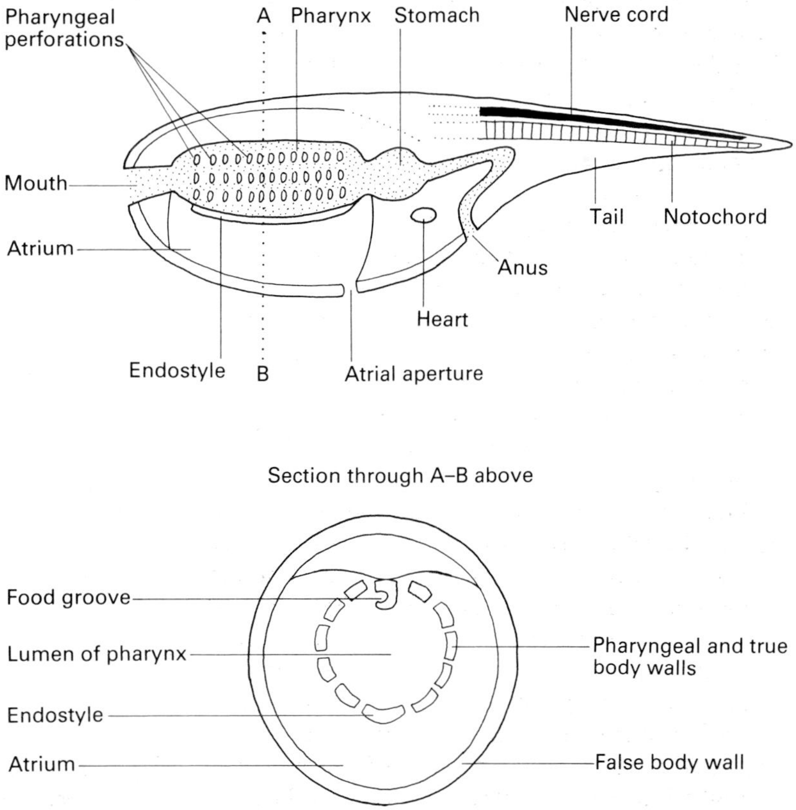

Fig. 7.29. The basic body plan of a generalized invertebrate chordate.

6 Mucociliary suspension feeders (as described in Section 9.2.5), the feeding water current leaving the body through the pharyngeal slits.
7 Usually the pharyngeal region surrounded by a secondary body wall which encloses an atrial cavity connecting to the exterior by a single pore (see Fig. 9.14).
8 Epidermis not secreting an external cuticle and without cilia.
9 At some or all stages of the life history, an internal, dorsal skeletal rod, the notochord.
10 At some or all stages of the life history, a hollow nerve tube running dorsally to the notochord.
11 At some or all stages of the life history, a muscular post-anal tail serving as the organ effecting swimming.
12 Circulatory system partially open.
13 Eggs cleave radially.
14 Development deuterostomatous, usually indeterminate and indirect.
15 Marine.

7.4A Subphylum UROCHORDATA (tunicates)

7.4A.1 Diagnostic and special features

1 A notochord, hollow nerve cord and post-anal tail only in the larval stage (if present) and in one permanently larval group.
2 Without any coelomic body cavities.
3 Without excretory organs.
4 Without segmentation of muscular or other structures.
5 Body wholly enclosed within a secreted tunic, test or 'house', usually composed of cellulose and protein, and containing cells which have migrated from the body and, in some, with extracorporeal blood vessels.
6 A U-shaped gut and a large pharynx usually occupying most of the body volume.
7 Nervous system in the form of a ganglion between mouth and atrial apertures, from which individual nerves issue.
8 Hermaphrodite, usually with a single ovary and testis.

The urochordates are sessile or free-living, solitary or colonial filter feeders which display their chordate affinities only in the free-swimming larval stage.

7.4A.2 Classification

The 1400 species of urochordates are divided between three classes.

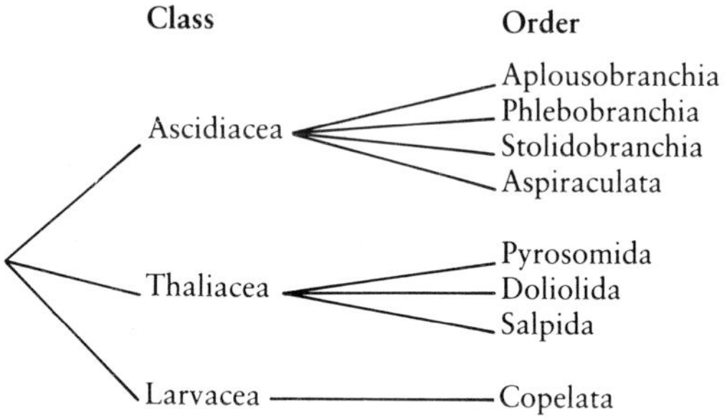

7.4A.2.1 Class Ascidiacea (sea-squirts)

This class contains the 1250 sessile, benthic species. A free-living tadpole larva (Fig. 7.30) is normally possessed and this eventually attaches itself by its head end to the substratum. Differential growth during metamorphosis then leads to an effective rotation of the anterior/posterior axis of the body through 180° so that the mouth and the associated branchial siphon of the tunic or test move to lie at the opposite end of the body to the point of attachment (Fig. 7.31). This branchial siphon leads via a tentaculated orifice into a huge pharynx, normally perforated by very many small slits (except in four deep-sea species, comprising the order Aspiraculata, in which the pharynx is reduced, the pharyngeal slits are absent, and the branchial siphon is modified into a prey-capturing series of prehensile lobes). The tentacles surrounding the opening to the pharynx serve to prevent the entry of large particles. The atrial aperture is also drawn out into a siphon which is sited near to the branchial one (Fig. 7.32).

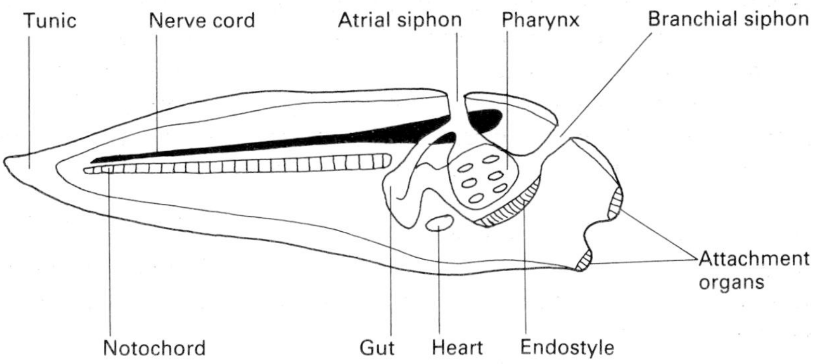

Fig. 7.30. Anatomy of the ascidian tadpole larva (after McFarland *et al.*, 1979 and others).

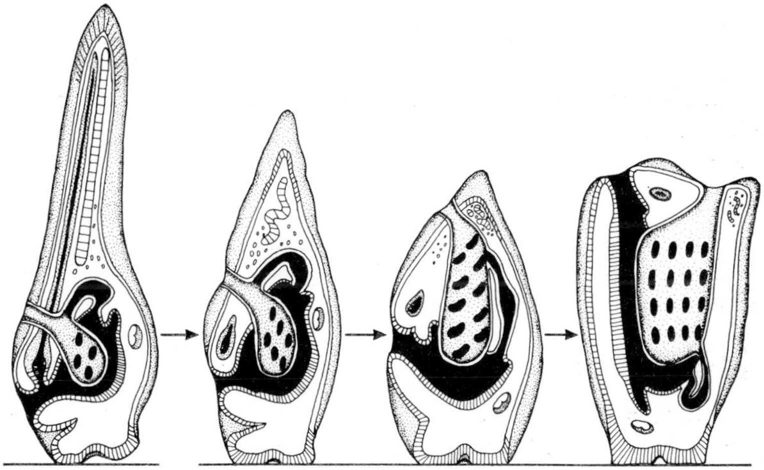

Fig. 7.31. Metamorphosis of the tadpole larva (after Barnes, 1980).

A basal heart is present which, unusually, pumps blood in two directions, the heartbeat reversing periodically. The blood is peculiar in that some of its cells contain high concentrations of heavy metals, particularly vanadium, niobium or iron, in association with sulphuric acid. Vanadium concentrations in these cells may attain 10^5–10^6 times the background levels in the external sea water.

Asexual multiplication by budding is common, proceeding in a variety of manners from different regions of the ascidian body. As in the unrelated cnidarians and bryozoans, budded individuals may remain associated with each other in colonies, although there is no zooidal polymorphism. Some colonial forms are embedded in a communal test, many sharing a single atrial opening; others are connected by basal stolons; and many colonies adopt characteristic shapes. In some, budding begins whilst the ascidian is still larval. The majority of ascidians, however, are solitary and these non-colonial forms are often much larger than the individual colonial zooids, up to 15 cm in height or more (Fig. 7.33).

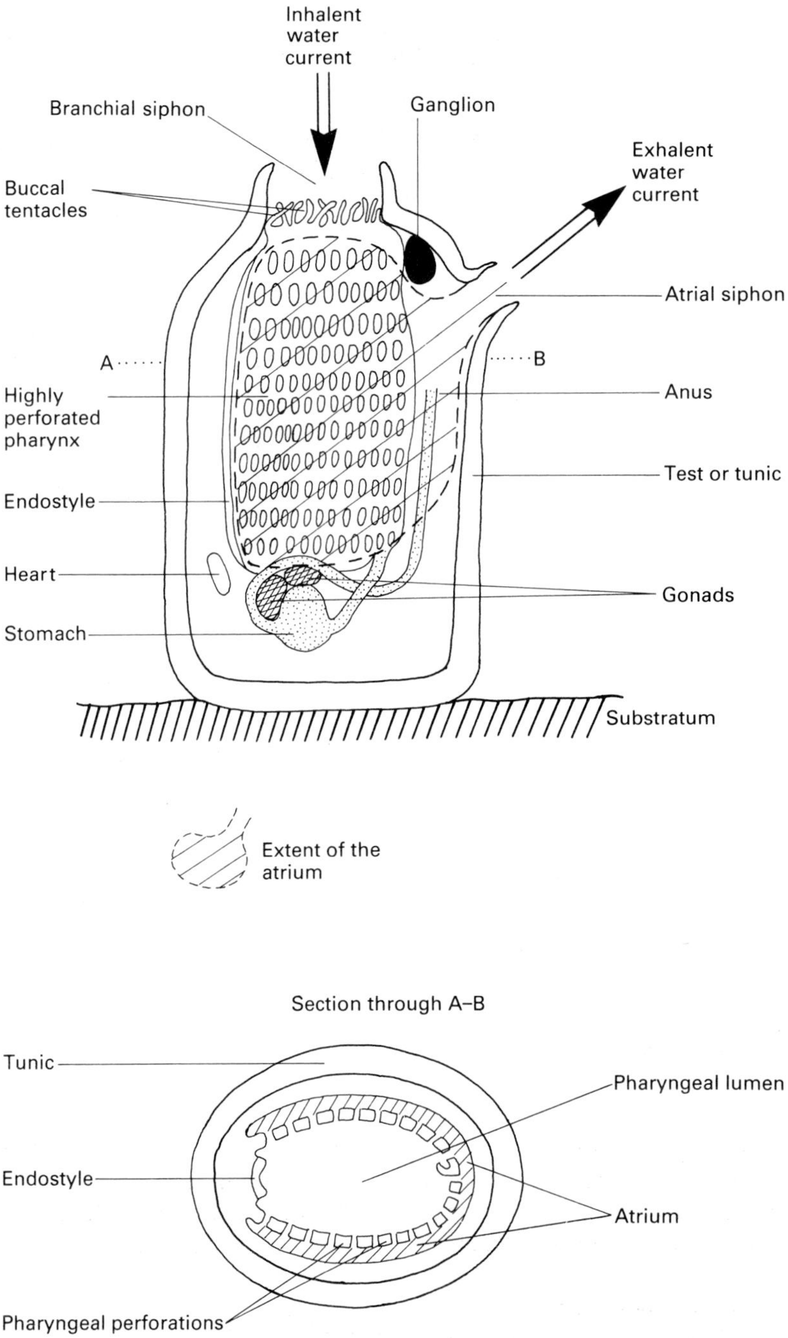

Fig. 7.32. Diagrammatic section through an ascidian (after several sources).

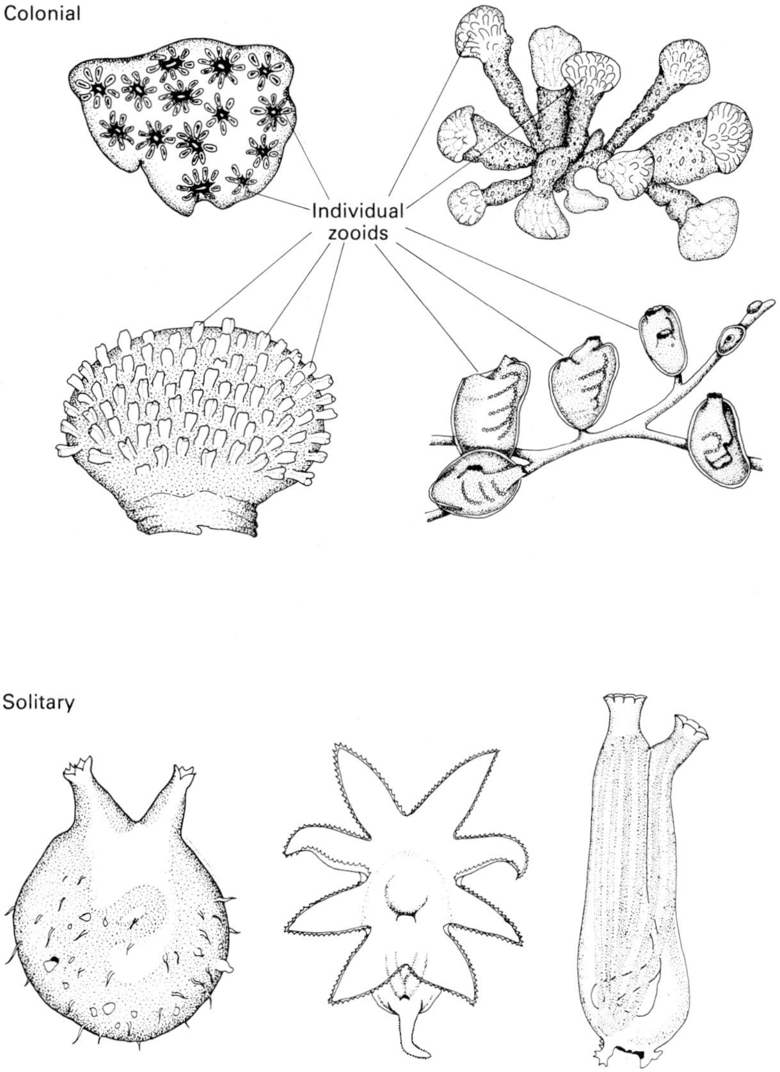

Fig. 7.33. Colony and body form of ascidians (after Millar, 1970 and others).

Besides the aberrant Aspiraculata, the sea-squirts are included in three orders dependent mainly on the location of the gonads and on the structure of the pharyngeal wall: in the Aplousobranchia, which are all colonial, the gonads are positioned in the loop of the intestine and the pharynx possesses a simple wall; the mainly solitary Stolidobranchia have their gonads embedded in the body wall alongside the pharynx which is folded longitudinally and bears internal bars; whilst the mostly solitary Phlebobranchia possess gonads sited as in the Aplousobranchia but their pharynx, although unfolded, has raised internal longitudinal bars formed by bifurcating papillae.

7.4A.2.2 Class Thaliacea (the pelagic tunicates)

The 70 thaliacean species are all planktonic and use their feeding current as a means of jet propulsion through the water, the branchial and atrial apertures being located at opposite ends of their fusiform or barrel-shaped bodies. Asexual budding also occurs in all forms, buds forming on a ventral stolon which arises immediately behind the pharyngeal endostyle.

Two different life styles occur. The order Pyrosomida includes colonial thaliaceans which occupy a common cylindrical test with a central lumen opening to the environment at only one end. Each zooid is positioned such that its branchial aperture is on the

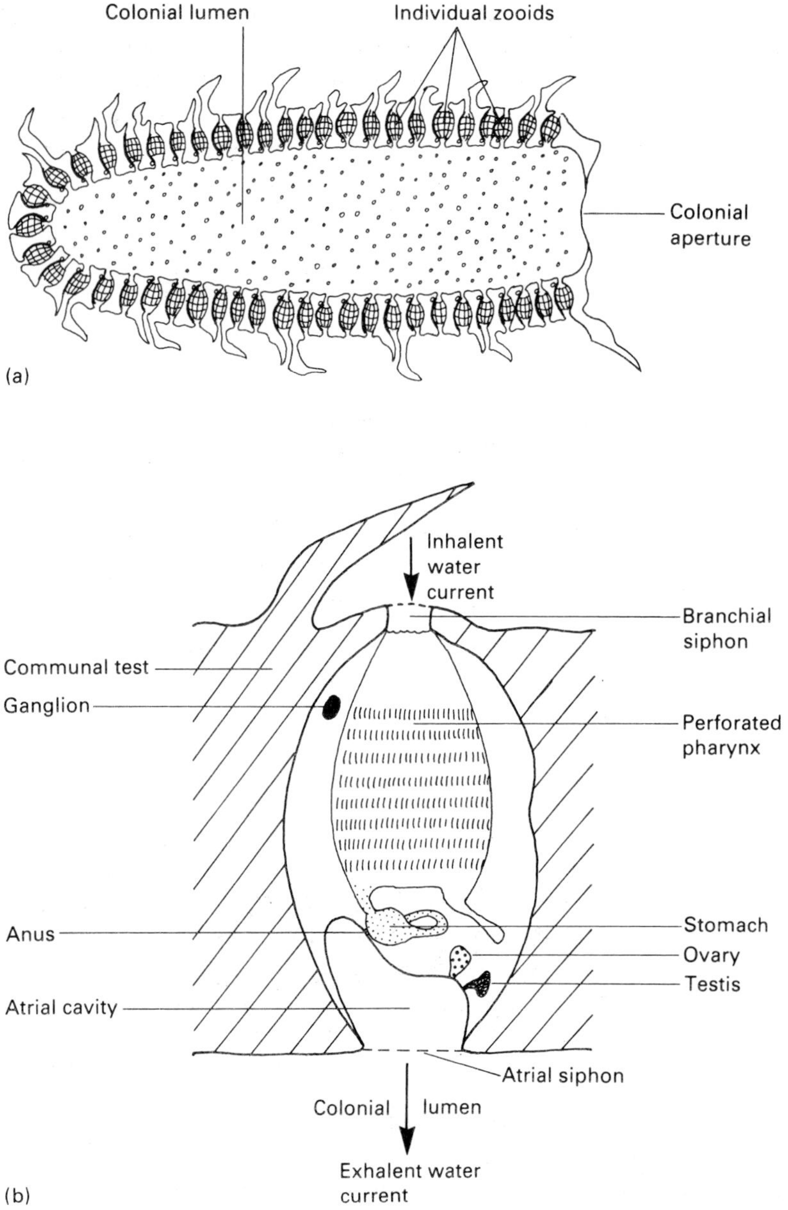

Fig. 7.34. Diagrammatic sections (a) through the colony and (b) through the individual zooid of a pyrosome (after Grassé, 1948 and Fraser, 1982).

outside face of the communal cylinder and its atrial aperture is on the inside face. All the feeding currents therefore discharge into the colonial lumen and the whole tubular colony, which may be several metres long (giving rise, it has been suggested, to several sightings of great sea-serpents), moves as a single unit through the water (Fig. 7.34). The individual zooids possess many pharyngeal perforations, superficially similar to those of the ascidians.

The orders Salpida and Doliolida, however, are mainly solitary, although they exhibit an alternation between solitary and aggregate forms of different sexuality. In the salps (Fig. 7.35), the solitary phase multiplies asexually, giving rise to chains of individuals which eventually separate to yield the solitary sexual 'generation'; in the doliolids, on the other hand, it is the solitary phase which reproduces sexually to produce the asexually multiplying aggregate 'generation' (Fig. 7.36). Both groups exhibit zooidal polymorphism (during different stages of their life history), possess bands or hoops of muscle around the body which drive water through (instead of ciliary power), have more or less transparent gelatinous tests, and have reduced numbers of individually

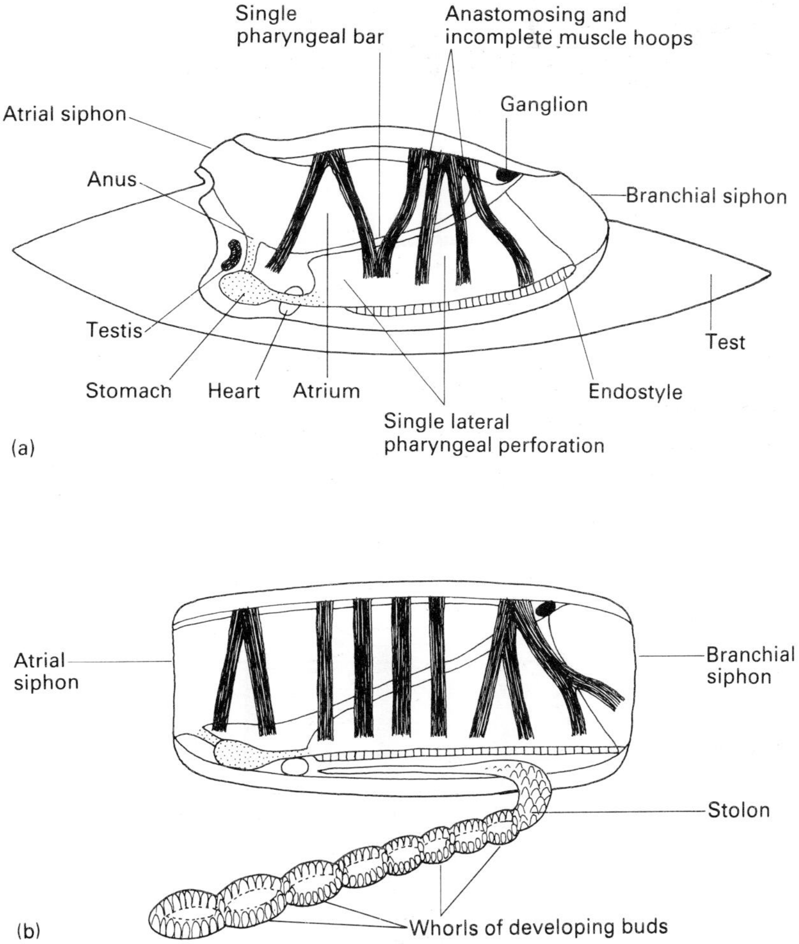

Fig. 7.35. The anatomy (diagrammatic) of the two stages in the life cycle of a salp: (a) the sexually reproducing aggregate 'generation';
(b) the asexually multiplying solitary 'generation' (after Berrill, 1950, Fraser, 1982 and others).

larger pharyngeal perforations—a development which reaches its extreme form in the salps which possess only two relatively enormous slits (see Fig. 9.15).

Only the doliolids have a larval stage in their life history, and this, it is widely believed, gave rise to the following class by paedomorphosis.

7.4A.2.3 Class Larvacea (appendicularians)

The planktonic larvaceans possess a morphology basically that of the characteristic urochordate larva (Fig. 7.37), the small body bearing a large persistent tail, complete with hollow nerve cord and notochord, and totalling only some 5 mm in length. Epidermal glands secrete not a cellulose test but a thin gelatinous 'house' which forms an external filtration apparatus (Fig. 7.38), water being drawn into this 'house' through screens by the beating of the tail and food particles, especially nanoplanktonic algae, being filtered by fine meshworks within the gelatinous structure (see Fig. 9.5). Filtered material is then

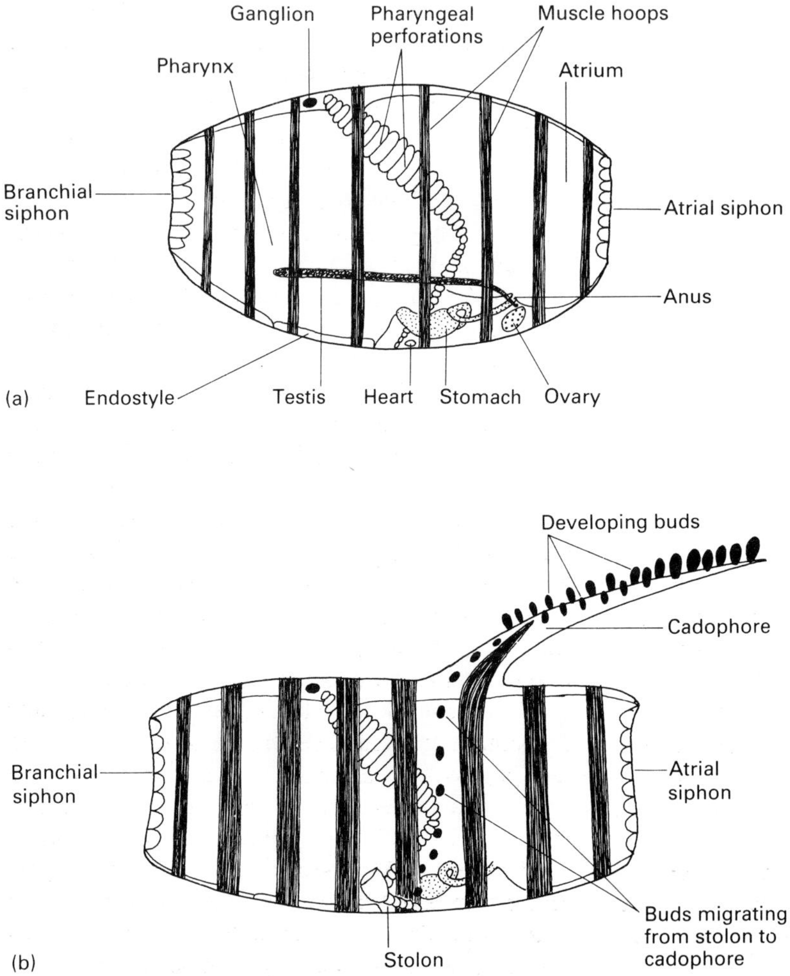

Fig. 7.36. The anatomy (diagrammatic) of the two stages in the life cycle of a doliolid: (a) the sexually reproducing solitary 'generation'; (b) the asexually multiplying aggregate 'generation' (after Berrill, 1950, Fraser, 1982 and others).

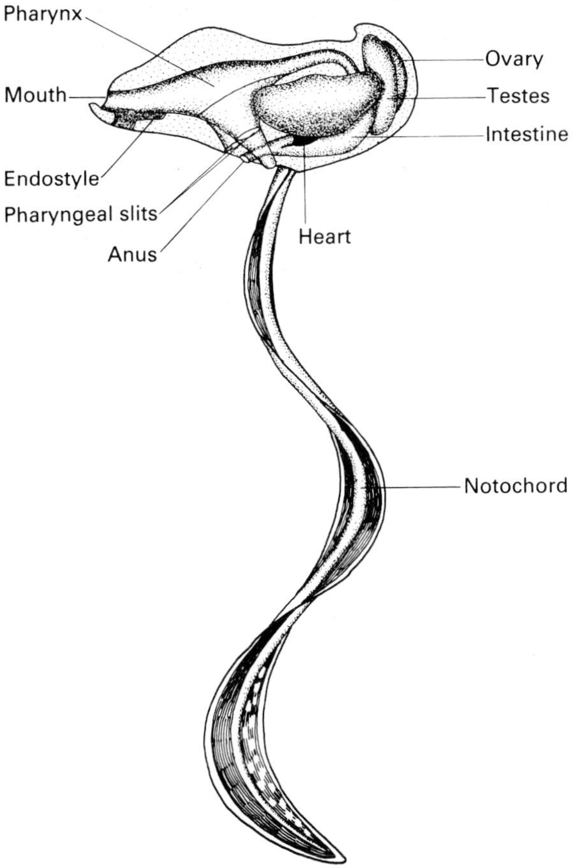

Fig. 7.37. The anatomy of a larvacean (after Alldredge, 1976).

ingested with the aid of a further water current, created by pharyngeal cilia, which leaves the gut through a single pair of pharyngeal slits opening individually on the body surface. The atrial cavity is therefore the area within the 'house'. When the screens and filters of the 'house' become irreversibly clogged, larvaceans can abandon their tests through a 'door' (which is normally in a closed position) and inflate a new pre-formed 'house'.

Larvaceans are all solitary, and reproduction is solely sexual. Unusually amongst urochordates, one species is gonochoristic. A single order is recognized for the 70 known species.

7.4B Subphylum CEPHALOCHORDATA (lancelets)

7.4B.1 Diagnostic and special features (Fig. 7.39)

1 Body laterally compressed, fish-like.
2 Notochord extends the whole length of the body.
3 The hollow dorsal nerve cord extends almost the whole length of the body, but is not dilated anteriorly to form a brain.
4 A persistent post-anal tail.
5 Body with serially repeated muscle blocks, nerves, excretory organs and gonads.
6 Body cavities formed enterocoelically from many serially repeated pouches; the ventral parts of the separate coelomic pouches merge together and the dorsal parts become obliterated by muscle.
7 Excretory system closely resembles protonephridia, but is formed by peritoneal cells.
8 Pharyngeal region covered by a secondary body wall formed by a pair of metapleural folds growing ventrally and fusing along the mid-ventral line.
9 Pharynx large, occupying half the body length.
10 Gonochoristic, with many single or paired gonads.

Cephalochordates are small (up to 10 cm long), free-living animals which are sedentary and benthic whilst feeding, but which are capable of swimming to change feeding location or to escape from predators, the notochord acting as an incompressible but flexible longitudinal strut. Contraction of the longitudinally arranged muscles therefore bends the body into a series of S-shaped wriggles, rather than shortening it (an equivalent system of movement to that seen in the chaetognaths and nematodes).

The cephalochordate feeding system is essentially the same as that employed by the urochordates, and in this group too the buccal region is guarded by a series of tentacles (the buccal cirri and velar tentacles—Fig. 7.40) which prevent the entry of undesired particles. The pharynx, however, is elongate rather than barrel shaped, and the atriopore is in the mid-ventral line just anterior to the anus. No surrounding test drawn out into siphons occurs.

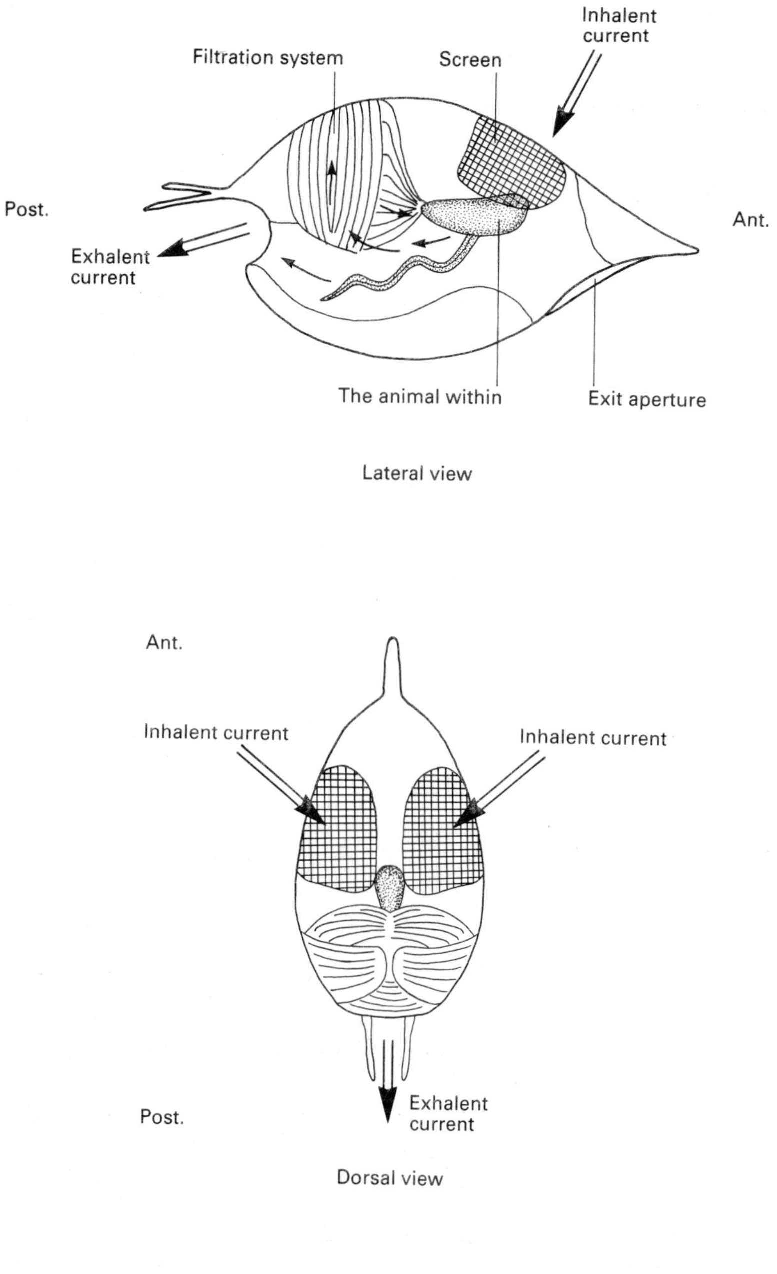

Fig. 7.38. The larvacean 'house' and the feeding currents induced through it (after Hardy, 1956 and others).

To some degree, the cephalochordates provide a link between the invertebrates and the third chordate subphylum, the Vertebrata, especially the agnathans which also lack appendages and jaws, and the larval stage of one group of which (the lampreys) possesses a mucociliary filtering pharynx similarly to the lancelets (although water is driven through its pharynx by muscular pumping). Most obviously, however,

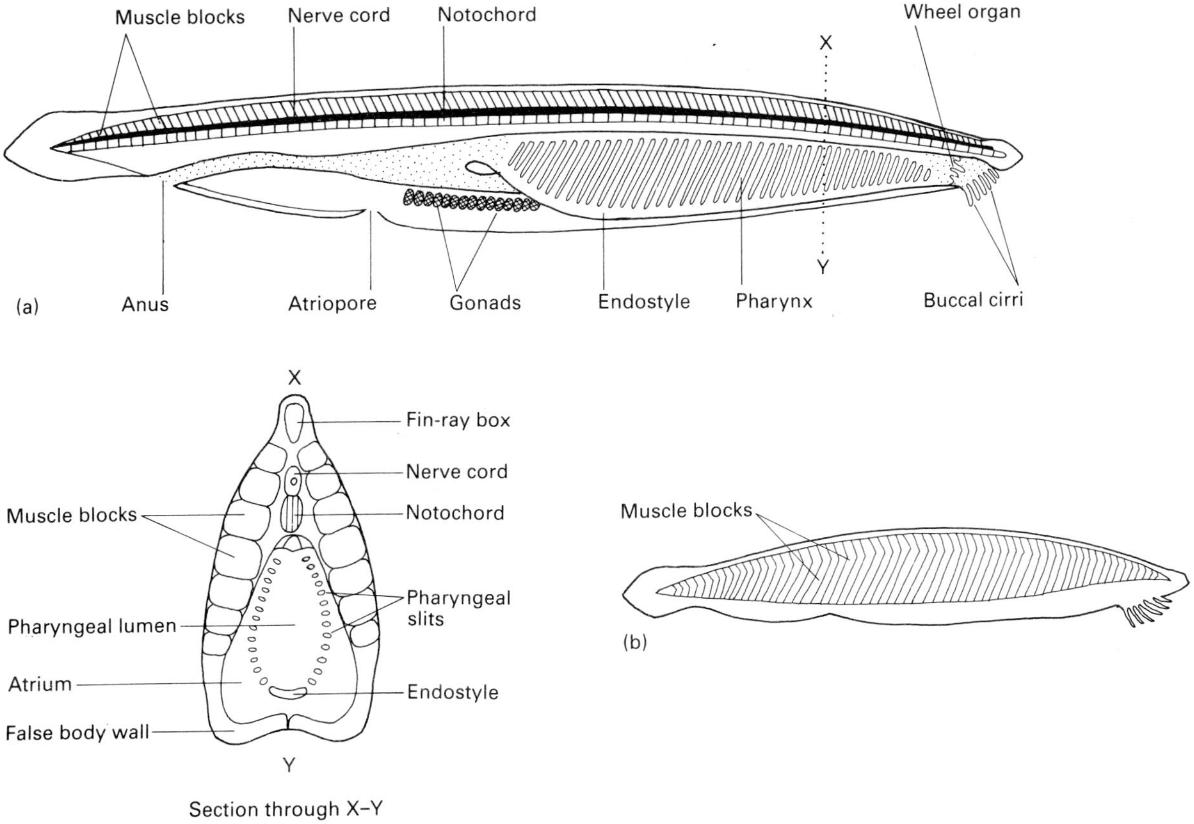

Fig. 7.39. Cephalochordates: (a) diagrammatic longitudinal section through the body; (b) appearance in life (after Young, 1962).

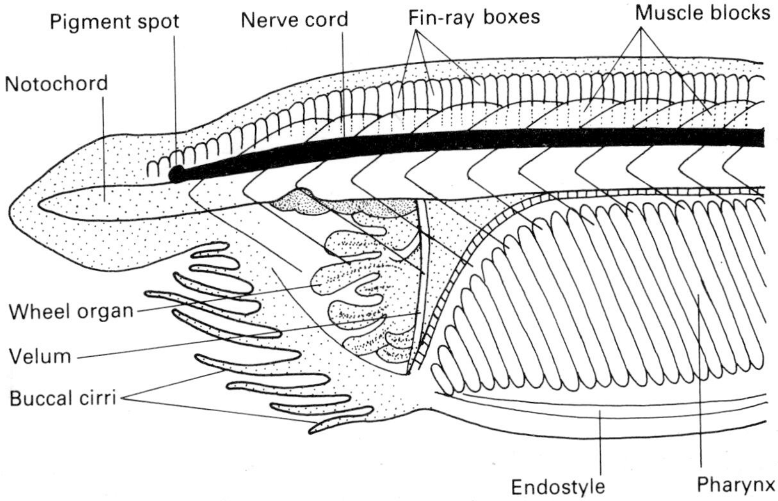

Fig. 7.40. The anterior end of a cephalochordate (after Young, 1962).

the cephalochordates differ from the vertebrates in their lack of a differentiated head with paired sense organs, a brain and a protective cranium, in possessing a nerve cord to which the muscles send processes (forming the apparent ventral roots) and peripheral nerve fibres which lack myelin sheaths, and in their very thin monolayered epidermis. Nevertheless, in outline fashion they do perhaps indicate the general nature of the likely bodily organization of the 'protovertebrate'. Equally, their basic similarity to an enlarged and neotenous version of the larval stage of the otherwise most unvertebrate-like urochordates is also apparent.

7.4B.2 Classification

The 25 known species are placed in a single class and order.

7.5 Further reading

Alvarino, A. 1965. Chaetognaths. *Oceanogr. mar. Biol., Ann. Rev.*, **3,** 115–94 [Chaetognatha].

Baker, A.N., Rowe, F.W.E. & Clark, H.E.S. 1986. A new class of Echinodermata from New Zealand. *Nature, Lond.*, **321,** 862–4.

Barrington, E.J.W. 1965. *The Biology of Hemichordata and Protochordata*. Oliver & Boyd, Edinburgh [Hemichordata and Chordata].

Berrill, N.J. 1950. *The Tunicata*. Ray Society, London [Urochordates].

Hyman, L.H. 1955. *The Invertebrates (Vol. 4) Echinodermata*. McGraw-Hill, New York [Echinodermata].

Hyman, L.H. 1959. *The Invertebrates (Vol. 5) Smaller Coelomate Groups*. McGraw-Hill, New York [Hemichordata and Chaetognatha].

Nichols, D. 1969. *Echinoderms*, 4th edn. Hutchinson, London [Echinodermata].

Young, J.Z. 1981. *The Life of Vertebrates*, 3rd edn. Clarendon Press, Oxford [Chordata].

8

INVERTEBRATES WITH LEGS: THE ARTHROPODS AND SIMILAR GROUPS

- Tardigrada
- Pentastoma
- Onychophora
- Chelicerata
- Uniramia
- Crustacea

Besides their general protostomatous condition, the six phyla included in this chapter really share only two characteristic anatomical features. They bear pairs of legs along all or part of the length of the body, each pair usually being served by ganglionic swellings on the longitudinal nerve cord/s; and they possess pseudocoelomic body cavities, often filled with blood and then termed haemocoels.

Three of the phyla (the Tardigrada, Pentastoma and Onychophora) are soft-bodied animals which use their body cavities as hydrostatic skeletons: in effect, these are worms with soft fleshy, unjointed, claw-bearing legs formed by finger-like outgrowths of the body and capable of being moved by extrinsic muscles. We suggested in Chapter 2 that such worms-with-legs might be regarded as 'protoarthropods' and as survivors of the original wave of arthropodization which took place in the Precambrian or very early Cambrian. The onychophorans do appear to be related in some way to some of the true arthropods (the uniramians), but the affinities of the tardigrades would appear to be as much—if not more—towards the (ancestral?) pseudocoelomate phyla as to the other protoarthropods, and the precise relationships of the pentastomans are obscure to say the least, largely as a result of their specialized parasitic life style.

In contrast, the 'true' arthropods (the Crustacea, Chelicerata and Uniramia) possess a hard, jointed, sclerotized cuticular exoskeleton composed of chitin and protein, sometimes impregnated with calcium carbonate (Fig. 8.1). This covers the whole body including the legs, which are therefore also jointed (the word arthropod is derived from the Greek: *arthron*, joint; *podos*, feet) (Fig. 8.2). As in various other animals, being covered by an external cuticular system imposes constraints on growth, and necessitates a series of moults. This problem is particularly acute in respect of the arthropods where the cuticle is also the skeleton. During moulting, therefore, the old skeleton is partly resorbed and then shed, and a new soft skeleton, which has been developed beneath the old, is inflated (by the intake of air or water into the body) and hardened. The animal is especially vulnerable during this period and moulting often takes place whilst in hiding. In origin, the exoskeleton was probably a series of protective plates or hoops of cuticles, as seen in the kinorhynchs (Section 4.6) and in some tardigrades (Section 8.1), for example, and this was later adapted to serve a skeletal function in partial replacement of the ancestral hydrostatic pseudocoel/haemocoel. In some, replacement has remained only partial since a number of arthropods still extend their legs by hydraulic pressure and only flex them using their exoskeletal–muscular system.

The body of an arthropod is in origin fundamentally monomeric, although extensive metamerism of the body wall, exoskeleton and some internal structures has occurred in association with each pair of legs (see, e.g., Fig. 2.10). In essence, therefore, the arthropod body plan is that of a small anterior region (acron) and an equivalent posterior

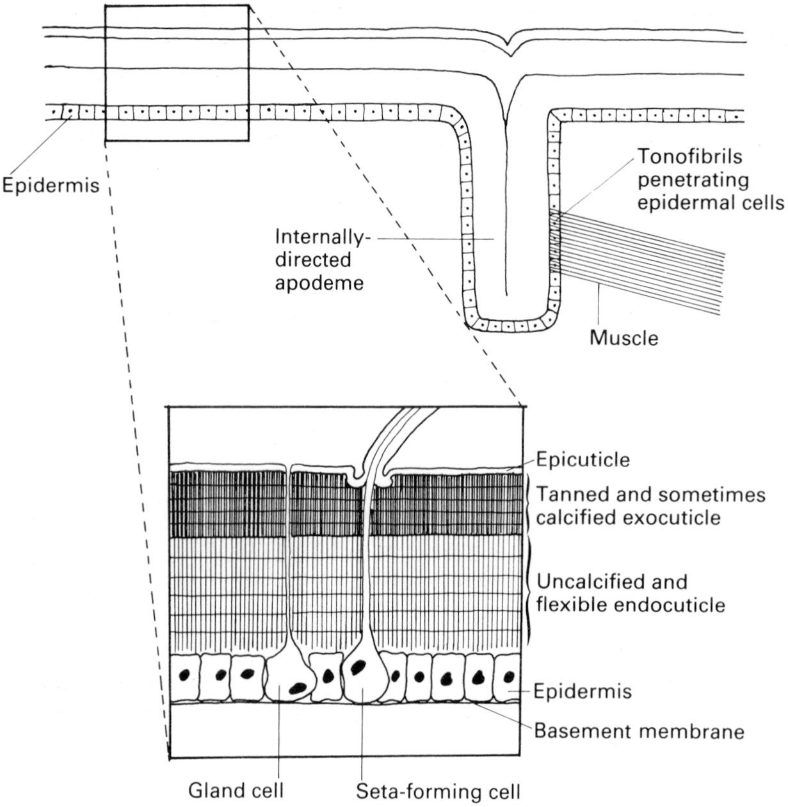

Fig. 8.1. Section through the cuticular exoskeleton of an arthropod showing the various layers and an internal projection (an apodeme) serving an internal skeletal function (after Hackman, 1971 and others).

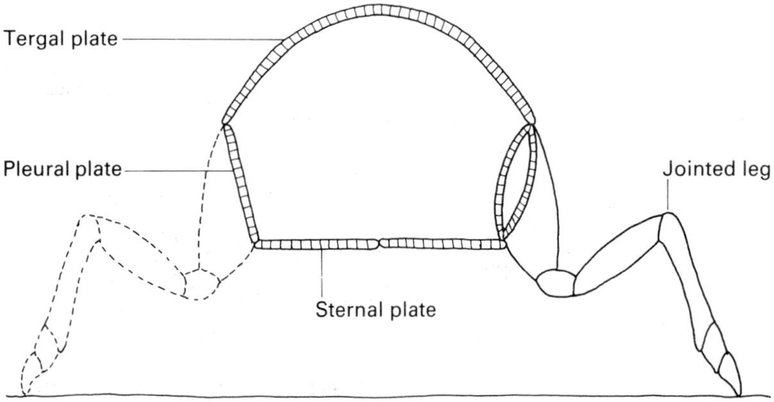

Fig. 8.2. Transverse section through the body of an arthropod showing the various exoskeletal plates encircling the body and a jointed limb. The number of separate articles of which the limb is composed varies from group to group, each phylum having its own nomenclature.

portion (telson) without legs, and number of intervening sections (the segments) each with one pair of legs, and with serial repetition of leg-based organs (muscular, nervous, skeletal, etc.). Non-leg-associated organs, for example the excretory and reproductive systems, are not serially repeated, however. In many lines of arthropods, there has been considerable fusion of leg-bearing segments, loss of appendages, and/or differentiation of various body regions; specialization of the anterior legs into feeding organs is particularly widespread. Apart from the common modification of their cuticle to serve a wholly or partially exoskeletal function (and any feature consequent on this, e.g. the jointed condition of the legs), the absence of cilia, and the tendency to develop compound eyes, the arthropods share few features, however, and hence it is most likely that the arthropod state represents a grade of organization rather than the distinguishing characteristic of a single phylogenetic line.

Together the arthropods comprise the large majority of animal species, mainly as a consequence of their successful conquest of the land and of the ease with which small terrestrial organisms can speciate. Their success as terrestrial animals, in marked contrast to most groups of invertebrates, probably owes much to the evolution of water-conserving excretory systems and gasous-exchange organs, and the development of a desiccation-resistant impermeable epicuticle.

8.1 Phylum TARDIGRADA (water bears)

8.1.1 Etymology

Latin: *tardus* slow; *gradu* step.

8.1.2 Diagnostic and special features (Fig. 8.3)

1 Bilaterally symmetrical; minute, squat.
2 Body more than two cell layers thick, with tissues and organs.
3 A through, straight gut.
4 Body monomeric, although with four pairs of short, unjointed, claw-bearing legs on which the

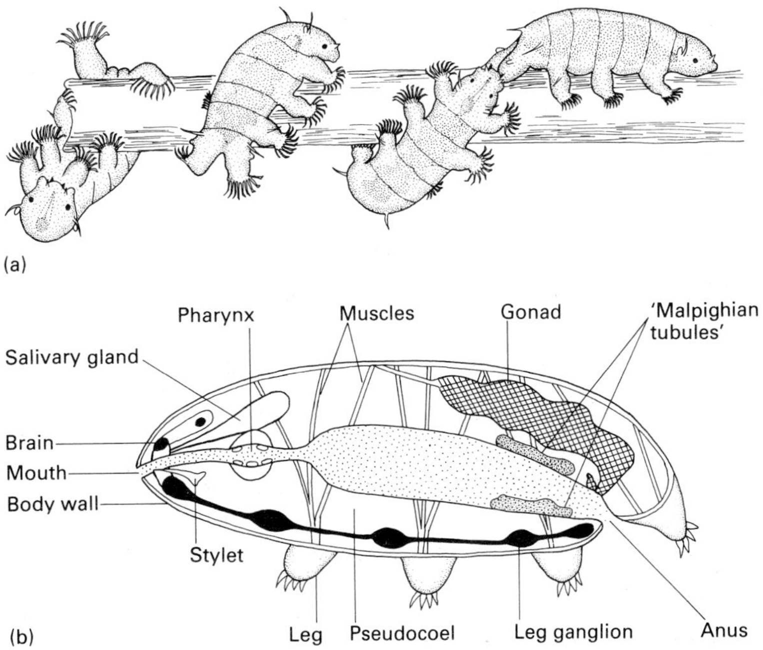

Fig. 8.3. Tardigrade structure: (a) external appearance of animals crawling on a filamentous alga (after Marcus, 1929); (b) a diagrammatic longitudinal section through a generalized tardigrade (after Cuénot, 1949).

animals crawl (Fig. 8.3a) using extrinsic muscles; leg pairs served by serially repeated nerve ganglia.

5 Well-developed pseudocoelomic body cavity forming the hydrostatic skeleton.

6 Body wall with cuticle-covered epidermis, but without muscle layers; essentially non-chitinous cuticle moulted and often bearing spines and/or thickened into plates; network of individual smooth muscle cells criss-cross the body.

7 Body has a fixed number of cells (eutelic).

8 A muscular, pumping pharynx bearing chitinous plates (placoids); a pair of buccal stylets can be protracted through the mouth to pierce the prey.

9 Without circulatory system or gaseous-exchange organs.

10 Three 'Malpighian tubules' possibly form an excretory system in some species.

11 Nervous system with a brain and paired longitudinal ventral cords bearing leg-associated ganglia.

12 Gonochoristic, sometimes parthenogenetic; with a single gonad.

13 Development direct.

14 Free-living, inhabiting water films interstitially or associated with vegetation on land, in fresh water and in the sea; cryptobiotic.

The tardigrades display a peculiar amalgam of features: like the other protoarthropods, they possess paired claw-bearing legs; like the lophophorates and deuterostomes, they show during their development, albeit only transitorily, paired enterocoelic pouches (five pairs); and with the pseudocoelomates they share a general level of bodily organization and life style.

Although several species occur in relatively permanent habitats, such as marine sands and shell gravels, many characterize temporary water films and labile water bodies. These latter tardigrades have evolved a variety of resistant stages. When the film of water around moss leaves evaporates, for example, the tardigrades too lose water through their permeable cuticles. Most of their bodily fluids can be lost as they shrivel to small barrel-shaped 'tuns' (Fig. 8.4). Tuns can survive for up to 10 years in the dry state (and probably for longer), their oxygen consumption falling to one six-hundredth of normal.

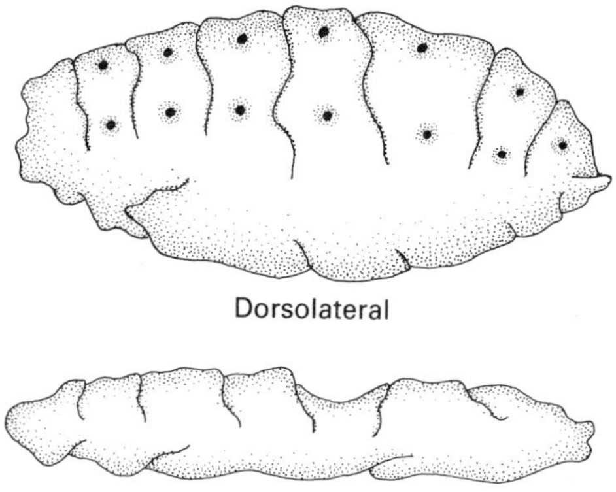

Fig. 8.4. A tardigrade tun (after Morgan, 1982).

In this state, they can withstand temperatures of −272°C for over 8 hours, and of up to 150°C. Less severe adverse circumstances can be avoided (a) by encystment, the animal pulling in its legs, detaching itself from its cuticle, as during moulting, and curling up into a ball within the cuticular shell, or (b) by the production of thick-walled resting eggs. It is possible that including such phases of suspended animation, the tardigrade lifespan may exceed 60 years.

All species are suctorial feeders, although the prey consumed may be algal or plant cells, or associated interstitial or cryptobiotic animals such as rotifers, nematodes and other tardigrades. A few are parasitic, one within the gut of gastropod molluscs. The mouth is applied to the food and the two stylets are then protracted by muscles through the mouth, piercing the prey. Fluids and organelles can thereafter be sucked into the gut by the pumping action of the pharynx, the pharyngeal placoids probably serving to macerate any solid particles ingested.

8.1.3 Classification

The 400 living species, which range in size from 0.05 to 1.2 mm in length, vary little in their anatomy, except in respect of ornamentation of the cuticle and in the form and number of their claws (Fig. 8.5); all are contained within a single class.

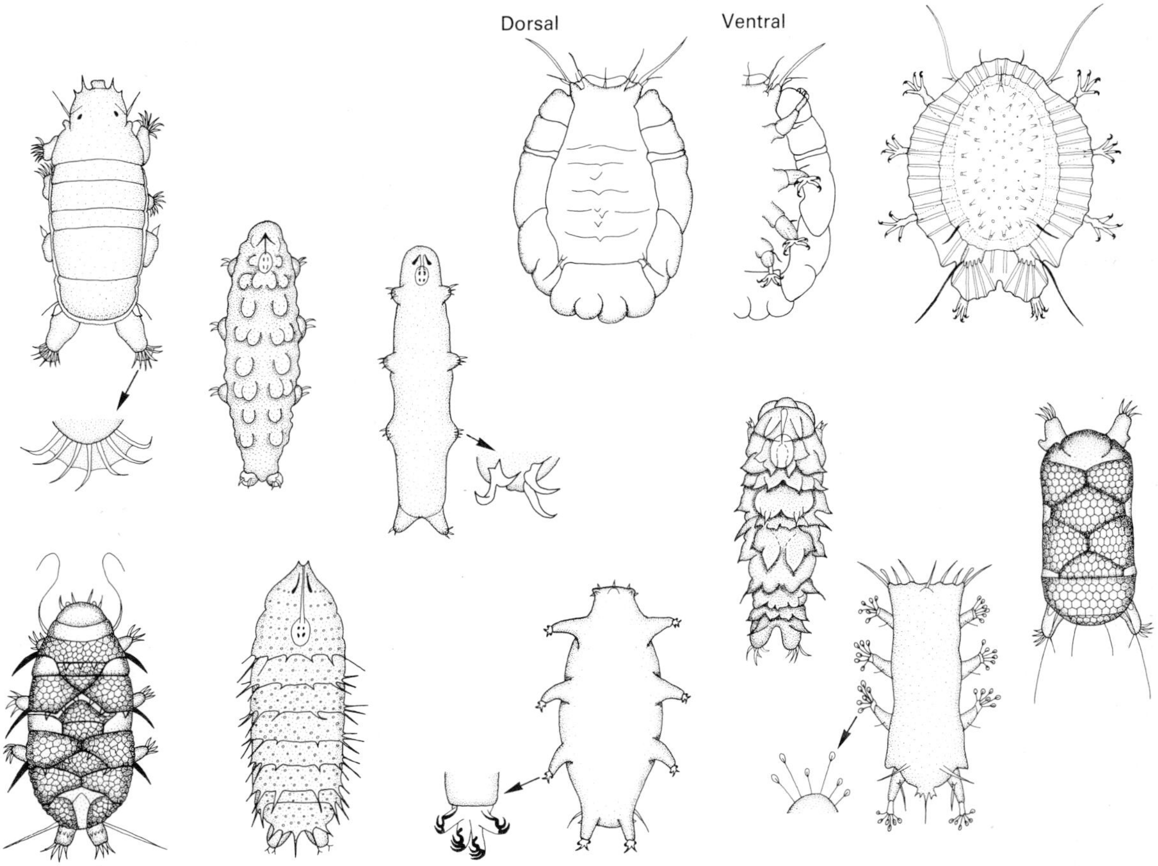

Fig. 8.5. Variation in cuticular form and ornamentation of tardigrades, and in shape and number of their claws (after Morgan & King, 1976 and others).

8.2 Phylum PENTASTOMA (tongue-worms)

8.2.1 Etymology

Greek: *pente*, five; *stoma*, mouth.

8.2.2 Diagnostic and special features (Fig. 8.6)

1 Bilaterally symmetrical; flattened, vermiform.
2 Body more than two cell layers thick, with tissues and organs;

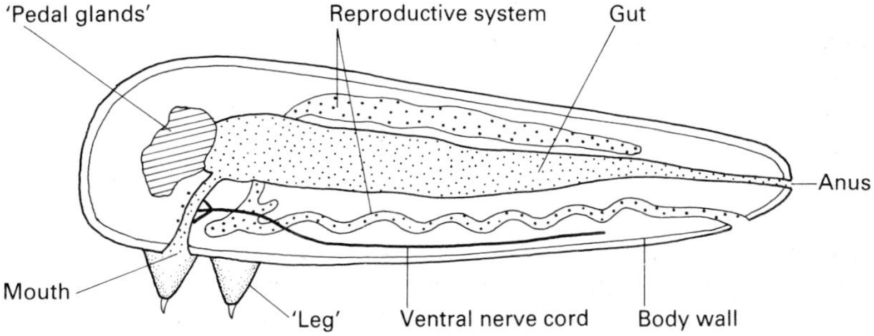

Fig. 8.6. Diagrammatic longitudinal section of a female pentastoman (principally after Cuénot, 1949).

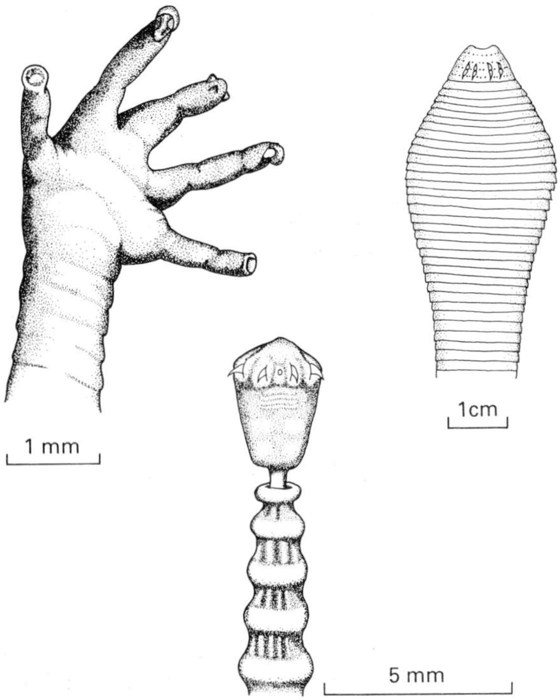

Fig. 8.7. Variation in morphology of the anterior end of the body of pentastomans, showing the extent of development or reduction of the 'legs' (ventral views) (after Kaestner, 1968).

3 A through, straight gut.
4 Body monomeric, although annular; with, anteriorly, two pairs of claw-bearing 'legs' or with only two pairs of claws (Fig. 8.7).
5 A pseudocoelomic hydrostatic skeleton.
6 Body wall with cuticle-covered epidermis and layers of striated circular and longitudinal muscles; cuticle chitinous, porous and moulted.
7 Without excretory, circulatory or gaseous-exchange organs.
8 Nervous system with a brain, a ganglion associated with each 'leg' (or all five ganglia fused into a single mass), and a ventral nerve cord.
9 Gonochoristic; with one or two gonads.
10 Fertilization internal, via copulation.
11 Development via three 'larval stages', the first with three pairs of lobed, leg-like, unjointed appendages.
12 Blood-consuming parasites of the naso-pulmonary system of vertebrates.

Like many endoparasites, the bodies of pentastomes are dominated by the reproductive system and very many, small eggs are produced. Fertilization is internal, and the fertilized eggs, numbering up to half a million at a time, are retained within the maternal body for a period before being released. The uterus which accommodates them then effectively occupies the whole of the body of the female worm, enlarging over 100 times. Three 'larval' or juvenile stages are passed through sequentially, the first while still within the egg shell. This passes out of the host's body through its alimentary canal (all hosts are predatory terrestrial or freshwater vertebrates, 90% of pentastoman species parasitizing reptiles). If this larva, still within its egg capsule, is swallowed by an intermediate host (an omnivorous or herbivorous insect, fish or tetrapod), the larva emerges and bores its way into the intermediate host's tissues by means of three chitinous stylets, moving on its short, stumpy legs. On reaching a specific region (the host's liver, etc.), the larva encysts and develops into the secondary larval stage. If the infected intermediate host then becomes the prey of the definitive reptile, bird or mammal host, the tertiary larva, which resembles a small version of the adult, emerges and migrates to the lungs or nasal passages. Some species lose their legs during one of the final larval moults, retaining only the terminal hook-shaped claws; many larvae also originally possess a pair of claws per leg, this number reducing to one per leg by the adult stage.

8.2.3 Classification

Some 100 species of these small (2 – 16 cm long) worms have been described; all can be placed in the one class.

8.3 Phylum ONYCHOPHORA

8.3.1 Etymology

Greek: *onychos*, claws; *-phoros*, bearer.

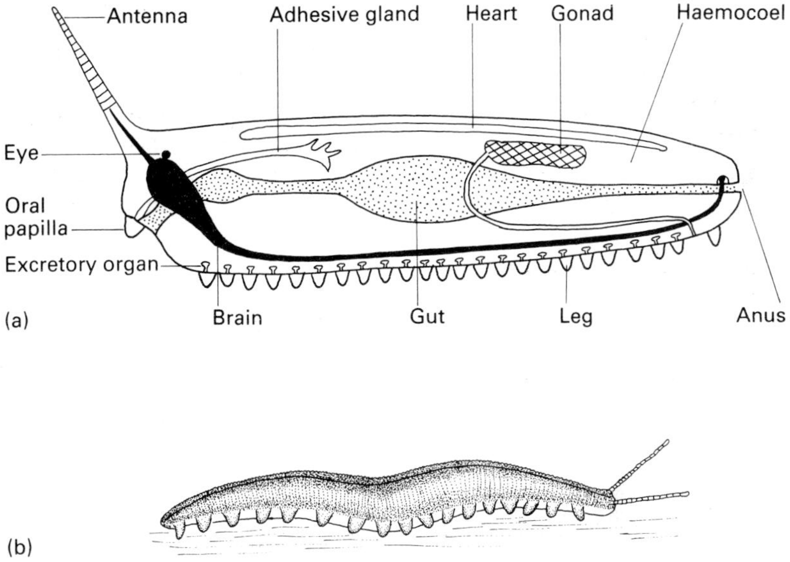

Fig. 8.8. Onychophoran structure: (a) diagrammatic longitudinal section; (b) external appearance (after Sedgwick, 1888 and Cuénot, 1949).

8.3.2 Diagnostic and special features (Fig. 8.8)

1 Bilaterally symmetrical; elongately and cylindrically vermiform.
2 Body more than two cell layers thick, with tissues and organs.
3 A through, straight gut, bearing, anteriorly, a pair of mouthparts, each with two claw-like mandibles (forming an inner and outer jaw blade); fore- and hind-guts lined with cuticle; without digestive diverticula.
4 Body with 14–43 pairs of short, unjointed, fleshy legs along its length; each leg a hollow evagination of the body bearing a terminal pad, pairs of claws and intrinsic muscles (although leg movements are effected by extrinsic muscles); each leg pair has associated with it a pair of heart ostia and of excretory organs.
5 A well-developed haemocoelic body cavity forming the hydrostatic skeleton; with a tubular heart but without other blood vessels.
6 Body wall with a cuticle-covered epidermis and layers of circular, oblique and longitudinal smooth muscle; cuticle very thin, flexible and chitinous.
7 Excretory organs serially repeated pairs of sac-like glands, the anterior ones forming salivary glands and the posterior ones gonoducts.
8 Gaseous-exchange organs simple tubular tracheae issuing in tufts from numerous small spiracles.
9 Nervous system with brain and a pair of very widely separated ventral cords joined by nine–ten rung-like cross-connectives in each leg 'segment' but without distinct ganglia; sense organs include a pair of annular antennae, each with a small, simple eye at its base.
10 Gonochoristic, with paired gonads; fertilization internal, via spermatophores.
11 Development direct.
12 Free-living, terrestrial.

For many years, the onychophorans have been of scientific interest chiefly as living examples of a half-way stage between the worm and arthropod grades of organization. Like worms, they are soft-bodied and possess hydrostatic skeletons, ciliated excretory ducts and smooth muscle layers in the body wall; and, similarly to the arthropods, they bear legs, tracheae, a heart with ostia, longitudinally partitioned blood sinuses, and jaws derived from the appendages, in this case from the claws that terminate the walking legs. The precise structure of their arthropod-like features, however, strongly indicates that they have achieved them in parallel and that whereas they may illustrate what the early uniramians, for example, may have looked like, the onychophorans cannot be ancestral to any known arthropod group. Their tracheae, for instance, are simple, mostly unbranched tubes issuing many at a time from the many (up to 75) spiracles that are scattered over each leg-bearing 'segment' (Fig. 8.9), and their jaws, which move in the anterior/posterior plane, act independently of each other and function as ripping organs by virtue of their pointed tips rather than as chewing appendages. Other onychophoran peculiarities include the structure of the ventral nerve cords, with their numerous connectives but without 'segmental' ganglia.

The onychophorans are terrestrial animals, largely confined to humid microhabitats and environments. Their cuticle is only 1 μm thick (thinner even than the thin arthropod epicuticle, see Fig. 8.1) and is permeable to water, whilst the spiracles have no closing mechanism. Lost water can, however, be replaced by evaginating thin-walled vesicles on to damp surfaces through slits in the cuticle. Onychophorans are mainly nocturnal predators, detecting prey with their antennae and capturing even quite active animals such as grasshoppers by spraying a mucus-like substance distances of up to 0.5 m from adhesive glands which open on paired oral papillae which flank the mouth. This mucus hardens almost immediately on exposure to the air to form an extremely sticky meshwork entangling potential prey. The same technique serves as a method of defence.

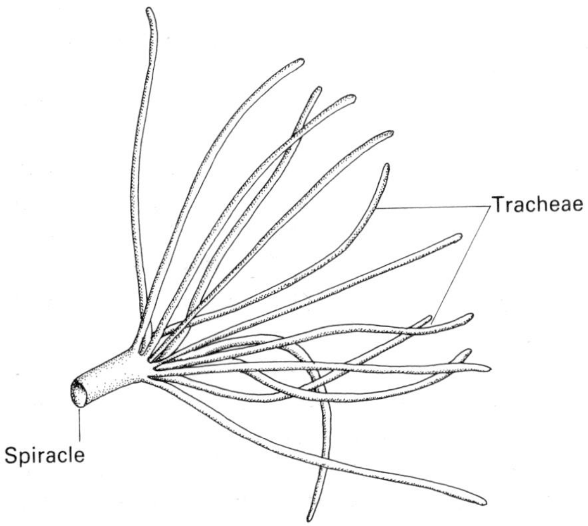

Fig. 8.9. A spiracle and tuft of tracheae (after Clarke, 1973).

8.3.3 Classification

The 70 species of onychophoran achieve lengths of up to 15 cm; all are placed in a single class and order.

8.4 Phylum CHELICERATA

8.4.1 Etymology

Greek: *chele*, talon; *cerata*, horns.

8.4.2 Diagnostic and special features (Fig. 8.10)

NB, the Pycnogona differ in numerous respects—including in many of those listed below—from the other chelicerates.

1 Bilaterally symmetrical, <1 mm–60 cm) long arthropods varying in body shape from elongate to almost spherical.

2 Body more than two cell layers thick, with tissues and organs.

3 A through, straight gut, from the mid-gut region of which issue from two to many pairs of digestive diverticula which secrete enzymes and intracellularly digest and absorb food (these diverticula arise not as outgrowths from the embryonic gut but by partitioning of the embryonic yolk masses before the gut becomes differentiated); mouth anteroventral.

4 Body divided into two regions, an anterior 'prosoma' formed by the acron and six appendage-bearing segments, and wholly or partly covered by a dorsal carapace, and a posterior 'opisthosoma' without legs and with only highly modified appendages, if any.

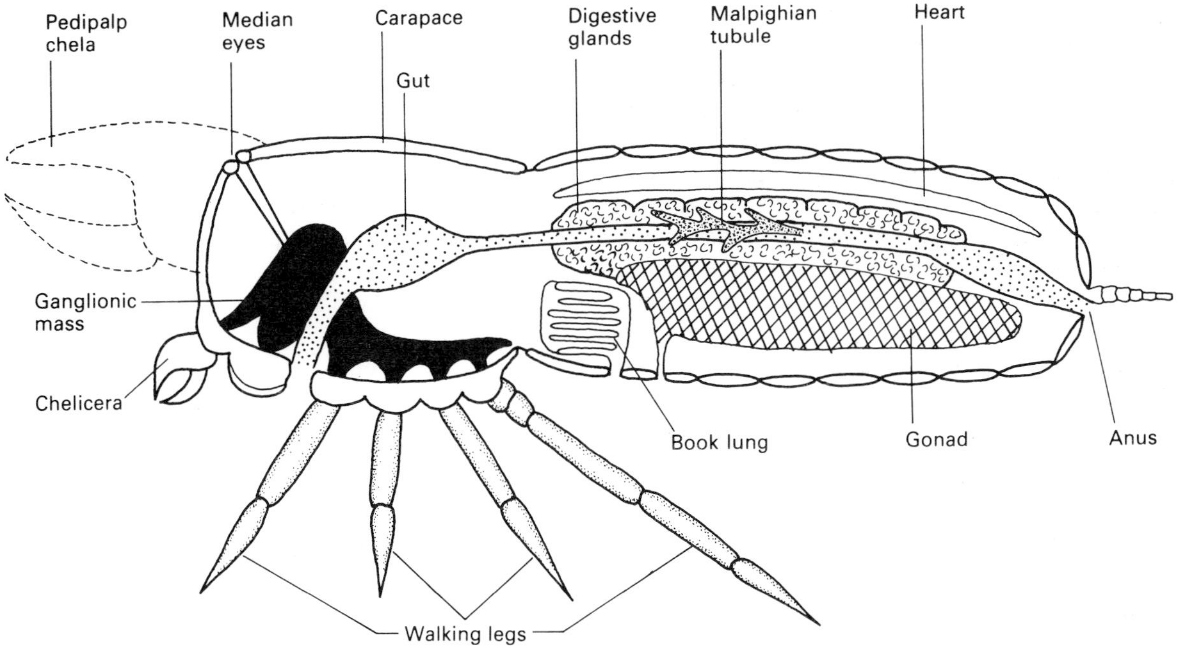

Fig. 8.10. Diagrammatic longitudinal section through a generalized chelicerate.

5 Appendages uniramous; prosomal appendages comprising one pair of chelate, subchelate or stylet-like 'chelicerae', one pair of chelate, leg-like or feeler-like 'pedipalps', and four pairs of walking legs, all attached near to the ventral mid-line and, in some, extended by haemocoelic pressure; without antennae or jaws.

6 Only one pair of appendages (the chelicerae) form mouthparts, although medially directed processes of the basal article of one or more other limbs ('coxal endites') may crush food or spoon it into the mouth.

7 Usually (unless secondarily lacking) with direct median and indirect lateral ocelli on the prosoma; in one group aggregations of the lateral ocelli form compound eyes.

8 Opisthosoma sometimes externally segmented and then with up to twelve segments, in some divided into a broad anterior 'mesosoma' and a narrow posterior 'metasoma', and in several with a projecting post-anal spine, sting or flagellum.

9 A prosomal excretory system of blind-ending coxal glands, and/or an opisthosomal system of branched endodermal Malpighian tubules arising from the mid-gut and discharging mainly guanine.

10 A non-calcareous exoskeleton and sometimes also with a plate-like mesodermal endoskeleton in the prosoma.

11 Gaseous-exchange organs associated with the opisthosomal appendages or with their embryological primordia; in marine forms, these are external gill-books, in terrestrial forms, the internal lung-books and the sieve- or tube-tracheae derived from them.

12 Blood system involved in the circulation of respiratory gases and usually containing haemocyanin.

13 Nervous system with separate ganglia along the length of the body or, more usually, concentrated into a single prosomal mass.

14 Gonochoristic, with external fertilization in solely marine classes (although the two partners associate closely in pseudocopulation during mating) and internal fertilization via copulation or sperma-

tophores in the primarily terrestrial class; gonopores on the second opisthosomal segment.
15 Juvenile stages small versions of the adult, usually hatching with the full complement of limbs.
16 Originally benthic marine, one class has colonized the land and fresh water highly successfully.

Chelicerates differ from the two other arthropod phyla in a number of major respects, as can be seen from the listing above, paralleling the uniramians in their evolution of Malpighian tubules and tracheae. The most obvious distinctive feature concerns their appendages. All lack jaws or any other limbs which are capable of working against each other to bite or chew. Food may be caught and ripped apart by chelate or subchelate limbs (the pedipalps and/or chelicerae); nevertheless only very finely particulate or, characteristically, liquid food can be ingested. Indeed, the mouth itself is usually screened by setae to prevent the entry of large particles. Even so, the chelicerates are almost entirely predatory. Their feeding speciality is to hold the prey close to the mouth, pour digestive enzymes into it, and then imbibe the products of this external predigestion. If enzymes are not actually injected into the prey, digestion or mechanical breakdown occurs in a preoral space enclosed by the coxal endites of some or all of the prosomal limbs, which are often arranged almost radially around the mouth. In association with this method of feeding, the fore-gut is adapted to form a pump or pumps, and the mid-gut, especially its many diverticula (which can occupy most of the body volume), is the site of final digestion and absorption. Further, the chelicerates lack antennae, although the pedipalps or first or second pairs of walking legs may be modified to serve a similar function.

Within the chelicerates, there is a morphological series indicating how a tracheal system could evolve from external gills. In the merostomatans, which are marine and almost certainly the ancestral stock of the chelicerates, the organs of gaseous-exchange are—as in most other marine animals—external gills. These arise from the posterior margin of the flap-like opisthosomal appendages, the beating of which drives water over the gill lamellae. Appropriately, the terrestrial arachnids are air-breathing but, in contrast to the uniramians (see Section 8.5), their characteristic gaseous-exchange organs have retained an embryological association with the posterior margin of the opisthosomal limb primordia (in the arachnids, these opisthosomal limb primordia do not develop into legs, although the spinnerets of spiders and the pectines of scorpions are highly modified opisthosomal appendages). The lung-books (Fig. 8.11), for example, form in this way and are equivalent to a series of gill lamellae housed in pockets sunk into the body. Each lung-book is an invagination with, dangling into it, a series of parallel lamellae held apart by struts, between which air moves by diffusion and into which blood flows within a sinus. In some lung-books, the lamellae are elongate and tube- rather than plate-like, and dependent on the number of such tubes these are termed sieve-tracheae (many closely associated tubes) or tube-tracheae (few tubes). These

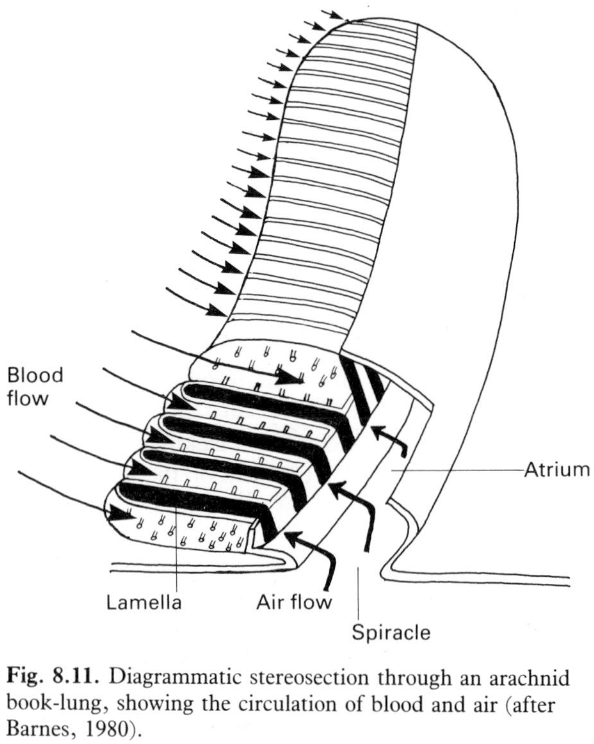

Fig. 8.11. Diagrammatic stereosection through an arachnid book-lung, showing the circulation of blood and air (after Barnes, 1980).

tracheal systems are therefore essentially elongate, internalized, leg-associated gills. In some spiders, additional secondary tracheae have developed from an alternative origin: from hollow, internal projections of the exoskeleton ('apodemes'—see Fig. 8.1).

8.4.3 Classification

The 63 000 described species of chelicerates are placed in three rather disparate classes, of which one is arguably unrelated to the other two.

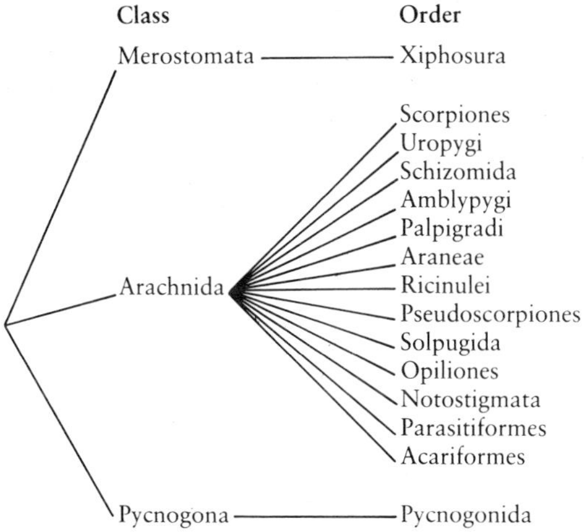

8.4.3.1 Class Merostomata (horseshoe crabs)

Although a dominant group of invertebrates until the Permian, the merostomatans are today represented by only four species in one order (Xiphosura). The horseshoe crabs are large marine chelicerates with a thick, horseshoe-shaped carapace covering the large prosoma and extending both anteriorly, so that the mouth is mid-ventral, and laterally, hiding the appendages from dorsal view. The small, hinged opisthosoma is a flat plate partly inset into a notch in the carapace and fringed laterally with stout spines; terminally, it bears a long, post-anal, caudal spine (Fig. 8.12). The chelicerae, pedipalps and all but the last pair of walking legs are chelate, this last pair bearing instead several terminal spines or lamellae used during burrowing (Fig. 8.13).

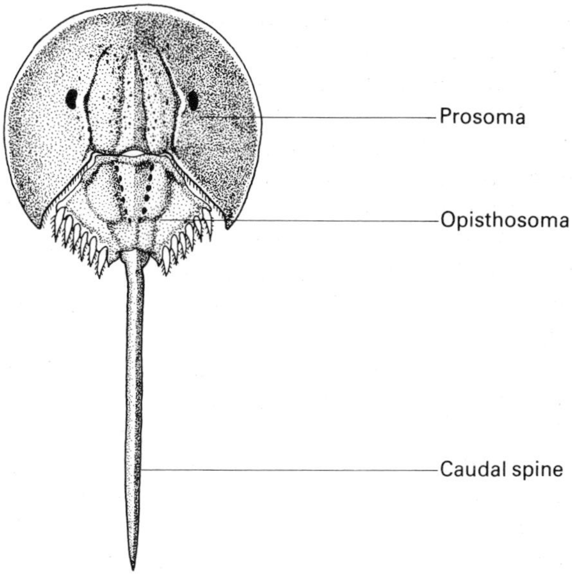

Fig. 8.12. A merostomatan in dorsal view (after Kaestner, 1968).

Unlike the other surviving chelicerates, horseshoe crabs possess almost a full complement of opisthosomal appendages: the most anterior pair are small and tubular, and the remaining six pairs form flat plates which propel the animals during their upside-down swimming. The first of these plates also serves as a genital operculum, whilst the posterior five in addition bear the external gills. Other features not inappropriate in a marine animal but atypical amongst the chelicerates in general include the lack of Malpighian tubules and the occurrence of external fertilization. Alone amongst living members of this phylum merostomatans also have large, diffuse, lateral compound eyes of a rather individual type; in contrast to the arachnids, they have their gonads and digestive diverticula in the prosoma.

The horseshoe crabs are nocturnal benthic animals found in shallow coastal waters, where they burrow in soft sediments and crawl over the surface, preying on large molluscs and polychaetes which are crushed

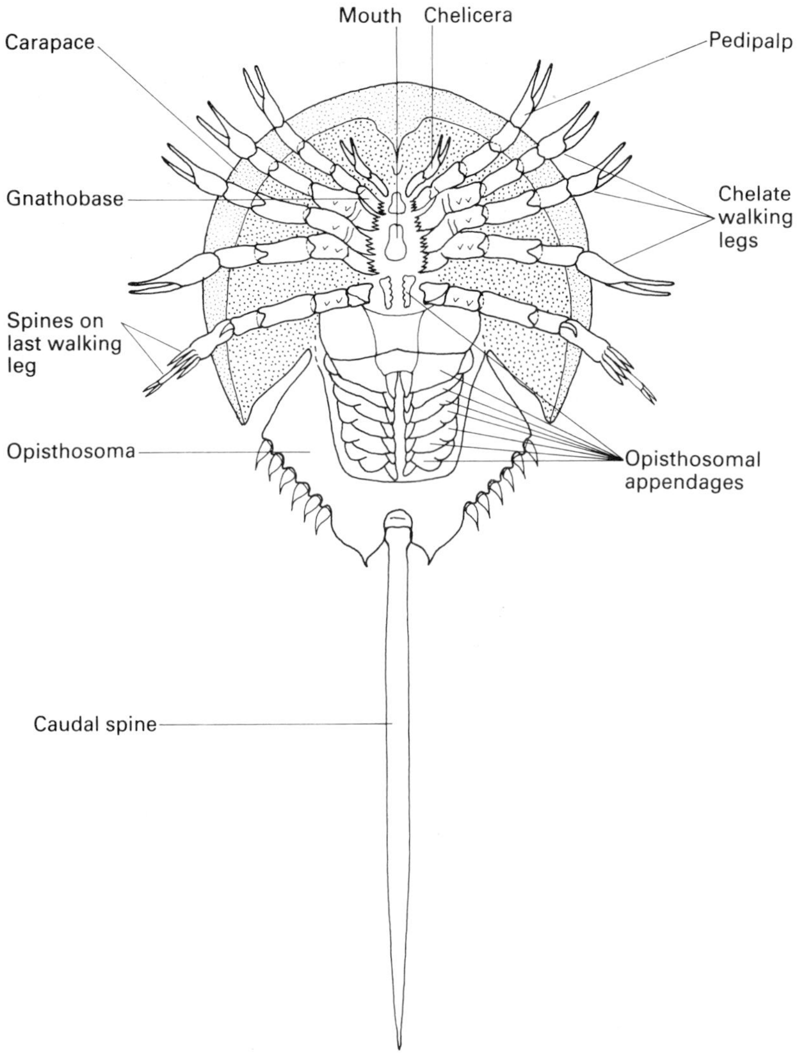

Fig. 8.13. Semi-diagrammatic ventral view of a merostomatan showing the various appendages (after Savory, 1935).

by the coxal endites ('gnathobases') and in a specialized gizzard.

8.4.3.2 Class Arachnida

The almost entirely land-based arachnids comprise over 98% of living chelicerate species and display a number of marked adaptations to terrestrial existence, e.g. a Malpighian-tubule excretory system, internal air-breathing gaseous-exchange organs, a cuticle waterproofed by a wax layer, and internal fertilization. In morphology, they range from elongate, well-armoured forms with large raptorial pedipalps, conspicuous external segmentation, and opisthosomas divided into meso- and metasomes (scorpions, whipscorpions, etc.) to almost spherical species with thin exoskeletons, no externally visible segmentation and less evident pedipalps (most spiders and mites, etc.); this diversity is reflected in their division into 13 orders (Fig. 8.14). The

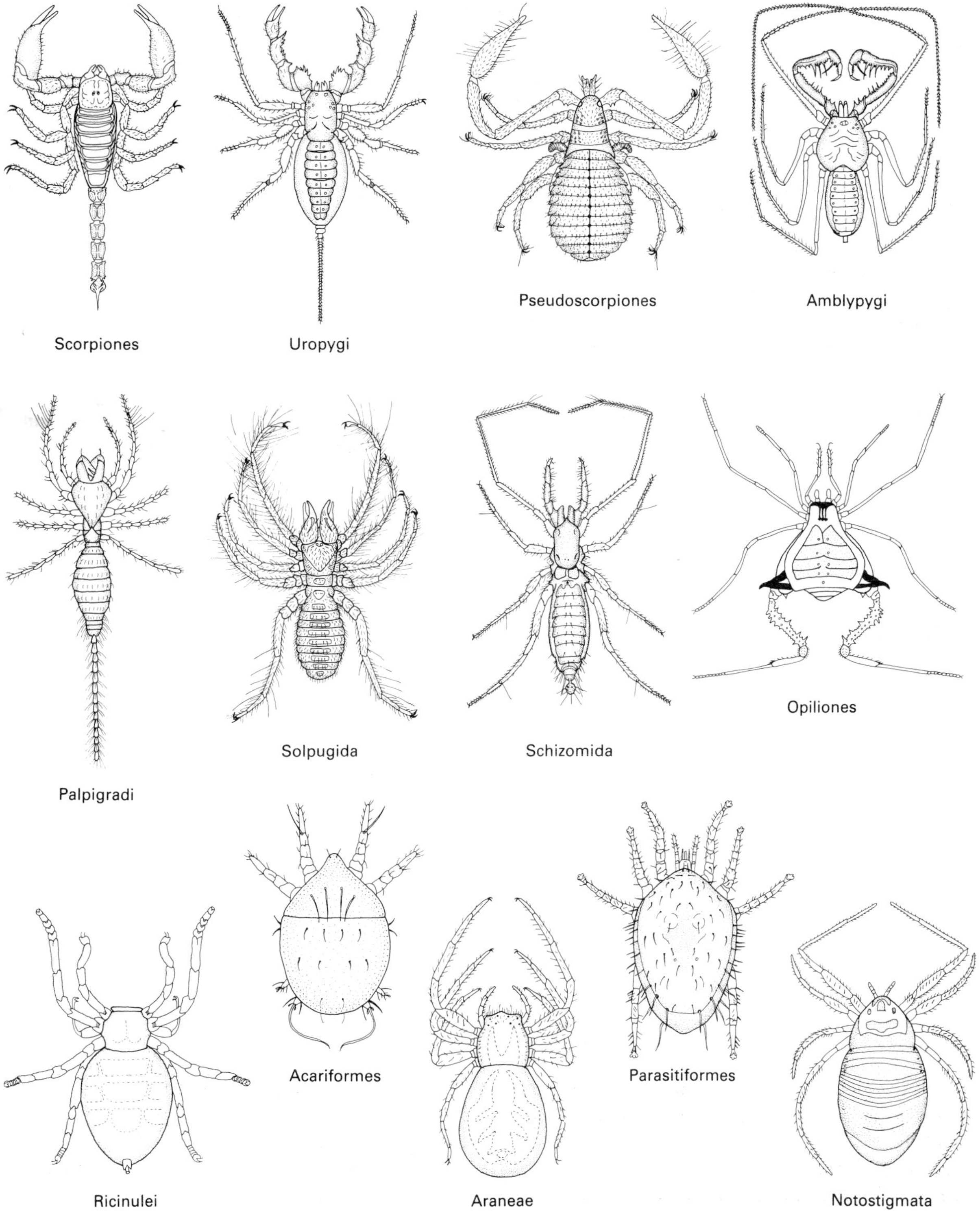

Fig. 8.14. Body form of the various arachnid orders (after Savory, 1935; Hughes, 1959 and others).

scorpion-like body form is almost certainly the primitive condition as it is little removed from that of some of the (now extinct) eurypterid merostomatans, which were freshwater and probably amphibious in habitat. Several modern arachnids have recolonized fresh waters, and some mites inhabit the sea.

In contrast to the xiphosurans, the arachnids are characterized by the absence of compound eyes and of opisthosomal ambulatory appendages, by the presence of long, non-chelate walking legs, and by the location of the gonads and digestive diverticula in the opisthosoma. In several groups of arachnids, the general reduction in massiveness of the exoskeleton has extended to the freeing of the last two prosomal segments from the carapace, although these may have separate dorsal plates, and in the evolution of a soft, flexible opisthosoma. In most mites these last two prosomal segments have been incorporated into the opisthosoma to form a two-regioned body differing from the standard chelicerate pattern. More generally, however, the prosoma and opisthosoma are retained as the two fundamental divisions of the body, the two being separated by an internal diaphragm or by a narrow stalk ('pedicel'); this separation may permit the prosomal body pressure to be raised (to extend the legs) without affecting that in the opisthosoma.

The arachnids are a most successful group, with a life style based mainly on preying on the equally

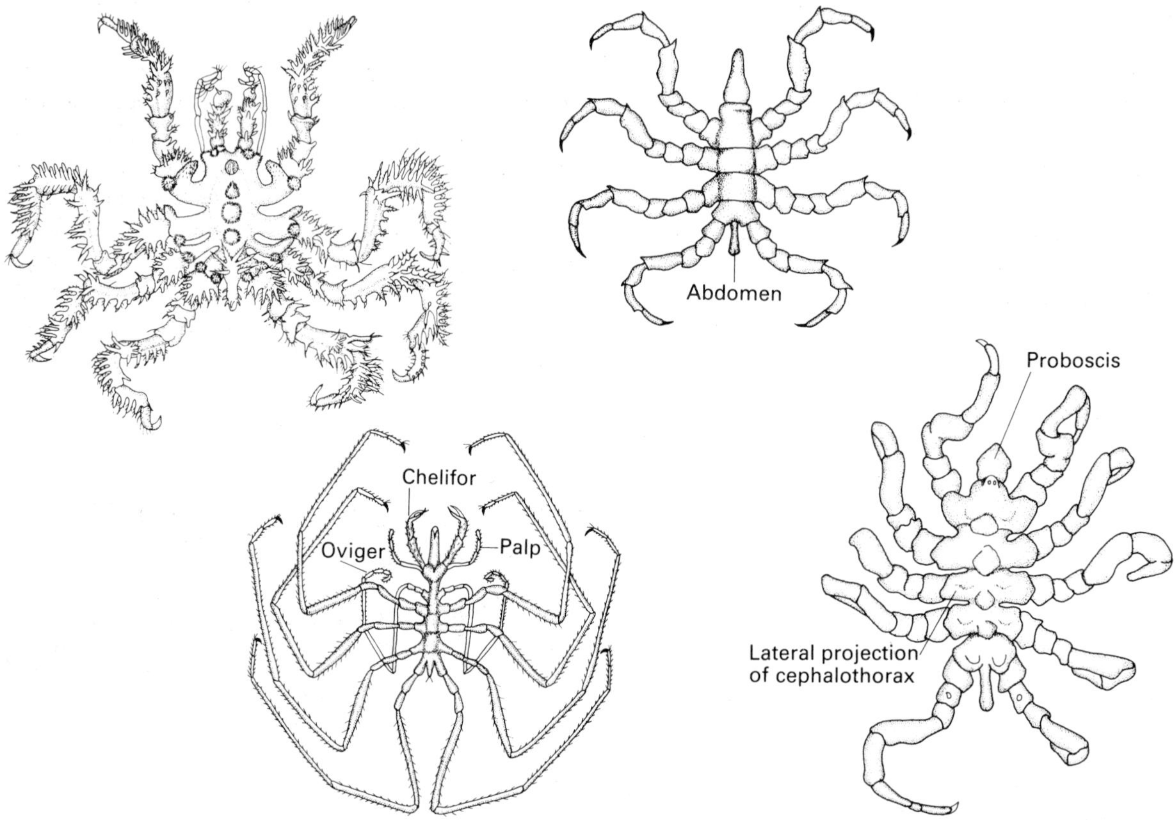

Fig. 8.15. Pycnogonan body forms (principally after Hedgpeth, 1982).

successful insects. Notable adaptations to permit the capture of such highly mobile organisms include the possession of poison glands, the secretions of which can be injected into their prey, and the production of silk, a phenomenon culminating in the construction of webs by spiders. They range in size from < 0.1 mm to 18 cm, and include over 62 000 species. Unusually, some mites are non-predatory, feeding on plant substances and on detritus; several are parasites.

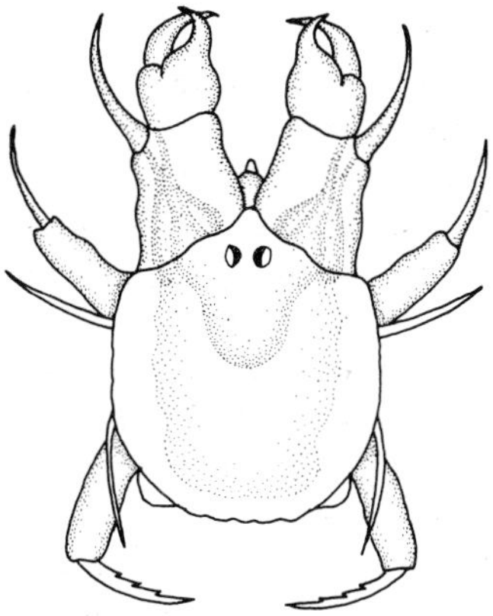

Fig. 8.16. A pycnogonan protonymphon (after Kaestner, 1968).

8.4.3.3 Class Pycnogona (sea spiders)

These small benthic marine arthropods (body length <6 cm) are considered by several zoologists to be unrelated to the other chelicerates, and possibly to the other arthropods. Their first appendage, the chelifor, *is* chelate like a chelicera; the second one, the palp, *may* be the homologue of the pedipalp; and most species *do* possess four pairs of walking legs. Some species, however, have five pairs of walking legs and a few, six pairs, whilst an additional appendage, the oviger, is located in between the palp and the first walking leg. This is attached ventrally, not laterally as in the other appendages—the walking legs articulating with stout lateral projections from the body (Fig. 8.15). In several species, the chelifores, palps and ovigers are vestigial or absent; the walking legs, however, are always well developed and may be very long (up to 75 cm span has been recorded).

The cephalothorax (prosoma?), which is not covered by any carapace, bears a central prominence, with two pairs of simple eyes, and anteriorly or ventrally, a large, tubular, non-retractable proboscis on which the mouth is terminal. The abdomen (opisthosoma?) is a minute, unsegmented papilla lacking any appendages. No excretory or gaseous-exchange organs occur. Peculiarly, the gonads extend well down into the legs, as indeed do the digestive diverticula, and the ova ripen in the legs. During

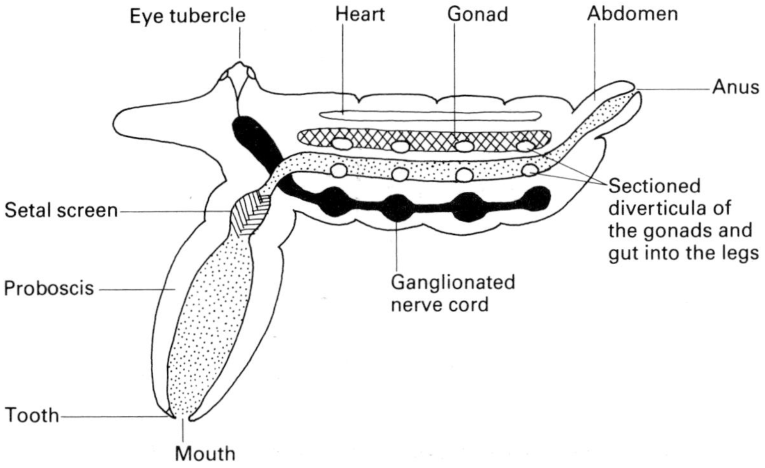

Fig. 8.17. Diagrammatic section through the body and proboscis of a pycnogonan (after King, 1973).

pseudocopulation, the gametes are discharged from multiple gonopores positioned at the base of each walking leg or at those of the last two pairs, and, after fertilization, the male gathers up the eggs and carries them around in balls attached to its ovigers. Eventually, a distinctive larval stage with three pairs of appendages, the protonymphon (Fig. 8.16), hatches and, most commonly, begins a semiparasitic life on or in cnidarians or molluscs.

The adults mostly feed on sponges, cnidarians or bryozoans, part of the prey being grasped by the chelifores (if present) whilst powerful suction generated by the pharynx results in ingestion of tissues, aided by the gnawing action of three teeth occurring at the tip of the proboscis. Within the pharynx, further teeth or strong setae macerate the food, and a collar of setae at the pharyngeal base prevents the intake of anything other than minute particles into the oesophagus (Fig. 8.17).

More than 1000 species have been described, all referable to a single order.

8.5 Phylum UNIRAMIA

8.5.1 Etymology

Latin: *unus*, one; *ramo*, branch.

8.5.2 Diagnostic and special features (Fig. 8.18)

1 Bilaterally symmetrical; <1 mm–35 cm long arthropods varying in body shape from extremely elongate to almost spherical.
2 Body more than two cell layers thick, with tissues and organs.
3 A through, straight gut lacking any digestive diverticula.
4 Body divided into two regions, a 'head' formed by the acron and three or four appendage-bearing segments, and a 'trunk' bearing pairs of walking legs; in one subphylum, the trunk comprises a series of up to 350 relatively uniform segments, the great majority of which bear walking legs; in the other subphylum, the trunk is differentiated into a 'thorax' with three pairs of legs, and an 'abdomen' of up to eleven segments with only highly modified appendages, if any.
5 Appendages uniramous, those of the head comprising one pair each of 'antennae', 'mandibles', and maxillae, and in some groups a second pair of maxillae, those of the trunk all form functional or modified walking legs; without chelicerae or chelate limbs.
6 Two or three pairs of mouthparts (the mandibles and maxillae), members of each pair working with or against the other; the basal articles of the maxillae, or second maxillae in those groups having two pairs, fuse to form a plate flooring the pre-oral cavity (the labium or 'lower lip').
7 Head with lateral ocelli, frequently organized into compound eyes; sometimes also with median ocelli.
8 Trunk, but not head, externally segmented.
9 Most members of one subphylum with one to two pairs of wings on the thorax.
10 With a fat body in the haemocoel, often closely associated with the gut.
11 Excretory system in the form of zero to two pairs of maxillary glands in the head, and one to 75 pairs of unbranched ectodermal Malpighian tubules arising from the hind-gut near its junction with the mid-gut and discharging mainly ammonia and/or uric acid.
12 Exoskeleton calcareous or, more commonly, non-calcareous.
13 Gaseous-exchange organs paired, branched tracheal tubes through which air diffuses; tracheae open (primitively) via one pair of spiracles on each leg-bearing segment and terminate, internally, in tracheoles; tracheal system not associated embryologically with the appendages.
14 Blood without any circulatory function in respect of the respiratory gases, and without respiratory pigments (except in a few larval stages).
15 Gonochoristic, with internal fertilization via spermatophores or copulation; ancestrally with gonopores on last trunk segment, modified in several forms.
16 Several groups with a distinct larval stage quite unlike the adult form; others hatching with less than

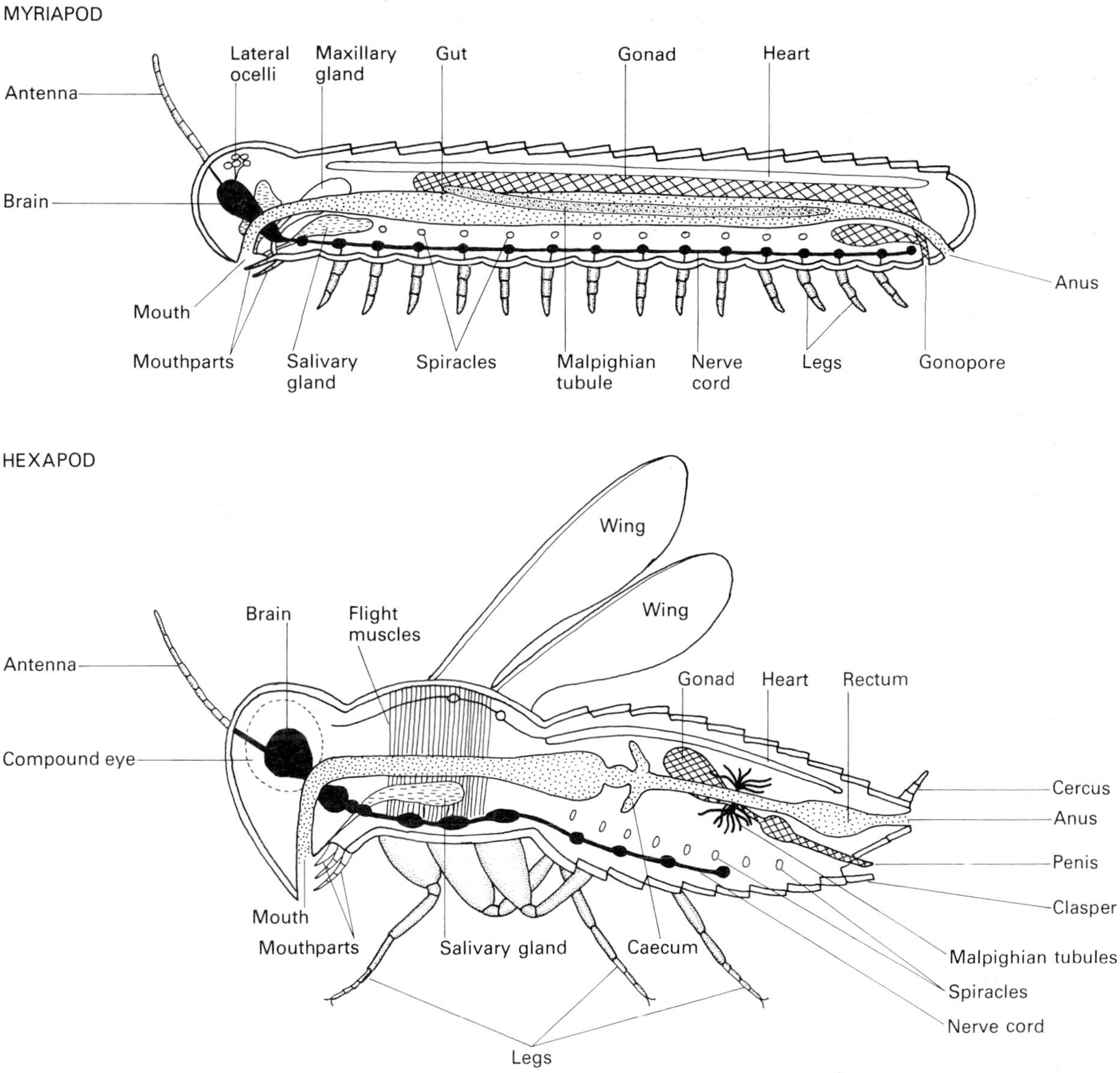

Fig. 8.18. Diagrammatic longitudinal sections through two generalized uniramians.

the full complement of segments (and legs), additional segments being added at each moult, even, in some, after reproductive maturity.

17 Primarily terrestrial, several in fresh water but very few in the sea.

The Uniramia probably originated on land and, with the arachnid chelicerates which they outnumber in both numbers of species and numbers of individuals, they are the dominant land invertebrates. Part of the reason for their terrestrial success is held in common with the arachnids. Both groups, independently, have acquired a gaseous-exchange ghsystem which relies on diffusion of the gases from the environment to the tissues and vice versa and is therefore more efficient in preventing water loss than, for example, is that of the tetrapod vertebrates, in which much water vapour is expired during forced ventilation of their lungs. The shared Malpighian-tubule excretory system is, by virtue of its association with the gut, also at least potentially capable of reducing water loss through the reabsorption of water from the nitrogenous wastes after their discharge into the hind-gut. And the presence of the cuticular exoskeleton around the external surface of the body is a greater barrier to water loss than are the soft, moist integuments of annelids and molluscs. Nevertheless, in the majority of the uniramian classes, such prevention of water loss is only partially successful, and they remain largely tied to moist microhabitats in and near the soil. Their spiracles cannot be closed, for example; little water is reclaimed from the nitrogenous waste, which is largely in the form of ammonia; and the cuticle is relatively permeable. It is in only one class that the water loss has been further, and greatly, reduced by the evolution of spiracular closing mechanisms, of an impermeable waxy epicuticle, and of a water reclamation system in the rectum. This class, that of the pterygote insects, is also by far the largest and most diverse of the Uniramia.

Although the above water-retaining mechanisms may explain why the uniramians are at least as successful as the arachnids, they do not appear able to account for their much greater apparent success. This is probably largely attributable to the possession by the uniramians of jaws capable of biting and chewing. They can take solid foods into the gut, and are not confined to a liquid diet and therefore to prey which can be externally predigested. In effect, this means that plant materials are available to the uniramians, and the land is, above all, characterized by an abundance of relatively tough plant tissues which need to be cropped and chewed before they can be ingested. In marked contrast to the almost entirely predatory arachnids, only one major group of the Uniramia, the centipedes, are exclusively carnivorous; although, for the reasons outlined in Chapter 9, most species are consumers of dead and decaying plant substances rather than the living plants themselves.

In origin, the uniramian jaw, the mandible, is the first walking leg, of which only the basal portion develops. Its cutting edge is a large, medially directed projection immovably attached to the remainder of

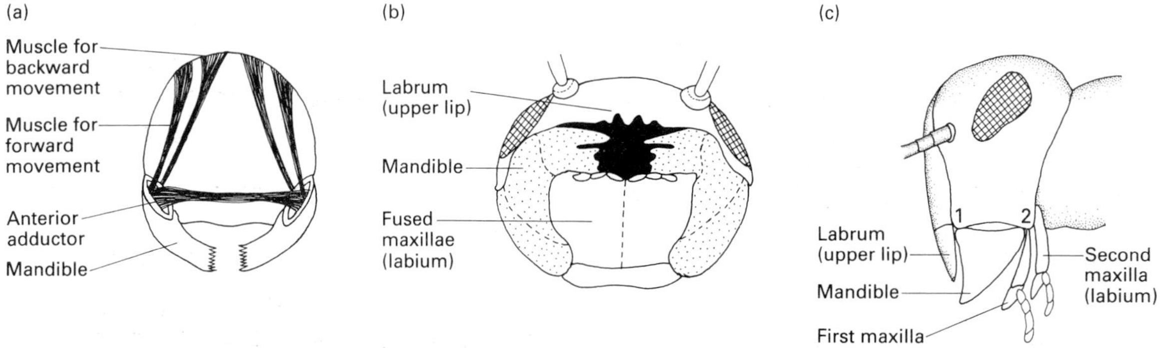

Fig. 8.19. Uniramian mouthparts: (a) a diagrammatic transverse section through the head showing the primitive state in which the mandibles are aligned vertically downwards and operated by a similar series of muscles to those of the walking legs; (b) an anteroventral view into the pre-oral cavity of a millipede, showing the upper lip (labrum) formed by sclerites of the head capsule, and the lower lip (labium) formed by fusion of the maxillae (note the horizontal orientation of the mandibles); (c) side view of an insect head with three pairs of mouthparts and a mandible articulating with the head capsule at two points (labelled 1 and 2). (After Kaestner, 1968.)

the jaw in all but the millipedes and symphylans. Primitively, the mandible articulates with the body in the same manner as the other walking legs, being oriented perpendicularly to the head capsule (Fig. 8.19a). In most uniramians, however, its alignment has changed so that it lies parallel to the ventral surface of the head, having undergone a rotation towards the anterior. Its erstwhile dorsal articulation with the body has therefore become a posterior one and, in some, a second point of articulation is developed anteriorly (see Fig. 8.19c). The freedom of movement of the jaws is then mainly limited to an opening and closing motion. The floor or posterior margin of the space in front of the mouth in which the mouthparts function is formed by the fused basal parts of the maxillae, or second maxillae in groups having two pairs, which have also been derived from walking legs (Fig. 8.19b). This contrasts with the position in the two other arthropod phyla, in which this labium is part of a ventral exoskeletal sternite of the body. In the myriapods, mouthpart anatomy varies little in its overall structure, but, in the pterygote insects, the morphology of the mandible and maxillae is extremely diverse, correlated with their radiation into numerous feeding modes (see Section 8.5,3B.2).

8.5.3 Classification

The more than one million uniramian species can be distributed between two subphyla, the one with four component classes and the other with six. The uniramian body plan is relatively conservative, however, and the differences between most of these classes are not nearly so marked as between those of many other large phyla, e.g. of the Mollusca, Chelicerata and Crustacea.

8.5.3A Subphylum MYRIAPODA

The myriapod classes probably represent the first major radiation of this phylum (or the surviving members of it), one of these classes then providing the starting point of the second major radiation in which the classes of the second subphylum arose.

Their primary distinguishing feature is that the trunk comprises a series of more-or-less identical segments, each of which, except for one or two terminal ones and sometimes for the first, bears a pair of walking legs; there is no differentiation into a thorax and abdomen. They do share a number of other common features, although these are not exclusively myriapod characters (since several are shared with the apterygote insects—see Section 8.5.3B.1). There is only a single pair of (usually elongate) Malpighian tubules, for example; the head lacks median ocelli but bears the 'organs of Tömösvary', which are probably sensitive to air-borne chemicals and maybe also to humidity; the individual articles of the antennae possess their own muscles; the nervous system is not concentrated, having ganglia in each trunk segment; and, as indicated above, their powers of water retention are relatively poor. The various classes differ mainly in the numbers of pairs of maxillae, the position of the gonopores, the numbers of dorsal tergal plates, and in which segments lack legs.

Although an exclusively terrestrial subphylum, several species in each class inhabit the intertidal zone of the sea.

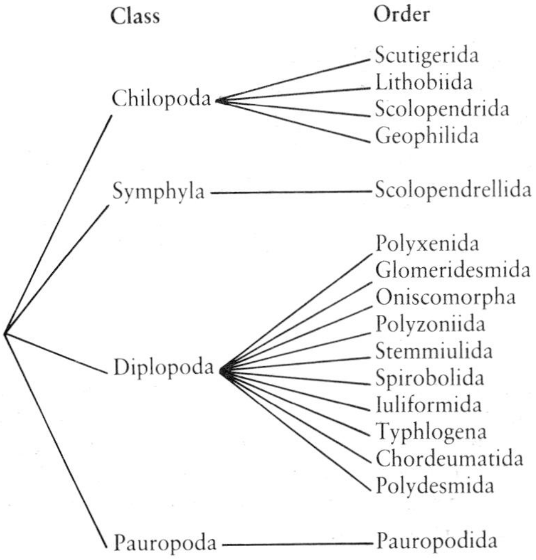

8.5.3A.1 Class Chilopoda (centipedes)

Centipedes are elongate to very elongate, dorso-ventrally flattened myriapods with a head possessing two pairs of maxillae, and a trunk comprising from 15 to more than 181 leg-bearing segments (always an odd number) together with two terminal legless ones, the pregenital and genital (Fig. 8.20). All their legs are similar, although sometimes increasing in length posteriorly, except for the first pair which, diagnostically, are modified into prey-catching organs—large, claw-like, poison-gland containing fangs (Fig. 8.21).

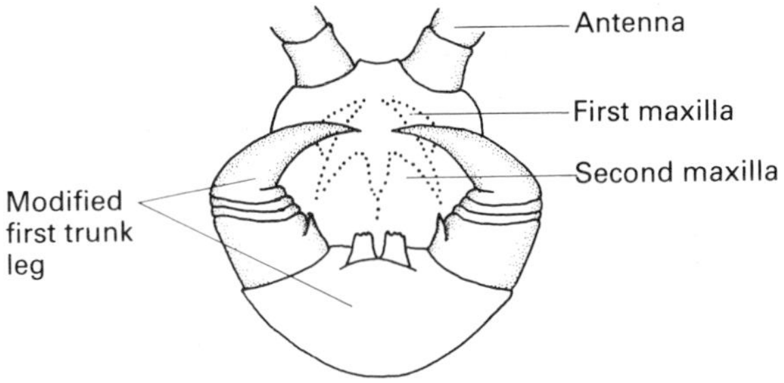

Fig. 8.21. The highly modified first pair of centipede trunk limbs (after Borror *et al.*, 1976).

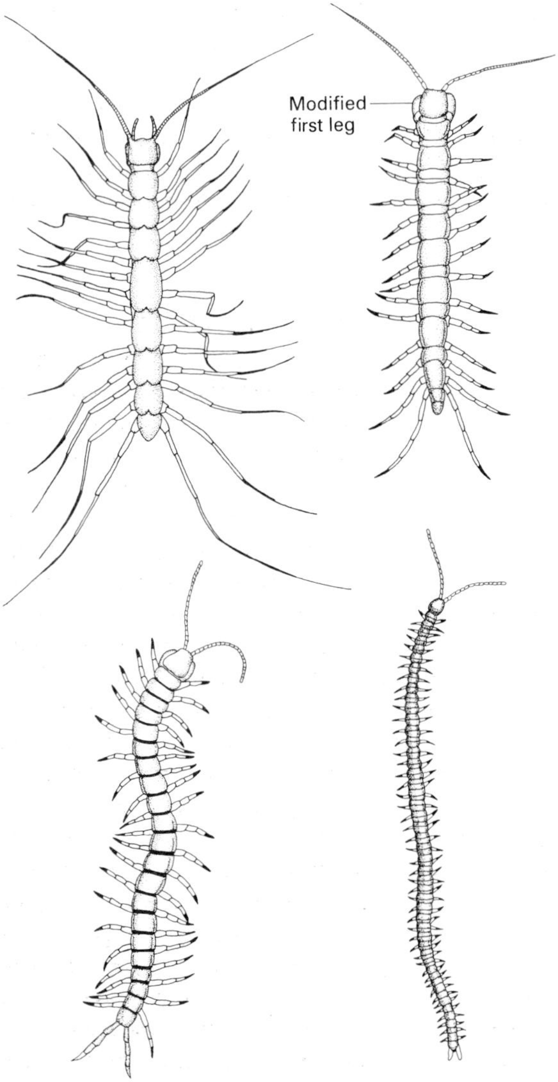

Fig. 8.20. The body form of centipedes (after Lewis, 1981).

Although most species have quite large numbers of lateral ocelli and one group (the order Scutigerida)—alone amongst the myriapods—possesses large compound eyes, the carnivorous centipedes are nocturnal or live beneath the soil surface and locate their arthropod and oligochaete prey with the antennae or, more rarely, with the legs. Some groups lack eyes.

As in most myriapods, sperm transfer is effected by means of an external spermatophore, which is extruded from the terminal gonopore only after considerable courtship behaviour. In all but one group, the male also protrudes a spinneret from the same orifice and lays down a series of silk threads in which the spermatophore may be suspended and which may guide the female to it. Several species guard the eggs and even the young after they have hatched.

Unusually in this subphylum, however, the young of many species hatch with the full complement of segments and legs (including, somewhat paradoxically, those centipedes with the largest number of segments), whilst, more typically, others hatch with less than the adult number, some with only four.

The 3000 species, which attain lengths of up to 27 cm, are divided between four orders.

8.5.3A.2 Class Symphyla

The symphylans are small myriapods (< 8 mm long) sharing the same general body plan as the centipedes, i.e. there is a head with two pairs of maxillae, and a trunk with, in this group, twelve leg-bearing segments and two terminal leg-less ones, the last fused with the telson (Fig. 8.22). Spinnerets are also located on the last free body segment. In contrast to the centipedes, however, there are many more dorsal tergal plates than there are segments—up to 24—permitting increased flexibility of the body; the gonopores secondarily open anteriorly (on the third segment); and the first pair of walking legs do not form fangs, although they are distinctive in being smaller than the following ones, sometimes only half the size. Also, unusually amongst myriapods, the single pair of short tracheae open via spiracles on the head.

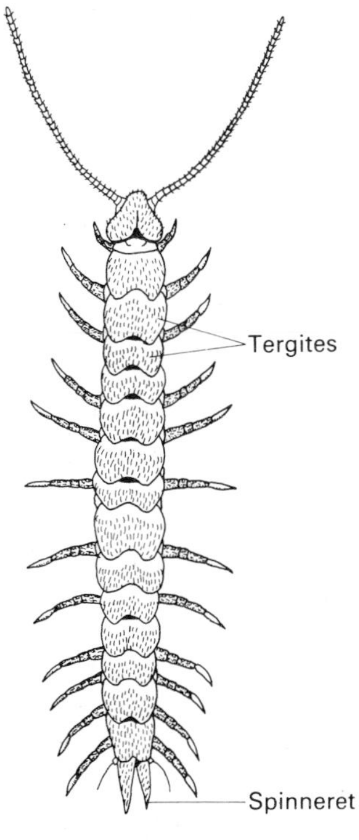

Fig. 8.22. The external appearance of symphylans (after Kaestner, 1968). Note that the number of tergites exceeds that of the leg-pairs.

In a number of other respects, however, this class resembles the members of the subphylum Hexapoda, particularly the apterygote insects. Their mouthparts are essentially similar, for example, and small styli and eversible coxal sacs are present in association with most legs; the coxal sacs can be everted by haemocoelic pressure and used to take up environmental water. Since, like most myriapods, symphylans hatch with less than the full complement of segments, it is possible that the hexapods evolved from symphylan-like ancestors by paedomorphosis (some millipedes and pauropodans also hatch with only three pairs of legs).

The 160 species of the small, blind, soft and pallid symphylans are mainly herbivorous, feeding especially on live rootlets and living in soil, leaf litter and rotting wood, under stones, etc. All species are placed in a single order.

8.5.3A.3 Class Diplopoda (millipedes)

The short to very elongate millipedes (Fig. 8.23) have as their most conspicuous distinguishing feature the occurrence of 'diplosegments'. The diplopod trunk is composed of a leg-less first segment, three following segments each with the typical myriapod single pair of legs, and then a series of from five to more than 85 'rings' each with two pairs of legs, ganglia, heart ostia, etc., in front of the one or more leg-less segments which comprise the terminal segment-proliferation zone. Each diplosegmental ring, which together comprise most of the almost cylindrical or, rarely, flattened trunk, is formed by the partial or complete fusion of the trunk segments in pairs; at the very least, such diplosegments share a common dorsal tergite. In many species, one or two leg pairs (those of the seventh apparent segment, or the posterior pair of the seventh and the anterior pair of the eighth) are, in the male, modified to varying extents to form copulatory 'gonopods', which collect sperm from the gonopores on the third segment and transfer it to the corresponding gonopores of the female. In other species, the mandibles are used to transfer the sperm, or the gonopores of the two sexes may be brought into close proximity, or, in classic myriapod fashion, sperm packets and silk guide threads are produced. The eggs of most millipedes hatch with only seven trunk 'segments', new ones being added throughout life, and long after the onset of sexual maturity.

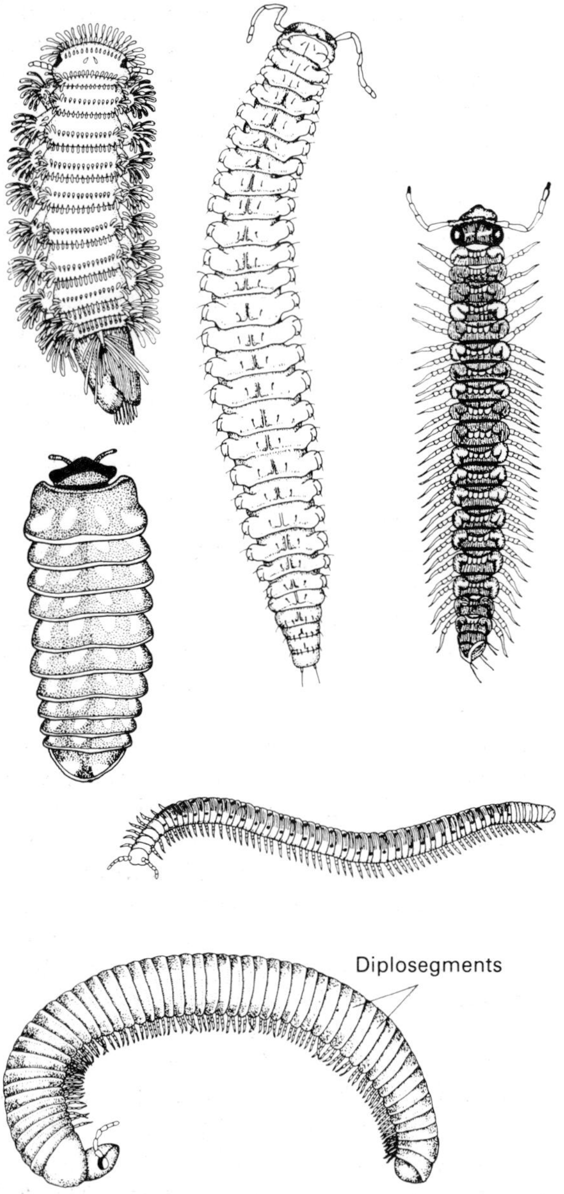

Fig. 8.23. Millipede diversity (after several authors, principally Blower, 1985).

Other millipede characteristics include the presence of only one pair of maxillae, of large numbers of lateral ocelli arranged in blocks which superficially resemble compound eyes (some species, however, lack eyes), and, except in one order, of a calcified exoskeleton—millipedes being the only uniramians to have such.

The 10000 species are disposed in ten orders. All are slow-moving feeders on plant material, usually only after it has started to decompose, which they obtain in the same microhabitat types as the symphylans (rotting logs, leaf litter, etc.). The largest achieve a length of 28 cm.

8.5.3A.4 Class Pauropoda

The minute pauropods (<2 mm length) live in the same habitats as the symphylans and millipedes, where they bite into fungal hyphae and suck out their contents by means of a pumping fore-gut. Although inconspicuous, they are often abundant. In their body plan, they bear a similar relationship to the millipedes as do the symphylans to the centipedes. The head, for example, bears a single pair of maxillae; the first trunk segment is leg-less, as are the last two; the gonopores are anterior, on the third segment; the trunk segments are arranged in incipient diplosegments, a single tergal plate partially or wholly covering segments 1 and 2, 3 and 4, 5 and 6, and so on (Fig. 8.24), up to the total of eleven or twelve segments, the last fused with the telson; and the young hatch with few segments and most with three pairs of legs.

Pauropods are not just incipient millipedes, however, as they show a number of distinctive features. They lack a heart, for instance, and most lack tracheae; even when present, the tracheae are very small. Further, they possess branched antennae.

The 500 species, all of which are blind, soft and colourless, are placed in a single order.

8.5.3B Subphylum HEXAPODA (insects)

Only one characteristic separates all the members of this subphylum from the Myriapoda: the hexapod

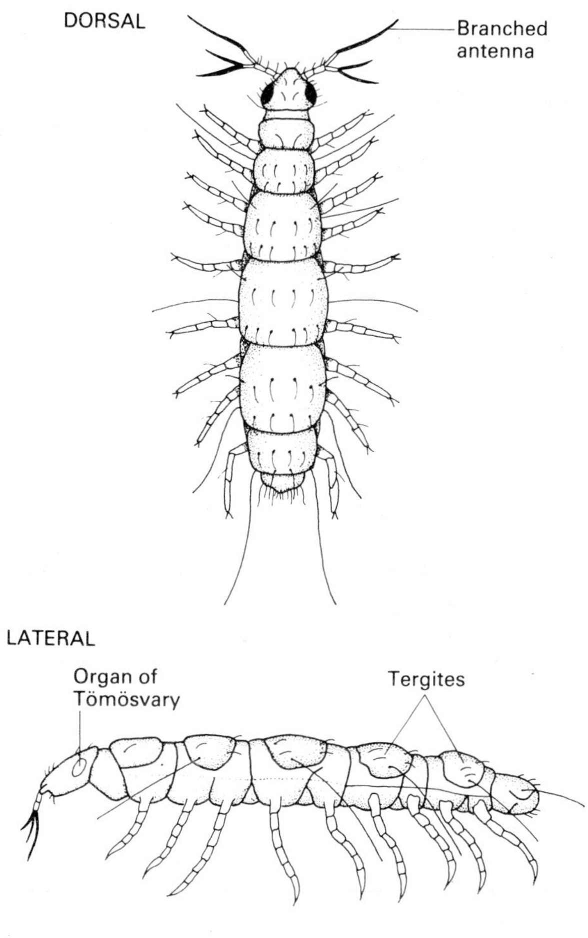

Fig. 8.24. The external appearance of pauropods (after Kaestner, 1968 and Borror *et al.*, 1976). Note that the tergites are less numerous than the leg-pairs.

trunk is subdivided into a thorax of three leg-bearing segments, and an abdomen of eleven segments without walking legs; although abdominal appendages of some form may be present and loss or fusion may reduce the number of abdominal segments to less than eleven. All hexapods also have two pairs of maxillae and terminal or subterminal gonopores, and most possess median ocelli as well as lateral ocelli or compound eyes.

8.5.3B.1 The Apterygote classes (wingless insects)

Five primitively wingless groups of insects are recognized, each of which must have diverged from the others at a very early stage of insect evolution (if indeed they did not derive separately from their myriapod ancestor or ancestors), and hence each can be regarded as constituting a separate class. Only one of these classes shows any clear affinity with the winged insects, which comprise the sixth hexapod class. All the apterygote classes are small, with only a single order each, and they are most conveniently treated together.

The five classes are broadly separable into three groups, of which the first (the diplurans, Class Diplurata; the spring-tails or collembolans, Class Oligoentomata; and the proturans, Class Myrientomata) all possess mouthparts partially enclosed within the head capsule, and share, between them, a number of other features in common with the myriapods rather than with the other insects. Their habits, including reproductive behaviour, and habitats are also very myriapod-like; not surprisingly, therefore, many entomologists would redefine the Myriapoda so as to include these groups within that subphylum. Amongst the presumed ancestral characters which they have retained are: mandibles with a single articulation with the head; a three-lobed hypopharynx (a median tongue-like organ associated with the salivary glands); eversible coxal sacs and abdominal styli; antennal articles with their own musculature; organs of Tömösvary (or structures very similar to them); and abdominal appendages of some form. All are small (generally with lengths of less than 7 mm) and, perhaps for that reason, lack Malpighian tubules, nitrogenous waste being eliminated through the mid-gut epithelium; two of the three classes also lack eyes.

The Diplurata (Fig. 8.25a) are without particular specializations such as distinguish the other two classes. The Myrientomata (Fig. 8.25b), however, have vestigial antennae, their function having been taken over by the first pair of thoracic legs, and piercing, stylet-like mandibles. Like most myriapods, the number of their trunk segments

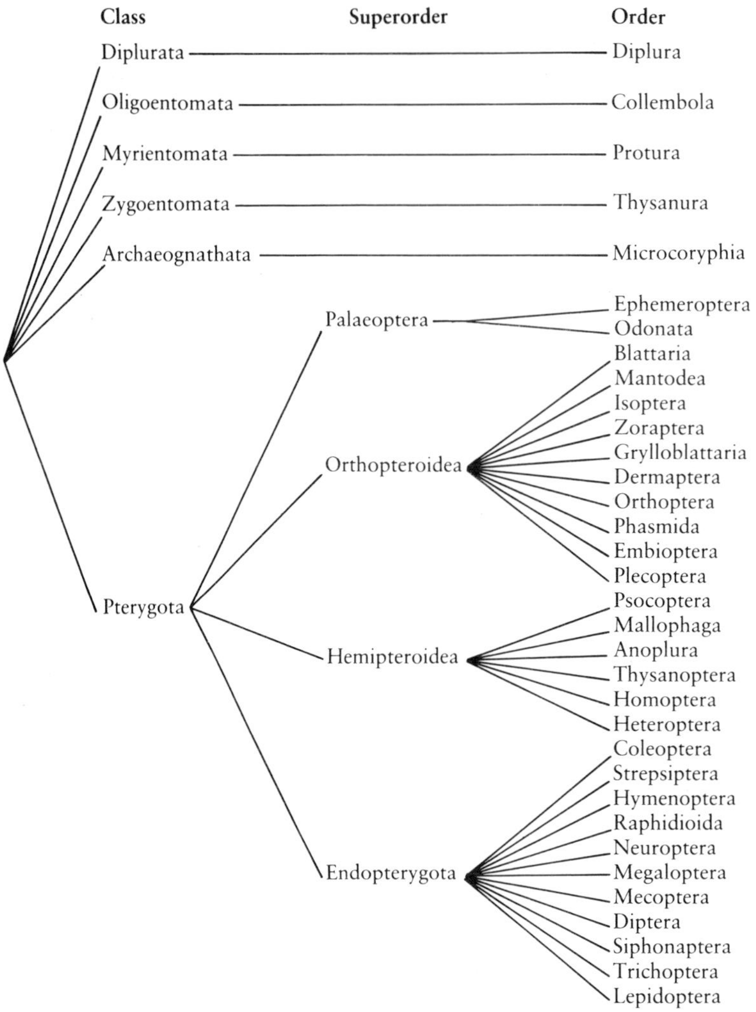

increases after hatching until the full adult complement of eleven is attained. The Oligoentomata (Fig. 8.25c), on the other hand, have retained only five abdominal segments (plus the telson) and have adapted two pairs of abdominal appendages to form the organ which gives them their common name of spring-tails. The limbs of the fourth segment form a spring which can be forcibly extended by haemocoelic pressure. When not in use it is held by a catch developed from the appendages of the third segment. Further, the limbs of the first abdominal segment are involved in the formation of a large ventral tube which contains eversible, coxal-sac-like vesicles and is of debated function. It may be used to take up environmental water. In none of these three classes are the gonopores on the eighth (female) and tenth (male) abdominal segments as in all other insects, nor do they display low rates of water loss.

The second grouping, containing the silverfish and firebrats (Class Zygoentomata) (Fig. 8.26a), differs from the pterygote insects to a much smaller extent. Apart from their primitive winglessness (and the thoracic structure associated with this state), they differ essentially only in the retention of three

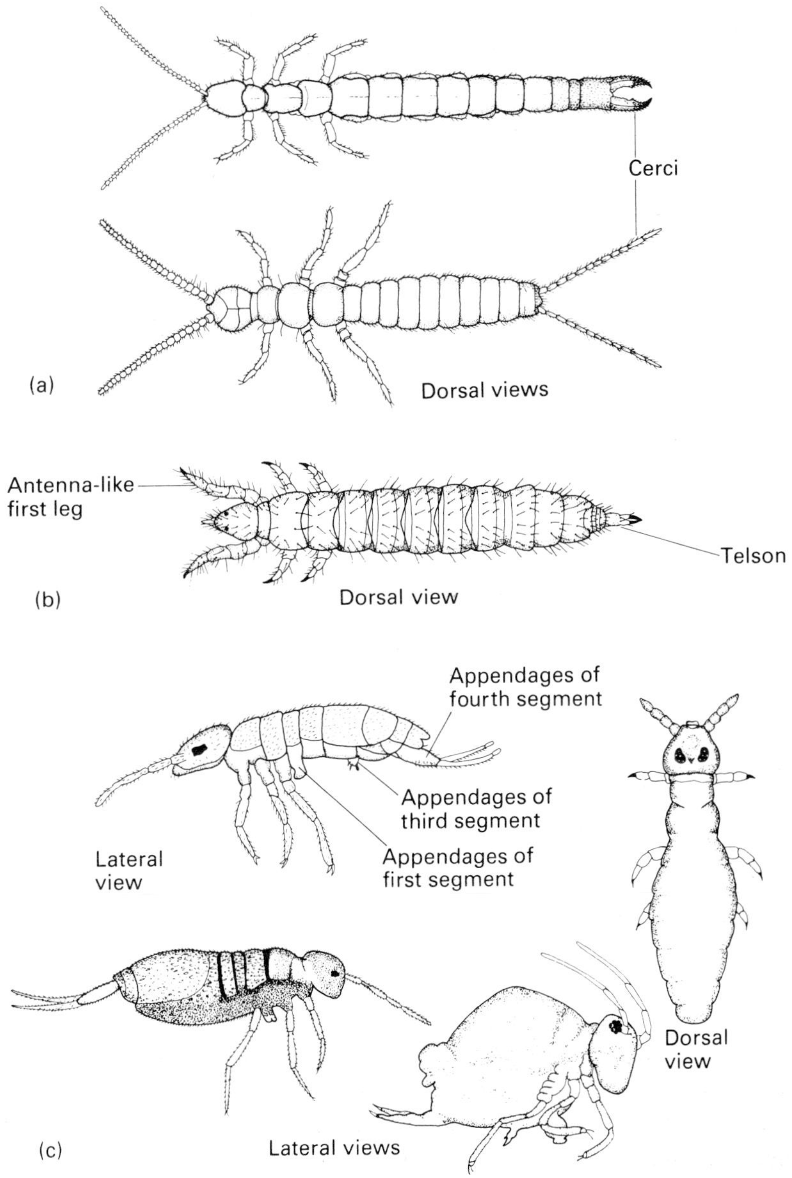

Fig. 8.25. The myriapod-like apterygote classes: (a) two diplurans; (b) a proturan; (c) various spring-tails, one showing the modified abdominal appendages (after Imms, 1964 and Wallace & Mackerras, 1970).

ancestral (and general apterygote) characteristics: the presence of abdominal styli; the absence of copulatory organs, sperm transfer being by externally deposited spermatophores; and in continuing to moult after attaining sexual maturity. For this reason, they are generally regarded as being close to the ancestry of the pterygotes.

The bristle-tails (Class Archaeognathata) (Fig. 8.26b), which form the third general group, straddle the interface between the Zygoentomata and the other

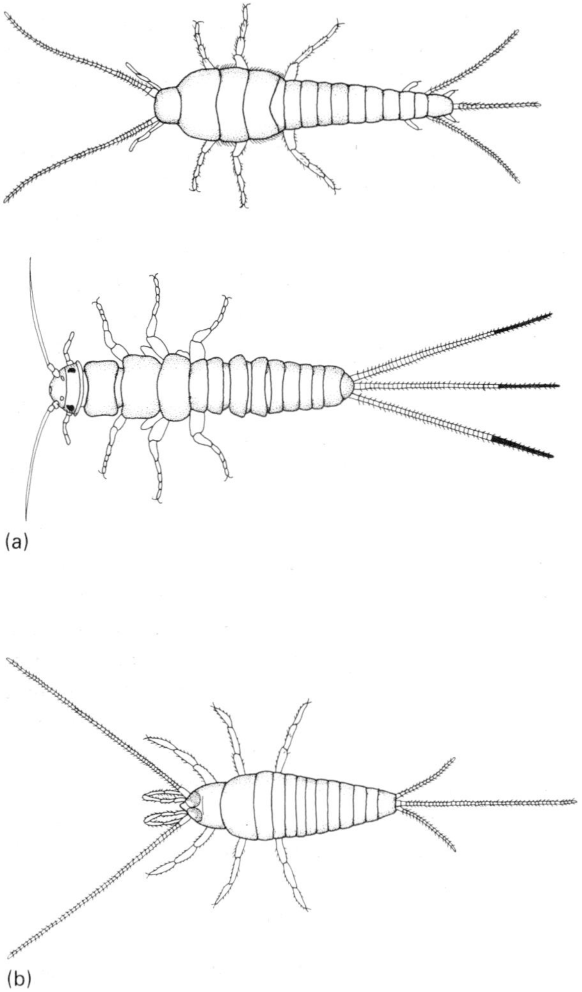

Fig. 8.26. The hexapod-like apterygote classes: (a) two zygoentomatans; (b) an archaeognathan (after Imms, 1964 and Daly *et al.*, 1978).

apterygotes, possessing some of the advanced features of the silverfish and firebrats but retaining many of the ancestral ones of the myriapod-like groups, often in respect of the same appendage or organ system. Thus the mouthparts are not enclosed within the head, but the mandibular articulation, hypopharynx and maxillae are all otherwise in the ancestral myriapod state. They also display all the primitive characters retained by the zygoentomatans. Just as it is possible that the winged insects were descended from the Zygoentomata, so it is possible that the latter were derived from the Archaeognathata, and so both classes are usually included within the same grouping as the 'true insects' by entomologists. They are both also generally larger (up to 2 cm in length) and less tied to humid litter and soil microhabitats than the other apterygotes.

The five classes together total somewhat over 3100 species, the majority of which are scavengers and feeders on decaying plant material, although some capture other small hexapods.

8.5.3B.2 Class Pterygota (winged insects)

The 29 orders of winged insects, including secondarily wingless forms, comprise the third major radiation of the Uniramia, and it is this class which currently dominates the phylum with some 98% of its living species. Above all else, their success has been made possible by emancipation from the humid types of microhabitats to which all the other uniramian classes are confined. And this, in turn, is largely a result of their ability to restrict the rate at which water leaves their bodies. The evolution of this state probably occurred in a number of stages, beginning with the properties of the cuticle and of the tracheae.

It is possible that an epicuticular wax layer was originally of adaptive value as a water-repelling hydrofuge layer: in environments liable to become waterlogged, small animals can become trapped in water films, and water uptake is a much more serious problem than water loss; the maxillary glands of apterygotes may have to pump out considerable volumes of surplus water. Some spring-tails have developed a series of wax-covered tubercles which serve a hydrofuge function, and the ancestral pterygote can be envisaged as being similarly equipped. It also may well have lacked a tracheal system, as do most of the myriapod-like apterygotes, this perhaps largely being a consequence of the small size of any animal originating by paedomorphosis, especially when its ancestors were already small. Once tracheae developed (were redeveloped), the waxy hydrofuge layer could spread to cover the whole body surface without impeding gaseous exchange.

Terrestrial arthropods may lose more water through their tracheal system, however, than through the cuticle, and so a waterproofed integument would by itself be ineffective.

Characteristically, pterygotes possess muscles which close the spiracles (and may have others to open them) and, for example, insects living in dry habitats may open the spiracles for only a small fraction of the time. Other means by which water loss through the spiracles is reduced may also be present. One such is to store the carbon dioxide generated in a non-gaseous phase for long periods. As oxygen in the tracheal air is consumed and not replaced by an equivalent volume of carbon dioxide, so a partial vacuum develops and serves to draw more air in through the spiracles in a one-way flow, counteracting any tendency of water vapour to diffuse out.

A final refinement of the water-conservation system is located in the hind-gut in the form of a mechanism whereby water can be resorbed from the faeces and excretory discharges as they pass through. Again, this may take several forms, of which one widespread one involves the active secretion of inorganic and/or organic solutes into a system of intercellular spaces within pads or papillae in the rectum wall. Water then diffuses passively from the gut lumen into these spaces along the osmotic gradient, and as the fluid is transported into the haemocoel, the solutes are absorbed and recycled (Fig. 8.27).

The colonization of non-humid habitats made possible by these heightened powers of water economy would have had many repercussions on the biology of the early pterygotes, and these account for many of the characteristic features of the group. The deposition of external sperm droplets or spermatophores, for example, is not a viable strategy in relatively dry conditions, and pterygotes transfer sperm directly, by copulating, although the spermatozoa are still usually contained within a spermatophore. Primitively, the females retain the ancestral paired gonopores and the male possesses paired copulatory organs, although in most groups only a single, median female pore occurs and correspondingly the two male organs fuse into a single structure. Males also possess a pair of claspers which hold the female whilst the eversible distal section of the male duct or distinct gonopod is inserted through her gonopore to deposit one or more spermatophores or sperm packets into the bursa copulatrix or some other section of the female reproductive tract. On release from the spermatophore, the sperm are usually stored—sometimes for considerable periods

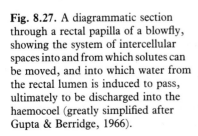

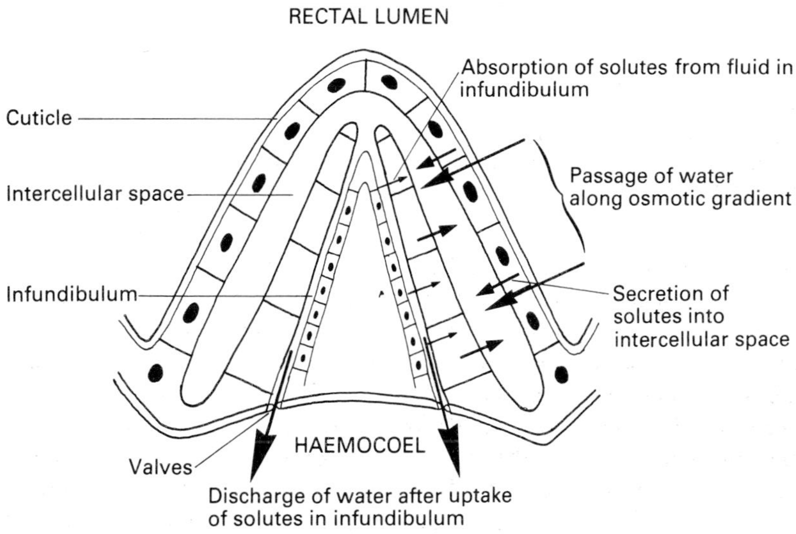

Fig. 8.27. A diagrammatic section through a rectal papilla of a blowfly, showing the system of intercellular spaces into and from which solutes can be moved, and into which water from the rectal lumen is induced to pass, ultimately to be discharged into the haemocoel (greatly simplified after Gupta & Berridge, 1966).

of time—in a spermatheca, until eventually they fertilize the eggs as these are laid, entering through minute pores in the shell which characteristically protects pterygote eggs. A number of female insects mate only once, although the sperm received may fertilize several batches of eggs; no species moults once it has achieved sexual maturity. Many possess an ovipositor, permitting the eggs to be laid in specific, usually concealed sites; in others, the terminal portion of the abdomen can be extended in the form of a tube to achieve the same end, whilst in several hymenopterans (bees, wasps, etc.) the ovipositor has been adapted to form a sting and is no longer associated with egg laying. The male claspers and gonopod, and the female ovipositor (Fig. 8.28), are derived from the ancestral abdominal appendages of the genital segments; apart from the widespread occurrence of a pair of terminal cerci,

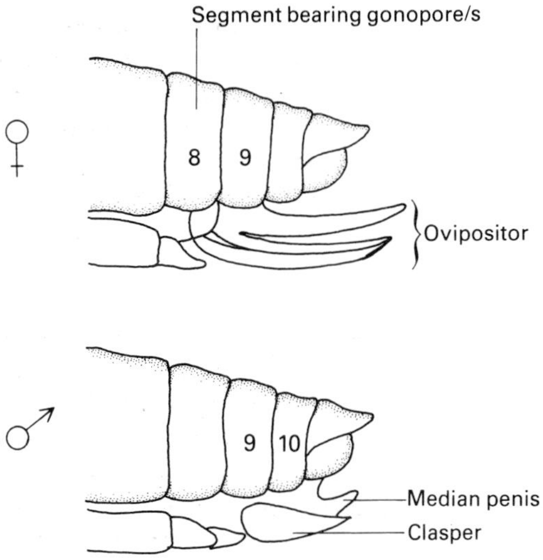

Fig. 8.28. Modification of the segmental abdominal appendages of the ancestral uniramians to serve egg-laying and copulatory functions in the pterygotes (lateral view) (after Snodgrass, 1935). The ovipositor of the female represents the basal portions of the appendages of the eighth and ninth abdominal segments; the male gonopore/s belongs to the tenth segment, although in most insects it is displaced anteriorly so as to appear to be on the ninth. Note, however, that many pterygotes have evolved secondary copulatory organs and means of oviposition.

which are the appendages of the eleventh segment, these copulatory and egg-laying organs are the only remnants of the series of these segmental appendages to remain functional in the adult pterygote.

More than any other feature, however, the Pterygota are diagnosed by their possession of wings—another development only possible after release from confinement to humid habitat systems. One pair of wings, in addition to the walking legs, occurs on each of the second and third thoracic segments of the adult, although one or both pairs may be reduced or lost. The origin of these wings, and the original selective advantages of the wing precursors, are unknown, although they may have developed from the trachea-rich, plate-like lateral expansions of the thoracic tergites that occur, for example, in some zygoentomatans and in various fossil hexapods (see Section 10.6.3). Certainly in living pterygotes, they form from evaginations of the body wall, four buds of epidermis growing out dorsolaterally. The upper and lower surfaces of these buds fuse together except along a series of channels in which blood flows and tracheae and sensory nerves are located. Eventually, the epidermal cells secrete a thin encasing cuticle, and when the moult into the adult occurs, the small, somewhat fleshy wing buds are inflated by haemocoelic pressure, the lining of the channels is reinforced with cuticle (to become the final 'wing veins'), the epidermis degenerates, and the wing comprises a thin double layer of cuticle. Of necessity, therefore, the wings are operated by extrinsic muscles, some, and only some, of which are attached to the wing base. The wings must also be hinged to permit them to describe the elliptical or figure-of-eight path required for flight. In small insects, the wings may beat at rates of up to one thousand times per second and, unusually amongst animals, there are several contractions of the flight muscles consequent on each received nerve impulse (see Section 16.10.5). The wings of primitive insects are usually relatively large and contain numerous wing veins, although the size of wing and number of veins have been greatly reduced in many evolutionary lines (Fig. 8.29); this correlates with a progression from long-bodied animals with a stable but

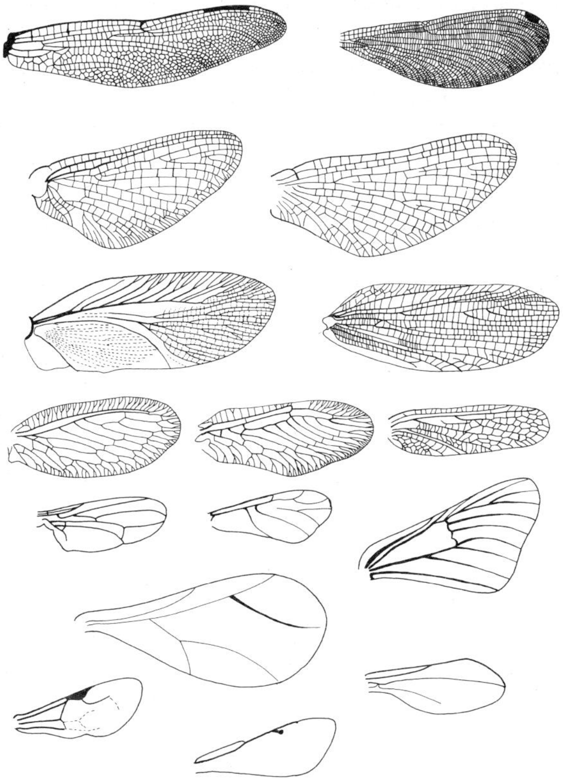

Fig. 8.29. Variation in the venation of pterygote wings (after several authors). Primitive wings with numerous veins are shown at the top of the figure, and down the figure is arranged a series displaying a progressive reduction in the number of veins.

unmanoeuvrable flight to short-bodied forms with unstable but highly manoeuvrable flight characteristics. These same lines have also evolved the ability completely to fold up the wings when they are not in use, a development which reaches a climax when all or part of the fore pair are modified to serve as protective coverings for the folded hind wings. Insect flight is considered in Section 10.6.3.

The final factor in the evolutionary success of the winged insects is probably the diversity of the food sources which they can exploit. This is reflected in an almost bewildering variety of mouthpart structure, different orders modifying different appendages, although the pterygotes are united, and contrast with most apterygotes (except the Zygoentomata), in having mouthparts which are not enclosed within the head capsule, a mandible with two points of articulation, and a single-lobed hypopharynx. The ancestral form of these appendages was almost certainly the standard myriapod system adapted for biting and chewing, and several modern species have retained this configuration (Fig. 8.30a). From this, however, have evolved many lines of piercing and sucking mouthparts, of which only a few examples can be given here.

The bugs, for example, can, dependent on group, feed on plant or animal fluids, by means of a beak formed from the labium (the second maxillae) within which are located two pairs of stylets derived from parts of the mandibles and maxillae (Fig. 8.30b). The mandibular stylets pierce the prey's tissues and allow the maxillary stylets to enter the wound. The structure of the maxillae is such as to enclose a pair of canals, down one of which passes saliva, and up the other the tapped fluids. The mosquitoes (Fig. 8.30c), on the other hand, have two additional stylets formed by the labrum (upper lip), which encloses the food canal, and the hypopharynx, which contains the salivary duct. In both cases, however, the labium serves solely as a guard and it telescopes or folds back as the other mouthparts enter the tissues.

Other insects suck more readily available liquids, such as nectar or fluids oozing from wounds. Butterflies and moths, for example, imbibe nectar through an elongate proboscis which can be extended by haemocoelic pressure (reduction in this pressure leads to re-coiling under the proboscis's own elasticity). This is formed from part of the maxillae; all the other mouthparts are reduced or missing (Fig. 8.30d). In some moths, the tip of the proboscis is sharp and barbed, and can be used as a piercing organ. In the houseflies, it is the labium which functions as the organ of uptake. Again, the mandibles are absent, but in this group the labrum and hypopharynx are enclosed within a groove in the large labium, which bears terminally a large, soft, bi-lobed 'labellum' which acts like a sponge (Fig. 8.30e).

In yet further species, the mandibles although huge are not used in feeding at all, but rather in defence or in fighting for mates (Fig. 8.30f); and some members of at least eight groups do not feed when adult and have vestigial mouthparts. In a number of pterygotes, especially those with a distinct larval stage

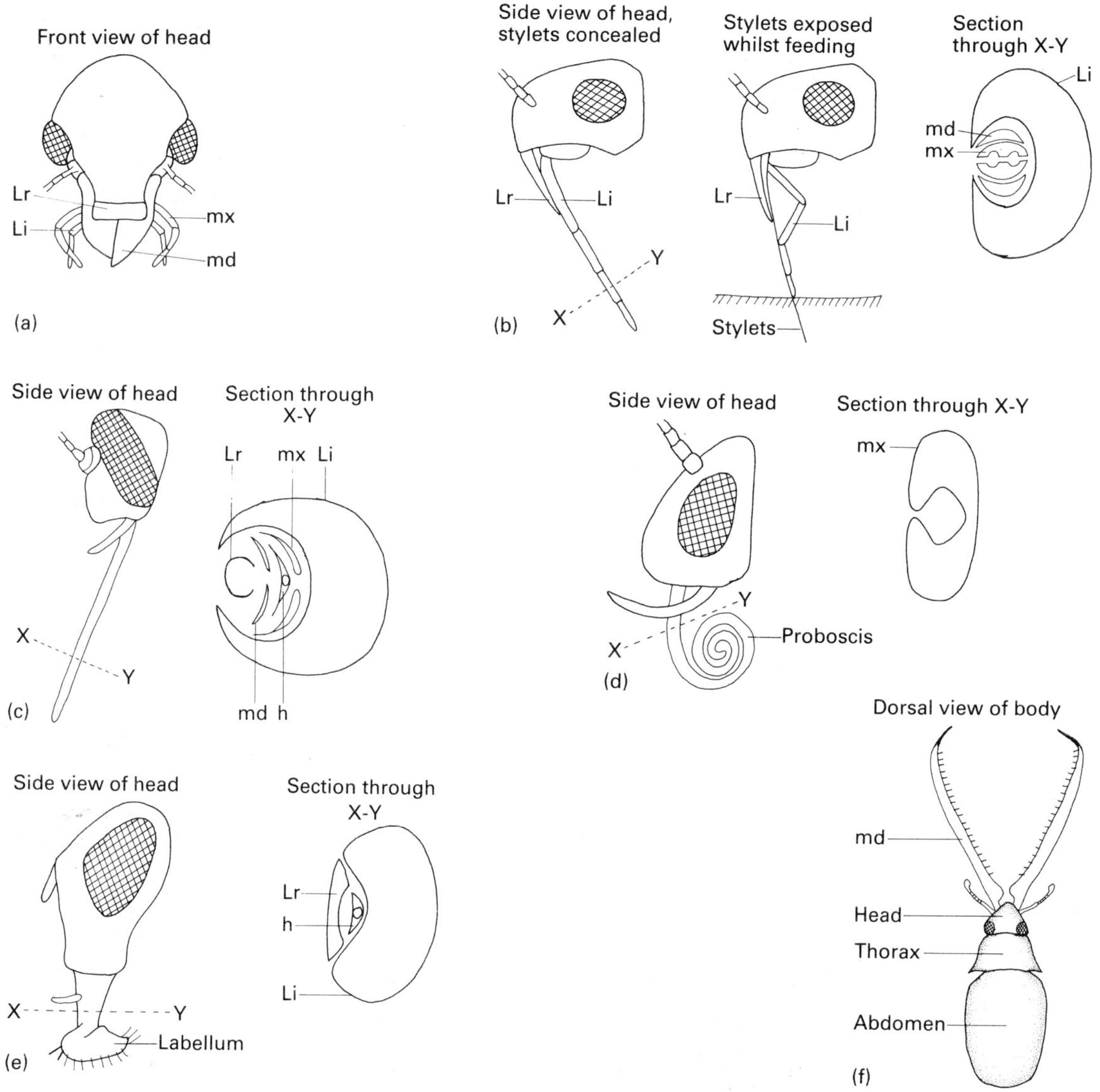

Fig. 8.30. Examples of the adaptive radiation of pterygote mouthparts (see text): (a) biting mandibles of a beetle (Coleoptera); (b) the piercing/sucking beak of a bug (Heteroptera); (c) the mouthparts of a mosquito (Diptera); (d) proboscis of a nectar-sucking butterfly (Lepidoptera); (e) sponging mouthparts of a housefly (Diptera); (f) the greatly enlarged mandibles of a male lucanid beetle (Coleoptera) used in intra-sexual combat and not for feeding (after several authors, principally Borror *et al.*, 1976). Key to labels: lr = labrum; md = mandible; mx = maxilla; li = labium; h = hypopharynx.

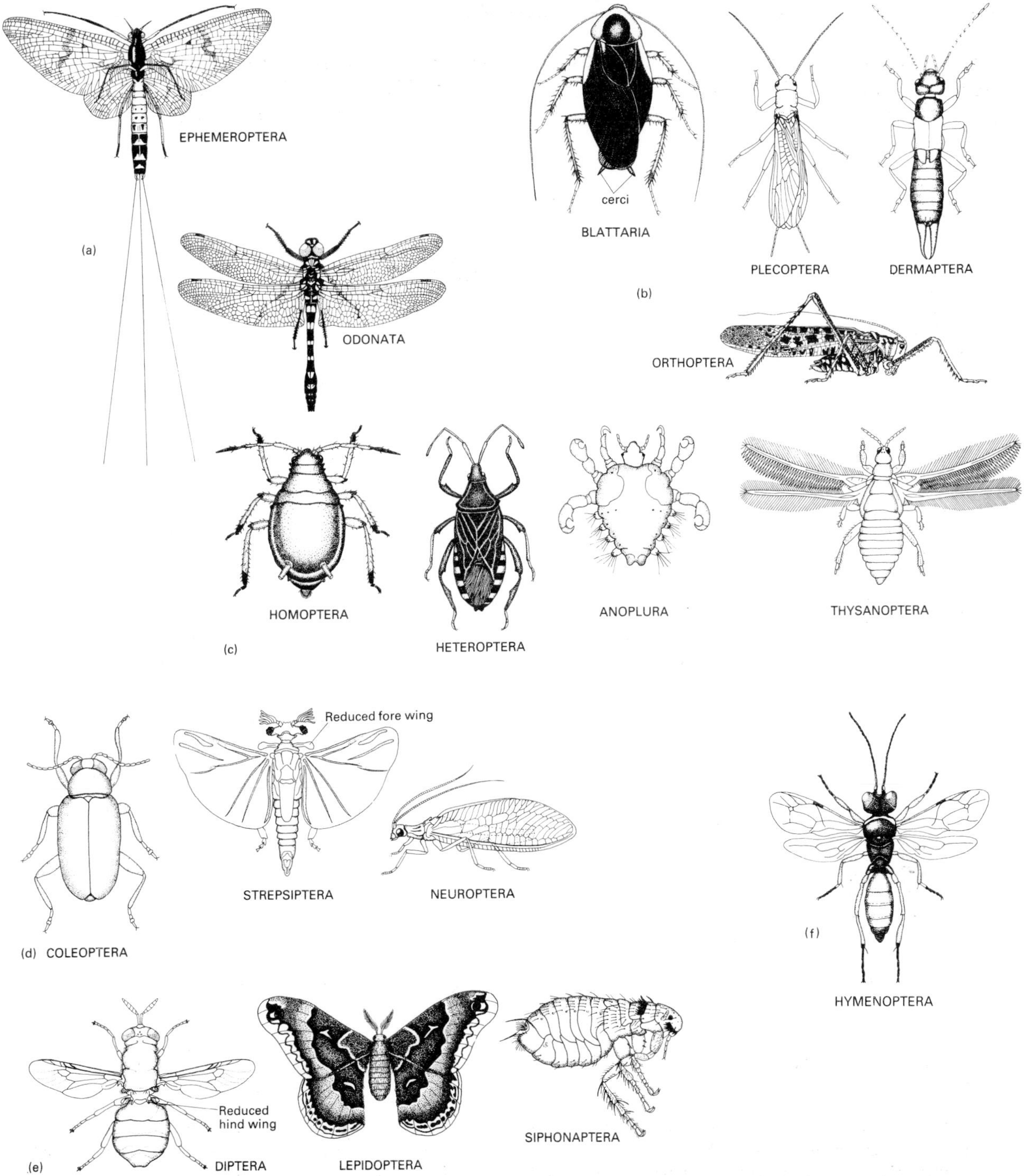

Fig. 8.31. Diversity of pterygote adult body form: (a) representative palaeopterans; (b) orthopteroids; (c) hemipteroids; (d) neuropteroid endopterygotes; (e) pancrpoid endopterygotes; (f) hymenopteroid endopterygotes (after various authors).

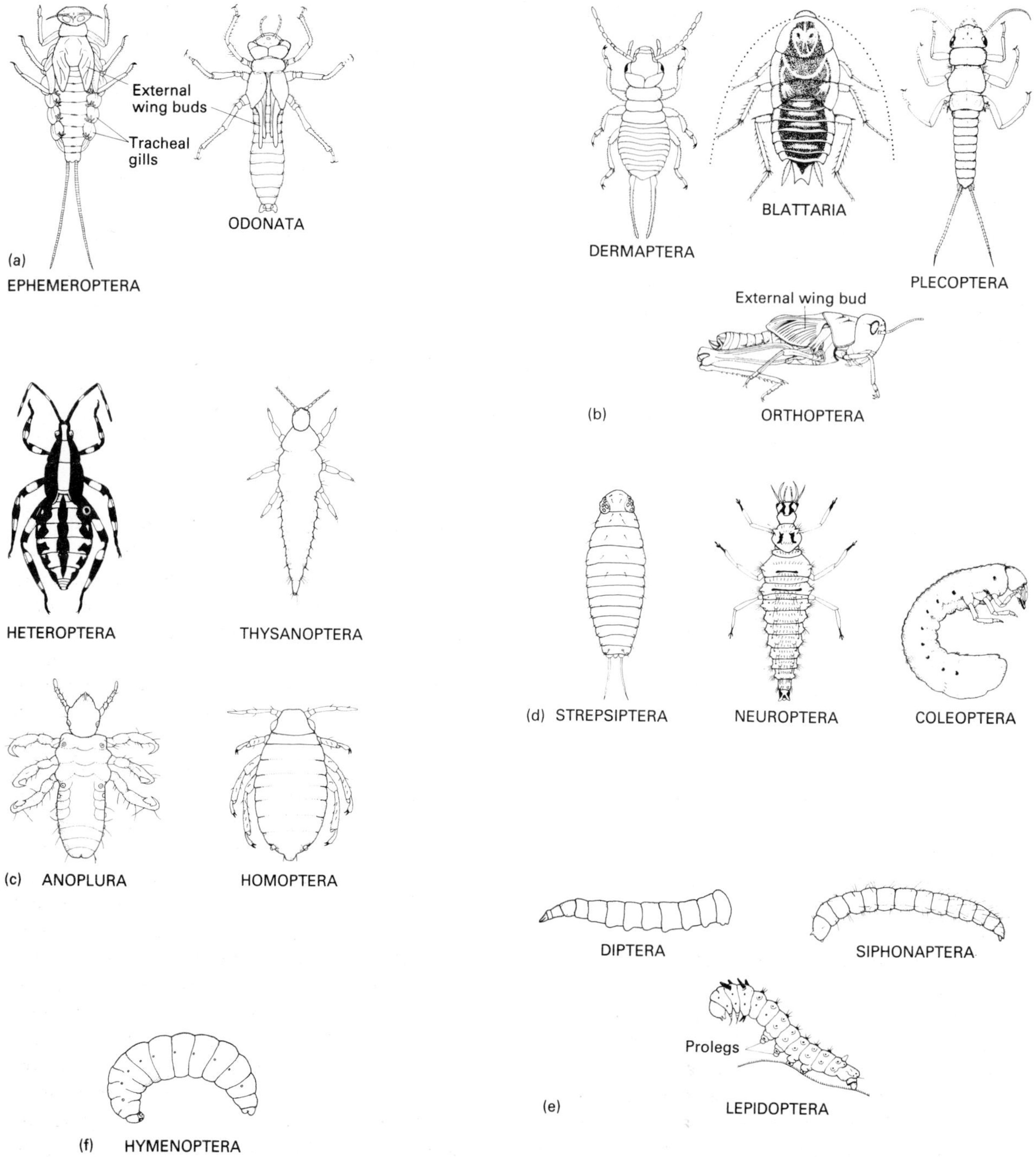

Fig. 8.32. The juvenile forms of the pterygote groups illustrated in Fig. 8.31: (a) palaeopteran naiads; (b) and (c) orthopteroid and hemipteroid nymphs; (d)–(f) endopterygote larvae (after various authors).

or with aquatic juveniles, the main part of the life cycle and the one concerned with feeding occurs before sexual maturity; the adult is a short-lived dispersal and reproductive 'machine'.

The one million species of winged insects are broadly divisible into four clusters of orders (Fig. 8.31). The palaeopterans (two orders) are chiefly distinguished by a (primitive) inability to flex their wings, which are large and have numerous longitudinal and cross veins, back over their abdomens. Their juvenile stages (naiads), which are long-lived and pass through many moults, are always aquatic and hence differ in their adaptations from the adults. Many, for example, have external pairs of tracheal gills: leaf-like abdominal plates which resemble true gills except that instead of containing a blood supply (the blood system of uniramians not being concerned with the distribution of respiratory gases), they possess closed tracheal tubes. These paired structures may be derivatives of the ancestral abdominal appendages. All other pterygotes are 'neopteran', i.e. can lay the wings flat over the abdomen when at rest.

The second cluster is the orthopteroids (10 orders). This basal group of the Neoptera, typically, have biting mouthparts, and an unconcentrated nervous system, simple elongate multiarticulate antennae, a pair of terminal cerci, numerous Malpighian tubules, eversible copulatory organs, and a series of several nymphal instars which change slowly and progressively towards the adult bodily form. All the adult organ systems, including wings, compound eyes, etc., gradually develop through the various juvenile stages. To a considerable extent, therefore, the ecology of the young is the same as that of the adult.

The six hemipteroid orders display a similar progressive change of the nymphs towards the adult form, although in most the changes are concentrated in the final instar or instars. The majority are fluid feeders and have parts of the maxillae adapted to form sclerotized stylets. Characteristically they possess few (four or less) Malpighian tubules, lack cerci, have a much-reduced wing venation, bear short antennae with few articles, and display a distinct gonopod. None of their nymphs possess ocelli.

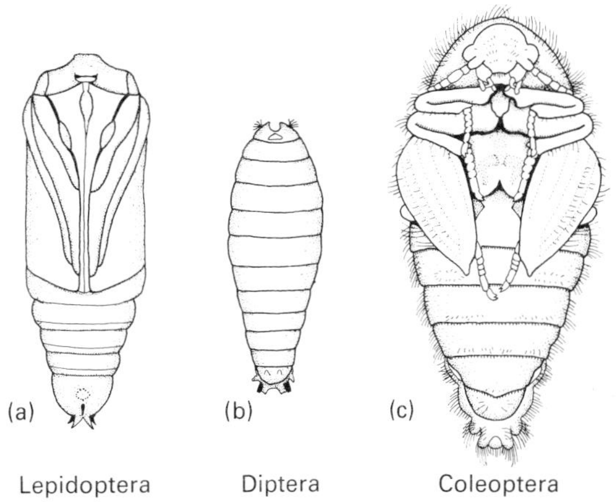

Fig. 8.33. Endopterygote pupae (ventral views), that of the dipteran is enclosed within a 'puparium', the cast integument of the last larval instar (after Jeannel, 1960 and Wallace & Mackerras, 1970).

In several respects, the remaining cluster, the endopterygotes (11 orders), resemble the hemipteroids, although they do have short cerci and may bear many Malpighian tubules. But whereas all the other neopterans display the gradual transition from the first nymphal stage to the adult referred to above, during which the wing buds develop externally to the body, in this group larval stages occur which differ profoundly from the adult form in their anatomy, diet and, often, habitat—a characteristic unique amongst the terrestrial arthropods. Many flies and hymenopterans, for example, have limbless worm-like larvae, whilst, conversely, lepidopterans and some other hymenopterans have larval stages with additional, secondary abdominal walking legs ('prolegs') (Fig. 8.32). In all these larvae, the wing buds develop beneath the integument and hence are not visible externally. Since the larvae differ so markedly from the adult, there is a complete metamorphosis separating the two (see Section 15.4.2). This occurs during a specialized inactive phase, the 'pupa' (Fig. 8.33); in it, all or many of the larval tissues are broken down by histolysis, the adult form is assembled *de novo*, and the developing wings appear outside the integument for the first time. Some pupae retain the moulted integument of the last larval instar as a protective casing around them (Fig. 8.33b). This strategy of separate larval and adult ecologies appears, evolutionarily, to have been highly successful in that 85% of all species of winged insects are endopterygote; some larval stages can even occur in what, to a basically terrestrial animal, are most inhospitable habitats, such as intertidal marine sediments and temporary pools of water—the larva of one African midge has powers of cryptobiosis, after dehydration, rivalling those of the tardigrades (see p. 222).

8.6 Phylum CRUSTACEA

8.6.1 Etymology

Latin: *crusta*, a rind or crust.

8.6.2 Diagnostic and special features (Fig. 8.34)

1 Bilaterally symmetrical, <0.1 mm–60 cm long arthropods (some achieving a leg span of up to 3.5 m, and others a weight of >20 kg) varying in body shape from elongate to spherical.

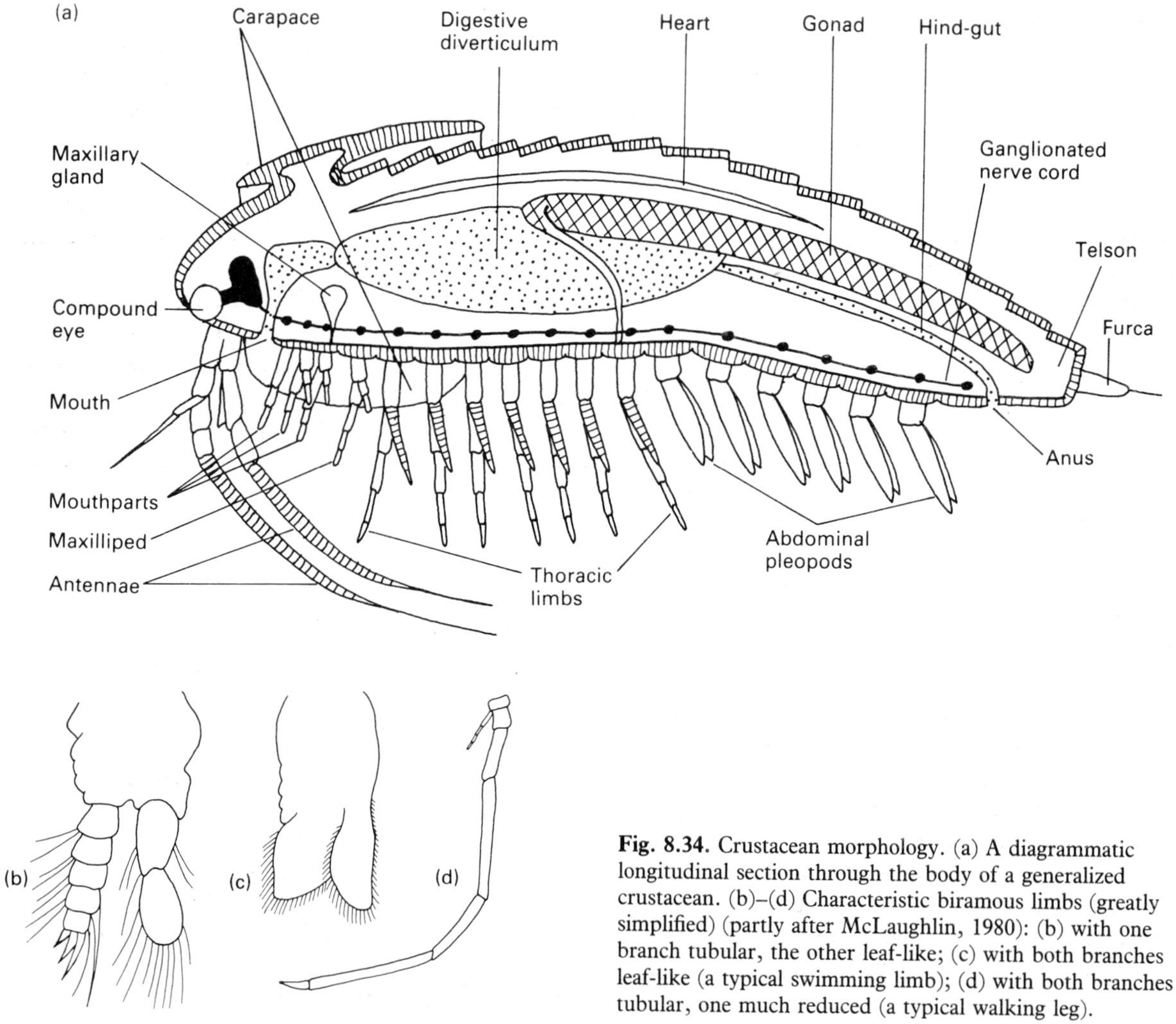

Fig. 8.34. Crustacean morphology. (a) A diagrammatic longitudinal section through the body of a generalized crustacean. (b)–(d) Characteristic biramous limbs (greatly simplified) (partly after McLaughlin, 1980): (b) with one branch tubular, the other leaf-like; (c) with both branches leaf-like (a typical swimming limb); (d) with both branches tubular, one much reduced (a typical walking leg).

2 Body more than two cell layers thick, with tissues and organs.
3 With a through, straight gut, from the small mid-gut region of which issue two digestive diverticula in which digestion and absorption of food take place; those species not already feeding on very small particles or fluids possess a grinding mechanism in the fore-gut; the digestive diverticula arise as outgrowths from the embryonic gut.
4 Different groups of crustaceans vary markedly in the manner in which the body is subdivided and in the number of segments comprising each division; nevertheless, basically a head, formed by the acron and five appendage-bearing segments, can be distinguished, as can a trunk of from two to >65 segments and a terminal telson which often bears a pair of processes, the 'furca'; the first and in various groups up to seven other trunk segments are often fused with the head to form a 'cephalothorax'; and the trunk is usually subdivided into a thorax and abdomen on the basis of its appendages, the thoracic segments not incorporated into the cephalothorax comprising the 'pereon' and the abdominal ones, the 'pleon'; the cephalothorax and, in some groups, most or the whole of the body is enclosed within an outgrowth of the head, the 'carapace', which extends laterally to overhang the sides of the body.
5 The cylindrical or leaf-shaped appendages are all basically biramous, the two branches normally being of different size and shape, and often bearing further secondary branches; the appendages of the head comprise two pairs of 'antennae' (the first pair being the 'antennules'), one pair of 'mandibles', and two pairs of 'maxillae'; those of the trunk vary greatly in their number, shape and regional differentiation; primitively each segment bears a pair of limbs, although those of the abdomen are often absent; without chelicerae but some limbs may be chelate.
6 With three pairs of primary mouthparts (mandibles and the two maxillae) and with, in many groups, one to three pairs of accessory mouthparts, 'maxillipeds', arising from those thoracic segments incorporated into the cephalothorax; members of each pair of mouthparts work with, or against, each other.
7 Head with median ocelli or lateral compound eyes, the latter sometimes located on movable stalks.
8 Trunk, but not head, normally with evident external segmentation, often concealed beneath the carapace, and sometimes lost.
9 Excretory system of blind-ending antennal and/or maxillary glands.
10 Exoskeleton often calcareous.
11 Gaseous exchange effected across the inner wall of the carapace, across the general body surface, or by means of gills developed from parts of the thoracic or abdominal limbs.
12 Blood system with haemocyanin and, rarely, other pigments.
13 Nervous system with paired ganglia in each segment; primitively with separate ventral cords and ganglia, more often fused together; all thoracic ganglia sometimes fused into a single mass.
14 Gonochoristic or, rarely, hermaphrodite, with internal fertilization via copulation by means of gonopods or penes; gonopore location variable, often thoracic.
15 Eggs usually carried by the female or brooded within specialized pouches; some hatch with the full adult complement of segments, most do so as a 'nauplius' larva with only three segments (Fig. 8.35) (some highly specialized parasitic forms are recognizable as crustaceans only by virtue of their nauplius larvae).

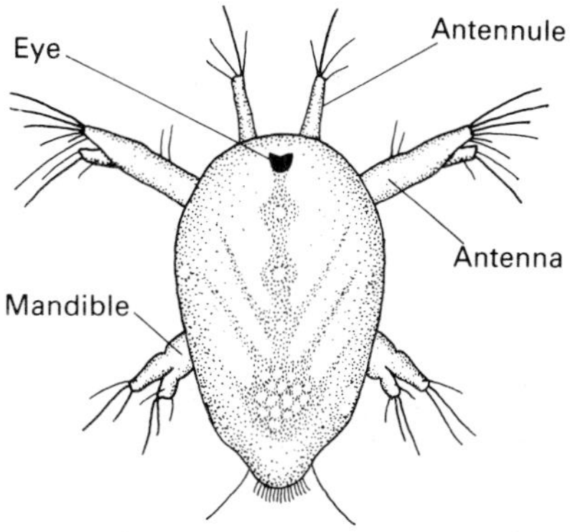

Fig. 8.35. A nauplius larva (after Green, 1961).

16 Essentially marine, although several (13%) are freshwater and a few (3%) are terrestrial.

The crustaceans are *the* marine arthropods, with all but 3% of the arthropod species known from the sea. They dominate the plankton (as they also do in fresh waters) and they are one of the three or four most important members of the benthos, in respect both of interstitial and of macroscopic species. In addition, several are parasitic.

To a considerable extent, their success must be attributable to the general arthropod characteristic of jointed limbs. Legs and/or swimming paddles permit rapid locomotion and provide effective means of moving from one habitat patch to another. Doubtless, the other arthropod feature of an exoskeleton also contributes; certainly the larger, relatively heavily armoured forms possess a significant measure of immunity from casual predation. But perhaps the most important reason for the virtual ubiquity of the aquatic Crustacea is the variety of their body forms, even though their internal anatomy is relatively uniform.

Whereas the chelicerate (Section 8.4) and hexapod (Section 8.5.3B) bodies conform to a standard pattern, and the myriapods (Section 8.5.3A), although displaying much variation in the number of segments, nevertheless possess a very conservative body form, such is far from the position in the crustaceans; there is no such thing as the typical crustacean body plan. Some have a head and trunk, others head, thorax and abdomen, several possess a cephalothorax, pereon and abdomen, and one major group only a cephalothorax and abdomen; in a few, the abdomen is missing, and, in some, a differentiated thorax or head is effectively absent too—further, the number of segments comprising these different bodily blocks also varies from group to group, and even within a single class. Similarly, the form of the limbs can range from walking legs, equivalent to those of the chelicerates and uniramians, to foliose paddles; the antennae, for example, may be sensory, the organs of propulsion or of food collection, the means of attachment to a host, claspers used during copulation, and so on. An acorn barnacle, parasitic copepod, conchostracan, crab and anthurid isopod (Fig. 8.36) display few, if any, obvious signs that they belong to the same group of animals.

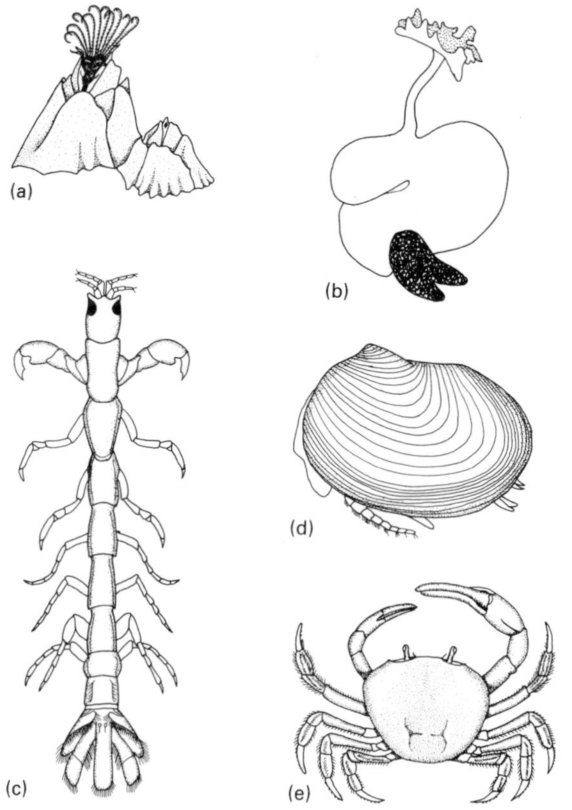

Fig. 8.36. The diversity of crustacean bodily form: (a) acorn barnacles (Cirripedia); (b) a parasitic copepod (Copepoda); (c) an isopod (Malacostraca); (d) a conchostracan (Branchiopoda); (e) a crab (Malacostraca) (after various authors).

This structural plasticity has permitted crustaceans to swim, burrow, crawl, bore into wood, live cemented to rock, hunt, browse, suspension and deposit feed (see Fig. 9.4), parasitize most phyla of animals including their own, and it is probably no exaggeration to conclude, occupy every possible type of marine niche. They are, however, principally aquatic animals. Members of several groups are terrestrial, but usually only marginally so. In particular, their gaseous-exchange systems remain those of their aquatic relatives, restricting them to

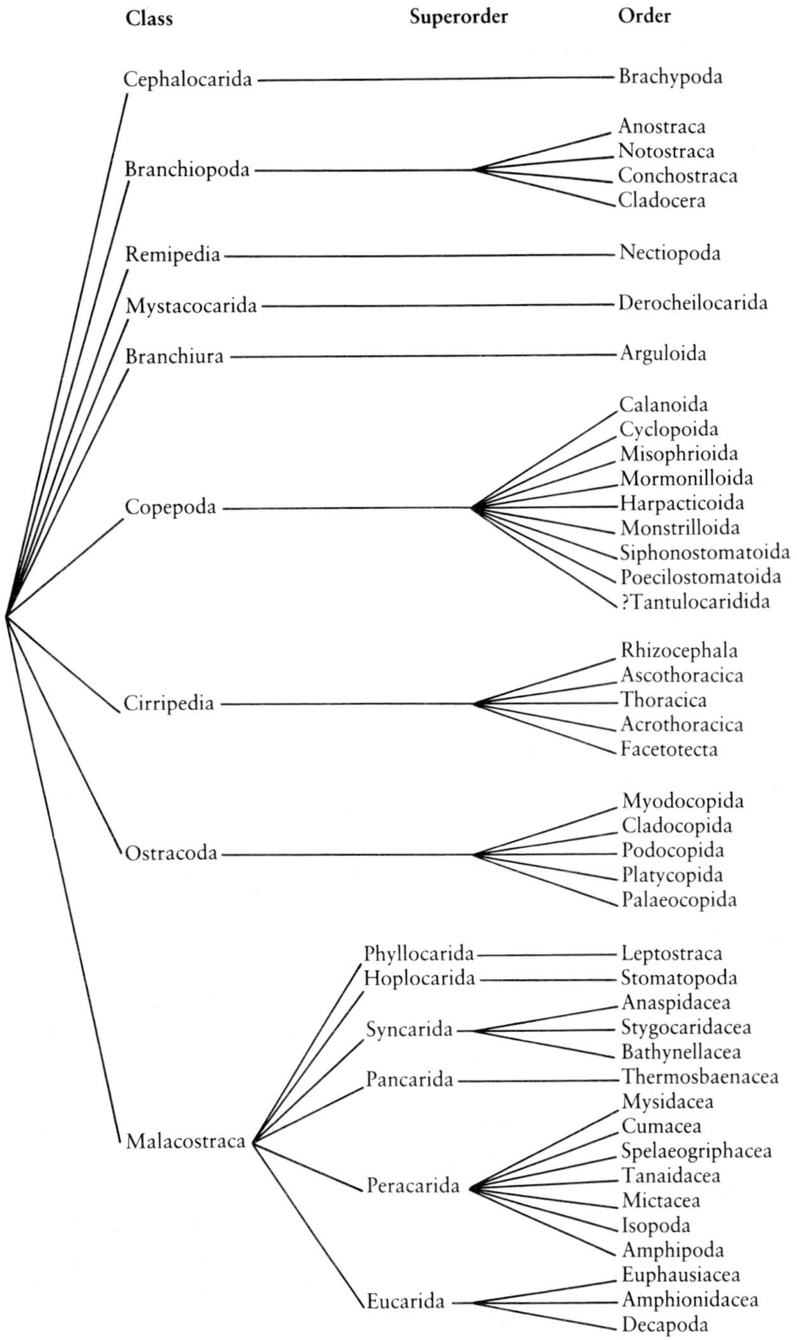
Class
Superorder
Order
Cephalocarida
Brachypoda
Branchiopoda
Anostraca
Notostraca
Conchostraca
Cladocera
Remipedia
Nectiopoda
Mystacocarida
Derocheilocarida
Branchiura
Arguloida
Copepoda
Calanoida
Cyclopoida
Misophrioida
Mormonilloida
Harpacticoida
Monstrilloida
Siphonostomatoida
Poecilostomatoida
?Tantulocaridida
Cirripedia
Rhizocephala
Ascothoracica
Thoracica
Acrothoracica
Facetotecta
Ostracoda
Myodocopida
Cladocopida
Podocopida
Platycopida
Palaeocopida
Malacostraca
Phyllocarida
Leptostraca
Hoplocarida
Stomatopoda
Syncarida
Anaspidacea
Stygocaridacea
Bathynellacea
Pancarida
Thermosbaenacea
Peracarida
Mysidacea
Cumacea
Spelaeogriphacea
Tanaidacea
Mictacea
Isopoda
Amphipoda
Eucarida
Euphausiacea
Amphionidacea
Decapoda

humid habitats. Moreover, some, e.g. the land crabs, must return to water to breed since their pattern of reproduction and development still includes an aquatic larval phase. Only the sowbugs or woodlice include species with specific terrestrial adaptations, and they are the most widely distributed of the land Crustacea. In addition to the pleopodal gills typifying their order (Isopoda), several woodlice have developed trachea-like invaginations of the integument of the pleopods (the abdominal appendages) which extend into the haemocoel within these limbs. Other woodlice, however, have only evolved mechanisms for keeping the surfaces of their gills moist, for example by channelling water droplets from the body surface on to the pleopods.

8.6.3 Classification

The almost 40000 species of this phylum are divided between nine classes. It is becoming increasingly customary to apportion these nine to four larger groupings (e.g. subphyla), of which the relatively unspecialized classes Remipedia and Malacostraca are placed in separate monotypic subphyla, and the two remaining subphyla, both of which probably originated by paedomorphosis, contain the Cephalocarida and Branchiopoda (subphylum Phyllopoda) and Mystacocarida, Branchiura, Copepoda, Cirripedia and Ostracoda (subphylum Maxillopoda). Although the Malacostraca are often termed the 'higher Crustacea', they are a basal, rather primitive group; it is the Phyllopoda and Maxillopoda which show the most advanced features.

8.6.3.1 Class Cephalocarida

The small, blind cephalocarids (<4 mm in length) are detritus-feeding, bottom-dwelling marine animals which were only discovered in 1955 although, locally, they can be abundant. They display some features which are usually regarded as being primitive. The body (Fig. 8.37) is divided into a head, thorax and abdomen, without any development of a cephalothorax or carapace, and all eight pairs of thoracic limbs are similar and effectively identical to the second maxillae, each one having both leaf-like and tubular elements. The eleven-segmented abdomen lacks appendages, however, except for the first segment which retains reduced limbs to which the egg sacs are attached. Unusually, cephalocarids are hermaphrodite with the paired ovaries and testes sharing a common duct, and development is a rather gradual process through the many larval stages.

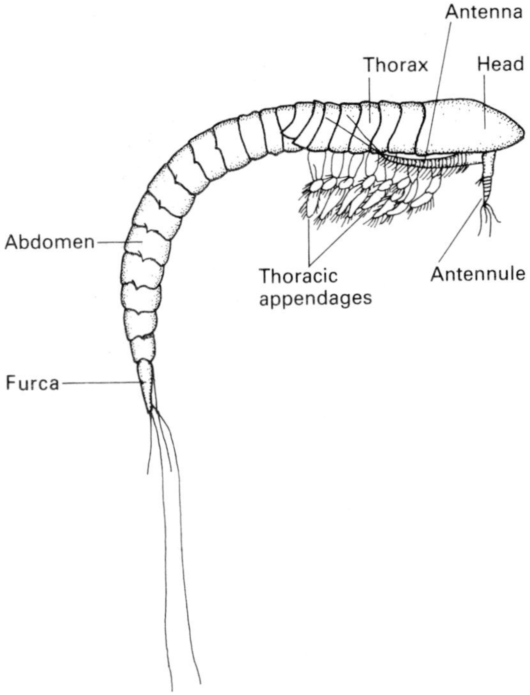

Fig. 8.37. A cephalocarid (lateral view) (after Sanders, 1957).

The ten known species are all elongate and cylindrical, their bodies terminating in a telson with a long furca of which each ramus bears a bristle. Only one order is included.

8.6.3.2 Class Branchiopoda

The branchiopods are a diverse group (Fig. 8.38) of mainly freshwater crustaceans characterized by small to vestigial head appendages (except, usually, the antennae), by not having any trunk segments fused

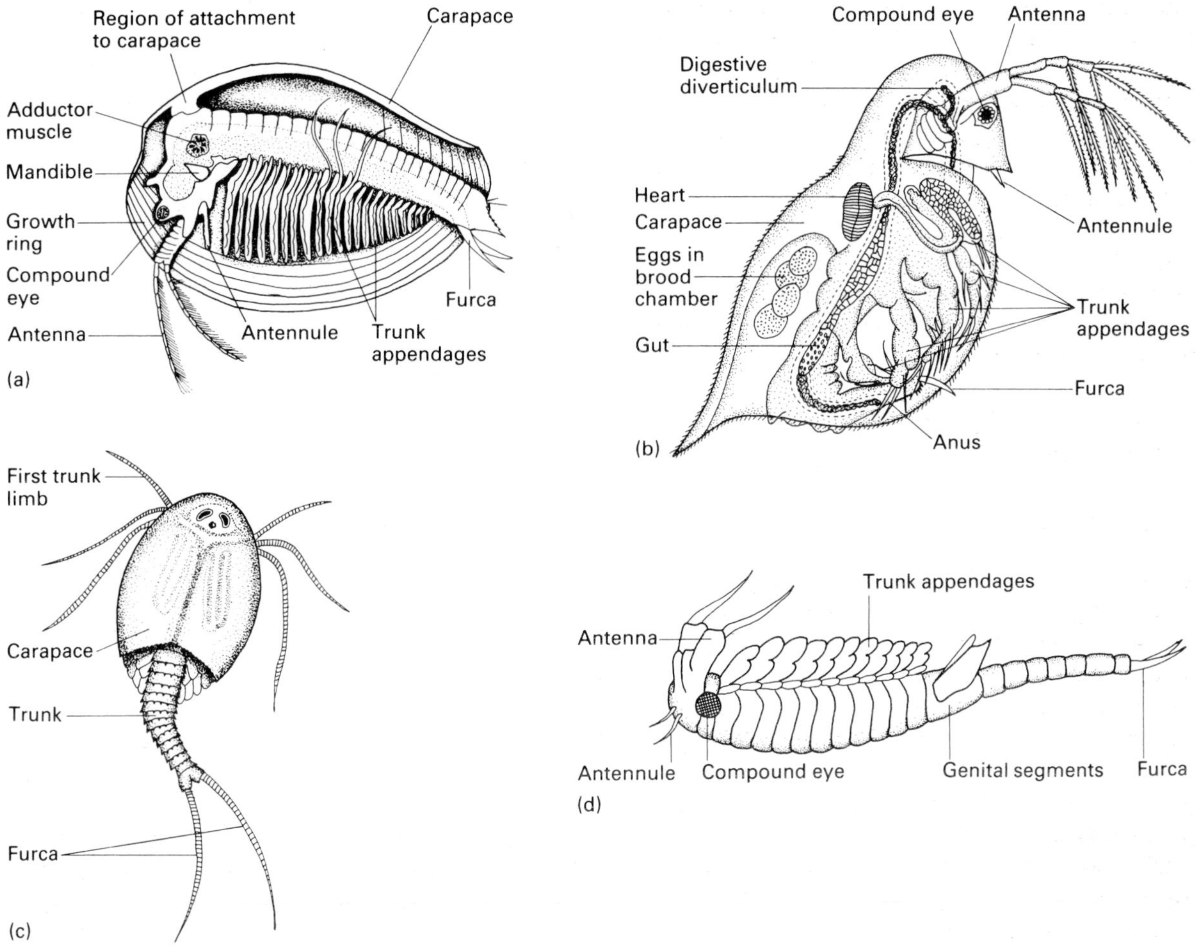

Fig. 8.38. Branchiopod morphology: (a) Conchostraca, lateral view with the left-hand valve of the carapace removed (see also Fig. 8.36d) (after Kaestner, 1970); (b) Cladocera, lateral view as seen through the transparent carapace (after Belk, 1982); (c) Notostraca, external appearance in dorsal view (after Kaestner, 1970); (d) Anostraca, lateral view of its upside-down swimming position (after McLaughlin, 1980).

to the head, and by the trunk bearing a series of similar limbs which usually decrease in size posteriorly, the last few segments lacking limbs altogether. These limbs are typically leaf-shaped swimming and/or filter-feeding organs supported more by haemocoelic pressure than by the cuticle; they also bear gills. Many species reproduce parthenogenetically and brood their eggs; several produce resting stages.

The four orders differ considerably in their body forms. The conchostracans (clam shrimps) and cladocerans (water fleas) both have short, sometimes almost circular, bodies, locomotory antennae, a claw-like furca, and a dorsal brood chamber within the laterally compressed carapace. But whereas the clam shrimps possess a series of up to 30 or more trunk segments and their carapace encloses the whole body including the head, the carapace of the water fleas, although often large, never encloses the head and is, in some, reduced to a small dorsal brood chamber, whilst there are never more than six pairs of trunk limbs (Fig. 8.38a and b). The conchostracan carapace is not moulted but grows by the addition of concentric rings, like those of the bivalve molluscs, and like them it is held shut by an adductor muscle which acts against an elastic hinge ligament.

In the Notostraca (tadpole shrimps), on the other hand, the carapace is a wide, dorsoventrally flattened, somewhat horseshoe-shaped shield from out of which projects the narrow cylindrical posterior end of the body with its two long, annulate furcal rami (Fig. 8.38c). Notably, the posterior trunk segments are only partially differentiated, so that one apparent segment may bear up to six pairs of limbs. Up to 70 pairs of trunk limbs can occur, of which the eleventh carry the brood chambers and the first are larger than the rest. The fourth group, the Anostraca (brine or fairy shrimps), lack any carapace (Fig. 8.38d) and form their brood chamber within the body, from the dilated vagina, although when full of eggs the genital segments form a large projecting bulge, the ovisac. Both the notostracans and anostracans characteristically occur in harsh environments, e.g. temporary pools or saline lakes, and have developed resting stages in extreme form. Their eggs are tolerant of temperature extremes and desiccation; some can remain dormant for up to 10 years.

None of the 850 living species exceeds 10 cm in length, most are less than 3 cm.

8.6.3.3 Class Remipedia

This class is represented by a single blind species described from a marine cave in 1980 (Fig. 8.39), and as yet very little is known of its biology. The smallish, elongate, translucent body (<3 cm in length) comprises a short, carapace-less cephalothorax of the head and first trunk segment, and a long trunk of over 30 similar segments each with a pair of leaf-like, lateral limbs used in upside-down swimming. The head appendages include a pair of peculiar rod-like processes in front of the antennules and prehensile mouthparts, including the maxilliped. A further eight species, all from marine cave-like habitats in the Atlantic, are in the process of being described.

8.6.3.4 Class Mystacocarida

Mystacocarids (Fig. 8.40) are minute (<1 mm in length), elongate, pigmentless, interstitial marine crustaceans distinguished chiefly by a head which is divided into a small anterior and a large posterior portion, and a trunk of ten segments of which the first bears a maxilliped even though it is not fused to the head. Although the head appendages are large (and are used in locomotion), those of the trunk are either reduced to small, single-articled structures, as on segments 2–5, or are missing; the telson, however, bears a large, pincer-like furca. A primitive feature of this group is that the members of each opposing pair of trunk ganglia touch but do not fuse; neither

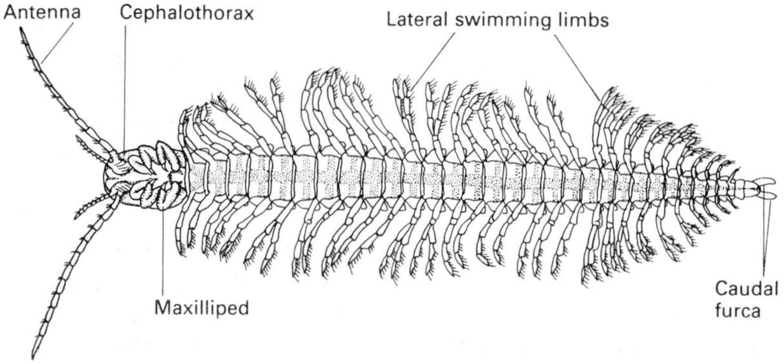

Fig. 8.39. A remipedian (ventral view) (after Yager & Schram, 1986).

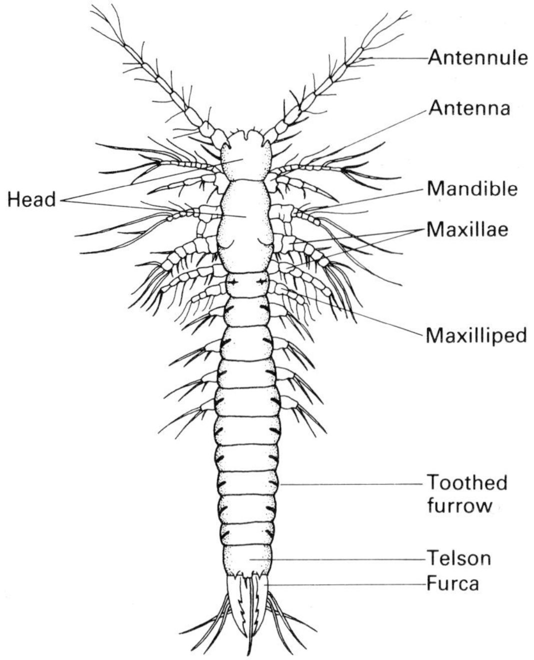

Fig. 8.40. A mystacocarid (dorsal view) (after Kaestner, 1970).

are compound eyes or digestive diverticula present, possibly as a consequence of their small size. One peculiar feature, otherwise unknown in the Crustacea, is that the posterior portion of the head and each trunk segment bears a pair of lateral, toothed grooves of which the function remains to be discovered.

The ten species are all placed in a single order.

8.6.3.5 Class Branchiura (fish lice)

The branchiurans are smallish (<3 cm in length) periodic ectoparasites of marine and freshwater fish, piercing their much larger prey with the small mandibles and ingesting a fluid meal, usually blood, like marine fleas or mosquitoes. Their markedly dorsoventrally flattened body comprises a cephalothorax of head and first thoracic segment, a pereon of three segments, and a bi-lobed unsegmented abdomen; the cephalothorax and, in some, much of the pereon is covered by a large, flat carapace, circular, bi-lobed or arrowhead in shape, which extends laterally or posteriolaterally (Fig. 8.41), and bears a pair of compound eyes.

Their head appendages are either minute or modified into organs of attachment to the fish, ending in hooks or, as in the case of the first maxillae, often in large, stalked suckers. All four pairs of thoracic appendages, including that of the segment wholly or partially incorporated into the cephalothorax, form swimming limbs; the abdomen, however, lacks any appendages. Somewhat unusually, the eggs are not brooded or carried by the female but are attached to the substratum or to benthic vegetation.

The 150 species are contained within a single order.

8.6.3.6 Class Copepoda

The copepods are the dominant members of the marine plankton and to a slightly lesser extent of that in fresh water as well; there are numerous interstitial benthic species; and about one-quarter of all species are parasitic, attacking animals ranging from sponges to whales. Although most species are small (<2 mm in length), exceptionally one free-living species approaches a length of 2 cm and an ectoparasitic form achieves 0.3 m.

Basically, the copepod body comprises a head, with well-developed mouthparts and antennae, a six-segmented thorax bearing swimming limbs, and a five-segmented appendage-less abdomen, but various segments may be fused together in a wide range of fashions (one thoracic segment, at least, always being fused to the head, and, in many, a second is also incorporated into the cephalothorax), and the primary division into cephalothorax, pereon and abdomen does not necessarily reflect the manner in which, in practice, the body is subdivided, if indeed there is any regional differentiation. The parasitic species, for example, show various degrees of bodily degeneration, in extreme form including loss of all apparent segmentation and of appendages. In the free-living (and some parasitic) species, there is a major functional subdivision of the body, marked by a point at which the posterior region articulates with the anterior. Except in the cylindrical interstitial

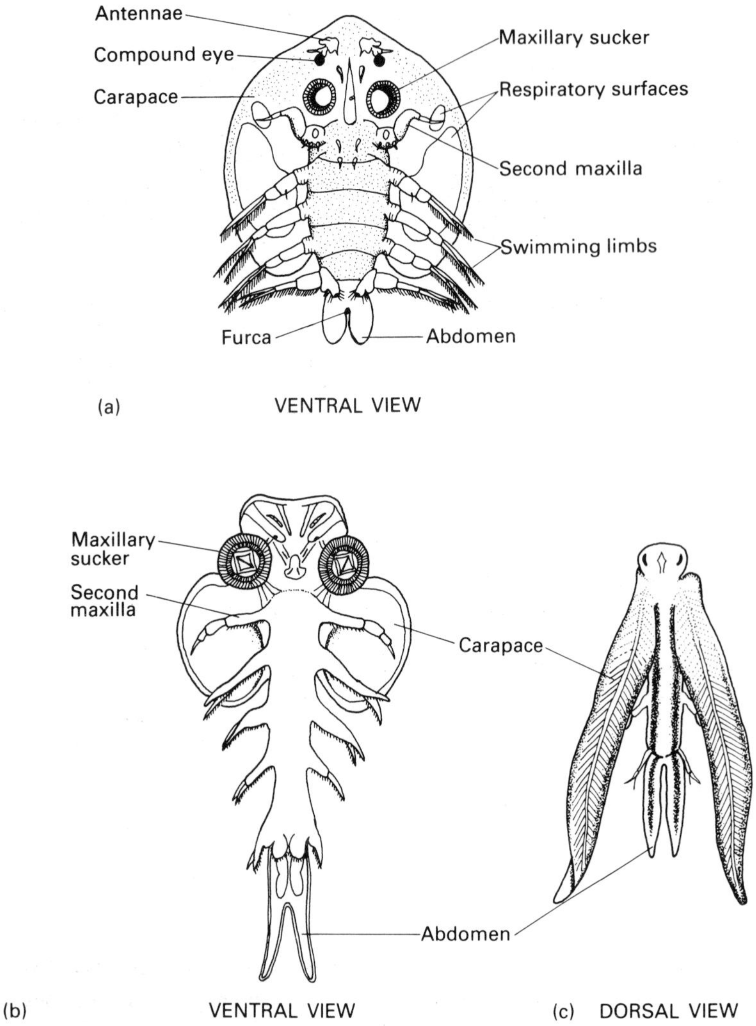

Fig. 8.41. Branchiuran morphology and diversity (after Kaestner, 1970).

species, the portion anterior to this articulation (the 'prosome') is oval, sometimes elongately so, whilst the posterior part of the body (the 'urosome') is narrowly tubular (Fig. 8.42). Although in one group, this articulation is indeed sited at the division between cephalothorax + pereon and abdomen, in many it occurs between the third and fourth segments of the pereon (in these groups corresponding to the fifth and sixth thoracic segments) so that the last segment of the pereon (and thorax) forms part of the urosome.

Copepods always lack a carapace and compound eyes.

Parasitic species display a wide range of body forms (Fig. 8.43), the sac- or worm-like bodies of the most degenerate types being unrecognizably crustacean except during their development. These often carry their eggs in long strings, as opposed to the one or two oval egg sacs of the free-living species.

The 8400 species can be divided between nine orders, one of which is only tentatively placed here. These, the poorly known tantulocarids, are minute

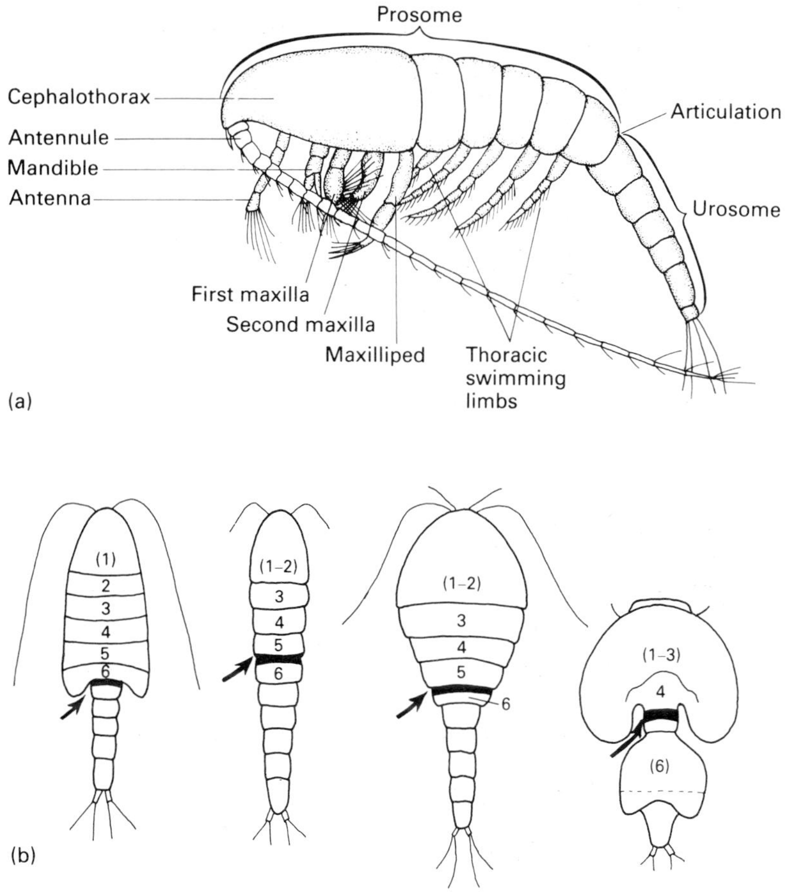

Fig. 8.42. Copepod morphology: (a) external appearance of a free-living copepod, in lateral view, showing the appendages; (b) the location of the articulation between the prosome and the urosome (arrowed) in various types of copepod (the thoracic segments are numbered, those fused to other regions being in parentheses) (after Kaestner, 1970).

ectoparasites of other marine crustaceans, which live permanently attached to their host by an oral disc: they lack cephalic appendages and hatch as advanced infective tantalus larvae. The female is little more than an egg sac which remains attached to the host by the larval head, whilst the free-living male develops from a mass of dedifferentiated tissue within the larva. It has been suggested that they should constitute a separate class, the Tantulocarida, related perhaps to the Copepoda and perhaps to the Cirripedia.

8.6.3.7 Class Cirripedia

As a group, the cirripedes are the most highly modified of the Crustacea, being either sessile or dwellers in other organisms in a parasitic manner. They are effectively headless, most lack an abdomen, and there is little or no evident segmentation. In the extreme form of the order Rhizocephala, they resemble nothing so much as a bracket fungus, with a network of fine tubes spreading through all the tissues of the host (almost invariably a decapod crustacean) and an external sac containing the gonads (Fig. 8.44a).

The order Ascothoracica (Fig. 8.44b), which parasitize cnidarians and echinoderms, are the least specialized anatomically. In a few of these, there are the rudiments of a head bearing chelate antennules (a most bizarre feature for a crustacean), and a thorax

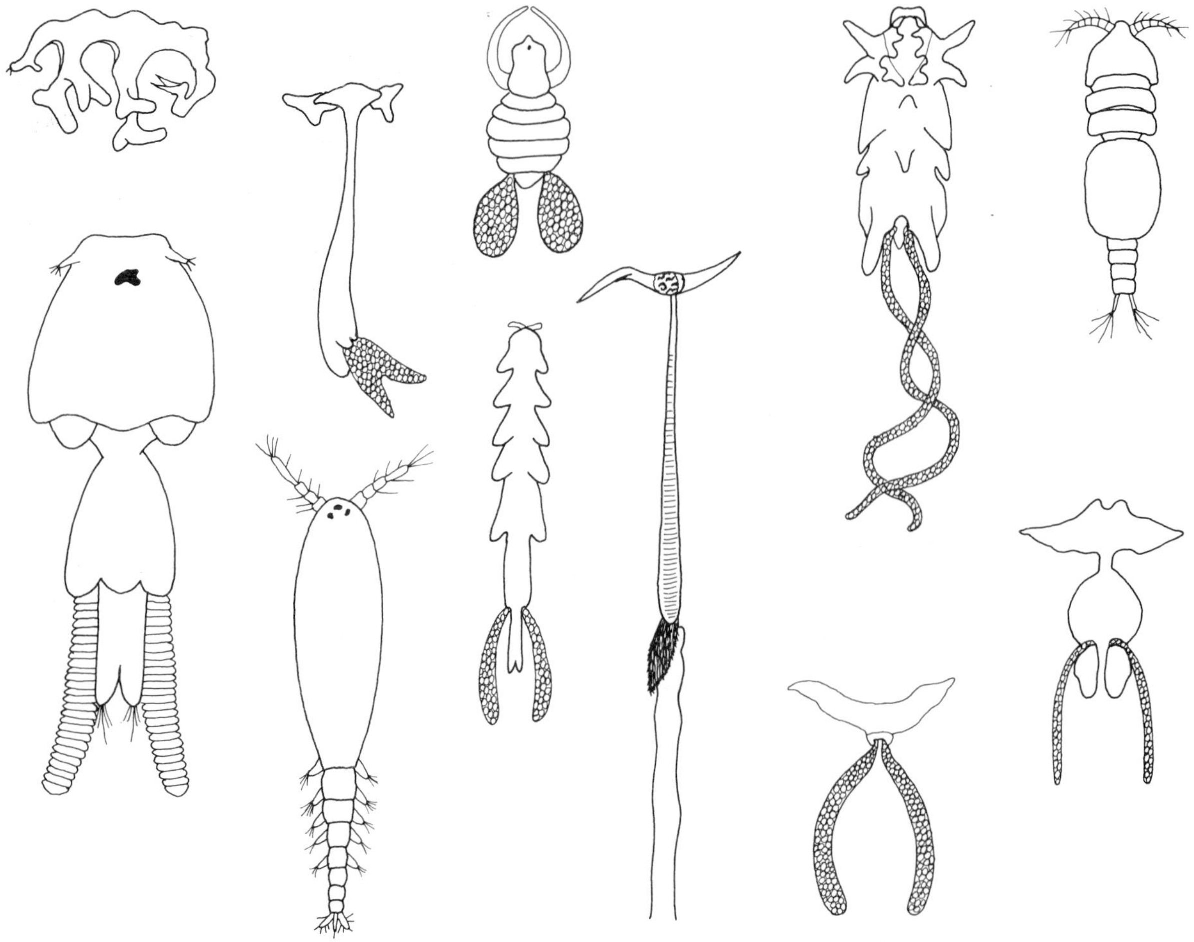

Fig. 8.43. The range of bodily form and degeneration in parasitic copepods (after various authors).

with six pairs of swimming limbs, all enclosed within a bivalved carapace, from out of which projects a free abdomen of five segments. The more familiar barnacles (order Thoracica) can be visualized as ascothoracicans which have become sessile, the six pairs of thoracic legs—the 'cirri'—forming filtering or food-collecting organs, and the bag-like carapace being reinforced with from few to many calcareous plates (Fig. 8.44c and d) that are not moulted but grow around their margins. Like the other cirripedes, initial attachment to, in this case, the substratum (rock, shell or seaweeds) is effected by the antennules; after successful attachment, this pre-oral region may enlarge to form an elongate column, as in the goose- or stalked barnacles, or it may form only a thin attachment disc, as in the acorn forms. The fourth group, the order Acrothoracica, are essentially similar to the barnacles, although they lack calcareous plates and bore into corals or, more rarely, mollusc shells. In both barnacle-like groups, the penis may be very long in order to reach into nearby attached (and therefore sessile) individuals. A fifth order, the Facetotecta, is known only from larvae (the so-called Y-nauplii and Y-cyprids).

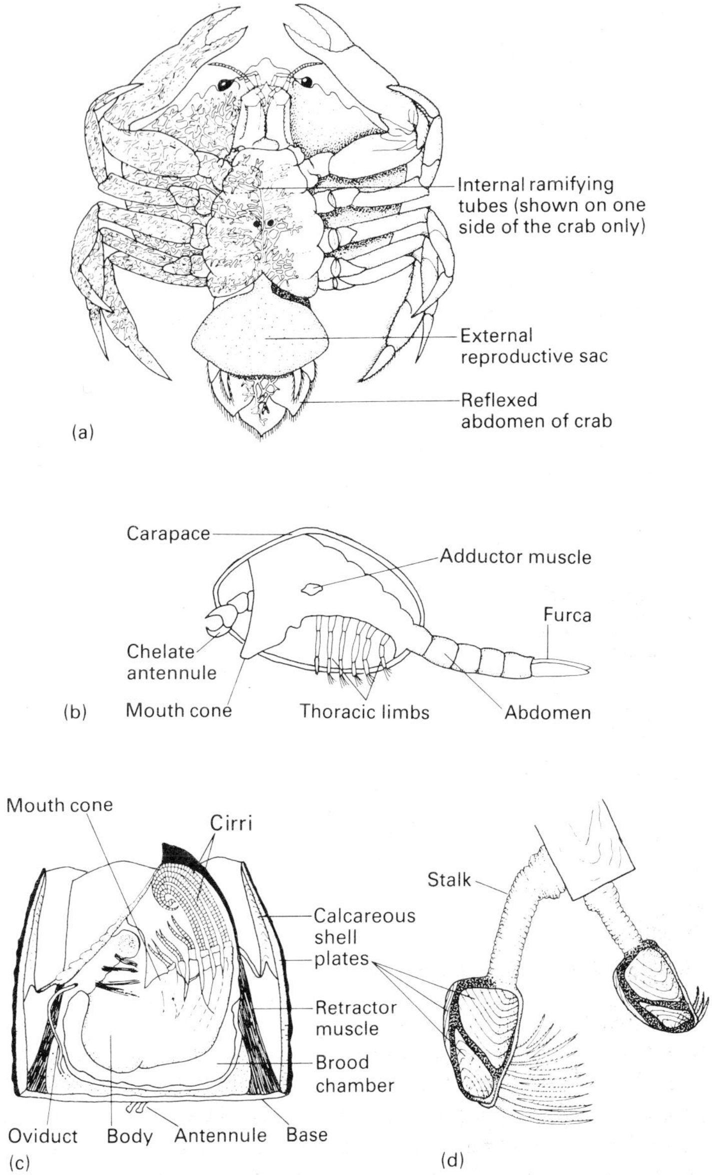

Fig. 8.44. Cirripedia: (a) a rhizocephalan infesting a crab, the latter shown in ventral view as if transparent; (b) a diagrammatic lateral view of an ascothoracican shown with the left-hand carapace valve removed; (c) an acorn barnacle in lateral view shown with some of the carapace and its plates removed (see also Fig. 8.36); (d) external appearance of two stalked barnacles attached to a block of wood. (a and b after Kaestner, 1970; c and d after Zullo, 1982.)

Within the class as a whole, there are clear trends towards hermaphroditism or reduction of the male sex to minute proportions, and towards reduction of the gut—in many, it is blind-ending and, in the rhizocephalans, absent. All have an individual type of second larval stage after the nauplius, the 'cypris', which locates the host or settlement site and thereafter metamorphoses into the young adult stage. The 1000 species are all marine.

8.6.3.8 Class Ostracoda

Ostracods are very small crustaceans (mostly <1 mm long, although rarely approaching 2 cm) with a short oval body enclosed within the bivalved and often calcareous shell formed by the carapace (Fig. 8.45). Like the conchostracans (and the bivalve molluscs), the two carapace valves are provided with transverse adductor muscles which act against an elastic hinge ligament; hinge teeth may even be present. Unlike those groups, however, the shell is shed and reconstituted at each moult.

Their sac-like bodies have no visible signs of segmentation but, judging from the appendages, what was ancestrally the head forms half of the body volume. Only a total of five, six or seven pairs of

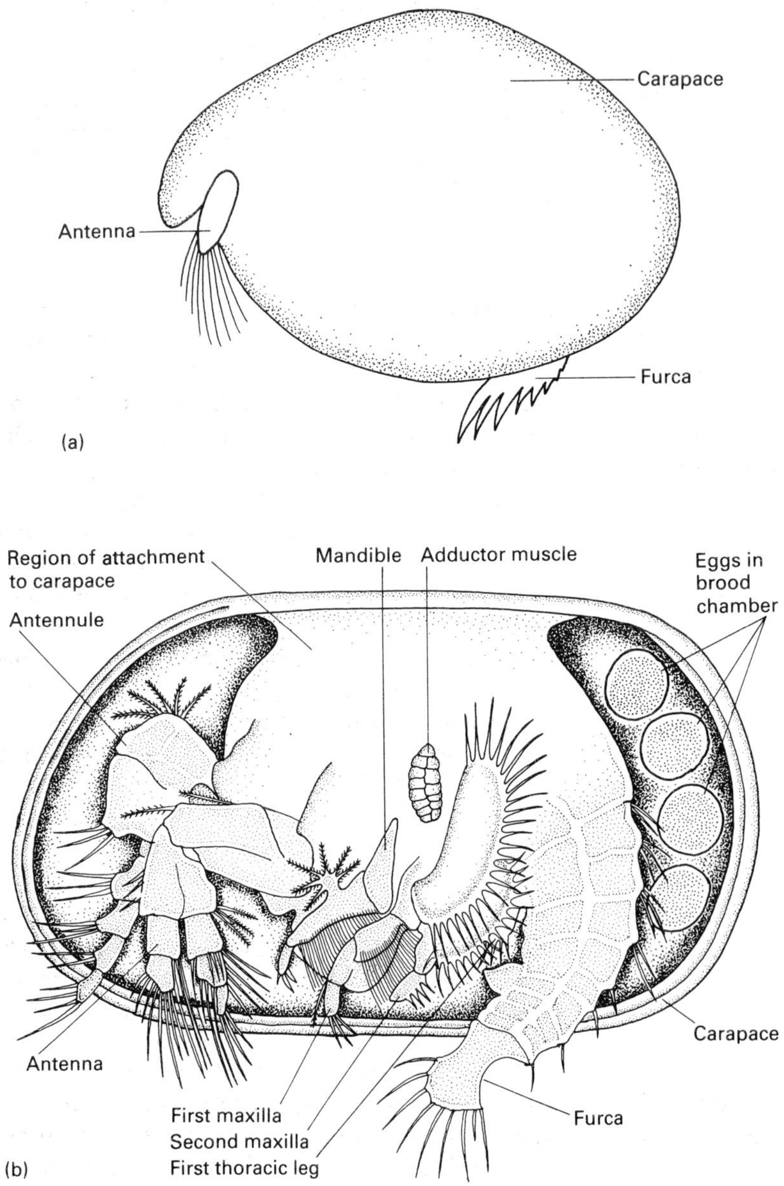

Fig. 8.45. Ostracoda: (a) external appearance in lateral view; (b) shown with the left-hand valve of the carapace removed (after Cohen, 1982).

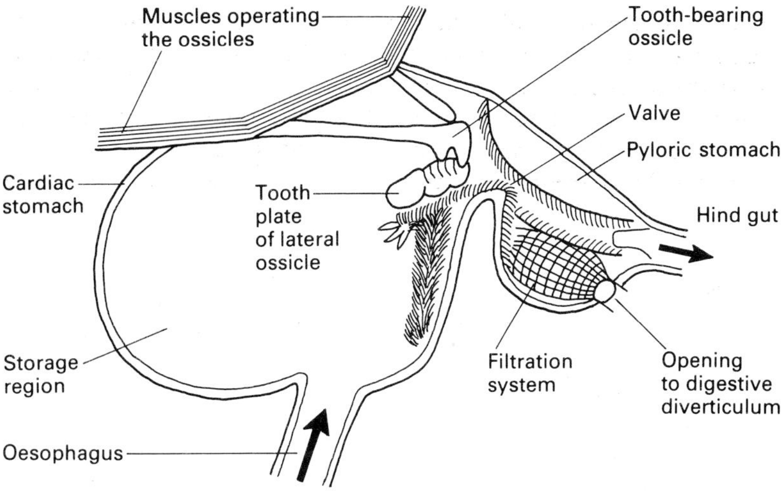

Fig. 8.46. A diagrammatic longitudinal section through the fore-gut stomach of a decapod malacostracan showing the ossicles of the 'gastric mill' which grind the food ingested, and the filtration system in the pylorus guarding the opening into one of the digestive diverticula (after Warner, 1977).

appendages are present, of which the first four certainly belong to the 'head' (the antennules, antennae, mandibles and first maxillae), and the last two, one or both of which may be missing, are 'thoracic'; but authorities differ on the nature of the intervening fifth pair. Most regard it as the cephalic second maxilla, but some consider it to be thoracic and term it either the maxilliped or first thoracic leg. If it is the second maxilla, then some ostracods are the only crustaceans to have retained only the appendages of the head. The function of these various limbs varies from group to group, but the antennae are usually the main swimming organs, and the first thoracic limbs, if present, are often the walking legs. The eggs are usually brooded beneath the carapace, although they may simply be attached to the substratum or to submerged vegetation.

The five orders (one of them as yet known only from empty shells) contain a total of 5700 species, most of them marine. A few species are terrestrial, having invaded the humus/litter layer in forests via fresh water.

8.6.3.9 Class Malacostraca

The Malacostraca is by far the largest class of the Crustacea, with some 23000 species, and, arguably, it contains a greater diversity of body forms than any other class in the animal kingdom; just one of its 16 orders, the Decapoda, includes such varied organisms as crabs, crayfish, shrimps and hermit-crabs. The main feature which unites the class is that the body fundamentally comprises a head, an eight-segmented thorax, and a six- (or rarely seven-) segmented abdomen, all these regions being equipped with a full complement of segmental appendages, including the abdomen. Their diversity, however, can be gauged from the fact that from none to all eight thoracic segments may be incorporated with the head into a cephalothorax; that from none to three pairs of thoracic appendages may form maxillipeds; and that a carapace may be present or absent (either primitively or secondarily), whilst, if present, it may cover some or all of the anterior region of the body (from only the first two thoracic segments up to all the thoracic and several abdominal ones).

Typical malacostracan specializations are the presence in the fore-gut of a 'stomach', in which the food is ground into finely particulate form and any coarse particles remaining are filtered out of the material which then passes into the digestive diverticula (Fig. 8.46), and the development of the last pair of abdominal appendages (the 'uropods') into a tail fan with the telson, which terminates the body instead of the more usual furca. Unless secondarily lost, a well-developed pair of compound eyes are also

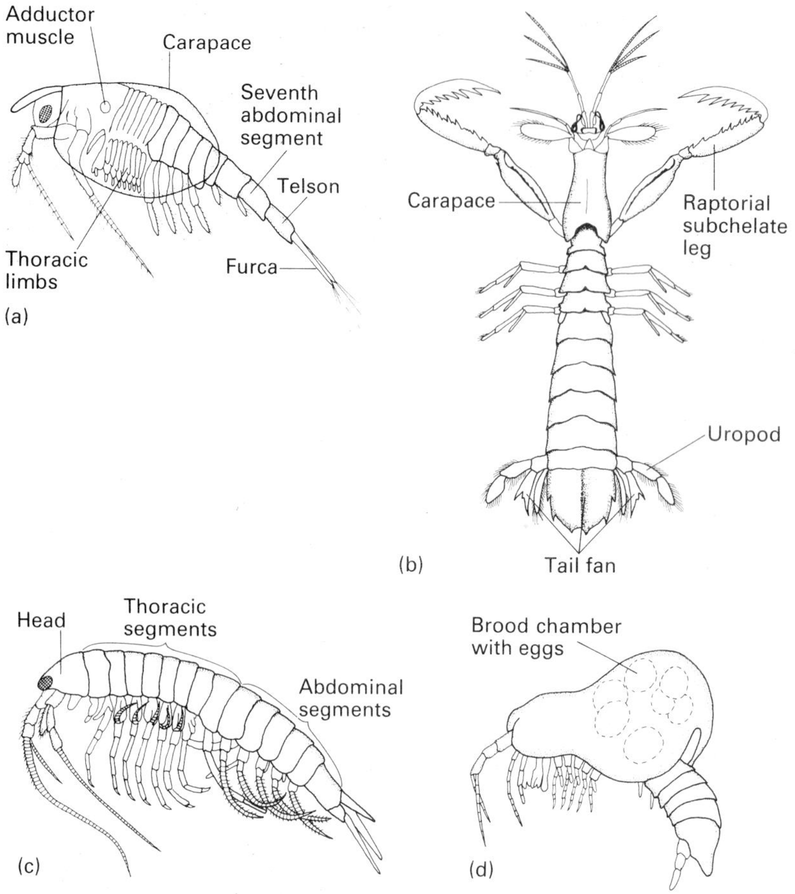

Fig. 8.47. Malacostracan diversity I. (a) Phyllocarida, seen in lateral view as if the carapace was transparent; (b) Hoplocarida, dorsal view; (c) Syncarida, lateral view; (d) Pancarida, lateral view (the carapace of the female depicted is greatly enlarged and swollen to act as the brood chamber) (after various authors).

characteristic. Many species are large and well calcified, and several show marked degrees of concentration of the nervous system and complex behaviour patterns. They are important members of the marine nekton and benthos; many occur in freshwater streams, rivers and lakes; and several are terrestrial, including all those crustaceans which can survive in other than permanently humid habitats.

Six major groups of orders can be distinguished. Two of them have retained the basic body plan of head, thorax and abdomen, i.e. without any development of a cephalothorax or of accessory mouthparts. One, the Phyllocarida (Fig. 8.47a), are the only malacostracan group to have retained a seventh abdominal segment and the furca, and to lack uropods. Their most obvious feature is a large, laterally compressed, bivalved carapace which covers the thorax and its eight pairs of leaf-like limbs, setae on which enclose the brood cavity. The second group, the Hoplocarida (mantis shrimps) (Fig. 8.47b), also have a carapace but it is much smaller, covering only half of the dorsoventrally flattened thorax. Particularly characteristically, the first five pairs of their thoracic limbs are subchelate and are not concerned with locomotion; the second pair are large and raptorial, whilst the next three pairs, in females, carry the egg masses. In some species, the second pair form 'fists' which can be shot out at 1 cm ms^{-1} to deliver an impact equivalent to a ·22-calibre bullet!

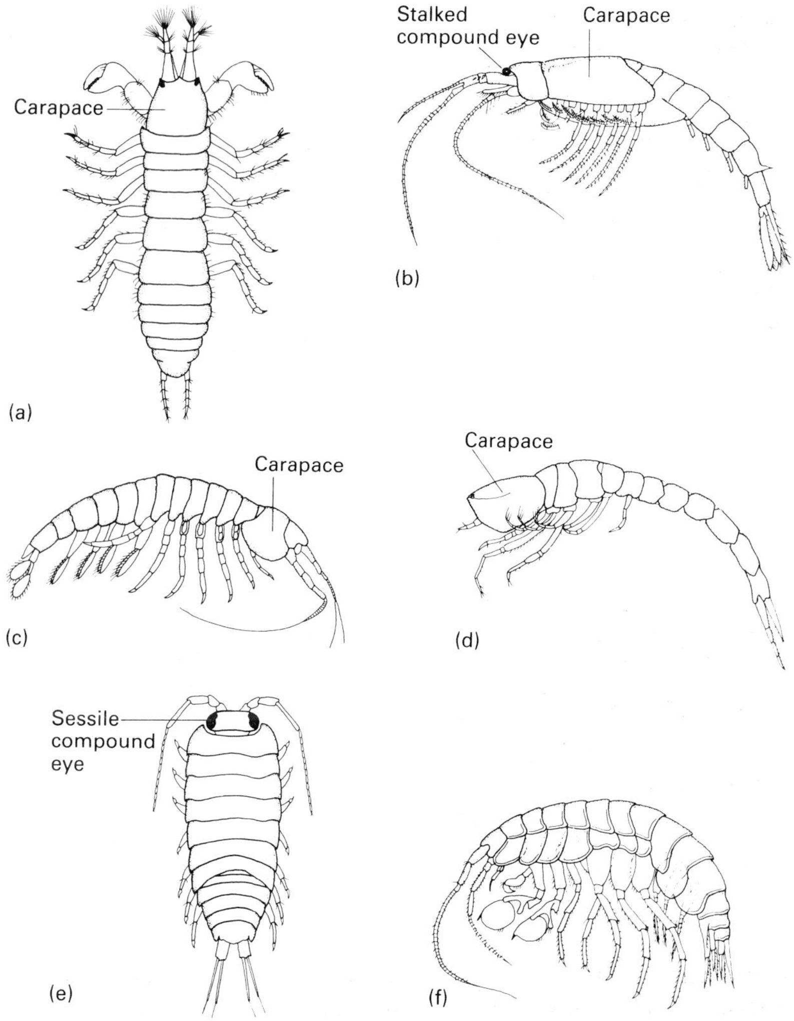

Fig. 8.48. Malacostracan diversity II. The Peracarida: (a) a tanaid; (b) a mysid; (c) a spelaeogriphacean; (d) a cumacean; (e) an isopod (see also Fig. 8.36c) and (f) an amphipod (a and e in dorsal view; the others lateral) (after various authors). Note the differential development of the carapace and the various shapes of the appendages.

In contrast to these two groups, the next three superorders have at least one (and usually only one, but three at the most) thoracic segment fused to the head in a cephalothorax, and therefore possess a pereon of seven segments and an abdomen of six. Carapace development, however, is very variable. The solely freshwater Syncarida (Fig. 8.47c) lack any carapace (primitively, it is generally thought) and tend towards the loss of appendages from the rather uniform trunk segments. The blind Pancarida (Fig. 8.47d), which inhabit caves and interstitial habitats, sometimes in hot springs of up to 45°C, are also mainly freshwater; they are distinguished by a short carapace which serves as the brood chamber (as well as the centre of gaseous exchange). The much larger third grouping, the Peracarida (Fig. 8.48), however, form their brood chamber from outgrowths of the pereon limbs. A carapace and stalked compound eyes were probably present in the ancestral members of this superorder, but both are subject to marked reduction in various lines so that their compound eyes are mostly sessile and in the members of two

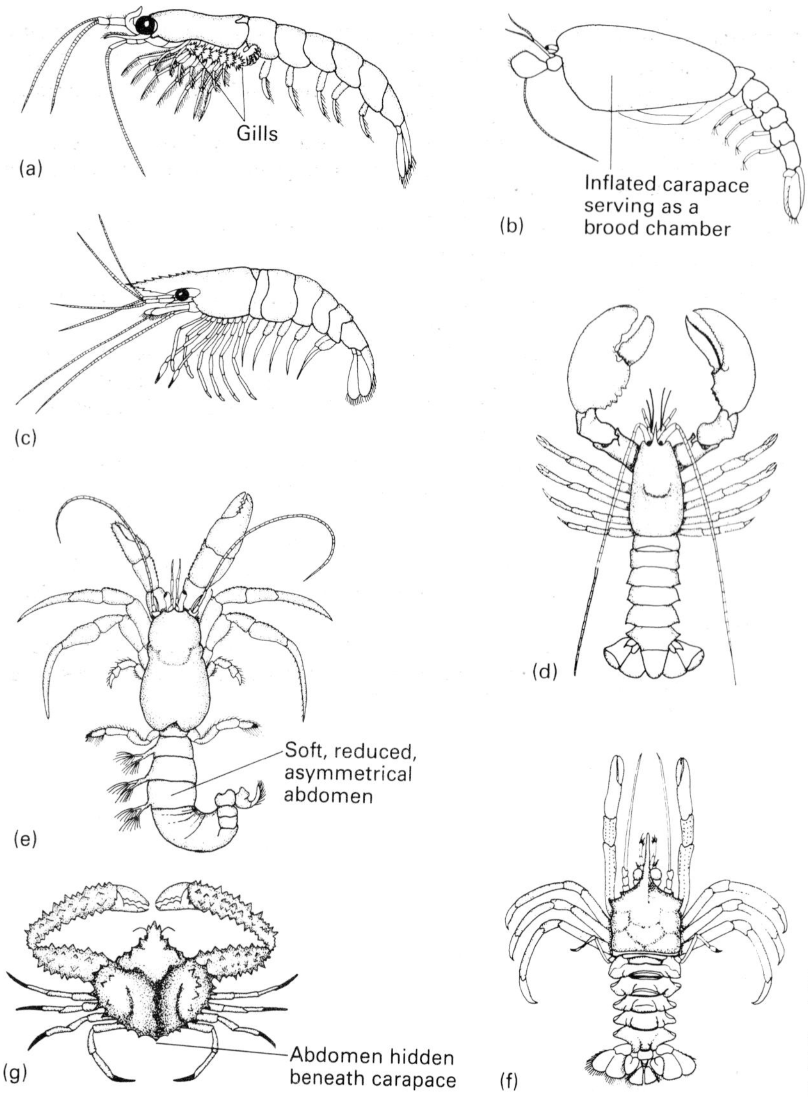

Fig. 8.49. Malacostracan diversity III. The Eucarida: (a) a euphausiid; (b) an amphionid; (c)–(g) decapods; (a–c in lateral view; d–g dorsal) (after various authors).

important orders (the Amphipoda and Isopoda) a carapace is absent. Peracarids are very variable in form and habitat: some swim, several burrow, and many crawl using the appendages of the pereon as walking legs, a number are parasitic; their bodies may be elongate or squat, dorsoventrally or laterally compressed or shrimp-like. The woodlice have successfully colonized the land (see Section 8.6.2).

The final group, the Eucarida (Fig. 8.49), have all their thoracic segments fused and incorporated into the cephalothorax, so that the body comprises only a cephalothorax and abdomen; and the whole cephalothorax is enclosed by the carapace which usually extends down laterally to cover the thoracic gills within a protective chamber. In general, the abdominal pleopods are used in swimming, and, in decapod females, for carrying the egg masses, whilst the thoracic limbs either serve as feeding appendages, or for prey-capture and walking. In the largest order, the Decapoda, there are three pairs of maxillipeds and an evolutionary trend towards reduction of the abdomen, which culminates in the crabs in which the small abdomen is folded underneath the large and often broad carapace, so as not to be visible from above. Alone amongst crustaceans, some crabs are much broader than they are long.

8.7 Further reading

Borror, D.J., De Long, D.M. & Triplehorn, C.A. 1976. *An Introduction to the Study of Insects*, 4th edn. Holt, Reinhart & Winston, New York.

Boudreaux, H.B. 1979. *Arthropod Phylogeny with Special Reference to Insects*. Wiley, New York.

Chapman, R.F. 1969. *The Insects: Structure and Function*. English Universities Press, London.

Clarke, K.U. 1973. *The Biology of the Arthropoda*. Edward Arnold, London.

Cloudsley-Thompson, J.L. 1968. *Spiders, Scorpions, Centipedes and Mites*. Pergamon Press, Oxford [Chelicerata & Myriapoda].

Daly, H.V., Doyen, J.T. & Ehrlich, P.R. 1978. *Introduction to Insect Biology and Diversity*. McGraw-Hill, New York.

Gupta, A.P. (Ed.). 1979. *Arthropod Phylogeny*. Van Nostrand Reinhold, New York.

Kaestner, A. 1968. *Invertebrate Zoology*, (Vol. 2). Wiley, New York [Onychophora, Tardigrada, Pentastoma, Chelicerata & Myriapoda].

Kaestner, A. 1970. *Invertebrate Zoology*, (Vol. 3). Wiley, New York [Crustacea].

King, P.E. 1973. *Pycnogonids*. Hutchinson, London.

Lewis, J.G.E. 1981. *The Biology of Centipedes*. Cambridge University Press, Cambridge.

Little, C. 1983. *The Colonisation of Land*. Cambridge University Press, Cambridge [All terrestrial arthropods].

Manton, S.M. 1977. *The Arthropoda*. Oxford University Press, Oxford.

McLaughlin, P.A. 1980. *Comparative Morphology of Recent Crustacea*. Freeman, San Francisco.

Ramazzotti, G. 1972. *Il Phylum Tardigrada*, 2nd edn. Istituto Italiano di Idrobiologia, Pallanza.

Savory, T.H. 1977. *Arachnida*. Academic Press, New York.

Schram, F.R. 1986. *Crustacea*. Oxford University Press, New York.

Sedgwick, A. 1888. *A Monograph of the Development of* Peripatus capensis, *and of the Species and Distribution of the genus* Peripatus. Clay, London.

III

INVERTEBRATE FUNCTIONAL BIOLOGY

Whereas Section II described the diversity of invertebrate body plans and biologies, this section concentrates on the unifying features of their functional anatomy, physiology and behaviour. It is clear that, whatever their structure, evolutionary history or ecology, all animals have certain common requirements needed to achieve, at least potentially, their own individual survival and that, in the longer term, of their genes. Hence they possess equivalent suites of functional systems permitting the acquisition of the necessary resources and information, and the processing and ordering of these inputs. The selective advantages associated with different body plans, life styles and habitats, however, have favoured different solutions to many of these common problems; and, as in Section II, the following chapters present invertebrate functional biology against a background of the various selection pressures and of optimal solutions to interacting pressures.

Some requirements are necessary for immediate survival, and selection here may often act powerfully on living individuals. Animals, for example, need to find, consume and assimilate energy- and chemical-containing food materials, often in the face of considerable competition for these resources (Chapter 9), whilst at the same time avoiding becoming the food of other consumers (Chapter 13). Other necessities—those permitting any animal to be capable of functioning at all—can be envisaged as having been faced and overcome relatively early in the evolutionary history of most lineages, and thereafter as being subject largely to stabilizing selection. Thus most animals require locomotory systems and all need some parts of their bodies to be capable of movement to obtain food, escape from consumers, avoid unfavourable environmental conditions, etc. (Chapter 10); they must also exchange respiratory gases with their environment and carry out energy-yielding metabolic reactions (Chapter 11); being of a different chemical composition to their environment, regulation of their internal composition and/or concentration, including the elimination of metabolic waste products, will be required (Chapter 12); and information on both the internal and external environments must be obtained, evaluated and, if appropriate, acted on, and the various levels of development or activity of the different functional systems of the body have to be timed and co-ordinated if an individual multicellular organism is to act as a unitary whole in its behaviour and physiology (Chapter 16). Finally (Chapters 14 and 15), animals must adopt reproductive and life history strategies that will maximize their genetic contribution to future generations; and the zygotes or other propagules formed must, in turn, develop into organisms capable themselves of reproducing and acquiring and processing their own resources, sometimes via distinct larval stages adapted for dispersal, the finding of specific host organisms, or feeding before metamorphosis into the reproductive adult form.

Since research into all these various fields has been carried out on only a very limited number of invertebrates, the animals upon which this section is based will be a minor fraction of those described in Section II. For understandable reasons, this experimental material has been drawn mainly from the more numerous and larger invertebrate groups (the arthropod phyla, the molluscs, etc.). This, however, will serve to redress the apparent bias against these so-called 'major phyla' created by the approach adopted in Section II.

9
FEEDING

To some extent, animal feeding is a neglected field of enquiry. True, there is a wealth of information available on, for example, the mechanics of suspension feeding and on how predators catch their prey, and much is known of gut anatomy and digestive physiology, but why animals feed on what they do consume has largely been assumed to be self evident: that is what they are adapted to capture and digest. But why are the most primitive types of animals almost exclusively carnivorous? And why is so much plant material left unconsumed? Why are some species more generalist in their diets than others? And why should some wading birds prefer mussels to worms?

In this chapter, we look at the phylogenetic constraints which have channelled animal feeding (the evolutionary past), at the different types of feeding mechanism possessed by animals (the heritage of the past), and at the rapidly growing field which investigates the pros and cons of taking different individual items within the broad range of a diet and how the prey can influence consumer choice (the ecological present).

9.1 Introduction: the evolution of animal modes of feeding

All the phyla of animals, with the possible exception of the Uniramia, evolved in the sea, a very stable and uniform habitat. Except in the marginal intertidal and immediately coastal zones, physical variables such as temperature, salinity, ionic composition, oxygen saturation, etc., do not pose any threats to the survival of marine animals, and physiological limitation of biological activity does not occur. Marine species therefore only require two—or possibly three—requisites for survival, both or all of them relating to feeding. They need to obtain sufficient food; they must avoid becoming the food of other organisms; and they may require an exclusive area of space in which to achieve the other two. Successful survival of the individual requires only these. If, of course, their genes are to survive through a longer period of time than the lifespan of any individual organism, they will also need to produce the maximum possible number of surviving and reproducing offspring (see Chapter 14). All other biological attributes, whether anatomical, physiological, biochemical or developmental, are simply the mechanistic ways of maximizing the chances of obtaining and processing these fundamental requisites.

The first animals inhabited the surface of the Precambrian sea bed (Chapter 2): what would they have had available to them as potential food? The answer can only be colonial or unicellular bacteria and protists, which, in all but the shallowest of waters, would mainly have also been heterotrophic. Sufficient light to permit photosynthesis penetrates only some 100 m into open sea water, and usually much less than this in shallow regions (some 20–30 m into coastal seas, and sometimes only a few centimetres in highly turbid silt-laden inshore waters). Most of the continental shelf and all of the ocean bed receives no sunlight. The primary production of organic materials by the photosynthetic protists of the sea is therefore a surface phenomenon, far removed from the vast majority of the benthic habitat. Today, most benthic animals are almost entirely dependent, ultimately, on the rain of dead and already partially decayed material from above, and on the organisms responsible for its decomposition both during its descent through the water and after it has settled on the sea bed.

Only in the marginal shallow waters would living photosynthesizers have been available to any animal which could either filter them from suspension in the water or else graze or browse the attached forms. Apart from the phylogenetically-isolated sponges, filter feeding was a much later specialization in animal evolution, as was browsing and grazing (Chapter 2). Bacterial chemosynthesis on the sea bed was, however, and in some areas still is, an important source of primary production; both bacterial photosynthesis and chemosynthesis are characteristic of anoxic or microaerobic subhabitats or reduced substrata.

It is therefore not surprising to find that the simplest surviving flatworms, and presumably the ancestral forms, are and were essentially consumers of bacteria and protists associated in some manner with the bottom sediments and rocks. What is perhaps more surprising is that in overt or covert ways this ancestral diet has dominated the nutritional lives of all the flatworm descendants, even including the terrestrial ones. Bacteria and protists are abundant in marine sediments, but they are individually small and are often relatively widely dispersed in space. They rarely occur in dense clumps, although we will consider some exceptions later. This has had two fundamental repercussions on animal life styles.

First, they must be mobile to find new supplies when local stocks have been exhausted—and mobility is one of the hallmarks of the animal condition—and, second, a consumer of small, widely scattered organisms must itself be relatively small. The larger an animal is, the greater will be its metabolic requirements and the more food in total it must obtain per unit time. This can be illustrated by a plot of body size against the corresponding energy expenditure on maintenance (Fig. 9.1). A large animal could not subsist on a diet of bacteria or small protists because it could not find and consume enough of them per unit time.

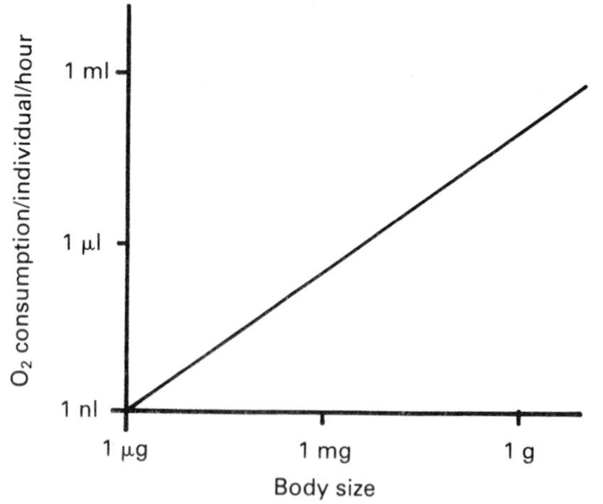

Fig 9.1. The relationship between body size and metabolic rate (after Phillipson, 1981). (See also Fig. 11.16.)

But to increase in size is itself likely to be selectively advantageous, for three reasons:

1 Larger animals tend to be able to produce more offspring than smaller animals (i.e. differential reproductive output).

2 Larger size can confer greater immunity from consumption by other organisms (i.e. differential survival).

3 Larger animals tend to be able to displace smaller animals from limited shared resources (i.e. differential survival again).

In addition, assuming equal digestibility, it is always energetically more efficient to ingest larger individual items of any given type of food than it is smaller ones, and to achieve this in turn requires larger size.

So considering these selective benefits of increase in size, it is also not surprising to find that most surviving flatworms are larger than the bacteria-consuming species and that they have turned to the consumption of larger food items—other animals—rather than bacteria and protists. Today, the two more significant basal stocks of animals, the bilateral flatworms and the radial coelenterates, are both essentially carnivorous. Nevertheless, there is a limit imposed on flatworm body size by the constraint of diffusion distance (Section 11.4.1), and so the continued selective advantage of increased size would render advantageous any morphological changes which permitted this to occur. This, together with escape from flatworm predation, probably provided two of the main impetuses behind the wide diversification of vermiform animals which occurred early in animal evolution.

One notable concentration of protist tissue does occur, however, in the shallowest of marine regions, in the form of filamentous or thallose, colonial or multicellular algae: seaweeds. Although the larger seaweeds pose particular problems (see below), the filamentous forms, and the juvenile stages of the macroscopic types, can relatively easily be removed from their attachment to the substratum by a rasping organ (see Fig. 5.3) making available for digestion this abundant and concentrated source of photosynthetic tissue. One of the early flatworm derivatives, the Mollusca, evolved in association with this specialized feeding mode, safe both from the vigorous water movements typical of shallow waters and from predation beneath their dorsal protective shells. Radular rasping is also an effective means of obtaining food from sessile or sedentary animal prey otherwise protected with external shells, sheaths or boxes, and indeed from terrestrial plants, and during their subsequent evolution the molluscs extended their basic feeding method to utilize most of the other available types of food materials.

The other early flatworm derivatives either retained the ancestral flatworm diet or evolved mechanisms for concentrating the rain of particles carried by, or sedimenting out of, the overlying water. Under many marine conditions, the fall-out rate and the suspended organic load of sea water are effectively constant, and so, provided that a consumer positions itself appropriately to intercept and concentrate this supply, it need not move much, if at all, or only one organ system need be moved, and being sessile or sedentary will reduce the basal metabolic requirements for energy and therefore the minimum quantity of food needed per unit of body weight. A diet largely based on bacteria, protists and dead organic matter is then still possible in a relatively large animal.

Several of these other early flatworm descendants extracted particles from suspension in the water by creating a current through some form of filtration device ('suspension feeding'); others collected those particles and their microbial associations which had already sedimented out on to the substratum surface ('deposit feeding'); and yet further forms intercepted the passive fall-out before it reached the bottom ('sedimentation interceptors').

Suspension-feeding groups include the sponges, the lophophorate phyla and the invertebrate chordates, and several members of the Mollusca, the Annelida and the arthropod groups, including some which later in phylogeny exported suspension feeding into the water column and up into the zone of photosynthetic production. The alimentary adaptations of suspension feeders will be covered in a later section of this chapter (Section 9.2.5) but here we can note some generalities of the feeding process. All suspension feeders have a filter of some form and either a feeding location positioned so as to intercept a natural (and persistent) water current or else means of creating their own current through the filter.

In sponges, the same cells—the choanocytes—both create the water current by the beating of their flagella and trap the food particles contained in it on the collar of microvilli surrounding the flagellum (Fig. 9.2). The small mesh size of this microvillar

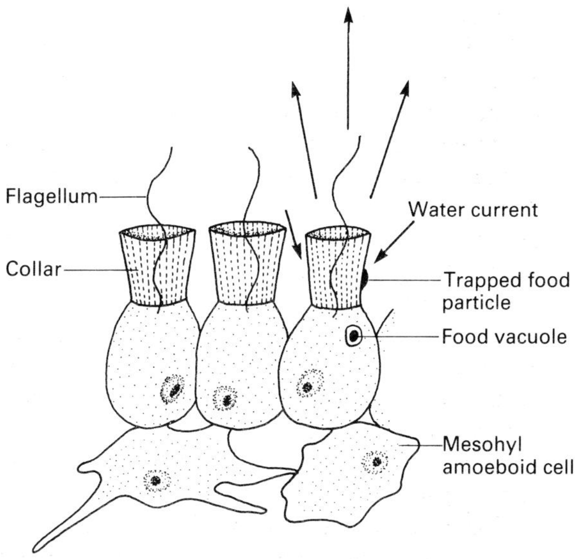

Fig. 9.2. Sponge choanocytes (after Brill, 1973).

filter enables particles down to the size of bacteria to be retained. In all suspension feeders, it would clearly be advantageous not to filter the same volume of water more than once, and sponges avoid this by drawing water in through many small pores scattered over the general body surface, and by expelling it through only one or a few larger exit apertures (see Section 3.2). The more powerful water current which results drives the filtered water well away from the sponge body. (The colonial and planktonic pyrosome tunicates have evolved an equivalent system and the single colonial exhalent current further serves to drive the colony through the water by jet propulsion; Fig. 7.34.)

In contrast, the suspension-feeding products of the radiation of flatworm descendants have adapted or

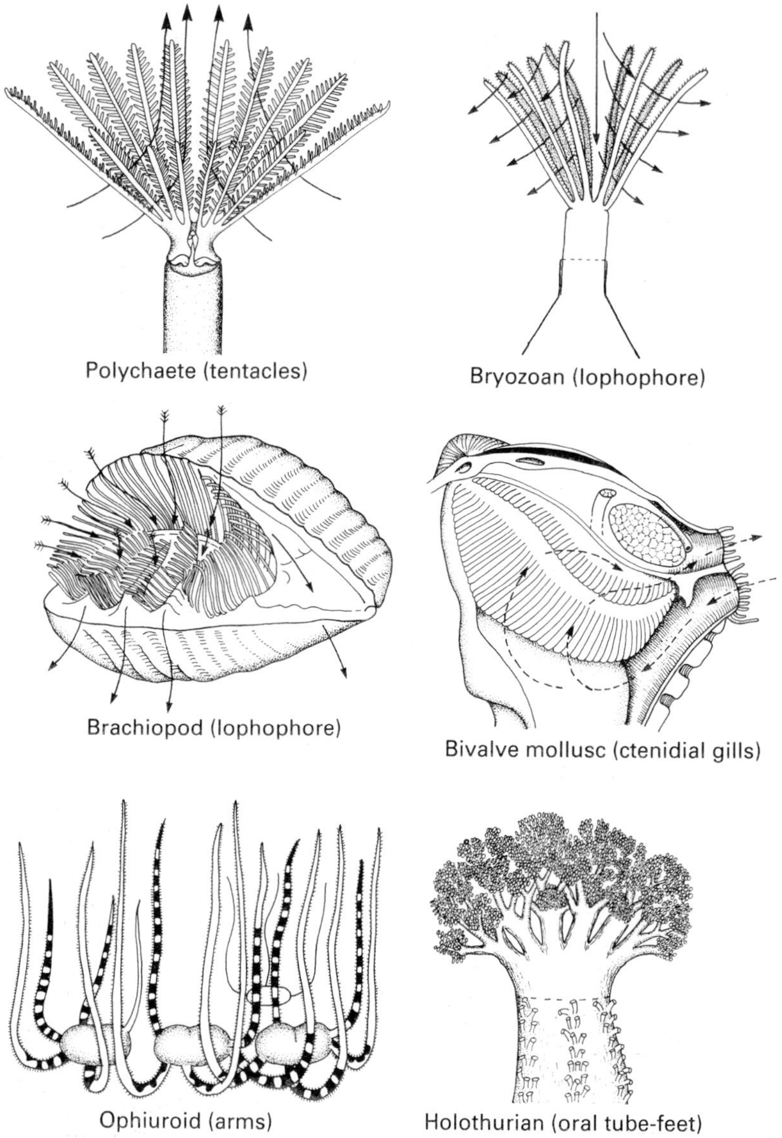

9.3. External organs used for muco-ciliary suspension feeding (after several sources).

evolved specific organs for filter feeding: tentacles on or near the head in the annelids, the mesosomal lophophore in the phoronans and their derivatives, greatly enlarged ctenidial gills in bivalve molluscs, and so on (Fig. 9.3). All, however, operate in the same basic fashion. The beating of cilia in tracts on some region of the filtration organ create the water current, whilst other ciliary tracts trap and convey the food particles, with the aid of mucus, to the mouth. Particle capture is largely size specific, but yet further tracts of cilia, either on the filter itself or on some accessory organ, sort the trapped particles into those to be ingested and those to be rejected, often in the form of a pseudofaecal pellet. The invertebrate chordate system is exactly equivalent (Section 9.2.5).

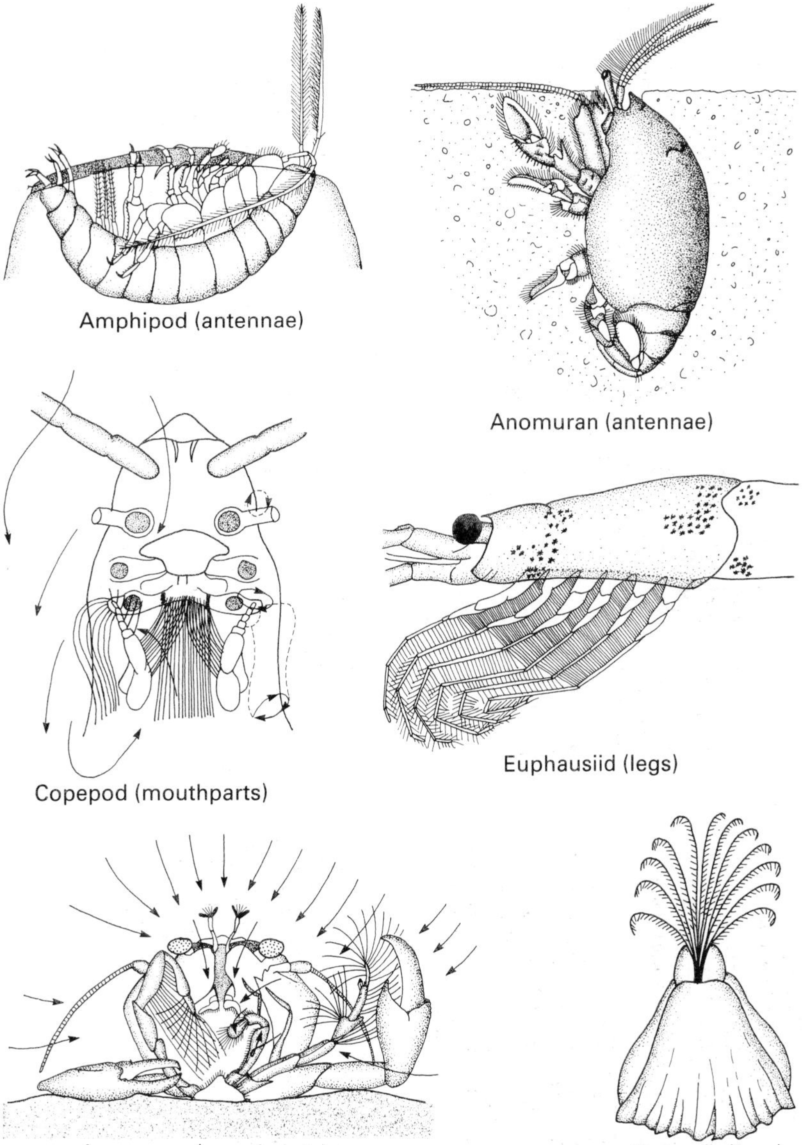

Fig. 9.4. Setose filtration systems in crustaceans (after several sources).

The arthropod filter feeders, however, lack cilia by virtue of their possession of an exoskeleton, and instead use setae on various of their appendages to form the filter (Fig. 9.4). Their feeding current is created by swimming-like movements of the filter-bearing or of other limbs. Setose filters are generally coarser than ciliary ones and therefore can trap only larger particles.

All the animals using tentacle-like structures to suspension feed, and many other filter feeders as well, are sessile or sedentary, and the tentacular feeding organ is effectively radially symmetrical with the mouth usually being located in the centre of the feeding apparatus—the same functional system as shown by the primarily radial cnidarians (see Section 3.4). The body is generally protected by an external shell or tube or is situated within a burrow in the substratum. The animal and/or its delicate feeding apparatus can therefore be withdrawn for safety from predators into cover. Under such semi-enclosed circumstances, a terminal anus would of necessity result in the passage of faecal material along the whole length of the external body surface before it could escape to the environment. Accordingly, there is a marked tendency in tubiculous and equivalent species for the gut to be U-shaped, with the anus, and often the excretory organs, discharging near the anterior end of the body in the path of the exhalent water current (e.g. Fig. 7.8). The same anatomical system is also found in many sedentary deposit feeders for the same reasons (e.g. Fig. 4.42). (Alternatively, the burrow may possess (at least) two openings at the surface and the occupant maintain a unidirectional current of water through it, or, more rarely, the animal may live doubled up within its tube so that both mouth and anus are at the single aperture.)

Some annelids, including the ragworm *Nereis diversicolor* and *Chaetopterus*, and the appendicularian tunicates, amongst others (some gastropod molluscs, echiurans, etc.), have evolved a filtration apparatus which is not part of the body at all, but is a secreted structure. A meshwork of mucous threads is formed in the shape of a net or bag, and a current of water is drawn through it by means of movements of all or part of the body equivalent to those used for locomotion. After an interval, the mucous net is consumed, together with such material as it has collected, and a new filter is secreted (Fig. 9.5). Some corals, other gastropod molluscs and insect larvae have independently evolved a similar feeding system, although largely dependent on natural water currents rather than self-induced ones.

The collection of surface material by deposit feeding requires no such elaborate organs as those of suspension feeders, and indeed some worms simply consume the surface layer with a completely unspecialized mouth region. Nevertheless, in a number of cases, notably in the sipunculans, holothurians and echiurans, and in several annelids and some hemichordates, a specialized series of lobes or tentacles around the mouth or an extensible proboscis, all covered by ciliary tracts, may be moved through or across the sediment, the more effectively to collect food particles in the vicinity of the burrow system (Fig. 9.6). Deposit feeders with lobes or tentacles around the mouth share several features in common with suspension feeders (as we noted above in respect of the U-shaped gut) and several of them can feed in both modes. Two problems peculiar to deposit feeders, however, are (a) that organic material may comprise only a small proportion of the background sediment, and (b) that much of the organic matter present may be the relatively refractory and indigestible residues which have accumulated precisely because they cannot be used by animal consumers, only by bacteria (see Section 9.2.6).

Therefore, except in shallow regions where benthic photosynthetic protists can live on the sediment surface, deposit feeders may be dependent on such bacteria as can convert these refractory organic residues into digestible materials, and on those protists and interstitial animals as are themselves dependent on the bacteria. In any event, large quantities of sediment may have to be ingested (and deposit feeders may rework benthic sediments and act as powerful agents of bioturbation) or large quantities must be processed by sorting organs in order to obtain sufficient digestible food. The tentacles of many deposit-feeding worms are capable, for example, of extending large distances away from the burrow in which the body is housed. Except in particularly rich areas, deposit feeders cannot therefore achieve the densities attained by suspension feeders, for which space rather than food is often the limiting resource. Because many are ultimately reliant on bacterial productivity rather than on the rate of sedimentation of detrital particles, and because bacterial productivity may itself be limited by factors other than the supply of carbon (such as nutrient shortage in the interstitial water), the growth rate of deposit feeders is also often slower than that of suspension feeders.

The final category of consumers of materials

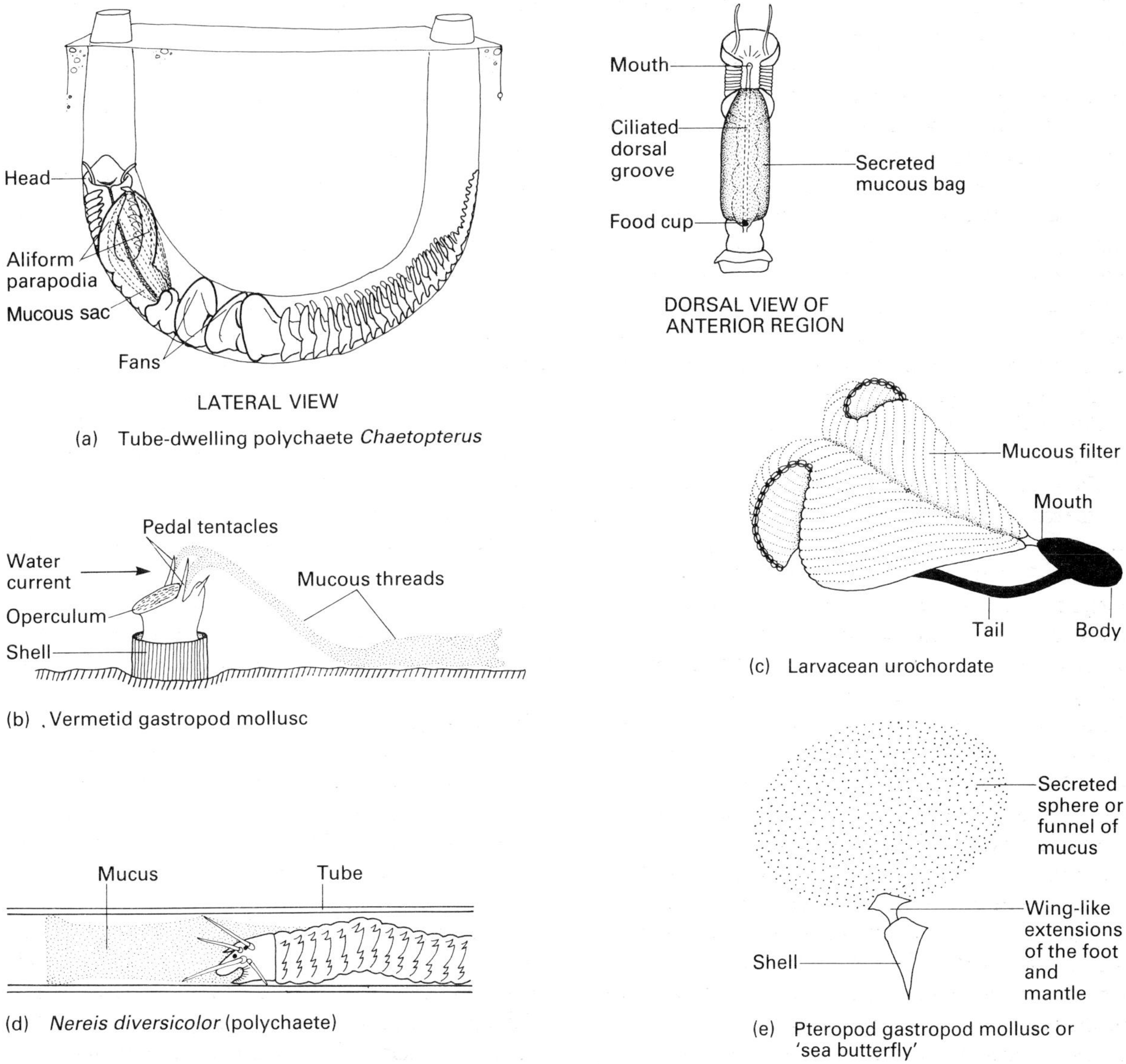

Fig. 9.5. Filter feeding by the use of a secreted mucous meshwork (after several sources).

originating in the overlying water, the sedimentation interceptors, such as several of the stalked echinoderms, have a body attached to the sea bed but positioned well above it so as to intercept the supply of detritus before it becomes diluted by incorporation into the sediment (Fig. 9.7). A series of radially arranged arms bearing hydraulically operated and mucus-laden papillae collect the particles, which are then conveyed to the mouth, located in the centre of the circle of arms, along ciliated grooves.

Judging from the surviving members of the ancestral animal groups, there was probably one other, and completely different, mode of nutrition present very early in the history of animal multicellularity. Several marine animals living in very shallow waters contain in their surface tissues symbiotic

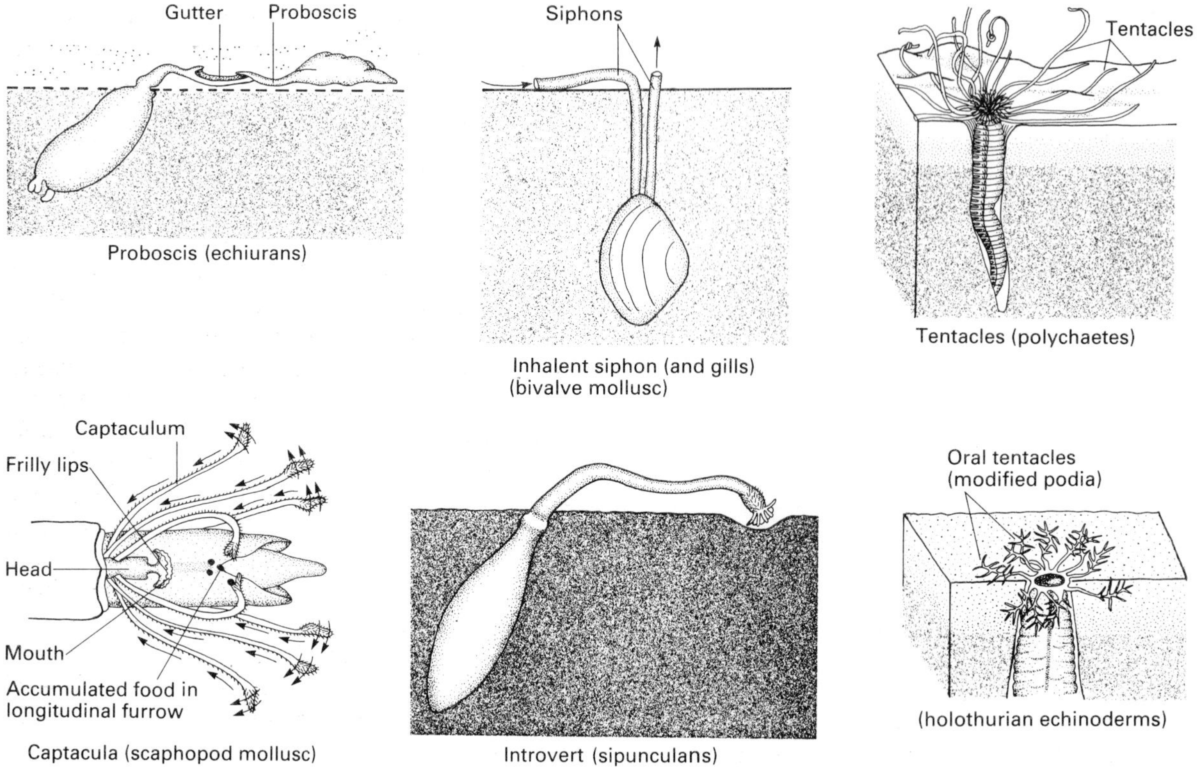

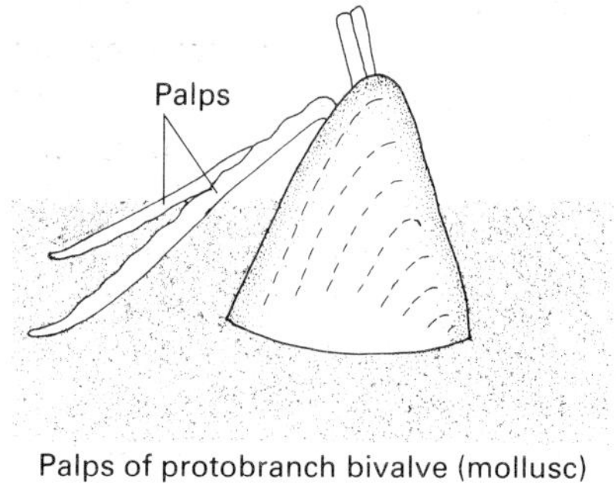

Fig. 9.6. Organs of mucociliary deposit feeding in burrowing invertebrates (after several sources).

Chapter 9

...micrograph of algal ...of a cnidarian ...with

Attachment
holdfast

Substratum

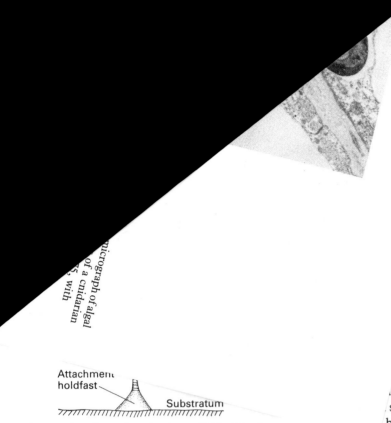

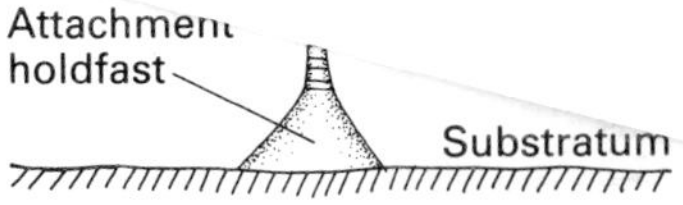

Fig. 9.7. A sediment-intercepting stalked echinoderm (crinoid) (after Clark, 1915).

oxyphotobacteria (prochlorophytes and cyanobacteria) or photosynthetic protists (unicellular dinoflagellates or chlorophytes), and some flatworms and cnidarians appear to be entirely dependent on these symbionts for their nutrition, although most are not capable of digesting them. Such acoels, soft corals and a jellyfish live with their bodies permanently (in the case of sessile species) or temporarily (in mobile forms) exposed to full sunlight. Their symbionts photosynthesize in the normal manner, but in part using inorganic nitrogen and phosphorus, stripped of their organic binding during the animal's metabolism, as a source of nutrients, and using the host's respiratory carbon dioxide as a carbon source. The animal partner appears in some way to render the symbiont's cell walls thinner and/or more leaky

...the photosynthesized ...lipids and amino acids, ...he animal's tissues where ...the breakdown products ...symbionts. In effect, the ...ve populations of photo- ...algae, which can attain ...$^{-3}$ of host tissue (Fig. 9.8). ...re at least partially dependent ...nthesis.

...gonophorans are equivalently ...le or in part, on symbiotic ...acteria which utilize the reduced ...s and methane that issue from ...les in the sea floor; and recently ...tes and bivalve molluscs associated ...ide-rich marine habitats have also ...obtain their nutrition in a similar ...ng some 50% of the carbon fixed by ...oxidizing bacteria. That this symbiotic ...o animal nutrition occurs in represen- ...st animal phyla, and particularly widely ...s thought to be closest to the ancestral ...sts that although multicellularity itself ...ve had a symbiotic origin (Chapter 2), the ...success of the early groups of animals may been aided by it, especially perhaps where food ...vailability was limiting. (In fact, the chloroplasts of some chromophytan algae may have been derived from other, one-time endosymbiotic, enkaryote algae, and, if this is so, then some eukaryotes have evolved from symbiotic unions of separate organisms which were themselves eukaryote.)

The feeding methods and food items reviewed above must for hundreds of millions of years have been the only ones found in the animal kingdom amongst free-living species, and marine animals today still basically subsist on a diet of living or dead bacteria, protists and each other. Eventually, however, animals of marine ancestry colonized the land and although terrestrial bacteria, protists and other animals could continue to permit essentially unchanged ancestral diets for many (except for the ineffectiveness of suspension feeding on land), in

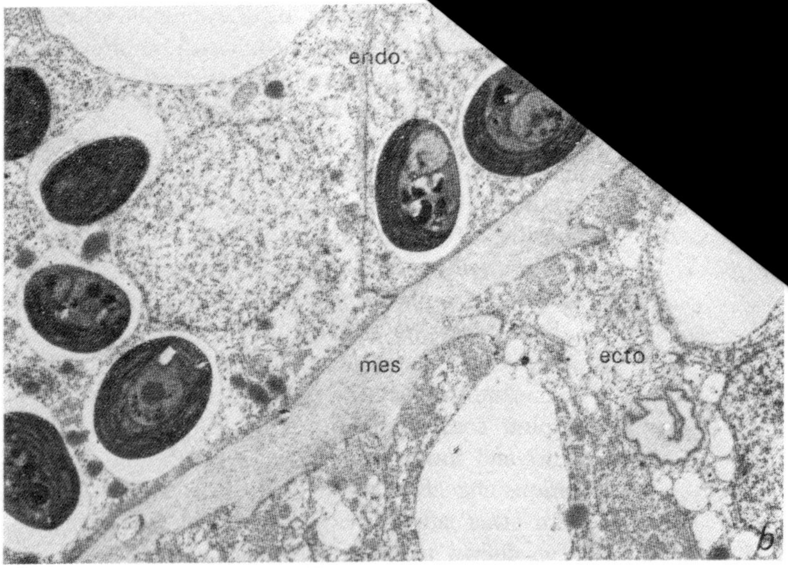

Fig. 9.8. An electro
symbionts in the tissue
(from Muscatine *et al.*, 19
permission).

their new habitat they would have encountered a new, plentiful, but very different, food source—the terrestrial bryophytes and tracheophytes. These pose a considerable problem to consumers in that a high proportion of their biomass is in the form of tough, indigestible cell walls of cellulose and almost inert, supporting, structural polymers such as lignin. Ancestrally, most animals had never needed cellulases or similar enzymes to digest their food, and the early terrestrial species must have found themselves completely ill-equipped to tackle this abundant but refractory source of potential food. Not until the Cretaceous did plant feeding become widespread on the land.

To some extent, however, a few marine species had already overcome a very similar problem. The larger seaweeds of shallow marine habitats also contain complex carbohydrates, in this case to provide the strength required to resist vigorous water movements. Once their macroscopic form has developed, they are as potentially usable as is the majority of terrestrial plant tissue. Only two groups of marine animals have evolved the ability to cope with this generally unattractive material whilst it is still living.

One group is certain sea-urchins. Most consumers of seaweed material can do so only after bacteria have converted refractory polysaccharides into digestible form; these include the classic deposit feeders. Bacteria, especially the anaerobic fermenters, can break down a very wide range of organic molecules, including those in petrol and many plastics, but instead of relying on this activity taking place in the external environment, some sea-urchins have internalized the process. They have a special region in their gut in which a culture of anaerobic fermenting bacteria is maintained. In effect, the sea-urchins take into their guts the coarse seaweed tissue and by so doing feed the bacteria; the bacteria digest this material, and the echinoderms then subsist on a diet of the products of bacterial digestion and on the bacteria themselves. A further problem associated with seaweed material is its low nitrogen status per unit weight, but, under anaerobic conditions, several bacteria can fix atmospheric nitrogen dissolved in the sea water and hence the gut bacteria can boost the

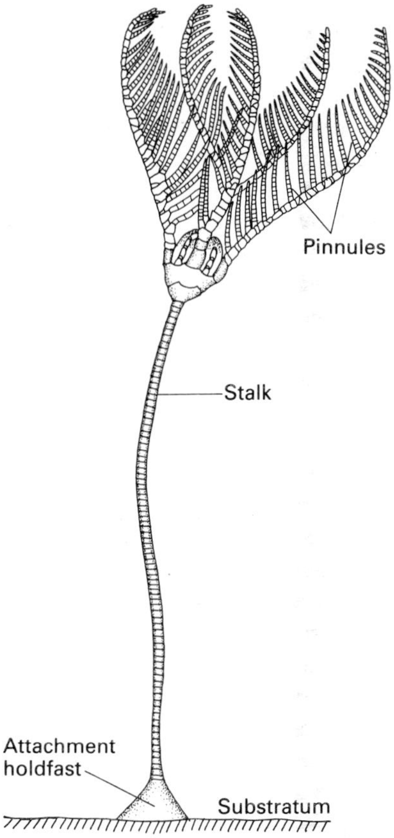

Fig. 9.7. A sediment-intercepting stalked echinoderm (crinoid) (after Clark, 1915).

oxyphotobacteria (prochlorophytes and cyanobacteria) or photosynthetic protists (unicellular dinoflagellates or chlorophytes), and some flatworms and cnidarians appear to be entirely dependent on these symbionts for their nutrition, although most are not capable of digesting them. Such acoels, soft corals and a jellyfish live with their bodies permanently (in the case of sessile species) or temporarily (in mobile forms) exposed to full sunlight. Their symbionts photosynthesize in the normal manner, but in part using inorganic nitrogen and phosphorus, stripped of their organic binding during the animal's metabolism, as a source of nutrients, and using the host's respiratory carbon dioxide as a carbon source. The animal partner appears in some way to render the symbiont's cell walls thinner and/or more leaky than normal, and some of the photosynthesized products, including sugars, lipids and amino acids, therefore diffuse out into the animal's tissues where they are assimilated, and the breakdown products then diffuse back to the symbionts. In effect, the animal milks its captive populations of photosynthetic bacteria and algae, which can attain densities of 30 000 mm^{-3} of host tissue (Fig. 9.8). Other marine species are at least partially dependent on symbiotic photosynthesis.

Some deep-sea pogonophorans are equivalently dependent, in whole or in part, on symbiotic chemoautotrophic bacteria which utilize the reduced sulphur compounds and methane that issue from vents and fumaroles in the sea floor; and recently various oligochaetes and bivalve molluscs associated with other sulphide-rich marine habitats have also been shown to obtain their nutrition in a similar fashion, receiving some 50% of the carbon fixed by their sulphur-oxidizing bacteria. That this symbiotic contribution to animal nutrition occurs in representatives of most animal phyla, and particularly widely in the groups thought to be closest to the ancestral ones, suggests that although multicellularity itself may not have had a symbiotic origin (Chapter 2), the origin and success of the early groups of animals may well have been aided by it, especially perhaps where food availability was limiting. (In fact, the chloroplasts of some chromophytan algae may have been derived from other, one-time endosymbiotic, enkaryote algae, and, if this is so, then some eukaryotes have evolved from symbiotic unions of separate organisms which were themselves eukaryote.)

The feeding methods and food items reviewed above must for hundreds of millions of years have been the only ones found in the animal kingdom amongst free-living species, and marine animals today still basically subsist on a diet of living or dead bacteria, protists and each other. Eventually, however, animals of marine ancestry colonized the land and although terrestrial bacteria, protists and other animals could continue to permit essentially unchanged ancestral diets for many (except for the ineffectiveness of suspension feeding on land), in

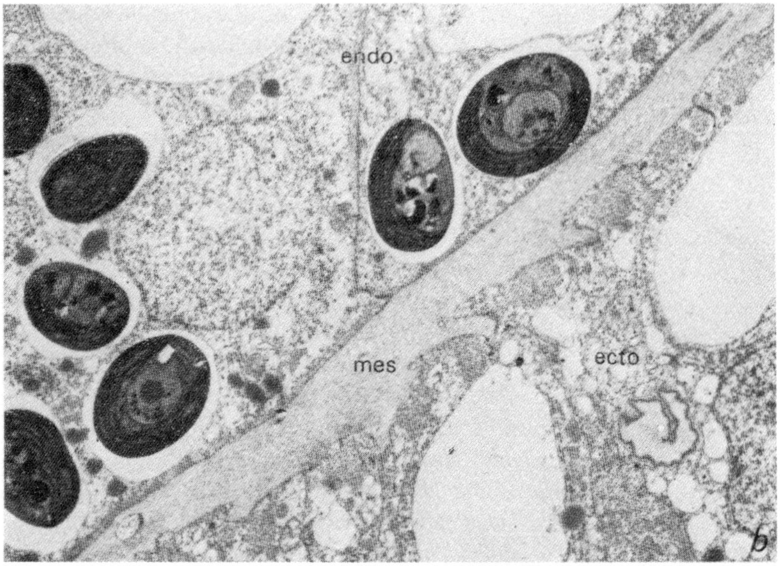

Fig. 9.8. An electronmicrograph of algal symbionts in the tissues of a cnidarian (from Muscatine *et al.*, 1975, with permission).

their new habitat they would have encountered a new, plentiful, but very different, food source—the terrestrial bryophytes and tracheophytes. These pose a considerable problem to consumers in that a high proportion of their biomass is in the form of tough, indigestible cell walls of cellulose and almost inert, supporting, structural polymers such as lignin. Ancestrally, most animals had never needed cellulases or similar enzymes to digest their food, and the early terrestrial species must have found themselves completely ill-equipped to tackle this abundant but refractory source of potential food. Not until the Cretaceous did plant feeding become widespread on the land.

To some extent, however, a few marine species had already overcome a very similar problem. The larger seaweeds of shallow marine habitats also contain complex carbohydrates, in this case to provide the strength required to resist vigorous water movements. Once their macroscopic form has developed, they are as potentially usable as is the majority of terrestrial plant tissue. Only two groups of marine animals have evolved the ability to cope with this generally unattractive material whilst it is still living.

One group is certain sea-urchins. Most consumers of seaweed material can do so only after bacteria have converted refractory polysaccharides into digestible form; these include the classic deposit feeders. Bacteria, especially the anaerobic fermenters, can break down a very wide range of organic molecules, including those in petrol and many plastics, but instead of relying on this activity taking place in the external environment, some sea-urchins have internalized the process. They have a special region in their gut in which a culture of anaerobic fermenting bacteria is maintained. In effect, the sea-urchins take into their guts the coarse seaweed tissue and by so doing feed the bacteria; the bacteria digest this material, and the echinoderms then subsist on a diet of the products of bacterial digestion and on the bacteria themselves. A further problem associated with seaweed material is its low nitrogen status per unit weight, but, under anaerobic conditions, several bacteria can fix atmospheric nitrogen dissolved in the sea water and hence the gut bacteria can boost the nitrogen status of the overwhelmingly carbohydrate sea-urchin diet. Although in this sense pre-adapted to terrestrial herbivorous diets, no echinoderm has become terrestrial, but many of the most successful terrestrial herbivores have solved the problems in a parallel manner.

The other group is the gastropod Mollusca, and not surprisingly in view of their browsing and grazing origins in shallow waters, they are one of the few animal groups to have evolved the required enzymes to break down a large number of carbohydrate polymers including, in some, cellulases. In contrast to the echinoderms, the gastropods have proved a successful group of land herbivores, as any gardener will testify.

Although a few terrestrial arthropods have also evolved cellulases, the strategy of a symbiotic gut microbiota is the one adopted by many of the more successful terrestrial herbivores, including cockroaches, termites, several beetles (and the mammalian vertebrates). These can only subsist on their plant diet through the intermediary of their symbiotic bacteria and protists, either via the fermentation of the carbohydrate chains and/or via the provision of the required additional levels of organic nitrogen. Such so-called herbivores are as dependent on bacteria and protists as are the acoel flatworms, the sipunculans and the deposit-feeding annelids.

Not all parts of a given plant are equally refractory, however; the young photosynthetic leaves, in particular, pose less problems to animal digestion. Even so, few types of consumer are able to break down unbroken cell walls to get at the cell contents by their own enzymatic repertoire. Land invertebrates lacking the ability to decompose cellulose, symbiotically or enzymatically, have evolved two techniques for releasing the contents of plant cells or otherwise obtaining plant fluids. Caterpillars, grasshoppers and various other insect consumers of leaves bite small pieces or thin strips from the plant and utilize the contents of such cells as were ruptured during detachment and subsequent chewing of the leaf fragment. Intact cells, however, are unavailable and hence relatively little food (one-third, on average) is obtained from each fragment ingested. Large quantities of material must therefore be consumed to offset the inefficiency of utilization, and this inefficient system can only be maintained because of the large biomass of the raw material available. The second technique is to tap into the plant's fluid-transport systems by means of piercing and sucking mouthparts (Fig. 9.9; see also Fig 8.30), as seen in all homopteran bugs, e.g. aphids. They avoid the refractory structural carbohydrates but may still require symbiotic gut bacteria to boost the nitrogen status of this dilute and protein-deficient liquid.

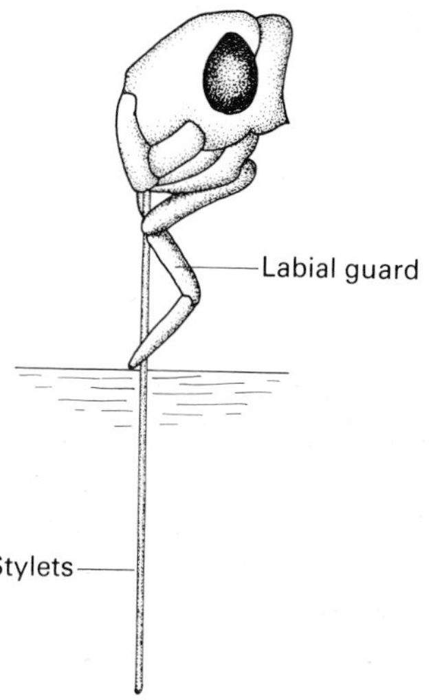

Fig. 9.9. The head of a plant-sucking bug, with its stylet inserted into the tissues of the plant (after Barnes, 1980).

The exploited plants, however, would clearly be advantaged if they could deter consumption of their photosynthetic tissues and in common with other sessile organisms (colonial marine animals, for example), i.e. those least able to defend themselves by other means, they have evolved mechanical and chemical defences of great subtlety (Section 9.2.4). As is usual in such systems, this has led to arms races between the consumed and consumer in terms of toxin production and detoxification or resistance.

Only one category of plant consumer efficiently utilizes plant materials without the aid of symbiotic bacteria and protists. These are species feeding on substances or structures produced by the plants specifically to attract animal consumption—nectar, fruits, nuts, etc.—in order to achieve animal-effected pollination and/or dispersal. Although fruits and nuts are generally targeted on vertebrates, nevertheless arthropods and molluscs may pirate this resource before it is consumed by birds or mammals.

Fungi, with their chitinous cell walls and lack of refractory supporting tissues, present fewer basic problems for animal digestive systems and are widely consumed. Indeed, some insects go so far as to collect plant material, chew it into a pulp, and then use that pulp as a substrate for fungal growth, ultimately consuming the fungi. To a degree, fungi replace the ancestral algal component in terrestrial animal diets.

Despite the widespread popular notion that terrestrial food chains are predominantly of the form plant → herbivore → carnivore, an impression which is reinforced by the amount of scientific attention which has been devoted to herbivorous animals (especially grazing mammals), the base of most terrestrial food webs is not living plant tissue and the grazing down of this by herbivores, but the decomposer food chain. It is through the detritus- and litter-feeding animals that most energy flows; a pathway mediated by bacteria, protists and fungi. Terrestrial animals have therefore not escaped from the consequences of their marine ancestry in terms of what they can efficiently process, and even today less than 3% of forest productivity is consumed, whilst it is still alive, by herbivores.

9.2 Types of animal feeding: patterns of acquisition and processing

9.2.1 Classification of feeding type

The influence of early assumptions of the overriding importance of the grazing food chain in ecological interactions is also reflected in the names given to the classical trophic levels. For many years, animal feeding was classified on the basis of this food chain and of the systematic affinities of the food species consumed, such that 'herbivores', 'carnivores' and the intermediate 'omnivores' were recognized. This classification of feeding possesses many disadvantages: relatively few species are exclusively herbivorous or carnivorous, especially when all the stages in their life history are taken into account, so that most animals fall into the catch-all omnivorous category; secondly, being based on the old two-kingdom approach to classification (organisms being either 'plant' or 'animal'), it leaves open to question the placing of animals dependent on bacteria and protists, which as we have seen is an important group of consumers; and thirdly, the phylogenetic relationships of the prey species are not necessarily relevant to the feeding ecology of the consumer.

More recently, it has become customary to distinguish categories of consumer on the basis of their general feeding methods. Thus we have hunters, parasites, grazers and browsers, suspension feeders, deposit feeders, and those deriving their nutrition symbiotically. These divisions cut right across the boundaries of the systematic position of the prey species: grazers and browsers, suspension feeders and deposit feeders, for example, may consume any or all of bacteria, protists, fungi, plants and animals. If the feeding techniques of the animal consumers do not necessarily distinguish between the various kingdoms of organisms, then neither should we when analysing feeding biology, and these categories are those which were introduced in the foregoing section. Here we will cover them in rather more detail, including with reference to the alimentary and digestive features of each type.

9.2.2 Common features

Most invertebrates possess a through gut with a more or less anterior opening to the exterior, the mouth, into which food is taken, and a more posterior opening, the anus, from which indigestible residues and excretory products discharged into the gut can be voided. Some, however, principally the cnidarians and the flatworms but also a number of other types most of which have secondarily evolved this condition, have a blind-ending gut with a single opening serving both for ingestion and egestion; and others, including several parasitic forms but also a few free-living species, lack an alimentary canal and absorb food materials from internal symbionts or directly across the outer body surface. Variations on these themes are numerous. In sessile species, for example, the mouth is often located in the centre of the upper surface, rather than being anterior; in some free-living forms with a recent sessile ancestry, e.g. the free-moving echinoderms, the mouth is located in the centre of the lower surface; and the chordates and some others have evolved a gut with multiple openings to the external environment (Section 9.2.5).

Developmentally, the gut comprises three regions: an anterior fore-gut of ectodermal origin and lining (an inpushing of the outer body surface); a similarly ectodermal hind-gut; and an endodermal digestive and absorptive mid-gut, often bearing blind-ending

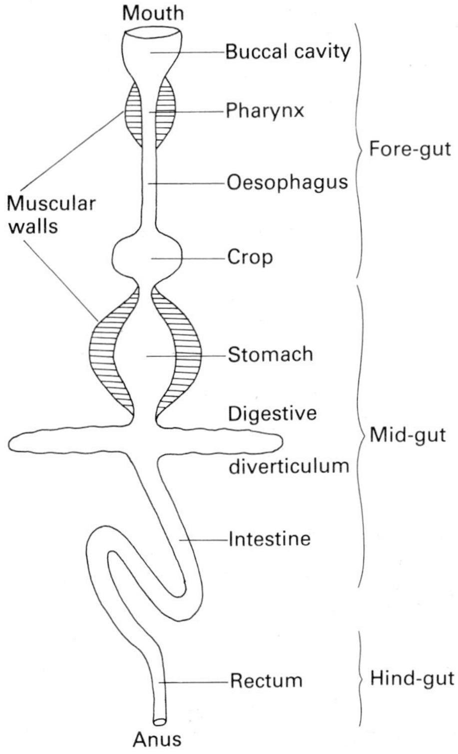

Fig. 9.10. The regions of a generalized invertebrate gut (diagrammatic).

diverticula. These three basic regions are further subdivided into different functional sections (Fig. 9.10). The fore-gut may comprise (a) a buccal cavity into which the mouth opens and into which 'salivary glands' discharge (sometimes specialized to produce sticky secretions, anticoagulants or toxins), (b) a muscular pharynx which may aid in food ingestion by acting as a pump, by forming an eversible organ of prey capture, or, in chordates, by serving as a filter for suspension feeding, (c) a short region of conduction, the oesophagus, and/or (d) a storage organ, the crop, especially well developed in consumers taking large meals at infrequent intervals which are released only gradually into the mid-gut.

The mid-gut is typically differentiated into (a) a muscular stomach, (b) various blind-ending secretory and/or absorptive outgrowths, and (c) an intestine. The stomach is the site of mechanical breakdown and sorting of the ingested material, sometimes possessing a distinct region—the gizzard—concerned with trituration, and it may also be partially the region of digestion. More often, however, either or both of digestion and absorption are effected in the various diverticula or caeca which issue from the gut just posterior to the stomach. In molluscs and crustaceans, these caeca may be elaborated into large complex organs—the hepatopancreas.

Primitively, digestion is largely intracellular, food particles being taken up by phagocytosis by the cells lining the absorptive region and being digested within vacuoles. This is prevalent in the sponges, coelenterates, flatworms and in various other animals in which the food is in finely particulate form by the time that it reaches the mid-gut. In more complex animals, and particularly in those species ingesting large individual masses of food, much digestion proceeds via enzymes secreted into the lumen of the gut and the products of digestion are then absorbed by the cells of the gut wall. In such species, the diverticula of the gut are normally purely secretory, and the intestine is the site of absorption. Extracellular digestion permits enzymatic specialization both of the individual cells and of different regions of the gut, but it also necessitates the production of a greater total quantity of enzymes since the lumen of the gut is a large open system in which it is difficult to maintain optimum concentrations of digestive activity.

Finally, the hind-gut (if present) is formed by the rectum, in which water may be absorbed (in some terrestrial animals; see Fig. 8.27) and faecal pellets may be formed prior to discharge through the anus.

From their food, all animals require energy-yielding compounds for immediate and later use, amino acids for synthesis into structural and metabolic proteins, and various other elements and compounds, such as vitamins, for use as catalysts or in other biochemical reactions. That is to say that they require to ingest carbohydrates, lipids, proteins and vitamins, and to digest them into a form suitable for absorption, if they cannot be absorbed directly. Animals with a diet low in proteins and/or vitamins normally require symbiotic gut bacteria for the synthesis of these compounds, and others ingesting complex carbohydrate polymers may also require symbiotic bacteria and protists to ferment them into simpler organic molecules. Much of the mass of animal faeces is composed of gut bacteria.

9.2.3 Hunters and parasites

Hunters are mobile animals which attack, kill and consume individual prey items one at a time, almost invariably other mobile animals. Three broad types can be distinguished: pursuit hunters, such as squid, which chase after, catch and subdue highly mobile prey; searchers, such as many gastropods and arthropods, which actively forage, seeking out prey items with less-developed powers of movement than themselves; and ambushers, such as spiders and preying mantises, which, apart from the final dart or spring, may be relatively sedentary.

Pursuers and ambushers (and to a lesser extent searchers) possess weapons of prey capture and immobilization. Most characteristically, these are either organs surrounding the mouth region, e.g. chelate or subchelate appendages in arthropods, sucker- or hook-bearing arms in cephalopods, and spines in chaetognaths (Fig. 9.11), or are powerful jaws associated with the anterior gut, in annelids mounted in an eversible pharyngeal region which can be forcibly and rapidly shot out to catch a prey item. Some ambushers attract their prey to them by mimicry: a number of siphonophores, for example, possess tentacles bearing copepod mimics, and they prey largely on copepod feeders (mostly other crustaceans). Once caught, the prey can be ingested whole, torn apart by appendages and consumed piecemeal, or have its body fluids sucked out. If ingested whole, the anterior region of the gut is capable of being distended to accommodate the meal.

Many searchers, on the other hand, feed on relatively sedentary prey which may protect themselves from attack by external coverings of calcium carbonate, cellulose, chitin, etc. (Sections 9.3.3 and 13.2.1). Searching hunters, if they have specialized predatory techniques, are therefore adapted: to bore through protective casings (e.g. by use of the radula in molluscs); to prize apart elements of the casing sufficiently to evert the stomach through the gap, then to secrete enzymes on to the unprotected tissues and to absorb the products of the extracorporeal digestion; to swallow the prey whole and crush the shell in the gizzard; or to suck out individual polyps or zooids from the communal matrix by means of a proboscis or stylets and a pharyngeal pump (e.g. several opisthobranch molluscs, pycnogonids, etc.).

Several types of hunter are suctorial feeders. In addition to the groups just mentioned, several predatory insects and above all the spiders either suck the fluids directly from their prey or inject salivary proteolytic enzymes into captured individuals (together with paralysing toxins) which liquefy the tissues so that they may be pumped into the predator. It is clearly but a small step from suctorial feeding of this type to an ectoparasitic life style, feeding on a host's fluids without killing it. The categories of hunter and parasite merge cleanly into one another, and in large measure their differentiation is simply a matter of relative sizes of consumed and consumer. A species of leech sucking the blood of a large mammal is unlikely to cause the death of the prey attacked by withdrawing what to the mammal would be an insignificant volume of blood; a different species of leech, on the other hand, sucking the blood

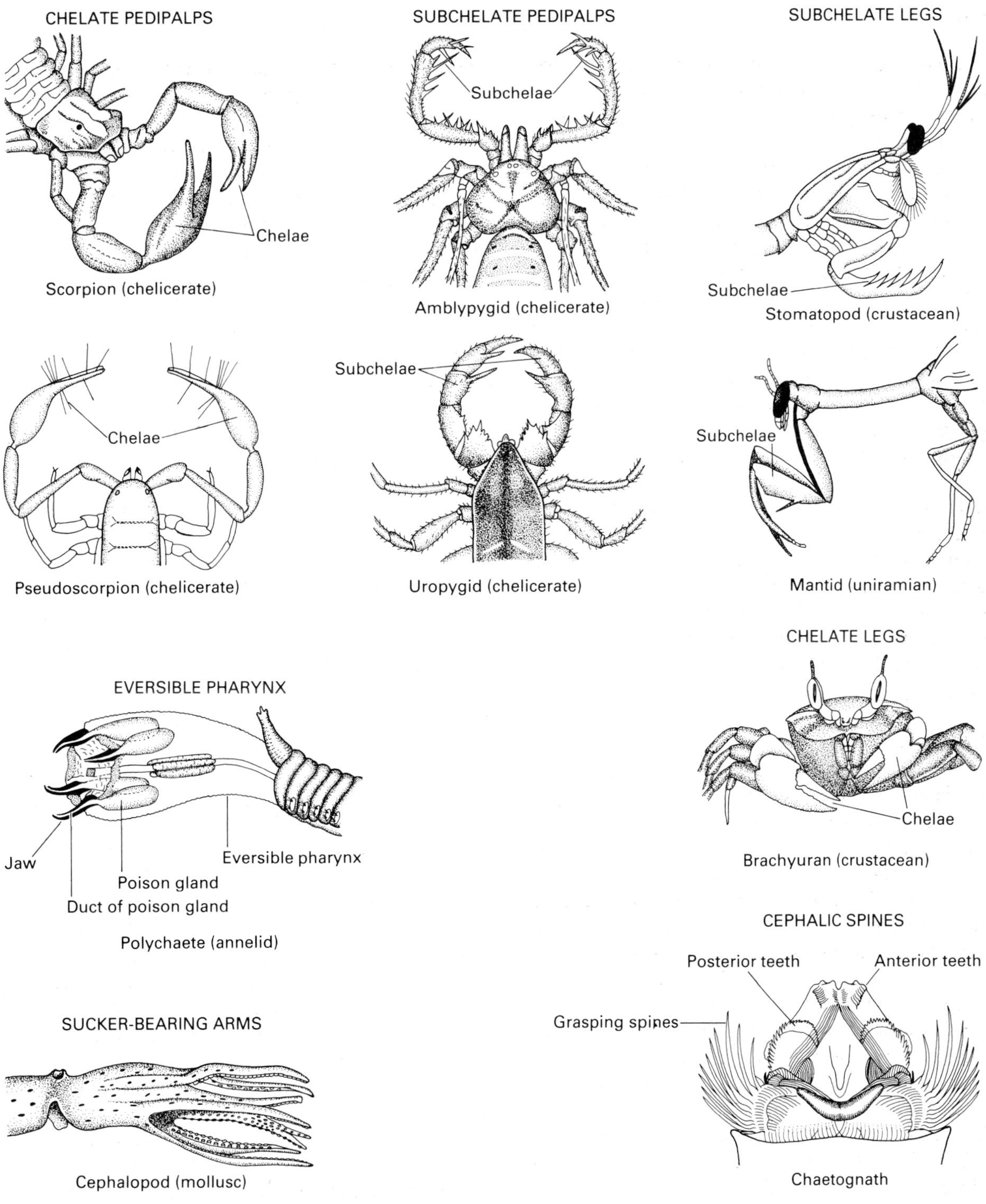
CHELATE PEDIPALPS
Chelae
Scorpion (chelicerate)
SUBCHELATE PEDIPALPS
Subchelae
Amblypygid (chelicerate)
SUBCHELATE LEGS
Subchelae
Stomatopod (crustacean)
Chelae
Pseudoscorpion (chelicerate)
Subchelae
Uropygid (chelicerate)
Subchelae
Mantid (uniramian)
CHELATE LEGS
Chelae
Brachyuran (crustacean)
EVERSIBLE PHARYNX
Jaw
Eversible pharynx
Poison gland
Duct of poison gland
Polychaete (annelid)
CEPHALIC SPINES
Posterior teeth
Anterior teeth
Grasping spines
Chaetognath
SUCKER-BEARING ARMS
Cephalopod (mollusc)

Fig. 9.11. Organs of prey capture (from many sources).

of a small pond-snail would, by so doing, kill it and would therefore qualify as a hunter. Even amongst equally sized animals there are problems of categorization. Some planktonic polychaetes attack and consume the head ends of arrow-worms, an attack which the prey survives, later regenerating its missing head. Is the polychaete a predator or a parasite?

A similar argument applies to many endoparasites. Several hymenopteran insects, for example, complete part of their life cycle within other animals and kill them in the process. Such 'parasitoids' consume their prey slowly from the inside towards the surface, rather than more rapidly the other way around as in the classic hunter. In these insects, the adult injects an egg or eggs into the prey individual (usually another insect), and the developing larval stage/s consume the host's tissues before pupating and metamorphosing. The adult stage is a typical hunter except in the sense that it does not do the consuming, only the attacking; its progeny are the consumers. Again, however, it is really only a question of relative size. A hymenopteran or dipteran larva is large relative to the size of the parasitized insect prey and therefore the host may only be just large enough to support from one to a few larvae through to pupation. But the smaller adult stage of, say, a trematode fluke or a nematode, although feeding on the host's tissues in an essentially identical fashion, is very small compared to a mammal host, for example, and it does not cause its host's death. Because the two feeding patterns are not distinguishable, except in extreme form, it is not surprising that many groups of small predatory invertebrates also contain parasitic species.

A rather clearer distinction can be drawn, at least in respect of feeding biology, between hunters and parasites which consume the fluids and/or tissues of their prey on the one hand, and gutless endoparasites inhabiting a host's alimentary tract on the other, although again the evolutionary transition between an endoparasitic consumer and an endoparasitic absorber is not a very major one. Several parasitic worms, most notably the tapeworms and the acanthocephalans, lack guts but possess an outer cuticle which is disposed in a series of microvilli (termed microtriches) equivalent to those possessed by the absorptive cells of the guts of other animals. Within the alimentary canal of their host, these worms absorb the products of their host's digestive processes, their cuticle also protecting them from this enzymatic activity. Whereas the other types of 'parasite' considered above are in effect micropredators, tapeworms and acanthocephalans are genuinely parasitic in the normal English usage of that word in that they subsist on the resources obtained by another species. The same is true of some other gut parasites (including several nematodes which do have fully functional alimentary systems) that consume only the intestinal contents of their host. Yet other gut-inhabiting animals, however, consume the tissues of the gut wall and the host's blood and are therefore predatory.

Feeders on animal tissues and fluids ingest a readily digestible, protein-rich material, and accordingly typically they secrete proteases and possess relatively short, simple guts, specialized only—if at all—anteriorly, in respect of a crushing gizzard (if whole prey are ingested) or a pumping or bulk-storage crop. The high protein content of animal material also makes it attractive to some essentially non-carnivorous animals at certain critical stages of their life histories. Some female dipteran insects, mosquitoes for example, require to take a blood meal to provide the protein to invest in eggs, although the males and the larval stages may not consume any animal food.

9.2.3 Grazers and browsers

Grazers and browsers are mobile consumers of sessile prey, cropping exposed tissues without, usually, killing the prey individual or colony. On land, the food sources are plants and fungi, but, in the sea, colonial animals (such as cnidarians, bryozoans or tunicates), bacterial colonies and multicellular algae can be grazed in an equivalent manner. Removal of the food materials requires the possession of hard biting or rasping mouthparts—the radular ribbon of molluscs (Fig. 5.3), Aristotle's lantern in sea-urchins (Fig. 7.24), the sclerotized jaws of insects (Fig. 8.30), etc.—although finding and acquiring the resources are not the major problems faced by grazers and browsers (in marked contrast to most types of hunter), since consumable material is often plentiful. The difficulties encountered by this category of consumer are (a) the chemical defence systems evolved by the otherwise defenceless sessile prey species, and (b) the small proportion of digestible material in every unit weight ingested as a result of the abundant structural or protective refractory compounds in their prey, and, often, the protein-deficient nature of the utilizable organics.

The problem of the high proportion of indigestible carbohydrate polymers in the bodies of seaweeds and plants, e.g. agar, algin, laminarin, cellulose, etc., has, as we have seen (Section 9.1), partly been solved enzymatically, in association with the evolution of elongate alimentary canals (especially the mid-gut region) to increase the area of those regions responsible for the digestion of refractory materials, and partly with the aid of symbiotic bacteria and protists housed in specialized gut compartments, often the hind-gut but sometimes the crop or stomach. A large storage crop is often also present.

The alimentary symbionts ferment the polysaccharides anaerobically, releasing fatty acids and other simple carbohydrates which can be absorbed. This system reaches its greatest development in those termites which consume the most refractory of all natural organic materials, wood. In these, the hind-gut is large, larger than the whole of the rest of the gut (Fig. 9.12), and it contains a dense culture of hypermastiginan flagellates. These ingest the wood particles phagocytically and themselves contain symbiotic bacteria which are probably mainly responsible for digesting some of the cellulose content of the wood. The lignin component is probably indigestible. The isopod crustacean *Limnoria* (the gribble) is another wood feeder, but it appears to lack gut symbionts, digesting some of the cellulose and hemicellulose components of the wood with its own enzymes. Structural carbohydrates, if they can be digested, provide an abundant source of energy-yielding materials, but relatively little else. In *Limnoria*, the required proteins must be derived from fungi infecting the wood consumed; wood without any such decomposer organisms already colonizing it cannot sustain the animal.

Scale effects also operate in the consumption of relatively non-nutritious plant materials. Large animals, i.e. vertebrates of rabbit size or larger, have a low specific metabolic rate, reducing the requirement for energy per unit body weight, and are able to store larger quantities of herbage in their gut fermentation chamber because of their large body volume; being homoiothermic, mammals are also able to maintain efficient fermentation temperatures. They are therefore able to subsist on coarse grasses, etc., ingested in bulk. Small animals, such as herbivorous invertebrates, must, however, feed on higher quality material and must gain entry to the plant cell contents by piercing the cell, rasping away the cellulose wall with a radula, or biting through it with their mouthparts. Bulk ingestion is generally not possible, simply by virtue of their small size. Even small birds and mammals cannot subsist on low-grade materials: if herbivorous, the diet must be restricted to energy-rich seeds and similar items.

The chemical defences of plants are many and varied, including alkaloids (e.g. nicotine, cocaine, quinine, morphine and caffeine), glucosinolates, cyanogenic glycosides and tannins, and these substances would appear to be deployed specifically to deter consumption of their susceptible tissues. Some of the chemicals are straightforwardly toxic—natural insecticides, for example derris and pyrethrum. Others, e.g. the tannins, bind with proteins when released, rendering them indigestible and further reducing the protein status of the food ingested. Yet others mimic closely the consumer's own hormones or pheromones, adversely affecting growth, development or reproduction, and initiating inappropriate behavioural reactions. In some plants, these chemical defences are mobilized only when grazing pressure starts, so as to avoid the diversion of resources away from growth and reproduction when this is unnecessary, and it has recently been shown that in at least one case adjacent plant individuals of the same species can react, by chemical mobilization, when a nearby individual is attacked by grazers, even though they themselves have not yet been grazed.

Other plant defences are structural, for example hair-like projections which release sticky secretions or noxious chemicals when triggered, and even anatomical mimetic resemblances to leaves already infested with consumers (Fig. 9.13), or symbiotic, especially as seen in the associations of a variety of angiosperms and other tracheophytes with ants. The ants remove browsing insects from the plant by interference competition 'in exchange' for the provision of nectar and of nest sites. Comparable structural, chemical and other defence systems are displayed by marine animals subject to grazing pressure, although as yet these have been little studied.

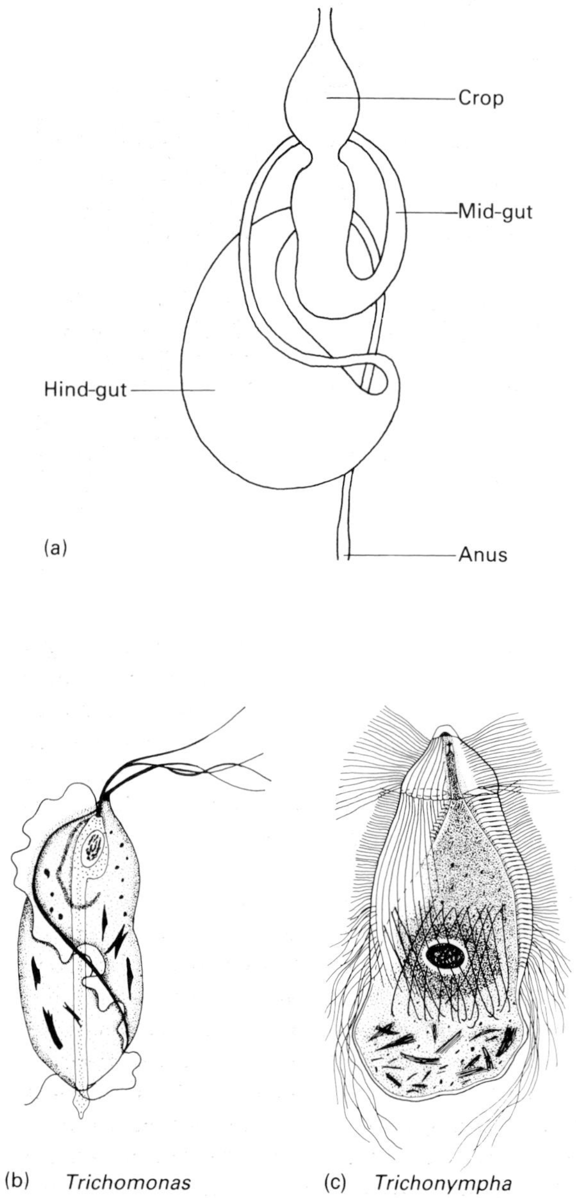

Fig. 9.12. The gut of a wood-eating termite (a), with two of the symbiotic flagellates (b) and (c) found in its hind-gut (a after Morton, 1979; b and c after Mackinnon & Hawes, 1961).

Nevertheless, consumers have been coevolving with their sessile prey for many millennia and individual grazers and browsers have evolved the ability to detoxify, avoid, sequester or excrete the specific defensive chemicals of their particular prey species, to bypass the structural defences, and to escape the attention of the defending ants by chemical mimicry of these insects. In spite of this, however, the huge biomass of living plant tissue present in most terrestrial habitats suitable for plant growth, and of seaweeds in kelp-forests and algal beds, must indicate that much macrophyte tissue remains effectively unavailable and/or unusable by grazers and browsers, and is only an adequate diet after death and decomposition.

Grazing and browsing are not the only feeding techniques appropriate to the consumption of macrophyte material. Some herbivores in both the sea (e.g. some opisthobranch molluscs) and the land (e.g. many hemipteroid insects) have circumvented the structural carbohydrate problem by sucking out the contents of individual cells or by inserting cannula-like mouthparts into the xylem or phloem vessels of tracheophytes (see Figs 9.9 and 8.30). In the latter case, hydraulic pressure in the transport vessel may be sufficient to pump the fluid directly into the gut of, for example, an aphid, and aphids parasitize the host plant in much the same way as does a tick or a female mosquito its animal host. Indeed, as noted above with respect to hunting carnivores, it is but a small step from ectoparasitically feeding on a larger host to becoming endoparasitic, and several nematodes, for example, are endoparasites of plants in the same manner as are other nematodes of animals, and several insect larvae

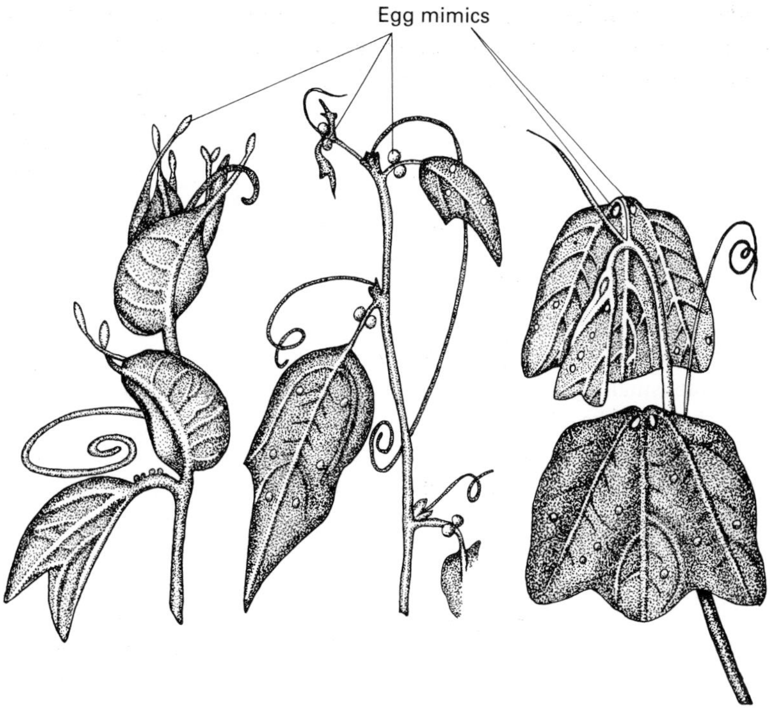

Fig. 9.13. The leaves and stalks of various passion-flower vines bear egg mimics to deter female *Heliconius* butterflies from depositing eggs on them (leaves already bearing eggs are avoided by the butterflies, probably because newly hatched larvae are cannibalistic). (After Gilbert, 1982.)

live within plants, there feeding on plant tissues by endoparasitic grazing.

Even when equipped with cellulases and equivalent enzymes or with symbiotic micro-organisms in the gut—and even more so in the absence of both—digestion and assimilation of macrophyte tissue are typically very inefficient, and in the case of liquid feeders, through-put of carbohydrate may be greatly in excess of requirements. Faecal production by these categories of consumer is therefore copious, and the faeces contain much unassimilated organic matter. They therefore provide an important ecological pathway by which materials photosynthetically fixed by living macrophytes are made available to other categories of animal consumer, especially to the deposit feeder.

9.2.5 Suspension feeding

The essential features of the means by which suspension feeders filter their food from water have been outlined in Section 9.1. In groups utilizing a non-anatomical external secretion (Fig. 9.5), there appear to be no further specifically suspension-feeding adaptations. Such animals are perhaps more nearly equivalent to ambushing hunters trapping relatively small individual prey in nets, at least in so far as their gut anatomy is concerned. Certainly the two feeding modes do intergrade: a web-building spider could equally well be argued to be an aerial suspension feeder or an ambushing hunter; and the cnidarians straddle the interface between the two in the sea. They possess radially symmetrical rings of tentacles around the centrally located mouth and the polyp phase, at least, is sessile—both characteristic suspension-feeding adaptations—and indeed their prey are mainly zooplanktonic animals suspended in the water. The separate prey items are not so much passively trapped, however, as attacked individually with nematocysts, albeit that the prey themselves discharge the nematocysts by accidentally touching them rather than the cnidarian initiating the attack.

In contrast, groups with a mucociliary filter-feeding system also possess a distinct series of alimentary specializations. In most such species, those in which the filtration apparatus is either freely projecting into the water or else is protected within an external shell (as in the brachiopods and the bivalve molluscs) (Fig. 9.3), the fore-gut is simply a short region connecting the external filter to the stomach. Most unusually, however, in the chordates, the fore-gut is the site of the filtration process itself and hence is highly specialized. The sides of the pharyngeal wall of these animals are perforated by numerous small openings, 'stigmata', which extend right through the body wall to open on the body surface. Water taken in through the mouth can then pass into the pharynx, through the stigmata and back to the environment in a unidirectional flow. Although unusual in that the pharynx is the site of filtration, this system has some parallels amongst other animals which probably indicate its evolutionary origins. Feeders on small particles often use a water current to drive collected particles into the mouth and this transport stream has to be expelled in some way. In the lophophorates, it appears that the water is simply 'regurgitated' at intervals through the mouth, and the ingestion of particles is then temporarily suspended. In the macrodasyid gastrotrichs (Fig. 4.9) and the cephalodiscid hemichordates (Section 7.2.3.2), however, the pharynx bears a pair of perforations extending through to the body surface through which such ingested water can be discharged without interrupting food intake. The early chordates would appear to have adapted a similar current for direct filter feeding, thereby doing away with the necessity of withdrawing an external lophophore-like organ at times of danger from predation, etc. (which also interrupts feeding), whilst the enteropneust hemichordates developed the same current for gaseous-exchange purposes, as indeed did the later aquatic chordates.

The pharynx of suspension-feeding chordates is very large, comprising the majority of the body volume. Being perforated by thousands of stigmata in many species, the body wall in the pharyngeal region is almost non-existent and would be completely ineffective. Accordingly, it has been replaced by a secondary or false body wall, formed by folds of tissue in the cephalochordates and by a secreted cellulose test in the tunicates. The body proper is therefore wholly or partially surrounded by a morphologically external cavity, the atrium, between the true and false body walls, into which

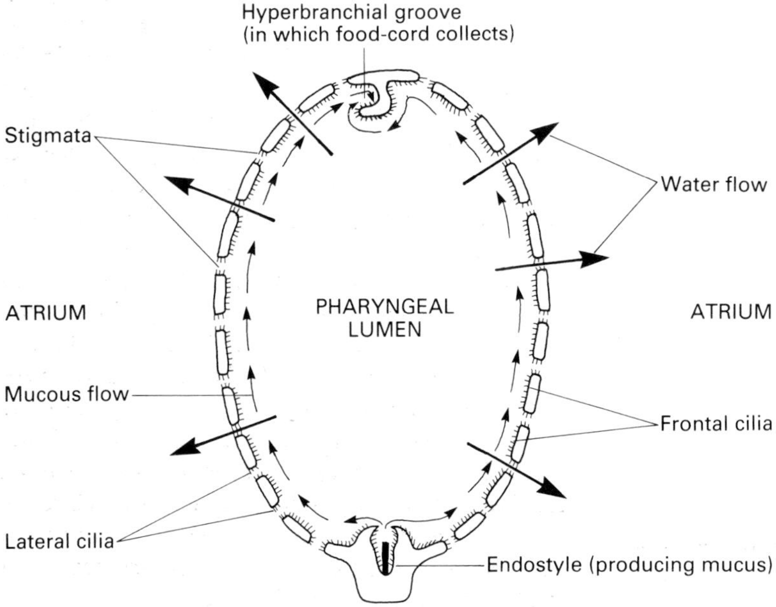

Fig. 9.14. Diagrammatic transverse section through the pharynx of the invertebrate chordates showing the nature of the filter-feeding system (after several sources).

water passes after flowing through the stigmata, and from which water is discharged to the environment via a single aperture, the atriopore or atrial siphon (Fig. 9.14). (Essentially the same system is, of course, present in vertebrate chordates. In fish, the gill slits (≡stigmata) perforating the body wall from gut to exterior are fewer, however, and the body wall is thicker, so that an additional false body wall is unnecessary, and the water current is used for gaseous- exchange purposes, not primarily for suspension feeding.)

Mucus is produced by a gland, the endostyle, running along the ventral mid-line of the pharynx,

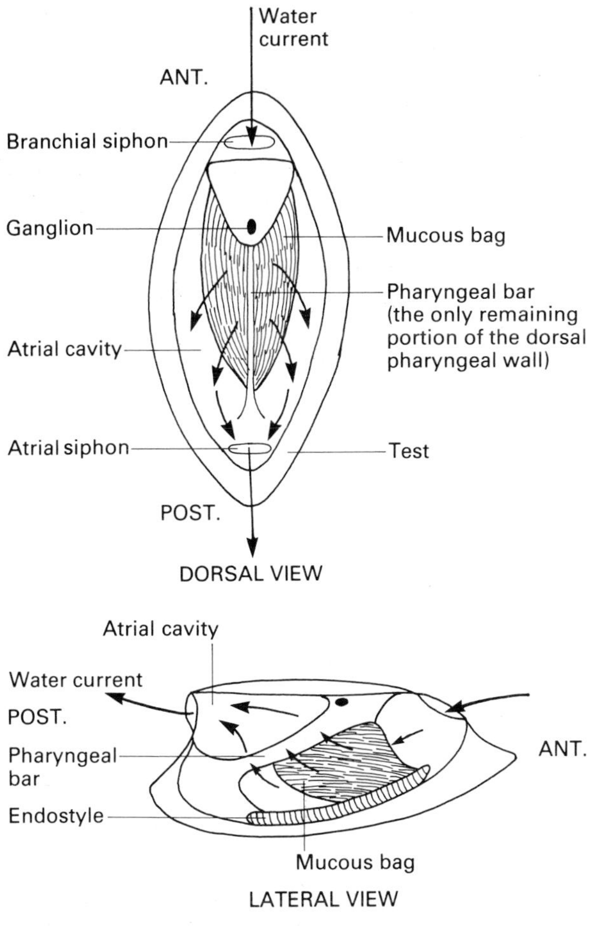

Fig. 9.15. A pelagic tunicate (for example a salp) showing the mucous bag suspended across the two large pharyngeal openings. (After Berrill, 1950 and others.)

and this mucus is induced to move up the perforated sides of the pharynx in thin sheets. The mucous sheets intercept potential food particles, in some cases down to 0.5μm in size, as the water flows out of the pharynx through the stigmata. Eventually, the food-laden sheets meet dorsally where they are formed into a longitudinal cord in the mid-dorsal hyper-branchial groove. All the motive power in this system, whether of water or of mucus, is provided by tracts of cilia in the pharynx, except in the pelagic salps.

In the pelagic tunicates generally, the feeding current also provides the means of propulsion, the oral and atrial apertures being located at opposite ends of the body. In association with this jet propulsion, the numbers of stigmata have been reduced, down to only two in the salps, the mucous sheets forming a single, internal, conical net slung across the lumen of the pharynx (Fig. 9.15) with the motive power of the water being provided by hoops of muscle around the body (Figs 7.35 and 7.36).

Not only in these invertebrate chordates, but also in the lophophorates and in the filter-feeding molluscs, food passes from the filtration organ (of whatever type) into the stomach in the form of a cord of food-laden mucus. In all mucociliary feeders, the pH of the stomach lumen is acid. This reduces the viscosity of the mucous cord, permitting the trapped food particles to be released. Digestion of the food and absorption of the products takes place within the stomach and/or in the mid-gut diverticula issuing near it. Undigested material passes through into a generally short intestine, in which the pH is alkaline, increasing the viscosity of the mucus again and rendering more easy the production of faecal pellets or strings.

The motive power to transport the mucous cord into the stomach is also provided by cilia, acting directly on the cord in the chordates, but indirectly in the lophophorates and molluscs. In both these latter groups, a rotating rod projecting into the stomach winches in the cord. The rod of the lophophorates is formed of mucus and faecal material and is housed in the pylorus, the cilia of which rotate it (Fig. 9.16). Every so often, the used rod is passed into the intestine, converted into a faecal pellet and voided, a new rod being then formed.

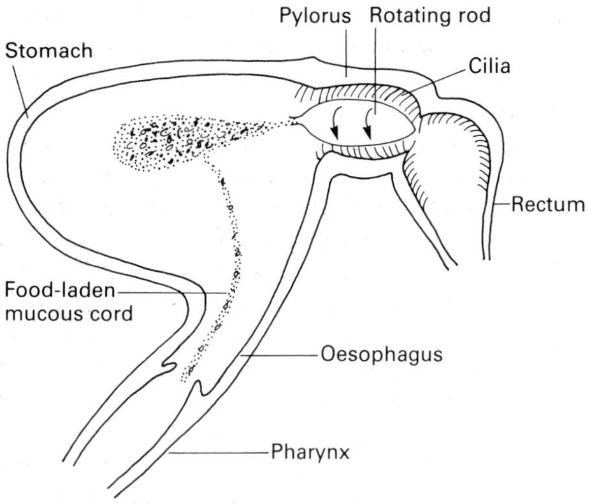

Fig. 9.16. The stomach of a bryozoan lophophorate showing the mucofaecal rod used to winch food into the gut (after Gordon, 1975).

In the molluscs, the rod, known as the crystalline style, is much larger, more permanent, and is composed of a hyaline mucoprotein. It also projects into the stomach, but in the molluscs from a special diverticulum, the style sac, the lining sac cilia rotating the style. Not only does the tip of the style act as a winch (Fig. 9.17), but the tip gradually dissolves in the stomach releasing the enzyme amylase. As the tip is lost, the style is slowly moved forward into the stomach (again by ciliary action) and more style material is secreted proximally so as to maintain constant stylar length. The tip is therefore maintained in contact with the stomach wall against which it rotates: this has the effect of a rotary pestle and serves to free particles from the now less viscous cord and distribute them over a gastric sorting region in which cilia can further subdivide the particles collected.

9.2.6 Deposit feeding

Probably the majority of living animal species consume detritus (organic material of small particle size) or litter (material of larger dimensions), but, in spite of this, the precise nature of their diet is still uncertain. A leaf ingested by, say, an earthworm is not just a dead leaf; it is a whole ecosystem in microcosm. On and in the leaf tissue will occur the bacteria and fungi responsible for its decay, together with various protistan consumers of the decomposer organisms (e.g. amoebae, ciliates, heterotrophic flagellates) and some only slightly less microscopic animals such as nematodes and mites feeding on the smaller organisms. On the surface of the leaf may well occur photosynthetic algae and cyanobacteria, and the dead and decomposing remains of other organisms, not least those derived from animal faeces. All of these living and dead components may be ingested by the consumer with the leaf, and the problem has been to determine which are digested and assimilated, and which meet most of the metabolic requirements of the consumer.

Nor is it necessarily the case that the consumer simply ingests background material and digests out whatever it can. Several deposit feeders are now known to be capable of much greater degrees of selective ingestion than was once thought to be possible; although few species in total have yet been re-examined in this light.

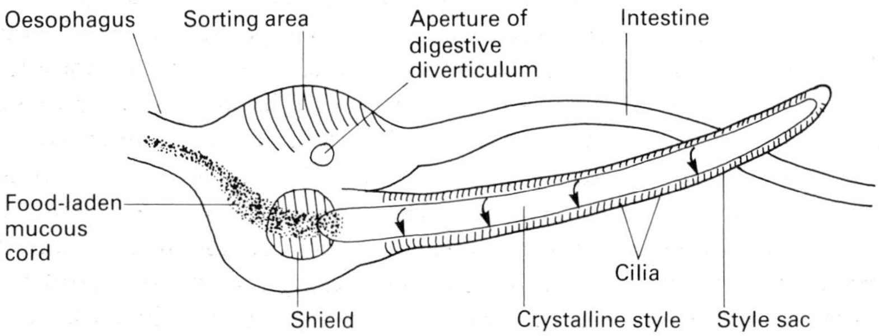

Fig. 9.17. The stomach of a filter-feeding mollusc showing the crystalline style used to winch food into the gut (after Morton, 1979).

Selective ingestion would certainly appear to be advantageous, if possible, since the nutritive value of different elements in the aggregate 'detritus' vary very widely. Comparison of the organic content of the sediment ingested by some unselective marine deposit feeders and of the faeces resulting after consumption, and exposure of samples of the sediment to enzymes known to be present in their guts, have both indicated that in such species the majority of what little organic matter is present in the sediment is not available to them. Less than 5–10% of the organic detrital pool may be digestible. (In clean sandy beaches, organic matter of *any* sort may comprise a very small proportion—$<1\%$—of the ingestible material. The litter layer in a forest may be richer in organic matter, but still relatively little of it may be nutritious.)

By the time any piece of organic debris becomes incorporated into the surface layer of litter or detritus on the substratum, it will have already lost most of its original food value. If a leaf, then the parent plant may have translocated all the soluble compounds into its perenniating tissues before shedding that leaf. Leaching will rapidly cause the loss of other organics in the first few hours after shedding. Any fragments of organic matter may only arrive on the substratum after passage through the gut of a consumer, during which most of the utilizable substances may have been removed: faecal material is one of the major sources of organic matter in sediments or soils. Therefore, the organic material remaining in an aged item of organic debris is likely to be in the form of those refractory structural, skeletal or protective substances that defy most animal digestive systems—detritus feeders do not generally possess cellulases, for example.

If, however, an item of debris is recent in origin, deposit feeders may be able to digest some of its contained organics, although even so its protein content is liable to be either very low or unavailable as a result of the presence of tannins (see Section 9.2.4). In general, though, it would appear most likely that consumers of dead organic matter (excepting scavengers of animal carcases) are reliant on the living associates of the debris, not on the debris itself which may merely be a suitable vehicle to transport the micro-organisms into the gut. On the land and in the fresh water, those fungi which are responsible for litter degradation are probably of primary importance as the real food source of deposit feeders, whilst, in the shallow fringes of the sea, photosynthetic unicellular or mat-forming protists are probably the elements of greatest nutritional import. Selective feeders capable of ingesting solely diatoms have been shown to be able to achieve a 70% assimilation efficiency, in comparison to less than 4% in related species feeding unselectively on the background organic pool. Throughout most of the sea, however, under circumstances in which photosynthesis is impossible, in which the fallout is sparse, and in which by the time debris reaches the sea bed only refractory substances remain, animal consumers must be dependent on bacteria, both for energy-yielding and proteinaceous materials.

The guts of deposit feeders are typically unspecialized, with a tendency towards elongate intestines (see Figs 4.42 and 4.45), for the same reason as in the grazing and browsing category. Like the latter too, faecal production may be copious although of much less nutritive value to other consumers.

9.2.7 Food from symbionts

Although relatively few animal species are totally dependent on the photosynthesis of endosymbionts for their nutrition in the manner displayed in Fig. 9.18 (see Section 9.1), many species, ranging phylogenetically from sponges to molluscs and from cnidarians to chordates, derive some nutritional benefit from the symbiotic photosynthesis of cyanobacteria (e.g. some sponges and echiurans), prochlorophytes (tunicates), or the more widespread dinoflagellates (present in the form of zooxanthellae in many marine invertebrates) or chlorophytes (in the form of zoochlorellae in mostly freshwater species). Some stony corals, for example, obtain two-thirds of their metabolic requirements from the intracellular zooxanthellae, and only one-third from external sources including the typically coelenterate

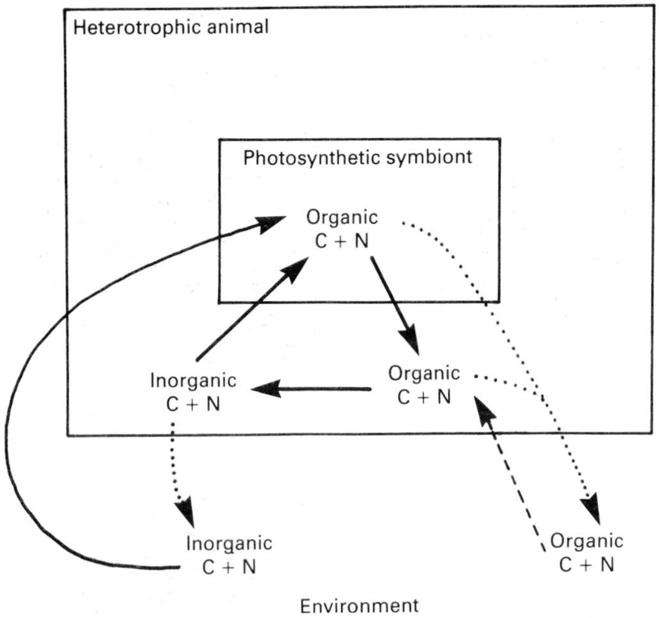

Fig. 9.18. Diagrammatic representation of carbon (C) and nutrient (N) fluxes in symbiotic interactions (after Barnes & Hughes, 1982).

capture of zooplankton. In a number of sea-anemones, the symbiotic zooxanthellae and the organelles of prey capture are located in different parts of the body: the symbiont-bearing 'organs' are extended during the hours of daylight and the nematocyst-bearing tentacles at night (Fig. 9.19). Symbiosis can also supplement herbivory. Sacoglossan molluscs feed suctorially on the cells of macroscopic green (and other) seaweeds, and some species can sequester intact and still functional chloroplasts from their prey into their digestive mid-gut diverticulum (Fig. 9.20). The chloroplasts become engulfed by the phagocytic digestive cells but in them continue to function photosynthetically for more than 2 months in some cases. Up to half of the carbon fixed passes to the mollusc and this may be sufficient to satisfy its respiratory requirements. (The chloroplasts of some *Euglena* may have had an evolutionary origin in a similar process.)

A rather different symbiotic relationship occurs between plants which secrete nectar and insects which gather to consume this sugar-rich fluid and thereby effect pollination. Several adult insects consume only nectar (to provide the energy required for flight); their larval stages are the feeding phase of the life cycle and they accumulate sufficient resources to permit the assembly of the adult body which then requires only energy-yielding substances during its short existence.

Finally, some relatively large, free-living animals lack guts at all stages of their life history (e.g. a few bivalve molluscs, oligochaete and polychaete annelids and all pogonophorans) and their mode of feeding was until recently somewhat mysterious, although symbiotic chemoautotrophic bacteria have now been demonstrated in the tissues of most such species. It is known, however, that regions of the body surface of at least some pogonophorans may also be drawn out into microvilli equivalent to those on the surface of gutless parasites living in the alimentary canals of other animals. It is therefore possible that some of these free-living worms are able to take up dissolved organic compounds from the surrounding water. Several other soft-bodied free-living animals have been suggested to be partially dependent on the uptake of external dissolved organics, although this

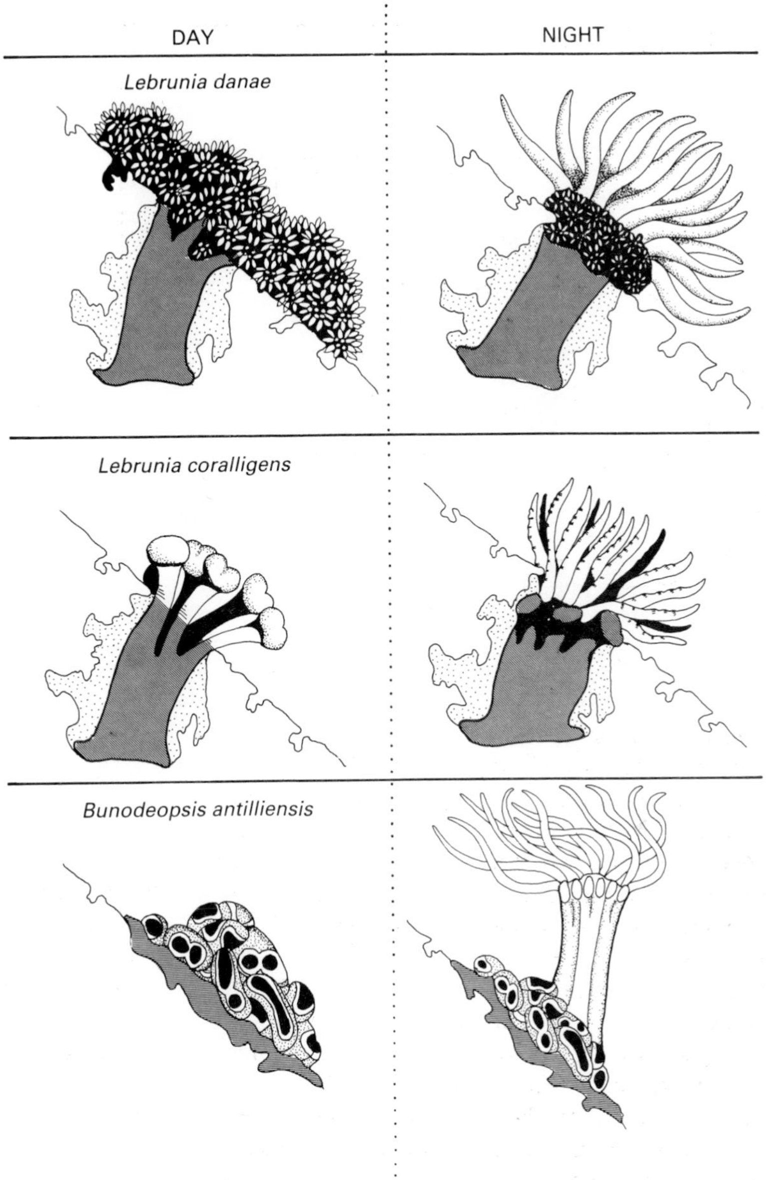

Fig.9.19. The use of different parts of the body at different times of the day by three coral-reef anemones: zooxanthellae-containing structures during daylight (left-hand column); and nematocyst-bearing tentacles at night (right-hand column) (after Sebens & De Riemer, 1977).

is far from being generally accepted, in part because of the infrequent occurrence of sufficiently high external organic concentrations to permit this mode of nutrition. External digestion with subsequent absorption of the digestion products across the body wall is another possibility, but no appropriate gland cells have so far been found.

9.3 Costs and benefits of feeding: optimal foraging

9.3.1 Introduction

We have seen that animals are clearly constrained in their types of diet by their evolutionary pasts: an

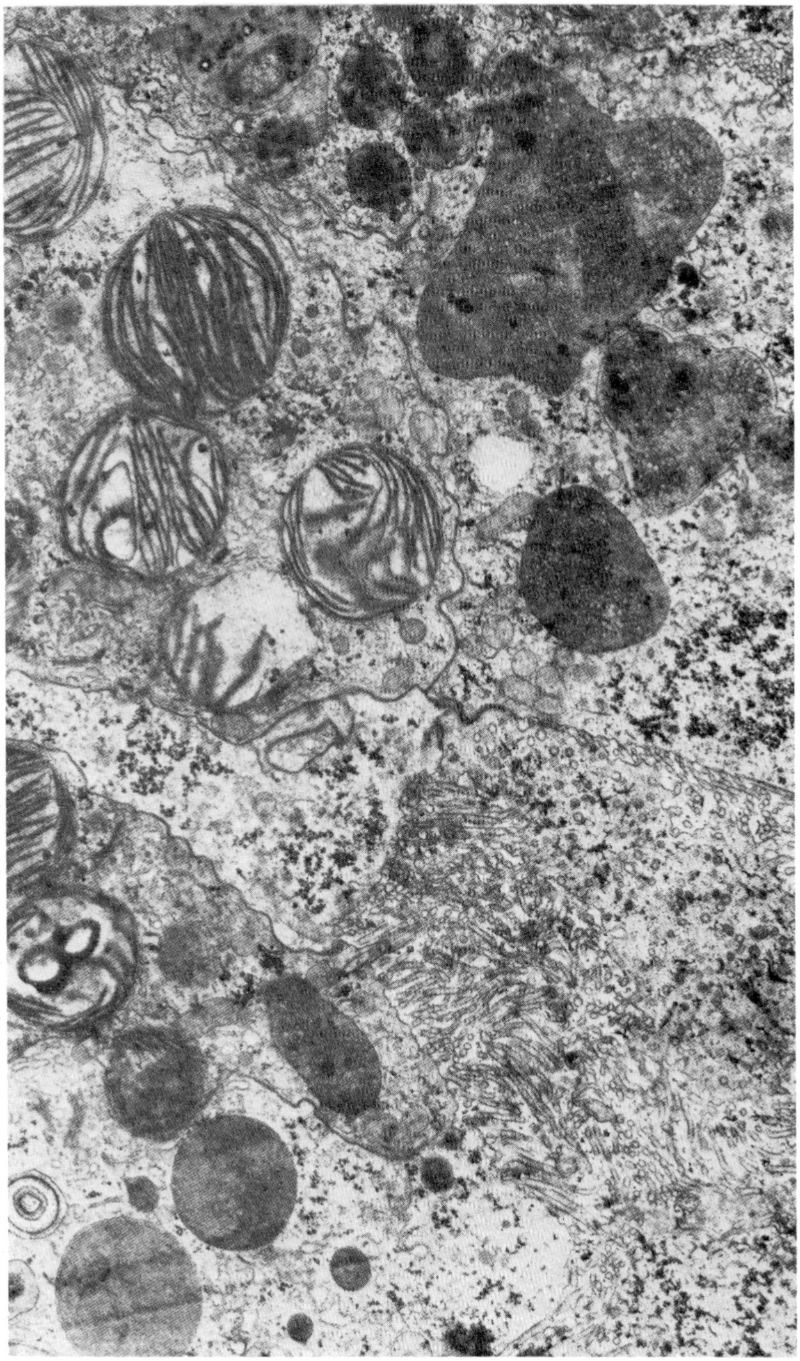

Fig. 9.20. An electronmicrograph of functional chloroplasts in the digestive diverticulum of a sacoglossan mollusc (from Trench, 1975, with permission).

animal in a lineage of suspension feeders could not easily adapt to pursuit hunting nor vice versa. Nevertheless, in many species there is a degree of flexibility possible, and in some generalist feeders this may even permit switching from one feeding mode to another, and from one prey type to a completely different sort of food. The estuarine polychaete *Nereis diversicolor*, for example, can suspension feed (Fig. 9.5d), deposit feed, behave as a hunter or a scavenger, or browse pieces from macroscopic algae (see Table 9.1).

All animals encounter a variety of potential food items during their daily lives and they are therefore faced with alternative food sources. For the most specialist of consumer, this may only be different individuals of the single prey species taken, but more generally the alternatives will at least include individuals or material of more than one food species. Hence, within limits set by their evolutionary pasts, animals face alternative courses of action, or 'choices', in ecological time—from minute to minute, hour to hour or day to day—in terms of whether to catch and consume a given item or to reject it in favour of another in space or time. Such 'decisions' are not without important potential consequences, since the benefit to be derived from consuming different items will vary. Some will be more nutritious than others. The most nutritious are also likely to be favoured by many consumers, and this is likely to give rise to intense utilization and resultant scarcity of that resource. Others are low in energy and nutrient content, and may therefore be generally less favoured foods but correspondingly relatively abundant and more readily available in any given habitat.

Nutritive value (per unit weight consumed) is only one element in any potential choice of food item, however, since besides yielding the obvious benefits, feeding activity also incurs costs. Whatever the form of the feeding process, time and energy have to be expended on finding, consuming, processing and digesting the food materials, and this time and energy could have been devoted to other purposes. The precise nature of feeding costs will vary from feeding type to feeding type, but in all cases one would expect selection to act in favour of a feeding strategy that maximizes the net gain (benefits minus costs) obtained. Food is one of the prime requirements for survival and hence any individual consumer maximizing this net gain would be at a selective advantage over individuals behaving differently, in respect of both survival and reproduction, since it would be more likely: (a) to be able to devote less time in total to foraging (thereby decreasing predation risk); (b) to grow faster or to a larger overall size (with consequent reproductive advantages); (c) to be healthier and therefore the more able to withstand parasitic infection, to escape from predators, etc.; and (d) to derive a greater benefit from fewer resources at times of food shortage.

How then can consumers maximize their net gain per unit time? This has been investigated via simple models of optimal foraging strategy and by testing the predictions of these models through experiment and observation. As is usual in the design of experiments, this has proceeded by isolating one particular variable, in this case prey choice. It should be remembered, however, that in the real world, an animal's activity at any one time is a compromise between many conflicting pressures. Maximization of net gain from feeding will in itself be advantageous, but so will be avoidance of being eaten by predators, breeding behaviour, and so on. These other desiderata may well place constraints on feeding and render consumption of prey less than optimally efficient: a consumer may have little option but to consume anything it finds in the brief period in which feeding is possible.

Further, the living food of animal consumers does not simply wait for its chance to nourish consuming species! Indeed, living prey have a greater vested interest in not being consumed than the consumer has in feeding on them; this has been termed the 'life/dinner principle'. In any encounter between, say, a cuttlefish and a shrimp, the cuttlefish has only its dinner at stake whilst the shrimp is fighting for its life, and selective pressures will vary accordingly.

In the following paragraphs, we will discuss how consumers could best maximize their net gains and how the potentially consumed could minimize that gain, thereby decreasing their own risk of being eaten. Mobile consumers, free to accept or reject discrete individual prey items, form the simplest case and we will devote most attention to that category; suspension and deposit feeders face rather different problems and so we will treat them separately.

9.3.2 The theory of optimal foraging

Any potential food item will have a particular 'food value' to a consumer. Animals require both energy-yielding compounds from their diet and those that will permit growth of somatic and reproductive tissues and cells, and so this overall food value should take both of these requirements into account. Unfortunately, the two may be required at different times of the year or during different stages of the life history, so that at some point energy may be the overriding factor, at another point organic nitrogen may be in particular demand, whilst for some species (e.g. land snails) there may be periods in which inorganic elements or compounds (such as calcium) may override all other requirements.

In practice, most work has used the energy content of the food as a convenient measure of its food value, and, in these simplified terms, the food value can be expressed as the energy which would be gained by consuming a given item (E_g) minus all the energetic costs associated with catching, subduing, consuming and digesting it (the 'handling cost', E_h). In addition, different potential food items will occur in any given habitat at different frequencies, and, if a consumer has a preference for some particular type of item, it may have actively to search for it and thereby incur a 'searching cost' (E_s). The net energy gain will therefore be:

$$E_g - E_h - E_s.$$

Alternatively, time, which can be more easily measured, may be substituted for energy on the debit side of this expression, and the food value given as energy gain per unit time. This will be E/T_h, if a prey item has already been encountered, or $E/(T_h+T_s)$ if a search-time element is included (where T_h is the 'handling time', T_s is the 'search time', and E is the available energy content $\equiv E_g$ above).

Let us take the simpler case first, in which there is no search-time element. A consumer encounters different potential food items at random in its habitat, and we are asking which of them it should eat in order to maximize its energy gain per unit time. The answer is that we would expect it to take preferentially those with the highest food value as measured by E/T_h (or, should nutrients, N, be the dietary requirement at that time, those with the highest value of N/T_h): consumers should choose the most profitable prey. This simple model gives rise to a number of testable and tested predictions:

1 A consumer should eat *only* the prey type with the highest food value, if the encounter rate with that prey type is sufficiently high that inclusion of other prey types (with lower food values) would decrease the average rate of energy intake.

2 If, however, the prey type with the highest food value is encountered at less than this rate, a consumer should expand its range of prey types consumed to include the next most valuable prey items, and so on.

3 If it takes a considerable time to distinguish different prey types (as, for example, when discrimination is by touch), lower value prey should be eaten if they are encountered frequently, even though higher prey may be still plentiful; whereas if prey recognition is instantaneous (as when discrimination is by sight), lower value prey should not be eaten if higher value foods are available instead. (Recognition

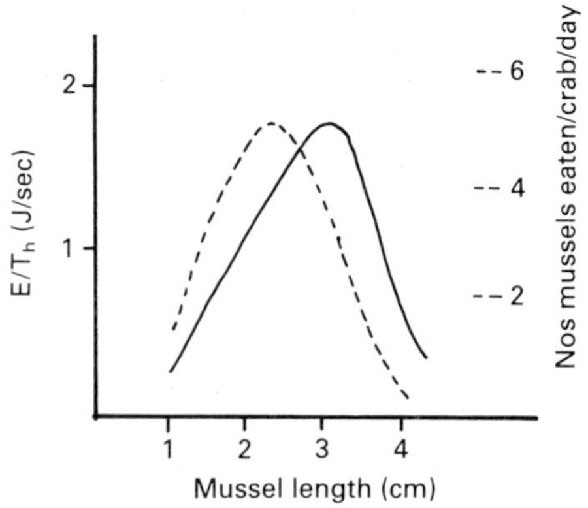

Fig. 9.21. The relationship between food value (E/T_h) and prey choice in 6.0–6.5 cm broad crabs (*Carcinus maenas*) feeding on mussels (*Mytilus edulis*)—see text. (After Elner & Hughes, 1978).

of prey food value is simply an additional component of handling time.)

All these predictions have been found to hold true, for example, in the green, harbour or shore crab, *Carcinus*, when feeding on mussels, *Mytilus*, of different individual sizes (see Figs 9.21–9.23; Box 9.1).

Tests involving a consumer and a variety of different prey species have been less frequently undertaken, but the polychaete *Nereis diversicolor*, for example, cited above as consuming a very wide range of different foodstuffs, does exhibit a preference hierarchy which corresponds well with the amount of assimilable energy in each food type (Table 9.1).

The basic model can now be made slightly more complex by the incorporation of different search times for different types of food. Let us assume that T_h is constant for items of given prey type and size, and vary the magnitude of T_s dependent on the frequency of various potential prey items in a habitat. Let there also be two types of prey, x and y, such that the food value of x is greater than that of y, i.e.

$$\frac{E(x)}{T_h(x)} > \frac{E(y)}{T_h(y)}$$

Clearly, if a consumer encounters prey of type x it should always eat it; it would never do better by

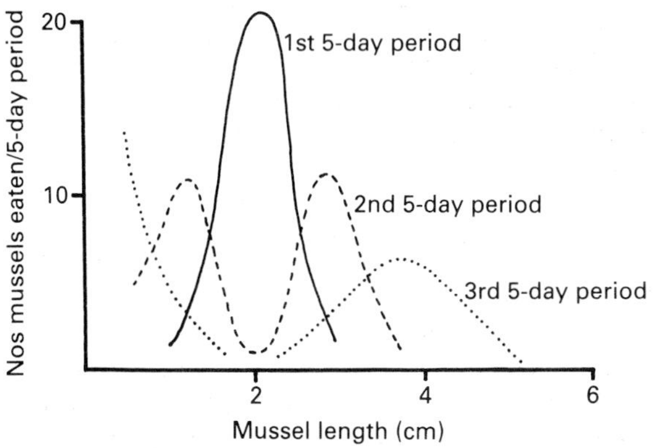

Fig. 9.22. Prey choice when the preferred size range of mussels (*Mytilus edulis*) could be depleted as a result of consumption by crabs (*Carcinus maenas*)—see text. (From data in Elner & Hughes, 1978.)

Box 9.1 Choice of mussel prey by crab consumers

By observation, it is relatively easy to quantify the time taken for a crab to break open a mussel and consume the flesh (T_h); and the energetic content of the flesh obtained (E) can be determined by bomb calorimetry. Crabs crush mussel shells with their chelae to open them, and there is therefore an appreciable handling time which increases more or less exponentially with mussel size. Large mussels thus take a very long time to open but have large values of E, whereas small mussels can be opened very quickly but contain little energy. For a given size of crab, E/T_h varies with mussel size as shown in Fig. 9.21; the same figure also displays the size range of mussels actually consumed by that size of crab. Clearly, the agreement between the mussels consumed and those with the greatest predicted food value is good, and prediction **1** (p. 300) holds true.

In this experiment any mussel eaten was replaced immediately by one of the same size, so that the highest value mussels were always maintained at a frequency high enough to result in depression of the average gain rate should a lower value prey be included in the crab diet. In a second experiment, however, consumed mussels were not replaced and the highest value mussels became reduced in frequency. The reaction of the crabs is shown in Fig. 9.22: as in prediction **2** (p. 300), they expanded their intake range to include the next most profitable sizes of mussels.

Crabs distinguish the sizes of different mussels by touch, and in a third experiment the relative proportions of different sized mussels were varied. Figure 9.23 shows that, in conformity with prediction **3** (p. 300), less valuable mussels were taken in small numbers when they were abundant, even though optimally sized, mussels were also plentiful.

When presented with different sized prey individuals of one prey type, *Carcinus* therefore appears able to select those sizes of mussels which maximize its intake of energy per unit time. In some way it must be able to assess the food value of different sized mussels.

Table 9.1 Feeding preferences of *Nereis diversicolor* and rank order of assimilable energy per food item

Feeding mode and food type	Preference order	Rank order of assimilable energy per item
Scavenging on dead *Macoma* (bivalve mollusc)	1	1
Hunting live *Tubifex* (oligochaete annelid)	2	2
Hunting live *Corophium* (amphipod crustacean)	3	4
Hunting live *Erioptera* larvae (dipteran insect)	4	3
Deposit feeding on surface sediment particles	5	5
Suspension feeding on particles in the overlying water	6	6
Browsing live *Enteromorpha* (green alga)	7	7
Browsing live *Ulva* (green alga)	8	8
Hunting live *Hydrobia* (gastropod mollusc)	9	9

rejecting it in favour of prey of type y (all other things being equal). But what if the first prey encountered by the consumer is of type y: should it eat it, or reject it and continue to search for the more profitable prey of type x? To maximize the net rate of energy gain, the decision to accept or reject y is dependent solely on the frequency of type x prey, i.e. on the magnitude of $T_s(x)$. Prey of type y encountered should be consumed if

$$\frac{E(y)}{T_h(y)} > \frac{E(x)}{T_h(x)+T_s(x)}$$

and it should be ignored if

$$\frac{E(y)}{T_h(y)} < \frac{E(x)}{T_h(x)+T_s(x)}$$

In other words, a given prey item encountered should be consumed if, during the handling time concerned, the consumer would not be able to obtain a more profitable item (see Figs 9.24 and 9.25; Box 9.2).

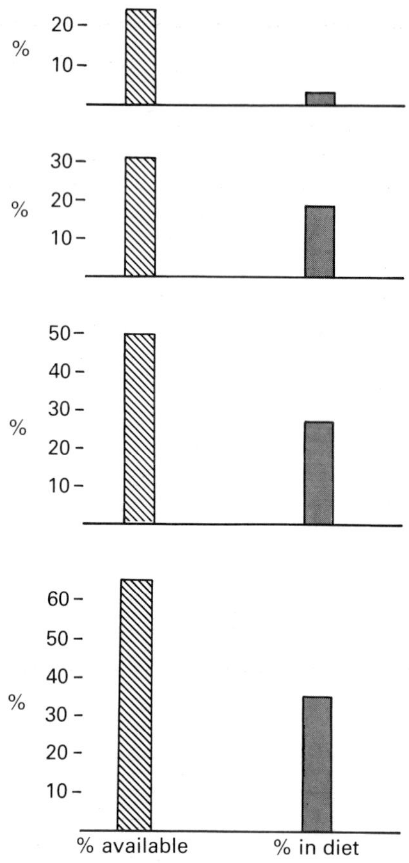

Fig. 9.23. The effect of different proportions of less valuable mussels (*Mytilus edulis*) on the preference displayed by crabs (*Carcinus maenas*)—see text. (From data in Elner & Hughes, 1978.)

A further complication is that many potential foods occur in highly clumped distribution patterns—in the form of discrete patches—for example, encrusting organisms on marine rocky substrata, ants in anthills, mysids in swarms, nettles in clumps, etc. Hence consumers face another type of problem: any given patch, region or area of local concentration will contain a finite amount of food, and consumption of that food may result in a decrease in its abundance and diminishing returns to the consumer. For how long, then, should a consumer remain in a patch before moving on to the next? Should it exhaust the

Box 9.2 'Search time' and prey choice: tits and mealworms

One of the difficulties associated with testing this conclusion is the problem of adequately quantifying search times: when a given consumer is moving through its habitat, how does an observer know whether it is searching for a given type of prey or carrying out some other activities? This analytical problem was solved in a neat manner by researchers at Oxford, by presenting a range of food items to a stationary consumer at different frequencies (rather than allow the consumer to move amongst the different items). The effect is the same, the encounter rate (equivalent to the search time) being adjusted by moving prey past at different rates. In this case the consumer was a bird (a great tit, *Parus*) and the prey were pieces of mealworm (larvae of tenebrionid beetles) of two different sizes such that the energy content of the larger was twice that of the smaller.

The mealworms moved past the captive bird on a conveyor belt (Fig. 9.24) and the bird was able to eat or ignore any potential food item passing in front of it. If it pecked at and consumed a given item, however, it missed the opportunity to take any other items which may have moved past it during the 5–15 seconds handling time involved. Different experiments were run with different search-time equivalents for the more profitable large mealworms and with different frequencies of more and less profitable items (Fig. 9.25).

The energy maximization model predicts selection in favour of large mealworms in experiments (c), (d) and (e), but not in (a) or (b). Such was indeed the case. Even though small mealworms were twice as abundant as large ones in experiment (e), as search time for the more profitable pieces decreased, so the consumer ignored the less profitable small mealworms and took mainly the large profitable items, the switch to selectivity occurring very close to the predicted point.

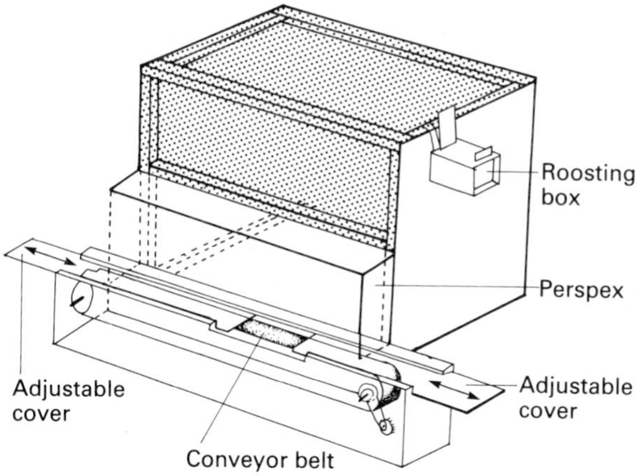

Fig. 9.24. The apparatus used to present various food items (mealworms) to a stationary consumer (a great tit) at different encounter rates—see text. (After Krebs *et al.*, 1977.)

patch completely first, or would it be advantageous to move before that point, and, if so, when?

Although superficially dissimilar, exactly similar problems are faced by other types of consumer. In high latitudes, for example, the amount of food available is dependent on the seasonal climate, and consumers have to decide when to migrate out of a geographical area of declining resources. When should the swallows leave? Several individual predator/prey interactions also fall into a similar category, especially where the consumer takes a considerable period of time to consume all the tissues of a captured item, and including those cases in which the consumer sucks the juices from its prey. With bees imbibing nectar from a flower, water-bugs sucking the fluids from mosquito larvae, and even lions feeding on an antelope, at first the rate of food intake from the prey is high but eventually it will fall when much of the easily obtainable and more nutritious materials have been ingested. When should the consumer leave the old item and obtain a new one? When all the food materials have been extracted, or before this point?

When a consumer begins to exploit a hitherto unexploited patch, geographical region or captured prey individual, the rate of energy gain per unit time will show a relationship with time of the form displayed in Fig. 9.26. Sooner or later the consumer

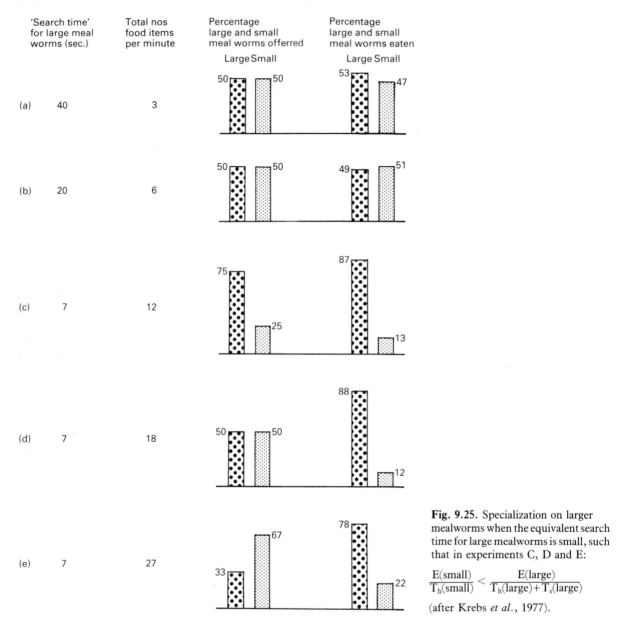

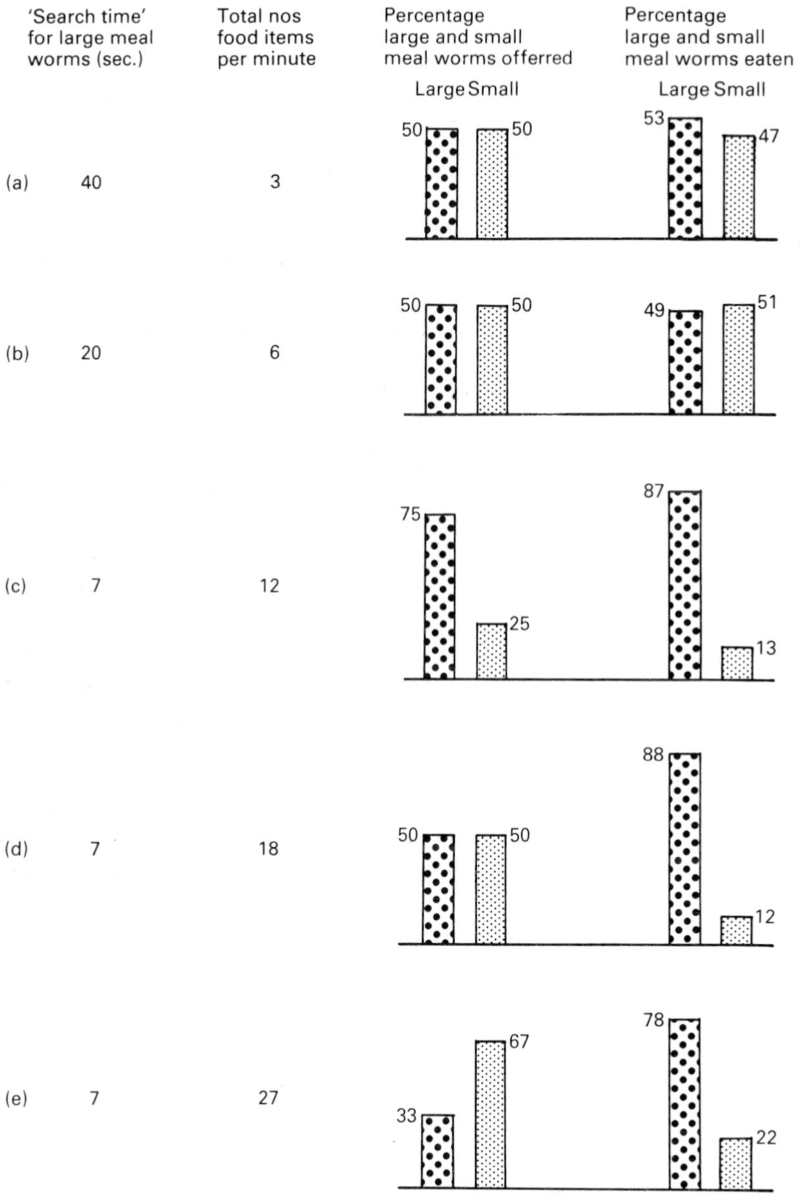

Fig. 9.25. Specialization on larger mealworms when the equivalent search time for large mealworms is small, such that in experiments C, D and E:

$$\frac{E(\text{small})}{T_h(\text{small})} < \frac{E(\text{large})}{T_h(\text{large}) + T_s(\text{large})}$$

(after Krebs *et al.*, 1977).

will have to leave and move elsewhere or catch another item, and movement between patches may be expensive in time and energy. Even such a slow moving animal as the winkle, *Littorina*, uses twelve times more energy per unit time when crawling than when grazing in one spot, and clearly long distance migration is very expensive indeed. So it is necessary to allow for the time (corrected for the different energy demands) taken to travel between, as well as within, patches.

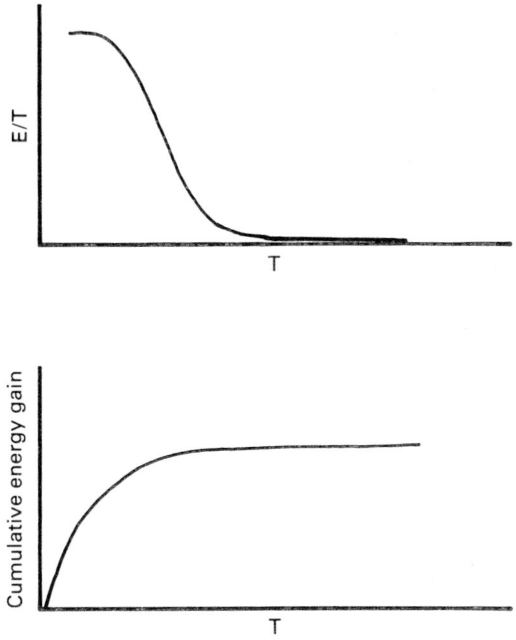

Fig. 9.26. Variation with time in the rate of energy gain from a finite pool—see text.

If then, we plot the average net rate of cumulative gain in energy from all the patches in a given habitat against time, and include the average time required to travel between these patches (Fig. 9.27), we can

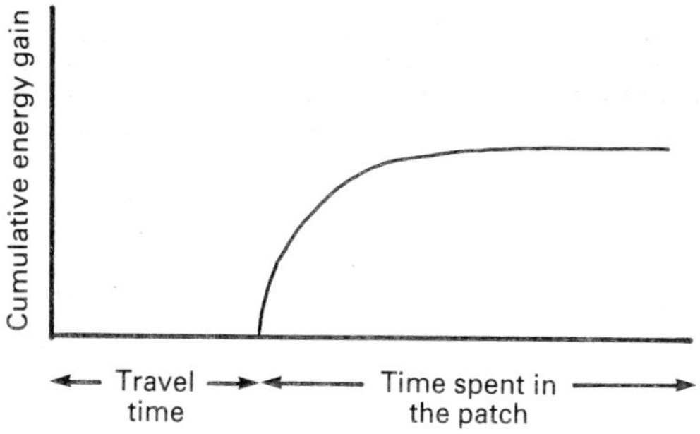

9.27. Energy gain from an exploited patch in relation to the time spent within each patch and the time taken to travel to that patch—see text.

ask the question 'what period of time spent in a patch will yield the maximum cumulative gain per unit total time (i.e. travelling time plus time within the patches)?' The answer will be given by the tangent from the origin to the cumulative gain curve (Fig. 9.28)—the line from origin to curve which has the steepest slope. Hence a consumer which leaves a patch or ceases consuming a given prey item when its rate of gain falls to this average value will maximize its energy intake per unit total foraging time. Consumers should give up feeding on a diminishing resource when they would do better by leaving it!

This being the case, then it follows that: (a) where patches vary in quality within any given habitat,

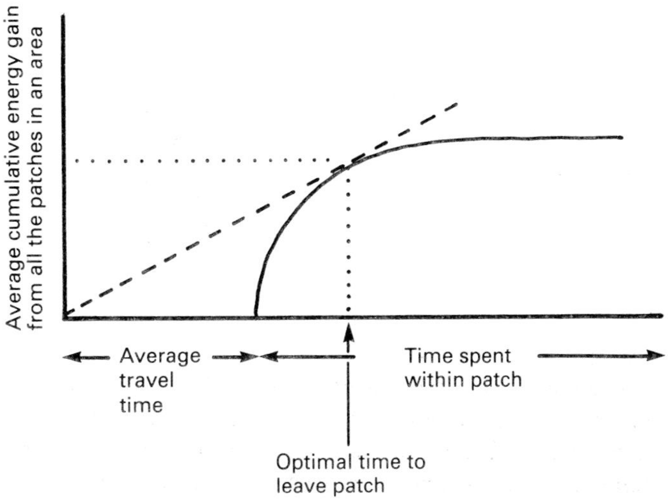

Fig. 9.28. The time spent within a patch that yields the maximum cumulative gain per unit total foraging time is given by the tangent from the origin to the gain curve—see text.

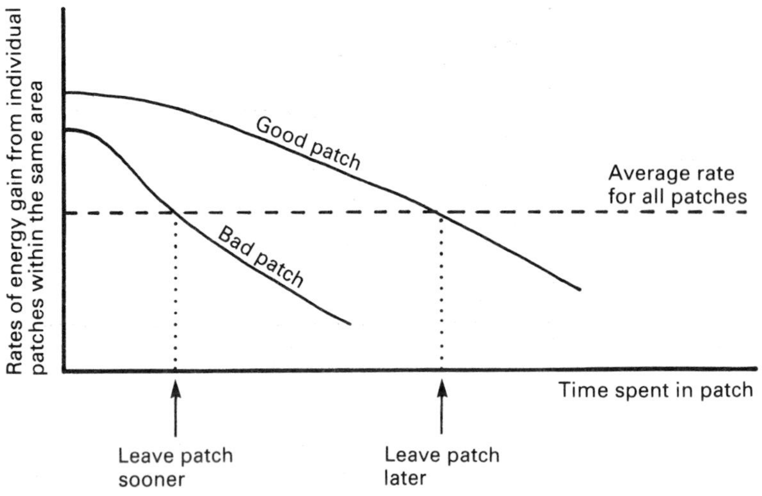

Fig. 9.29. The effect of patch quality on the optimal time to be spent in different patches within a given area—see text.

consumers should remain longer in the better patches (Fig. 9.29); and (b) where patches are of approximately the same quality but vary in their distances apart, it would be advantageous to remain in a patch longer if it took a longer travel time to reach it (Fig. 9.30) (i.e. when travel costs are higher, the returns from moving will be lower, and hence a consumer should remain longer in each patch).

9.3.3 Mobile consumers and their prey

How might the general considerations outlined above affect the feeding biology of mobile consumers feeding on individual items, and how can their prey minimize the chance of being consumed?

A given quantity of food can clearly be composed of many small prey items or a few large ones, and these two extreme cases bear different costs and different benefits. The gain from a single large item is great, but large prey are likely to be highly mobile and it may therefore require the expenditure of much time and energy to capture any single such item. Conversely, a diet of many small items may not require much time to be devoted to each individual capture but there will probably be a considerable search-time element. These differences will have repercussions on consumer and prey alike.

Pursuit hunters typically spend a large proportion of their total feeding time chasing large individual prey items, not always successfully. There is much energy expended on each potential prey individual, therefore, and correspondingly the gain from it must be great. It will be to the advantage of the hunter to minimize this pursuit cost as much as possible, and to the advantage of the prey to maximize it. This is likely to lead to an arms race between predator and prey, with the predator concentrating on that limited range of potential prey species that are most susceptible to its form of pursuit and capture whilst still being maximally rewarding energetically, and adapting more and more closely to the flight patterns of these prey species. The effect of maximizing net gain is for pursuit hunters to become specialist in their diet.

Besides the obvious pressure in favour of increased speed and agility, the most general means by which the prey species can decrease capture success rates of pursuit (and other) hunters is living in groups (swarms, schools, clumps, herds, etc.). This derives from a number of effects.

1 One commonly recorded observation is that a large group can detect and react to a predator's approach more quickly than can a small group or a solitary individual. The small, flightless, marine hemipteroid *Halobates*, for example, skates on the

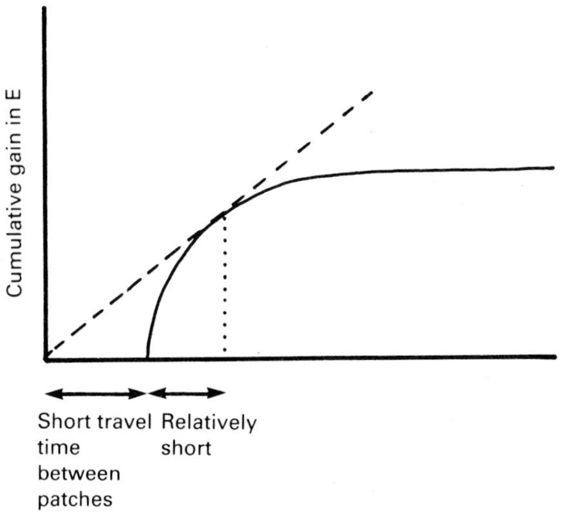

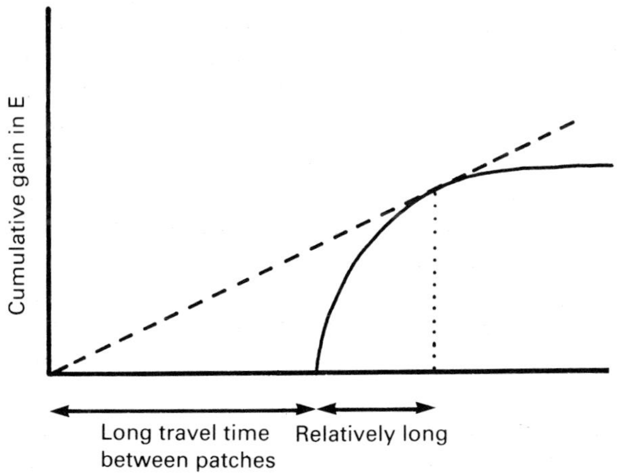

Fig. 9.30. The effect of differing average travel times on the optimal time to be spent in patches of the same quality—see text.

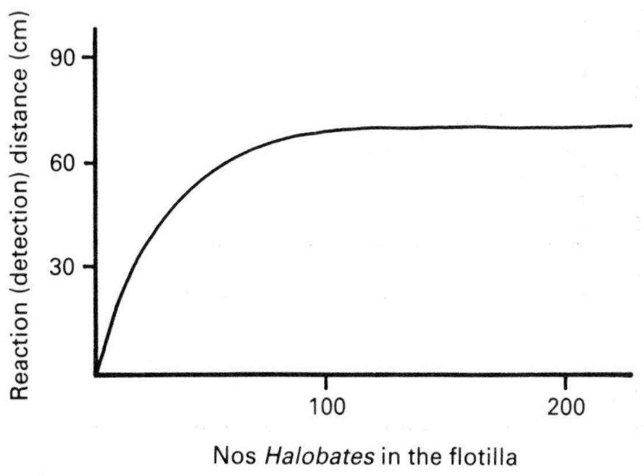

9.31. Variation in the distance away at which an experimental predator model evoked a behavioural response in the sea-skater *Halobates* in relation to the number of individuals in the flotilla (after Treherne & Foster, 1980).

surface film of the sea in a similar fashion to pond-skaters on inland waters. When not feeding, it aggregates in 'flotillas', and the flotillas display clearly defined responses to the approach of predators (birds, fish, predator models). The distance away at which an experimental predator model produced a behavioural avoidance response in the flotilla varied with flotilla size as shown in Fig. 9.31: detection distance was greater in larger flotillas, although clearly there is a critical size of flotilla after which there is no further increase in detection range (the maximum having been reached).

2 A second effect of living in a group is confusion of the potential predator. Pursuit hunters attack one potential prey item at a time, and we have already seen that it will be to the advantage of the consumer to capture a prey individual of high food value. Groups may react to the close proximity of a predator by the individual members scattering in all directions. This increases the difficulty of 'homing in' on any one target individual and results in considerable target switching as different prey move across the sensory field.

3 Predators may be deterred from attacking a group by group-defence behaviour, whereas they suffer no such inhibition when encountering a solitary member of the same species. Most of the classic examples of group defence are from within the vertebrates, but invertebrate examples are also known, most obviously in the social hymenopterans. In addition, some sawfly larvae exude a sticky resin from their mouthparts when attacked, e.g. by predatory insects. This defence does not deter pentatomid bugs from attacking solitary larvae, but bugs which attack larvae

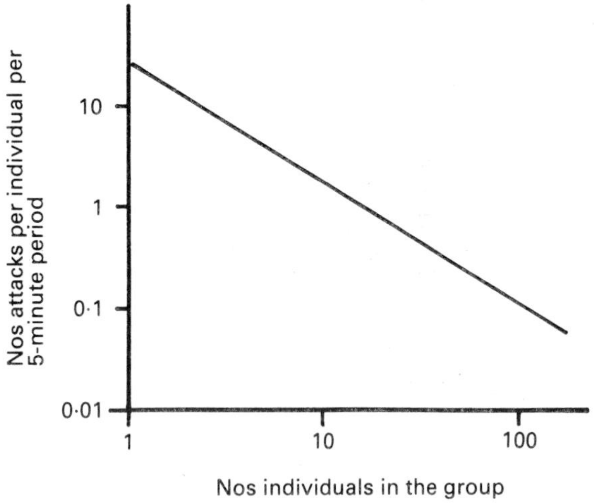

Fig. 9.32. Number of attacks by fish on individual sea-skaters, *Halobates*, in relation to the number of individuals in the group (after Treherne & Foster, 1980).

in aggregations rapidly become covered in the resin and may barely be able to move afterwards.

4 The three effects of group living described above all have the effect of increasing the time taken for a predator to make a successful capture; it may, on average, be advantageous for the predator to switch its attention to another, non-social prey. A fourth effect relates to the probability of any given individual within a group being the subject of a successful predatory attack, the larger the group, the smaller the probability that any particular individual will be the one caught during a given attack. This has been shown in the bug *Halobates*, cited above in respect of the detection effect, when preyed on by fish (Fig. 9.32). An individual in a group of ten receives only one-tenth of the number of attacks directed at a solitary individual, and a bug in a group of 100 receives only one-hundredth of that number.

The effect is not necessarily purely statistical. The probability of being the individual taken from a group is often highest for those on the periphery, and least for those in the centre of the aggregation. In a number of instances, individuals have been described as constantly moving from the periphery to the centre, leaving at the margins of the group those individuals least able to maintain a relatively safe position within the 'selfish herd'. In this way, individuals are actively minimizing their own chances of being consumed.

In contrast, hunters of the searching and ambushing categories prey on individuals which are easily caught by virtue of their small size (relative to the consumer). All and any easily captured prey may be consumed in order to ingest sufficient food in total, and searchers can most readily maximize their net gain by being generalist in diet. Further, because search time will be inversely proportional to the overall abundance of suitable prey, searchers are likely to have particularly generalist diets in food-poor habitats, and at times of food scarcity. At times of peak abundance of any one particular prey species, however, they are likely to spend a disproportionate amount of time utilizing that prey type (become temporarily specialist), switching to other food species if and when their numbers also increase.

Prey individuals subject to predation by searching hunters will be advantaged by maximizing a consumer's search time. This can be achieved in many different ways, which increase the difficulty of either finding the prey or of recognizing the prey as potential food. These include (Fig. 9.33):

1 Inhabiting microhabitats which cannot easily be investigated by most potential predators, e.g. dwelling in crevices, under stones, in burrows, etc.

2 Being cryptic, i.e. by a combination of shape, posture and surface pattern (or odour) blend in with the background and thereby increase recognition time (some species can change colour or pattern to merge with more than one background type; others can modify their local environment to conceal themselves more effectively).

3 Being mimetic, i.e. similarly to crypsis, increase recognition time by mimicking an object of no or low food value—a dead leaf, twig, faecal pellet, etc.—or a distasteful species or one capable of aggressive retaliatory defence.

Increase in the necessary search time is not the only means of decreasing an individual's food value to a consumer, handling time can also be increased and the energy content per unit weight can be

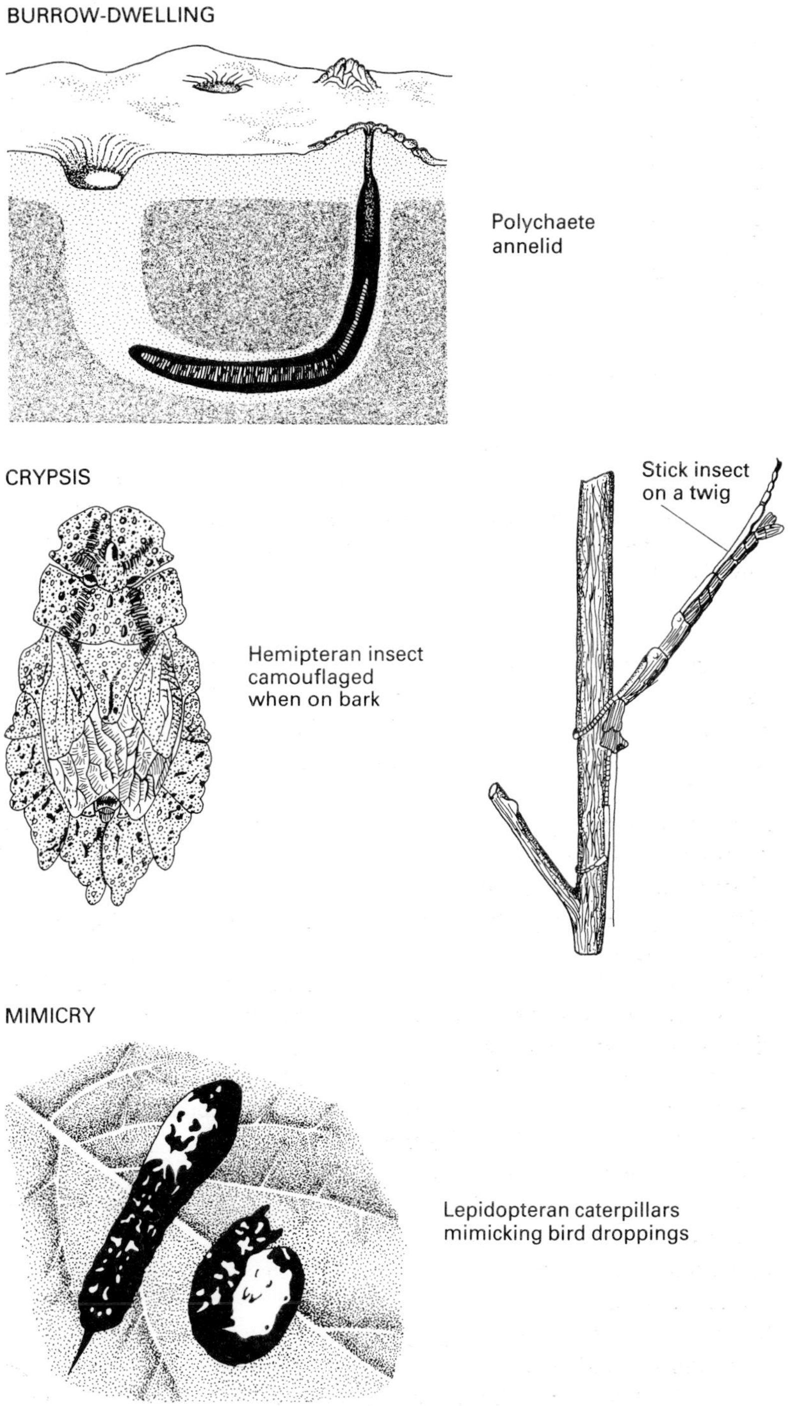

Fig. 9.33. Mechanisms for increasing the search times of potential predators (from various sources).

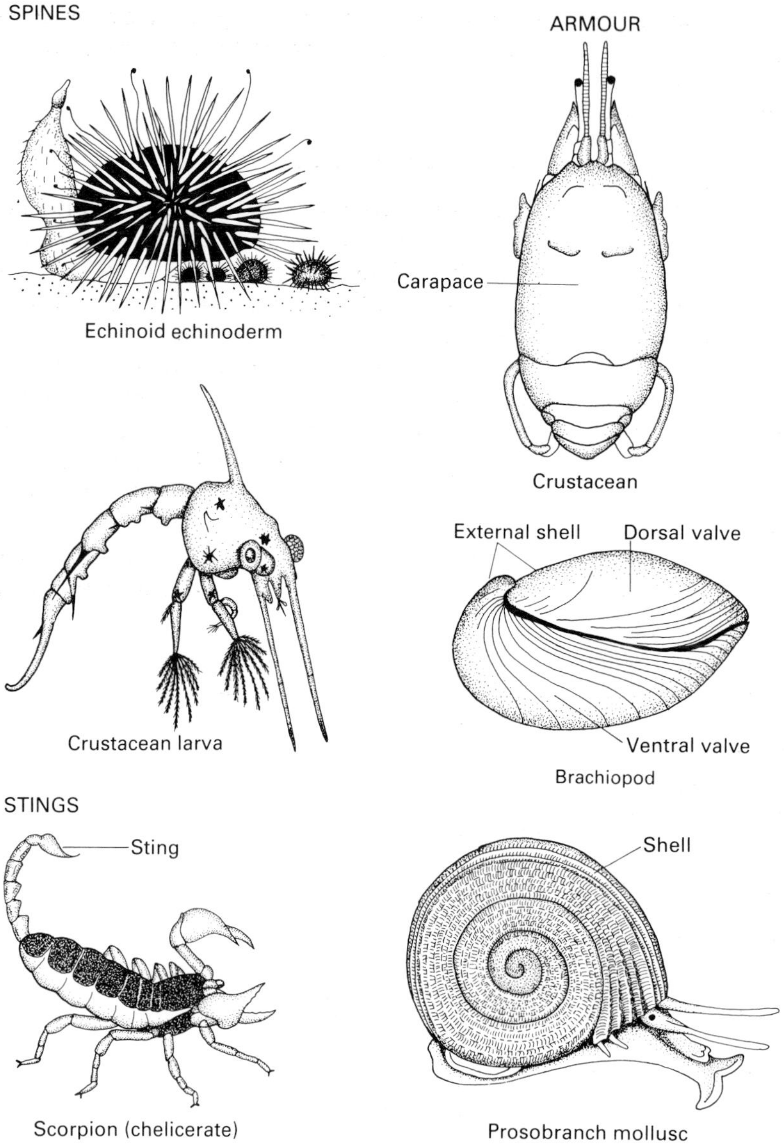

Fig. 9.34. Structures increasing the handling times of potential predators (from various sources).

decreased. Handling time is often increased by the evolution of (Fig. 9.34):

1 Defensive weaponry, e.g. stings, jaws, etc.

2 Defensive armour, e.g. shells, calcareous or chitinous plates, etc.

3 Shapes which increase the difficulty of grasping, manipulating and/or ingesting the food item, e.g. spines. Spines and similar protuberances on the body surface are particularly effective in this respect, for not only do they make handling time-consuming, but they also increase the effective body size—thereby taking the prey out of the size range of at least some predators—for little increase in body tissue.

Energy and/or nutrient content per unit body

weight or volume can be reduced by diluting the quantity of living tissue with inert, nutrient-poor or inedible substances. Palatability can be decreased by the presence of noxious substances in specific organs or generally in the tissues. In this case, a level of trial and error learning is necessary on the part of the consumer, and the prey often make this process as rapid and effective as possible by coupling their distastefulness with bright warning colouration.

Increase in handling time, decrease in energy content per unit mass and decrease in palatability (or outright toxicity) are also precisely the defensive tactics utilized by the prey of grazing and browsing consumers, which functionally are equivalent to searching hunters except in respect of the sessile and frequently photosynthetic nature of their prey. Searchers are normally larger and longer lived than their animal prey, but the relationship is usually the other way around in grazers/browsers and their food species: the prey are the larger and longer lived. Hence whereas it is difficult for the prey of searching hunters to do anything other than 'teach' their long-lived consumer to avoid them by use of noxious chemicals, it is easier and more advantageous for long-lived plants to kill their short-lived insect browsers with toxins and thereby cause selection for individuals which ignore that plant as food.

All of these defensive systems (and see also Section 13.2.1) make it advantageous for a generalist consumer to search for other prey types—more obvious, less protected, more palatable species—but by virtue of their abundance resulting from successful avoidance of generalist consumption, sessile or sedentary species relying mainly on passive mechanical or chemical defences are open to attack from specialist consumers which have managed evolutionarily to crack the defensive codes. Grazers and browsers, and specialist consumers of cryptic, mimetic and noxious animal prey have participated in similar evolutionary arms races with the prey species as have pursuit hunters and their highly mobile prey. Indeed, once specialist consumers have evolved the mechanisms whereby the toxic or noxious defences of their prey organisms can be rendered ineffective, the consumer may even use the chemicals to their own advantage by sequestering them in their bodies as a defence against their own predators.

9.3.4 Suspension and deposit feeders

Consumers which capture particles by filtering a current of water clearly have no pursuit or search costs. Instead they bear the costs of filtration and those associated with the rejection of unwanted particles. The mesh sizes of their filters can, in general, only be altered in evolutionary time and hence there is little or no possibility of differentially selecting out particles of high food value from the water: filters are size not edibility specific. A sessile filter feeder can only maximize its net rate of energy gain by altering:

1 The filtration rate in relation to the relative abundance of different types of particles in the water.

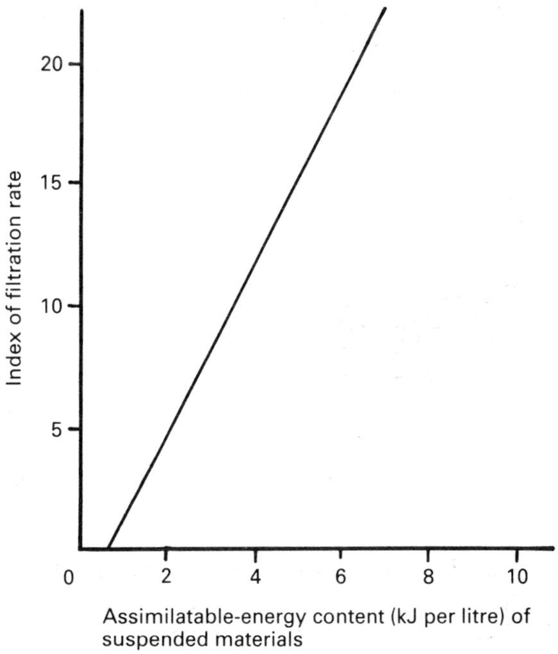

Fig. 9.35. The relationship between concentration of edible particles in the overlying water and the filtration rate of the polychaete *Nereis diversicolor* when behaving as a suspension feeder (after Pashley, 1985). (See Fig. 9.5d and Table 9.1.)

2 The rejection rate in relation to the relative nutritional values of the particles and to the costs of rejection.

Filtration rate should increase as the concentration of high-value particles in the water increases (Fig. 9.35), but there may be a limit determined by the ability of the gut to process the collected material. In mucociliary feeders, this could be the ciliary movement of the mucous cord within the gut; and in the arthropods sieving out particles with their setae, it could be the packing of the gut lumen with ingested particles. Once the gut becomes filled, the rate of energy intake will be limited by the passage time through the gut and the speed with which digestion can proceed. Hence filtration rate should decrease, once the gut is full, to that value which maintains the gut in a full state. Equivalently to the position with predators consuming only the highest value parts of a prey item before moving to the next (Section 9.3.2), it may pay suspension feeders, dependent on the quality and abundance of their food, to pass food rapidly through the gut digesting only some of it, rather than pass it through more slowly and digesting a greater proportion.

The cost of rejecting unwanted particles may be high, and therefore as the cost of rejecting a given particle type increases so the ingestion rate of that particle type may also be expected to increase. When rejection costs are very high, it may be less costly to reject even completely indigestible particles through the gut than by means of sorting organs. Under such circumstances, a suspension feeder could be in energetic deficit and it would then be advantageous to cease feeding completely until the suspended load of the water changed.

With such a food-collection system, there is very little, if any, opportunity to become a specialist consumer except on an evolutionary time scale and then only on certain size ranges of particles. Pelagic tunicates, for example, possess very fine filters capable of retaining bacteria; whales, on the other hand, sieve out the large planktonic euphausiids. These differences are simply functions of the mesh sizes of their filters, and also of the relative sizes of the filterers. Whales could not operate a filter capable of collecting bacteria and protists: they would be unable both to move forward through the water and to filter, for exactly the same reason that it is not possible to tow a fine-meshed plankton net through water rapidly and still permit it to catch anything. Water is simply pushed ahead of the mesh and little passes through. Such selectivity as suspension feeders show is therefore not exercised at the level of the diet, but is displayed mainly at the stage of the free-living larval phase possessed by most species, when it settles in an area likely to provide the sessile adult with an adequate supply of appropriate suspended food particles.

Deposit feeders are somewhat intermediate in character between the suspension-feeding category and those of the hunters and grazers/browsers considered above. In as much as they are consuming non-living organic matter, they are feeding on material of low nutritional status which is correspondingly plentiful and which, being dead, does not protect itself from being consumed. Some may simply pass the inorganic sediment, together with such organic matter as it may contain, through their guts, and they will therefore be subject to the same gut-packing restrictions as apply to some suspension feeders. When the potential ingesta contains less than a certain threshold digestible content, it may also be energetically advantageous not to process it at all but to move and resume feeding elsewhere. Many deposit feeders, however, are capable of selective ingestion (Section 9.2.6), and such species may behave like searching hunters, except, instead of the whole animal moving, the 'searching' is conducted by highly mobile feeding tentacles or similar organs.

In general, the living components of the diet of suspension and deposit feeders are minute in relation to the size of the consumer, and, although some of the consumed species do have chemical defence systems, most appear to lack any form of defence against consumption. Instead, they have rapid rates of sexual and/or asexual multiplication which are more than capable of making good predatory losses. In this they are somewhat equivalent to the prey of grazers and browsers, which characteristically possess sufficient potential for growth or asexual division to replace lost structures, such as leaves, polyps or zooids, so that only exceptionally severe rates of consumption cause the death of the individual or of the colony.

9.4 Conclusions

We have seen that animals originated as small consumers of bacteria, protists and each other and that these ancestral diets not only persisted relatively unchanged for millions of years but have also influenced animal feeding ever since. The selective advantage of large size favoured the evolution of mechanisms for increasing the quantities of these small and widely scattered items that could be ingested, but did not lead to any radically different new diets. Consumption of macrophyte material, whether algal or plant, evolved relatively late and, unless mediated by intestinal micro-organisms, remains an inefficient process, not least because of the poor nutritional quality of structural carbohydrate tissues, but also because of the defensive repertoires of these otherwise easily available materials.

The categorization of animals as 'herbivores', 'carnivores' or 'omnivores' is not helpful in understanding their feeding biology: on such a basis, most species are omnivores. Of more value is classification by feeding mechanism—hunters/parasites, grazers/browsers, deposit feeders, suspension feeders, absorbers of the products of internal symbionts, etc.—a system which cuts right across the taxonomic position of the prey species, and is concerned more with the consumer than with the consumed.

Although constrained to varying extents by their evolutionary pasts, all animals nevertheless face choices in ecological time: whether to attempt to consume this individual prey item rather than another (in space or time); whether to remain feeding in this area rather than move to another; and so on. Consumers do not appear to react to such 'decisions' randomly, and optimality theory provides one model with some success in predicting foraging behaviour. Within the constraints imposed by factors other than food quality and supply (e.g. predation, breeding biology), several animals do feed in such a manner as to maximize their gain (benefits derived minus costs borne). Prey species on the other hand have been selected so as to maximize the relative cost of consuming them, by their behaviour, their morphology or their biochemistry. As might be expected, the different feeding modes will have inherently different potential solutions to the maximization of gain, with consequent selective advantages in favour of a generalist or a specialist diet.

Feeding behaviour would seem to be just as much under a day-by-day selective control as other, more familiar aspects of animal biology.

9.5 Further reading

Barnard, C.J. (Ed.) 1985. *Producers and Scroungers*. Croom Helm, London.

Calow, P. 1981. *Invertebrate Biology: A Functional Approach*. Croom Helm, London.

Crawley, M.J. 1983. *Herbivory*. Blackwell Scientific Publications, Oxford.

Hodkinson, I.D. & Hughes, M.K. 1982. *Insect Herbivory*. Chapman & Hall, New York.

Jennings, D.H. & Lee, D.L. (Ed.) 1975. *Symbiosis*. Cambridge University Press, Cambridge.

Jennings, J.B. 1972. *Feeding, Digestion and Assimilation in Animals*, 2nd edn. Macmillan, London.

Jorgenson, C.B. 1966. *Biology of Suspension Feeding*. Pergamon Press, Oxford.

Mason, C.F. 1977. *Decomposition*. Edward Arnold, London.

Morton, J. 1979. *Guts*, 2nd edn. Edward Arnold, London.

Owen, J. 1980. *Feeding Strategy*. Oxford University Press, Oxford.

Smith, D.C. & Douglas, A.E. 1987. *The Biology of Symbiosis*. Edward Arnold, London.

Taylor, R.J. 1984. *Predation*. Chapman & Hall, New York.

Townsend, C.R. & Calow, P. (Ed.) 1981. *Physiological Ecology*. Blackwell Scientific Publications, Oxford.

Vermeij, G. 1987. *Evolution and Escalation. An Ecological History of Life*. Princeton University Press, Princeton.

10
MECHANICS AND MOVEMENT (LOCOMOTION)

In this chapter we will learn how invertebrates are able to do mechanical work on their environment. Most of the discussion will be about the ways in which animals move, but the mechanical principles are the same for sessile animals which expend energy and move the environment past themselves.

The opening section deals with basic mechanics and defines terms. This is followed by a discussion of the biological aspects of the generation of forces by animal cells. A common feature is the utilization of respiratory energy as discussed in Chapter 11 in the formation of cross-linkages between protein fibrils.

A comparative study of animal locomotion follows. This will discuss the influence of scale or relative size on locomotory systems and will review the potentialities of the grades of structural organization exhibited by the various groups of invertebrates. The chapter describes separately the locomotory systems of soft-bodied animals, which have what is often called a hydrostatic skeleton, and animals with hardened skeletal parts such as crustaceans and insects.

10.1 The principles of locomotion

Invertebrates show a diverse range of structure; many are soft-bodied but others have hardened tissues or coverings. Virtually all are able to perform mechanical work with the result that the body moves with respect to the environment or the environment is moved past the fixed body of the organism. The energy which is invested in such activities must result in a return to the animal. This return on the energy invested can be thought of as being an enhanced rate of feeding or other resource acquisition, a reduced rate of predation, the avoidance of harmful environmental changes, enhanced dispersal and/or contact with potential partners in sexual reproduction.

When animals move they obey the same basic principles as all moving objects (Newton's laws of motion). These principles state that:

1 If a body is at rest relative to its environment, it can be set in motion only by the application of an external force.

2 The application of an unbalanced force to a mass in motion results in an acceleration or deceleration of the mass in the direction of the force.

3 For every action there must be an equal and opposite reaction.

4 The energy within a closed system remains constant, though it can be changed from one form to another.

These concepts apply universally and will be illustrated in general terms before the movement and activities of the different groups of invertebrates are explained. Box 10.1 illustrates the general principles governing the locomotion of animals and defines some units.

When an animal moves it will have overcome forces which tend to resist its movement; this is generally referred to as friction or drag. Since the resistance of the medium through which an object moves tends to change its rate of movement (by causing deceleration), the resistance is a force. It can be expressed in the units defined in Box 10.1. In any system energy is conserved and mechanical energy can be converted into kinetic energy which is stored in the movement of a mass, but which will eventually be lost as heat. This is shown in Box 10.2.

Box 10.1 Mechanical terms and definitions

1 A **force** is detected by its effect on a **mass**.
A **force** tends to change the direction or rate of movement of a **mass**.
Force is the **mass×acceleration**.

$F = M \cdot a$

In SI units a force of 1 **newton** (N) is required to impart to a mass of 1 kilogram (1 kg) an acceleration of 1 metre per second per second.

2 When one body exerts a force on a second body, it is as though the second exerts a force of equal magnitude but in the opposite direction. This is called the **reaction force**.

3 When a force moves a mass, mechanical **work** is done.
Work is **force×distance**.

$W = F \cdot d$

$W = (M \cdot a)d$

The SI unit of work is the joule (J).
The rate of doing work is **power** and therefore power is equal to force×velocity. The S1 unit is the **watt** (W) where 1 watt = 1 joule per second.

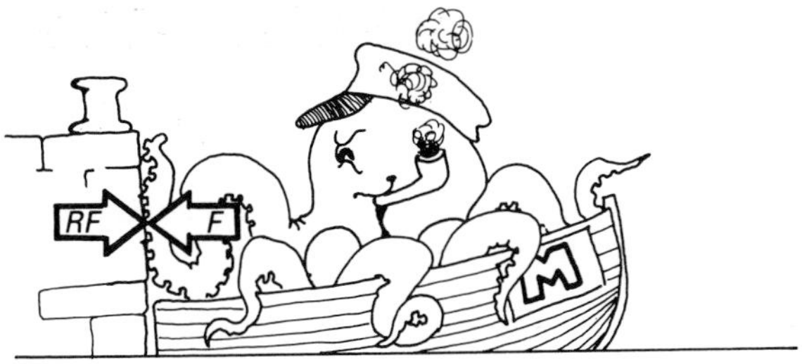

An illustration of the concept of force and reaction force.

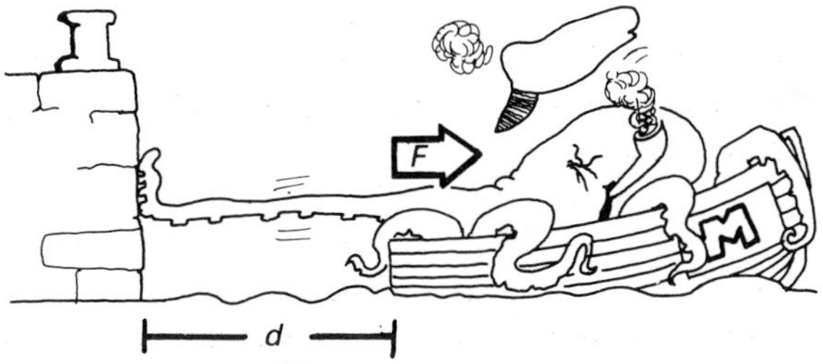

An illustration of work = force×distance.

The mechanical work performed in locomotion is ultimately derived from energy generated by chemical reactions in the cells of which the animal is composed (see Chapter 11), but the chemical work is not necessarily equal to mechanical work.

A weight-lifter may become exhausted in trying to lift a weight which does not move, but he has not, by these definitions, performed mechanical work. He has, of course, done metabolic work, his muscles use ATP at metabolic cost, and through his efforts he has produced heat. We therefore have to define the efficiency of locomotion in recognition of the fact that energy may be expended which does not contribute to the movement of the body. We can determine the mechanical work done by measuring mass, and distance moved. This work we could describe as the *useful work*. The total energy output of the animal in motion is more difficult to measure (but can be estimated by measuring the rate of oxygen consumption), and the efficiency of the animal (like any machine) can then be expressed as the ratio:

$$\frac{\text{output of useful work}}{\text{input of energy}}.$$

We shall see that some systems of animal locomotion may be more efficient than others, but that efficient systems may have limited power, i.e. the rate at which they can work may be limited. As elsewhere in biology we find that there are potential trade-offs, one important one being that between efficiency and absolute power.

It is shown in Box 10.2 that organisms in motion have inertia and will experience drag, that is, forces acting contrary to the direction of locomotion due to the resistance of the medium through which they are moving. The extent to which such forces are important is profoundly influenced by the relative size of the organism, its absolute velocity, and the viscosity

Box 10.2 Kinetic energy and friction

1 In Box 10.1 the force F imparts to our model (the octopus in the boat) an acceleration such that:

$$F = M \cdot a$$

The work done would depend on the distance moved by the mass while the force was applied and:

$$W = F \cdot d$$

2 If the boat was without friction the octopus would move forever with the energy stored in the form of kinetic energy, where

$$\mathrm{KE} = \tfrac{1}{2}\, m \cdot V^2$$

and V is the velocity in metres per second.
The boat comes to a rest because there is a frictional force which causes **deceleration**. This force is often called **drag**.
3 The kinetic energy will have been dissipated as **heat energy**. For the octopus and boat to continue to move at a constant velocity, the octopus must continue to exert a force equal to the frictional force.

1
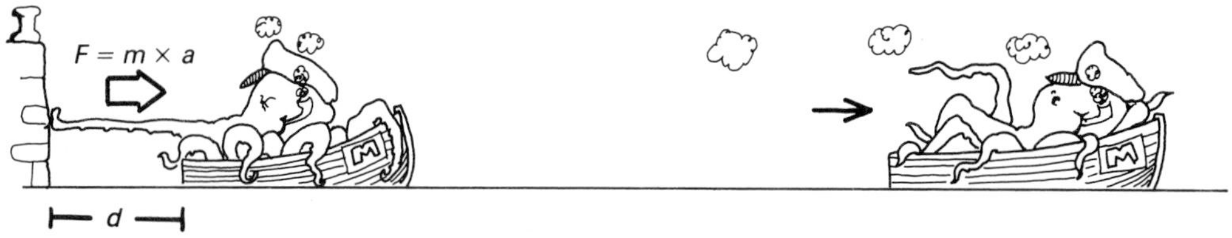

2
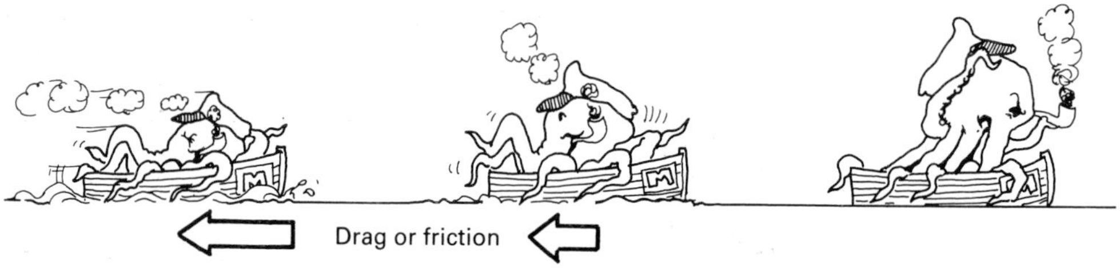

3
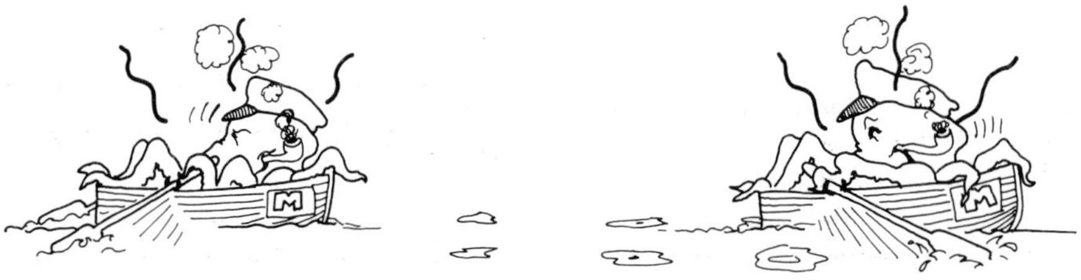

(or friction) of the medium over or through which it is moving. These interrelationships can be expressed as the **Reynold's number** (Re).

$$\mathrm{Re} = \frac{1\,V_{max}}{v}$$

where 1 is the length of the system,
V_{max} is the maximum velocity achieved by the system and
v is the viscosity of the medium.
(For water $v = 1 \times 10^{-6}\ \mathrm{m^2\,s^{-1}}$.)

Reynold's number is a measure of the relative importance of inertial and viscous forces. For minute locomotory systems, such as spermatozoa, flagellates, ciliates, ciliated larvae and the smallest invertebrates, e.g. the Rotifera (Section 4.9), it is very small, being in the range 10^{-3} to 10^{-1}. In these circumstances locomotion is almost entirely explained by reference to viscous forces alone. Some larger invertebrates achieve velocities such that Reynold's number is much greater than unity, and their locomotion cannot be explained by viscous forces alone. This is

Box 10.3 The resolution of forces

1 A force acts in a specific direction; it is a **vector**.
2 Two forces acting on a single mass but in different directions can be added to give a single **resultant force** which acts in a different direction.

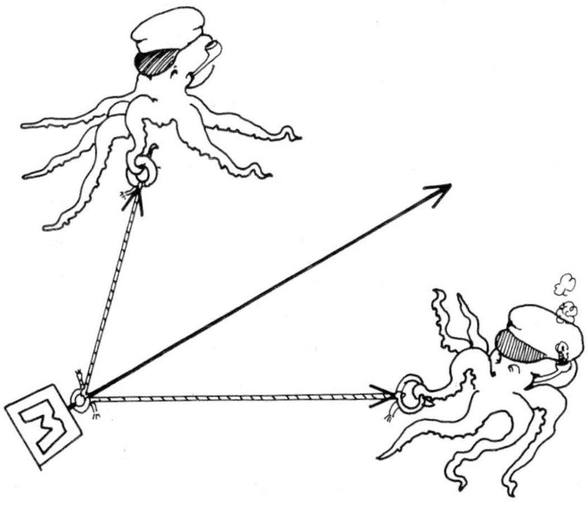

Forces acting in different directions can combine to produce a resultant force acting in some other direction.

3 a We can represent the lines of force from a mass to represent the **direction** in which they act.
b The length of the line is proportional to the **magnitude** of the force.
c The **direction** and relative **magnitude** of the **resultant force** can be estimated geometrically.

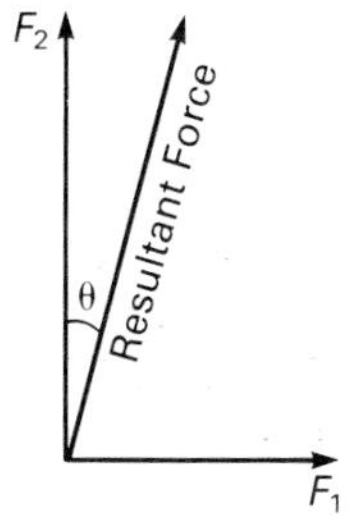

Geometrical resolution of forces. Sin $\theta = F_1/F$, cos $\theta = F_2/F$.

4 A single force can be resolved into two components acting at right angles to each other. The direction of one of the components is usually the direction of forward motion. The component of force at right angles to the line of motion is not relevant to the motion, but it may represent a metabolic cost.

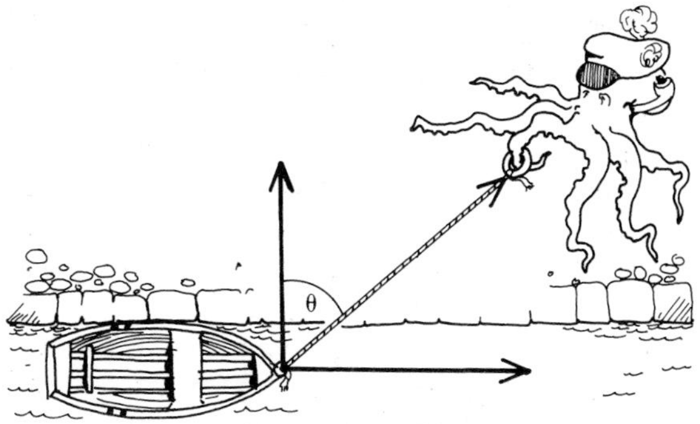

Resolution of the force acting in the line of motion given a force acting in some other direction. Force in direction of motion = F cos θ.

particularly so in the case of the insects and the cephalopod molluscs.

In Boxes 10.1 and 10.2 the forces generated were supposed to act along the line of motion, but the forces may be applied in other directions. The principle is explained in Box 10.3.

10.2 The generation of forces by animal cells

The locomotion of all animals is due to change in cell shape occurring against environmental resistance by the utilization of metabolic energy. Three types of cell contribute to animal locomotion—flagellated cells, ciliated cells and muscle cells (Fig. 10.1). Despite the differences in morphology of these cells, the cellular mechanisms which bring about the changes in shape are similar; they all involve ATP-dependent formation of cross-linkages between protein fibrils.

A true flagellum is a tubular projection of the cell surface bounded by a cell membrane continuous with that of the cell. It contains a bundle of protein fibrils, the axoneme, which, as the electron microscope shows, consists of a central core of two microtubular fibrils surrounded by a ring of nine double microtubular fibrils (Fig. 10.2). Changes in shape of the flagellum are brought about by sliding of the outer doublets along each other as a consequence of the formation of cross-linkages between their adjacent strands. The formation of each cross-linkage requires phosphate-bound energy provided by the ATP molecule (see Chapter 11). The same mechanism operates in cilia; the main differences between these structures relate to their absolute length and mode of deployment. The wavelength of a flagellum is characteristically less than its total length, consequently a single flagellum can generate a more or less constant force parallel to its longitudinal axis (Fig. 10.3a).

In contrast to a flagellum, the cilium is relatively short and it describes a fixed cycle of movements often involving a rather stiff power stroke and a more flexed recovery stroke (Fig. 10.3b). Constant propulsive forces are generated by the carefully sequenced and co-ordinated movement of large numbers of cilia which are deployed over large parts of the surface area, or in girdles or bands, each containing many thousands of these organelles.

Muscle fibres can generate forces because they have the intrinsic ability to become shorter against resistance. Muscle fibres are composed of tubular contractile elements, the myofibrils, each of which is approximately 1 μm in diameter and is itself composed of protein filaments arranged in a highly ordered fashion. In many muscles a functional unit called a sarcomere can be recognized; it consists of an arrangement of thick and thin filaments composed of two different proteins, *myosin* and *actin* (Fig. 10.4).

Cross-bridges can be formed between the myosin and actin filaments and the formation of these bridges and changes in the shape of the myosin molecule head cause the thin filaments to slide over the thick filaments unless the movement is opposed by an equal and opposite force. Continued contraction of the muscle fibre requires a repeated cycle of formation

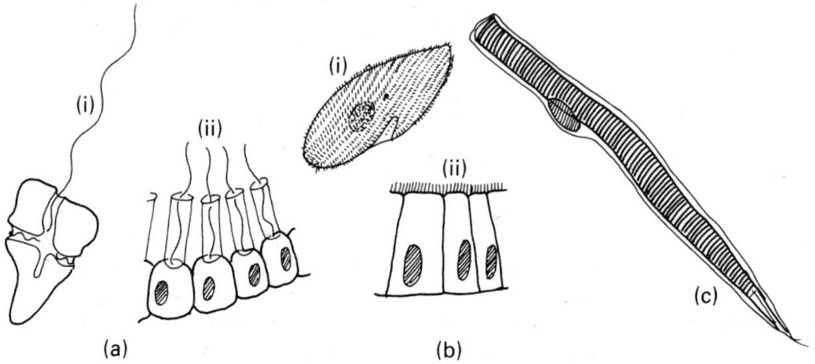

Fig. 10.1. Different types of locomotory cell. (a) Flagellated cells: (i) free-living flagellate; (ii) choanocytes of Porifera. (b) Ciliated cells: (i) free-living ciliate; (ii) ciliated cells of an animal. (c) Muscle cells.

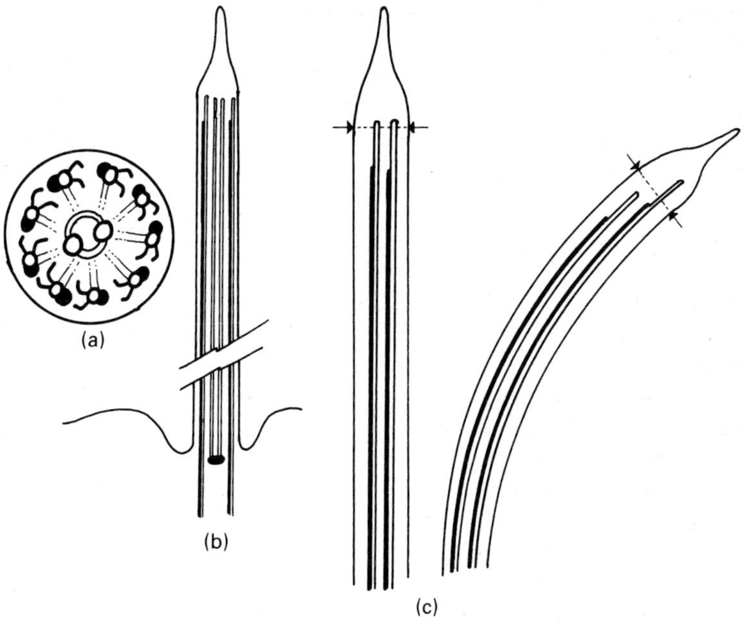

Fig. 10.2. Diagramatic representation of a cilium. (a) Transverse section; (b) longitudinal section; (c) the bending movement.

and detachment of the cross-bridges, and at each cycle a pulse of high energy phosphate must be supplied by ATP. The properties of a muscular system are determined by the rate at which sarcomeres shorten and the number of sarcomeres in series; this determines the intrinsic rate of shortening.

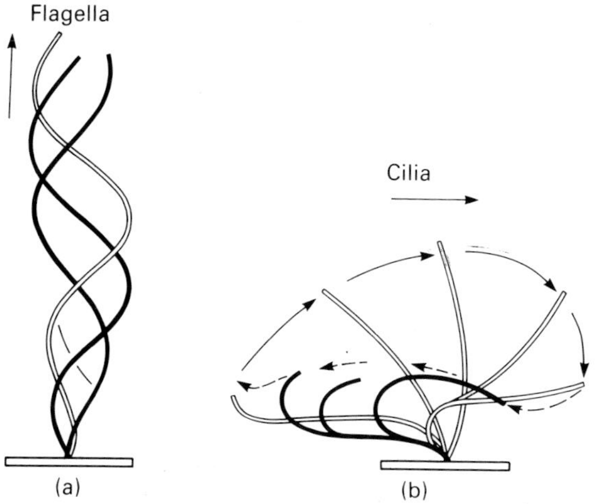

Fig. 10.3. Comparison of the direction of force generated by (a) flagella and (b) cilia.

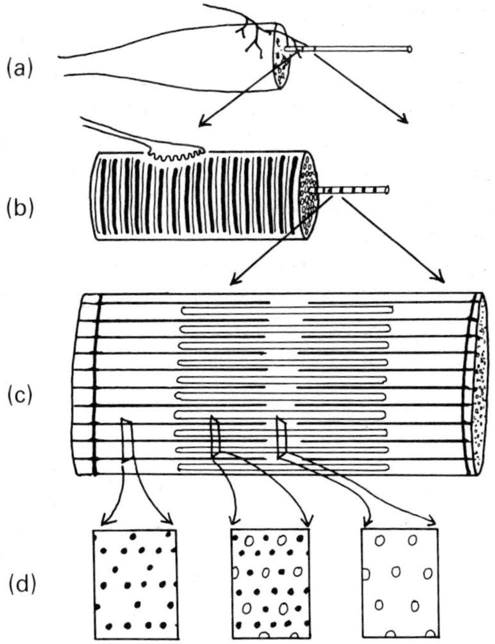

Fig. 10.4. The structure of a muscle at different levels of magnification. (a) The entire muscle with innervation; (b) a single muscle fibril with nerve ending (synapse); (c) the functional unit, the sarcomere; (d) electron-microscopic image of a fibril cross-section at different levels.

The force per unit cross-sectional area depends on the number of filaments in parallel. As the degree of overlapping between fibrils increases, the free cross-linkage sites become used up and the power of the muscle falls to zero (Fig. 10.5).

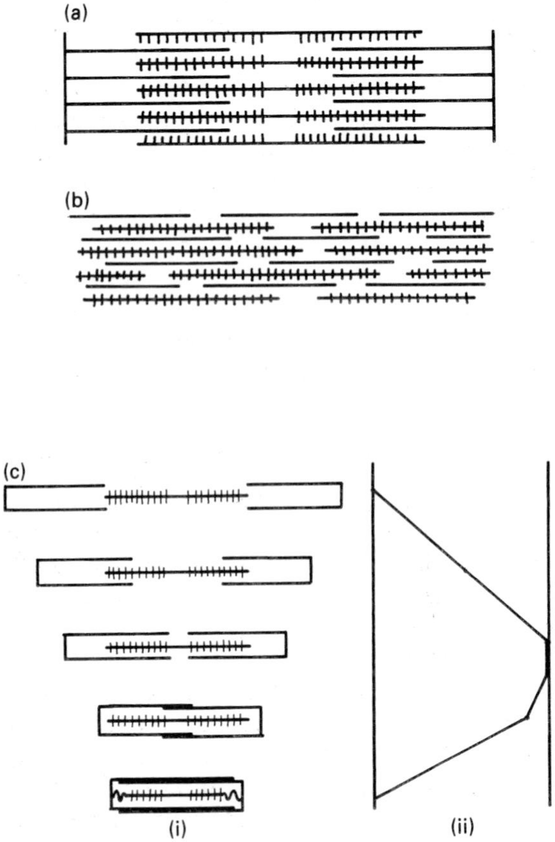

Fig. 10.5. The structure of muscles and production of force by muscle cells. (a) Diagrammatic representation of the structure of striated muscle; (b) diagrammatic representation of the structure of non-striated muscle; (c) the generation of force: (i) the structure at different levels of contraction; (ii) the force generated as a function of relative length.

10.3 Ciliary locomotion

Invertebrates (other than the very simple placozoans; Section 3.3) do not utilize flagella in their locomotory activities, although this method of locomotion is retained as the almost universal mechanism for the locomotion of their spermatozoa. Flagellated choanocyte cells are, however, used by the Porifera to drive water through their chambers and canals.

The deployment of large numbers of cilia over the entire body surface or in distinct ciliated bands or girdles, however, is a common locomotory device, but it is generally restricted to animals less than 10^3 μm in length. This is because the individual cilium has a fixed mode of beating and the only way in which the sum of the locomotory forces can be increased is by an increase in the number of cilia. This is limited, however, by the surface area of the organism, whereas the mass of the organism increases with the volume. Moreover, hydrodynamic efficiency of ciliated organisms decreases with increasing size; consequently larger organisms moving in this way are required to expend much greater energies to maintain the same relative velocities.

The larvae of many invertebrates have their cilia deployed in bands or girdles which either increase in length or number as growth proceeds (Fig. 10.6). Ctenophores have compound cilia and they can achieve speeds of up to 15 mm s^{-1}.

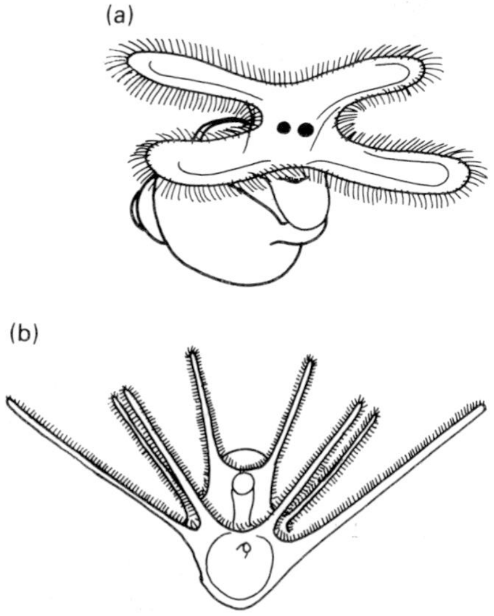

Fig. 10.6. Ciliated larvae with elongated ciliary bands. (a) Veliger larva of a gastropod mollusc; (b) pluteus larva of an ophiuroid echinoderm.

The activity of cilia is co-ordinated in such a way that each cilium normally has the same pattern and frequency of beating but individual cilia are at a different phase of the cycle. A wave of co-ordinated activity is seen to pass from cilium to cilium. This wave of activity may pass in the same direction as the direction of the power stroke or in the opposite direction so that each cilium makes its power stroke slightly after the one immediately in front of it. There will then be no interference between cilia during the execution of the power stroke. Cilia co-ordinated in such a way are said to exhibit metachronal rhythmicity (Fig. 10.7).

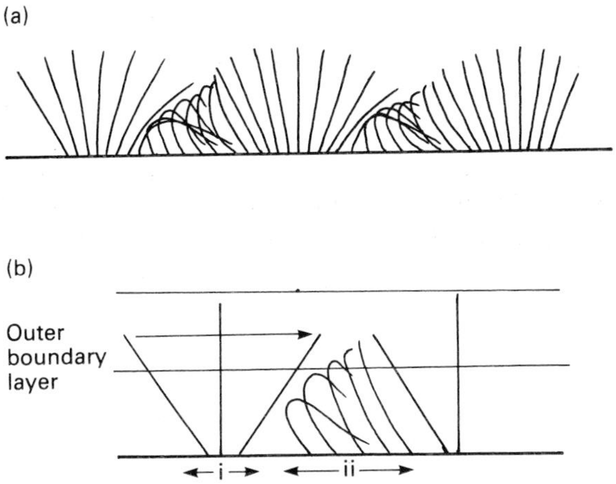

Fig. 10.7. Metachronal ciliary activity. (a) Individual cilia at a different phase of the cycle of activity; (b) the power stroke (i) and recovery stroke (ii) produce a net force in the outer boundary layer.

The transition from ciliary to muscular locomotion is likely to have taken place among the flatworm-like ancestors of the Bilateria, and the two modes of locomotion are found to coexist in some free-living flatworms and nemertines.

The epidermis of free-living flatworms (turbellarians) and nemertines is abundantly ciliated and the smallest specimens, which are about 1 mm long, lie at the upper end of the size range for efficient locomotion using ciliary mechanisms. Larger flatworms (triclads and polyclads), which retain ciliary creeping as the principal means of locomotion, do so by becoming flat; consequently their surface area increases by more than the square of the linear dimensions.

The largest organisms to move by ciliary creeping are probably nemertines, such as *Lineus longissimus*, which can achieve lengths of many metres yet have a maximum width of only *c.* 1–2 mm. This gives the required high surface area, but it is a solution open only to aquatic species. In terrestrial environments the high rates of water loss which would be associated with such a high surface area are not acceptable. The muscular activities of flatworms and nemertines are varied and involve pedal locomotory waves, peristalsis or looping movements with anterior and posterior adhesion. These have often been developed to a much greater degree by other phyla and the principles will be described in the following sections.

10.4 Muscular activity and skeletal systems

Chemical energy can be utilized in muscle cells to cause shortening of the muscle fibres, but such changes in cell shape can only be part of a **locomotory system** if there is a skeletal system to transmit the forces generated.

The skeletons of animals are of two fundamental types: *fluid skeletons*, which are utilized by a wide variety of soft-bodied invertebrates, and *rigid skeletons*, which are employed by the arthropod phyla and some echinoderms. The distinction between animals with rigid skeletons and fluid skeletons is far from being absolute. Many of the non-arthropod phyla have some rigid skeletal parts, the spines of echinoid echinoderms, the 'vertebrae' of ophiuroids and the aciculae in the parapodia of polychaete worms are examples.

Similarly in the arthropods a hydrostatic skeleton is often used for the transmission of forces as in the movements of the gut and in arachnids for the extension of jointed limbs. The rigid skeletons of the arthropods are exoskeletons in which the hardened skeletal elements enclose the soft tissues. In the chordates an elastic skeletal rod, the notochord, has been developed as a simple internal skeleton. From

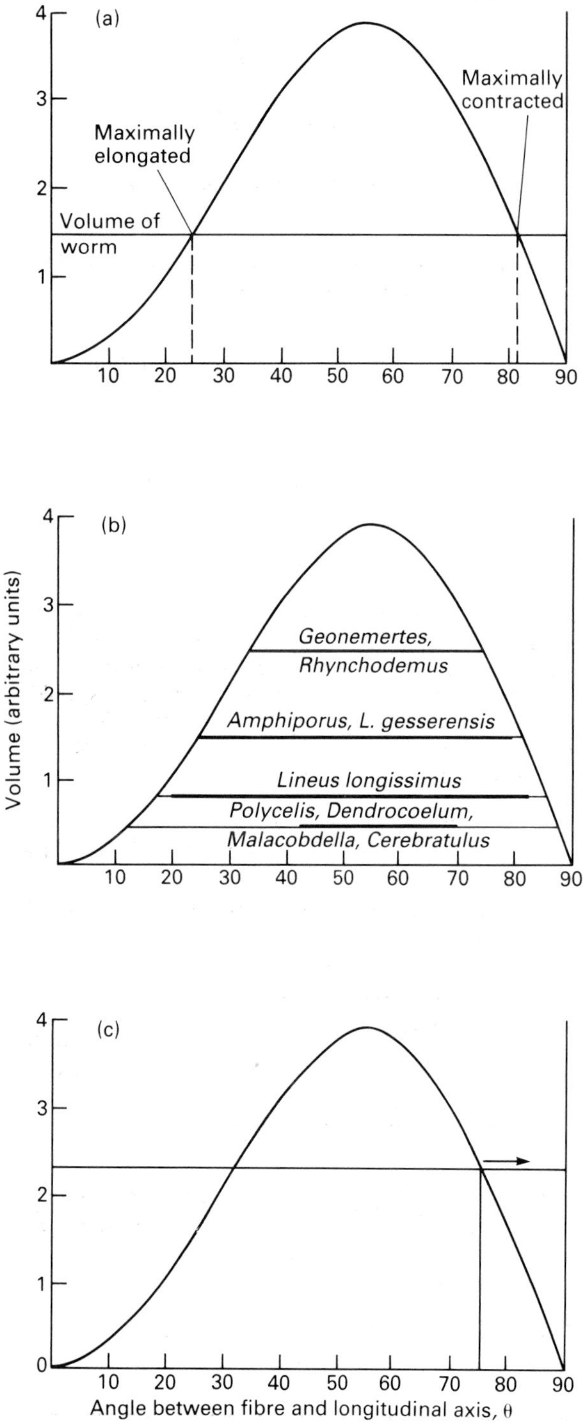

this origin the vertebrates with their characteristic cartilaginous or bony skeletons have evolved.

In all locomotory systems involving muscular activity the muscles can only generate power when they contract, and a force must be applied to restore them to their original length. This is most frequently accomplished by antagonism between different sets of muscle fibres, but the restorative force can be provided by the action of cilia (as in some cnidarians), or by energy stored in elastic tissues. The fluid skeletons of soft-bodied invertebrates can be employed in a wide variety of ways, but the basic principles are always the same. These are explained in Box 10.4.

Many soft-bodied animals are capable of very great changes of shape. Some of the most deformable are the nemertines but there are physical limits to deformability since the volume of the tissues and usually of the body cavities of animals are fixed. In the nemertines there are inelastic fibres in the body wall inclined at an angle to the logitudinal axis. If the fibres were to be stretched out until they were parallel to the longitudinal axis the enclosed volume would be zero, similarly if the fibres had an angle of 90° to the axis the volume would again be zero. Between these two impossible extremes is a position where the enclosed volume is maximal; it occurs when the fibres have an angle of approximately 55° to the longitudinal axis. The relationship between the volume enclosed by such a geodesic fibre system, and the angle between the fibres and the longitudinal axis is shown in Fig. 10.8a–c. The volume of a worm such as a nemertine is usually less than the maximum

Fig. 10.8. The physical properties of an inelastic spiral fibrous system and the changes in shape of worm-like animals. (a) The relationship between enclosed volume and the inclination of an inelastic fibre to the longitudinal axis of a cylinder around which it spirals. (b) The limits of deformability of some nemertine worms which have volumes less than the theoretical maximum volumes of the fibre system. (c) Relationship between fibre angle and volume in Nematoda. Because the animals have a high internal pressure the cross-sectional shape is circular. Contraction of the longitudinal muscles tends to reduce the volume. The consequent increase in pressure acts against the longitudinal muscles. These worms usually move with sinusoidal movements in which contralateral fibres contract alternately. (Redrawn from Clark, 1964.)

Box 10.4. The hydrostatic skeleton

1 The contraction of muscles surrounding a fluid-filled cavity will increase the pressure of the fluid:

$$\textbf{Pressure} = \frac{\textbf{force}}{\textbf{area}}$$

The SI unit of pressure is Newtons per square metre $\frac{N}{m^2}$

2 The **force** acting on any surface is therefore F = pressure×area.

3 Pressure acts equally at all points of the surface and acts at right angles to the surface (see below).

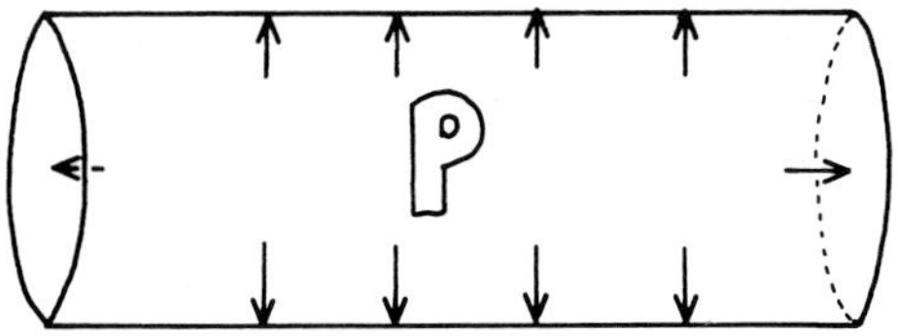

Pressure acts in all directions in a fluid.

4 The internal pressure in the hydraulic system can be resisted by the boundary surface.

If the resistance of the boundary surface is lower in any region than that necessary to contain the pressure, fluid movements will cause a change in shape. This is the basis of locomotion in all animals that use a fluid skeleton.

As shown below, muscles in the right- and left-hand parts of a cylinder oppose each other.

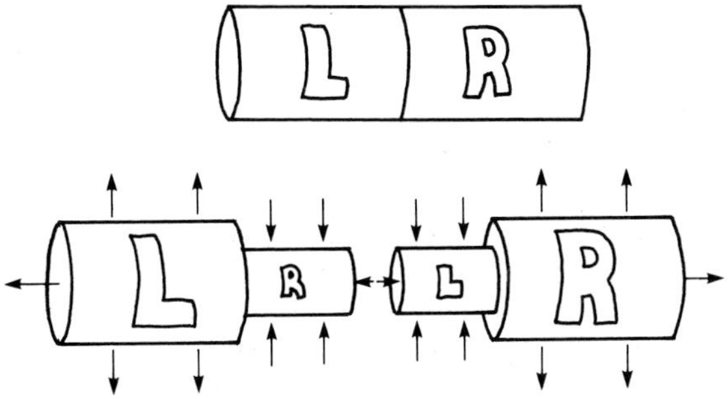

Antagonism between connected compartments in the system, in this case left and right.

5 A very versatile system results if muscles are arranged around the body wall as circular and longitudinal sets.

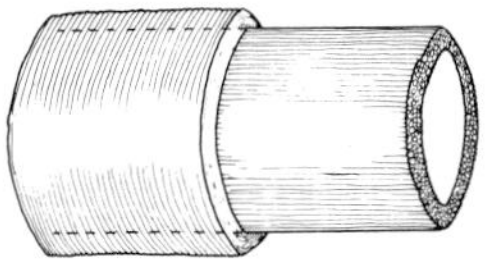

Opposing circular and longitudinal muscles, circular muscles on the outside.

6 Any section of an undivided cylindrical animal can then perform a wide variety of changes in shape.

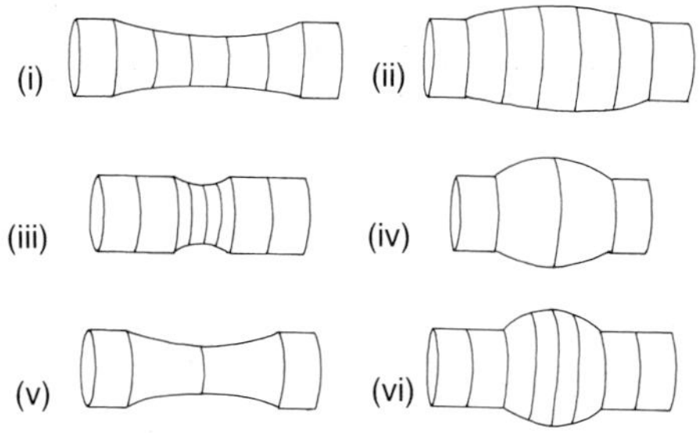

Body regions not of fixed volume. (i) Circular muscles contract, fluid exported; (ii) circular muscles relax, fluid imported; (iii) longitudinal and circular muscles contract, fluid exported; (iv) longitudinal and circular muscles relax, fluid imported; (v) (vi) body regions of fixed volume; (v) longitudinal muscles relax, volume constant; (vi) circular muscles relax, volume constant.

7 Changes in shape can perform mechanical work. The work done will be the product of the force applied (= pressure×area) and the distance moved.

In the example below a proboscis is extended while much of the body contracts.

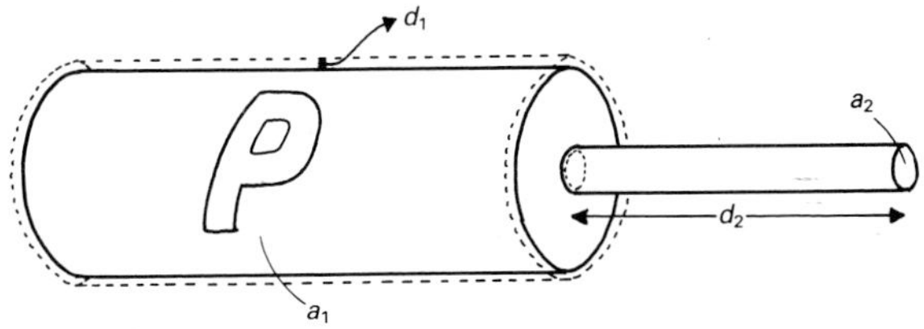

$$\frac{\text{total force of body wall}}{\text{area of contracting } a_1} = \text{pressure} = \frac{\text{force at proboscis end}}{\text{area of proboscis } a_2}$$

$$\frac{F'}{a_1} = P = \frac{F''}{a_2}$$

This is a hydrostatic lever, see also Box 10.5.

The work done J = $P \times a_2 \times d_2$.

volume of the fibre system as shown in Fig. 10.8a. When maximally contracted or elongated the worm will have a circular cross-section and its volume will be equal to the volume of the fibre system, at other lengths the cross-section of the animal will be elliptical. The performance of a number of nemertines in relationship to these theoretical limits is summarized in Fig. 10.8b. The mechanics of nematode worms is also determined by the properties of a fibre system in the body wall. Unlike the nemertines, the nematodes are relatively undeformable worms; they have a complex arrangement of fibres in the body wall (see Fig. 4.13) inclined at angles greater than 55° to the longitudinal axis. These worms also have high-pressure body fluids.

Contraction of the longitudinal muscles will tend to cause the angle of the fibres to increase and so tend to cause the enclosed volume to decrease (Fig. 10.8c). This in turn will cause an increase in internal pressure, and this increase will act against the longitudinal muscles and return them to their original length. In nematodes, therefore, unlike the model represented in Box 10.4, 5, there are no circular muscles.

Soft-bodied animals with oblique muscle fibres such as the leeches are able to set the fibre angle at 55° and to become in effect rigid. Muscles acting on rigid or jointed skeletons are also able to transfer forces from the site of generation to the point where they are applied to the environment. A rigid skeletal element can act as a mechanical lever, the principles of which are explained in Box 10.5. Note, however,

Box 10.5 Rigid skeletons—the principle of a lever

1 A force F can be made to move a load M when it causes rotation about a fulcrum. If F is the force applied and F_R is the force of reaction, they are related by the relationships

$$F \times d_1 = F_R \times d_2$$

where d_1 is the distance moved by F and d_2 is the distance moved by F_R.
Note that the pivot reverses the direction of the work done.

2 A small force can move a large load but energy will be conserved such that:

$$\frac{F_R}{F} = \textbf{mechanical advantage} = \frac{d_1}{d_2} = \textbf{velocity ratio}$$

3 A large effort can also be made to move a small load over a large distance.

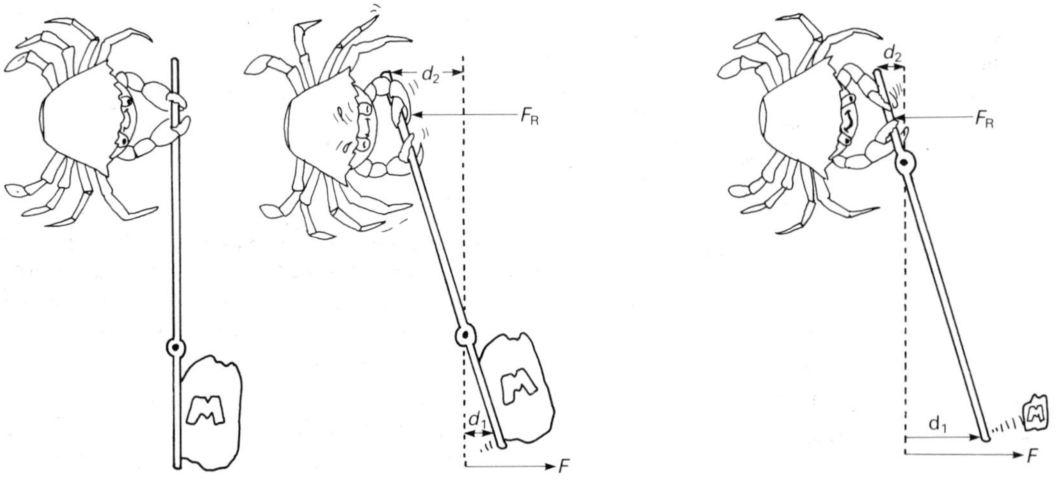

that the example illustrated in Box 10.4, 7, is also a lever system which depends not on relative length but on relative area.

10.5 The locomotion of soft-bodied animals

10.5.1 Locomotion over a firm surface

Many soft-bodied invertebrates are able to move over a firm substratum. A firm substratum may be defined as one where the surface close to the body of an animal does not move relative to parts of the substratum at some distance from the body when a force is applied to it by the animal. In order to do so, a force must be transmitted to the substratum through a fixed point, often referred to as the French term 'point d'appui'.

Many systems of locomotion involve the propagation of waves of contraction and relaxation in muscles with their longitudinal axis parallel to the direction of locomotion. Locomotion can be either in the same direction as the propagation of the wave or in the opposite direction. This is illustrated in Box 10.6. Flatworms, some cnidarians, and above all the gastropod molluscs move by means of waves of activity in the muscular surface applied to the substratum (pedal locomotory waves).

Pedal locomotory waves can be seen quite easily by examining the under-surface of a planarian or a snail while it crawls along a glass plate inclined at a shallow angle. In species of the land snail *Helix* several waves crossing the whole of the foot will be seen simultaneously, each moving in the same direction as the locomotion of the snail, but at a greater rate. In other species, for example, marine limpets such as *Patella*, a relatively small number of waves passing in the opposite direction to the movement of the animal can be seen. Moreover, in the limpets there is usually a half wavelength difference between the right and left sides of the foot. The waves are therefore of a number of different types, described as: *direct, retrograde, monotaxic* or *ditaxic*.

The different types of activity have different characteristics. The monotaxic, direct locomotory waves of *Helix*, in which the wavelength is much less than the length of the foot, can be considered to be a low-geared system, giving relatively low speed and poor manoeuvrability but the ability to move a relatively large mass or overcome large resistive forces. Ditaxic sytems with a wavelength as long as or longer than the foot give higher relative speeds and much greater manoeuvrability.

The molluscan foot is an organ of considerable complexity. Its principal skeletal element is a complex system of blood sinuses. Although the animals are acoelomate, there is a hydrostatic skeletal system within the foot. The principal muscle systems are not longitudinal muscle fibres, but opposing systems of anterior and posterior oblique muscle fibres (Fig. 10.9). It used to be thought that each

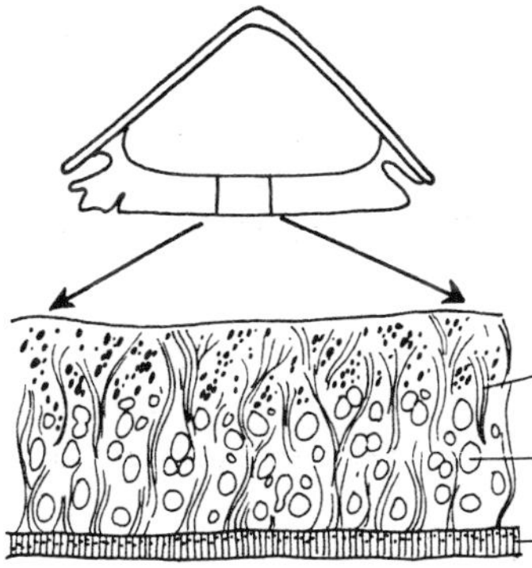

Fig. 10.9. The structure of the molluscan foot.

segment of the foot actually became detached (as shown in Box 10.6 a and b) during the passage of a locomotory wave, but this is unlikely due to the huge forces which would be necessary to separate two closely adhering wet surfaces. The formation of 'points d'appui' and zones of movement depend on the properties of the mucous layer beneath the foot. The mucus acts as an elastic solid under conditions of low lateral stress but as a viscous liquid under higher lateral stress. At the 'points d'appui' it acts

Box 10.6

The illustrations show the body of an animal in which five regions are identified (they may or may not be true segments) and which are attached to the ground when extended in a, when contracted in b. The waves of muscular activity can be in a muscular foot as shown in a(i) and b(i) or in a cylindrical worm as shown in a(ii) and b(ii).

The result is always the same. If the 'points d'appui' are formed in regions where the longitudinal dimensions of the body are maximal, the animal will move in the same direction as the wave—the wave is direct. If, on the other hand, the 'points d'appui' are in regions of longitudinal muscle contraction so that the fixed body regions are at their shortest, the animal will move in the opposite direction to the wave, the wave is said to be retrograde.

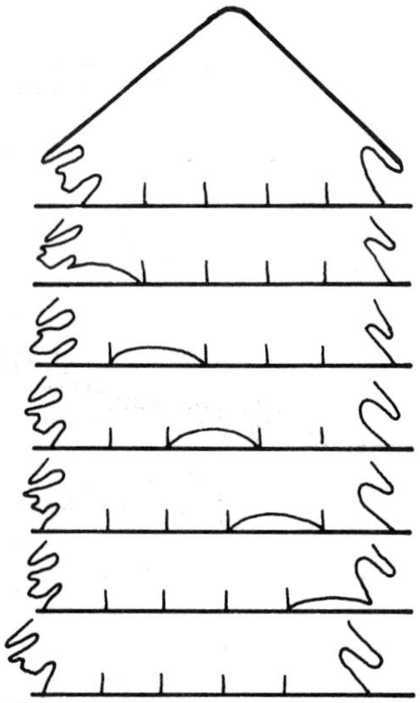

a(i) Points d'appui: short segments wave retrograde

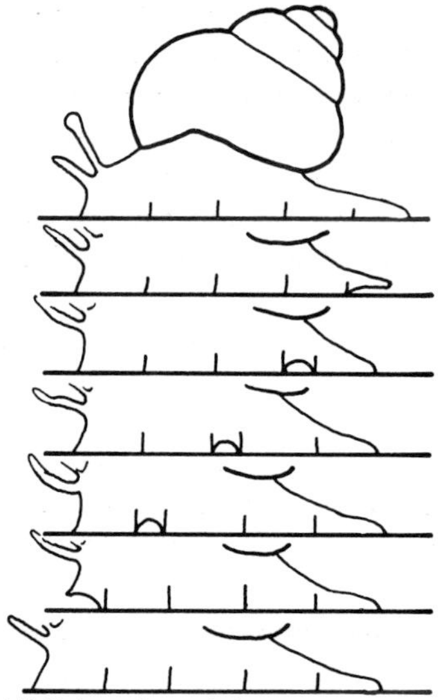

b(i) Points d'appui: long segments wave direct

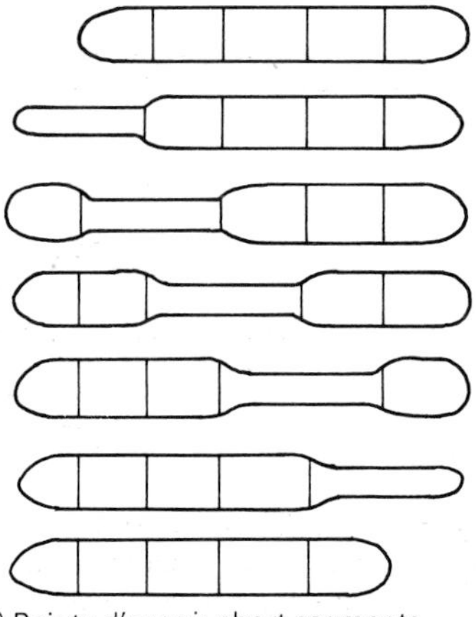

a(ii) Points d'appui: short segments wave retrograde (c.f. earthworms)

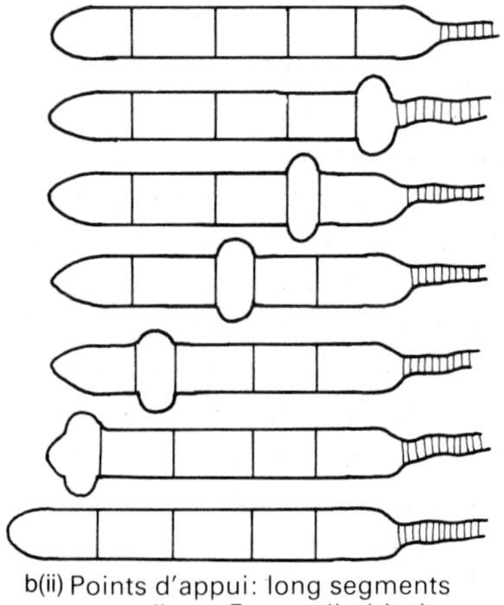

b(ii) Points d'appui: long segments wave direct. For a cylindrical worm this will require the erection of chaetae, but see also nereidiform locomotion as an elastic solid but is fluid under the wavefront where the lateral forces are greater.

Some snails, including *Helix*, can exhibit an alternative 'galloping' movement, in which retrograde waves of muscular activity, with a wavelength about equal to the length of the foot, are generated. This type of locomotion can further be modified to the stage where there is attachment to the substratum alternately at anterior and posterior extremities—a type of locomotion which is more characteristic of the looping movements of leeches described below.

Many large flatworms and most nemertine worms exhibit a muscular component to their locomotion in which alternating waves of contraction of the circular and longitudinal muscles generate retrograde peristaltic waves which enhance the locomotory abilities derived from the activity of the surface cilia. This system is most highly developed in the septate coelomate worms and is particularly characteristic of the earthworms.

Figure 10.10a illustrates the movement of an earthworm. Note that segments are stationary when the circular muscles are relaxed and longitudinal muscles contracted (see Box 10.6). The segments in front of the shortest segment are elongating and exert a backthrust against the ground through the 'points d'appui' while the posterior segments are contracting and exert tension through this point (Fig. 10.10b). There will naturally be a tendency for slip to occur, but this is prevented by the erection of stout chaetae.

The pressure waves associated with the contractions of circular and longitudinal muscles are separated (Fig. 10.11), and the inter-segmental septa effectively isolate pressure changes in individual segments. The highest pressures are recorded when the circular muscles contract and the segments would be penetrating the substratum. During the anchoring phase the pressure is lower. Isolation of pressures in this way is not possible in aseptate animals. This is thought to be one of the major advantages of the septate condition responsible for the evolution of metamerism in annelid worms. Each segment is a separate hydrostatic element with its own fixed volume (see Box 10.5). Although we have considered earthworms in the context of movement over a flat surface, there is little doubt that they are primarily adapted for movement between the interstices of the soil.

A dramatically different means of locomotion based on that of the earthworms has been developed by the leeches. These animals have posterior and anterior suckers which provide particularly effective 'points d'appui'. The whole body acts as a single hydrostatic system, the gut providing a single fluid-

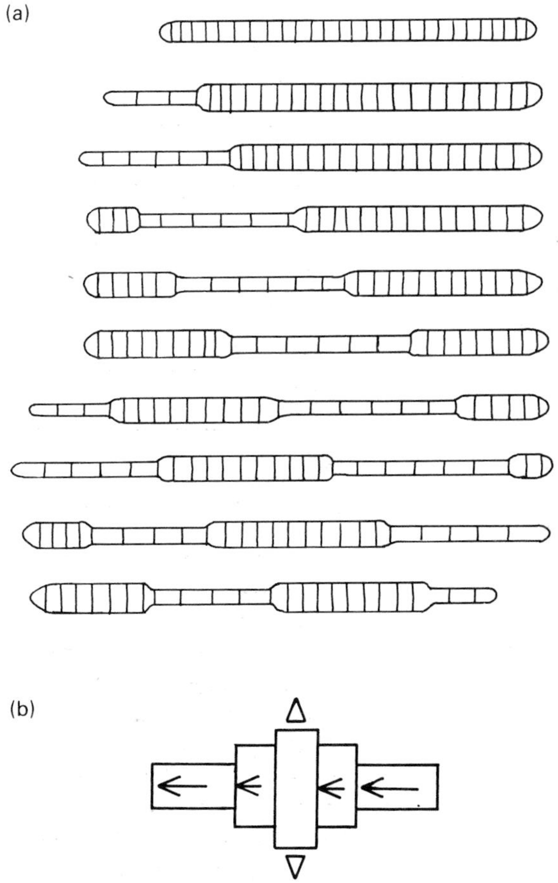

Fig. 10.10. (a) Successive stages in the movement of an earthworm. Segment volume is fixed as in Box 10.6, Fig. a(ii). The segments are stationary with respect to the ground when the longitudinal muscles are contracted, and circular muscles relaxed; segment volume is fixed and the locomotory wave is retrograde. (b) Forces acting on the fixed segments with erected chaetae.

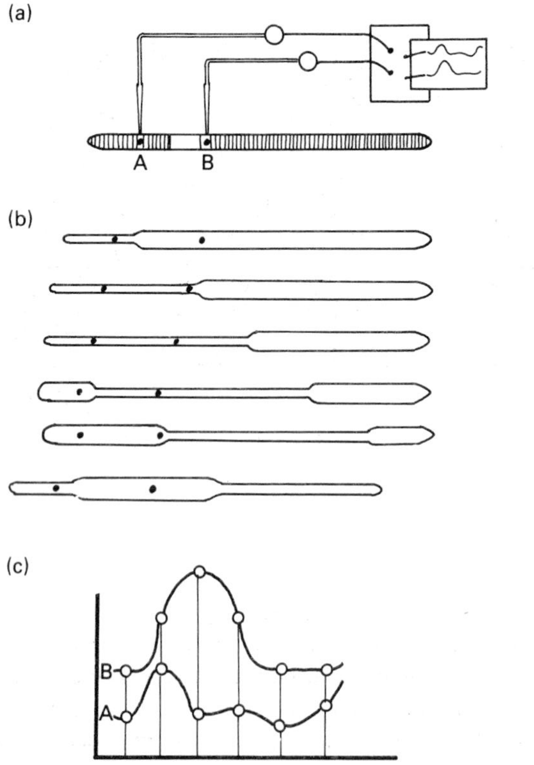

Fig. 10.11. Studies of locomotion in the earthworm. (a) System in which pressures in individual segments A and B can be recorded by the insertion of plastic tubes directly linked to a pressure transducer. (b) Successive position of the two segments A and B during recording. (c) The recorded pressure waves. (After Seymour, 1969.)

filled cavity, and attachment occurs alternatively by anterior and posterior suckers (Fig. 10.12a).

Similar movements are also exhibited by some insect larvae, e.g. lepidopteran caterpillars, in which arching movements are equivalent to the contraction of longitudinal muscles (Fig. 10.12b). In leeches the body cavity (coelom) is virtually obliterated by a deformable tissue, the botryoidal tissue, and the animals are effectively aseptate and acoelomate from a design point of view, although they are undoubtedly annelid worms which betray their coelomate and segmented origins in the details of their anatomy, especially in the nervous system.

The errant polychaete worms exhibit a mode of locomotion which is very different from both that of the oligochaetes and leeches. It involves the movement of multiple limbs, the tips of which are made to move backwards relative to the body but which, because the tips are attached to the ground, causes the body to move forwards (see Box 10.5).

When a *Nereis*, for example, is crawling slowly (Fig. 10.13a) there is a metachronal wave of activity in the parapodia passing forwards from the tail to the head and with the left and right parapodia being exactly one half wavelength out of phase. This direct wave ensures that each limb executes its power stroke without risk of interference with the limb immediately posterior, which will have executed its power stroke just beforehand. As a *Nereis* crawls more

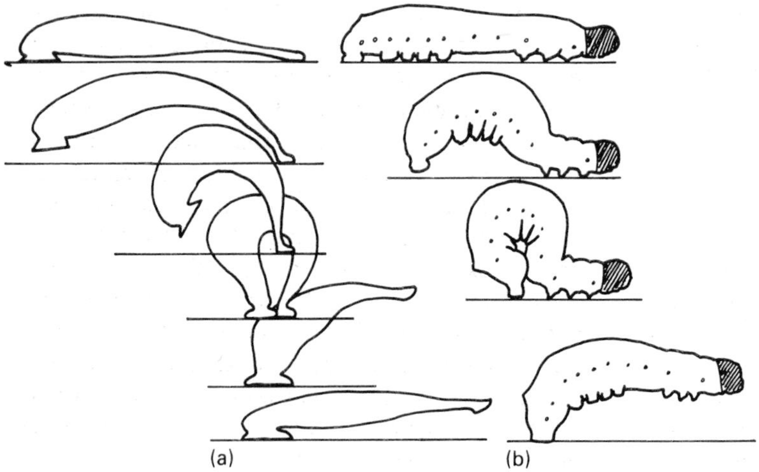

Fig. 10.12. Looping movements in which 'points d'appui' are alternately at anterior and posterior ends. (a) A leech; (b) a caterpillar (lepidopteran larva).

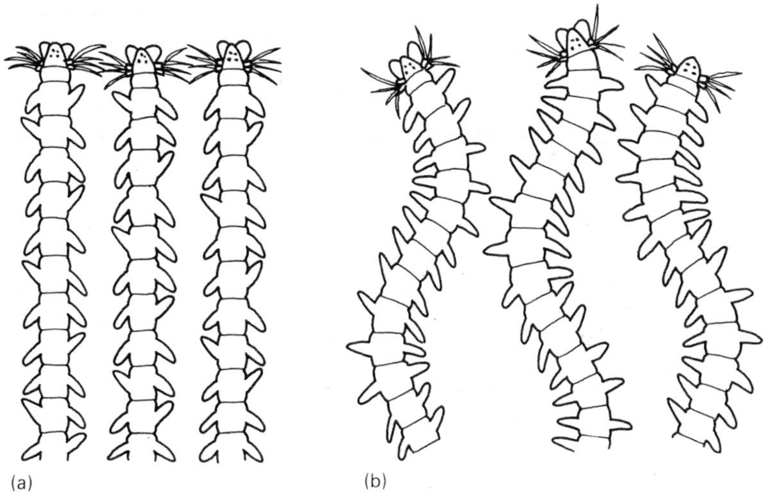

Fig. 10.13. Diagrammatic representation of locomotion in the polychaete *Nereis*. (a) Slow crawling; (b) rapid crawling. For reasons of simplification the movements of the chaetae are not shown. Chaetae can be protruded and retracted to increase step length.

quickly or swims, the same ditactic direct locomotory wave is exhibited but the longitudinal muscles of the body wall are also used to generate propulsive force (Fig. 10.12b). A synchronous wave of activity in the left and right longitudinal muscle is initiated so that each parapodium executes its power stroke as it is passed by the crest of the sinusoidal wave in the body wall which moves forwards synchronously with the wave of parapodial activity.

The forces transmitted to the ground are now not only those generated by the parapodial musculature, but also ones generated by the contraction of the contralateral longitudinal muscles. This locomotory pattern follows the general rule that direct locomotory waves should be observed in circumstances where 'points d'appui' are formed in regions of maximum length of the longitudinal muscles.

In many errant polychaetes the role of the longitudinal muscles in crawling is slight, most of the tractive power being developed by the parapodial muscles, which cross segmental boundaries. This tendency is especially marked in the scale worms and their relatives, and in these conditions the transverse septa are reduced or lost. The locomotion of these animals involves intrinsic and extrinsic muscles of the parapodium. The chaetae can be protracted and retracted by intrinsic muscles in the chaetal sac, and both intrinsic and extrinsic muscles operate the parapodium which can be lifted and depressed, and moved forwards and backwards causing stepping movements. The movement of some polychaetes therefore resembles the complex walking of myriapods.

The water vascular system of the echinoderms provides a unique ambulatory system. In starfish, for example, there are typically five arms and in each one there is a water vascular canal radiating from the circum-oral ring. Along each radial water vascular canal is a series of reservoir-like ampullae and tube feet (Fig. 10.14a,b). The ampullae and tube feet have muscular components which work against each other. Contraction of the muscles compressing the ampullae drives water into the tube feet whereas contraction of the tube feet must involve a mass flow of water into the ampullae. Each ampulla does not act as a simple reservoir for its own tube foot, as there is mass movement of fluid between regions of the system, and a tube foot can be extended to have a volume greater than that of its ampulla when fully dilated. In general, however, muscles which compress the ampullae act antagonistically to those which shorten the tube feet.

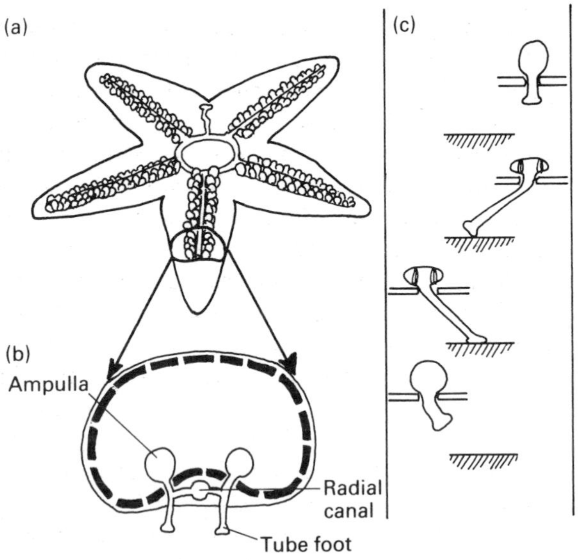

Fig. 10.14. The tube foot and ampullae in the water vascular system of an echinoderm: (a) the general arrangement; (b) a diagrammatic cross-section of an arm to show the water vascular system; (c) stepping cycle of a single tube foot (retractor muscles in the tube foot are not shown).

The tube feet are extended by hydraulic pressure, can perform simple step-like motions and may be provided with adhesive suckers at their tips (Fig. 10.14c).

A peculiar feature of starfish locomotion is that the tube feet do not exhibit any detectable metachronal rhythmicity.

10.5.2 Burrowing and movements within tubes

Burrowing worms tend to have large body cavities and to be at least partially aseptate. Animals with this structure can exhibit a wide variety of movements within the burrow (Fig. 10.15); four types of peristaltic locomotion are possible because any region of the body can simultaneously increase in length and girth. These movements may be used for irrigation, locomotion, or irrigation and locomotion together.

Unsegmented, and segmented but aseptate, animals with large body cavities can also perform major changes in shape which can be used when burrowing into the substratum; and the sipunculans,

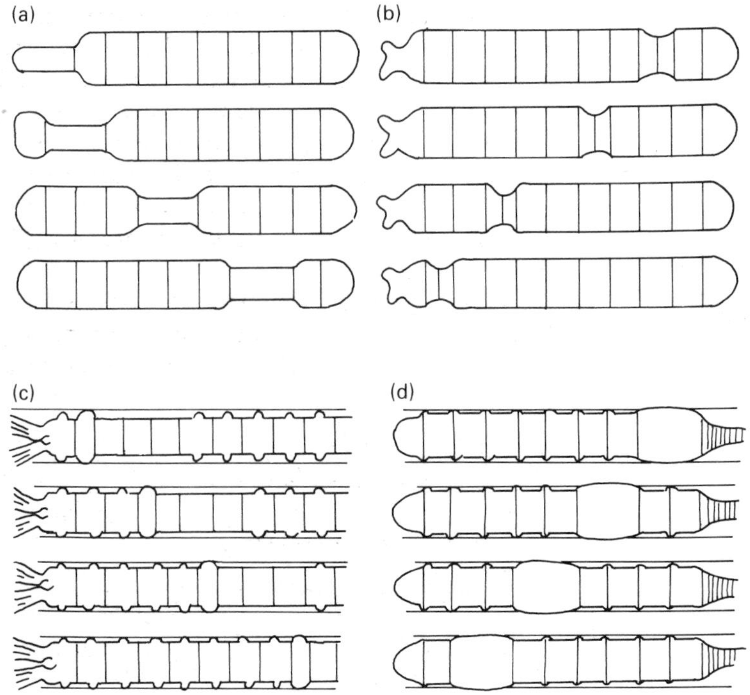

Fig. 10.15. Four modes of peristalsis which may be exhibited by tubiculous worms. In (a), (b) and (d) segment volume is not fixed. In (c), which represents the fanworm *Sabella*, segment volume is fixed due to complete septa. Locomotion during peristalsis is prevented by the formation of 'points d'appui' in segments of intermediate length.

for example, are capable of rapid re-entry into the substratum. The associated movements are illustrated in Fig. 10.16, together with a trace of the pressures generated internally. These animals have the capacity to exhibit a high rate of work, but such activity would not normally be maintained for long periods of time. During phases of high coelomic pressure all the muscles of the body must do metabolic work to maintain constant length. Mechanical work is performed by the controlled relaxation of specific muscles allowing certain regions of the body to be extended (see Box 10.4). The rate of work can be high, but efficiency is low due to the requirements for continuous muscle tone at all times of high coelomic pressure. It could be argued that the evolution of transverse septa in the annelids occurred at least partly as a means of resisting the outward movement of the body wall without muscular activity, and which thus increased the mechanical efficiency (cf. Section 10.1).

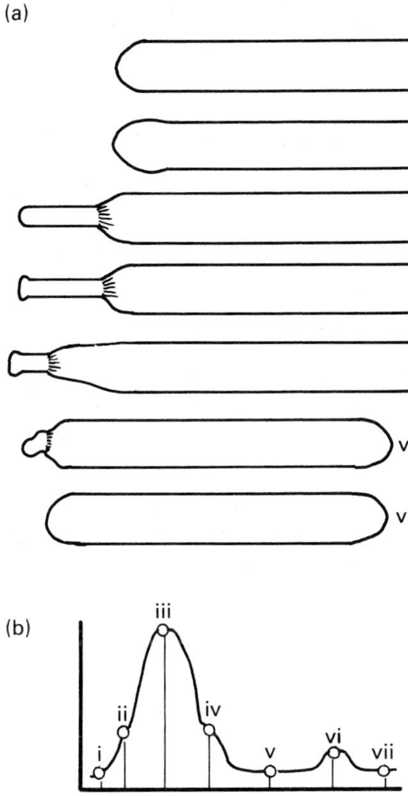

Fig. 10.16. (a) Diagrammatic representation of successive movements of a sipunculan during burrowing into or through the substratum. (b) Internal pressures recorded during the burrowing cycle. Note that highest pressures are recorded during penetration into the substratum (as in earthworms, cf. Fig. 10.11) but all regions of the body have the same pressure. (After Trueman & Foster Smith, 1976.)

The polychaete worm *Arenicola marina* is a segmented worm which is perhaps secondarily adapted to a burrowing existence. Its segmented structure confers a number of advantages. Nervous co-ordination is enhanced by the segmented ganglia; there are chaetae with parapodia for formation of 'points d'appui' and a well-developed vascular system with segmental gills. Mechanically *Arenicola* has an undivided trunk coelom which gives it the wide repertoire of locomotory movements needed and it is capable of high rates of working. This gives it the ability to re-enter the substratum by the alternate formation of penetration and terminal anchor points (Fig. 10.17), which are also to be found in burrowing cnidarians and in the later stages of burrowing by bivalves (Fig. 10.18).

Arenicola normally lives in an open J-shaped tube which is part of a U-shaped burrow system (Fig. 9.32). The animal excavates and eats sand at the base

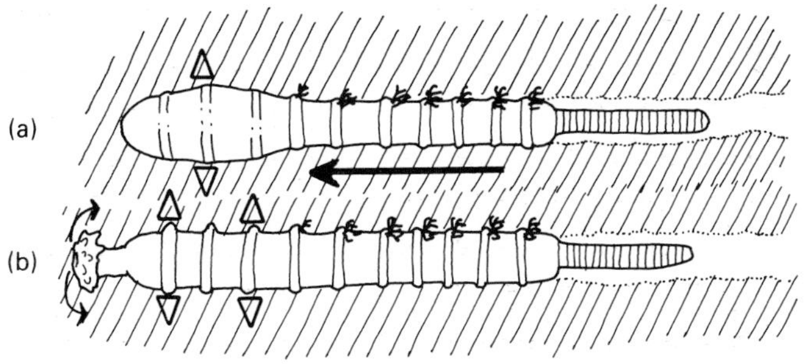

Fig. 10.17. Different stages in the burrowing of the lugworm. (a) Terminal anchor. The anterior part of the worm is dilated while the posterior is drawn forwards. (b) Penetration anchor. The anterior segments form 'flanged' anchor points while the proboscis excavates the sand and the head is thrust forwards. (After Trueman, 1975.)

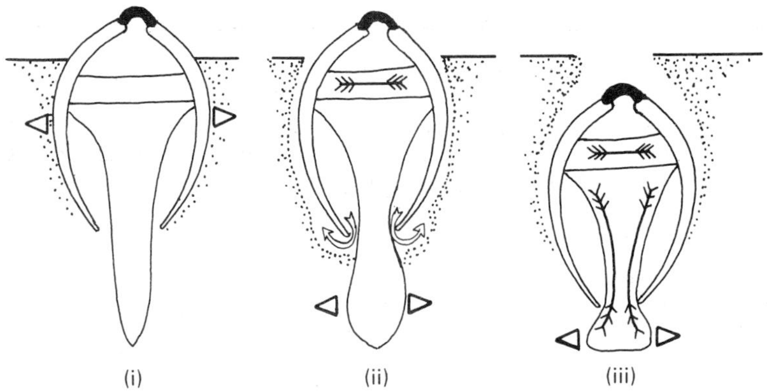

Fig. 10.18. Burrowing of a bivalve mollusc. (i) The shell gapes and forms an anchor during penetration of the foot. (ii), (iii) The tip of the foot is dilated while the shell is drawn down into the sand. (After Trueman, 1975.)

of the head shaft using scraping and ingesting movements by the proboscis. The rate of working is low, and for this activity the anterior coelom is isolated by the pharyngeal septa. These have valves and can either isolate the anterior coelom for sustained activity or, when open, allow the high pressures generated in the trunk coelom to do work via the extruded proboscis when burrowing more actively.

Tubiculous worms which retain the intersegmental septa, such as the fanworm *Sabella*, have a special problem. It is necessary for them to irrigate the tube by peristalsis. With fixed segment volume they can only do so by employing a peristaltic wave of type (c) in Fig. 10.15; consequently they tend to move in a direction opposite to the waves. This is prevented in *Sabella* by the formation of 'points of d'appui' when the segments have intermediate length through the erection of chaetae. Other segments slip inside the very smooth lining of the animal's tube.

10.5.3 Swimming

The smallest soft-bodied invertebrates swim using the locomotive power of bands of cilia as described in Section 10.3. Apart from the Ctenophora, however, which have ciliated comb-rows, larger invertebrates use muscular power in swimming. In order to do so, the forces generated by the muscles must be transmitted to the medium (usually water) through which the animal is moving. For larger animals the Reynold's numbers (see Section 10.1) will be greater than unity and the inertia of the body cannot be ignored. A large animal whose $Re > 1$ will, when swimming, set in motion a mass of water which will normally move in a series of vortices. Swimming of this kind is particularly characteristic of vertebrate animals, but there are many soft-bodied invertebrates which are not neutrally buoyant in water but are able to generate lift and propel themselves forward. The principal mechanisms are: (a) the backward propagation of waves in smooth-bodied animals; (b) paddling and rowing; and (c) jet propulsion.

Smaller animals for which Re is less than 1 cannot exploit vorticity. They may still be able to swim, but the forces that result in locomotion are due to differential drag.

Much of the mechanical analysis of undulatory swimming is based on a study of the movements of the common eel, but comparable locomotory waves can be generated by smooth-bodied worm-like invertebrates; a good example is the leech *Hirudo*. When it swims, *Hirudo* is flattened dorsoventrally and retrograde waves are generated. The wave passes backwards at a velocity (U) relative to the ground which is greater than the forward velocity (V) of the body. The principles of this type of swimming are outlined in Box 10.7.

The swimming of the rough-bodied polychaete worms is rather different. The wave passes in the same direction as the locomotion. This is partly due

Box 10.7 The swimming of smooth-bodied worms

1 The swimming of smooth-bodied animals can be analysed by considering the motions of a single length of the body.

2 Such a unit describes a figure-of-eight motion in which it has an angle θ to the transverse axis of motion as it crosses the axis of motion. In this analysis only the forces acting at the mid-line are considered although this is a simplification.

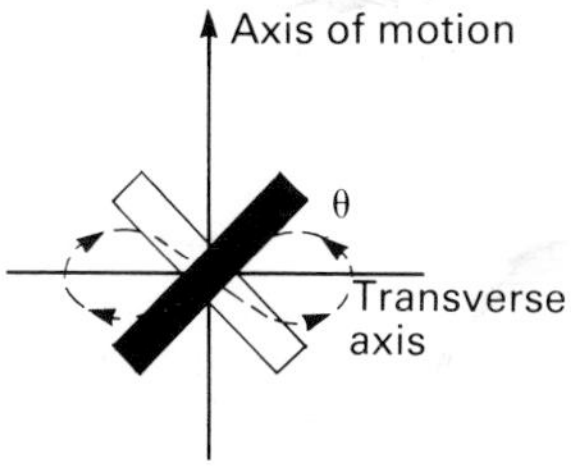

An understanding of the forces involved requires a knowledge of the resolution of forces (see Box 10.3)

3 If the body were at rest the motion of body as it crosses the longitudinal axis of motion would exert a force and set in motion a mass of water (see Box 10.2—kinetic energy and friction).

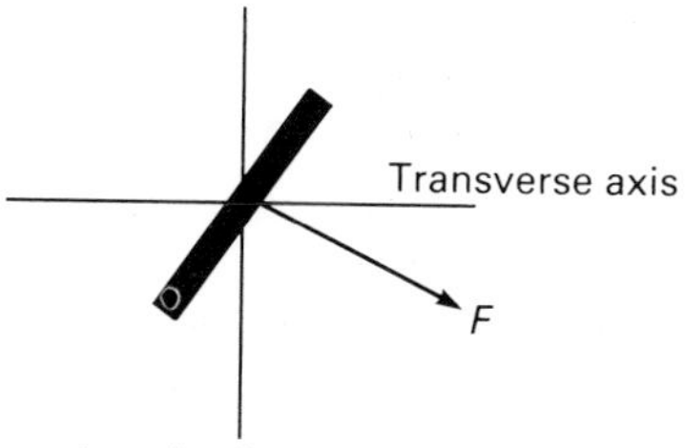

One component of the force F, F'', will cause water to move along the surface of the body, but the Component F' acts at right angles to the body and causes water to be set in motion.

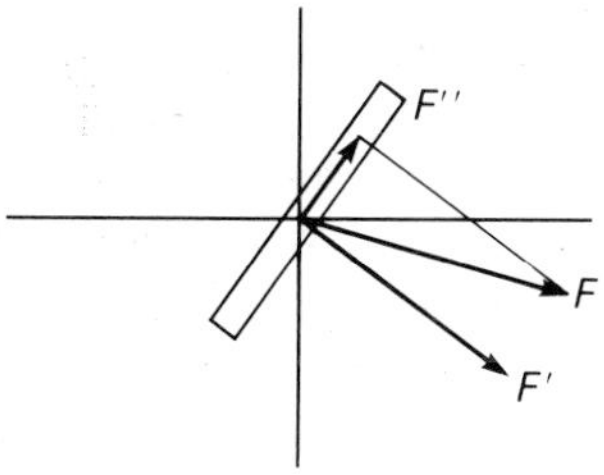

The reaction Force RF' equal and opposite to F' can be resolved into components parallel to and normal to the axis of motion of the animal.

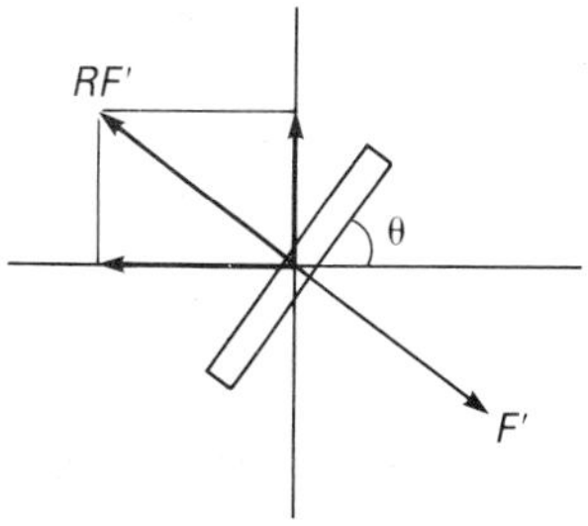

During the passage of a complete wave the transverse forces cancel each other out but the components acting along the axis of motions are summed in effect.

As the animal gathers speed the motion of the length of body is such that the angle between the motion of the body and the axis of movement is reduced. This angle is called the angle of attack (α).

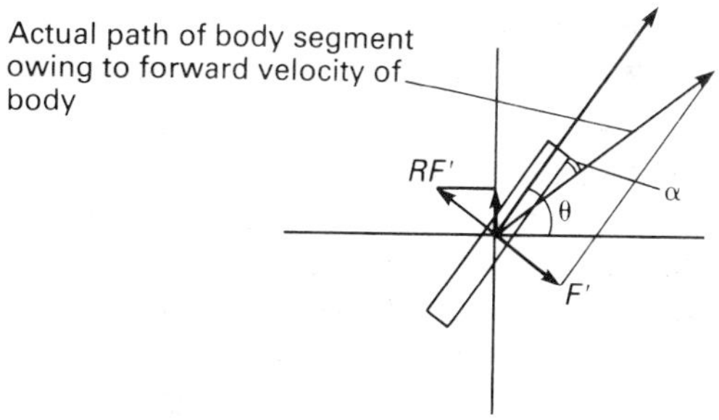

This effectively reduces the Reaction Force acting along the axis of motion. The animal will accelerate until this force is equal to the drag.

to the complex fluid dynamics that result from the oscillation of a flanged body, but is also due to the actions of the parapodia which execute a power stroke as they are passed by the crest of the wave and thus exert a backward force on the water (see Fig. 10.13b above).

Many large vertebrates exhibit a thrust and glide mode of swimming in which kinetic energy is stored in the body following intermittent power strokes. The cephalopod molluscs are also able to achieve this by jet propulsion. Water in the mantle cavity is forcibly extruded by contractions of the muscles lining the mantle cavity. Since this water has mass and is accelerated backwards there must be reaction force which drives the animal forwards. The principal is further explained in Box 10.8.

10.6 The locomotion of invertebrates with jointed limbs

The Crustacea, Chelicerata and Uniramia are three diverse and successful groups of invertebrates. They are all characterized by their hardened exoskeletons; this enables them to have jointed limbs, and the success of the three groups must in part be attributed to the locomotory abilities which this confers. In addition, pterygote Hexapoda (insects) are the only invertebrates to have evolved wings; this gives them the power of true flight and is one of the factors that has allowed their dominance of terrestrial environments. Despite their polyphyletic origins, the walking limbs of the most highly evolved members of the Crustacea, Chelicerata and Uniramia are remarkably uniform in structure as a result of convergent evolution. The limbs are composed of a series of jointed elements becoming progressively less massive towards the tip (Fig. 10.19a). Each joint is articulated to allow movement in only one plane. These limb joints allow extension and flexion of the limb; rotation of the limb plane at the basal joint with the body is also possible and this is often responsible for forward movement. The body is typically carried slung between the laterally projecting limbs (Fig 10.19b) and walking movements do not involve any raising or lowering of the centre of gravity.

Box 10.8 Jet propulsion

1 In jet propulsion the force against the water and the reaction force act along the line of movement.

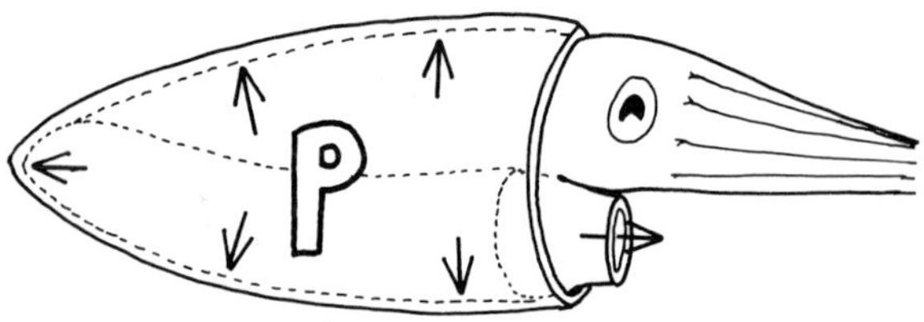

2 In cephalopods the pressure in the mantle cavity is raised by the contraction of muscles surrounding it. There is therefore a force acting at right angles to the body wall in all places.

Force = pressure × area.

3 At the opening of the siphon the same force results in the acceleration of a mass of water which moves out of the mantle cavity.

Work done = $\frac{1}{2} mV^2$ (= kinetic energy)

Because the mass of water expelled is smaller than the mass of the body of the animal the jet is expelled at higher speed than the animal is propelled in the opposite direction Jet engines with this characteristic have low efficiency.

4 Part of the work done by the muscles in contracting the mantle cavity is used to distort the elastic body wall but there are also antagonistic muscles which restore the mantle cavity to its original volume.

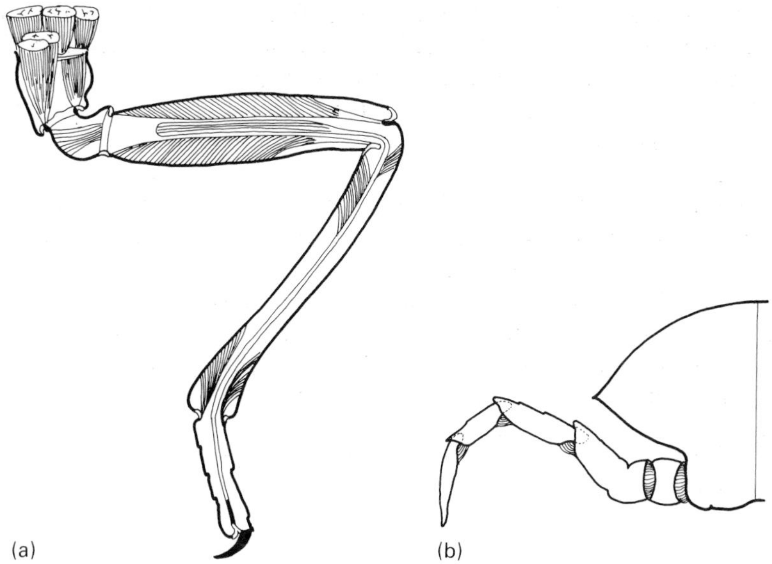

Fig. 10.19. (a) A typical arthropod limb. Note the bulk of muscles in basal sections, tendons and joints. (b) The characteristic attitude of the arthropod limb. Protraction involves rotation of the proximate joint in a horizontal plane and extension of the more distal joints.

As the base of the limb is rotated, the tip of the foot can be made to trace a linear path parallel to the axis of motion by extension and flexion of the limb joints. The limb acts as a mechanical lever (see Box 10.5) and in long-legged arthropods the velocity ratio is such that a large force can be made to move the relatively small load through a large distance, giving a rapid walking gait.

The bulkiest of the muscles providing the locomotory power are located not within the limb but in the body, and since the animals do not bob up and down during forward locomotion, only the limb undergoes major changes in momentum. This is kept to a minimum by the common structural design, in which the mass of the limb is reduced from base to tip. Lateral undulations of the body such as those exhibited by polychaete worms (see Fig. 10.13) and some centipedes (see below) do involve changes in momentum, and suppression of this is likely to have been a powerful influence in arthropod evolution.

Although the walking limbs of insects and advanced crustaceans, such as crabs, are similar, the evolutionary history of the limbs are quite different. The limbs of the crustaceans are derived from an ancestral type which served many different functions including feeding and respiration, in addition to locomotion.

10.6.1 Locomotion in crustaceans

The crustacean limb is based on a plan which has several component parts: two unpaired basal segments, a foliaceous epipod, and paired distal parts—the exopod and the endopod (see Fig. 8.34b–d). In the living phyllopod Crustacea, such as the Cephalocarida and the Branchiopoda (fairy shrimps)

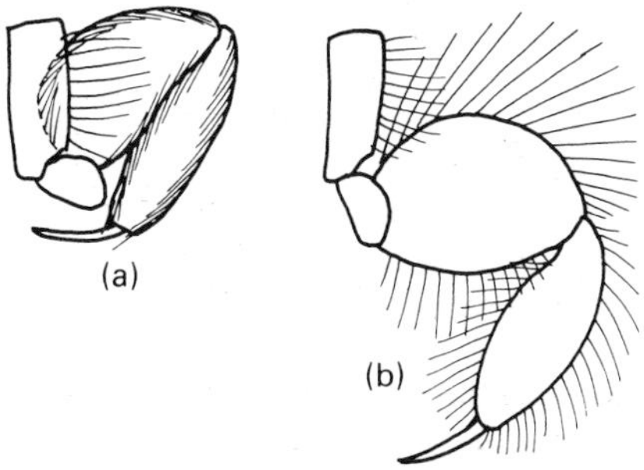

Fig. 10.20. The role of setae in the generation of differential drag forces in the swimming of Crustacea with relatively low Re numbers. (a) Attitude during recovery stroke; (b) attitude during effector stroke. (After Hesseid & Fowtner, 1981.)

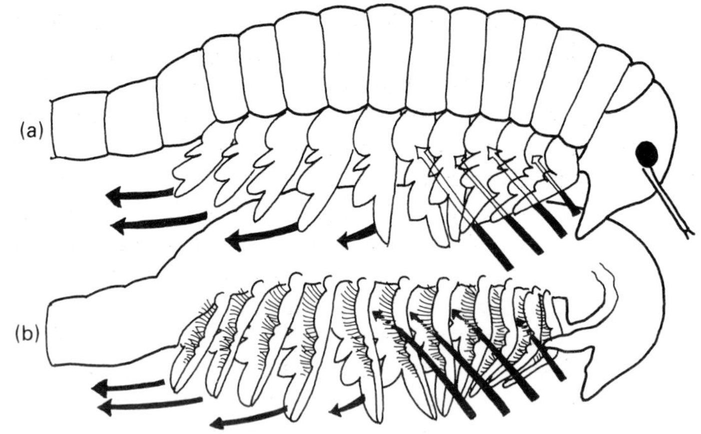

Fig. 10.21. The swimming/feeding currents produced by the limbs of phyllopod Crustacea. (a) Outer view of the right-hand side; (b) inner view of the left-hand side, showing the bristles which filter the food particles which are then passed forwards to the mouth. Water is drawn into the mid-ventral line and then forced backwards when it passes through the filter on the endopod or inner limb part.

(see Figs 8.37–8.39), there are similar limbs on each non-cephalized thoracic segment, which serve the functions of both locomotion and feeding.

The crustacean limb can act as a paddle in which the surface area presented to the water is at a maximum during the power stroke, but is much reduced due to folding and the movements of the setae during the recovery stroke (Fig. 10.20). There is differential drag between effector and recovery strokes. The swimming of the primitive crustaceans however is also due to the expansion and contraction of the spaces formed between the limbs. Water is drawn into the mid-line between the limbs (Fig. 10.21) and passes into the inter-limb space on each side and exits laterally. The unidirectional flow is due to the valve-like action of the outer exopod. Food particles are trapped by setae on the endopod and are passed from seta to seta along a mid-ventral food groove forwards towards the mouth. The metachronal rhythm of the limb movement is therefore responsible for both locomotion and feeding and probably also for respiration.

A special feature of crustacean locomotion is the relative ease with which the transition between walking and swimming may be made. Many of the shrimp-like malacostracans have thoracic walking limbs and abdominal swimming limbs (pleopods) (Fig. 10.22). There are many lines of evolution leading to the loss of walking ability (see Fig. 8.49a) or reduction in the role of the swimming pleopods, as in the lobsters and crabs (see Fig. 8.49d–g). The lobsters also have a well-developed escape reaction involving the last pair of abdominal appendages, the uropods. The powerful muscles of the abdomen can be contracted to result in its sudden flexion, which causes the uropods to exert a force on the water and the lobster moves rapidly backwards (see Section 16.10.4).

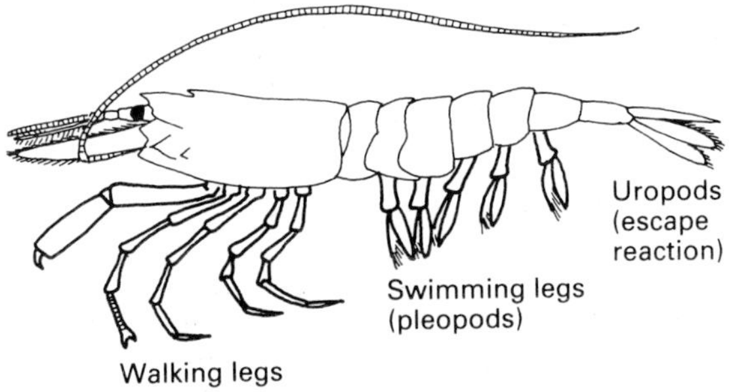

Fig. 10.22. Shrimp-like malacostracan with both walking and swimming limbs.

In the true crabs (Brachyura) the abdomen is reduced and most walk on five pairs of thoracic limbs. A transition to swimming can still be made, however; the last thoracic limbs can be flattened and modified for sustained swimming as in the portunid *Callinectes* (Fig. 10.23). This crab swims by rotational movements of the legs which act as hydrofoils.

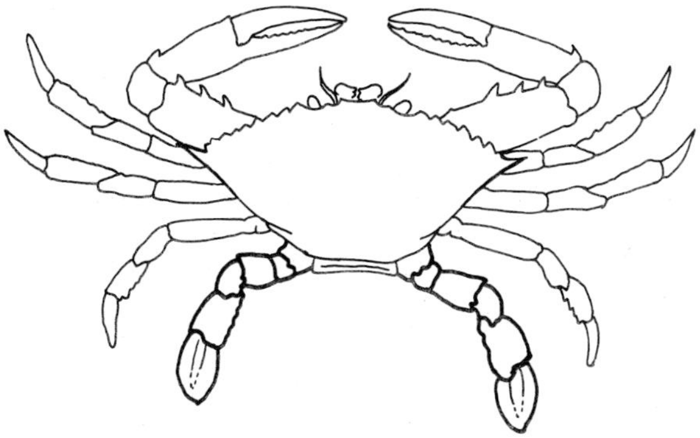

Fig. 10.23. The swimming crab *Callinectes*. The hind limbs act as hydrofoils during swimming.

The transition between swimming and walking in crustacean evolution is a consequence of the aquatic environments to which they have largely been restricted. In sea water, invertebrates can achieve near neutral buoyancy and the high viscosity of water (as compared to air) allows low terminal velocities. The structural design of the Crustacea is capable of supporting walking activity in a terrestrial environment, but other aspects of the physiology of the Crustacea have prevented a truly terrestrial or aerial radiation. This has, however, been a feature of the Chelicerata and Uniramia.

10.6.2 Terrestrial locomotion: walking

Animals living in terrestrial environments are much denser than the air in which they live; they therefore require structural support, and those that move frequently or quickly make use of rigid skeletal elements which interact with the ground. The skeletal elements may have to resist considerable deformative forces. These include bending forces, normal to the axis of the limb, and torsion forces acting about the axis of the limb.

The structure which gives this property with the minimum mass is a hollow cylinder. Rapid locomotion also requires flexible links (joints) between the rigid elements, tendons for the transmission of the forces generated by muscles and energy-storing devices. The arthropod cuticle is ideally suited to meeting all of these demands.

The basic material is a carbohydrate/protein complex, chitin. This is a tough and flexible substance but is not itself suitable for forming the rigid skeletal elements which are required. The cuticle can be hardened however by the formation of linking between the protein elements. This tanned protein is called sclerotin. It can give to the cuticle a very high degree of rigidity comparable with bone or it can be relatively soft. This enables the cuticle to form flexible hinges characteristic of the articulation of arthropod limbs. Each limb joint usually permits movement in one plane only, this being achieved by the development of internal buttresses.

The simple flexible hinge joint allows for flexion by muscular activity while extension of the joint often involves the elasticity of the joint membrane or involves a hydraulic mechanism in which the forces causing extension of the joint are transmitted by the haemocoelic fluid in the limb.

The stepping locomotion of terrestrial arthropods involves rotation of the laterally directed limb axis. In the Diplopoda (millipedes) there are a very large number of short limbs (Fig. 8.23). A metachronal rhythm of movements passes forwards as each limb makes its power stroke shortly after the limb immediately behind. This could be referred to as a 'low-geared' locomotory system in which a relatively large force can be applied to the anterior end of the animal, but maximum speed is low. Diplopoda are herbivorous animals mostly living in rotten wood and similar substrata, and this is a suitable mode of movement for that mode of life.

Many of the Chilopoda (centipedes), however, are more active predatory animals capable of greater speed. Figure 10.24 shows tracings of a centipede running progressively faster. The movements are very similar to those of the nereid polychaetes (Fig. 10.12). There is, however, a tendency for the number of limbs in contact with the ground to be fewer and the flexion of the body to be greater with increasing speed.

Lateral undulations of the body serve to increase step length in arthropods with short limbs, but they

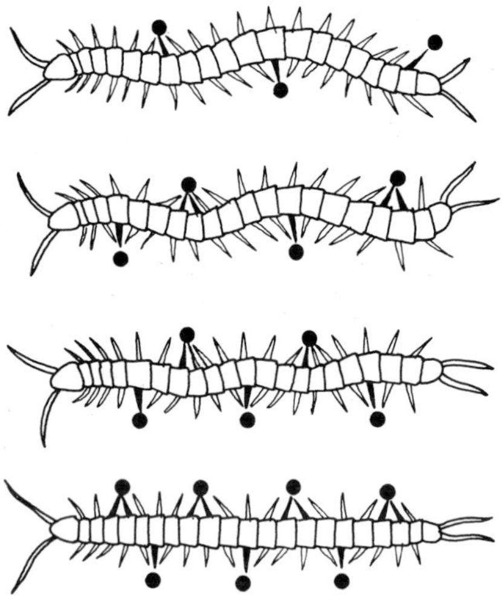

Fig. 10.24. Tracings of a centipede running progressively faster. The black dots mark limb tips which are stationary and in contact with the ground. (After Manton, 1965.)

introduce significant changes in momentum. This can be circumvented by increasing the length of each limb and reducing the number of limbs, and in the evolutionary history of the arthropod phyla there has been a marked tendency for a reduction in the number of walking limbs. In Crustacea, the Decapoda have five pairs, but often only three or four pairs are actually used in walking. The arachnids have four pairs and the insects three pairs of limbs. Walking requires the lifting of the limb (elevation) and its forward movement (protraction) followed by lowering of the limb (depression) and its backward movement relative to thge body (retraction). During retraction the limb tip will be anchored to the ground and the centre of gravity of the animal moved forward relative to that of the limb tip. Movements in the various hinges in the limb can maintain the body at a constant height above the ground during such a step.

Figure 10.25 shows the side view and trajectories of the limbs of a scorpion as it walks forwards. Interference between the limbs is reduced by the different lateral position of each limb tip during the movement. The trajectory of each limb is rather different and this must involve complex co-ordination of the movements of the joints in each limb. The trajectories of individual limbs are generally non-overlapping in this way as indicated in Fig. 10.26. Most arthropods walk forwards rotating the basal joint of the limb relative to the body (Fig. 10.25b–d) but crabs walk in a sideways fashion, protraction being achieved by extension of the lower limb joints.

With a small number of limbs there are problems of stability if too many limbs are in motion at once. In insects which have only three pairs of limbs stability requires that the limbs are moved in such a sequence that at least three limbs are in contact with the ground at all times, and that the centre of gravity falls within a triangle drawn between the limb tips. The most frequently observed stepping pattern in insects meets this requirement. Two sets of limbs are moved alternatively, each set forming a stable base while they are in contact with the ground. The limbs move in metachronal sequence from rear to front with legs on opposite sides of a single segment being completely out of phase with each other. These features are also exhibited by multi-limbed animals such as nereid worms (Fig. 10.13) and centipedes (Fig. 10.24).

Different walking speeds can be achieved within the alternating triangle gait by alterations of the relative duration of promotion and retraction, there being a marked tendency for the p/r ratio to increase as the rate of forward progression is increased.

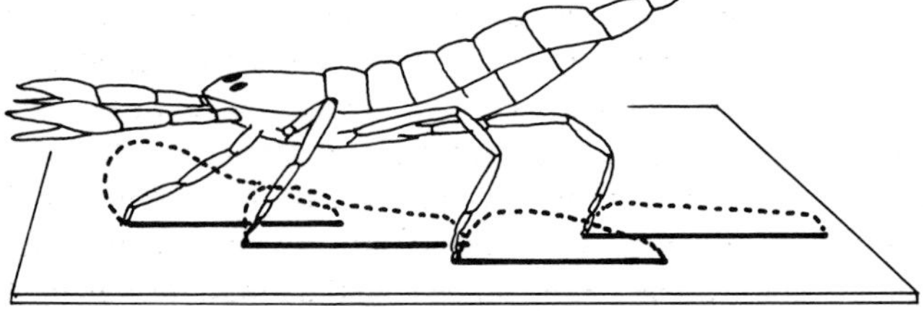

Fig. 10.25. Side view and trajectories of the limb movements of a scorpion. Note that the paths of the individual limbs do not cross each other. (After Hesseid & Fowtner, 1981.)

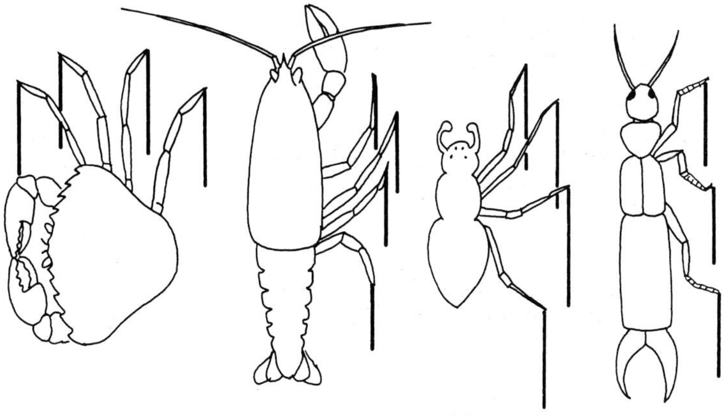

Fig. 10.26. The limb trajectories of a variety of arthropods: (a) crab; (b) lobster; (c) arachnid; and (d) insect. (After Manton, 1952.)

Some insects show variation on the alternating triangle pattern, especially where one pair of limbs is specialized for a different function, as in the mantids where the first pair of limbs rarely support the body.

10.6.3 Terrestrial locomotion: flight

The physical properties of the sclerotin-tanned arthropod cuticle are also such that the evolution of true flight was possible for the pterygote insects.

The cuticle has almost ideal properties for an aerofoil. It can produce a thin, light, flexible, lamellar membrane which can be supported and stiffened by veins and struts of sclerotized cuticle. The aerofoil is, in fact, a corrugated plane but it behaves in a manner which is equivalent to a solid section of similar chord and depth although it is far lighter. Furthermore, elastic proteins which store energy can be used in the generation of rapid wing movements while power is generated progressively by the muscles.

Flying insects first evolved some 200 million years ago. There has therefore been a long time during which the basic mechanism has been modified and refined. Consequently present-day insects exhibit a very wide range of structural and aerodynamic adaptations, much more diverse for instance than those of the birds or bats. Figure 10.27 shows a selection of wing shapes for insects which are here drawn all the same size but which, in reality, differ markedly in absolute size (see also Fig. 8.29).

The structural modifications which first made possible the evolution of flight in insects were probably involved in temperature regulation. Indeed, many living insects use their wings as do the Lepidoptera (e.g. butterflies) to raise the body temperature by absorbing solar energy, while others such as the Hymenoptera (e.g. bees and wasps) will raise body temperature prior to flight by flapping movements. It has been suggested that changes in relative scale could bring about a changed situation in which wings, which were primarily adapted for temperature regulation, acquired aerodynamic properties.

Precise temperature regulation, for instance, requires wing movement so that the development of wing-like extensions for temperature regulation would have been associated with the appropriate musculature. These primitive flying insects probably used relatively slow wing movements and would have been capable of gliding flight during which stable or steady state aerodynamic forces operated. The aerodynamics of many of the living insects, however, can best be understood if non-steady state conditions and unstable air flows are taken into account.

Living insects exhibit two different structural adaptations for the generation of wing movements. The Odonata (dragonflies) are thought to be amongst

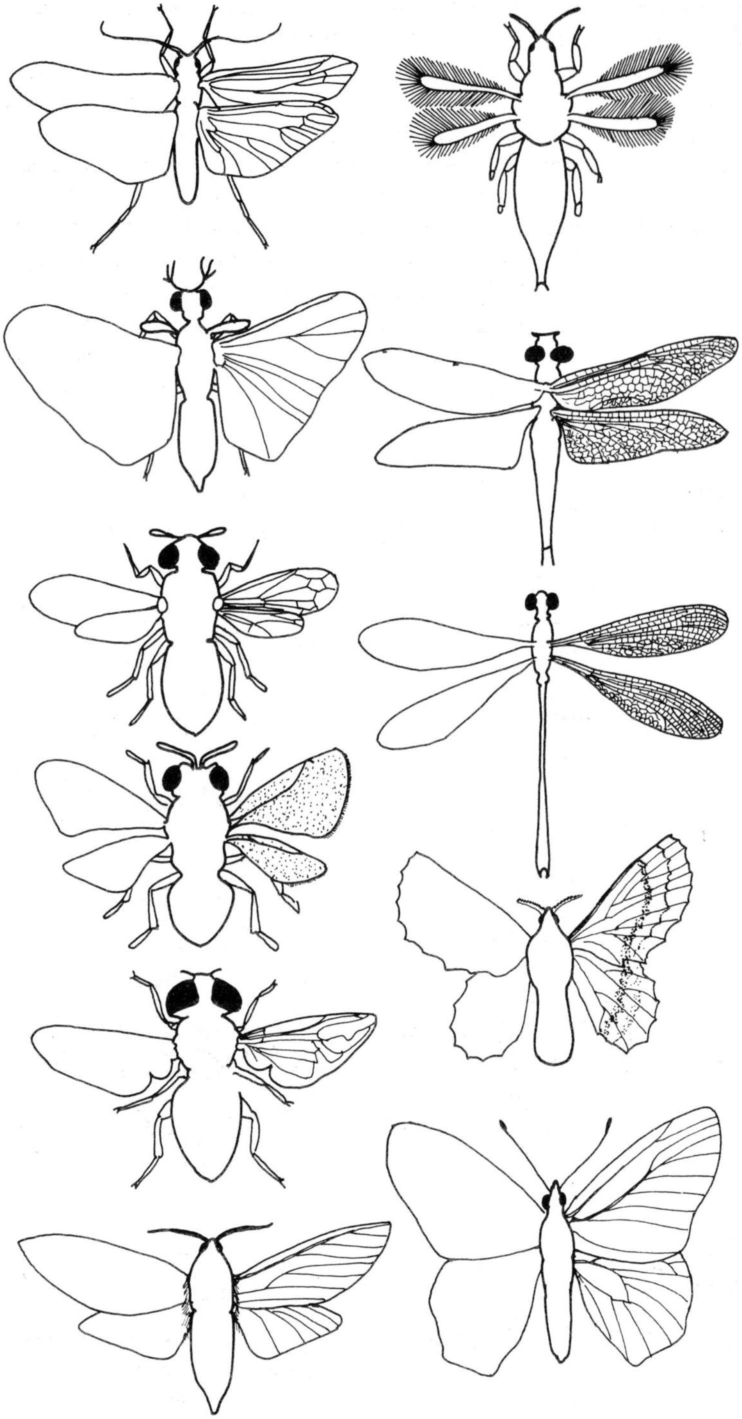

Fig. 10.27. Wing shapes in insects. These animals are here drawn as approximately the same size; in reality they have markedly different dimensions. (See also Fig. 8.29.)

the most primitive of the living pterygotes (winged insects) and in these, the wings are operated by direct flight muscles which attach to the wing bases as shown in Fig. 10.28a. In the dragonflies there is a direct one-to-one relationship between the nerve impulses stimulating the direct flight muscles and the movements of the wings and thorax, as shown in Fig. 10.28a (iii). Such insects are said to have synchronous flight muscle innervation (see Section 16.10.5).

The wing movements of the majority of insects, however, are caused by the action of indirect flight muscles which cause changes in the shape of the thorax which, through the articulation of the wing bases, causes the upward and downward movement of the wings. This structure is shown in Fig. 10.28b. The nerve impulses which stimulate the indirect flight muscles are not synchronized with the observed frequency of movements in the thorax which cause the wing movement (see Section 16.10.5). These insects are said to have asynchronous flight muscle innervation. The nerve impulses shown in Fig. 10.28b (iii) maintain the flight muscles in a state of excitation which is characterized by high levels of intracellular Ca^{2+} ions. In this state the muscles can be made to contract by stimulation of their stretch

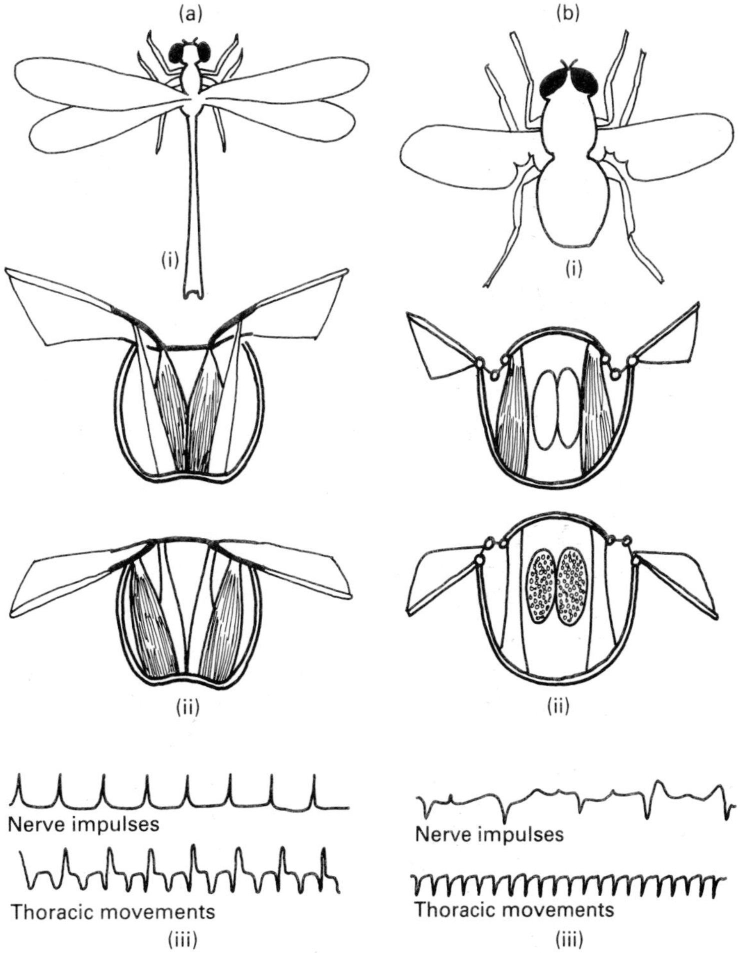

Fig. 10.28. Structural and neuromuscular adaptations for flight in insects. (a) Direct flight musculature, e.g. Odonata: (i) a representative dragonfly; (ii) cross-section of the thorax of a dragonfly showing the direct flight muscles attaching to the wing bases; (iii) traces of the synchronous nerve impulses and thoracic movements recorded in insects of this type. (b) Indirect flight musculature, e.g. Diptera: (i) a representative fly; (ii) cross-section of the thorax showing the main antogonistic flight muscles; (iii) traces of the asynchronous nerve impulses and thoracic movements recorded in insects of this type. (After Pringle, 1975.)

receptors. The frequency of wing beating is then a function of the physical properties of the thoracic box. One of the effects of the contraction of one set of flight muscles is to cause a stretching of the opposing set which sets in train a sequence of alternate contractions which will be maintained so long as the flight muscles are in the excited state.

It was formerly thought that under the influence of the indirect flight muscles the thorax of a dipteran fly would 'click' from one stable position to another

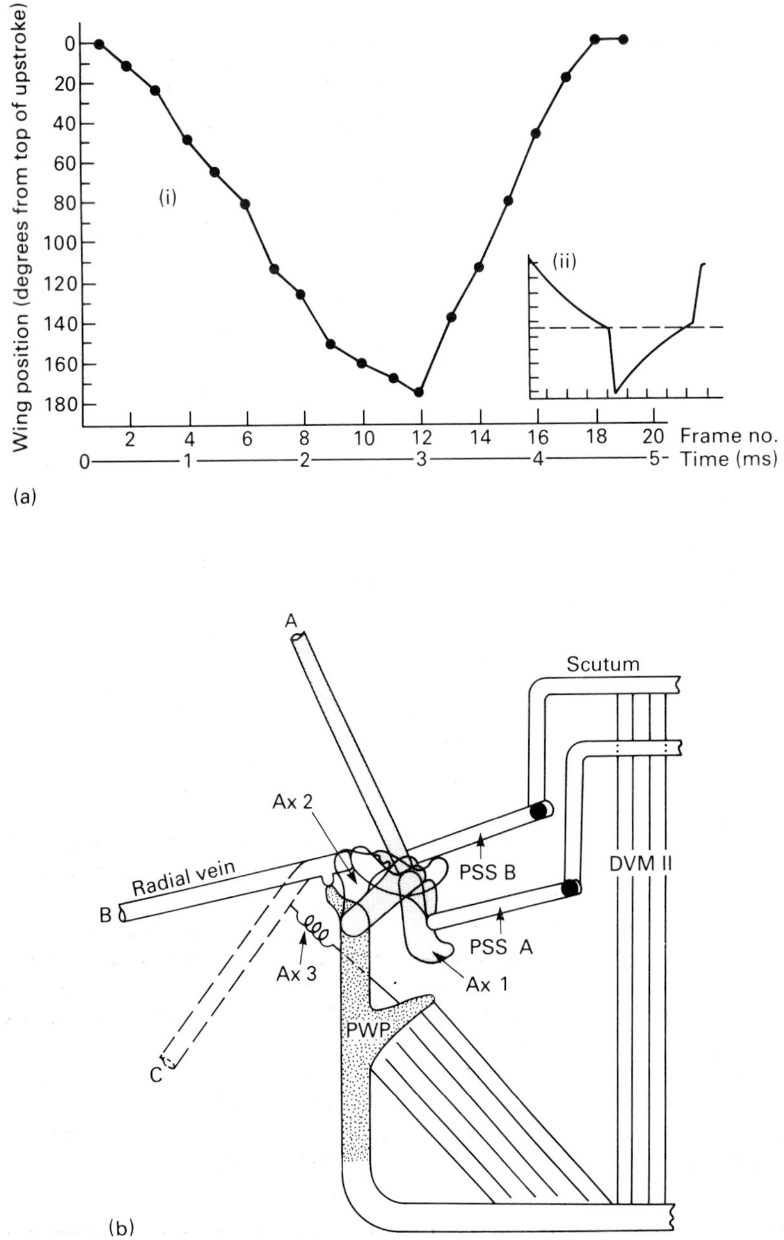

Fig. 10.29. (a) (i) Progressive position of the wing tip of a dipteran fly obtained by cinematography at 4000 frames per second; (ii) the predicted plot based on the idea of a simple 'click' mechanism. (b) A semi-diagrammatic model of the wing articulation of a dipteran. Lowering the wing from position A to position B raises the dorsal arch of the thorax, the scutum, and thereby activates the dorsoventral muscles (DVM II) which contract and raise the wing. Axillary muscles (Ax 3) can increase the amplitude of the wing stroke and this increases the energy stored in the elastic proteins of the thorax by deformation. This contributes to the rapid rise of the wing as observed in Fig. 10.29a. (After Miyan & Ewing, 1986.)

Box 10.9 Vortex generation and flight

1 The flight of insects is a consequence of non-steady state airflow generated by the movements of the wings. An important feature is the production and shedding of vortices of moving air.

In a vortex a mass of fluid rotates in a ring or ellipse about a central line, as shown below. The vortex ring possesses momentum and energy.

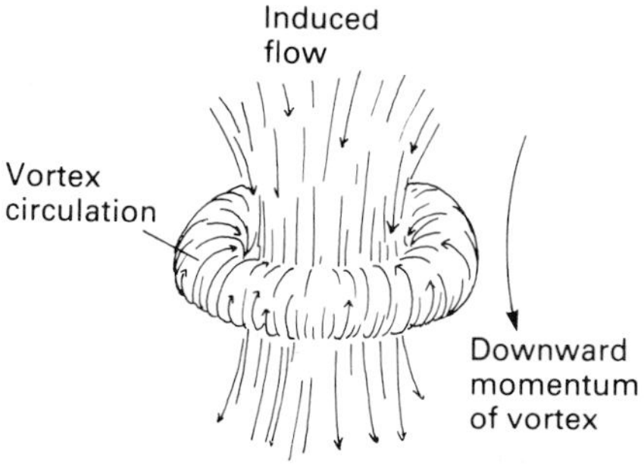

A diagrammatic representation of a vortex ring. In order to generate a vortex in a fluid, work must be done.

2 In flight the movements of the wings generate circulation about the aerofoil surface which is shed from the wing as a vortex when the wings reverse direction or change shape and inclination (see below).

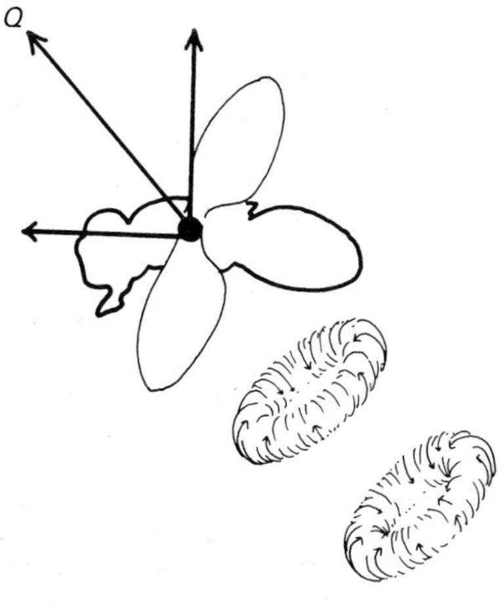

A flying insect can be supposed to be supported and propelled forwards by the reaction force Q resulting from the downward-shed vortex rings.

3 Circulation is often induced when wings are flung apart after having been clapped together (see also Fig. 10.30). This is often called a clap/fling mechanism.

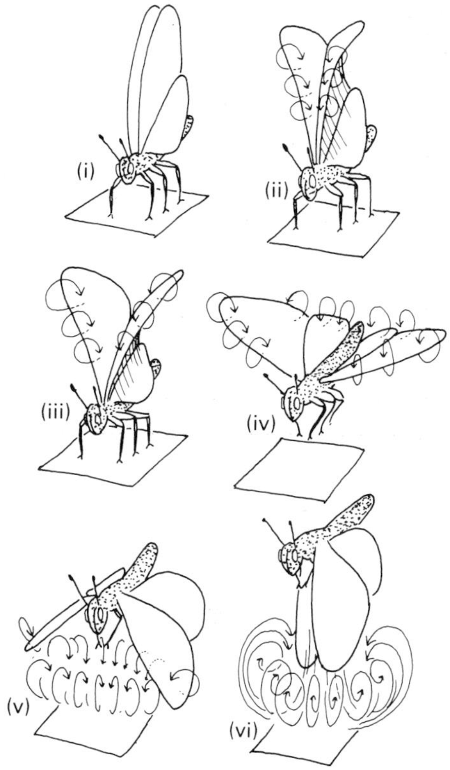

The near-vertical take-off of a butterfly (above) results from the inducted and downward shedding of a vortex (after Kingsolver, 1985).
(i) The wings held together in the 'up' position;
(ii), (iii) As they move apart, circulation is induced over the leading edges of the wings (see also the venation patterns on the wings in Fig. 10.27 which stiffen the leading edge of most insect wings).
(iv) Acceleration of the mass of circulating air induces lift and the butterfly loses contact with the ground.
(v), (vi) As the wings reach the bottom of the stroke, the circulating air is shed from the wing surface as a vortex ring with downward momentum.

with the wings in the up or down position. The kinematics of the wing beat, however, do not accord with this model, which predicts a slowing of the wing as tension is developed in the thorax towards the mid point of the beat followed by rapid acceleration to the stable position (Fig. 10.29a).

A diagramatic representation of an alternative model is given in Fig. 10.29b. It is proposed that the movements of the muscles which lower the wing cause a raising of the dorsal arch of the thoracic box through its locked articulation with the wing bases. This causes stretching of the dorsoventral muscles (stimulating their contraction) and stores energy in an elastic system which is released during the upstroke.

In order to fly, an insect must be subject to a reaction force which has components equal to its weight and equal to or greater than its drag when moving through air. The two components of the reaction force are lift and thrust. They can be generated by an aerofoil as it is moved through air.

The flight of locusts can be understood in this way and painstaking observations of the wings of other insects suggest that as an aerofoil the wing is moved in such a way as to generate lift and thrust during *both* the upward and downward stroke. The wing tip traces a roughly elliptical closed path relative to the insect, which is a saw-toothed pattern relative to a stationary point of reference given the forward velocity of the insect.

The coefficient of lift generated by some insects, however, is greater than could be expected from a conventional aerofoil at the recorded Reynold's numbers. This is especially striking in hoverflies and others which are capable of stationary flight. The key to the understanding of the flight of these and most other insects is through the recognition of the non-steady state, unstable forces that are generated by the wing movements. The flying insect leaves in its wake vortices of moving air and the flight of the insect is due to the reaction force resulting from the generation of vortex movements about the wing surface. The principle is further explained in Box 10.9.

When the wings of an insect are clapped together then flung apart, the anterior stiffened edges are the first to be pulled away from each other and air rushes in to the low pressure space developing between the wings (Fig. 10.30). Work is done in accelerating the mass of circulating air and the reaction force has components of lift and thrust. Once the moving air has maximum velocity no further work can be done by the wings until the vortex is shed, which may occur when the wings change their direction of movement at the bottom of the wing stroke. The take-off of a butterfly is also described in similar terms in Box 10.9.

The insects are such a diverse group of animals that it is unlikely that a single theory of flight adequately describes them all. The very smallest

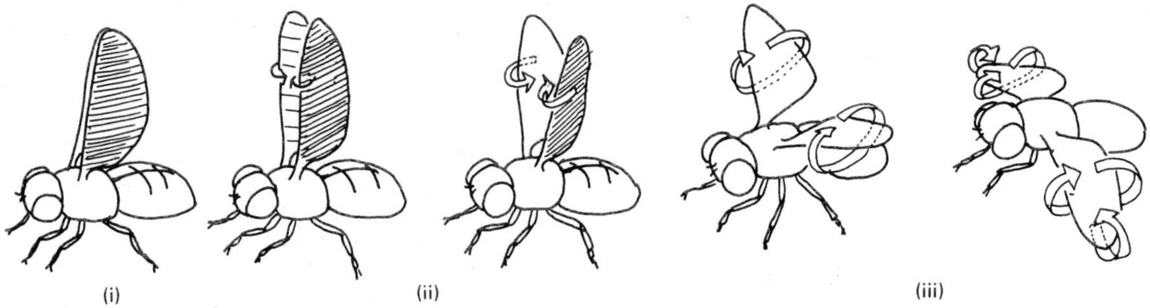

Fig. 10.30. The clap/fling mechanism for vortex generation. (i) At the top of the stroke the wings are clapped together. (ii) The wings are flung apart, the stiffened leading edges being separated first, thus establishing a circulation of air around the wing. (iii) Downward movement of the wing accelerates the air; work is done and the reaction force generated has components giving lift and thrust. (After Weis-Fogh, 1975.)

insects such as the 'smoke midges' have very low Re values and they can be considered to row themselves through the air as if it were a viscous fluid. These insects have feathery wings in which the aerofoil characteristics have been lost (see Fig. 10.27). In contrast are the larger Coleoptera which achieve flight characteristics to give Re values as high as 23 000. The stiffened forewings of these animals may therefore behave as conventional aerofoils.

10.6.4 Terrestrial locomotion: jumping insects

Some insects have the ability to jump; in most of them it is an important escape reaction, and anyone who has tried to catch a flea will be aware how effective this unpredictable behaviour is. The ability to jump is particularly well developed in the fleas, grasshoppers and leaf hoppers.

In order to jump, an insect must exert a force against the ground sufficient to impart a take-off velocity to its mass. The height of the jump will be defined by the relationship:

$$\text{kinetic energy } (\tfrac{1}{2}mV^2) = \begin{array}{l}\text{potential energy at the}\\ \text{end of the jump } (mgh)\end{array}$$

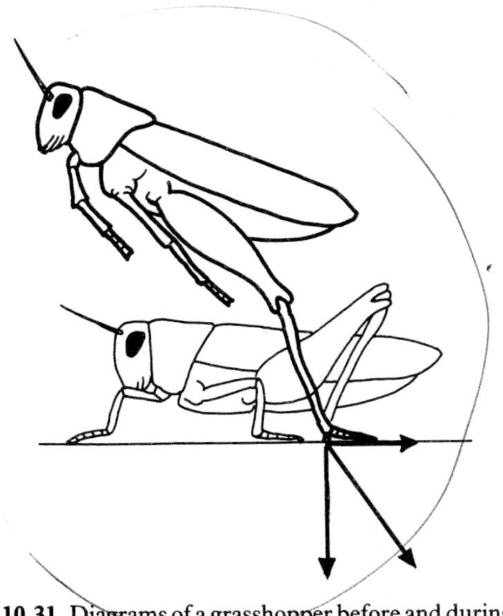

Fig. 10.31. Diagrams of a grasshopper before and during its jump. A force is transmitted to the ground via the articulation of the hind feet. The long legs increase the time during which the force can act and so contribute to the acceleration achieved.

where m is the mass of the insect,
V is the velocity at take-off,
g is the acceleration due to gravity and
h is the height of the jump.

It follows that $h = \dfrac{V^2}{2g}$ and that $h = \dfrac{\text{kinetic energy}}{mg}$

The force exerted against the ground by the tip of the leg of a jumping insect will have vertical and horizontal components as shown for a grasshopper in Fig. 10.31, and the vertical component is equal to $F \sin \theta$.

A jumping insect will only continue to accelerate while its feet are in contact with the ground and the take-off velocity will be determined by the scale of the force and the time over which it acts (see also Box 10.2). This in turn will be determined by the length of the legs, and insects which jump all tend to have relatively long legs. It was explained in Section 10.2 that the rate at which muscle cells can contract is limited by their physiology; in flying insects this constraint is overcome by the evolution of indirect flight muscles and the ability to store energy in elastic components of the body wall by deforming the thorax (see Section 10.6.3). This is also true of most jumping insects. The separate elements of a generalized insect leg are identified in Figure 10.19a and reference will be made to this in order to describe the jump of some insects. The leg of the flea is characterized by elongation of both coxa and femur. When it prepares to jump the femora are raised to the position shown in Figure 10.32a. A force is then loaded into a patch of very elastic cuticle in

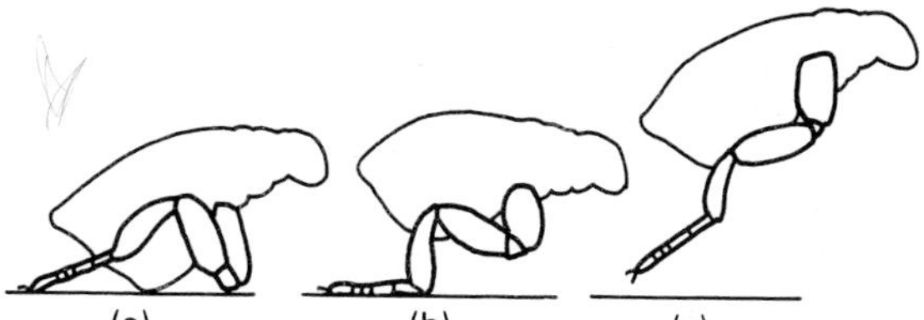

Fig. 10.32. The jump of a flea. (a) The femora are raised and energy is stored in deformed elastic proteins of the cuticle. (b) Locked femora are released by the relaxation of muscles. The force exerted against the ground by the tibia gives the flea a specific velocity which will determine the height of its jump.

the thoracic wall. Although the fleas are flightless insects it is thought that this patch of very elastic cuticle has been developed from that of the flying insects and the former flight muscles are used to distort the thoracic box. The thoracic segments are held in their new position by a ratchet-like mechanism of pegs and sockets in the integument and the flea can remain in this position cocked ready for a sudden release of stored energy during the jump. This is initiated by relaxation of the muscles which raised the femora, the pegs and sockets disengage, and a large force is exerted by the tibia against the ground as the thoracic segments return to their original position.

10.7 Conclusions

This chapter has provided a general introduction to the locomotory systems of invertebrates. You should understand the effects of scale and the reasons why larger animals use muscle cells to develop force whereas the smallest animals are able to use cilia or flagella.

You should understand that all systems of animal movement obey the same mechanical laws and that a skeletal system is always involved in the application of a force. The skeletal system may transmit forces through a liquid in an enclosed space or through a rigid system of levers.

Some invertebrates are sluggish, slow-moving animals, others are highly mobile and agile; the cephalopod molluscs and the insects, for instance, illustrate very advanced locomotory techniques involving on the one hand jet propulsion and on the other, flapping flight. The analyses given here are expressed in simple terms, but many readers may wish to proceed to a more rigorous analysis. These readers are referred to the list of further reading below.

An understanding of the mechanics of animal locomotion is crucial to an understanding of the evolutionary origins of animal groups, since any proposed ancestor must be structurally sound. It must 'work' and obey fundamental physical laws just as all the living invertebrates must do.

10.8 Further reading

Alexander, R. McN. 1982. *Locomotion of Animals.* Tertiary Level Biology. Blackie, Glasgow.

Clark, R.B. 1964. *Dynamics in Metazoan Evolution.* Clarendon Press, Oxford.

Elder, H.Y. & Trueman, E.R. 1980. *Aspects of Animal Movement.* Society for Experimental Biology Seminar Series. Cambridge University Press, Cambridge.

Hesseid, C.F. & Fowtner, C.R. (Eds) 1981. *Locomotion and Energetics in Arthropods.* Plenum Press, New York.

Kingsolver, J.G. 1985. Butterfly engineering. *Scient. Am.*, **253** (2), 90–7.

Rainey, R.C. (Ed) 1984. *Insect Flight.* Blackwell Scientific Publications, Oxford.

Trueman, E.R. 1975. *The Locomotion of Soft-Bodied Animals.* Edward Arnold, London.

Weis-Fogh, T. 1975. Unusual mechanisms for the generation of lift in flying animals. *Scient. Am.*, **233** (5), 81–7.

11

RESPIRATION

The need for oxygen used to be thought of as a fundamental property of all living things. Oxygen is taken up over respiratory surfaces such as gills and lungs, and once inspired it is used to oxidize organic substances—largely but not exclusively carbohydrates—to yield the energy needed to power all active body processes. Life originated without oxygen, though, so aerobic respiration is not an essential feature of organisms and indeed some can and do still function anaerobically. This chapter reviews processes of aerobic and anaerobic respiration in invertebrates. We begin by considering the biochemical basis of respiratory processes and then, focusing on aerobic respiration, we consider how O_2 is obtained from the environment and transferred to the tissues, and how the uptake of O_2 is influenced by a myriad of environmental factors.

11.1 Central importance of ATP in respiration

Phosphorylated nucleotides, and particularly *adenosine triphosphate*, play an important part as intermediaries in the transfer of energy from the fuel (foodstuffs) to the power-consuming processes of metabolism. Potential energy in the food is transferred to so-called high-energy phosphate bonds (designated ~P) viz:

$$\text{energy from absorbed food} + \text{A}-\text{P}\sim\text{P}+\text{Pi} \rightarrow \text{A}-\text{P}\sim\text{P}\sim\text{P}$$

where Pi is inorganic phosphate, and this stored energy can then be yielded up to metabolism giving ADP+Pi. Note, however, that the term 'high-energy phosphate bond' is used rather loosely. The energy is *not* stored in the covalent linkage between the phosphate and the rest of the molecule. Rather, the phosphate-bond energy is a reflection of the energy content of the whole triphosphate molecule, before and after its conversion to the diphosphate.

11.2 Backbone of catabolism

Glycolysis and the tricarboxylic acid (TCA) cycle form the backbone for catabolism in all the invertebrate phyla. These pathways are familiar and are well covered in other textbooks and so only a brief description will be given here.

Figure 11.1 gives a *very stylized* summary of these two pathways. The fuel is glucose, derived directly from food or by enzyme-mediated transformations of other molecules from food or from body stores.

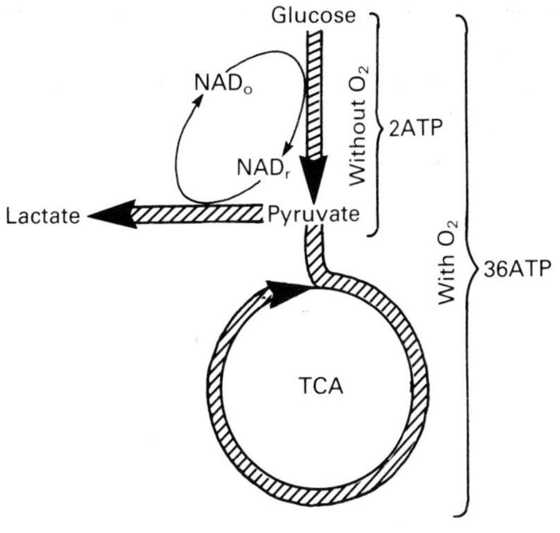

Fig. 11.1. An extremely stylized version of the metabolic pathways concerned with generating ATP.

Glycolysis takes place in the cytoplasm of cells and can occur without oxygen. It generates ATP from the direct involvement of substrate passing down this pathway—*substrate level phosphorylation.*

The *TCA cycle* occurs in mitochondria and requires oxygen. It generates reduced nicotinamide adenine dinucleotide (NAD_r) which is used, in turn, to generate ATP by donating electrons (becoming oxidized; NAD_O) to an electron transport system of cytochromes in which the final electron acceptor is oxygen. This is *electron transport phosphorylation.* Notice from Fig. 11.1 that the involvement of oxygen allows the generation of more molecules of ATP per molecule of glucose than the part of the process that does not involve oxygen.

11.3 Generating ATP without oxygen

Before the evolutionary origin of photoautotrophs, organisms had to function without oxygen. Moreover, environmental anoxia (no O_2) can occur now: for littoral animals during exposure to air at low tides; during burrowing in reducing substrata; in many parasitic habitats. Additionally, specific tissues may become anoxic though the organism as a whole has ample oxygen. For example, the muscles concerned with escape reactions in bivalve molluscs typically undergo anaerobic metabolism.

Hence the 'first' organisms depended upon anaerobiosis, and some continue to do so. In theory there are a large number of possible anaerobic pathways but four main ones have been used by invertebrates. Different pathways have evolved to meet particular energy requirements.

Lactate pathway—the best known pathway. It is illustrated in Fig. 11.1; the NAD_r, from glycolysis, being reoxidized by the reduction of the terminal pyruvate to lactate under the catalytic action of lactate dehydrogenase. As already noted, it is not efficient in ATP production, but it can generate ATP at a high rate without oxygen and is commonly used to support burst work in vertebrates when muscle tissue can temporarily run out of oxygen. It is not of universal occurrence in invertebrates. It probably occurs in insect leg muscles but not in flight muscles which are so well served by tracheae and tracheoles (see Fig. 11.6b) that they rarely become anoxic.

Opine pathway—this is similar to the lactate pathway and is adapted for burst work—the rapid but not necessarily efficient generation of ATP. In it the carbohydrate is catabolized by glycolysis but the reduction of pyruvate is replaced by its reductive condensation with an amino acid to form an opine—an amino acid derivative:

$$\text{glucose unit} + 2 \text{ amino acid} + 3 \text{ ADP} + 3 \text{ Pi} \rightarrow 2 \text{ H}_2\text{O} + 2 \text{ opine} + 3 \text{ ATP}$$

Several pathways can be identified depending upon the amino acid and enzyme used. For example, a substance called octopine is formed in the mantle muscles of cephalopods during burst work. Strombine, another opine, has been recorded from some bivalve molluscs.

Succinate pathway—organisms such as bivalves inhabiting anoxic muds and endoparasites inhabiting anaerobic sites in their hosts, like the vertebrate gut, have evolved anaerobic pathways that, whilst not being capable of generating ATP rapidly, can generate more per glucose input than the lactate and opine pathways. The succinate pathway retrieves energy from the NAD_r, generated in the initial glycolytic pathway, by an electron transport phosphorylation with fumarate rather than oxygen as a final electron acceptor. The basic framework of the process is illustrated in Fig. 11.2. Phosphoenolpyruvate (PEP), the molecule before pyruvate in glycolysis, is converted to oxaloacetate (by the addition of CO_2—carboxylation—catalysed by PEP carboxykinase) then to fumarate. Fumarate is oxidized to succinate by an electron transport system and succinate may be further metabolized to proprionate and other volatile fatty acids. Oxaloacetate, fumarate and succinate are all intermediaries in the TCA cycle, but occur there in exactly the reverse sequence to the succinate system just described. Hence this pathway can be considered as the TCA put in reverse. A part of the pyruvate in this succinate system can also be converted to lactate, acetate, alanine and ethanol. A range of end products is therefore produced. Overall, the process leads to

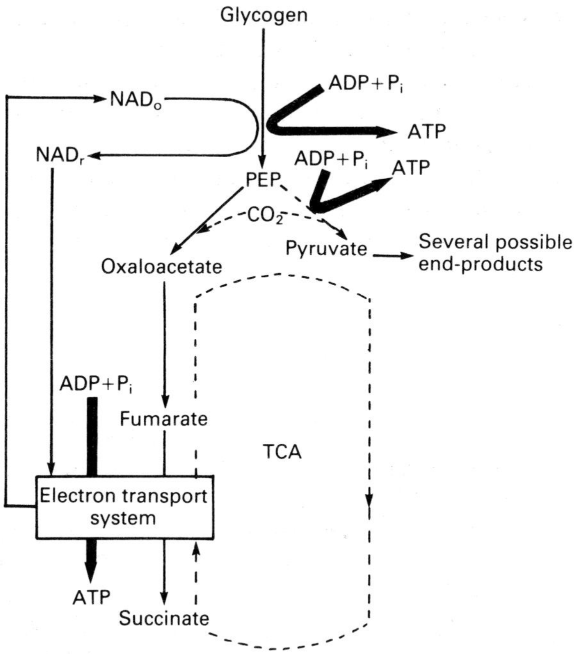

Fig. 11.2. Modified anaerobic pathway used by endoparasites and bivalves living in conditions of hypoxia for long periods (after Calow & Townsend 1981).

the production of about 4 to 6 ATP molecules per glucose-unit input.

A very similar reaction pathway, described for bivalves, produces succinate by reduction of oxaloacetate; however, the latter is not produced from PEP but from an amino acid, aspartate, by transamination reactions.

Phosphagens—these are important in burst work and act as stores, accepting ~P from ATP in periods of relaxation and delivering it up under periods of hard work and anoxia:

Phosphoarginine+ADP = Arginine+ATP

The above system is common in the invertebrates whereas phosphocreatine is used by vertebrates. Exceptions to this rule are echinoderms that may have both phosphagens and the annelids that contain four other phosphagens in addition to phosphoarginine.

Figure 11.3 maps the phyletic distribution of the major pathways (not including phosphagens). All pathways are widespread. Thus the succinate pathway might either have evolved from the glycolysis–TCA system or vice versa, i.e. by the reversal of one or the other (above). It is generally accepted that amino acids were prominent components of the early biotic environment and it has been suggested that the earliest system for the generation of ATP involved amino acids as both electron donors and acceptors. Hence, opine pathways could be primitive. Clearly all anaerobic pathways were present at an early stage in the evolution of the invertebrates and their distribution now is probably attributable to selection pressures—specific adaptations to the ecological circumstances in which they have to work. Some systems have evolved to meet the physiological needs of burst work (produce lots of ATP but not necessarily very efficiently because it is not sustained), others have evolved to supply ATP more efficiently, if less quickly, in response to long-term and sustained anoxia.

11.4 Uptake of oxygen

11.4.1 Diffusion is paramount

Aerobic metabolism depends on oxygen being made available to respiring tissues. Fundamentally, this depends upon diffusion—oxygen molecules moving from high to low partial pressure (Po_2) according to the dictates of Fick's Law. The rate of diffusion of oxygen through tissues depends on Po_2 gradients and also on the properties of the tissues—the latter are often expressed as a diffusion coefficients. On the basis of reasonable assumptions about these coefficients and the oxygen demands of tissues it can be calculated that the distance between metabolizing tissue and a respiratory surface can be no more than 1 mm. This is one of the reasons why large, solid-bodied turbellarians have to be flat whereas smaller acoels and rhabdocoels do not (Fig. 11.4). Many jellyfish and anemones, however, do grow very large despite the fact that they are also solid. Here, though, the outer and inner layers of tissue are thin and in

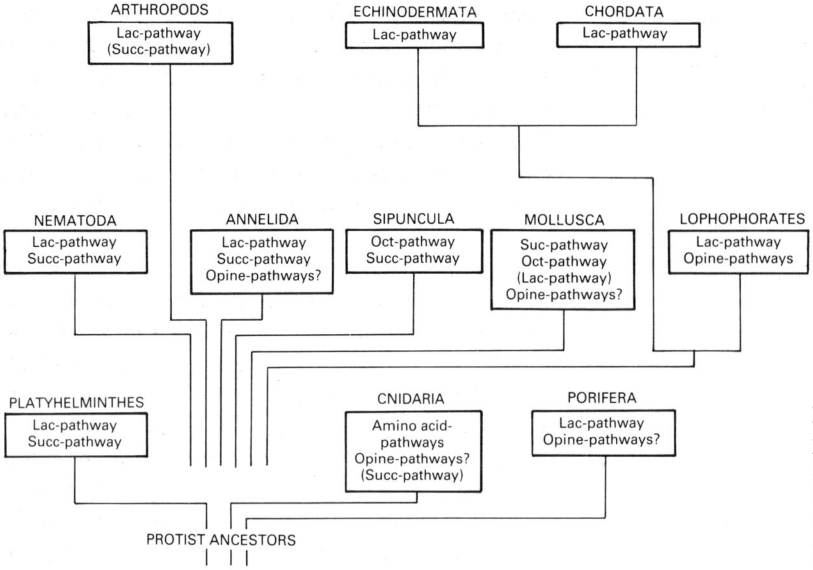

Fig. 11.3. Phyletic distribution of major pathways involved in ATP production. Modified from Livingstone (1983). Key: Lac = lactate; Succ = succinate; Oct = octopine; pathways in parentheses represent special circumstances.

direct contact with water in the external environment or the gastrovascular cavity. The thicker mesoglea is largely devoid of cells and has a low metabolic demand. In anemones, the more cellular mesoglea is usually less thick and, because of the complex folding of the gastrodermis (p. 67), always within 1 mm of water circulating in the gastrovascular cavity.

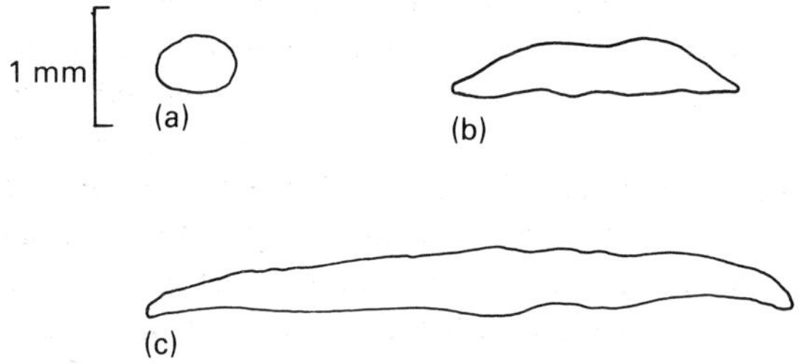

Fig. 11.4. Transverse sections, drawn to the same scale, through various turbellarians: (a) rhabdocoel, (b) triclad, (c) polyclad. (After Alexander, 1971.)

11.4.2 Circulatory systems

A solution to the limitations imposed by diffusion is the evolution of circulatory systems. These increase the capacity for transporting O_2, so that the 1 mm limitation is escaped, and by rapidly removing oxygen from respiratory surfaces maintain a steep Po_2 gradient and so increase the rate of O_2 intake. Two main kinds of system have evolved (Fig. 11.5): the open system with large haemocoel (persistent blastocoel or expanded blood vessel) typical of arthropods and molluscs and the closed system of arteries and veins well developed in the Annelida. Both require muscular pumps, e.g. contractile tubes in arthropods, accessed by pairs of valved ostia (Fig. 11.5a), and muscular, pulsating blood vessels (lateral 'hearts') in annelids. Echinoderms and hemichordates have somewhat intermediate systems consisting of small vessels connected to larger sinuses. Holothurian echinoderms, though, have well-developed closed systems. In association with the invasion of land, isopod crustaceans have evolved larger, more muscular hearts than other crustaceans and vessel-like lacunae that form an almost closed system. Insects, though having an open system, do not make much use of it for oxygen transport. Instead they have evolved an elaborate and extensive system of tubes, tracheae and tracheoles (Fig. 11.6), that ensure all actively metabolizing tissues are provided for by

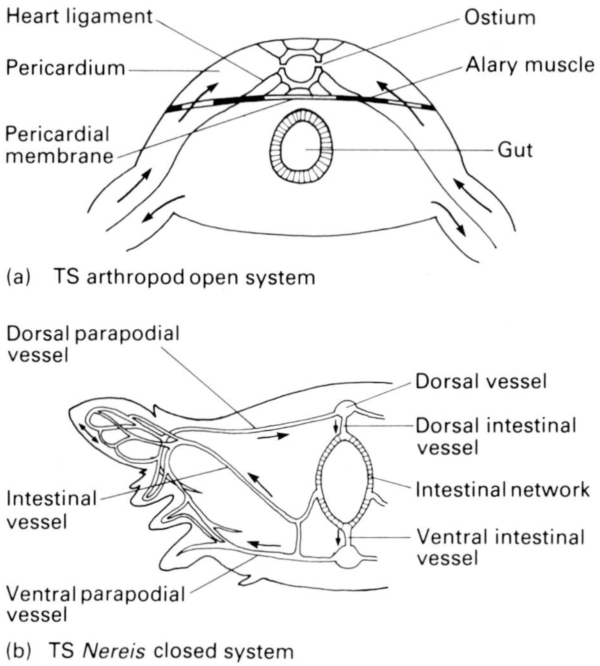

Fig. 11.5. Two kinds of blood system: (a) open system of an arthropod; (b) closed system of a polychaete annelid (*Nereis*).

almost direct supply of gaseous oxygen. (The way these work is explained in the legend to the figure.)

11.4.3 Blood

Primitively, bloods were probably colourless and similar in composition to sea water, as in many present-day gastropod and bivalve molluscs. Such saline media can carry comparatively small amounts of oxygen and the bloods of many invertebrates contain oxygen-carriers, i.e. specialized proteins capable of binding reversibly with oxygen, so-called respiratory pigments. In the bloods of annelids, for example, oxygen in solution accounts for < 1 vol % (ml O_2/100 ml blood) as compared with 1–10 vol % associated with respiratory pigments. These pigments resemble one another in that they consist of conjugated proteins linked to prosthetic groups containing a metal. Haemoglobin and chlorocruorin have a similar prosthetic group, called haem, which is a porphyrin, linked to one atom of ferrous iron. In haemocyanin the prosthetic group is not a porphyrin but a polypeptide and is attached to copper and sulphur. Similarly in haemerythrin the prosthetic group is not a porphyrin though it is attached to iron. Some properties and the distributions of these pigments are summarized in Table 11.1.

The amount of O_2 carried by a blood (oxygen capacity) depends upon the quantity of pigment it contains and the extent to which it can take up and yield O_2 (oxygen affinity). An important consideration is the amount of oxygen that a pigment can take up relative to oxygen availability in the environment (ambient Po_2). This can be expressed in a so-called *oxygen dissociation curve,* obtained by exposing blood to oxygen at a series of Po_2s and measuring uptake by changes in the optical property of the pigment. These curves are often, though not invariably (see below), sigmoid (Fig. 11.7) and a useful index of the mode of functioning of a pigment is given by the Po_2 at which it becomes half saturated (known as P_{50}). These P_{50} values vary with (a) temperature, lower temperatures shifting the curves to the left (Fig. 11.8a) and (b) pH, lower pHs most often resulting in lower oxygen affinity and shifting the dissociation curve to the right (Fig. 11.8b)—known as the (normal or negative) Bohr effect. Since high CO_2 causes low pH, this will have a Bohr effect and since CO_2 is likely to be highest in tissues with high metabolic rates, the result will be to yield O_2 when and where it is most needed. Reverse (abnormal or positive) Bohr effects have been recorded for some invertebrates, e.g. *Limulus,* the chelicerate king crab, that crawls on the surfaces of mudflats, where periodic oxygen deficiency with increased ambient levels of CO_2 may favour this response for increased uptake despite a disadvantage for unloading O_2 at the tissues. Not all cases of reverse Bohr effects can be so easily explained and this is why they are referred to as 'abnormal'.

There is variation both within and between pigments in terms of their dissociation curves and P_{50} values and particular pigments are also certainly adapted to the environment, both internal and external, in which they have to operate (see, for example, Fig. 11.7). For instance, the following correlations can be noted:

1 P_{50}s are high in animals that live in a high ambient Po_2, have respiratory surfaces that present modest diffusion barriers and therefore have blood Po_2s not far removed from ambient. This situation prevails in *Sabella* (Fig. 11.7) with a blood pigment of chlorocruorin ($P_{50} \simeq 30$ mmHg). This subtidal animal is tubiculous but is able to extend its crown of tentacles into well-oxygenated water. Moreover, though enclosed in a tube of mud and mucus, the body surface

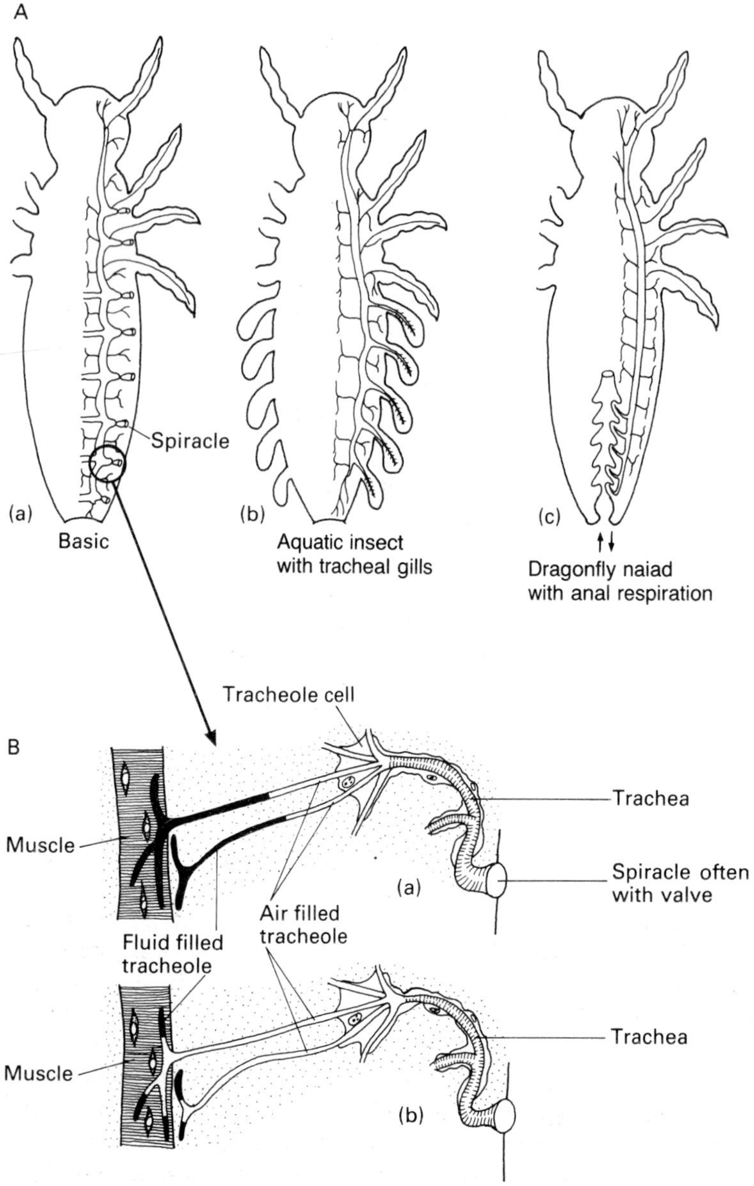

Fig. 11.6. A Tracheal systems: (a) basic; (b) aquatic insect with gills; (c) dragonfly naiad with anal respiration **B** Enlargement of tracheal system (enlarged from **A** a)—air enters through spiracles, which in most insects have valves. Large tubes (tracheae) lead to small (tracheoles). The latter can contain fluid (a) but during active metabolism the osmotic pressure of surrounding tissues increases and the fluid is withdrawn (b). Air is then drawn into the tracheoles, and passes by diffusion through their walls into the tissues. (Redrawn from various sources.)

Table 11.1 Respiratory pigments (after Schmidt-Nielsen, 1979)

Pigment	Description	Molecular weight	Occurrence in animals
Haemocyanin	Copper-containing protein, carried in solution, blue when oxygenated but colourless when deoxygenated	300 000–9 000 000	Molluscs: chitons, cephalopods prosobranch and pulmonate gastropods; not bivalves. Crustaceans: crabs, lobsters. Chelicerates: *Limulus, Euscorpius*
Haemerythrin	Iron-containing protein, always in cells, non-porphyrin structure, violet in colour		Sipunculans: all species examined. Polychaetes: *Magelona* Priapulans: *Halicryptus, Priapulus* Brachiopods: *Lingula*
Chlorocruorin	Iron–porphyrin protein, carried in solution, dichromic —red in concentrated solution, green in dilute solution	2 750 000	Restricted to four families of Polychaetes: Sabellidae, Serpulidae, Chlorhaemidae, Ampharetidae Prosthetic group alone found in starfish (*Luidia, Astropecten*)
Haemoglobin	Iron–porphyrin protein, carried in solution or in cells; most extensively distributed pigment, red in colour	17 000–3 000 000	Vertebrates Echinoderms: sea cucumbers Molluscs: *Planorbis*, Pismo clam (*Tivela*) Hexapods: *Chironomus, Gastrophilus* Crustaceans: *Daphnia, Artemia* Annelids: *Lumbricus, Tubifex, Arenicola, Spirorbis* (some species have haemoglobin; some chlorocruorin; others no blood pigment), *Serpula* (both haemoglobin and chlorocruorin) Nematodes: *Ascaris* Flatworms: parasitic trematodes Ciliates: *Paramecium, Tetrahymena*

is irrigated by peristaltic action of body wall musculature, and this appears to occur continuously even when the crown is withdrawn. Another low-affinity pigment is the haemocyanin of the squid *Loligo* ($P_{50} \simeq 50$ mmHg). Yet another animal included in this category is the errant burrowing polychaete *Nephtys* with haemoglobin as pigment. As will be seen from Fig. 11.7 this has a relatively low P_{50} ($\simeq$5–7 mmHg) but it will also be noted that it has a hyperbolic not a sigmoid dissociation curve—and compared with a sigmoid curve of similar P_{50}, this would become less saturated at higher P_{O_2}s—or to put it another way, the pigment with a sigmoid curve becomes fully saturated at lower P_{O_2}s than one with a hyperbolic curve with equivalent P_{50}. This animal is intertidal and lives in impermanent burrows in sand; water being drawn through these by ciliary action when submerged. Under these conditions the pigment acts as a transport mechanism capable of taking up and yielding O_2 at high P_{O_2}. When the tide ebbs the burrow collapses and there is an abrupt change from high to low P_{O_2}—probably occurring so rapidly that

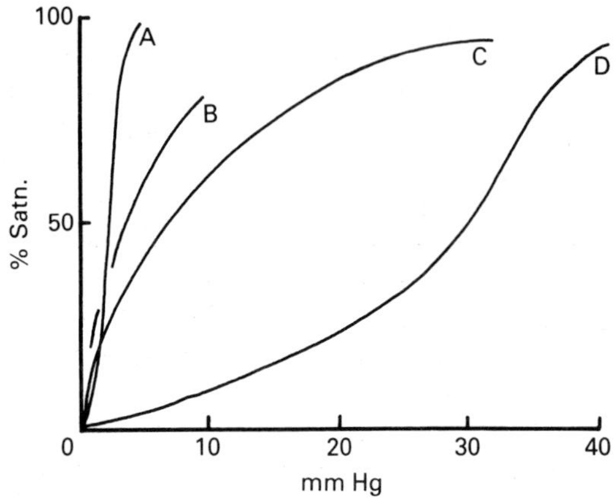

Fig. 11.7. Oxygen dissociation curves—plot of uptake of O_2 from the atmosphere (% complete saturation) versus oxygen availability in the atmosphere (Po_2). A = *Arenicola marina* (pH 7.5, 19°C); B = *Nephtys hombergii* (pH 7.4, 15°C)—vascular; C = as B but coelomic fluid; D = *Sabella spallanzanii* (pH 7.35, 26°C). (Data from Jones, 1955, 1972.)

it offers no possibility for uptake and transport at reduced Po_2—and the animal very likely becomes anaerobic in the intertidal period (c. 6 h). Note also that the vascular haemoglobin of *Nephtys* has a higher affinity than the coelomic haemoglobin—possibly because uptake of O_2 is over the whole body surface directly into the coelomic haemoglobin from which it is transferred to the vascular system. The oxygen dissociation curves of the vascular and coelomic haemerythrins of the sipunculan *Dendrostomum zostericolum* are exactly the reverse of those shown by *Nephtys*; coelomic pigment has a higher affinity than vascular. In this animal, tentacles are the respiratory organs. Oxygen diffuses through these and is transported from them by a vascular system, then passing from this into coelomic fluid.

2 P_{50}s are low in animals that live at high ambient Po_2 but in which the respiratory surfaces present a considerable diffusion barrier, for here blood Po_2 is low relative to ambient. Many decapod crustaceans fit this category and have P_{50}s ranging from 1 to 20.

3 P_{50}s can be very low (< 1) in animals that are subjected to low ambient Po_2s. Tubificid oligochaetes and chironomid larvae (both with haemoglobin) that are common in organically rich or organically polluted habitats fit this category. The chironomids live in burrows that they actively irrigate. However, they do not do this continuously and even when the surrounding water is fully saturated, conditions within the burrows can become hypoxic (= low O_2), when a high-affinity pigment is of value for uptake and possibly also for storage of O_2 over short periods of anoxia. *Arenicola*, a marine intertidal annelid, that often burrows in deoxygenated muds also has a low P_{50} (Fig. 11.7). As with chironomids, this animal can irrigate its burrow but not continuously and the high-affinity pigment again may be of importance as a short-term O_2 store. It is unlikely, though, that the amount of O_2 stored could sustain *Arenicola* for the whole time that it is exposed between tides and, any-

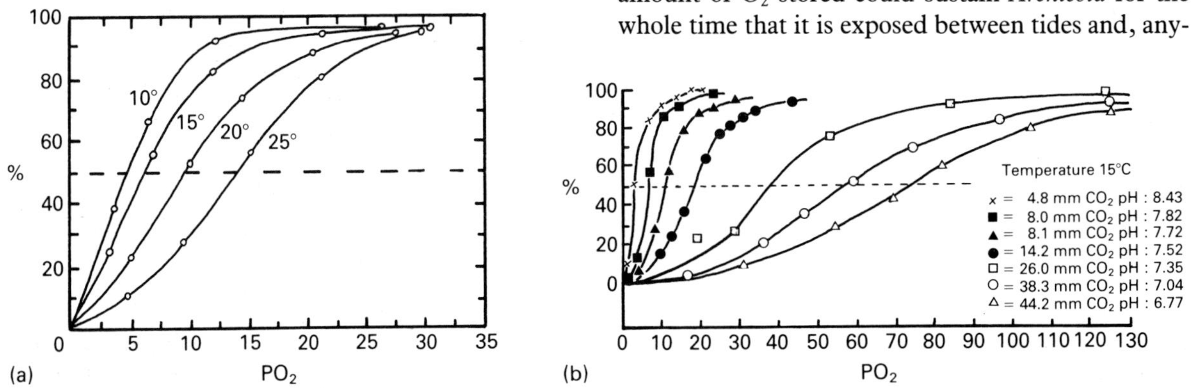

Fig. 11.8. Effect of temperature (a) and CO_2 and pH (b) on dissociation curves. Both results are for haemocyanin from crustaceans. (After Waterman 1960.)

way, under these conditions it is probably capable of some aerial respiration. Finally, it can be noted that the dissociation curve of *Arenicola* is relatively unaffected by low pH, and this again may be an adaptation to life in anoxic muds, which are often acidic.

11.4.4 Respiratory organs

Another limitation on oxygen uptake is the size of the respiratory surface. An unspecialized surface is adequate for flatworms and long, thin nemertines, nematodes and some annelids. However, further increases in size or activity and the evolution of outer protective and relatively impermeable coverings, or shells, required the evolution of vascularized respiratory surfaces. Evaginated surfaces are common in aquatic forms, e.g. 'gills' of some polychaetes and arthropods (Fig. 11.6), ctenidia of molluscs, gill books of *Limulus* and podia of some echinoderms. Figure 11.9 shows the ctenidial system, in diagrammatic form, of an aquatic gastropod. Notice that blood is pumped through the 'gill', by the heart, in the opposite direction to the flow of water (effected mainly by lateral cilia). This countercurrent flow is typical of gills (though not those of *Arenicola* or cephalopods) and ensures that water with lowest P_{O_2} is in contact with the least oxygenated blood, hence maximizing the efficiency of transfer of O_2. By contrast, invaginations are common in terrestrial groups, e.g. tracheae (Fig. 11.6), lung books and sieve and tube tracheae of arachnids (Fig. 11.10) and 'lungs' of pulmonate molluscs. Some aquatic forms have invaginated surfaces—the respiratory trees of holothurian echinoderms (Fig. 11.11) and the anal structures of some insect larvae (Fig. 11.6). Some aquatic insect larvae have physical gills and plastrons (Fig. 11.12), and the way these work is explained in the legend to Fig. 11.12.

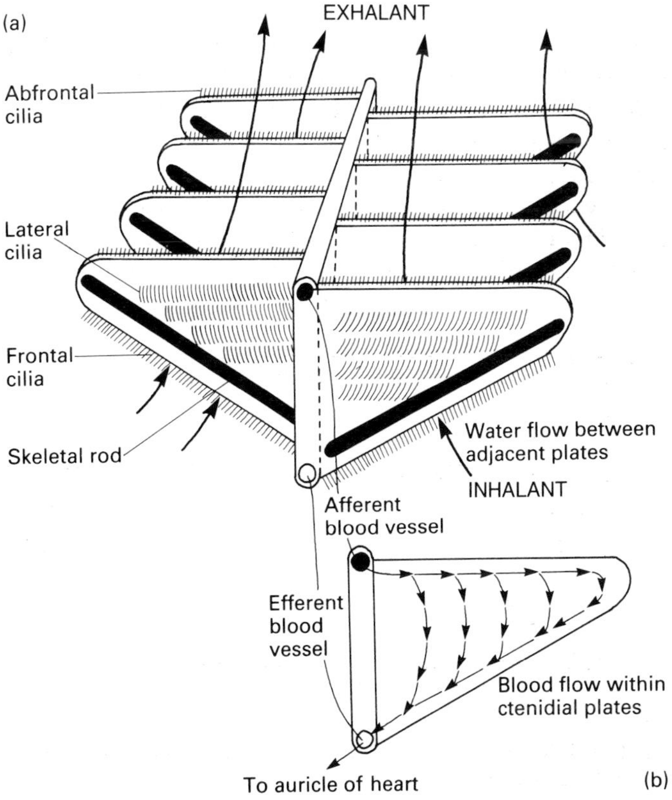

Fig. 11.9. Ctenidial system of a gastropod: (a) shows flow of water over 'gills'; (b) shows flow of blood through them. (After Russell-Hunter, 1979.) See also Fig. 5.18.

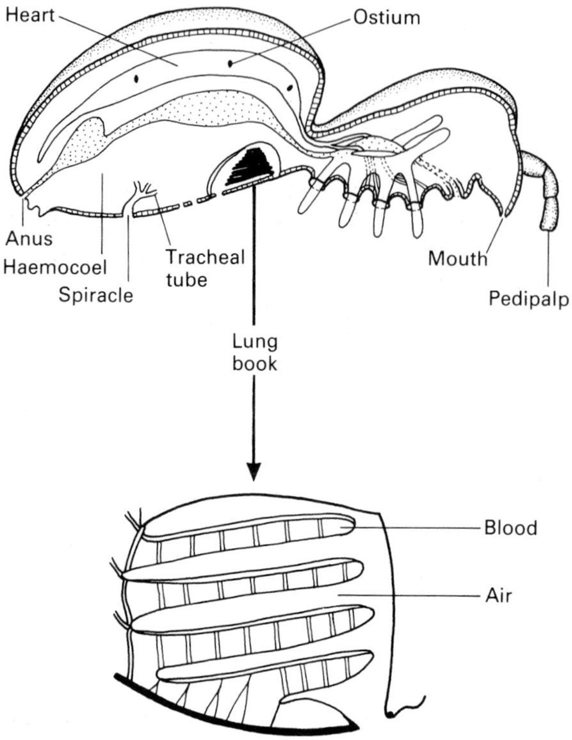

Fig. 11.10. Lung-book of a spider (after Snow, 1970). See also Fig. 8.11.

11.4.5 Ventilation

One other mechanism that assists in oxygen uptake is the ventilation of the respiratory surfaces. Again this ensures that the supply of oxygen to the respiratory surface is continuously replenished so that the gradient across the surface is maximized. This occurs in animals both with and without modified respiratory surfaces, but not in the tracheal systems of insects. Arrangements range from lateral cilia on the gill lamellae of bivalves to the muscular pumps servicing the respiratory trees of holothurians and anal structures of dragonfly naiads. Tubiculous polychaetes use ciliary currents and/or peristaltic movements (above). Crustacea commonly depend upon undulating appendages. Ventilation often increases as environmental Po_2 is reduced. Figure 11.13 shows apparatus for simultaneously measuring activity patterns (by measuring changes in electrical impedance on a respirometer) and oxygen consumption (by electrodes) in aquatic leeches. As ambient Po_2 falls, increased activity brings about increased O_2 uptake relative to resting leeches. However, random movements can be as effective as regular ventilation movements in improving the supply of O_2.

11.5 Measuring metabolism

Most of the energy that goes into respiration ultimately appears as heat (Fig. 11.14). Hence, if the heat emanating from animals could be measured, it would provide a useful and accurate assessment of metabolism. Devices that are sensitive enough to record the small quantities of heat emanating from

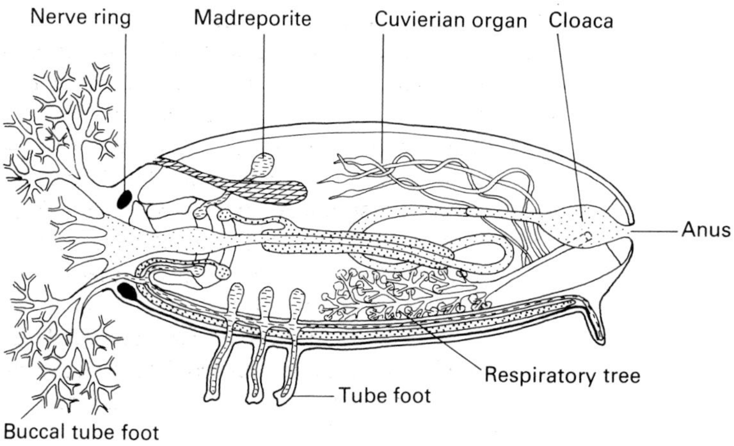

Fig. 11.11. Diagrammatic longitudinal section of an holothurian echinoderm showing respiratory trees (after Nichols, 1969). See also Fig. 7.27.

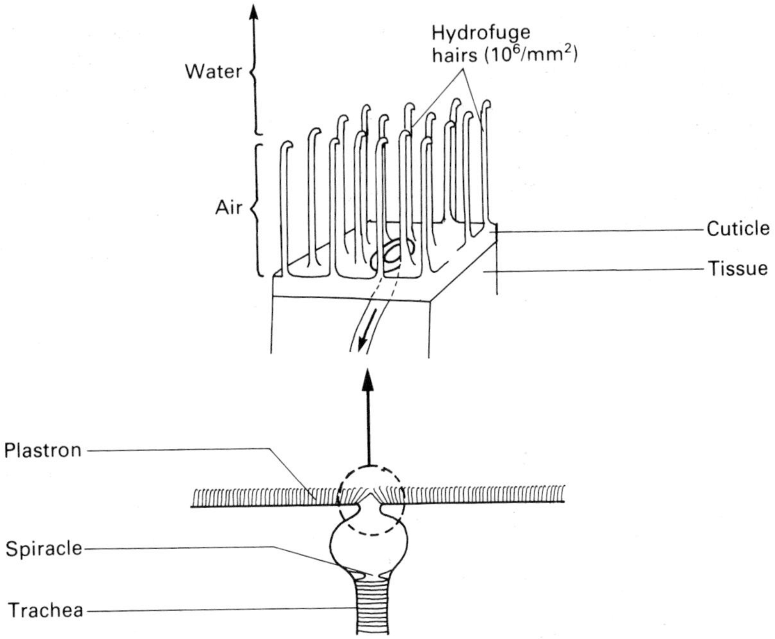

Fig. 11.12. A plastron. Part of the body surface is covered with hydrophobic hairs providing a non-wettable surface, where air remains permanently. It thus acts as a non-compressible gill into which oxygen diffuses from the water; i.e. as O_2 is withdrawn the thick pile prevents intrusion of water, so the volume remains constant and PO_2 must fall causing O_2 to diffuse in from the aquatic environment. In the aquatic insect *Aphelocheirus* the plastron can withstand pressures of several atmospheres before collapse. After Ramsay (1962) and Eckert & Randall (1983).

invertebrates, involving very sensitive thermopiles, are available—they are called direct microcalorimeters—but they are not commonly used. A trace from a calorimetric experiment, involving one of them, with an aquatic oligochaete, *Lumbriculus variegatus*, as experimental subject, is illustrated in Fig. 11.15. The traces show: (I) the aerobic state; (II) the anoxic state; (III) what happens after the animals are poisoned. Notice that aerobic metabolism occurs at about four times the rate of anoxic.

It is easier to measure oxygen consumption than heat production. Various techniques are available; these range from chemical titration of oxygen in aqueous solution to physical measures of the volume or pressure of the gas. Several kinds of very accurate and sensitive electrodes are also available. Hence oxygen levels can be measured in the environment of sealed systems containing animals, or in the inflows and outflows of open systems. By definition, this kind of technique will only monitor aerobic processes and so is likely to underestimate respiratory metabolism because it is possible that invertebrates use anaerobic processes in conjunction with aerobic ones even when there is plenty of oxygen available. Just when, how and under what circumstances of reducing PO_2, metabolism shifts from an aerobic to an anaerobic basis is not exactly clear and must await some more results from direct calorimetry.

Nevertheless, most studies of respiration have used oxygen consumption as a measure of metabolism. This kind of information is used below to describe how various factors influence respiratory metabolism. However, there are still complications since the metabolism of invertebrates is so sensitive to the condition of the animal or the surroundings in which it is being kept that what is being measured in one experiment need not be comparable with that being measured in others. The following classification is helpful. Total respiratory metabolism comprises: *standard metabolism*, recorded when the organism is at rest; *routine metabolism*, recorded in routinely active animals; *feeding metabolism*, recorded in animals that have just fed; and *active metabolism*, recorded in animals undergoing substantial activity. Experimentalists often aim at standard metabolism as a repeatable measure of metabolism.

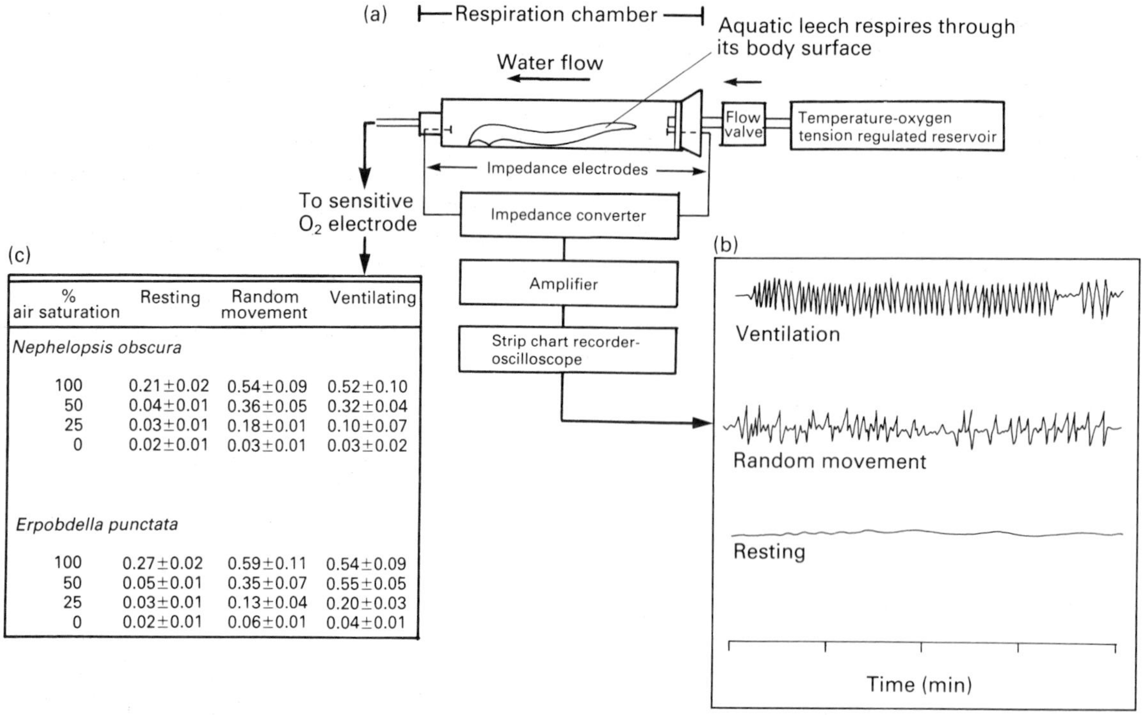

% air saturation	Resting	Random movement	Ventilating
Nephelopsis obscura			
100	0.21±0.02	0.54±0.09	0.52±0.10
50	0.04±0.01	0.36±0.05	0.32±0.04
25	0.03±0.01	0.18±0.01	0.10±0.07
0	0.02±0.01	0.03±0.01	0.03±0.02
Erpobdella punctata			
100	0.27±0.02	0.59±0.11	0.54±0.09
50	0.05±0.01	0.35±0.07	0.55±0.05
25	0.03±0.01	0.13±0.04	0.20±0.03
0	0.02±0.01	0.06±0.01	0.04±0.01

Fig. 11.13. Schematic representation of apparatus described in the text (a) and some results from it (b & c). (b) shows patterns of movements picked up by the impedance system and (c) shows levels of O_2 consumption (μl mg dry wt^{-1} h^{-1} ± SE) of two species of leech at different ambient O_2 concentrations when individuals were carrying out particular patterns of movement. Note that movement is correlated with increased O_2 consumption at all concentrations except the least, but there is little difference between ventilatory and random patterns of movement. (After Wrona & Davies, 1984.)

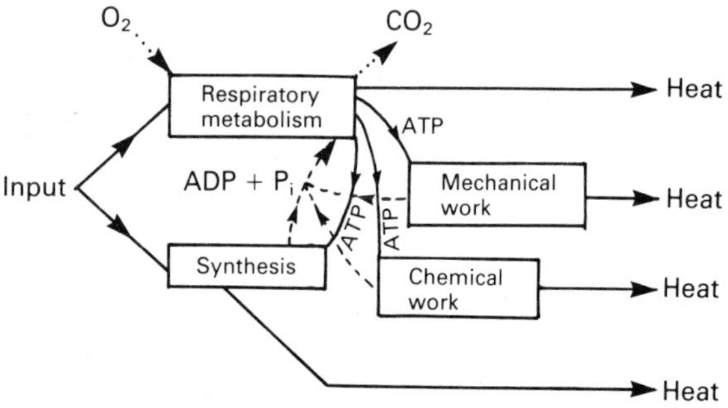

Fig. 11.14. Flow diagram of use of food energy. Respiratory processes generate ATP inefficiently, with heat loss, and most of the energy carried by ATP appears as heat after doing work.

11.6 Factors influencing respiration

The following is a list of some of the better studied factors that influence the respiration of invertebrates. They are ordered in decreasing intimacy of association with organisms, i.e. from what have been described as intrinsic to extrinsic factors.

11.6.1 Body size

It is reasonable to expect big invertebrates to respire more than small, but, since oxygen uptake is surface dependent (above) and body surfaces increase in two dimensions whereas body mass increases in three dimensions, it is also expected that the relationship between body mass and oxygen consumption will not be one of simple proportionality. Comparisons of different-sized animals within species and of species with different body sizes indicates that standard respiratory rate increases, but at a reducing rate, with body size, represented by the following equation:

Resp. rate = a $(\text{weight})^b$ 11.1

where a and b are constants and b is usually less than 1. Taking logarithms gives:

log. Resp. rate = K+b(log weight) 11.2

where K = log a. Hence plotting the logarithms of oxygen consumption against the logarithms of body mass should give a linear relationship with a slope equivalent to b (Fig. 11.16).

Dividing equations 11.1 and 11.2 throughout by weight gives:

Resp. rate per unit weight = $a(\text{weight})^{b-1}$

log (Resp. rate per unit weight) = $K+(b-1)$ log weight

Since b is less than 1, $b-1$ will be negative, so plotting log Resp. rate per unit weight against log weight should give a straight line with a negative slope (Fig. 11.17). Hence, as expected above, respiratory rate per unit weight reduces with weight. However, if these relationships were simply a matter of surface area not keeping pace with body mass as mass increases, then b should be 0.67 and $b-1$ should be -0.33. The rationale behind this is straightforward.

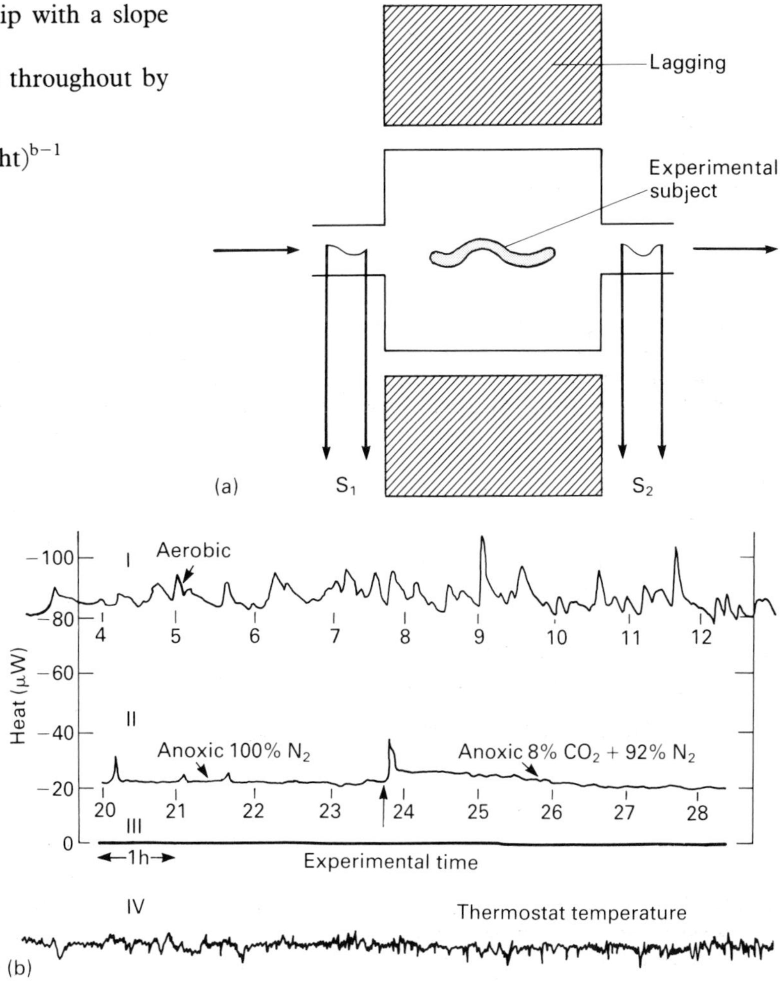

Fig. 11.15. Direct microcalorimeter. (a) The apparatus: very sensitive sensors (S_1 and S_2) measure the temperature of water flowing into and out of a well-lagged chamber containing the experimental subject. (b) Some results described more fully in the text. (After Gnaiger, 1983.)

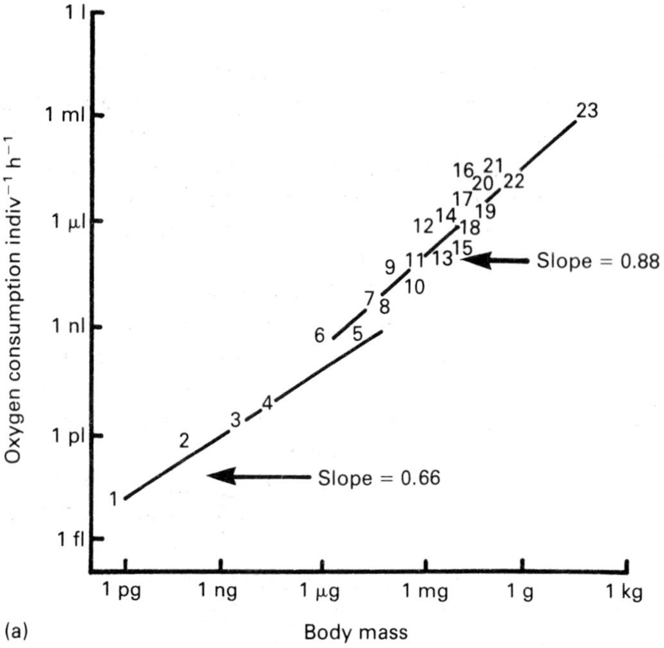

(a)

Code number	Group	*n*	*b*
	Unicellular ectotherms		
1	Bacteria	5	0.68 (1–2)
2	Fungi	2	
3	Flagellates	4	1.33
4	Ciliates	5	0.28
5	Rhizopoda	5	0.93
	Multicellular ectotherms		
6	Nematoda	24	0.82
7	Microcrustacea	12	0.91
8	Mites	71	0.61
9	Collembola	29	0.74
10	Isoptera (larvae)	4	0.75
11	Enchytraeidae	61	0.87
12	Coleoptera (larvae)	17	0.67
13	Isoptera (larvae)	21	0.94
14	Formicidae (workers)	23	1.14
15	Lumbricidae (cocoons)	3	1.00
16	Opiliones	30	0.69
	Multicellular ectotherms cont.		
17	Diplopoda	77	0.79
18	Aranea	6	0.81
19	Isopoda	40	0.69
20	Mollusca	6	0.76
21	Coleoptera (adults)	14	0.81
22	Lumbricidae (adults)	18	0.76
23	Macrocrustacea	3	0.81

(b)

Fig. 11.16. (a) Double-logarithmic plots of O_2 consumption against body mass for a variety of taxa. (b) List of taxa used in compiling a, each with the slopes *b* of log–log plots of the data for the taxon. (After Phillipson, 1981.)

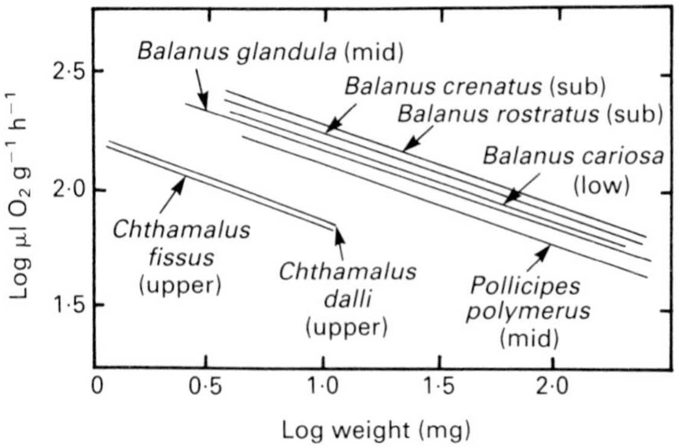

Fig. 11.17. Relationship between log weight-specific O_2 consumption and log body weight, for several species of barnacles. (After Newell & Branch, 1979.) See text for details.

Respiration, being surface dependent (above), should be proportional to body surface area which, in geometrically similar bodies, is in turn proportional to body length squared (l^2). Mass is equivalent to volume and hence, again assuming geometrical similarities, to l^3. So respiration should be proportional to the cubed root of mass ($\sqrt[3]{l^3} = l$) squared ($= l^2$), i.e. to ($\sqrt[3]{M^2} = M^{0.67}$). However, rarely are values of b precisely 0.67 and often they approximate more to a value between 0.67 and 1 (Fig. 11.16). Hence, it is usually assumed that though geometrical factors play a part in determining the size dependency of respiration, they cannot be the only or perhaps even the main basis for this relationship. Just exactly what factors are involved remains to be elucidated.

11.6.2 Activity

A mobile limpet consumes about 1.4 times more oxygen than an inactive one; this figure is about 2 for *Gammarus*, an amphipod, 3 to 4 for *Palaemonetes*, an estuarine shrimp and more than 100 for a flying locust. Again these observations are not unexpected; movement involves the beating of cilia and flagella and the contraction of muscles and so leads to an enhancement of metabolism over the standard rate.

11.6.3 Feeding

Oxygen consumption often increases immediately after a meal, only to subside again shortly afterwards, e.g. by between *c.* 10 and 40% in the crustacean *Macrobrachium rosenbergii*. This postprandial respiratory response is variously referred to as *specific dynamic action* (SDA) or a *specific dynamic effect* (SDE) or a calorigenic effect.

Explanations for SDE include:

1 the cost of processing the food in the gut;

2 the cost of degrading and excreting proteins absorbed from food that are in excess of requirements;

3 the cost of using raw materials obtained from the food to synthesize new tissues.

In a careful piece of work on the mussel, *Mytilus edulis*, Bayne & Scullard (1977) recorded a 25% increase in the postprandial oxygen uptake of this animal, of which more than 80% was attributable to the cost of filtering the food and less than 20% to the cost of ammonia excretion.

Continuous exposure to low rations is often associated with a reduced oxygen consumption. Part of this is due to the absence of SDE, but part can be due to reductions in levels of activity and even economization in maintenance metabolism. Certainly, different species of barnacles that occur at different levels on the shore and experience different levels of food supply have different rates of oxygen consumption (Fig. 11.17). *Balanus crenatus* and *B. rostratus* have an uninterrupted food supply for much of the tidal cycle and have high metabolic rates. Lower and mid-shore species, such as *Balanus glandula* and *B. cariosa*, have intermediate levels whereas upper shore *Chthamalus* species, that experience a substantially reduced feeding time with each tidal cycle and also for several days during neap tides, have very low rates of oxygen consumption.

11.6.4 Temperature

Changes in ambient temperatures can only influence metabolism if they influence body temperatures—something which is *generally* true for invertebrates but not for mammals and birds. Invertebrates are therefore described as *poikilothermic* (*poikilo* = Gk varied) and mammals and birds as *homeothermic* or *homoiothermic* (*homoio* = Gk keeping same). However, many invertebrates, such as those in the tropical open oceans and the deep seas, live at constant temperatures and are therefore not strictly poikilothermic. Hence a more general term to describe the metabolic properties of invertebrates is *ectothermic*, referring to the source of the heat, and mammals and birds are *endothermic*. But even this classification is not truly general, for some insects generate large amounts of heat in their flight muscles (by endothermy) and this can be used to maintain a constant thoracic temperature (i.e. homoiothermy). For example, bumblebees cannot fly if their flight muscle temperature drops below 30°C or rises above 44°C. At least 90% of energy expended by a flying bee is released as heat within the thorax and, during vigorous flight, the temperature here can rise several degrees within seconds. Moreover, there is evidence that in free flight bumblebees can maintain a fixed internal temperature at different external temperatures. Large bees (primarily queens) can remain in free flight at 0°C, and under these conditions maintain a thoracic temperature at 30°C and above. Such regulation is maintained in a small part passively—by the thick pile of insulating hair on the thorax—but mainly by regulating flight effort.

In conclusion, then, no one term is strictly applicable to all the thermal characteristics of invertebrates but most are poikilothermic and ectothermic.

Within the normal temperature range of a poikilotherm, respiration is increased by increasing ambient temperature. However, as might be expected from the way chemical reactions respond to temperature, the relationship is non-linear—each additive increment of temperature causing a multiplicative increment in respiration. A widely used index of the effect of temperature on metabolism, that takes this into account, is the Q_{10} value, defined as follows:

$$Q_{10} = \frac{R_2^{10/t_2-t_1}}{R_1}$$

$$\log Q_{10} = \frac{(\log R_2 - \log R_1)\,10}{t_2 - t_1}$$

where R_1 and R_2 are metabolic rates (e.g. of oxygen consumption) at temperatures t_1 and t_2 °C respectively. Note that if $t_2 - t_1 = 10$°C then Q_{10} is given by the ratio of R_2 and R_1, so Q_{10} indicates the factor by which R is multiplied up for each 10°C increase in temperature. Following an acute change in temperature Q_{10}s of 2 and more are often recorded, but Q_{10} can change dependent on the range of temperatures over which it is measured.

Moreover, the immediate response following a temperature change need not be a lasting one. Figure 11.18 shows what happens when mussels, *Mytilus edulis*, are transferred from an ambient temperature (10°C) to 5 and 15°C. Oxygen consumption increases dramatically for the 15°C group and then reduces to a lower, steady level, whereas it first reduces for the 5°C group and then shifts with time to a higher steady level. This process is known as *acclimation*; note that Q_{10} values for acclimated values (after the adjustments) have to be lower than for the acute response and for perfect acclimation tend to 1. Such a response is thought to be adaptive because it allows con-

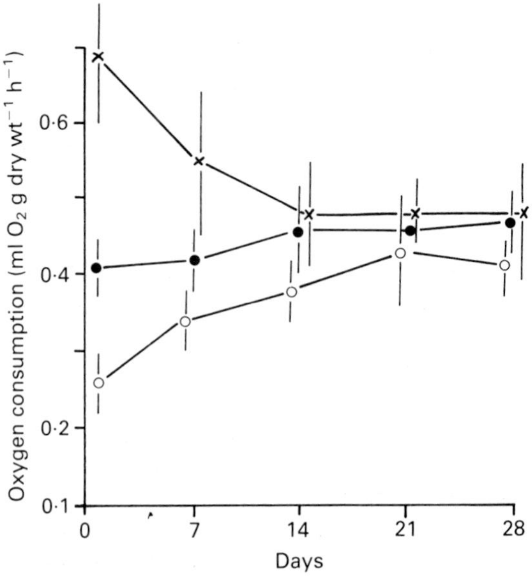

Fig. 11.18. Acclimation by *Mytilus*. ● = continuously at 10°C; × = 10°C to 15°C; ○ 10°C to 5°C, see text. (After Widdows & Bayne, 1971.)

servation of energy at high temperatures and the maintenance of a high level of ATP production at low temperatures so that body maintenance and vital activity can be kept going. This kind of response has been recorded in various platyhelminths, molluscs, annelids and echinoderms.

Acclimation can also operate in the reverse direction; i.e. oxygen consumption reduces further at low temperatures and increases further at high temperatures with acclimation time, so Q_{10}s between acclimated rates increase from the acute rates. This occurs in some limpets, both freshwater (*Ancylus fluviatilis*) and marine (*Patella aspera*). At low temperature this kind of acclimation may be adaptive, operating like aestivation, and allowing the conservation of energy under winter food shortage or oxygen depletion, perhaps associated with ice cover, in freshwater habitats. Elevated metabolism at high temperature is more difficult to explain and subtle rate compensations are probably involved.

The acclimatory response illustrated in Fig. 11.18 is called *positive acclimation*; that of the limpets is called *negative/reverse acclimation*. The extent of these processes varies from species to species and might also depend upon the state of the animal. Thus in some intertidal invertebrates standard metabolism acclimates (positively) quickly and almost completely to temperature change ($Q_{10} \simeq 1$) whereas the Q_{10} values of routine metabolism are always greater than 1. This response is possibly adaptive because littoral animals are subject to considerable fluctuations in temperature between tides when they are inactive and cannot feed. Some ecological patterns are discernible in the occurrence of low Q_{10}s for the oxygen consumption of quiescent organisms: subtidal organisms, such as *Strongylocentrotus franciscanus* and *Anemonia natalensis*, as well as lower-shore, burrowing animals such as *Diopatia cuprea* and some bivalves, that do not experience marked, short-term variations in ambient temperature, have $Q_{10}s > 1$, whereas intertidal organisms such as *Littorina littorea*, *Strongylocentrotus purpuratus*, *Macoma balthica*, *Actinia equina* and *Bullia digitalis* have $Q_{10}s \simeq 1$. But there are also many contradictions, with some intertidal organisms such as *Patella vulgata* showing no suppression of Q_{10} and some subtidal organisms, such as the polychaete *Hyalinoecia*, having low Q_{10}s.

11.6.5 Oxygen concentration (PO_2)

Aquatic invertebrates often have to suffer low PO_2; for example, reduced PO_2 is symptomatic of the organic loading associated with pollution in fresh waters and can occur during air exposure at low tides, below a thermocline in lakes and during burrowing in reducing substrata. Some invertebrates respond by adjusting ventilation and blood circulation rates such as to maintain a fixed level of oxygen consumption despite reductions in ambient PO_2 (e.g. Fig. 11.19). These are known as *regulators*. On the other

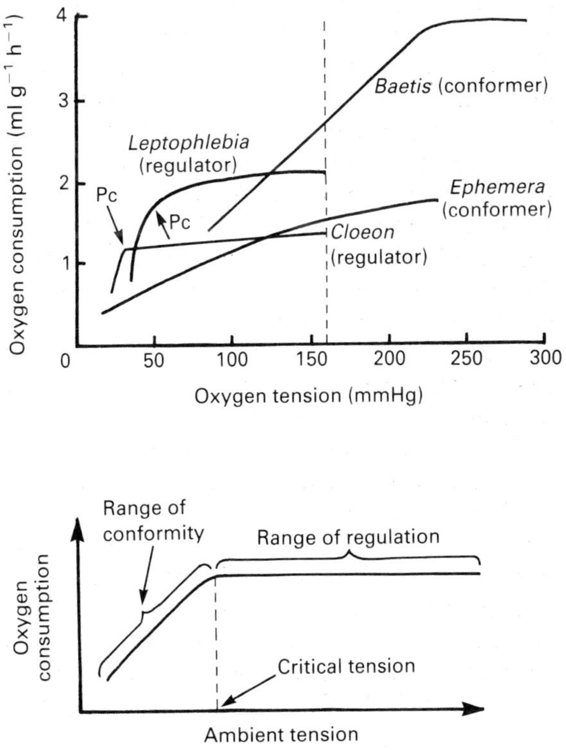

Fig. 11.19. Oxygen consumption as a function of ambient O_2 tension. The vertical dashed line depicts the O_2 content of water in equilibrium with the atmosphere. (After Fox *et al.*, 1937.)

Table 11.2 Examples of regulators and conformers. Pc is the critical Po_2 at which regulation breaks down and regulators become conformers

Regulators	Pc*	Conformers
Paramecium (ciliate)	50	*Spirostomum*
Tetrahymena (ciliate)	2.5	
Trypanosoma (flagellate)		
Pelmatohydra (hydrozoan)	60	Various sea anemones
Aurelia (scyphozoan)	120	*Cassiopea* (scyphozoan)
Tubifex (oligochaete)	25	*Nereis* (polychaete)
Lumbricus (oligochaete)	76	*Erpobdella* (leech)
Mytilus (bivalve)	75	*Sipunculus* (sipunculan)
Helix (pulmonate)	75	*Limax* (pulmonate)
Loligo (squid)	45	
Cambarus (crayfish)	40	*Limulus* (chelicerate)
Uca (crab)	4	*Homarus* (lobster)
Cloeon (ephemerid nymph)†	30	*Baetis* (ephemerid nymph)†
Hyalophora (moth larva)	25	*Tanytarsus* (chironomid larva)

*The values of Pc must be accepted with caution since the transition from regulation to conformity is seldom sharply defined. Furthermore the shape of the oxygen uptake/tension curve is often influenced by degree of acclimation to abnormal Po_2s.
†See also Fig. 11.19.

hand, the oxygen consumption of some species changes in concert with environmental Po_2. These are known as *conformers* (Fig. 11.19). Regulators cannot regulate without limit with reducing Po_2 and the critical Po_2 at which a regulator becomes a conformer is often denoted as Pc. Lists of conformers, regulators and Pc values are given in Table 11.2. Often animals naturally exposed to low Po_2 are adapted as regulators and have low Pc values. Thus, tubificid worms survive in organically polluted stretches of rivers and have a low Pc value of 25 mmHg. Similarly *Uca*, the fiddler crab, is a common inhabitant of marshes and mudflats, and is a good regulator. On the other hand *Baetis* inhabits fast-flowing, well-oxygenated streams.

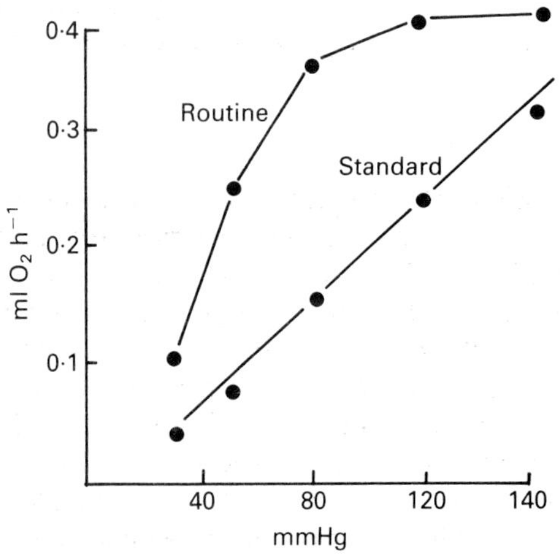

Fig. 11.20. Effect of ambient O_2 tensions on the metabolism of *Mytilus* (after Bayne *et al.*, 1976). See text for further discussion.

Again different components of metabolism can respond differently to Po_2; for *Mytilus*, for example, standard metabolism conforms whereas routine metabolism regulates down to about 80 mmHg (Fig. 11.20).

Following exposure to hypoxia or anoxia, oxygen consumption often becomes increased relative to what it was before exposure. This is the paying of the so-called oxygen debt. It includes recharging of the phosphagen and ATP pools and the disposal of anaerobic end products. However the process is not straightforward, for there is often a lack of correlation between end-product accumulation and the size of the oxygen debt and typically ATP and phosphagen levels are recharged rapidly in the initial part of the process. Other adjustments must be involved; e.g.

intracellular pH, reduced by accumulation of acidic end products, must return to its preanoxic state, and extra oxygen may be required to resaturate body fluids and respiratory pigments.

In conclusion it should be noted that too much oxygen can be as harmful as too little. Free oxygen (superoxide) radicals can denature macromolecules, particularly proteins. This is important when oxygen concentrations are very high, as they can be in the tissues of invertebrates harbouring symbiotic algae (such as many coral-reef dwelling organisms), for oxygen is an end product of photosynthesis. In response, these organisms have evolved cellular defences against oxygen toxicity, i.e. the enzyme superoxide dismutase (SOD), which can eliminate damaging superoxide radicals, and catalase which eliminates the H_2O_2 generated by SOD. Figure 11.21 shows that for a number of invertebrate groups, SOD activity increases with chlorophyll concentration in their tissues, i.e. the potential for the photosynthetic production of oxygen by the symbionts. However, the relationship is not the same for all groups and these differences can be ascribed to differences in the localization of the symbionts: SOD activity is highest in Cnidaria where zooxanthellae are exclusively or predominantly intracellular (cytoplasm is directly exposed to oxygen) whereas for the urochordate *Didemnum*, where the symbiotic prochlorophytes are not only extracellular but also extraorganismal (occurring in the lining of the common cloacal chamber), oxygen can be quickly removed in the outgoing current without coming into contact with animal cytoplasm. Some of the lower values in the cnidarian range are from species that are shaded in caves and under ledges.

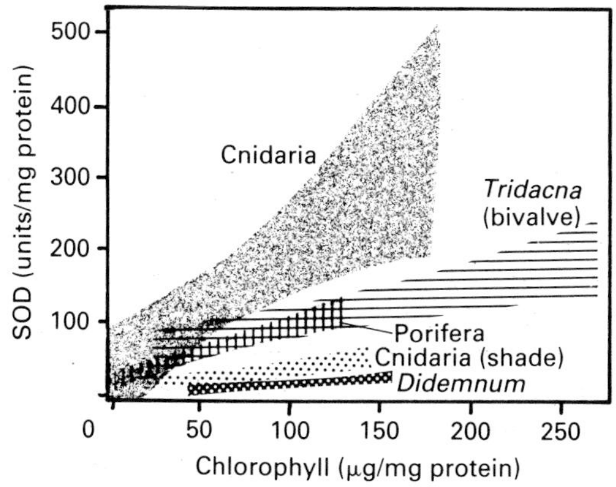

Fig. 11.21. Relationship between SOD activity and chlorophyll concentration for major groups of coral-reef invertebrates (after Shick & Dykens, 1985).

11.6.6 Salinity

Because changes in salinity can involve active behavioural and physiological responses in marine invertebrates it is to be expected that they are accompanied by changes in oxygen consumption. *Metapenaeus monocercus* is a prawn that occurs in both fully marine and brackish situations. Oxygen consumption of individuals from a brackish population was lowest in 50% sea water and increased towards hypo- and hyperosmotic conditions. By contrast the oxygen consumption of individuals from a marine population was lowest in 100% sea water and increased to a maximum in 25% sea water.

The actual costs of osmoregulation, and active transport, are probably very slight—so most of these differences are probably due to whole-organism responses, such as differences in activity patterns. For a more detailed discussion of this see Newell (1979).

11.7 Conclusions

'Respiration' refers to processes that release biologically useful energy from food materials, particularly carbohydrates. These processes are largely, but not exclusively, aerobic, and so 'respiration' is also used to refer to the process of obtaining O_2 and getting rid of waste CO_2. The biochemical aspects of the process are sometimes called 'internal' or 'tissue respiration' and the physiological aspects of O_2 uptake as 'external respiration'. The reader should now be familiar with the mechanism behind both, the systems that obtain O_2 from the environment and deliver it to the tissues, and the intrinsic and extrinsic factors that influence metabolic rates.

11.8 Further reading

Alexander, R. McN. 1971. *Size and Shape*. Edward Arnold, London.

Bayne, B.L. & Scullard, C. 1977. An apparent specific dynamic action in *Mytilus edulis*. *J. mar. biol. Assoc., UK,* **57**, 371–8.

Calow, P. & Townsend, C.R. 1981. Resource utilization in growth. In: C.R. Townsend & P. Calow (Eds) *Physiological Ecology: an Evolutionary Approach to Resource Utilization.* Blackwell Scientific Publications, Oxford.

Eckert, R. & Randall, D. 1983. *Animal Physiology,* 2nd edn. W.H. Freeman, New York.

Heinrich, B. 1979. *Bumble-bee Economics.* Harvard University Press, Cambridge, Massachusetts.

Jones, J.D. 1972. *Comparative Physiology of Respiratory Mechanisms.* University of Pennsylvania Press, Philadelphia.

Livingstone, D.R. 1983. Invertebrate and vertebrate pathways of anaerobic metabolism: evolutionary considerations. *J. Geol. Soc.* **140,** 27–38.

Newell, R.C. 1979. *Biology of Intertidal Animals,* 3rd edn. Marine Ecological Surveys Ltd., Kent.

Newell, R.C. & Branch, G. 1979. The influence of temperature on the maintenance of energy balance in marine invertebrates. *Adv. mar. Biol.,* **17,** 329–46.

Phillipson, J. 1981. Bioenergetic options and phylogeny. In: C.R. Townsend & P. Calow (Eds) *Physiological Ecology: an Evolutionary Approach to Resource Utilization.* Blackwell Scientific Publications, Oxford.

Ramsay, J.A. 1962. *A Physiological Approach to the Lower Animals.* Cambridge University Press, London.

Schmidt-Nielsen, K. 1979. *Animal Physiology: Adaptation and Environment.* Cambridge University Press, Cambridge.

12

EXCRETION, IONIC AND OSMOTIC REGULATION AND BUOYANCY

Materials in excess of metabolic requirements have to be removed from the bodies of animals. They include indigestible residues that are lost in the faeces (Chapter 9) and CO_2 that is lost in respiration (Chapter 11). There are other excess materials, though, and these include water, various ions and the breakdown products of excess proteins and amino acids. It is the latter, nitrogen-containing substances, that are normally referred to by physiologists as excretory products. Nevertheless, the processes involved in the removal of excess ions, water and nitrogen-containing substances from the bodies of invertebrates are often so intimately related that it is sensible to treat them together. In this chapter we therefore start with an account of true excretion, then consider ion–water problems, and finally describe the structure and functioning of so-called excretory systems that are invariably associated with ionic and osmotic regulation but not always with nitrogen excretion.

12.1 Excretion

Waste amino acids derive from excesses absorbed across the gut wall and from the catabolism of proteins. These are most usually broken down further, in an oxidative process that yields ketoacids and ammonia:

$$\underset{\text{amino acid}}{NH_2\text{–}\overset{\overset{\displaystyle R}{|}}{C}HCOOH} + \tfrac{1}{2}O_2 = \underset{\text{ketoacid}}{O{=}\overset{\overset{\displaystyle R}{|}}{C}COOH} + \underset{\text{ammonia}}{NH_3}$$

The ketoacids can easily be used in other metabolic pathways, but because ammonia is extremely toxic it must either be removed rapidly from the body or be stored in a harmless form (below). Ammonia is very soluble and is easily lost from aquatic animals by a process of diffusion across the total-body and respiratory surfaces. Excretion dominated by ammonia is known as *ammoniotelism* and is very common in aquatic invertebrates, including aquatic insect larvae (cf. terrestrial adults, below).

In some types, ammonia can, however, be converted to form less toxic urea:

$$\begin{array}{c} NH_2 \\ | \\ C=O \\ | \\ NH_2 \end{array}$$

The use of this substance in excretion is known as *ureotelism* and is typical of mammals. Its use is not widespread in invertebrates, possibly because it has the disadvantage of being toxic whilst still entailing the loss of considerable amounts of water in its excretion. Ureotelism is, however, practised by some Platyhelminthes, Annelida and Mollusca.

A dominant excretory product of terrestrial invertebrates (and non-mammalian, terrestrial vertebrates) is uric acid—*uricotelism*—and this is a member of a general class of molecules, the purines, others of which are used by some invertebrates—*purinotelism*.

The main reason for the evolution of these excretory products is that they have low toxicity and because of their low solubilities are excreted as solids that require little water for removal. Uricotelism is of importance in the Onychophora, Uniramia and to a lesser extent in the Crustacea and Mollusca. The main excretory product of Arachnida is guanine, another purine.

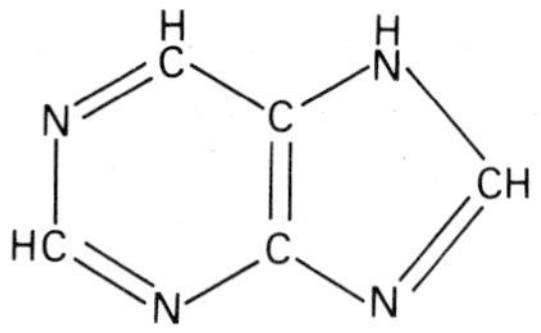

General purine

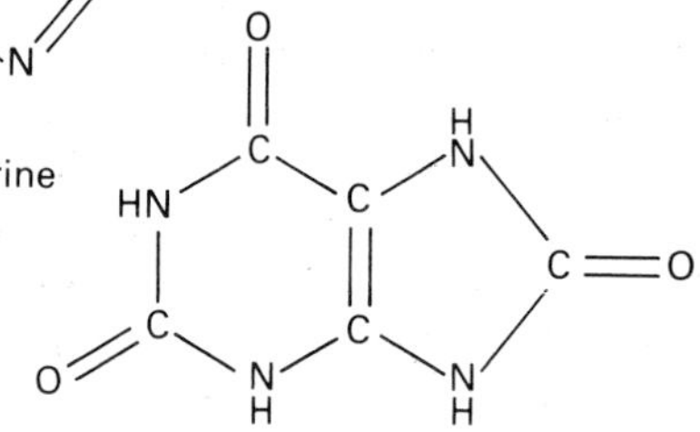

Uric acid

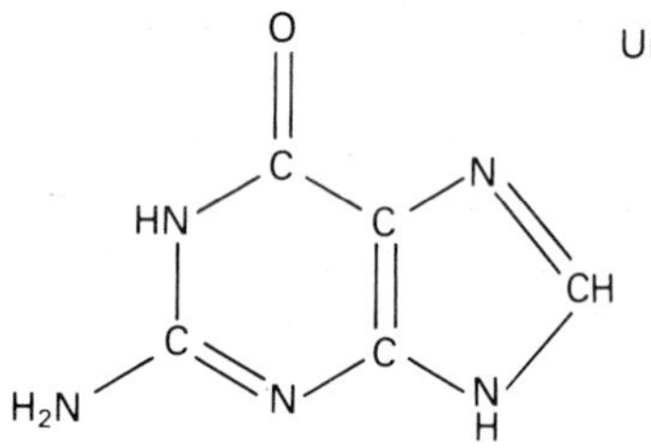

Guanine

An exception to these trends is found in the Crustacea. Here, even in terrestrial isopods, ammoniotelism predominates. The cuticular surfaces of these animals are permeable, being largely devoid of the epicuticular wax that occurs in insect cuticles (Section 12.2.5), and ammonia can be readily excreted by gaseous diffusion. The tissues are also apparently less susceptible to ammonia than those of most animals. Moreover, the emission of ammonia not only serves an excretory role but it also assists in exoskeleton formation, i.e. the alkaline conditions created at the surface by ammonia provide carbonate (from dissolved CO_2) for $CaCO_3$ precipitation. Molluscs might also make use of ammonia in the hardening of shells. Other terrestrial invertebrates that excrete ammonia directly are some oligochaetes and myriapods, but these, like the isopods, live in humid microhabitats where the risks of desiccation are not serious.

Metabolic economy must also be another factor that has played a part in the evolution of excretory systems. Thus purines might be less toxic than ammonia, but they are associated with carbon and hence potential energy. Urea is intermediate in these respects. Table 12.1 compares and contrasts the toxicity, water requirements and energy/carbon losses associated with each of the three main products. Just how important these losses are in the economy of the whole organism is, however, problematical, and of more importance might be the energy lost (mostly as heat) in the metabolic processes leading to the formation of the excretory products.

It should also be realized that excretory substances, particularly less toxic ones such as the purines, need not leave the body at all. This is certainly true in terrestrial and aquatic pulmonate snails in which uric acid can accumulate to considerable extents as snails get larger and older. Urea has also been found to accumulate in the tissues of the tropical pulmonate, *Bulimulus*, during aestivation. The significance of this is uncertain but, as well as serving an excretory function, the enhanced urea concentration in the tissues and body fluids of this snail may cause an increased osmotic pressure and thus a reduced evaporative water loss. And this might increase survival during prolonged drought. Uric acid also accumulates during development in cleidoic eggs. Again pulmonate snails provide good examples—the uric acid content of the egg of *Lymnaea* increases from about 0.5% fresh weight at cleavage to about 4.5% fresh weight at hatching. And it has been argued that the evolution of the cleidoic condition might have been the prime impetus for the evolution of

Table 12.1 'Economics' of excretory products. * Rough quantitative index. Data from Pilgrim (1954)

		Heat combustion			
	C/N	kJ mol^{-1}	kJ mol^{-1} N^{-1}	Toxicity	Water requirements
Ammonia	0	378	378	***	***
Urea	0.5	638	319	**	**
Uric acid	1.25	1932	483	*	*

purinotelism. Certainly both traits have been of considerable importance in the conquest of land.

12.2 Osmotic and ionic regulation

The body fluids of animals are dilute saline solutions with sodium chloride as the predominant electrolyte—in other words they resemble sea water. Indeed it is generally considered that this reflects an origin of life within the sea. Yet there are appreciable differences between the body fluids and sea water (Table 12.2) and Macallum in the 1920s saw in this evidence for a gradual change in the composition of the oceans from what it was when life forms originated to what it is now. However, palaeochemical research has since shown that early oceans did not differ appreciably in composition from present-day ones.

An implication of Table 12.2 is, therefore, that even animals living in sea water have to regulate the internal composition of their body fluids. This intensifies in more dilute aquatic environments—estuaries and fresh waters—where the inhabitants have to maintain body fluids more concentrated than their surroundings. The opposite is the case in animals (a) that having occurred in fresh waters evolved more dilute body fluids than sea water but then returned to the sea, and (b) that live in very concentrated media, such as salt lakes. In terrestrial situations, of course, the main problem is the conservation of water.

We now consider the ionic and osmotic challenges associated with these four main ecological circumstances in more detail. A brief definition of various terms, important in understanding ionic and osmotic regulation, is given in Box 12.1.

Table 12.2. Concentrations of ions in plasma or body-cavity fluid as a percentage of the concentration in body fluid dialysed against sea water. After Barrington (1979)

	Na	K	Ca	Mg	Cl	SO_4
Cnidaria						
Aurelia aurita	99	106	96	97	104	47
Echinodermata						
Marthasterias glacialis	100	111	101	98	101	100
Urochordata						
Salpa maxima	100	113	96	95	102	65
Annelida						
Arenicola marina	100	104	100	100	100	92
Sipuncula						
Phascolosoma vulgare	104	110	104	69	99	91
Crustacea						
Maia squinado	100	125	122	81	102	66
Dromia vulgaris	97	120	84	99	103	53
Carcinus maenas	110	118	108	34	104	61
*Pachygrapsus marmoratus**	94	95	92	24	87	46
Nephrops norvegicus	113	77	124	17	99	69
Mollusca						
Pecten maximus	100	130	103	97	100	97
Neptunea antiqua	101	114	102	101	101	98
Sepia officinalis	93	205	91	98	105	22

*This grapsoid crab is the only animal in the table which is hypo-osmotic (ionic concentration 86% that of sea water).

Box 12.1 Some definitions

When solutions are separated by a barrier (membrane) permeable to both solvent and solute:

1 Solutes (ions) pass from the more concentrated to the less concentrated solution by *diffusion* (in fact these ionic movements can be complicated by differences in electrical potential across the membrane and this will be considered in more detail in Chapter 16).

2 Solvent passes from the less to the more concentrated solution by *osmosis*, and *osmotic pressure* is said to force solvent to pass from the low to the high concentration—osmotic pressure is a colligative property, i.e. dependent on the number of solute particles and not on their kind.

Processes 1 and 2 continue until the concentration gradient is abolished.

Membranes associated with organisms are usually permeable to both water and solutes *to some extent* and are subject to simultaneous movement of both, for example:

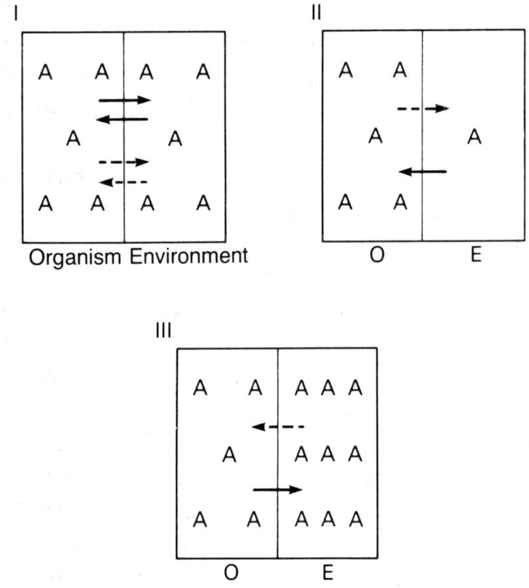

A = ion; - - ► net movement of ions; → movement of water

In I the organism is *iso*-osmotic with its environment

In II the organism is *hyper*osmotic with its environment

In III the organism is *hypo*-osmotic with its environment

Isotonic is often used synonymously with iso-osmotic, etc., but this is not strictly correct—since tonicity describes the response of the cell/organism (in terms of volume) and organisms iso-osmotic with surrounding ions are usually, but not always, *isotonic*, e.g. sea urchin eggs are isotonic in iso-osmotic NaCl but not in iso-osmotic $CaCl_2$. For more explanation see Potts & Parry (1964).

Osmoconformers are animals in which osmotic concentrations of body fluids conform with the environment, i.e. remain iso-osmotic with it.

Osmoregulators are animals that maintain osmotic concentrations of body medium different from the environment.

Ionic regulation is the regulation of concentration of solutes in body fluids that usually differ markedly from their concentration in the environment.

Euryhaline: animals tolerate wide variation in salt concentration of water in which they live.

Stenohaline: animals have limited tolerance to variations in salt concentrations in water in which they live.

Concentrations: it is easy to understand what is meant by a weight of substance but *moles* (= molecular weight in grams or the Avogadro's number [6.023×10^{23}] of molecules of an element or compound) are more elusive. They are, nevertheless, more useful since they measure the number of solute particles—and the osmotic pressure of a solution depends directly upon the number of particles that it contains.

Molarity = moles/litre of solution.

Molality = moles/kg of pure solvent.

12.2.1 Marine environment

Most marine invertebrates have body fluids iso-osmotic to sea water and are osmoconformers. Nevertheless, as we have already seen, the ionic composition of body fluids can differ quite appreciably from that of sea water and so there has to be ionic regulation. Something of this is illustrated in the data in Table 12.2. The following points are worthy of note: (a) *Aurelia* regulates mainly sulphate ions, keeping them below the concentration in sea water—this may be to do with buoyancy, since sulphate ions are heavy and their replacement by chloride might decrease the density of this jellyfish and prevent it from sinking—see also the cuttlefish, *Sepia* (and Section 12.2.2); (b) apart from potassium, echinoderms do not regulate their body fluids appreciably and this may be one feature of their physiology that has largely restricted them to the sea; (c) magnesium levels are generally low in the body fluids of arthropods and this might be to do with the fact that the arthropods are generally very active and magnesium can act as an anaesthetic; note in contrast, though, that magnesium concentration is not particularly depressed in the fast-moving *Sepia*.

12.2.2 Aside on buoyancy

Aquatic animals denser than water would tend to sink, so it is advantageous for swimming animals to maintain a density equal to or less than that of water otherwise they spend energy in keeping themselves from sinking. We have already seen that this can be achieved, at least in part, by ionic regulation. Here we illustrate this further but also draw attention to other adaptations involved with buoyancy.

Heliocranchia is a deep-sea squid. Here the fluid-filled pericardial cavity is very large. The fluid is less dense than sea water, containing far less sodium but a very high concentration of ammonium ions. The latter form as end products of protein metabolism and diffuse into the acidic pericardial fluid and become trapped there. Finally, as noted for *Aurelia* and *Sepia*, the anions in the pericardial fluid of this squid are almost exclusively chloride, with the heavy sulphate ions being excluded.

Other methods of increasing buoyancy involve: (a) the removal of ions without replacement, but then the body fluids would be hypo-osmotic to sea water and this therefore involves osmotic costs; (b) a reduction in heavy substances, for example, some pelagic gastropods such as heteropods and pteropods have reduced or no shells; (c) an increase in substances such as fats and oils, that are less dense than water, is very common in planktonic crustaceans from both marine and freshwater environments; (d) use of gas floats, such as the soft-walled floats of *Physalia* and the rigid-walled floats formed from the gas-filled chambers of nautiloid cephalopods and the cuttlebone of the cuttlefish (Fig. 12.1).

12.2.3 Freshwater environment

If marine animals are immersed in dilute sea water, their body fluids either follow the osmotic conditions of the environment faithfully (osmoconformers) or they resist the dilution of their body fluids (osmoregulators). Examples are given in Fig. 12.2. Estuarine, and to some extent littoral, animals are naturally subjected to periodic dilution, between tides and after heavy rain, and show both patterns of response.

Osmoconformers, though, are not necessarily more stenohaline than osmoregulators. Thus the osmotic pressure of the blood of the mussel, *Mytilus edulis*, closely follows that of the surrounding water whether it be in the North Sea (normal salinity) or in the Baltic ($<$ half normal salinity). One reason for this tolerance is that though it does not regulate extracellular body fluids it does regulate intracellular ones and amino acids might play an important part in this. These can increase in concentration during dilution, so increasing the osmotic pressure of the intracellular fluids. Certainly, excretion of ammonia is accelerated in *M. edulis* when transferred from high to low salinity. This presumably signals the greater catabolism of proteins and hence amino acid production, intracellularly, in conditions of osmotic stress. Indeed, there appears to be genetic variation in leucine aminopeptidase (LAP), an important enzyme for protein catabolism, and variants with higher catabolic rates are found in higher frequencies in low-salinity, estuarine conditions.

Osmoregulators can control water influx and ionic efflux by evolving a less permeable body surface and/or by actively pumping water out and regulating the influx and efflux of ions to body fluids. Figure 12.3 indicates how the permeability of the exoskeletons of crustaceans is lower in animals from more dilute or littoral habitats.

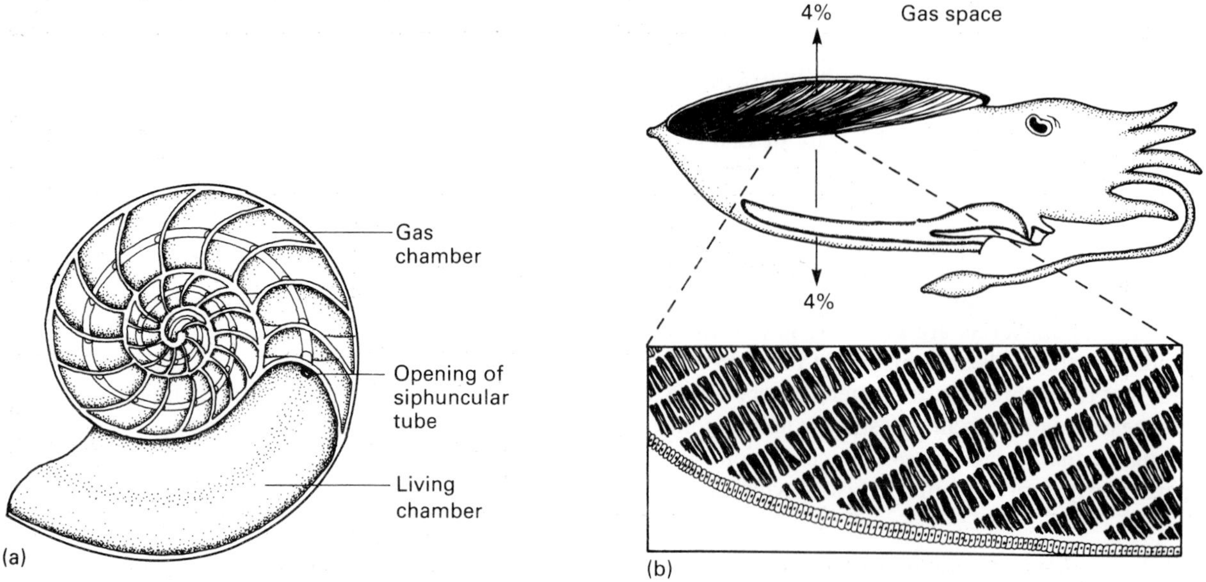

Fig. 12.1. (a) The chambered shell of *Nautilus*. As it grows this animal adds new gas chambers to its shell, one by one. Initially these are filled with a liquid in which NaCl is the major solute. Sodium is removed from this by active transport. The fluid is therefore hypo-osmotic to sea water and water is removed leaving a gas space. Gas diffuses in: N_2 reaches 0.8 atm, the same as in the water and the animal's own tissues, but O_2 is at lower pressure. Because of the rigid system the gas pressure remains at *c.* 0.9 atm, irrespective of the depth at which the animal swims. (b) Cuttlebone is a laminar structure containing gas of approximately the same composition and pressure as in *Nautilus*. However, the oldest and most posterior chambers (marked black) contain fluid — again a solution of largely NaCl. The quantity of gas present can be varied by altering the volume of liquid. In this way the cuttlefish varies its own density and hence can vary its buoyancy and thus depth. This is done by altering the concentration of the NaCl solution by ionic movements; e.g. near the surface the fluid approximates to sea water in its osmotic concentration, but at greater depths the concentration is lowered, the fluid becomes hypo-osmotic to sea water and the animal's own blood, and this generates an osmotic force that balances the tendency of the increased hydrostatic pressure to push water into the cuttlebone (after Schmidt-Nielsen, 1979).

Animals living in fresh water are similar to the osmoregulators in brackish waters, but have to regulate throughout life. Figure 12.4 indicates how the body fluids of some of these animals respond to increasing salinity. Notice that there are great differences in the concentration at which they maintain the body fluids in low salinity (i.e. fresh water)—thus the bivalve, *Anodonta*, and the water flea, *Daphnia*, have lower concentrations than the amphipod, *Gammarus*, and the corixid bug, *Sigara*—but these are all lower than the concentration of the body fluids of marine equivalents, presumably to reduce the costs of maintaining a large differential in osmotic pressure between internal and external environments.

Most freshwater animals regulate by having an epidermis of low permeability (Fig. 12.3) and by producing copious amounts of urine. Thus if urine production per day is expressed as a percentage of the body weight of the producer, it is usually considerably less than 10% for marine invertebrates, but considerably more than this for freshwater invertebrates. For example, *Astacus* and *Gammarus* produce around 40% body weight per day, *Daphnia* $> 200\%$ and the freshwater bivalve *Anodonta*, $> 400\%$. Usually the urine of these freshwater invertebrates is hypo-osmotic relative to body fluids, useful ions having been selectively removed in appropriate parts of 'excretory' organs (p. 382). Nevertheless, despite this and the impermeability of the epidermis, some

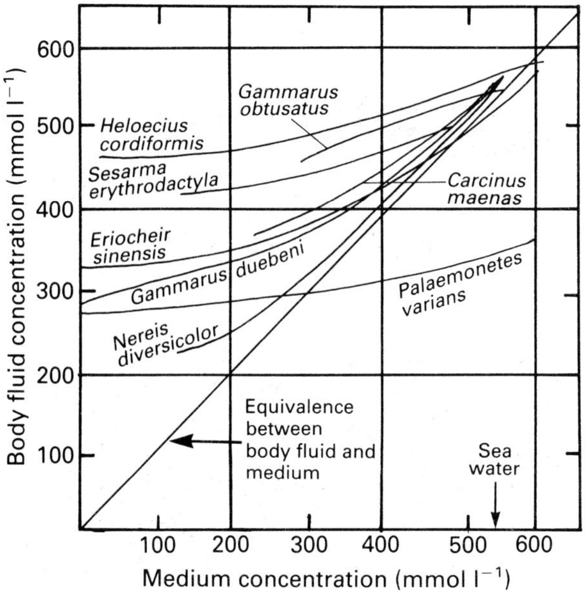

Fig. 12.2. Relation between concentrations of body fluids and medium for several brackish-water animals (after Schmidt-Nielsen, 1979).

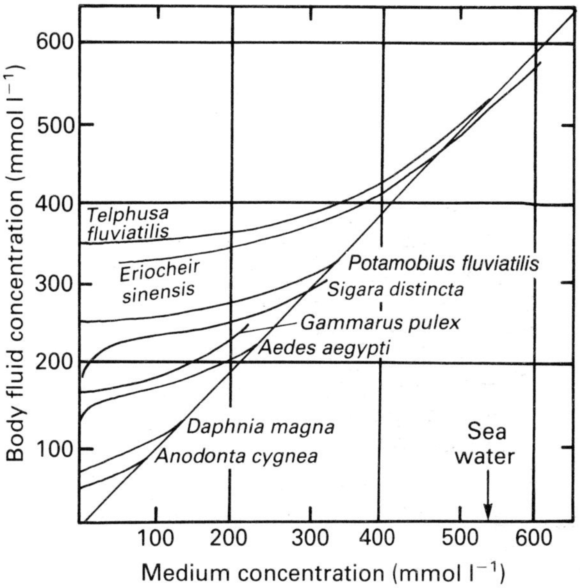

Fig. 12.4. As Fig. 12.2 but for various freshwater animals (after Schmidt-Nielsen, 1979).

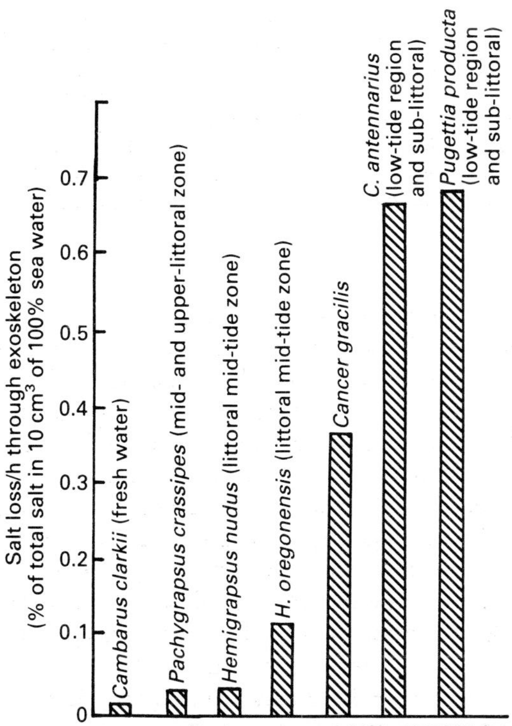

Fig. 12.3. Permeability of the exoskeleton of crustaceans from different environments (after Hoar, 1966).

ions are lost and must be replaced. Some of this occurs from the food, but many freshwater invertebrates are also capable of the direct, active uptake of ions from the surrounding medium. This can be demonstrated by first soaking them in distilled water, to reduce the concentration of ions in their body fluids, and then returning them to ordinary fresh water. After a short time, the ionic composition of the body fluids returns to normal even though at all times the freshwater medium is more dilute than the body fluids. This implies uptake by active transport. These processes have been most thoroughly researched in freshwater crustaceans such as the crayfish, *Astacus*, the freshwater isopod, *Asellus*, and freshwater species of the amphipod, *Gammarus*. It is probable that the general body surface plays some part in this process, but there is conclusive evidence that for crustaceans the gills are the organs of active ion transport.

Larvae of the mosquito *Aedes aegypti*, with blood containing abnormally small amounts of sodium ($<$ 30% normal), can be produced by rearing them in distilled water. This can be rectified within several hours of transfer to normal fresh water. Experiments that involve damaging the anal papillae and blocking the guts of these animals have shown that 90% of this uptake occurs through the papillae and most of the rest through the gut. In terms of ion uptake, the rectal gills of dragonfly larvae play a similar role as the anal papillae of *Aedes*.

12.2.4 Hyperosmotic environments

A few marine invertebrates, such as the shrimp *Palaemonetes*, are hypo-osmotic with respect to sea water (Fig. 12.2). It is generally agreed that these animals have secondarily returned to the sea from an original freshwater existence. There are also a number of grapsoid crabs that have body fluids hypo-osmotic with respect to sea water (see footnote† to Table 12.2) but there is no question that the ancestors of these ever spent time in fresh water and the reason for the hypo-osmotic state here is unclear.

The general problem for organisms in hyperosmotic circumstances, though, is to keep salts out and water in. Invertebrates living in saline waters more concentrated than sea water face the same problem. *Artemia*, the brine shrimp from salt lakes, is an excellent example. This animal maintains its body fluids at a lower osmotic concentration than its surroundings by active regulation; water is taken in by drinking and excess ions are removed through the gills in adults and a specialized neck organ in larvae.

12.2.5 Terrestrial environment

Potentially, the greatest physiological problem for terrestrial animals is dehydration. However, some

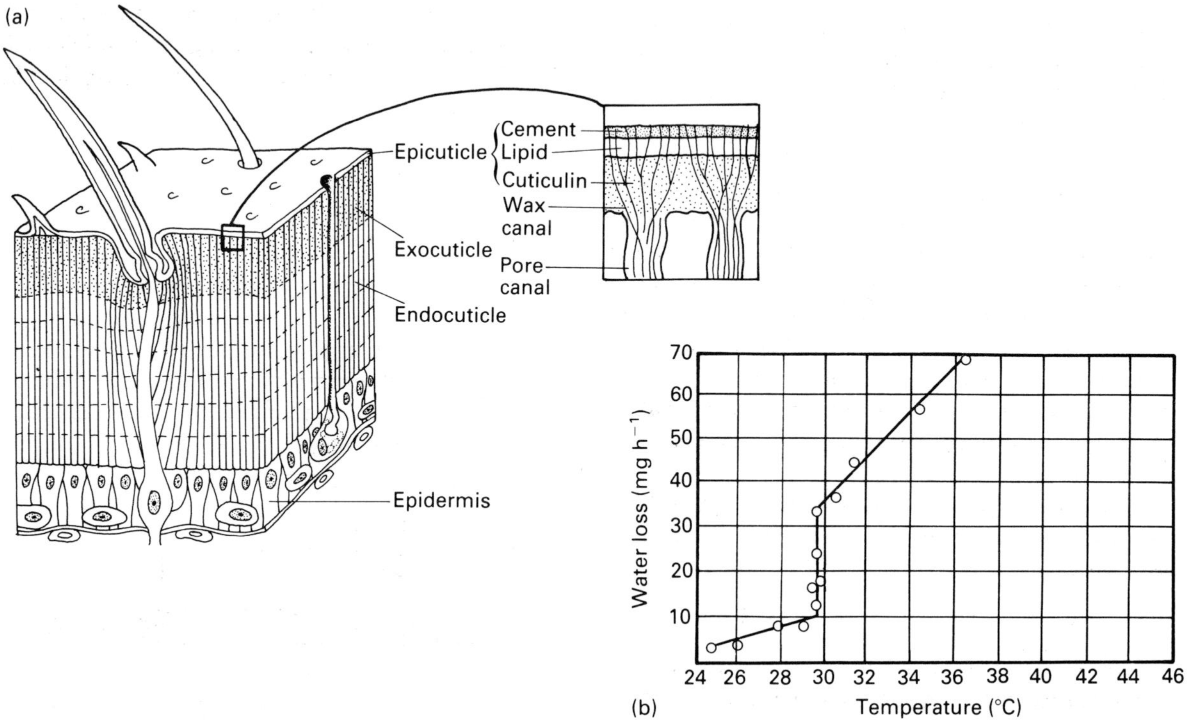

Fig. 12.5.(a) Section of the cuticle of an insect, showing the wax layer (after Edney, 1974); (b) water loss from the cuticle of the cockroach with increasing air temperature (after Beament, 1958).

terrestrial animals, such as earthworms, slugs and snails, have moist surfaces and are restricted to moist habitats in the soil and leaf litter and are not truly terrestrial. After rainfall, conditions in the soil might even become hypo-osmotic and under these conditions some nematodes actively remove water from their own tissues, possibly directly through the intestine. At the other extreme are the insects that can live in very dry environments and have an impervious exoskeleton. The latter is achieved not by the cuticle itself, but by the epicuticular layer of wax (Fig. 12.5a) (see also Section 8.5.3B.2). For example, if this is abraded the evaporative water loss from the body is increased greatly. Similarly, the importance of the wax layer for the retention of water has been demonstrated by measuring the rate of water loss at different temperatures. In this experiment there is a sudden jump in the rate of water loss at a temperature coincident with the melting point of the wax coating (Fig. 12.5b).

As already noted (p. 368), terrestrial crustaceans lack an epicuticular wax layer and are mostly restricted to moist habitats. Although terrestrial isopods are often found in microhabitats of high humidity they are sometimes exposed to drier air; water lost by evaporation can be replaced, in part, from the moist food these animals eat—decaying plant material—and some are capable of drinking and taking up water through the anus. The desert isopod, *Hemilepistus*, avoids excess desiccation behaviourally, by living in burrows during the heat of the day. These may be up to 30 cm deep, are much cooler than the surface desert and their relative humidity can be as high as 95%. (For a more detailed account see Edney, 1957.)

Insects also take up water by drinking, from food and also by the metabolic oxidation of organic molecules e.g.

$$C_6H_{12}O_6 + 6O_2 \longrightarrow 6CO_2 + 6H_2O$$

Moreover, some terrestrial insects and arachnids are able to absorb water vapour directly from the atmosphere. Just how this is achieved is not clear but, in different groups, rectal and buccal epithelia and possibly the tracheal system are involved.

12.2.6 'Invasions' of land and freshwater habitats

It seems indisputable that life evolved and radiated in the marine environment (Chapter 2). Invasions of other major habitats might either have occurred directly (sea to land; sea to fresh water) or indirectly (sea to land via fresh water; sea to fresh water via land). Additionally the sea/land transition might have occurred across the surface of the littoral region or through its interstices. These transitions are illustrated schematically in Fig. 12.6. We have already discussed and dismissed Macallum's theory (p. 368) but it is not unreasonable to presume that ancestral habits and habitats have left some mark on the composition of the body fluids of extant animals. Thus it is possible to hypothesize that terrestrial animals derived directly from marine forebears are likely to have body fluids with higher osmotic pressures than those with freshwater forebears (p. 371). And this is largely supported by the data summarized in Table 12.3. As might be anticipated, though, there are complications. For example, the

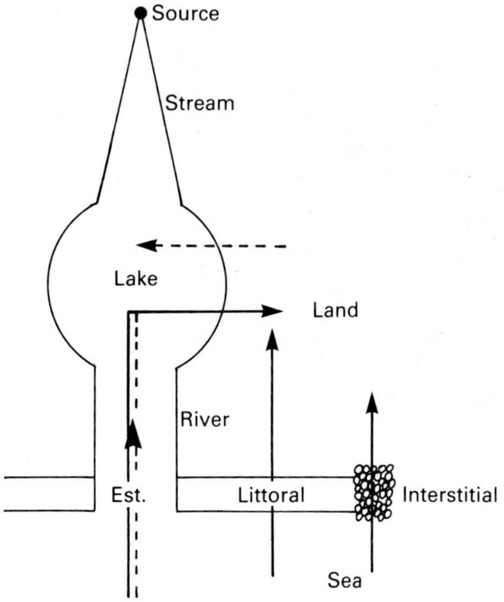

Fig. 12.6. Routes for the 'invasion' of land (—) and freshwater (– –) habitats.

terrestrial decapod, *Holthuisana transversa*, is thought to be derived from a freshwater ancestor, and yet has blood with a relatively high osmotic pressure. However, though small freshwater crustaceans do have body fluids with low osmotic pressures (< 300 mosmol), freshwater decapods, possibly because of their large sizes, have retained much higher values (> 500 mosmol), so the higher values in terrestrial species are not incompatible with freshwater origin.

Animals that followed the interstitial route to land, such as some nemertines, nematodes and annelids, are likely to have experienced conditions somewhat intermediate to those that pertain in fully terrestrial and freshwater situations. This is because sediments act as buffers to changes in salinity of the overlying water. Nevertheless, in the transition from sea to land, interstitial water will become more dilute so that animals adapted for life in the soil are likely to have evolved good mechanisms for ionic regulation. Hence, very possibly these animals, like those with freshwater ancestors, should have dilute body fluids and the terrestrial nemertine, *Argonemertes dendyi*, that is thought to have invaded land through the interstitial route, does indeed have body fluids with relatively low osmotic pressures (Table 12.3).

From the very limited survey in Table 12.3, it would appear that most invasions of land by invertebrates have occurred directly rather than indirectly. Also two major groups of terrestrial invertebrates, the insects and arachnids, are not mentioned in the table but very probably derived directly from marine ancestors. By contrast to this emphasis in the invertebrates, terrestrial vertebrates have evolved exclusively from freshwater ancestors. Both the terrestrial vertebrates and the terrestrial arthropods, particularly the insects (Section 8.5.3B.2), have body surfaces that are adapted to resist desiccation. More soft-bodied terrestrial invertebrates avoid desiccation by living in humid habitats, but probably have to suffer some desiccation, from time to time, and hence have to tolerate some osmotic concentration. The evolution of dilute body fluids, associated with freshwater origins, also appears to be associated with some inability to tolerate an increase in the concent-

Table 12.3 The osmotic pressures of the blood of some terrestrial animals (modified from Little, 1983)

Species	Taxonomic position	Osmotic pressure (mosmol)	Possible route to land
Nemertea			
Argonemertes dendyi	(Hoplonemertea)	145	Marine littoral
Annelida			
Lumbricus terrestris	(Oligochaeta)	165	Fresh water
Mollusca			
Eutrochatella tankervillei	(Prosobranchia)	67	Fresh water
Poteria lineata	(Prosobranchia)	74	Fresh water
Pseudocyclotus laetus	(Prosobranchia)	103	Brackish water
Helix pomatia	(Pulmonata)	183	Salt marshes
Pomatias elegans	(Prosobranchia)	254	Marine littoral
Agriolimax reticulatus	(Pulmonata)	345	Salt marshes
Crustacea			
Talitrus sp.	(Amphipoda)	400	Marine littoral
Holthuisana transversa	(Decapoda)	517	Fresh water
Porcellio scaber	(Isopoda)	700	Marine littoral
Cardisoma armatum	(Decapoda)	744	Marine littoral
Coenobita brevimanus	(Decapoda)	800	Marine littoral

All figures are averages for active animals in damp conditions on land, or for equilibrium with fresh water ration of body fluids—possibly proteins evolved for dilute conditions are more easily denatured at higher osmotic pressure. Hence, soft-bodied invertebrates with freshwater ancestors are less well-fitted for terrestrial life than invertebrates with direct marine ancestry.

Having invaded land, some invertebrates secondarily invaded freshwater habitats, for example, some nemertines, myriapods, insects and gastropod molluscs. The insects have been particularly successful in fresh water; out of *c*. 1 million described living species, 25 000 to 35 000 are freshwater for at least one stage of their life cycle. Osmotic problems are probably less important for adult stages because of their impervious cuticles. But for larvae, with hydrophilic and poorly chitinized cuticles, complex mechanisms of regulation have evolved, and ion pumps, located in special structures (see below), are important in this. These soft-bodied larvae and other

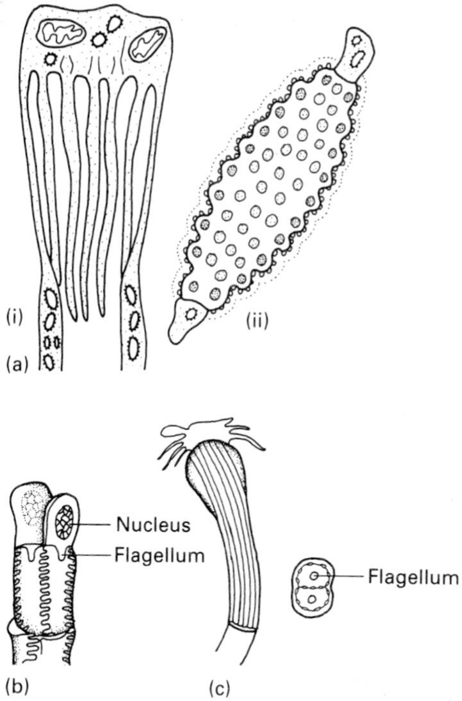

Fig. 12.7. (a) Flame bulb of a rotifer: (i) longitudinal section; (ii) transverse section. (b) and (c) solenocytes of respectively a priapulan and a gastrotrich (after Barrington, 1979).

soft-bodied invertebrates almost certainly experience the same selection pressures, in terms of osmotic conditions, as forms that have accessed fresh waters directly and hence are likely to have evolved body fluids with lower osmotic pressures. Because of this they are probably indistinguishable in these terms from invertebrates with direct origins.

12.3 Excretory systems

'Excretory systems' occur in all major invertebrate phyla, except the Cnidaria and Echinodermata. Porifera do not have 'excretory systems' as such, but freshwater species do possess contractile vacuoles. 'Excretory systems' occur in ammoniotelic animals and hence cannot all be concerned with nitrogen excretion. Instead, their prime function is almost certainly concerned with osmotic and ionic regulation, and occasionally with true excretion, and the same is probably the case for the contractile vacuoles of freshwater sponges (hence the use of quotation marks). Here we first describe the structure and then consider the function of 'excretory systems'.

12.3.1 Structure

Two major categories of system can be distinguished on the basis of development: nephridia, tubules of ectodermal origin that develop from external surfaces and grow in; coelomoducts, tubules of mesodermal origin that develop from internal tissues and grow out.

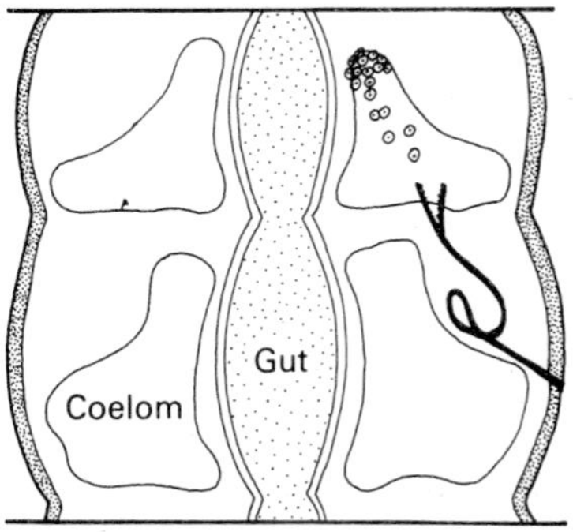

Fig. 12.8. A metanephridium, as in an annelid. (See also Fig. 12.15.)

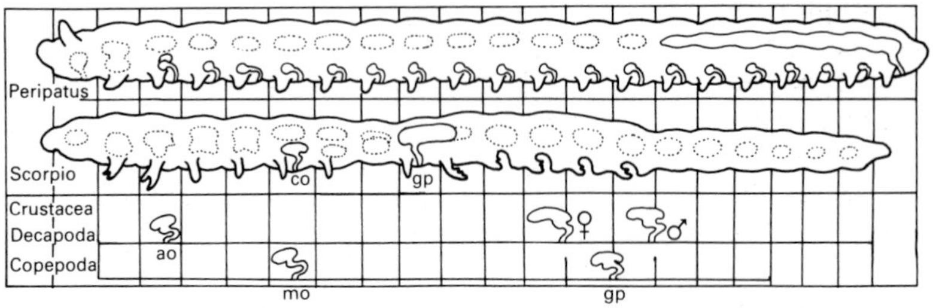

Fig. 12.9. Excretory and genital coelomoducts in the Onychophora and various arthropods (after Goodrich, 1945). ao = antennal gland; co = coxal gland; gp = genital pore; mo = maxillary gland.

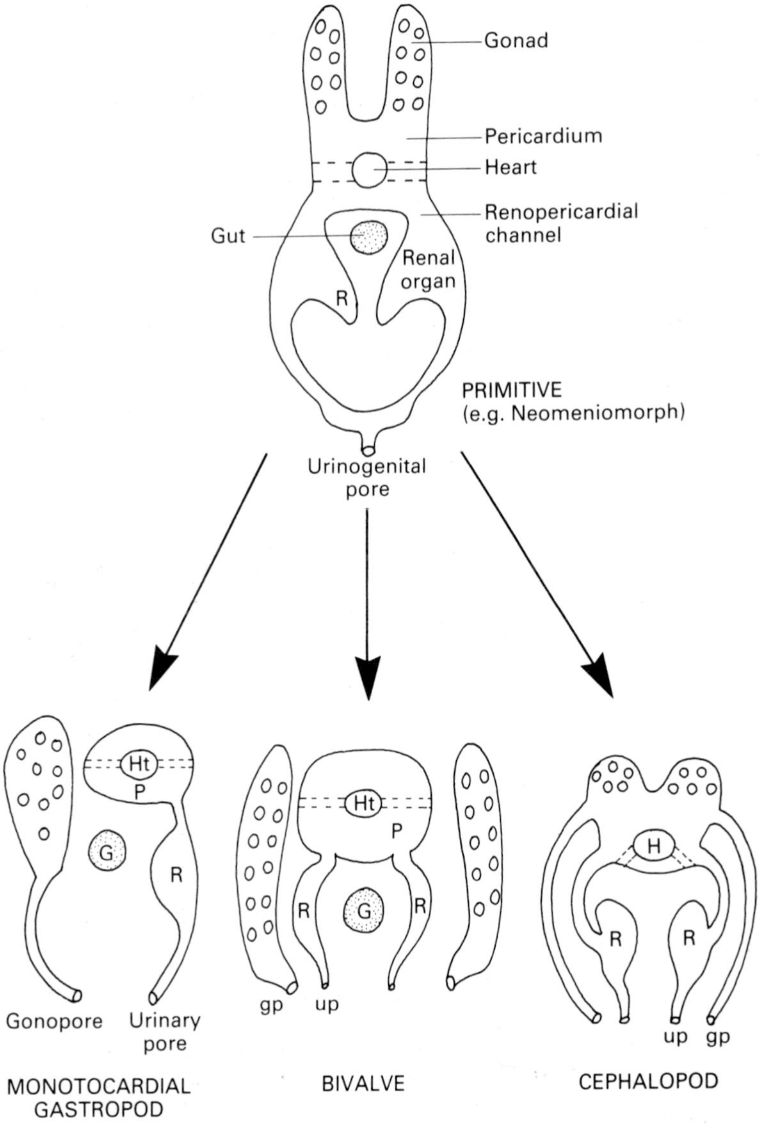

Fig. 12.10. Reproductive and excretory structures in molluscs (after Goodrich, 1945).

Nephridia consist of two main types: (i) protonephridia—the closed system of tubules that end in flame cells in platyhelminths (Section 3.6.2) and nemerteans, in flame bulbs in rotifers, and in solenocytes in *Priapulus*, some gastrotrichs, polychaetes and archiannelids (Fig. 12.7); and (ii) metanephridia with ciliated funnels, as in oligochaete annelids (Fig. 12.8) and, arguably, in some other groups.

Coelomoducts are tubular, sometimes ciliated, excretory structures which, before their embryological origin was fully understood, were often labelled 'nephridia'. They occur in onychophorans, arthropods and molluscs. *Peripatus* bears a pair, known as coxal glands, in almost every segment (Fig. 12.9); but they are fewer in number in the other groups. In the crustaceans, glands open to the base of the second antenna (= antennal gland; Fig. 12.9) and/or second maxilla (= maxillary or shell gland; Fig. 12.9), some uniramians have them associated with their maxillae, and in arachnids a pair open in the 6th segment (= coxal gland; Fig. 12.9). For molluscs the 'excretory' coelomoduct 'kidneys' have their origin in close association with the heart and gonads but in the course of evolution have become more or less distinct, in all the major classes, as illustrated in Fig. 12.10. [It should be noted that although these coelomoducts are by definition 'coelomic', in that their tubular cavities are bounded by a mesodermally derived membrane, they are *not* associated with coelomic body cavities. In fact, animals with excretory coelomoducts do *not* possess coelomic body cavities, and whilst coelomate invertebrates often possess 'coelomoduct' connections between their coeloms and the external environments, these are *not* used for excretion!]

Some systems have both ectodermal and mesodermal components and are referred to as mixonephridia or nephromixia (Fig. 12.11). They are particularly common in polychaetes, but the 'nephridia' of phoronans, sipunculans, echiurans and brachiopods are probably also of this type.

Not related to the above series of organs at all are the 'excretory systems' of nematodes and pterygote insects. *For nematodes*, the system invariably consists of a ventral gland cell, situated in the pseudocoelom,

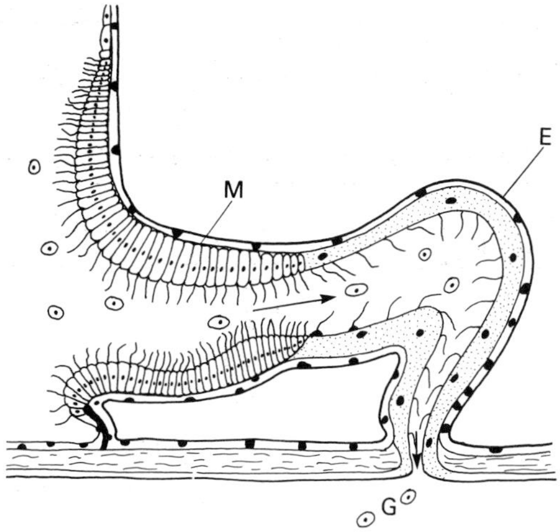

Fig. 12.11. Diagrammatic morphology of a nephromixium (= mixonephridium). M = mesodermal (= coelomoduct component); E = ectodermal (= nephridial component); G = gamete.

with a terminal ampulla opening to the exterior on the ventral surface by a pore. This may or may not be associated with a tubular system—the so-called H system (Fig. 12.12). The lateral canals are

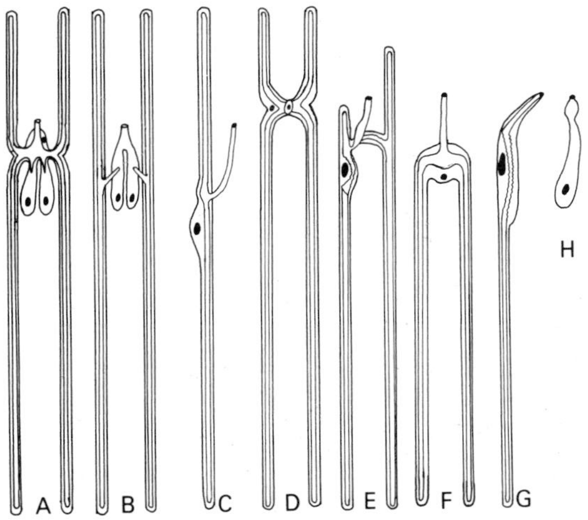

Fig. 12.12. Excretory systems in nematodes. A and B = H-type system with two ventral gland cells; C = asymmetric system, representing one arm of H system; D = H system without gland cells; E and F = shortened H system; G = shortened and reduced H system; H = single, ventral gland cell only present. (After Lee & Atkinson, 1976.)

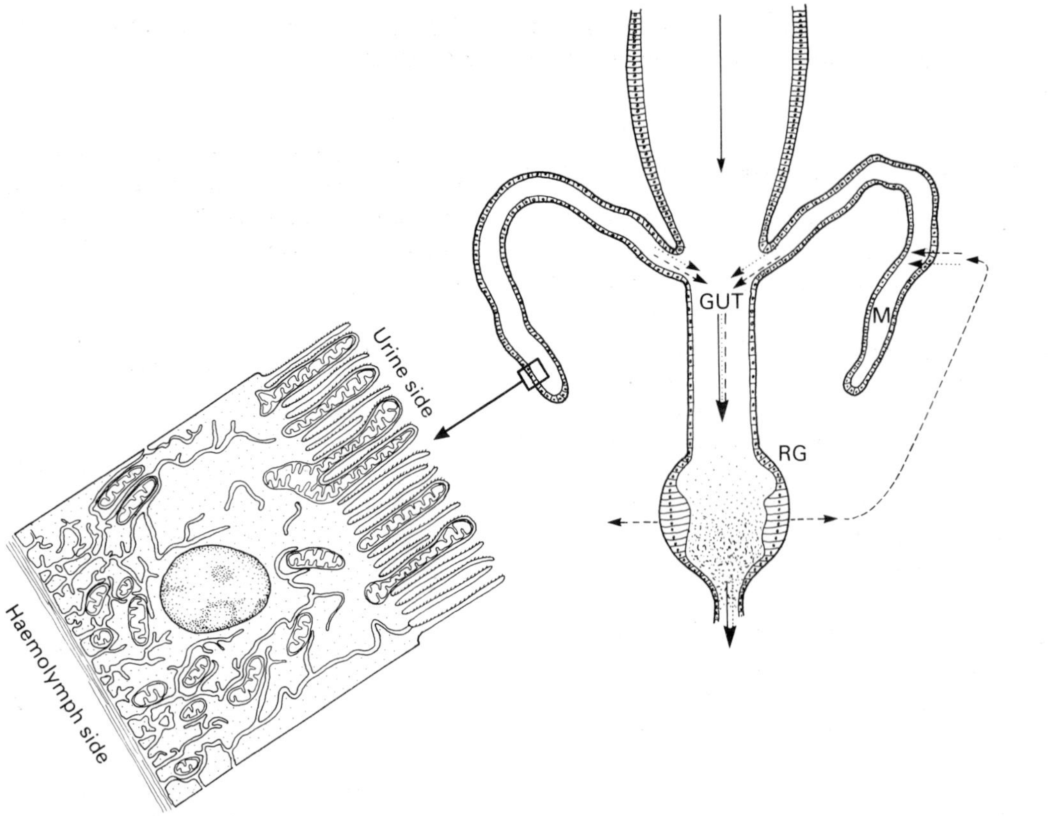

Fig. 12.13. Malpighian tubules (M) and rectal gland (RG) of insects. ——► =food; - - - -► = water and some ions (notice that the spatial organization of Malpighian tubules and rectal gland is such that water and ions can cycle); · · · ·► = uric acid. (After Potts & Parry, 1964). The ultrastructure of the Malpighian tube is also shown (after Oschman & Berridge, 1971.) Notice that it is rich in mitochondria suggesting that it is involved in active transport.

intracellular and the whole system has only one nucleus. *For insects*, Malpighian tubules are the characteristic excretory organs. Tubules, one to many in number, open into the intestine between the mid-gut and hind-gut as illustrated in Fig. 12.13. Tubules in the same general position as these, and probably with excretory functions, also occur in myriapods, some arachnids and even tardigrades, but have almost certainly evolved independently at least in the arachnids and tardigrades, so this is an example of convergence. Finally, another peculiar system is the so-called glomerulus of enteropneusts. This is formed by evagination of peritoneum into the proboscis coelom.

12.3.2 Function

The last section indicated that there are a large number of different types of 'excretory systems' with different embryological origins. Despite this diversity of structure, however, there is commonality of function in that only two basic processes are responsible for the formation of excreted fluid.

1 Ultrafiltration: fluid is passed, under pressure, through a semipermeable membrane that holds proteins and similar larger molecules back but allows water and smaller solutes to pass through.

2 Active transport: movement of solutes against a concentration gradient by a process that uses up energy. It can either occur into the excretory system (active secretion) or out of the excretory system (active reabsorption/resorption).

12.3.2.1 Systems thought to involve ultrafiltration

Pressure for ultrafiltration is thought to be generated by the beating flagella or cilia ('flame') in protonephridia, and by 'blood pressure' in coelomoducts, e.g. at the end sac of the antennal (green) gland of crustaceans (Fig. 12.14) and in the heart of the mollusc. Ultrafiltrates pass from these structures into tubules, and in the molluscs, via the pericardium, into the kidneys. All the low-molecular-weight constituents of the body fluids are filtered into the ultrafiltrate in proportion to their concentration in the fluids. Physiologically important molecules such as glucose and, in freshwater invertebrates, ions such as Na^+, K^+, Cl^- and Ca^{2+} are removed in the tubules of the system (see, for example, Fig. 12.14) leaving toxic substances and unimportant molecules behind to be excreted. As already noted, this selective reabsorption involves active transport.

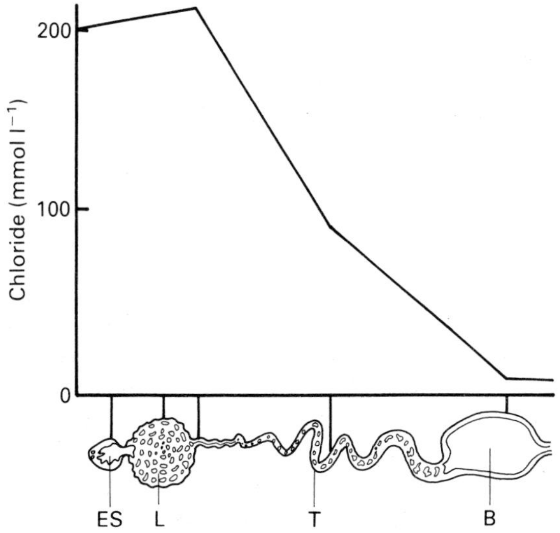

Fig. 12.14. Antennal (green) gland of crayfish with changes in chloride composition of urine as it passes along the organ. ES = end sac; L = labyrinth; T = renal tubule; B = bladder. (After Potts & Parry, 1964.)

12.3.2.2 Systems thought not to involve ultrafiltration

Ciliary filtration

Metanephridia and nephromixia direct fluids from the coelomic body cavity into the 'excretory' tubules by ciliary action. These fluids are again modified in composition by active transport in the tubule of the system (Fig. 12.15).

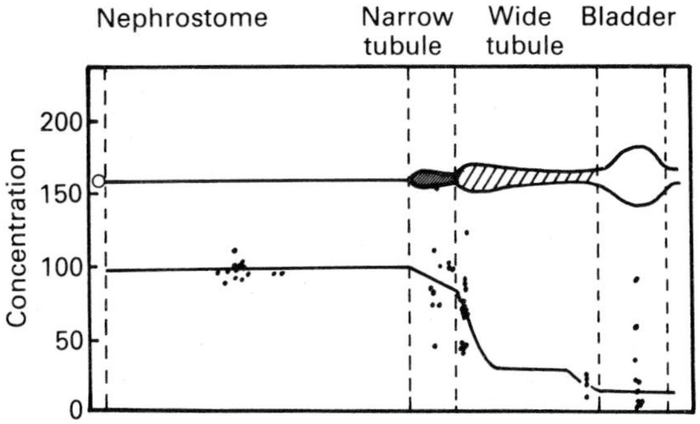

Fig. 12.15. Osmotic pressure along the length of a metanephridium of an earthworm (after Potts & Parry, 1964).

Excretory materials may also exit via this route. For example, in earthworms chloragogenous tissue is important in excretion. This tissue is of epithelial origin and lies in the coelom against the gut wall. Chloragogen cells store fat and carbohydrate, but they are also sites of deamination, and ammonia and urea pass from them into the coelom, and are swept into the nephridial funnels. Waste particles, released when the chloragogen cells disintegrate, may also be passed out through the nephridial system. In leeches, botryoidal tissue takes over the function of the chloragogenous tissue and in some groups, such as *Theromyzon*, waste products from it pass out through an open nephridial system. In other groups, though, such as *Glossiphonia* and *Hirudo*, the nephridial funnel is sealed from its tubules. Here cilia beat from the metanephridial funnels, but *into* coelomic cavities surrounding the funnels and serve to distribute amoebocytes that are actually formed in the funnels. The tubules still open to the outside and probably effect some ionic and osmotic regulation.

The Malpighian tubules of insects

The direct supply of O_2 to the tissues by a tracheal system has diminished the need for an efficient, pressurized blood system by insects. Hence the Malpighian tubules do not receive a pressurized blood supply, and are surrounded with blood at approximately the same pressure as the contents of the tubes. So ultrafiltration cannot be involved here. Instead potassium ions, and possibly other solutes, are actively secreted into the lumen of the tubule and water follows by osmosis. Urates are also secreted into the lumen and these are kept in solution by the relatively high pH at the distal end of the tubules. This fluid enters the hind-gut and water is resorbed, particularly by rectal glands (see Fig. 8.27), the pH falls and the urate is precipitated as uric acid. The latter is lost in the faeces. See Fig. 12.13.

12.3.2.3 The nematode system

Not much is known about the functioning of this system; however the tubes almost certainly differ in function from the gland cells. The former appear to be involved in ionic and osmotic regulation, but there are no cilia to create flows; it is usually assumed, by default, that ultrafiltration and active transport are important. It seems likely that the system plays a minor role in nitrogen excretion but some nematodes release secretions from the system that appear to have enzymatic activity. These probably derive from the gland cell.

12.4 Conclusions

You should now appreciate that, within the invertebrates, it is difficult, and even unhelpful, to disentangle an account of excretion (loss of nitrogenous wastes) from one of ionic and osmotic regulation and even of buoyancy. The formation, transport and removal of nitrogenous wastes, at some stage, inevitably involves the filtration and flow of body fluids, even if the excreta end up, like uric acid, to be dry. So excretion is intimately associated with the movement of liquids and ions and hence with processes of osmosis, diffusion, active transport and ultrafiltration—and the reader should now have a clear understanding of these. 'Excretory systems' are invariably associated with ionic and osmotic regulation but not always with excretion. Despite their diversity of structure, this chapter focused on the similarities in function of 'excretory systems' in terms of the key processes just noted and, again, it is these principles that the reader should take from the chapter. Finally, the replacement of ions of one mass with those of another, and the replacement of liquid with gas, influence the density of tissue and are, as will now be appreciated, important factors in the control of buoyancy in aquatic invertebrates.

12.5 Further reading

Barrington, E.J.W. 1979. *Invertebrate Structure and Function*, 2nd edn. Nelson, London.

Beament, J.W.L. 1958. The effect of temperature on the waterproofing mechanism of an insect. *J. exp. Biol.*, **35**, 494–519

Denton, E.J. & Gilpin-Brown, J.B. 1961. The distribution of gas and liquid within the cuttlebone. *J. mar. biol. Assoc., U.K.*, **41**, 365–381.

Denton, E.J. & Gilpin-Brown, J.B. 1966. On the buoyancy of the pearly *Nautilus*, *J. mar. biol. Assoc., U.K.*, **46**, 723–59.

Edney, E.B. 1957. *The Water Relations of Terrestrial Arthropods*. Cambridge University Press, Cambridge.

Edney, E.B. 1974. Desert arthropods. In: G.W. Brown (Ed.) *Desert Biology, (Vol. 2)*. Academic Press, New York.

Hoar, W.S. 1966. *General and Comparative Physiology*. Prentice Hall, New Jersey.

Horne, F.R. 1971. Accumulation of urea by a pulmonate snail during aestivation. *Comp. Biochem. Physiol.*, **38A**, 565–70.

Koehn, R.K. 1983. Biochemical genetics and adaptations in molluscs. In: P.W. Hochachka (Ed.) *The Mollusca (Vol. 2)*, pp. 305–30. Academic Press, New York.

Krogh, A. 1939. *Osmotic Regulation in Aquatic Animals*. Cambridge University Press, Cambridge.

Lee, D.L. & Atkinson, H.J. 1976. *Physiology of Nematodes*, 2nd edn. Macmillan, London.

Little, C. 1983. *The Colonisation of Land*. Cambridge University Press, Cambridge.

Needham, J. 1931. *Chemical Embryology* (3 vols). Cambridge University Press, Cambridge.

Pilgrim, R.L.C. 1954. Waste of carbon and energy in nitrogen excretion. *Nature (London)*, **173**, 491–2.

Potts, W.F.W. & Parry, G. 1964. *Osmotic and Ionic Regulation in Animals*. Oxford University Press, London.

Schmidt-Nielsen, K. 1972. *How Animals Work*. Cambridge University Press, Cambridge.

Schmidt-Nielsen, K. 1979. *Animal Physiology: Adaptation and Environment*. Cambridge University Press, Cambridge.

13
DEFENCE

This chapter first of all classifies the various threats to which invertebrates are exposed and then considers how these animals defend themselves against each type of threat. It therefore ranges widely from defence against predators, to defence against pathogens and even to defence, if a defence is possible, against ageing processes.

13.1 Classification of threats

13.1.1 There are two major classes

Figure 13.1 gives some examples of survivorship curves—numbers (or proportions) of individuals all born at roughly the same time (cohorts) alive at different ages thereafter. Curves in (a) are for field populations and here mortality is either focused on juveniles or occurs at a roughly constant rate at each age. Mortality in these natural populations is probably mainly due to ecological or extrinsic factors such as accidents, diseases and predation. Juveniles are often more vulnerable to these than adults, or all age classes are roughly equally susceptible.

Curves in (b) are for laboratory cultures. Here most extrinsic mortality factors can be excluded. However death still occurs but is focused on old-aged individuals. There is an increase in vulnerability with age, possibly due to intrinsic factors, i.e. to ageing processes or senescence.

13.1.2 Ecological (extrinsic) causes of mortality

Ecological causes of mortality are many and varied, but there are four main classes: accidents, diseases, predation and environmental stresses. The first three of these are self-explanatory. Environmental stress can either occur due to absence of an essential factor or to the presence of a stressor—a natural toxin or an artificial pollutant.

13.1.3 Ageing

Ageing apparently occurs when extrinsic mortality factors are excluded, and can therefore be ascribed to internal degeneration of systems, cells and molecules. These intrinsic effects can probably be traced ultimately to the denaturation of important biological molecules—nucleic acids and proteins—under the influences of such processes as thermal noise, cross-linkage of side-chains in macromolecules, autoxidation and so on. Yet it should not be imagined that these intrinsic processes cannot be influenced by extrinsic factors. Figure 13.2 shows that whole-body irradiation of *Drosophila* can shorten life to an extent that depends upon dose. However, despite these effects the shapes of the survivorship curves remain the same, and some gerontologists suggest that life-shortening here is due to accelerated ageing, possibly due in turn to increased damage to macromolecules—particularly DNA—by the high-energy radiation. Alternatively, addition of certain chemicals, such as vitamin E, to the culture media of nematodes or food of *Drosophila* extends their lives. These chemicals probably work by protecting macromolecules from damage; for example, vitamin E is probably an antioxidant.

13.1.4 Classification

All mortality can therefore be influenced by the external environment; it is just that some causes of

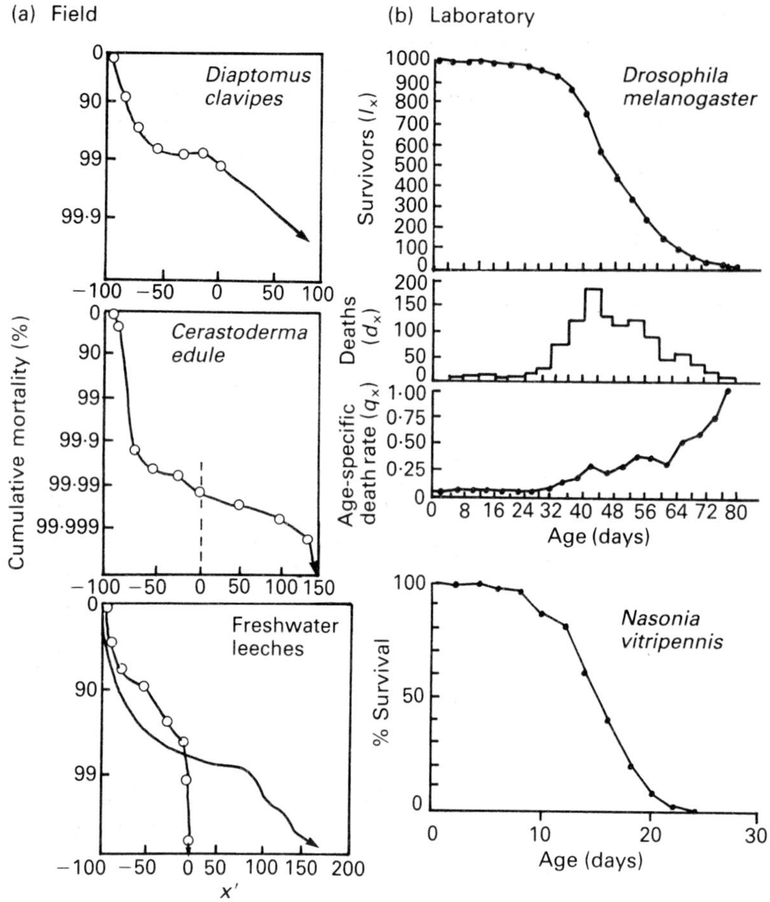

Fig. 13.1. (a) Survivorship curves for field populations of invertebrates. Note x' = % deviation from mean life-span. (After Ito, 1980.)
(b) Survivorship curves, distribution of ages at death and age-specific death rate from a laboratory population of *Drosophila melanogaster* (after Lamb, 1977) and survivorship curve for a laboratory population of the hymenopteran, *Nasonia vitripennis* (after Davies, 1983).

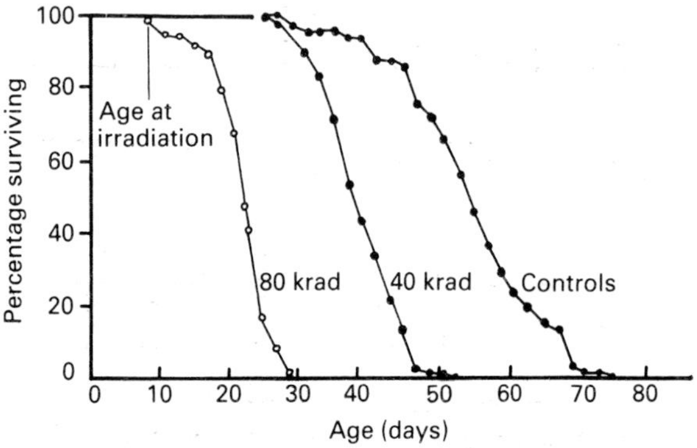

Fig. 13.2. Effect of whole-body gamma-radiation on male *Drosophila melanogaster* (after Lamb, 1977).

mortality are more intimately associated with organisms than others. Table 13.1 classifies mortality factors according to their intimacy of association with a recipient and results in a series ranging from predators at one end to ageing processes at the other. Conversely, the ease of exclusion and hence experimental manipulation of mortality factors reduces along the continuum in the reverse direction, from predators to ageing factors.

13.2 Defence

13.2.1 Against predators

All animals are potential food for other animals (see Chapter 9). They can protect themselves against

Table 13.1 Classification of mortality factors

Mortality agent	Accident	Predators	Disease	External stressor (e.g. pollutant)	Internal stressor (e.g. system degeneration)
Intimacy of association with recipient	X	X	XX	XX	XXX
Ease of exclusion by artificial means	XXX	XXX	XX	XX	X
Response of recipient		Defence	Immunological defence	Tolerance, resistance, repair	Repair

X = low, XXX = high.

being eaten in a variety of ways but these can be grouped into one of three main classes of responses: avoiding potential predators; dissuading them; actively repelling them.

13.2.1.1 Avoidance

This can involve keeping out of the way of predators and/or being inconspicuous. A possible example, involving both kinds of responses, is the extensive vertical migration exhibited by some marine and freshwater planktonic animals. Though these patterns are generally complex, they often involve downward migration, away from sunlight, during the day and upward migration during the night. Thus, animals avoid being conspicuous to predators in the light and only come to the surface at night when they are likely to be less conspicuous to visual predators (Fig. 13.3a). This behaviour is most marked in species and age classes that are particularly vulnerable to predation. Moreover, in a unique study of a copepod in a series of isolated mountain lakes, Gliwicz (1986) has shown that vertical migrations only occur in lakes that contain planktivorous fishes. Also he was able to make observation on the migratory patterns of the copepod in one lake at several different stages in stocking with planktivorous fishes. Twelve years after stocking there was little evidence of vertical migration, but some 23 years later the copepods showed strong migration away from the water surface

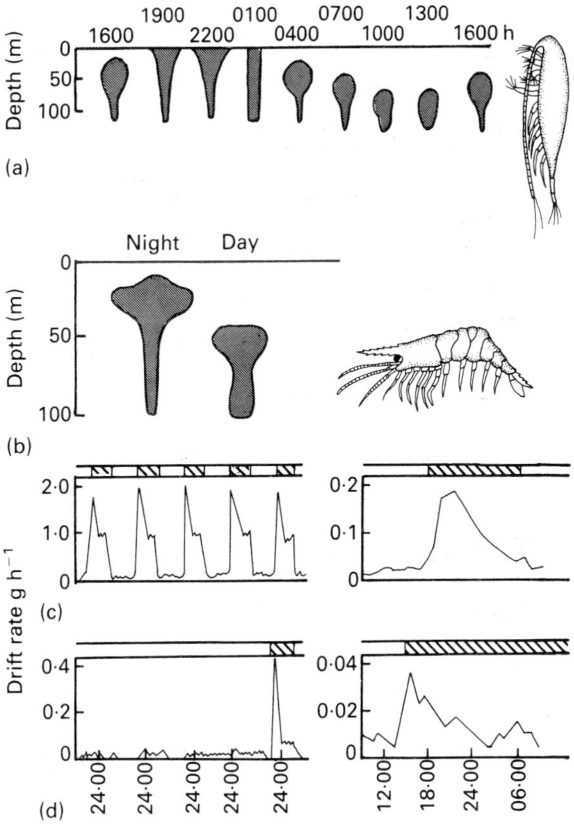

Fig. 13.3. Vertical migration of (a) the copepod *Calanus* and of (b) a deeper-dwelling acanthephyrid prawn (after Barnes & Hughes, 1982). The influence of light and dark on drift of some freshwater invertebrates: (c) open stream; (d) in experimentally manipulated systems (after Holt & Waters, 1967).

during the day. However, predation cannot be the whole explanation for the evolution of all vertical migration, for in some invertebrates the daytime descent is often to great depths, far in excess of what would be needed to avoid the light, and some migrating zooplankters luminesce at night so increasing their conspicuousness. Other possible explanations involve optimal exploitation of patchy food (Chapter 9), energy economies and improved horizontal migration.

Similar vertical migrations, but of a more limited kind, are practised by freshwater invertebrates that live on submerged stones: during the day they are under stones and are inactive, whereas during the night they often emerge on to upper surfaces and become more active. Large numbers of these normally benthic organisms can be collected floating free in running-water systems. This invertebrate drift, as it is called, is particularly abundant during nightfall (Fig. 13.3b), possibly because it is then that invertebrates crawl on to exposed, upper-stone surfaces and become vulnerable to being washed away.

Escape reactions are an extreme form of locomotory avoidance reactions. They involve use of either normal locomotory responses or the deployment of specialized behaviour. The cuttlefish, *Sepia,* has an 'ink sac' which contains fluid composed of granules of melanin. When attacked, it ejects an ink screen and immediately becomes very pale and swims at right angles to its original path of flight. Bivalve molluscs, such as the usually sessile cockle, *Cerastoderma*, can achieve escape reactions by sudden and rapid foot and shell contractions. Extremely powerful escape movements are made by *Cerastoderma* in response to the tube feet of starfish—probably evoked by a substance released into the water.

Cryptic (concealing) colouration is another widespread method of avoidance in invertebrates (see Fig. 9.33). Examples probably occur in all phyla but have been particularly thoroughly studied in some snails and insects.

1 Banding in *Cepaea nemoralis*. This species of land snail produces a wide range of shell colours and patterns by varying the underlying hue of its whole shell and the number, width and intensity of bands. These variations are genetically controlled. Thrushes feed on *Cepaea,* finding shells by sight and breaking them open on rocks known as thrush anvils. In the mid 1950s A.J. Cain and P.M. Sheppard showed that, on average, shells broken by thrushes have colours more easily seen than those carried by snails still living in the same area. Different patterns are more difficult to see in different places and at different times of the year. Banded light-coloured shells, for example, are difficult to see in lush vegetation (grass fields, and hedgerows) when sharp contrasts of light and dark are produced by the interplay of entering light and narrow shadows. In dark woodlands, uniformly dark and unbanded shells are more difficult to see (Fig. 13.4).

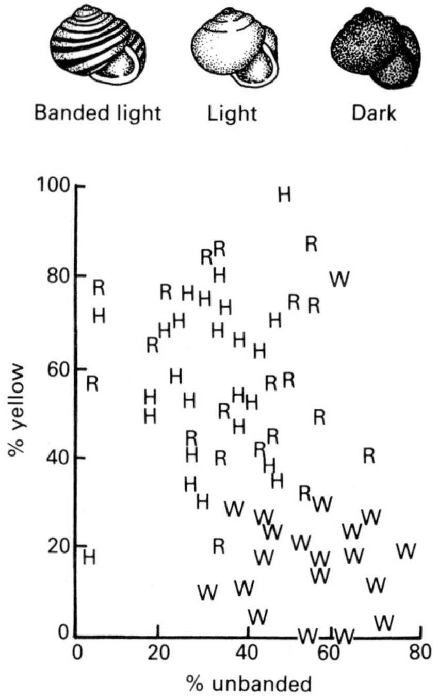

Fig. 13.4. Frequencies of morphs of *Cepaea* in various habitats. Banded light shells are difficult to see in low vegetation (herbage, R; hedgerows, H) whereas dark unbanded shells are more difficult to see in dark woodland (W) (after Calow, 1983; original work by Cain & Sheppard, 1954).

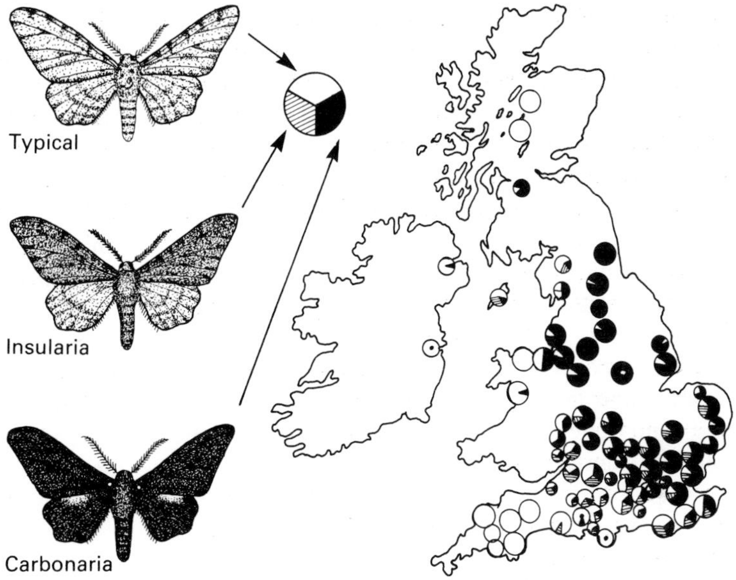

Fig. 13.5. Frequencies of three morphs of *Biston betularia*. Insularia and carbonaria are both melanics. Although looking intermediate between typical and carbonaria, insularia is in fact controlled at a different locus (after Sheppard, 1958).

2 Melanism in *Biston betularia*. A large number of moths and other insects have evolved wing and body patterns that make them inconspicuous on lichen-covered trees. Industrial pollutants have killed lichens and sooted up tree trunks. In these areas the previously adapted *typical* moths were replaced by black, *melanic*, forms. This is known as industrial melanism (Fig. 13.5). Again in the 1950s H.B.D. Kettlewell showed that birds could be responsible for this in the peppered moth (*Biston betularia*); by direct observation, typical forms were found to be more vulnerable to bird predation than melanics at polluted sites and vice versa at non-polluted sites. Kettlewell also released peppered moths into an old cider barrel lined with vertical black and white stripes. Sixty-five per cent of the moths took up a position on a background they matched (typicals on a white and melanics on a dark background), so moths appear to have appropriate behaviour to allow them to take up resting positions on backgrounds upon which they are least conspicuous.

(As might be expected there is more to the evolution of banding in shells and melanism in the peppered moth than indicated by the short descriptions given above. For a more thorough description the reader should refer to genetical texts, e.g. Berry, 1977).

Finally, camouflage need not make the bearers inconspicuous but can make them resemble objects not usually associated with food. Many insects resemble parts of plants: twigs and leaves. The young larvae of some of the swallowtail butterflies are conspicuous but escape attack because they are black with a white saddle on their backs and thus resemble bird droppings (see Fig. 9.33)! They change their colour pattern dramatically when they become too large for this method of concealment.

Erichsen *et al.* (1980) have carried out some novel experiments to test the influence of this kind of camouflage on predation by birds. They offered great tits (*Parus major*) a choice between large and small mealworms (*Tenebrio molitor*) in straws; but the large ones were in opaque straws that resembled artificial twigs whereas the small ones were in clear straws and were easily visible. The large worms gave more energy per mouthful than the small ones (about twice as much) and were therefore a more profitable choice, but the small worms were more instantaneously recognizable than the large ones that required the bird to take time inspecting and picking up 'twigs' in the search for worms. In choice experiments the birds consistently selected the small mealworms where 'twigs' were abundant, but switched to larger ones where 'twigs' were rare.

Other animals achieve camouflage by attaching to themselves material from the external environment. Spider crabs pick up pieces of algae and other materials which are attached to patches of hooks on their exoskeletons. The cases of caddis larvae might also serve a similar function. Thus larvae of *Potamophylax cingulatus* with leaf cases are less likely to be eaten by trout when on a leafy background than on a sandy one. During larval growth, this animal switches from a leaf to a sand case, but sandgrain cases are no more likely to be eaten when on a leafy as compared with a sandy background. This lack of difference might be due to the low palatability of sandgrain cases (see below). (For more details, see Hansell, 1984.)

13.2.1.2 Dissuasion

Animals can dissuade predators from eating them by either physical or chemical defences.

Calcareous fortifications are used widely—spicules of sponges, calcareous skeletons of corals, tubes of annelids, shells of molluscs and brachiopods, calcified boxes of some lophophorates and tests of echinoderms. Chitinized skeletons of arthropods also form fortifications and are sometimes reinforced with

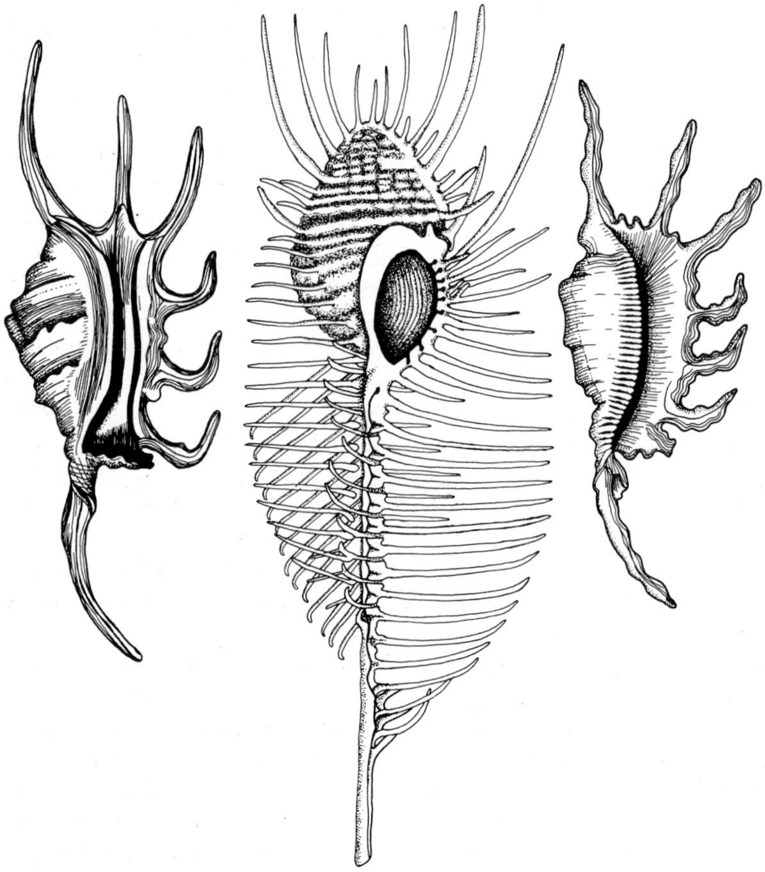

Fig. 13.6. Defensive sculpturing in shells.

calcium. The latter is particularly true of cirripedes, sessile crustaceans which become covered externally with thick, calcareous plates (see Fig. 8.36a).

Sometimes inorganic inclusions may simply so dilute the nutritious tissues that they make the animals poor quality food. Such may be the function, at least in part, of the calcareous or siliceous skeletons of sponges, the calcareous skeletons of cnidarian corals, the tubes of grit and sandgrains that surround some polychaete annelids and the stony cases of some freshwater insect larvae (see above). Surprisingly, a small number of marine turbellarians have been found to bear calcareous scales or rods embedded in their body walls (Chapter 3). These probably have a supportive function but it is not unreasonable to assume that they also operate as tissue diluents. Moreover, it is also plausible that the molluscan shell evolved from this kind of origin.

Shells are a particularly obvious form of physical fortification (see Fig. 9.34). Strong shell sculpture, occluded shell apertures and low spires coupled with a thick shell are effective devices amongst many marine, benthic gastropods in thwarting the attacks of fishes, crabs, lobsters and other shell-breaking predators (Fig. 13.6). Occluded teeth, inside the apertures of land snails, exclude shell-entering predacious beetles. Strong sculpture in the form of various shell protuberances, ribs, knobs and spines strengthen shells and make them effectively larger and hence more difficult or time-consuming for a predator to handle. Lack of specialized molluscivores may in part explain the general absence of thick shells and elaborate architecture in the freshwater molluscan fauna. Poor calcium availability in fresh water may also be partially responsible for these differences.

An unusual, protective body covering occurs in the ascidians. Their bodies are covered by an epithelium, one cell thick; but this is not the outer covering of the body. Instead it is surrounded by a tunic (the group are sometimes referred to as tunicates—Section 7.4A). This is usually quite thick but varies considerably in texture from being soft and delicate to being tough and similar to cartilage. It consists of a fibrous material, a principal constituent of which, in many, though not all species, is a kind of cellulose called tunicin. Also present are proteins and inorganic inclusions, such as calcium. The tunic can be vascularized and contains amoeboid cells, so it is not just a dead covering.

Not all physical fortifications are secreted by the defenders. Again particularly good examples of this are the cases of caddis larvae. These are built from materials from the surrounding environment. A similar example, but where the defences are formed from materials rejected by the defender, is the so-called 'faecal shield' of the beetle larvae *Cassida rubiginosa*. This consists of a compressed packet of cast skins and faeces carried on a fork-like organ held over its back. The shield is manoeuvrable and is used by larvae to protect themselves against attack from other insects such as ants. Hermit crabs move into empty gastropod shells and thus save on the need to invest in the production of a thick exoskeleton. Their uropods are modified and the larger, left one is used for hooking on to the columella of the shell (see Fig. 8.49e).

Chemical methods of dissuasion are also common in the invertebrates. Many invertebrates lace their tissues with toxins. As much as 0.3% of the body weight of a nemertine can be made up of neurotoxins. Numerous shell-less gastropod molluscs, e.g. opisthobranchs and pulmonate slugs, use various toxins, including sulphuric acid. Some sponges produce irritating substances and an extract, prepared from freshwater sponges, has been shown to be fatal when injected into mice. Other animals may also make use of these toxins. Some crabs decorate themselves with sponge, perhaps for camouflage or perhaps to 'cash in' on protection derived from the toxins produced by the sponge. Similarly, some hermit crabs occupy snail shells with attached sponges and anemones and derive similar protection.

Turbellarian flatworms probably do not secrete toxins directly into their tissues but produce rhabdoids—rod-shaped epidermal bodies, arranged at right angles to the surface and secreted by epidermal gland cells (Section 3.6.3.1). They are discharged when the worms are irritated and that they have a defensive role is suggested by the following easily carried out experiment. Stickleback fishes readily eat the oligochaete, *Tubifex tubifex*, and may take it from forceps. If, however, the *Tubifex* is first smeared with mucus produced by prodding a flatworm then it is rejected by the fish. If, finally, *Tubifex* is coated in mucus from the trails of normally moving worms it will be eaten. The mucus from the disturbed worms contained many rhabdoids whereas that from the non-disturbed worm contained few if any rhabdoids. Rhabdoids probably also have other functions such as in the rapid formation of mucus itself, or as antimicrobial agents.

Toxins are common in insects and they can 'borrow' toxic compounds from the plants upon which they feed. For example, the grasshopper *Poekilocerus bufonius* feeds on milkweeds that contain a number of complex toxins that can disrupt cardiac function—so-called cardenolides. The grasshopper extracts these from its food and stores them in a poison gland. When attacked by predators it defends itself by ejecting a poison spray rich in the plant-derived toxins. When these grasshoppers are maintained on a diet without milkweeds, the cardenolide content of the spray is reduced ten-fold. Monarch butterflies also feed on milkweeds and lace their tissues with cardenolides, making themselves distasteful to avian predators. Butterflies grown from larvae not fed on milkweeds again have no harmful effects on predators. Toxic chemicals such as this, not produced in the insects but received from plants, are sometimes called *kairomones* to contrast them with *allomones*, toxins that confer advantages on the organisms (i.e. plants) that produce them. (For more details see Nordlund and Lewis, 1976).

Chemical toxins are often associated with warning colouration. Correlations of this kind can be found in many invertebrate phyla, from vividly coloured nemertines and slugs to brightly coloured insects. Such colouration tends to be associated with simple patterns and the colours frequently include red, yellow, or black and white. Everyone will be familiar with the black and yellow bands of bees and wasps.

The evolution of toxins and warning colourations is not straightforward. Their only evolutionary virtue is if they inhibit predation, and yet the only way a predator can be aware of them is by 'having a go'. Kin selection is a possible explanation, where the sacrifice of one individual carrying a gene for the warning toxin can protect relatives in the same group that carries the same gene. Similarly, such a gene could also spread if the bearer were not easily damaged, so that it could survive detection by a predator or if predators were repelled by an obnoxious stimulus before attempting a strike. The majority of warningly coloured insects are tough and not easily damaged and, as with slugs, frequently emit strong odours. Nemertines can regenerate tissues lost to predators.

Warning colours can be mimicked by non-toxic animals and this resemblance is referred to as Batesian mimicry, after the man who first made it explicit. Since the warning colour of the *mimic* is false whereas that of the *model* is not, it follows that: (a) the model must be noxious and bright coloured; (b) the model must be more common than the mimic for if the model were rare the predator would not learn that it is protected and the whole relationship would fail; (c) mimics must occur in close association with models and closely resemble them. Mullerian mimicry is another form of mimicry whereby noxious species converge on the same pattern because they get advantages from each other. Criterion (b) does not apply here and resemblance, criterion (c), need not be as precise. Wasps and bee species carry the same pattern of banding and this is Mullerian mimicry. Many dipterans, particularly hoverflies and some lepidopterans, have evolved wasp/bee-like appearances and this is Batesian mimicry. Figure 13.7 gives some lepidopteran examples.

It is not too difficult to appreciate how Mullerian mimicry might have evolved, but a major problem with Batesian mimicry is that the mimic must resemble the model closely enough to gain protection from it (criterion (c)), so how could forms intermediate between non-mimetic ancestor and mimic have been favoured? One possibility is a two-phase evolutionary process: the establishment of an approximate but, nevertheless, adequate resemblance by means of a major mutation, followed by a gradual improvement by natural selection of more usual small-scale genetic variance.

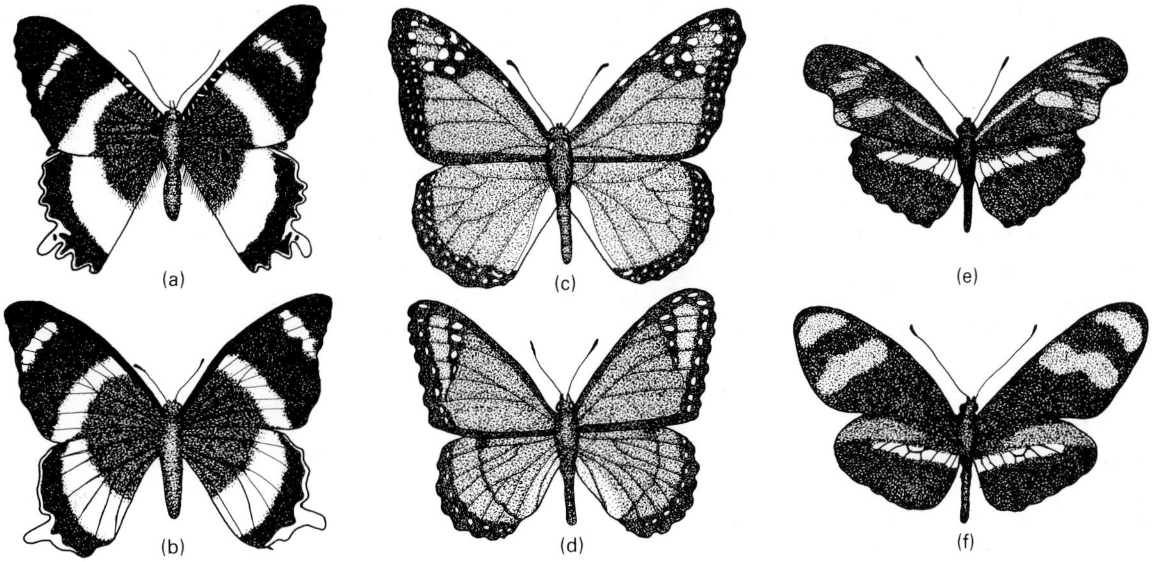

Fig. 13.7. Batesian mimicry between the moths *Alcidis agarthyrsus* (a) and its mimic *Papilio lag* (b) and between the North American viceroy *Limenitis archippus* (c) and its model a monarch *Danaus plexippus* (d) Mullerian mimicry between *Podotricha telesiphe* (e) and *Heliconius telesiphe* (f).

13.2.1.3 Repulsion

Organs that are used to capture and kill prey can often be used actively to repel predators; for example cnidoblasts of cnidarians are used for both attack and defence. Paradoxically, the cnidoblasts of cnidarians that fall prey to nudibranch molluscs can also be borrowed by the predators themselves as defences. Ciliary tracts in the stomachs of the nudibranchs carry undischarged nematocysts to projections on their dorsal surfaces (cerata) in which they are engulfed but not digested (Fig. 13.8). They are moved to the distal tips of the cerata, cnidosacs, which open to the exterior. Discharge from these may be effected by contraction of circular muscles around the cnidosacs. Nematocysts can be replaced in about 10 days and most nudibranchs use only certain types of nematocysts present in their prey. A small number of turbellarians utilize nematocysts of the hydroids they eat in the same way, as does the ctenophore, *E. rubra* (p. 71), those of its medusa prey.

The chitinized jaws and stings of arthropods are also examples of aggressive structures that are used defensively. On the other hand, some stings are specifically defensive, for example the sting of honey bees. This is formed from the modified ovipositor—no longer used for laying eggs—and consists of paired barbed lancets and the unpaired stylet (Fig. 13.9). At rest this lies in a pocket within the seventh abdominal segment. The mechanism of stinging is described in the legend of the figure. The poison is secreted by a pair of long glands in the abdomen and in the honey bee contains a poison and certain enzymes which cause the tissue of the victim to produce histamine.

Beetles of the genus *Brachinus,* commonly known as bombardier beetles, use a defensive spray to ward off predators, such as spiders, preying mantids and even frogs. When disturbed they release this from a pair of glands at the tips of their abdomens. The latter can be rotated so that they can spray accurately in virtually any direction. The active principals of

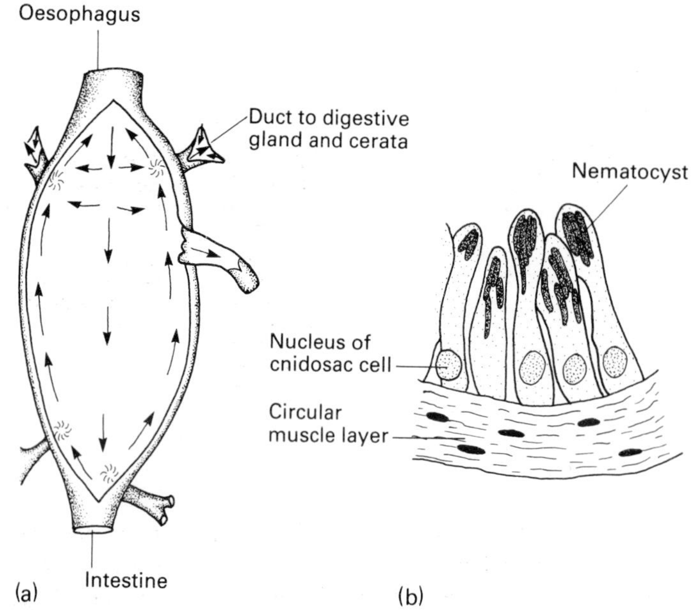

Fig. 13.8. Stomach of nudibranch opened dorsally (a), and cnidosacs (b) (after Barnes, 1980).

the secretion are benzoquinones, which are synthesized explosively by oxidation of phenols at the moment of discharge. An audible detonation accompanies the emission and the spray is ejected at 100°C!

Most millipedes are relatively slow moving, and as well as having thick calcareous exoskeletons for protection, also carry a battery of repugnatorial glands (Fig. 13.10). The openings are located on the sides of tergal plates or on the margins of tergal lobes—usually a pair per segment, though they are entirely absent from some segments. The composition of the secretion varies with species but can include aldehydes, quinones, phenols and hydrogen cyanide. The HCN is liberated just before use when a precursor and enzyme are mixed from a two-chambered gland.

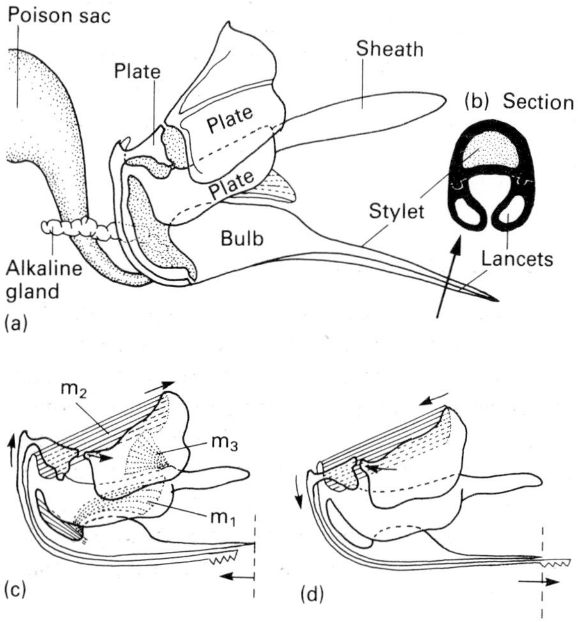

Fig. 13.9. Sting of bee, *Apis mellifera*. The shaft (stylet + barbed lancets) is depressed by contraction of m_1, then the contraction of the powerful m_2 muscles, running from quadrate plate to anterior part of the oblong plate, causes rotation of the triangular plate so as to push out the lancet. Retraction of the lancet is brought about by contraction of m_3. The muscles on the two sides of the sting work alternately, and by successive acts of protraction and retraction drive the lancets more deeply into the body of the victim. Poison is secreted by a pair of thread-like glands in the abdomen. Their secretion accumulates in the poison sac that opens at the base of the sting into the poison canal. The function of the so-called alkaline gland is uncertain. (After Imms, 1964.)

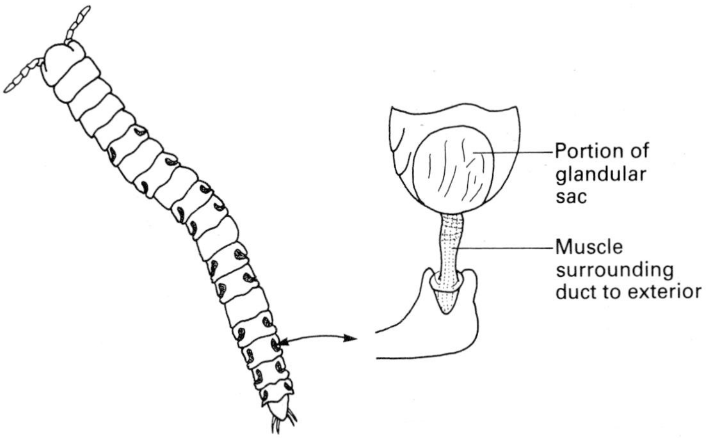

Fig. 13.10. Repugnatorial glands of a millipede (after Cloudsley-Thompson, 1958).

This fluid, which is toxic or repellent to other small animals and from large tropical species seems to be caustic to human skin, is usually released slowly but some species can eject it as a high-pressure jet or spray for 10 to 30 cm. Here, ejection is probably caused by the contraction of trunk muscles adjacent to the secretory sac. The carnivorous and faster running centipedes are less well endowed with repugnatorial mechanisms. They rely more for protection on speed and the use of poison claws that are also concerned with the capture of prey (Fig. 8.21). Some species do, nevertheless, carry repugnatorial glands and some lithobiomorphs bear large numbers of unicellular repugnatorial glands on the last four pairs of legs, that they can kick out at an aggressor, throwing off adhesive droplets.

The pedicellariae of echinoderms, found in Asteroidea and Echinoidea, are also organs that have evolved specifically for defensive purposes. These are specialized, jaw-like appendages and are used for protection, especially against larvae that might settle on the body surface (Fig. 7.18). There are three main kinds: pedunculate (stalked), sessile (attached directly to the test) and alveolar (somewhat insunk). One of the several kinds of echinoid pedicellaria contains glands that secrete a poison capable of rapidly paralysing small animals and driving larger predators away. The avicularia zooids of some bryozoans (Section 6.3.3.3) have the same function as pedicellariae.

An interesting form of repulsion is by startling potential predators. Some lepidopterans and other insects have large markings on their wings which appear to be imitations of the vertebrate eye. These animals normally rest with 'eyes' concealed but suddenly expose the spots when disturbed. Another possible explanation for them is that they deflect predators to less vulnerable parts of the body or even to defence organs; for example, some wasps have white abdominal spots near their stings. Experimental work with captive birds has produced evidence for both functions of eye-spots. Flash colouration also serves a startling function and the rapid colour changes that occur in cuttlefishes after disturbance provide a particularly vivid example of this.

13.2.2 Against internal invaders

Some pathogens can penetrate outer defences but then all organisms have an inner line of defence of one form or another to combat this. In the vertebrates this is achieved by an immune system that involves antibodies capable of neutralizing specific foreign agents. Invertebrates do not have such a specific immunological system, but they do have an inner line of defence—based generally on phagocytic cells capable of recognizing and eliminating foreign material.

13.2.2.1 Recognition of self and non-self are fundamental requirements

A classic example of self-recognition is given by sponge cell reaggregation. By squeezing cells through cloth or by putting them in a solution of ethylene diamine tetraacetic acid (EDTA) it is possible to dissociate whole sponges into slurries of individual cells. Mixtures of cells from different species and even clones within the same species will reaggregate in species-specific and clone-specific fashion.

Grafting experiments also make the same point. Gorgonians (colonial anthozoans), for example, reject grafts from different species (xenografts) and from genetically different individuals of the same species (allografts) but accept grafts from different parts of the same colony (autografts) which consistently fuse.

13.2.2.2 Phagocytic or amoeboid cells are of general importance in invertebrate self-defence

All the reactions noted above were for colonial and encrusting organisms. These are often found in situations where living space is limited and competition for it is keen, so that evolution of self-recognition might have been favoured as a means of maintaining self-integrity. Hence these kinds of systems of self-recognition may be a peculiar consequence of this

Table 13.2 The invertebrate phyla in which it has been recorded that amoeboid cells remove foreign material

Animal phylum	Particle or substance injected	Response: Phagocytosis	Response: Encapsulation
Porifera	India ink, carmine	+	
	Erythrocytes	+	
	Trematode redia, cercaria		+
Annelida	India ink, carmine, iron	+	
	particles, erythrocytes	+	
	Foreign spermatozoa	+	
Sipuncula	Latex beads, bacteria	+	
Mollusca	Carmine	+	
	India ink	+	
	Erythrocytes, yeast, bacteria	+	
	Thorium dioxide	+	
Crustacea	Bacteria, carmine	+	
Uniramia	Bacteria	+	
	Latex beads	+	+
	Iron, saccharide	+	
	Araldite implants		+
	Bacillus thuringiensis	+	
	India ink, carmine	+	
	Erythrocytes, bacteria	+	
Echinodermata	Bovine serum albumin	+	
	Bovine gamma globulin	+	
	Sea urchin cells (into a sea star)	+	
Urochordata	Carmine	+	
	Glass fragments		+
	Trypan blue	+	
	Thorium dioxide	+	

kind of evolutionary pressure and not a common basis for the evolution of an immunological system within the animal kingdom.

Immunological studies began when, in the early 1900s, Elie Metchnikoff introduced rose spines beneath the epidermis of bipinnaria larvae of starfish and found that within a short time these were 'attacked' by amoeboid cells. He obtained similar results when he injected anthrax bacilli into the larvae of the rhinoceros beetle, *Oryctes nasicornis*. From observations such as these, Metchnikoff proposed the idea that amoeboid cells, which are involved in intracellular digestion in many primitive invertebrates (Section 9.2.2), had been retained in the evolution of more advanced animals as an inner defence system. Phagocyte cells are certainly widespread throughout invertebrate animals and experiments involving the introduction of foreign material into living animals have indicated that they are capable of removing a variety of foreign particles (Table 13.2).

13.2.2.3 How do phagocytes discriminate between self and non-self?

Wandering phagocytes must be able to 'ignore' normal tissue of *self* but engulf *non-self* particles. They may also be involved in the removal of damaged 'self' due to the presence of xenobiotic, tissue-damaging substances such as pollutants.

Knowledge of the mechanism is limited. *A priori* it would seem likely that recognition occurs when a phagocyte makes contact with its target and that since foreign particles are unlikely to produce specific 'kill' signals, it is more likely that self cells produce specific 'don't kill' signals. There is no evidence for an intermediate antibody system as subtle as in the vertebrates, but there is evidence for opsonin (molecules that coat foreign particles so that they adhere to phagocytes and facilitate their phagocytosis) activity in the body fluids of those invertebrates that have fluid-filled cavities. Thus, amoebocytes from the chelicerate horseshoe crabs exhibit no significant bactericidal effect in the absence of serum, but killed *Escherichia* when serum was present. Similarly, phagocytosis of human erythrocytes by haemocytes (blood amoebocytes) from the octopus *Eledone cirrosa* occurred only after they had been exposed to *Eledone* serum. Extracts from many invertebrates act as agglutinins—cross-linking and binding various cells and bacteria *in vitro* (Table 13.2)— and these might have opsonic properties, binding foreign particles to the surfaces of phagocytes. These mechanisms are discussed further in Coombe *et al.* (1984).

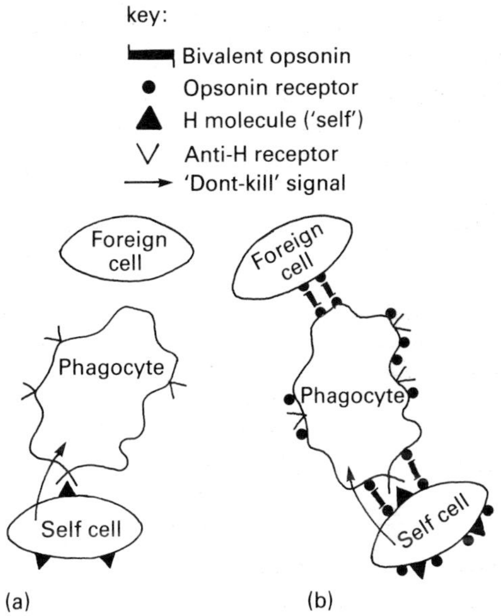

Fig. 13.11. Models of phagocyte immunological systems of invertebrates—see text for further explanation (after Coombe *et al.*, 1984).

Models showing how phagocytes might effect self-recognition are summarized in Fig. 13.11. Direct self-recognition (Fig. 13.11a) probably occurs in the phagocytes that 'patrol' the tissues of solid-bodied invertebrates, whereas the involvement of intermediate factors, opsonins (b), can occur in invertebrates with fluid-filled cavities—the fluids containing the opsonins.

13.2.2.4 Reproduction presents some complications

Where there is internal fertilization (Chapter 14), sperm with a foreign genotype are transferred to the tissues of another organism. Similarly, fertilized eggs and embryos, when they reside within the tissues of a mother (Chapter 15), are genetically semi-foreign. However, under normal circumstances these must not be destroyed by the immunological system of the parent—the 'mother' must be a willing host for the sperm or offspring. Just how the reproductive cells evade immunological destruction is not understood, but the following experiments are illuminating. Spermatozoa from allogenic donors are not phagocytosed when injected into the coelomic cavity of earthworms, whereas mammalian spermatozoa and spermatozoa from other species of earthworms are phagocytosed. Male sipunculan worms failed to encapsulate homologous eggs when they were injected into their coeloms, and this suggests that propagules have some generally evasive mechanism to deal with host immune systems, for eggs do not naturally occur within the male worm! Eggs damaged by staining, heating or sonication were rapidly encapsulated whereas frozen eggs, though apparently dead, were not. For more information, see Coombe *et al.* (1984).

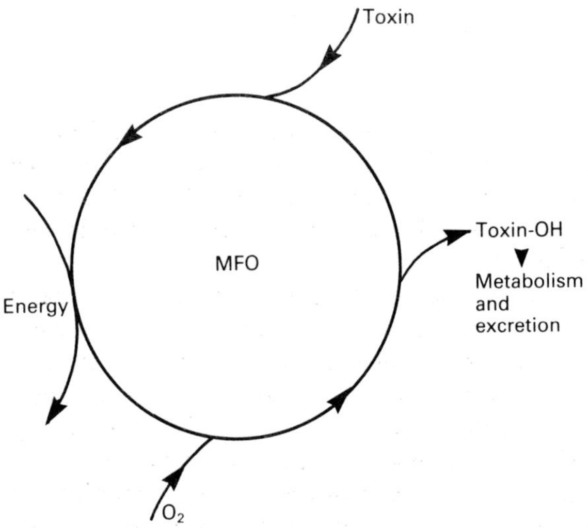

Fig. 13.12. Simplified representation of MFO system (after Calow, 1985).

13.2.2.5 Evasion of host response

Successful parasites have to be able to defend *themselves* against the host immune response. This has been carefully studied in some trematodes where a variety of defences has been discovered. Thus schistosomes coat themselves with antigens identical or similar to those of the host, these either being synthesized by the parasite (molecular mimicry) or derived from the host and bound to the surface of the parasite. Whatever their origin, these antigens mask parasite antigens so they are no longer recognized as foreign. Fasciolid flukes, on the other hand, produce substances that are toxic to lymphocytes and other immunological cells. Also, the glycocalyx of the parasite tegument appears to turn over at a high rate, and by this means flukes may be able to slough off host antibody.

13.2.3 Responses to pollutants

Environmental stressors and the responses that they provoke are so varied that it is not possible to treat them all in a comprehensive way. Some 'physiological defences', e.g. to stress from oxygen, salinity, etc., have already been treated in previous chapters (e.g. Sections 11.6.5 and 12.2). Here we describe some general responses that are elicited in invertebrates by general classes of pollutants: xenobiotics (organic toxins) and heavy metals.

13.2.3.1 Mixed function oxygenase (MFO) and xenobiotics

Organic pollutants, such as hydrocarbon compounds from oil spills, can penetrate the tissues of marine invertebrates. They are lipophilic and as such not easily metabolized. Instead they can accumulate in lipid depots and the lipid components of cell membranes until they reach concentrations at which they cause biochemical problems. However, some marine invertebrates, namely polychaetes and some molluscs and crustaceans, contain an enzyme system capable of oxidizing the toxin by literally adding oxygen atoms, making it more hydrophilic and thus more easily metabolized. The system is fairly non-specific in terms of the substrates that it will attack. It consists of several enzymes, and some cytochromes are present. It is primarily membrane-bound in the microsomal fraction of the endoplasmic reticulum of certain tissues (e.g. digestive gland or hepatopancreas). The process of oxygenation is energy expensive and is summarized in a very general way in Fig. 13.12; it involves expoxidation, hydroxylation and dealkylation. It is also an inducible system, i.e. the enzymes associated with it are being produced only when a xenobiotic challenge exists; for example, a cytochrome specifically associated with MFO increases in the tissue of mussels after 1 day of exposure to diesel oil and returns to control concentrations after 8 days' recovery.

Mixed function oxygenases also occur in herbivorous insects, where they are probably involved in dealing with natural organic toxins produced by plants as a defence against these herbivores. For example, polyphagous insects generally have higher MFO activities than stenophagous (more specialist) ones, probably because they are exposed to a wider variety of toxic compounds such as phenolics, quinones, terpenoids and alkaloids.

13.2.3.2 Metallothioneins

Heavy metals, such as mercury, cadmium, copper, silver and tin, can be extremely toxic for aquatic invertebrates. For example, they cause denaturation of enzymes by interacting with them and altering their tertiary configuration. However, some marine invertebrates—notably crabs, bivalves, limpets and some annelids—can detoxify heavy metals by binding them to specialized proteins called metallothioneins. These are low molecular weight compounds, rich in sulphhydryl (SH) groups due to high levels of the amino acid cysteine within them. The SH group is capable of combining with or chelating the metal and rendering it less toxic.

13.2.4 Repair—protection against ageing?

In Section 13.1.3 it was hinted that ageing of whole organisms is due to the accumulation of suborganismic damage. Evidence for this is as follows.

Tissue disruption—ageing in dipteran insects is associated with a declining flight capacity and this is correlated with degenerative changes in the structure of the flight muscle. Cell numbers in the brain of worker bees have been shown to decline from a mean of 522 at eclosion to 350 at 10 weeks (Fig. 13.13).

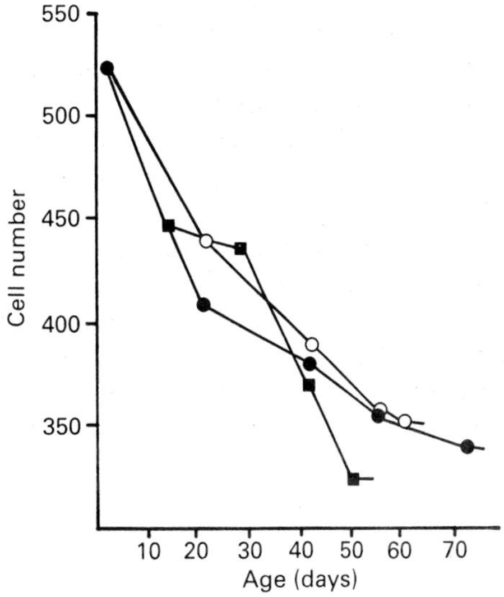

Fig. 13.13. Decline in number of cells in the brain of bees with age (after Rockstein, 1950).

Lipofuscin (known as age-pigment)—probably a product of lipid peroxidation and derived from the breakdown of membranes—has been found to accumulate with age in the tissues of nematodes and insects (Fig. 13.14).

Enzyme fidelity—there is evidence, particularly from studies of nematodes, that enzyme structure and function become impaired with age: enzymes become more sensitive to heat denaturation (indicative of changes in molecular organization), develop different immunological properties and have reduced catalytic capacities. However, these observations do not apply to all the enzymes that have been studied in nematodes nor to many enzymes studied from other animals.

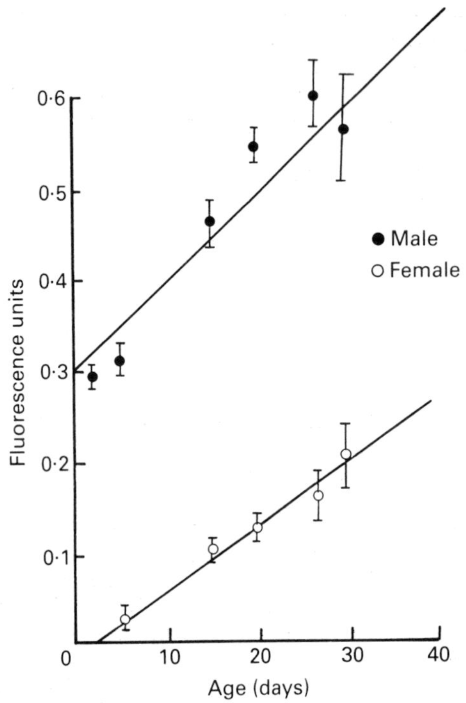

Fig. 13.14. Accumulation of lipofuscin (measured by its fluorescence) in *Drosophila melanogaster* (after Biscardi & Webster, 1977).

The root cause of all this damage is likely to emanate from molecular processes such as thermal noise, mistakes being made in the synthesis of proteins and a variety of other processes. At the same time, damaged protein molecules can, in principle, be replaced according to genetic instructions and whole cells can be replaced, again according to genetic instruction, by mitosis (Chapter 1). Indeed evidence from work on nematodes suggests that proteins containing abnormal amino acids have a more rapid turnover than normal proteins, but that this slows down as worms get older. Cell division occurs extensively in the Cnidaria, Platyhelminthes, Annelida and Mollusca but is very restricted in adult nematodes and insects. In freshwater triclad turbellarians there is a reduction in cell turnover with age (Fig. 13.15), which is accentuated with the onset of reproduction in semelparous forms (see Section 14.4). The latter is associated with accelerated ageing. The age-state, vitality, of an organism might therefore depend upon a balance between the generation of damage and its replacement or repair. Not surprisingly, therefore, ageing processes are least obvious in organisms such as cnidarians where tissue turnover is continuous and are most obvious in organisms such as nematodes

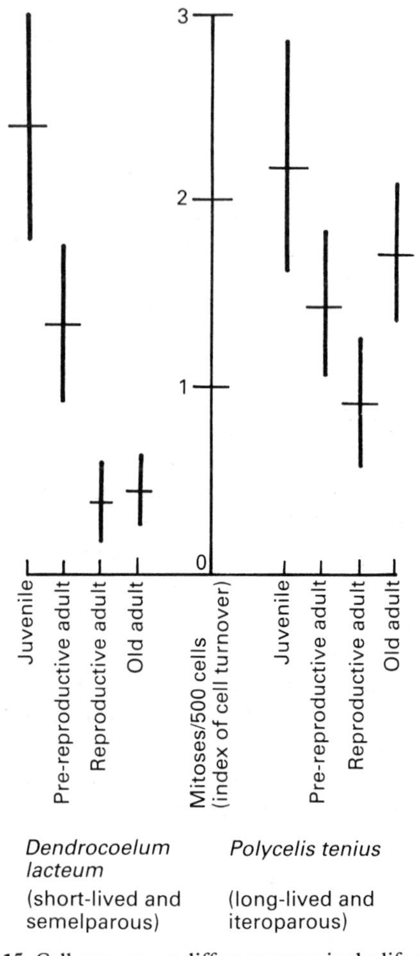

Fig. 13.15. Cell turnover at different stages in the life cycles of two triclads, one long-lived, the other short-lived. The latter invests considerably more in reproduction than the former. In both, cell division reduces from juvenile to adult. However, whereas there is a big further dip in this, following the onset of reproduction in the short-lived form, there is a smaller, non-significant dip in the long-lived form. It is thought that reduced turnover is *caused* by high investment in reproduction and shortens life by acceleration of senescence. (After Calow & Read, 1986.)

Table 13.3 Distribution of senescence

Animal phylum	Suspected lack of ageing in some species	Suspected presence of ageing in some species	Definite ageing in some species
Cnidaria	+	+	
Platyhelminthes	+	+	+
Mollusca	+	+	
Nematoda			+
Annelida	+	+	+
Rotifera			+
Arthropod phyla			+

and insects, where tissue turnover is limited (Table 13.3). Interestingly, mitosis is more extensive and persists for longer in those insects with the greatest longevity, namely coleopterans.

13.3 Conclusions

The threats to which organisms are exposed are diverse, as, not surprisingly, are the responses that they provoke. Yet it is possible to produce a coherent classification of threats that puts them on a continuum, varying from more to less intimacy of association with the organism that is threatened. Similarly, behind the diversity of defence responses it is possible to perceive some common features. For example, defence mechanisms, whether they be against predators, parasites, microbes, or wear and tear, are all expensive in material and energy. It is these general principles that readers should aim to take away from this and, indeed, all the chapters in this part of the book.

13.4 Further reading

Berry, R.J. 1977. *Inheritance And Natural Selection. New Naturalist* No. 61. William Collins & Co., Glasgow.

Blest, A.D. 1957. The function of eyespot patterns in the Lepidoptera. *Behaviour,* **11,** 209–56.

Calow, P. 1981. *Invertebrate Biology.* Croom Helm, London.

Calow, P. 1983. *Evolutionary Principles.* Blackie, Glasgow.

Coombe, D.R., Ey, P.L. & Jenkin, C.R. 1984. Self/non-self recognition. *Q. Rev. Biol.,* **59,** 231–55.

Davies, I. 1983. *Ageing.* Edward Arnold, London.

Eisner, T., Van Tassell, E. & Carrel, J.E. 1967. Defensive use of a "faecal shield" by a beetle larva. *Science, N.Y.,* **158,** 1471–3.

Erichsen, J.T., Krebs, J.R. & Houston, A.I. 1980. Optimal foraging and cryptic prey. *J. anim. Ecol.,* **49,** 271–6.

Gliwicz, M.Z. 1986. Predation and the evolution of vertical migration in zooplankton. *Nature (London),* **320,** 746–8.

Hansell, M.H. 1984. *Animal Architecture and Building Behaviour.* Longman, London.

Livingstone, D.R., Moore, M.N., Lowe, D.M., Nasci, C. & Farrar, S.V. 1985. Responses of the cytochrome P-450 monoxygenase system to diesel oil in the common mussel, *Mytilus edulis* L., and the periwinkle, *Littorina littorea* L. *Aquatic Toxicology,* **7,** 79–81.

Nordlund, D.A. & Lewis, W.J. 1976. Terminology of chemical releasing stimuli in intraspecific and interspecific interactions. *J. chem. Ecol.,* **2,** 211–20.

Rockstein, M. 1950. The relation of cholinesterase activity to change in cell number with age in the brain of the adult worker bee. *J. cell. comp. Physiol.,* **35,** 11–23.

Theodor, J.L. 1976. Histo-incompatibility in a natural population of gorgonians. *Zool. J. Linn. Soc.,* **58,** 173–6.

Turner, J.R.G. 1984. Darwin's coffin and Dr. Pangloss—do adaptationist models explain mimicry? In: B. Shorrocks (Ed.) *Evolutionary Ecology,* pp. 313–61. Blackwell Scientific Publications, Oxford.

14
REPRODUCTION AND LIFE CYCLES

Fundamentally, the multicellular eukaryotes are diploid organisms that reproduce sexually by the production of haploid germ cells, two such gametes fusing to form a new diploid zygote. The genetic programme in the germ cells usually differs from that of the parent because of the reassortment of genes that occurs during meiosis. Further, the new diploid organism is usually, but not always, formed by the union of gametes from different genotypes. Thus sexual reproduction and fertilization in multicellular organisms results in the creation of a single cell with a new and unique genetic constitution.

This chapter will be concerned with these processes of sexual reproduction, whilst the subsequent development and differentiation of the zygote and its division products will be treated in Chapter 15. Although the life cycles of animals all follow the same basic plan, there is much variation in the pattern of sexual reproduction, and in that of asexual multiplication which is also widespread in animals. We will therefore also review this diversity, its control, and the adaptive significance of the different variants, including the selective advantages of sexual reproduction itself. It will become apparent in the following pages that different combinations of reproductive and life-cycle traits are characteristic of different phylogenetic lines of animals, grades of animal organization, and ecological situations. This will make it possible to discuss invertebrate reproductive strategies in relation to current theories of life-cycle evolution.

14.1 Introduction

In Chapter 1 it was suggested that evolution is a natural consequence of systems that persist by a semi-conservative process of replication, and that all living things possess in common the same type of genetic programme, variation in which arises fom miscopying of base sequences during replication of the DNA molecule. The evolution of multicellular organisms, however, requires not just the semi-conservative replication of cells, but the replication of organisms, a process which we call *reproduction*, and in the vast majority of organisms this process also allows genetic recombination. This is referred to as sexual reproduction. Several systems of sexuality are possible; what they share in common is the creation of situations in which the exchange of genetic material between the programmes specific to different individuals is possible.

Three different systems are represented diagrammatically in Box 14.1. In the viruses, genetic exchange can occur when multiple invasion of a host cell by different strains of the same virus has taken place. In bacteria, part of the DNA molecule of one partner is transferred to another during conjugation and genetic exchange can occur between the complementary strands of DNA. All multicellular eukaryotic organisms have adopted the third of the systems of sexuality illustrated.

In animals, two very different types of germ cells are usually produced, the spermatozoan and the egg (Section 14.4.2). These are often derived from specialized cells set aside early in embryonic development, but this is not universal and the germ cells are sometimes derived from dedifferentiated somatic cells.

Multicellular organisms can have different life histories; many plants, for instance, have alternate generations of haploid and diploid individuals (Box 14.2), which may be morphologically different or identical. An alternation of generations of this type never occurs in animals, although there may be a striking alternation between diploid phases which reproduce sexually and asexually (Section 14.2).

Despite the essential similarity of the sexual process in animals, the conditions of sexuality and the organization of the life cycle are extremely varied. Invertebrate animals can differ from each other in a great many of what may be termed their 'life-cycle traits', often collectively and somewhat anthropomorphically described as the 'reproductive strategy' (Table 14.1).

14.2 The significance of sexual and asexual reproduction

14.2.1 Asexual reproduction in invertebrate life cycles

Sexual reproduction is widespread but many organisms can also reproduce asexually, that is produce offspring without recombination of genetic material. Offspring produced in this way will have a genetic constitution which is virtually identical to that of the parental organism. Asexual reproduction can take place either by subdivision of an existing body into two or more multicellular parts (*budding* and *fission*) or by the production of diploid eggs (*parthenogenesis*). Figure 14.1 illustrates these two basic mechanisms. Both are widespread among the invertebrates.

Fission is particularly common in the soft-bodied phyla such as the Porifera, Cnidaria, Platyhelminthes, Nemertea, Annelida and some Echinodermata. It is not often found in those that have an external casing and is unknown in the Mollusca and the arthropod phyla. Fission may involve simple transection into two fragments, each of which regenerates the missing parts (Fig. 14.1a) or it may give rise to multiple fragments each of which can reconstitute a complete animal (Fig. 14.2). Fission is usually combined with the capacity for sexual re-

Table 14.1. Reproductive traits of marine invertebrates

Trait			
Development	Pelagic	planktotrophic / lecithotrophic / mixed	Non-pelagic
Egg size	Small *c.* 50 μm ⟷		Large > 1000 μm
Fecundity	High 10^6 ⟷		Low 1
Brood frequency	Low 1 per annum ⟷		High Many per annum almost continuous
Broods per lifetime	One		Many
Longevity (Generation time)	Perennial Many years ⟷	Annual	Subannual A few days or weeks
Body size	Large Length > 1000 mm ⟷		Small < 1 mm
Spermatozoa	Simple		Advanced
Fertilization	External without sperm storage		Internal or with sperm transport or storage
Reproductive effort*	Large Made later		Small Made early

*Reproductive effort can be defined as $\frac{E_G}{E_S+E_G}$ or $\frac{\Delta E_G}{\Delta E_S+E_G}$ where E_G is the energy allocated to germinal tissues, E_S the energy allocated to somatic tissues, ΔE_G the annual increment in germinal tissues, and ΔE_S the annual increment in somatic tissues.

Box 14.1 Systems for sexual reproduction (Myxis)

Sexual reproduction is a system of self-replication which permits the exchange of genetic information (embodied in a DNA molecule) and the creation of an individual whose genetic information is derived by crossing over from different parental sequences.

1 Viruses. Bacteriophage viruses cause infection of a bacterial host by injecting their DNA core. The protein coat of the virus remains outside the cell. Inside the bacterium the viral DNA molecules provide the information for the creation of new protein coats and the lysis of the host.

(a) Mixed infections of different mutants for the same structural genes (A or B) are inactive in certain strains of bacterium, but mixed infections for different structural genes are active.

(b) The mutant virus particles are able to grow in strain k of the host bacterium.

(c) Genetic crossing over can occur from mixed infection of the non-complementary type in strain k and these

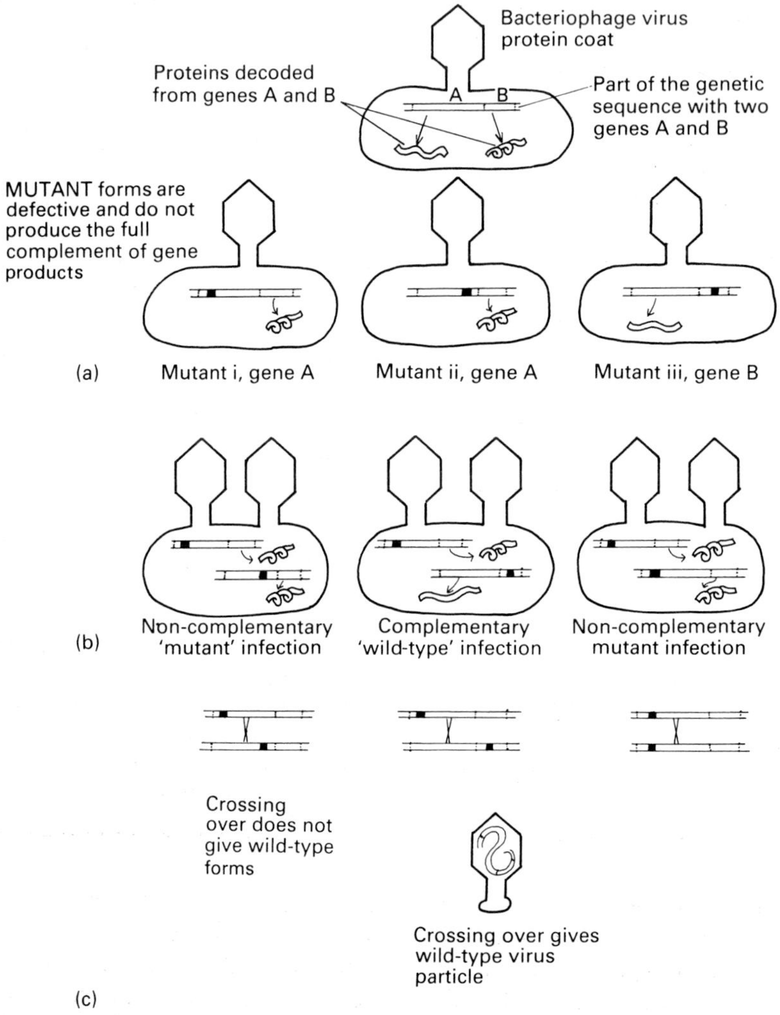

new virus particles can infect strain B, i.e. sexual reproduction has occurred.

2 Bacteria.

(i) Different strains of bacteria are able to conjugate. During the conjugation process an attachment stalk or *pilus* is formed and the DNA molecule (bacterial chromosome) from the donor (male-like) cell is passed to the recipient.

(ii), (iii) The length of DNA exchanged is time-dependent and can be interrupted by vigorous shaking of a culture.

(iv) Genetic exchanges can occur between the duplicated lengths of chromosome material. In this way bacteria can be produced which combine the genetic characteristics of the two parental individuals.

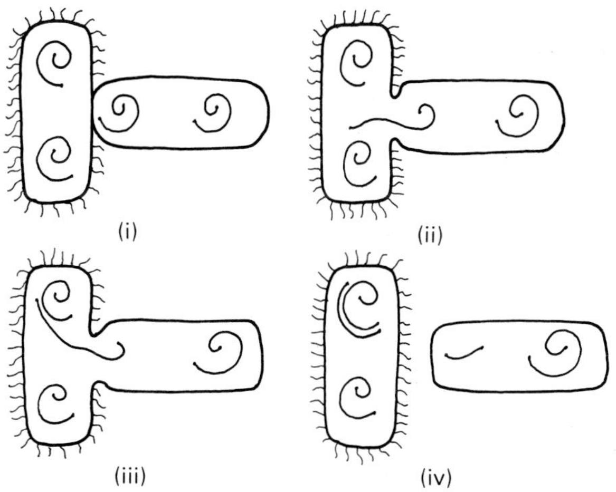

3 Sexual reproduction in higher organisms. Diploid cells have two sets of chromosomes, each containing coded genetic information from one parent. The diploid cell will give rise to haploid gametes by a special cell division called meiosis. The genetic information in the haploid germ cells may be identical to that of the parent or different due to crossing over.

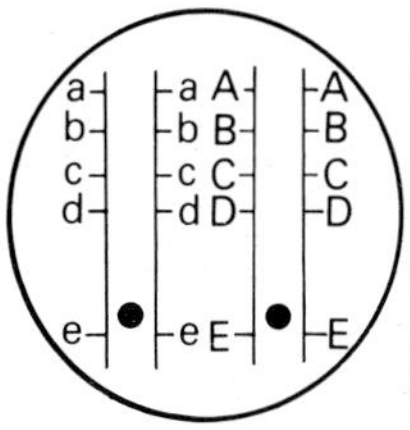

A diploid animal cell. The genes represented by the sequence a–e and A–E occur on chromosomes inherited from two parents. This cell is heterozygous

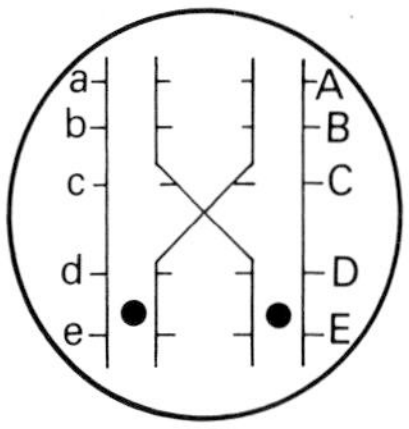

During MEIOSIS genetic exchange (crossing over) can occur between chromatids of the paired chromosomes

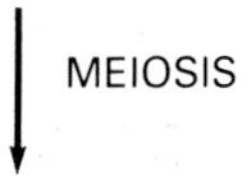

The gamete may have parental or recombinant sequences of genes

GERM CELLS

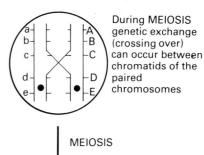

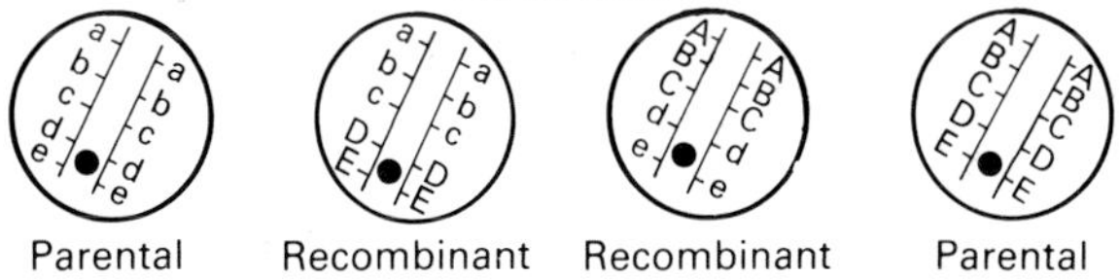

The frequency of crossing over is a function of the distance between the genes in the DNA sequence of the chromosome

Haploid germ cells from different parents (or sometimes from one parent) fuse (a process called fertilization) and form a diploid zygote. The genetic constitution of the zygote is not the same as that of the diploid cells which gave rise to the haploid gametes. It combines genetic traits from the two parental genomes in each set of chromosomes.

production in complex life cycles with asexual and sexual 'generations' (see Fig. 14.4a). The difficulties that fission presents to our concept of 'the individual' is underlined by those invertebrates in which fission is incomplete and thus gives rise to colonial organisms made up from a large number of structural units. Colonial structures are encountered particularly frequently in the Cnidaria, the bryozoans and the urochordates. Some compound or colonial invertebrates are illustrated in Fig. 14.3. These animals are

Box 14.2 Life cycles and sexual reproduction

1 Sexual reproduction in most protists. Example, *Paramecium*.

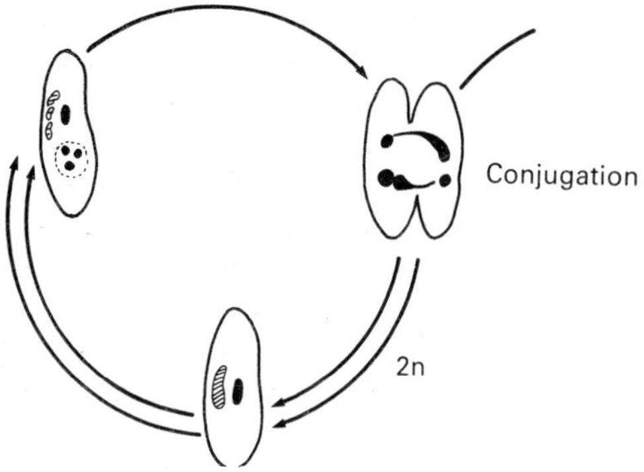

2 The life cycle of all multicellular animals (and some algal protists).
The multicellular adult is composed of several or many (often different) diploid cells.

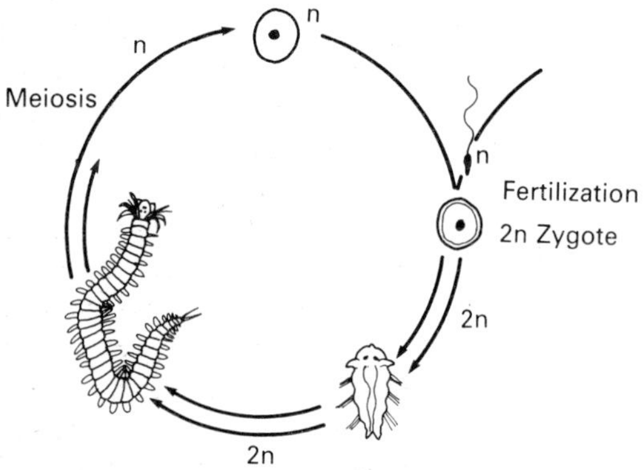

The multicellular body gives rise to haploid gametes.
Fusion creates a diploid zygote.
By a series of mitotic divisions a new diploid adult develops.

3 The life cycle of many algae and all higher plants involves an alternation between multicellular haploid and diploid phases. The haploid phase is called the gametophyte; it produces (by mitosis) sperm cells and/or egg cells.
Fusion of gametes creates a zygote.
By mitosis this develops into a multicellular body called the sporophyte.
Meiosis creates haploid spores.
Without fusion they proceed through mitotic divisions to create the haploid gametophyte.
The figure below illustrates this process in ferns.

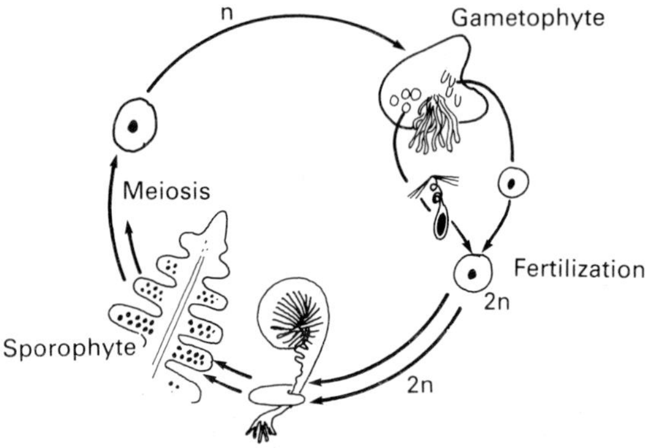

In higher plants the gametophyte is often reduced to a very small number of cells.
Animals never show this alternation of haploid and diploid generations. The alternation of sexually and asexually reproducing individuals (see Fig. 14.4) is not equivalent.

more realistically interpreted as being made up of a series of modular units, not individuals.

Parthenogenesis is also widespread among the invertebrates (in this book it is taken to mean asexual reproduction via eggs but some authors use the term parthenogenesis to include fission. It involves modification of meiosis so that the eggs are diploid and do not have to fuse with male germ cells in order to give rise to a diploid cell from which development can proceed. The term arrhenotoky is used to describe the related phenomenon in which unfertilized haploid eggs develop into males and fertilized diploid eggs give rise to females. Obligate parthenogenesis, in which sexual reproduction never occurs, is rare, but it is found in the bdelloid rotifers (see Section 4.9.3.1) in which males have never been observed. More frequently, parthenogenesis occurs cyclically together with episodes of sexual reproduction. One or several

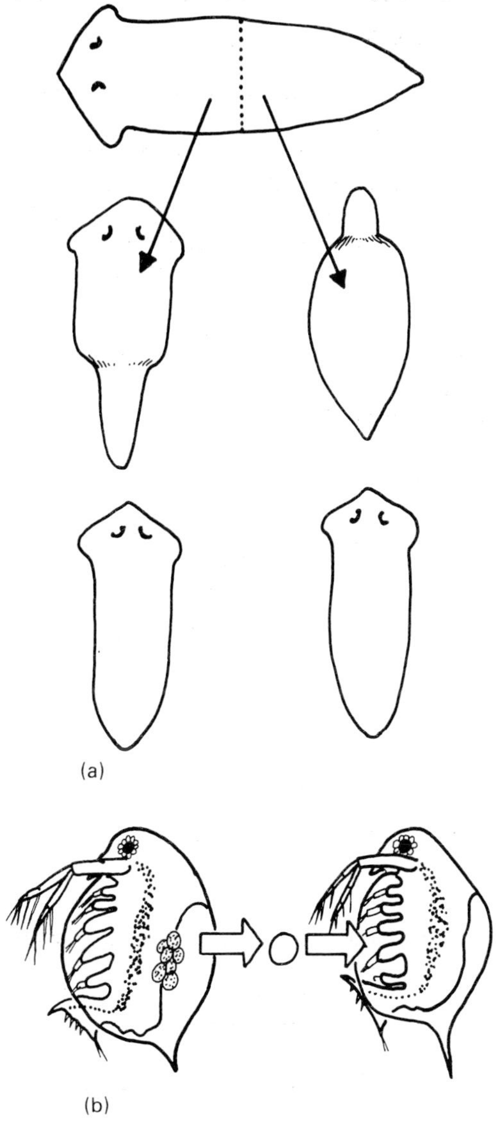

Fig. 14.1. Diagrammatic representation of two types of asexual (amiotic) reproduction. (a) Fission. (b) Parthogenesis.

generations of asexually reproducing individuals are followed by a generation of sexual individuals which usually give rise to resistant resting eggs (Fig. 14.4a). Monogont rotifers, many small freshwater Crustacea and aphids all characteristically exhibit this type of life history in which the diploid individual that emerges from the resting egg gives rise to a clone of genetically identical descendants. In these circumstances the growth rate of the population is at a maximum and the life cycle can be thought of as being a means for the maximum exploitation of a temporarily under-exploited food resource where other biological constraints limit body size. At some stage a new type of individual appears in the population which has two types of offspring, sexually reproducing females and males. The transition may be endogenously determined, but more often it is a response to changing environmental conditions such as crowding, food quality or shortening day lengths (see also Section 14.4.4). There may be morphological changes associated with this transition to sexual reproduction (Fig. 14.4b,c) and in the aphids complex changes of morphology also occur during the asexual phase (see Fig. 14.4d).

14.2.2 Patterns of sexuality

Over 99% of all invertebrates exhibit sexual reproduction at some stage in their lives and it is usually the only means by which reproduction can take place. Sexual reproduction in animals always involves fusion of relatively large, immobile female gametes, the ova, and minute, mobile male gametes, the spermatozoa, a phenomenon referred to as *anisogamy*. It is therefore possible to recognize male and female functions, but these are not necessarily assigned to separate individuals in the population. The phenomenon in which the sexes are separate in this way is described as *gonochorism* or alternatively *dioecy* (the corresponding adjectives being gonochoristic and dioecious); that in which the same individual can function both as male and female either simultaneously or sequentially is *hermaphoditism* (adj. hermaphroditic). The characteristics of these different systems are summarized in Table 14.2, and a summary of their distribution among the major groups of invertebrates is presented in Table 14.3. In the case of sequential hermaphrodites the sex change may occur at a species-specific size or age, no individuals being pure males or females; alternatively, the sex change may not be so

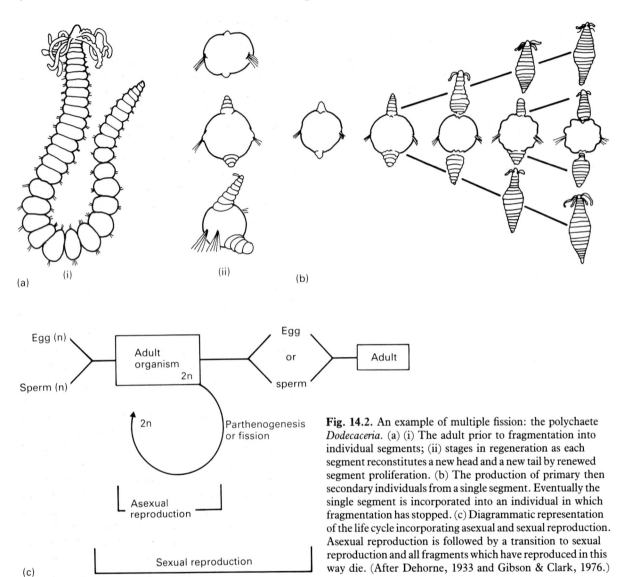

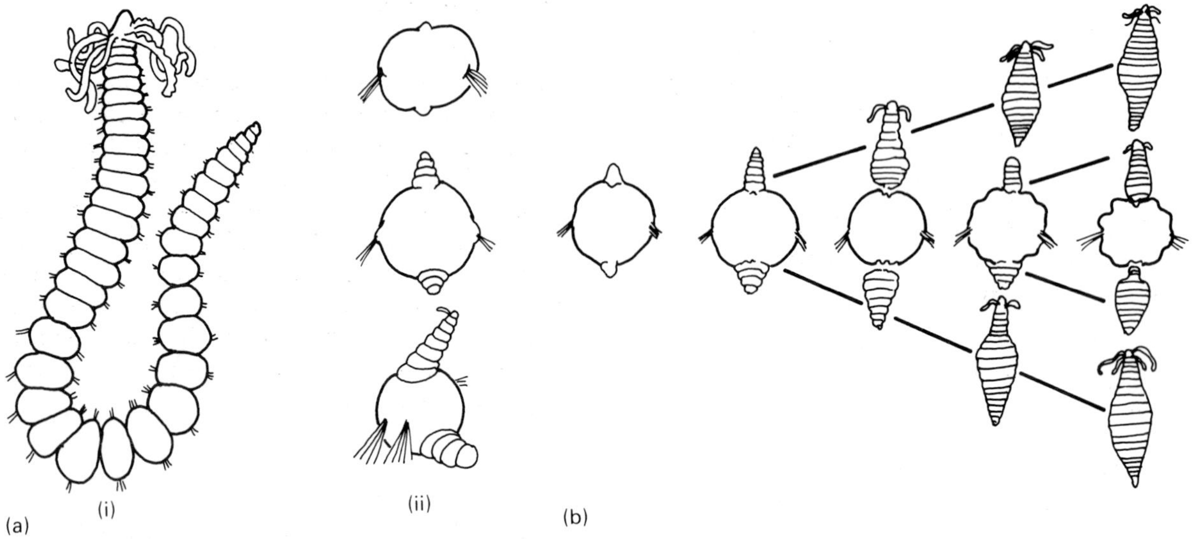

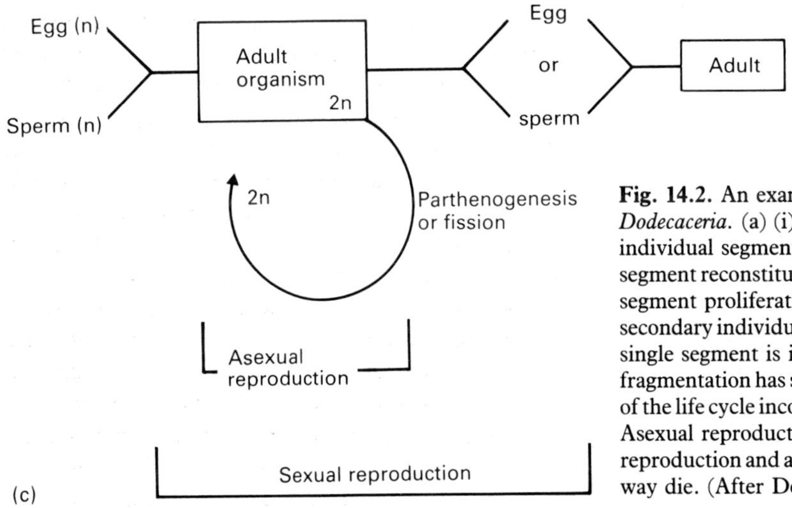

Fig. 14.2. An example of multiple fission: the polychaete *Dodecaceria*. (a) (i) The adult prior to fragmentation into individual segments; (ii) stages in regeneration as each segment reconstitutes a new head and a new tail by renewed segment proliferation. (b) The production of primary then secondary individuals from a single segment. Eventually the single segment is incorporated into an individual in which fragmentation has stopped. (c) Diagrammatic representation of the life cycle incorporating asexual and sexual reproduction. Asexual reproduction is followed by a transition to sexual reproduction and all fragments which have reproduced in this way die. (After Dehorne, 1933 and Gibson & Clark, 1976.)

precisely fixed, taking place at different ages or sizes in different members of the population. In the latter case some individuals may be purely male or female. The two situations are illustrated in Fig. 14.5a and b.

14.2.3 Mechanisms of sex determination

The sex determining mechanisms of invertebrates fall into three basic types: (a) maternal, (b) genetical and (c) environmental.

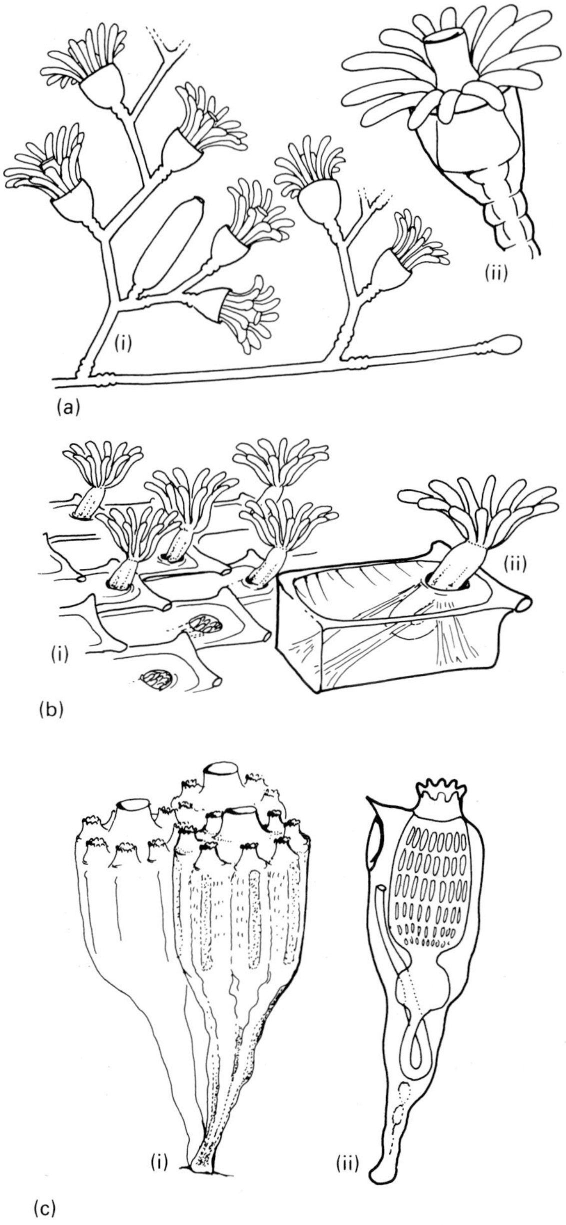

Fig. 14.3. Examples of colonial or 'modular' invertebrates: (a) the hydroid *Obelia*: (i) the branching colony, (ii) an individual polyp; (b) the bryozoan *Membranipora*: (i) part of the mat of zooids, (ii) an individual zooid; (c) the urochordate *Sydnium*: (i) three colonies showing individual inhalent siphons and a shared exhalent siphon, (ii) an individual zooid.

14.2.3.1 Maternal

In this case the sex of the offspring is determined by the mother through the production of different types of eggs. An example is illustrated in Box 14.3, **1**. In this system inbreeding is inevitable and all members of the adult population are females. In the Hymenoptera the queen honey bee is able to control the fertilization of her eggs; fertilized diploid eggs will normally develop as sterile female workers, but have the potential to become functional females if exposed to appropriate conditions during their development, whereas unfertilized eggs will develop parthenogenetically into haploid males.

14.2.3.2 Genetical

In some animals males and females have visibly different chromosomal complements in which females have two identical sex chromosomes, the so-called X chromosomes, and the males have unlike chromosomes, one X chromosome and one Y chromosome. The sex of a zygote is here determined by the sex chromosome complement of the fertilizing sperm. This phenomenon, described as heterogamy, is not as widespread as might be thought among the invertebrates. It occurs in several different forms among the insects and has been thoroughly investigated in the fruitfly *Drosophila melanogaster*. As shown in Box 14.3, female *Drosophila* have two X chromosomes and males have an odd pair: the X and Y chromosomes. At meiosis all the eggs receive an X chromosome but exactly half of the spermatozoa receive a Y chromosome. Random fusion of gametes therefore establishes a sex ratio of 1:1. In many invertebrates heterogamy has not been observed, nevertheless the sex ratio may be fixed.

14.2.3.3 Environmental sex determination

The sex of an individual is not always determined at or before fertilization; it sometimes depends on the environmental conditions experienced by the developing embryo or larva. One of the best known examples is that of the echiuran worm *Bonellia viridis*,

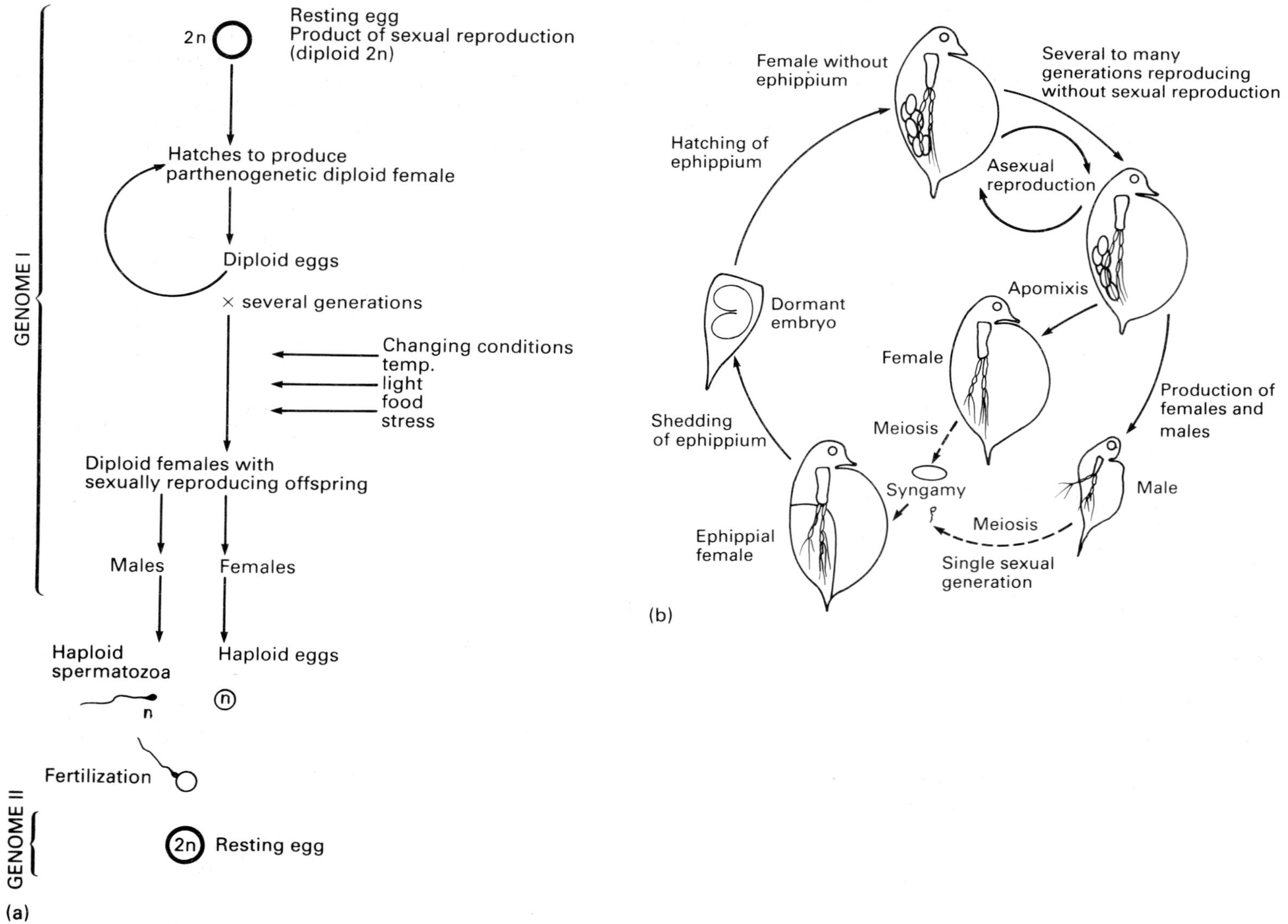
GENOME I
2n
Resting egg
Product of sexual reproduction
(diploid 2n)
Hatches to produce
parthenogenetic diploid female
Diploid eggs
× several generations
Changing conditions
temp.
light
food
stress
Diploid females with
sexually reproducing offspring
Males
Females
Haploid
spermatozoa
n
Haploid eggs
n
Fertilization
GENOME II
2n
Resting egg
(a)
Female without
ephippium
Several to many
generations reproducing
without sexual reproduction
Hatching of
ephippium
Asexual
reproduction
Apomixis
Dormant
embryo
Female
Production of
females and
males
Shedding
of ephippium
Meiosis
Syngamy
Male
Meiosis
Ephippial
female
Single sexual
generation
(b)

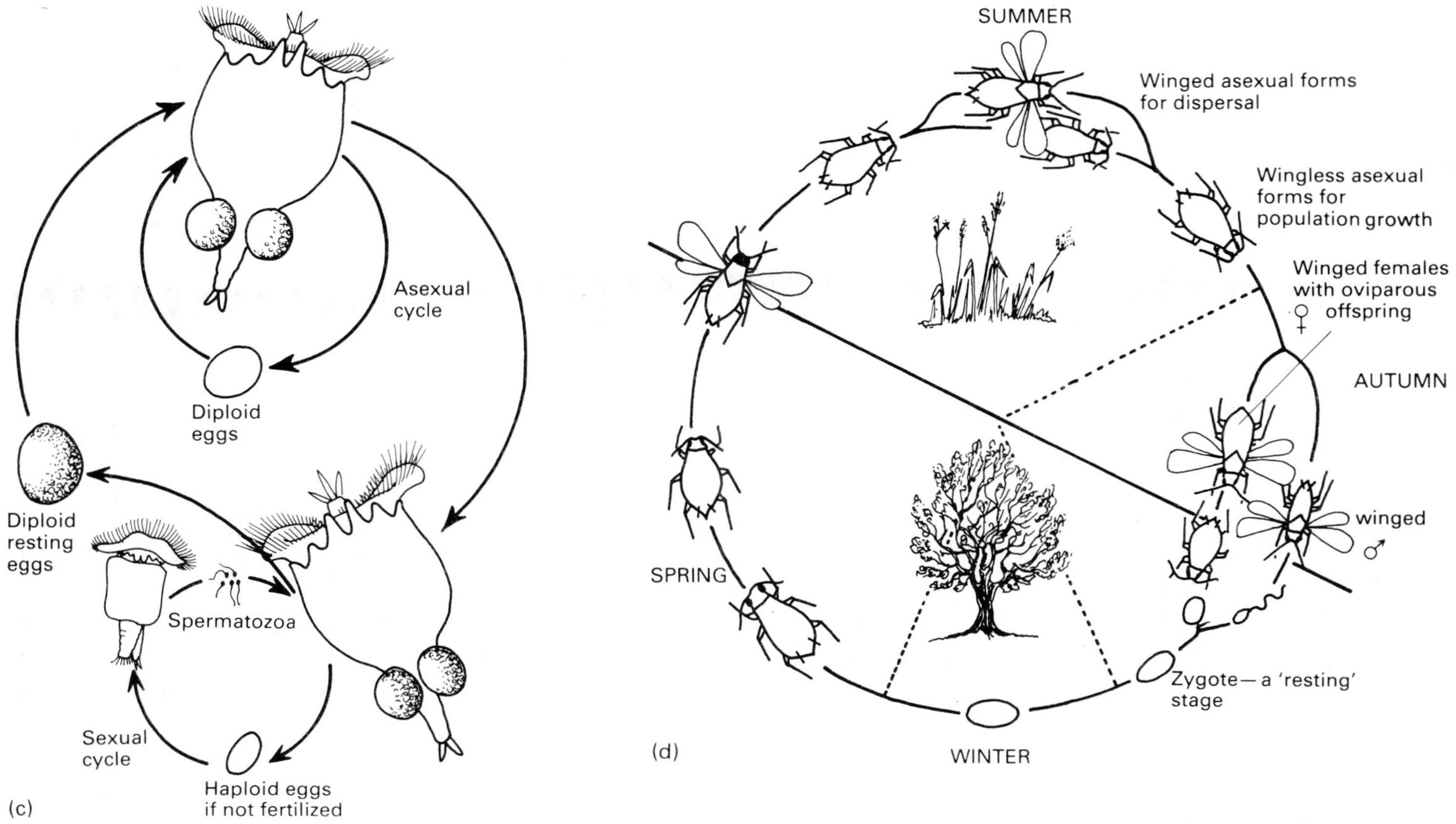

Fig. 14.4. Alternation of parthenogenesis and sexual reproduction in the life cycles of invertebrates. (a) The elements of a generalized life cycle; (b) the life cycle of the freshwater cladoceran *Daphnia*; (c) the life cycle of the rotifer *Branchionus*; the asexual phase is followed by a sexual one in which unfertilized females lay small eggs which hatch as males and fertilized females produce larger eggs which are the overwintering stage in the life cycle; (d) the life cycle of the bird-cherry aphid. All females, other than the oviparous females of autumn, are parthenogenetic and viviparous. There is marked polymorphism especially in the production of winged and wingless forms. (b after Bell, 1982; d after Dixon, 1973.)

Box 14.3 Systems of sex determination

1 An example of maternal sex determination; *Dinophilus gyrociliatus*, a minute polychaete worm.

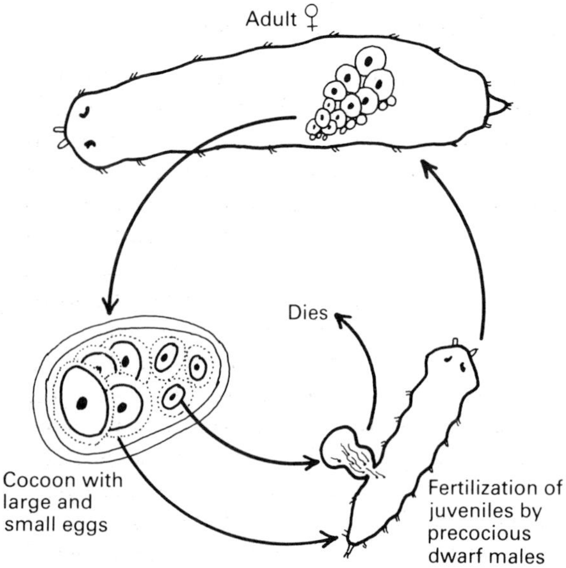

Two types of egg are produced in the ovaries.
Large eggs become females; smaller eggs become precociously mature dwarf males.
Insemination of the female embryos occurs in the cocoon.
All adults are inseminated females.

2 An example of genetic sex determination in the fruit fly *Drosophila*.

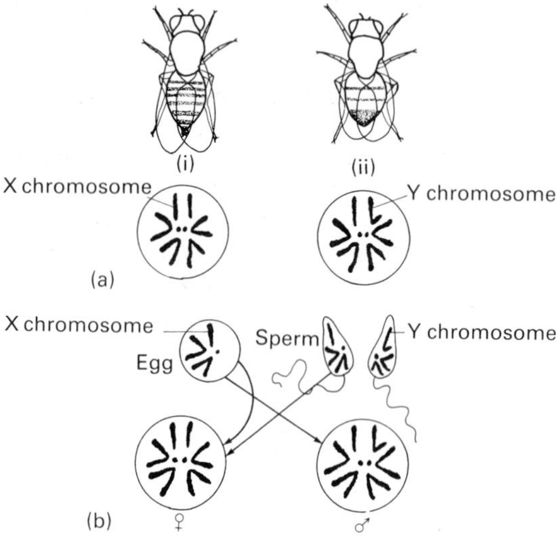

(a) (i) Females have a diploid genetic constitution with equal sex chromosomes XX. All the offspring have equal sex-forming capability.
(ii) Males have a diploid genetic constitution with unequal sex chromosomes, X and Y. Males produce sperm with two different sex-forming capabilities. Sperms with a Y chromosome will father males; sperms with an X chromosome will father females.
(b) The sex of the offspring is determined by which of the two possible sperm types happens to fertilize the eggs. The sex ratio will normally be 1 : 1 with this system of sex determination.

the males of which are dwarf and parasitic on the females (see Section 4.12). It was demonstrated at the beginning of the century that the sex of the free-swimming planktonic larvae is not fixed, those that settle on to mud become females and only those that settle on or very close to the large female proboscis will become males (Box 14.3).

Careful experiments have shown that the females release a substance which has a profoundly masculinizing influence on the developing larvae in the absence of which almost all larvae become females. In most other echiurans sex determination is genetical, but dwarf males have recently been discovered in some other families not closely related to *Bonellia*, and this may indicate independent evolution of a similar phenomenon in other members of the phylum.

3 Environmental sex determination.
In some organisms the sex of an individual is determined after fertilization. The example shows the echiuran worm *Bonellia viridis*.

The ciliated planktonic larvae have the potential to develop as a female (all the large adults are female) or as a dwarf male. Larvae settling on an uninhabited area of the sea floor tend to develop as females. Larvae settling in the vicinity of a female proboscis are induced by secretions of the female to develop into dwarf males. Careful experiments show that there is also some element of genetic determination of sex but this is normally overridden by the environmental factor.

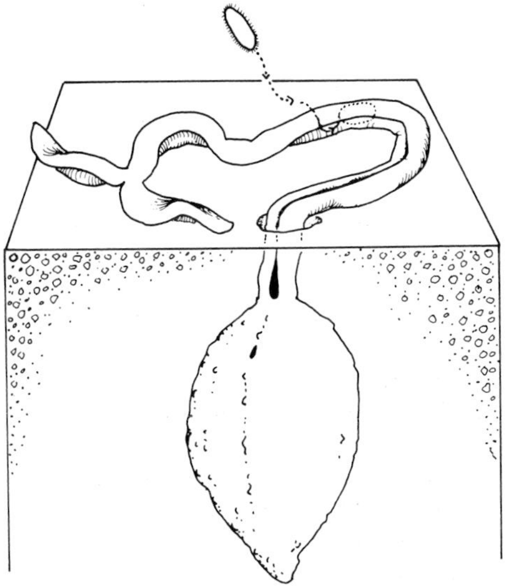

Echiurans are gonochoristic, but the process of sexualization in many sequential hermaphrodites is also profoundly influenced by environmental conditions. A pheromonal mechanism is thought to be involved in the sex determining mechanism of these animals and an example, the slipper limpet *Crepidula fornicata*, is illustrated in Fig. 14.6. These limpets form themselves into stacks: the lowermost individual is always a female and the upper ones are males, while intermediate animals in the stack may be hermaphrodites.

14.2.4 The adaptive significance of patterns of sexuality

14.2.4.1 Why reproduce sexually?

One of the outstanding challenges to evolutionary biology is to understand why different patterns of reproduction are observed. Why, for instance, is sexual reproduction so dominant? The answer is not as self-evident as might at first appear. Sexual reproduction creates diversity among offspring and it reduces the expression of potentially harmful genes, but these advantages must be offset against the disadvantages or costs. Sexual reproduction reduces the efficiency with which genes are transmitted between generations, since only half of the genome of the offspring of sexually reproducing animals is contributed by each parent. Since fitness is measured by the effective contribution of genes to subsequent generations this is a major cost. Other costs are associated with the need to devote resources to sexual display and courtship, and with the higher rates of mortality often incurred whilst finding a mate.

Many of the theories suggest that sexual reproduction has a long-term advantage to the species or group but concede a short-term advantage to individuals with the capacity for asexual reproduction; as such they are therefore group selection theories. Recently, more acceptable theories have been developed in which the selective advantage of sexual reproduction is perceived as a property of the individual, not of the group or species.

It has been suggested that sexually reproducing individuals have an advantage because their rather variable offspring, which together have many different genetic constitutions, will have an average fitness in the changed world of the future which is greater than that of the identical offspring of an asexual organism. Such a theory suggests that sexual reproduction will have its greatest advantage in unstable environments and will tend to lose its short-term ad-

Table 14.2 Systems of sexuality in animals

Gonochorism	All individuals in the population produce either male germ cells or female germ cells with the corresponding accessory glands and structures.
Hermaphroditism	Some individuals in the population produce both male and female germ cells and develop accessory glands and structures associated with male and female functions at some time during the entire lifetime. *Simultaneous* During the adult phase, individual animals produce male and female germ cells. They may have the capacity to donate male germ cells and to receive male germ cells (outbreeding) or may be self-fertilizing. *Sequential* Individual animals exhibit one sexual function prior to the onset of the second. **1** Protandric. Male functions are developed and active prior to the onset of female sexual functions. Male functions may be lost after the onset of female sexual function or male and female functions may then coexist (simultaneous hermaphroditism). **2** Protogynous. Female functions are developed and active prior to the onset of male sexual functions. May be protogynous or protogynous leading to simultaneous hermaphroditism. **3** Sequential alternating. As protogynous or protandrous but with subsequent sex reversal.

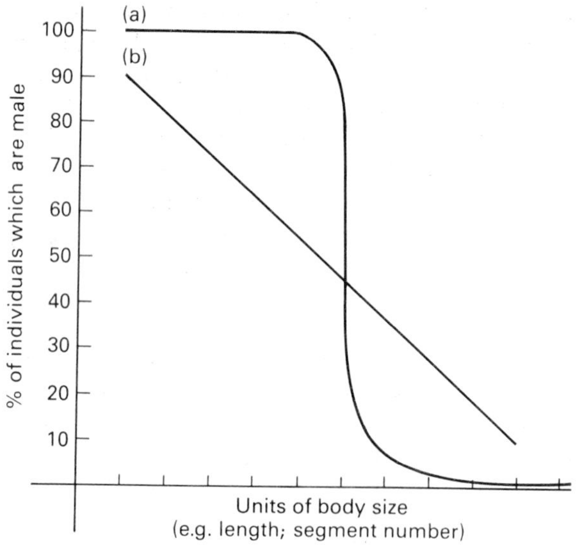

Fig. 14.5. Patterns of sex reversal in sequential hermaphrodites: (a) false gonochorism; (b) unbalanced hermaphroditism.

vantage in relatively stable ones. A survey of the distribution of asexual reproduction among invertebrates in different habitats, however, does not confirm that prediction. Asexual reproduction is particularly frequently encountered in organisms exploiting unstable, fluctuating resources while sexual reproduction is almost universal in those exploiting the most stable environments. Thus asexual reproduction is much more commonly encountered among freshwater than among marine representatives of the same taxonomic groups.

An alternative theory supposes that the selective advantage of sexual organisms arises from the ability of their diverse offspring to compete in the structurally complex world of a saturated environment where

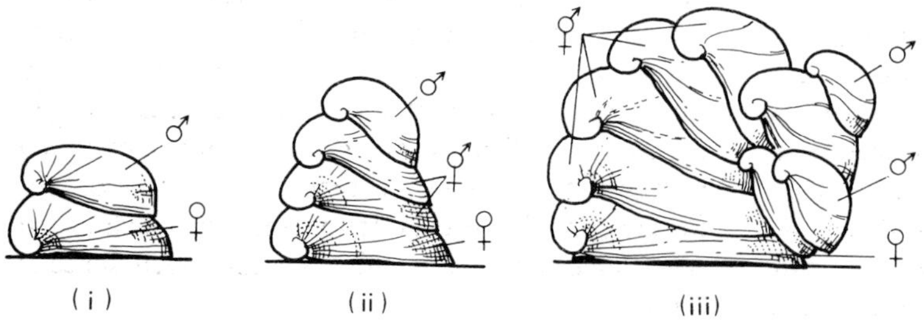

Fig. 14.6. Sex determination in the slipper limpet *Crepidula fornicata*: (i) early pair formation; (ii), (iii) later and more complex associations with males, females and intersexes.

there is intense competition. All sexually produced offspring are slightly different and the uniform offspring of an asexually reproducing animal will not be able to supplant them from all the different habitats which they can have come to occupy. The theory was called by its proposer (Bell, 1982) 'the theory of the tangled bank'. It suggests that sexual reproduction would predominate in stable complex environments, but that asexual reproduction might be expected to occur where species are maintained below the carrying capacity of the environment and opportunities for ecological opportunism exist. In these circumstances animals which have the highest possible rates of growth or growth in potential numbers of descendants are favoured.

There are other theories to account for the dominance of sexual reproduction (Fig. 14.7). Some, such as the suggestion that sexual reproduction is maintained as a mechanism which prevents potential parasites from successfully invading all the offspring of an organism, have predictions that are the same as more general theories such as that of the tangled bank.

14.2.4.2 Hermaphroditism versus gonochorism

Just as we can ask 'what are the selective advantages of sexual as opposed to asexual reproduction?'; we can also ask what determines whether animals which reproduce sexually should be gonochoristic or hermaphrodite. There are both taxonomic and ecological components of the variation; Table 14.3, for instance, shows that certain taxonomic groups are predominantly hermaphrodite, others almost exclusively gonochoristic, but it is also observed that hermaphroditism is more frequently encountered in certain environments. Freshwater and terrestrial annelids and molluscs, for instance, are hermaphrodite, whereas their marine relatives are predominantly gonochoristic. Similarly, deep-sea crustaceans are more often hermaphroditic than their shallow-water relatives. The variety of sexual condition among the invertebrates suggests that it is possible to seek functional explanations of the occurrence of hermaphroditism and that natural selection can determine the patterns of sexuality which are observed. Accordingly a number of models to account for the evolution of hermaphroditism have been proposed. Three of these are summarized below.

Low density model: when organisms exist in low densities or are immobile or sedentary then simultaneous hermaphroditism increases the probability that rare encounters between individuals will be fecund, and if other individuals are not encountered self-fertilization is possible.

Size advantage model: when one of the sexual functions has an advantage related to size but the other does not then sequential hermaphroditism will be adopted.

Gene dispersal model: when population numbers are low, inbreeding and random genetic drift may occur; in these circumstances hermaphroditism increases the effective population size.

A more general selectionist theory, which encompasses these different models, can be proposed. At the heart of the theory is the simple but profound observation that all sexually reproduced animals have precisely one mother and one father. Exactly half of the zygote is contributed by the mother and exactly half by the father, and both male and female reproductive functions are therefore equal means to reproductive success.

An organism has limited resources at its disposal and these must be allocated to maintenance and reproduction in such a way that the overall fitness of the organism is at a maximum. Some of the costs associated with sexual reproduction arise from the need to construct accessory structures—the glands and ducts which are necessary to discharge the gametes satisfactorily—and from increased mortality associated with the need to find a mate. Hermaphrodites must carry both types of cost, and their total fixed costs in relation to any finite resource to be invested in reproduction are likely to be higher. We would thus expect evolution to favour gonochorism.

This conclusion is modified, however, if the return on investment in either sexual function declines; in these circumstances it can be argued that hermaphroditism will be favoured. A declining return on energy allocated to female reproductive function, for instance, might be expected in animals which brood their offspring in a chamber within which space is limited.

Table 14.3 Conditions of sexuality in invertebrates

Phylum	Class	Notes
Porifera		All sponges have the ability to reproduce sexually although many also produce asexual fragments called gemmules
Mesozoa		Functional self-fertilizing hermaphrodites with alternation of sexual and asexual generations
Cnidaria		Gonochoristic but with frequent asexual reproduction by fission. Sometimes a complex life cycle with pelagic medusae producing gametes and asexually reproducing benthic hydroid phase
Ctenophora		Simultaneous hermaphrodites, probably not self-fertilizing
Platyhelminthes	Turbellaria Monogenea	Simultaneous hermaphrodites, probably cross-fertilizing Frequent asexual reproduction by fission Sexual reproduction sometimes unknown
	Trematoda	Simultaneous hermaphrodites, probably usually cross-fertilizing in multiple infections. Occasionally self-fertilizing Larval stages exhibit frequent asexual reproduction by multiple fission
	Cestoda	Simultaneous hermaphrodites, usually self-fertilizing. One genus gonochoristic
Gnathostomula		Simultaneous hermaphrodites
Nemertea		Virtually all gonochoristic, occasionally hermaphrodite in fresh waters
Gastrotricha		Simultaneous hermaphrodites, but parthenogenesis common
Mollusca	Chaetodermomorpha Monoplacophora Polyplacophora Scaphopoda	All gonochoristic
	Gastropoda	Highly variable sexual conditions Prosobranchia—mostly gonochoristic, but often sequential (protandric) hermaphrodites Opisthobranchia—mostly simultaneous hermaphrodites Pulmonata—all simultaneous hermaphrodites with cross-fertilization
	Neomeniomorpha	Hermaphroditic
	Bivalvia	Gonochoristic
	Cephalopoda	Gonochoristic
Rotifera	Bdelloidea	Obligate parthenogenesis, males unknown
	Monogonata	Cyclic parthenogenesis, gonochoristic at sexual phase with dwarf males
	Seisonidea	Gonochoristic
Kinorhyncha		Gonochoristic
Acanthocephala		Gonochoristic
Loricifera		Gonochoristic
Nematomorpha		Gonochoristic

Phylum	Class	Notes
Nematoda		Gonochoristic
Priapula		Gonochoristic
Sipuncula		Gonochoristic
Echiura		Gonochoristic, sometimes with dwarf males
Annelida	Polychaeta	Usually gonochoristic, sometimes sequential protandric hermaphrodites, occasionally simultaneous. Asexual reproduction by fission in some species
	Clitellata	Always cross-fertilizing simultaneous hermaphrodites
Pogonophora		Gonochoristic
Phorona Bryozoa		Simultaneous hermaphrodites, occasional gonochoristic species
Brachiopoda		Mostly gonochoristic, occasionally hermaphrodite
Entoprocta		Simultaneous or sequential hermaphrodites
Hemichordata		Gonochoristic
Echinodermata	Echinoidea Holothuroidea Crinoidea	Gonochoristic
	Asteroidea Ophiuroidea	Mostly gonochoristic, occasionally simultaneous hermaphrodites. Asexual reproduction by fission in some
Chordata	Urochordata	Simultaneous hermaphrodites
Crustacea	Branchiopoda Ostracoda	Gonochoristic, but with frequent parthenogenesis; males sometimes unknown
	Copepoda	Gonochoristic
	Cirripedia	Thoracicans usually functional hermaphrodites, but in some species gonochoristic with dwarf males. Others gonochoristic with dwarf males
	Malacostraca	Usually gonochoristic, but not infrequently sequential (protandric) hermaphrodites
Chelicerata		Almost always gonochoristic
Onychophora		Almost always gonochoristic
Tardigrada		Almost always gonochoristic
Pentastoma		Almost always gonochoristic
Uniramia		Almost always gonochoristic, occasionally with parthenogenesis or arrhenotoky. Males generally not known

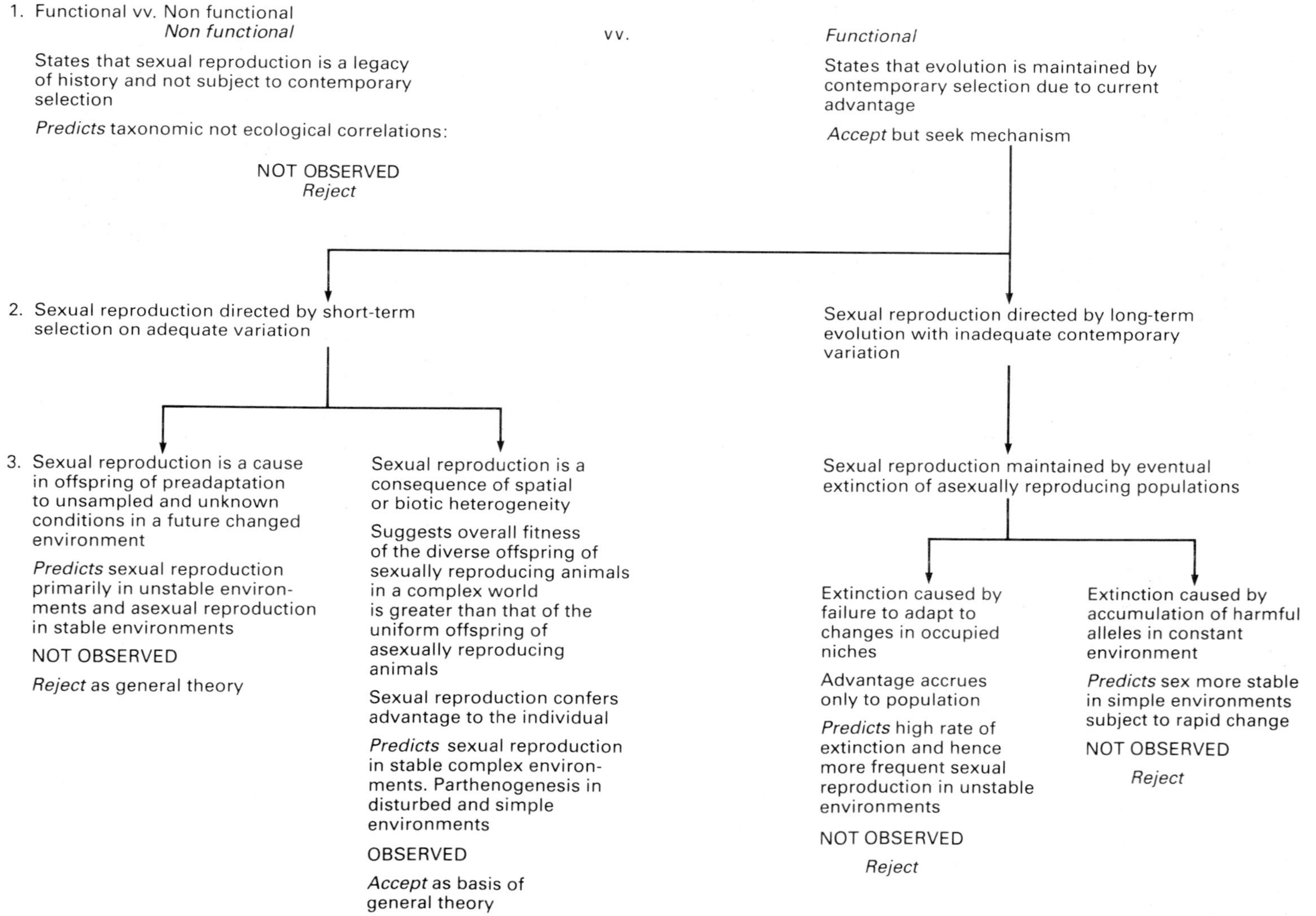

Fig. 14.7. A scenario for distinguishing between alternative theories of the selective advantage of sexual reproduction (after Bell, 1982)

14.3 The organization of sexual reproduction: life histories

14.3.1 Introduction

There are many different patterns of sexual reproduction, and the most important of the variable traits which together constitute the reproductive strategy are:

1 The maximum potential lifespan.
2 The number of breeding episodes per lifetime.
3 The pattern of gamete discharge which may involve storage and mass release or alternatively progressive discharge of gametes as they are produced.
4 The degree of synchronization between members of the population.
5 The pattern of mating and degree of outbreeding in the population.
6 The relative size and cost of the gametes.
7 The mode of development and the extent to which juveniles and adults are exposed to different selective pressures.
8 The relative proportion of total available resources allocated to reproduction.

The life cycles of animals are frequently classified according to the number of breeding episodes per lifetime and the duration of adult life. Several schemes, which are not mutually exclusive, are used; some of them are compared in Table 14.4. The different patterns of reproduction are also represented diagrammatically in Box 14.4.

14.3.2 Marine invertebrates

Patterns of reproduction are profoundly influenced by the environments in which animals live. Marine invertebrates are able to discharge naked gametes into the surrounding medium where fertilization may take place but freshwater and terrestrial invertebrates are not able to do this. The externally fertilized eggs of marine invertebrates frequently develop into mobile planktonic larvae (see the individual systematic sections and Chapter 15). These two factors have

Table 14.4 Systems for the classification of breeding patterns

Breeding occurs once per lifetime		
Nouns:	Semelparity/monotely	Virtually synonymous terms; monotely used primarily by annelid biologists
Adjectives:	Semelparous/monotelic	
All insects fall into this category; their life cycles are further described as follows:		
Univoltine	one generation per year, i.e. annuals	
Multivoltine	many generations per year	
Bivoltine	two generations per year	
Semivoltine	one generation every second year	
Breeding occurs several times per lifetime		
Nouns:	Iteroparity/polytely	
Adjectives:	Iteroparous/polytelic	
Annual iteroparity (polytely)	Breeding occurs in discrete episodes separated by periods of usually one year	
Continuous iteroparity (polytely)	Breeding more or less continuous during an extended breeding season. When total lifespan is one year or less may be indistinguishable from univoltine	

Box 14.4 Invertebrate life cycles

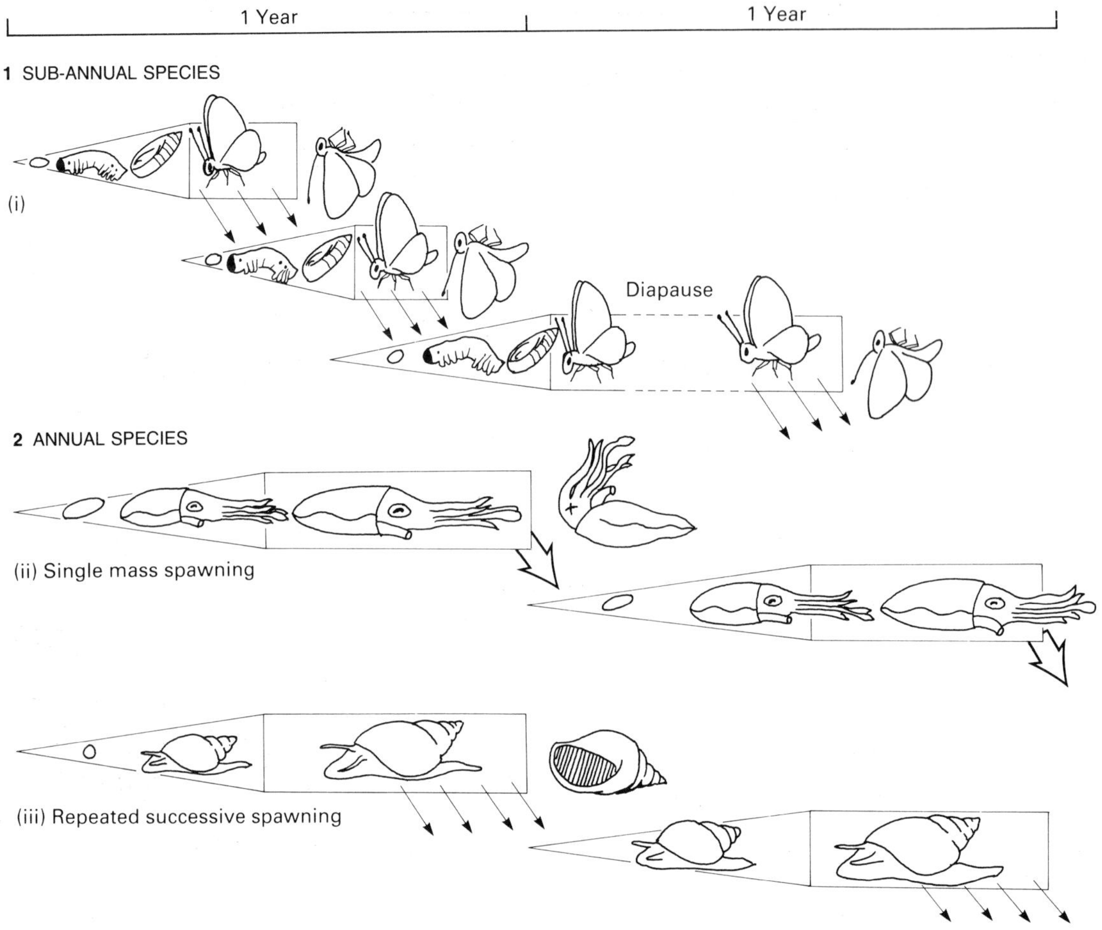

In each case a representative animal is illustrated. The band width represents quantity of resources: an increasing band width shows resource acquisition; utilization of resources is indicated by a decreasing band width.

1 Sub-annual

There are several generations in each year and in most examples in both freshwater and terrestrial temperate environments there is a reproductive diapause. In this example diapause is as the adult but it is frequently as the egg, the larva or the pupal stage.

2 Annual

There are many animals in temperate regions which live for precisely one year; there is then one generation per year. Breeding may take place as in example (ii) with a

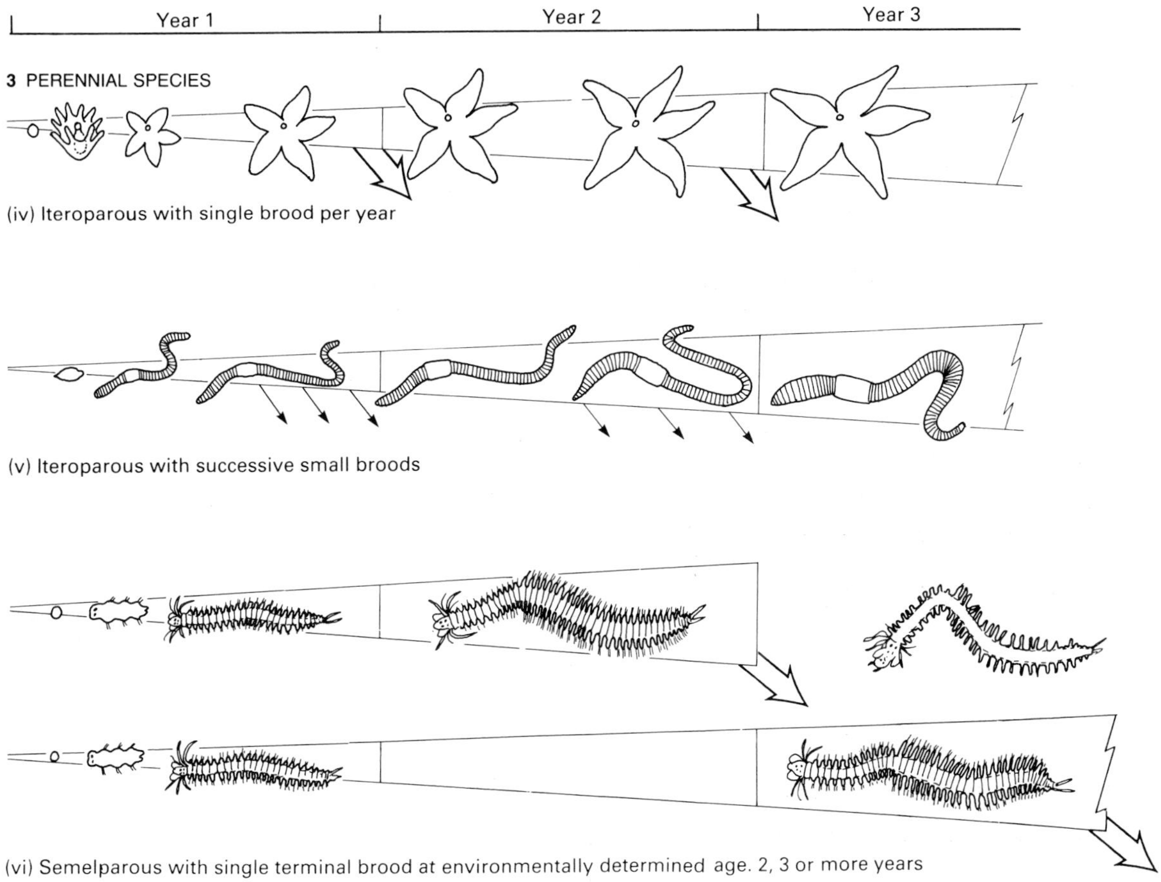

(iv) Iteroparous with single brood per year

(v) Iteroparous with successive small broods

(vi) Semelparous with single terminal brood at environmentally determined age. 2, 3 or more years

single mass spawning or, as in example (iii), spawning may be progressive over a period of several months.

3 Perennial

These have the potential to live for several years and to accumulate resources during this period. Two different patterns of resource allocation to reproduction are possible. In the commonest form, animals such as the starfish of example (iv) release gametes each year in an annual spawning episode. This may be synchronized such that all individuals spawn at the same time, but this is not necessarily the case as in example (v), an earthworm. Some perennial animals are also semelparous and spawn only once per lifetime but do so at any age from 1 to several years. The best known examples are cephalopod molluscs and polychaetes of the family Nereidae, as in example (vi).

profound influences on the whole pattern of reproduction.

The majority of the animal phyla with marine representatives have most or at least some species with pelagic larvae and external fertilization; they exhibit apelago-benthic life cycle as illustrated in Box 14.5. It is also true that all those phyla with pelagic larvae also have some or many representatives which have non-pelagic, benthic larvae, and this is usually interpreted as an advanced condition involving suppression or modification of the larval phase. A smaller number of species in most phyla have pelagic larvae which do not feed and are supplied with yolk by the parental organism. The different kinds of larvae are described as exhibiting planktotrophic pelagic, lecithotrophic pelagic or direct development. This subdivision of types is not absolute and many larvae exhibit mixed development. The possession of a pelagic larva is, however, usually interpreted as a primitive trait. Over 70% of all marine invertebrates in temperate zones have planktotrophic pelagic development which implies that in most circumstances this pattern of reproduction has clear advantages; these are thought to be due to some or all of the following:

1 Exploitation of the temporary food resource provided by phytoplanktonic blooms.
2 Colonization of new habitats.
3 Expansion of geographical range.
4 Avoidance of catastrophe associated with local habitat failure.
5 Avoidance of local and sib-competition.
6 Exposure of diverse offspring to the maximum degree of habitat diversity.

Pelagic development is frequently (but not always) associated with external fertilization, which in turn is associated with what is also thought to be a primitive trait—the possession of a simple round-headed spermatozoon of the type illustrated in Fig. 14.10. This type of spermatozoon recurs throughout the animal kingdom and these two traits help to define a pattern of reproduction which is peculiarly marine.

Associated with these two supposedly primitive traits are several others that together constitute a syndrome of covariable traits characteristic of the life histories of marine invertebrates (Table 14.5, 'original conditions'). Advanced or modified traits which depart from the supposed primitive condition are listed in Table 14.5 under 'altered conditions'; these are especially characteristic of minute marine invertebrates such as those inhabiting interstitial spaces in sands.

The evolution of a relatively large body size made it possible to adopt patterns of reproduction which it is suggested are primitive. Some phyla such as the Kinorhyncha have no large relatives, and these, together with the smallest representatives of phyla such as the molluscs, annelids and echinoderms (which do have large-bodied representatives), exhibit most of these advanced traits. Advanced or modified reproductive conditions are also to be found in marine invertebrates living in extreme or unusual environmental conditions such as deep-sea and polar regions, and in estuaries.

A major correlate of the 'primitive' reproductive traits is the mass discharge of gametocytes in an annual spawning crisis. It is very commonly found that marine invertebrates store the gametes in body cavities (often the coelom) and then release them in a single seasonal spawning crisis. One of the reasons why minute organisms have advanced traits may be because it is not possible for them to store sufficient quantities of gametocytes in this way. There are two quite different life cycles in which mass spawning occurs and in which a clear annual spawning season may be expressed at the population level (see Box 14.4,**3**). In the more common pattern, spawning occurs at annual intervals during a lifespan of 2 or more years, but in some species, most notably all members of the polychaete family Nereidae and all cephalopod molluscs other than the primitive nautiloids, spawning occurs only once per lifetime. Mass spawning in animals of this type is followed by the genetically determined death of the individual.

Advanced reproductive traits of marine invertebrates are also characteristic of soft-bodied invertebrates living in freshwater and terrestrial environments as described in the following section.

Table 14.5 Primitive and modified traits of marine invertebrates

Original conditions	Altered conditions
Eggs freely discharged into water, free pelagic development	Eggs not freely discharged, often brooded
External fertilization (sperm type primitive)	Internal fertilization (often with sperm storage) (sperm type modified)
Small quantity of yolk in the egg	Large quantities of yolk in the egg
Total equal cleavage	Unequal cleavage
Blastula with blastocoel	Blastocoel more or less obliterated
Gastrulation by invagination	Gastrulation by other types
Planktotrophy in the larva	Lecithotrophy in the larva
Autotrophic egg production	Heterotrophic egg production
Discrete, once per year reproduction with strong seasonality, between and within individual synchrony of gametogenesis	Repeated often more or less continuous reproduction with reduced seasonality, loss of between and within individual synchrony of gametogenesis
Long-term storage of accumulated germ cells	Repeated spawning of small batches of germ cells without accumulation
Large body size, often with capacious body cavity or intercellular spaces	Small body size, partially obliterated body cavity. Accumulation of gametes in intercellular spaces not possible

Some data from Jägersten, 1972.

14.3.3 Freshwater and terrestrial invertebrates

The choice between planktotrophic or lecithotrophic pelagic development and direct development, which is such a feature of reproduction in the sea, is not open to those animals inhabiting freshwater and terrestrial environments. The osmotic and other stresses in these habitats preclude release of naked unprotected spermatozoa and eggs, and fertilization must be internal. Similarly developing embryos must be protected against water loss or osmotic stress, e.g. they must be enveloped in a waterproof coat or in a cocoon. Consequently the primitive pelagic larval phase has been suppressed, and the larger, soft-bodied non-marine invertebrates exhibit the following reproductive traits:

1 Viviparity or deposition of eggs in impermeable membranes or cocoons.

2 Internal fertilization, requiring direct pairing between partners.

3 Structurally advanced, often filiform, spermatozoa.

4 Investment of relatively high levels of maternal resources in each egg, and consequently with lower fecundities.

5 Brood care or provisioning of offspring.

6 Repeated or episodic egg laying, exploiting the potential that an internal sperm store provides for continuous reproduction.

7 Hermaphroditism. (This is especially true for the soft-bodied invertebrates but is not the case for the arthropods.)

Some freshwater animals betray a recent origin from marine ancestors; the freshwater bivalves for instance, e.g. *Anodonta*, are very similar to their

Box 14.5. The life cycle of marine invertebrates

The eggs and spermatozoa of marine invertebrates may be released into sea water where fertilization can take place. This has profound implications for their reproductive biology.

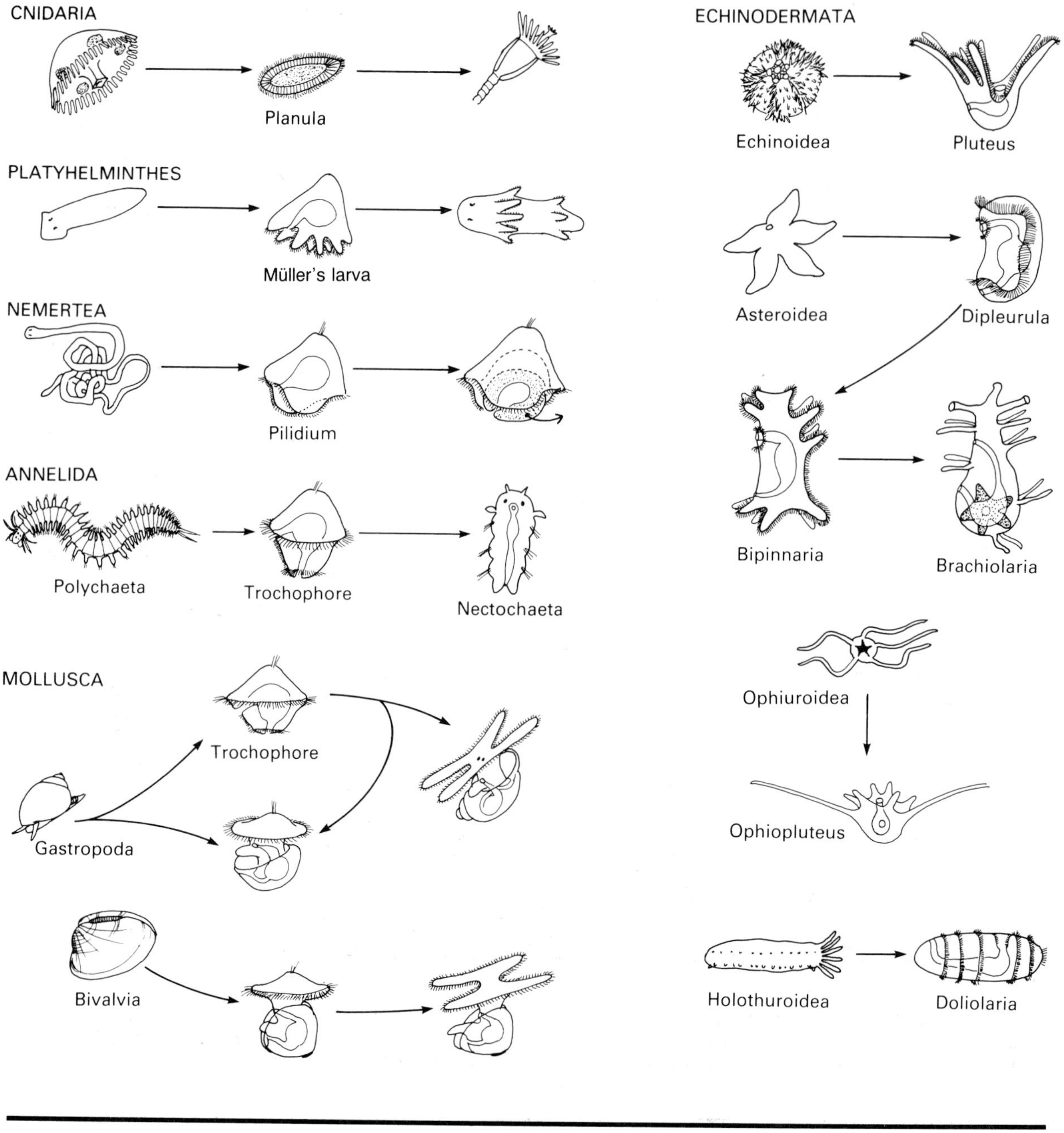

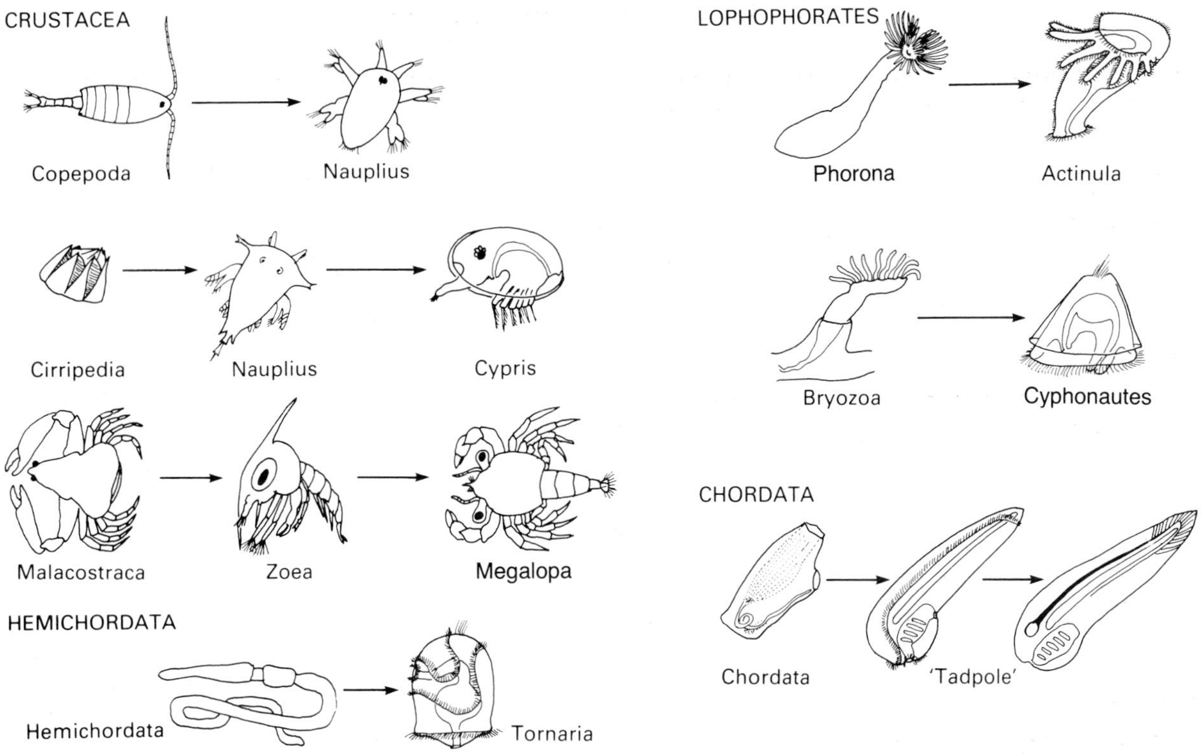

Fig. 1. Examples of the characteristic larvae of the major groups of marine invertebrates. The sketches show the adult form and the corresponding larvae. Note the adults are drawn to a much smaller scale than the larvae.

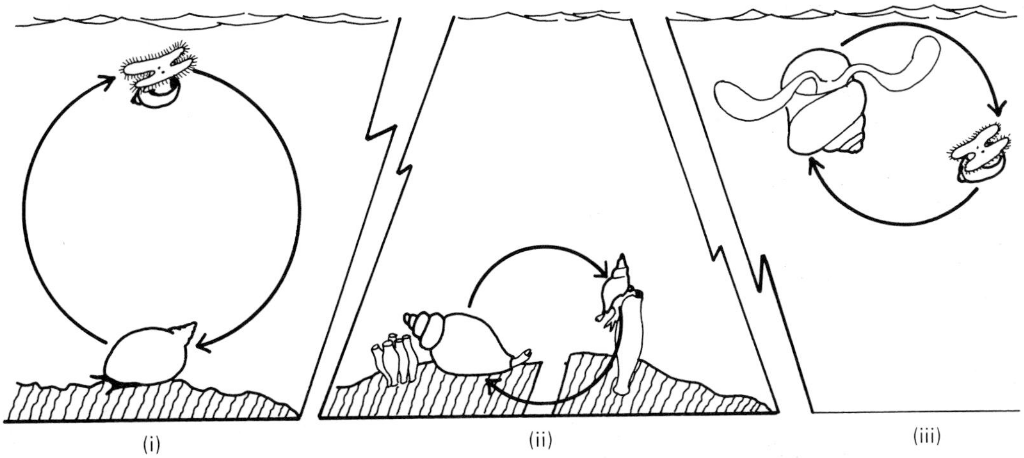

Fig. 2. Characteristic life cycles. (i) Pelago-benthic—with pelagic larval stage. (ii) Holo-benthic—suppression of the primitive pelagic larval stage. (iii) Holo-pelagic—with entirely pelagic history. All three patterns can be found in many groups. The examples given here are all gastropod molluscs.

marine relatives and they do release relatively large numbers of what are in effect modified veliger larvae called *glochidia*. The larvae are not free-living, however, but are external hitch-hikers on the gills or skin of freshwater fish.

The life cycles of the freshwater platyhelminths, clitellate annelids and pulmonate molluscs are remarkably similar. They are all simultaneous hermaphrodites (unless sexual reproduction has been suppressed) with complex sexual behaviour, highly specialized accessory glands and the capacity to protect their embryos. Their life cycles involve extended periods of egg-laying activity (see Box 14.4). The principal trade-offs that are made in such species are those between fecundity and longevity; some species live for only 1 year or less but others, usually those that achieve a larger adult body size, live for several years.

Freshwater and terrestrial environments are more extreme in physical variables than are marine ones, and most of the invertebrates inhabiting them have the capacity to enter a physiological resting state known as *diapause*. In this state the organism can lose water and withstand extreme conditions without harm; the metabolic demand is lowered and the requirement for an external energy source can be nil so that the periodic absence of food which occurs in temperate and polar regions can be withstood. Long-lived species such as some pulmonates and clitellates will enter the diapause state as adults, but many smaller forms do so as eggs. Thus many populations of leeches only exist as embryos in cocoons during the winter months. The habit of overwintering as resting eggs is particularly characteristic of minute forms such as the rotifers and several groups of crustaceans such as the cladoceran water-fleas. These organisms exhibit life histories in which there is an alternation of asexual and sexual generations (Fig. 14.4); the diapause eggs, which arise from sexual reproduction, hatch in the spring to give rise to new clones of offspring which can exploit the new resources. The bdelloid rotifers exhibit an extreme form of diapause; their encysted eggs can remain in what is virtually a state of suspended animation for years until they are blown into suitable conditions for growth. This capacity may be related to the absence of sexual reproduction; by avoiding unfavourable conditions they live to enjoy what are in effect permanently optimal conditions.

The insects (Hexapoda) and the spiders and mites (Arachnida) have most successfully adapted to terrestrial environments (see Sections 8.4.3.2 and 8.5.3B.2), and, particularly in the case of the insects, some have secondarily adapted to fresh waters as adults or as larval or nymphal forms. The reproductive biology of these groups differs from that of the soft-bodied invertebrates in one important respect. They are very rarely hermaphrodite; the vast majority of species reproduce only sexually and the sexes are always separate. Their success is due in large part to the development of a waterproof covering—the integument or cuticle—but is also due to their ability to lay waterproof eggs. Such eggs must be fertilized before they are laid; consequently all uniramians exhibit internal fertilization, often associated with complex copulatory behaviour. The prevalence of gonochorism seems to be a consequence of their high mobility.

It is traditional to subdivide the life cycles of insects according to modes of development—exopterygote (hemimetabolous) versus endopterygote (holometabolous) as described in Chapter 15, but a different, more functional approach will be used here. The life histories of insects can be regarded as different patterns of activities in pre-adult and adult stages. Insects grow through a series of moults or instars and it is possible therefore for the different instars to have different functions. The principal ones that they may have are:

1 Development and differentiation.
2 Food and other resource acquisition.
3 Dispersal and resource tracking.
4 Mating and mate selection.
5 Allocation of resources to offspring.
6 Selection of sites for offspring growth.
7 Oviposition.

A number of different insect life histories are analysed in this way in Box 14.6. Note the marked differences in the allocation of dispersal and resource acquisition functions and the contrasts which exist

Box 14.6 A functional analysis of the life histories of insects

1 Juveniles and adults exploit a similar food resource:
Juvenile functions—development and differentiation, resource acquisition.
Adult functions—resource acquisition, dispersal and resource tracking, mating and oviposition.
e.g. the Orthoptera (locusts and grasshoppers).

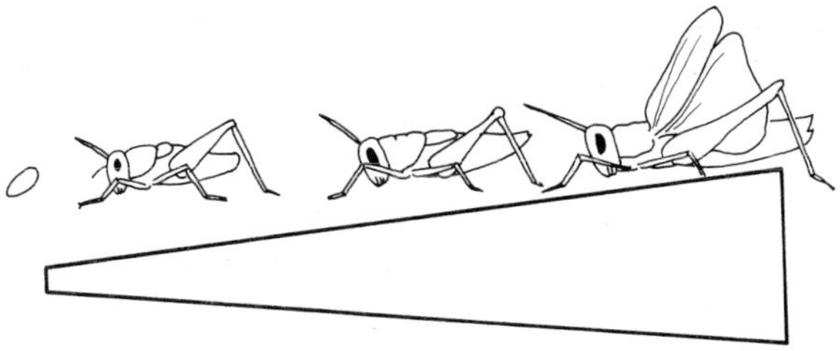

2 Juvenile and adults exploit a different food resource:
Juvenile functions—development and resource acquisition.
Pupal functions—differentiation and development.
Adult functions—resource acquisition, resource tracking, mating, and selection of sites for offspring.
e.g. blowflies and other Diptera.

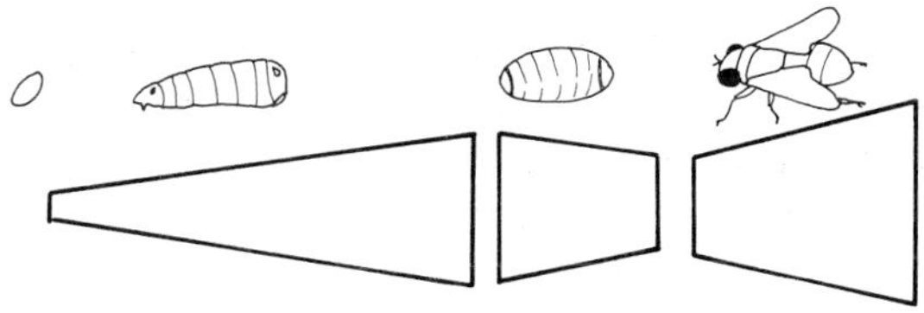

3 Adults ephemeral and not involved in resource acquisition; juvenile functions, development and resource acquisition, adult functions, mating, resource tracking and dispersal.

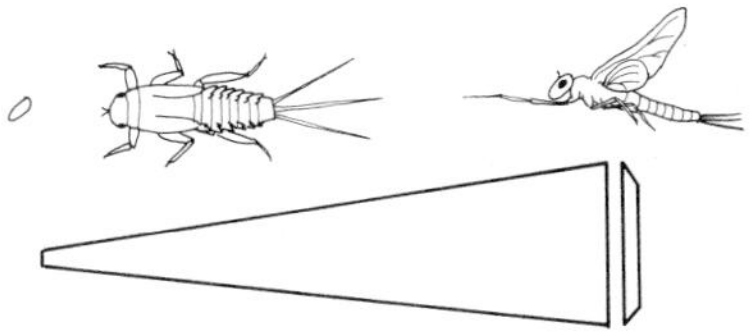

(i)

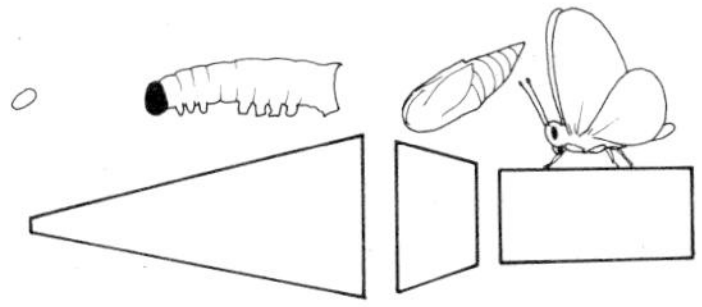

(ii)

(i) Larvae aquatic, e.g. Ephemeroptera.
(ii) Larvae terrestrial, e.g. Lepidoptera. In this example the adults are longer lived and have mouthparts specialized for nectar collection as a fuel for flight.

4 Adults solely responsible for resource acquisition—provisioning:
Larval roles—resource acquisition, dispersal and resource tracking, mating, allocation to offspring, selection of sites for offspring development, oviposition.
e.g. a solitary hymenopteran wasp.
The adult provides all resources available to the offspring.

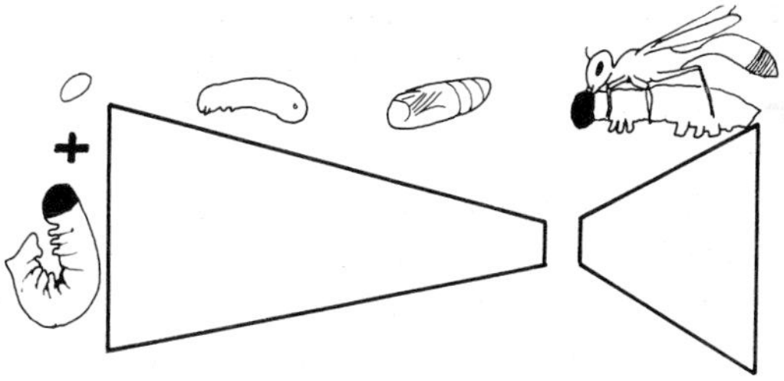

5 Resource acquisition, development, mating and oviposition separated by caste differentiation:
Juveniles—resource utilization and growth.
Sterile workers—resource acquisition.
Queens (fertile adults)—mating and mate selection, oviposition, dispersal and resource tracking.
Males (fertile adults)—mating—no role in resource acquisition.
e.g. honey bees.

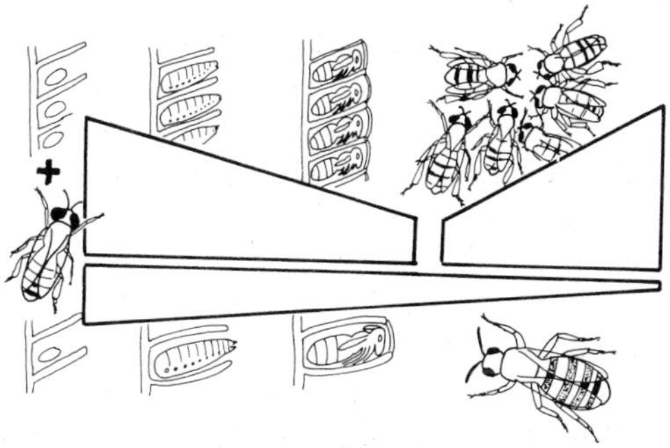

between most insects and the marine invertebrates where dispersal is often a function of the juvenile phase and resource acquisition a function of the adult phase.

The simplest life history is one in which there is a gradual transition during development to the adult condition and both adults and larvae exploit the same food resource. Both juveniles and adults are involved in resource acquisition but the adults have the additional roles of dispersal, resource tracking, mating and egg laying. This simple life history is exhibited by the Orthoptera (e.g. the locusts) and is illustrated in Box 14.6,**1**.

Not infrequently, however, adults and larvae feed in different ways, and are consequently subject to quite different selection pressures. Dipteran blowflies, for instance, have larvae feeding on a rich but temporary food resource, dead meat. This must be tracked by the parents which often feed in a quite different way (Box 14.6,**2**). Several insect larvae have an aquatic existence and the niche differentiation between adults and juveniles is then even more marked.

From this situation life histories may have arisen in which the roles of dispersal and resource tracking have been separated from those of resource acquisition, and the adults feed either very little or not at all. This separation of function has arisen independently many times in the evolution of the insects. The most extreme examples are the Ephemeroptera, Plecoptera and Trichoptera, the larvae of which are aquatic carnivores. As adults, they never feed and only live for a few hours, during which time they must mate and deposit eggs in an environment suitable for their subsequent survival and development (Box 14.6,**3**).

The well-known life history of the Lepidoptera is similar, although the adults do obtain some energy from nectar sipping. The caterpillar larvae have one basic function, which is to acquire as much of the available resource as quickly as possible. The transition to the adult phase, which has the roles of mating and resource tracking, involves a non-feeding pupal stage (Box 14.6,**3**). The food resource may be seasonal and it can be tracked through time as well as space by the intervention of diapause as egg, pupa or adult, or by migration.

The adults of some insects have acquired not only the functions of dispersal, mating and resource tracking, but also resource acquisition. They are solely responsible for finding the food and making it available to their offspring, a phenomenon described as provisioning; it is observed in some Orthoptera, Coleoptera and Diptera but it is most characteristic of the Hymenoptera (Box 14.6,**4**) and it is thought to have played a crucial role in the evolution of eusocial behaviour among the bees and wasps (Box 14.6,5).

14.4 The control of reproductive processes

14.4.1 Ultimate and proximate factors

The life cycles discussed in the previous sections involve complex sequences of cellular activity, and these must be co-ordinated in order to bring about an orderly progression of events and a proper relationship with outside factors, and where necessary to maintain the proper degree of synchrony between different members of the population. It is usually supposed that populations of animals that show clear cycles of reproductive activity do so in response to cycles of environmental change. Environmental conditions are not constant and it follows that certain times will be more favourable for reproductive activity than others; there are therefore evolutionary forces which select for reproduction to occur at the most favourable times. These forces are referred to as the *ultimate factors* controlling reproduction. They are not necessarily the same ones as those used to control the reproductive cycle. Gametogenesis and particularly oogenesis may take several months for completion, and quite different environmental signals may be used to regulate the progression of the cellular events that culminate in reproduction. Those environmental events used in this way are referred to as the *proximate factors*. The perceived environmental changes must be integrated within the central nervous system of the reacting individual and the information may be transduced in the form of a change in nervous, neuroendocrine or endocrine activity (Section 16.12). There is, in effect, a chain of command which results in a highly structured species-specific reproductive cycle (Fig. 14.8). This controls the differentiation of the germ cells, the flow of energy, and its relative allocation to reproductive processes, maintenance and growth.

14.4.2 Component processes: gametogenesis

Spermatogenesis results in the formation of the male germ cells and oogenesis in the formation of the female germ cells; they are rather different processes and are illustrated in Fig. 14.9 and described in a little more detail below.

14.4.2.1 Spermatogenesis

Spermatogenesis in most invertebrates is completed quickly; it often involves frequent mitotic divisions prior to the onset of meiosis to yield vast numbers of germ cells. Each spermatogonium which transforms to a primary spermatocyte will give rise to four spermatids which, by a process of differentiation, give rise to the spermatozoa (Fig. 14.9a). Many marine invertebrates have an essentially similar round-headed sperm type (Fig. 14.10a) but others have more advanced types (Fig. 14.10b–d). Advanced spermatozoa are associated with life cycles in which there is either sperm storage or internal fertilization or in which the eggs are highly protected (see above).

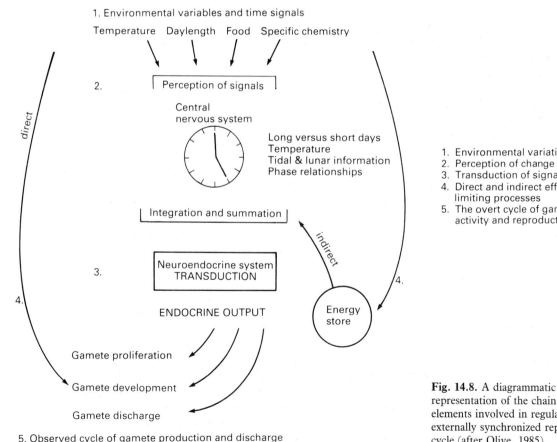

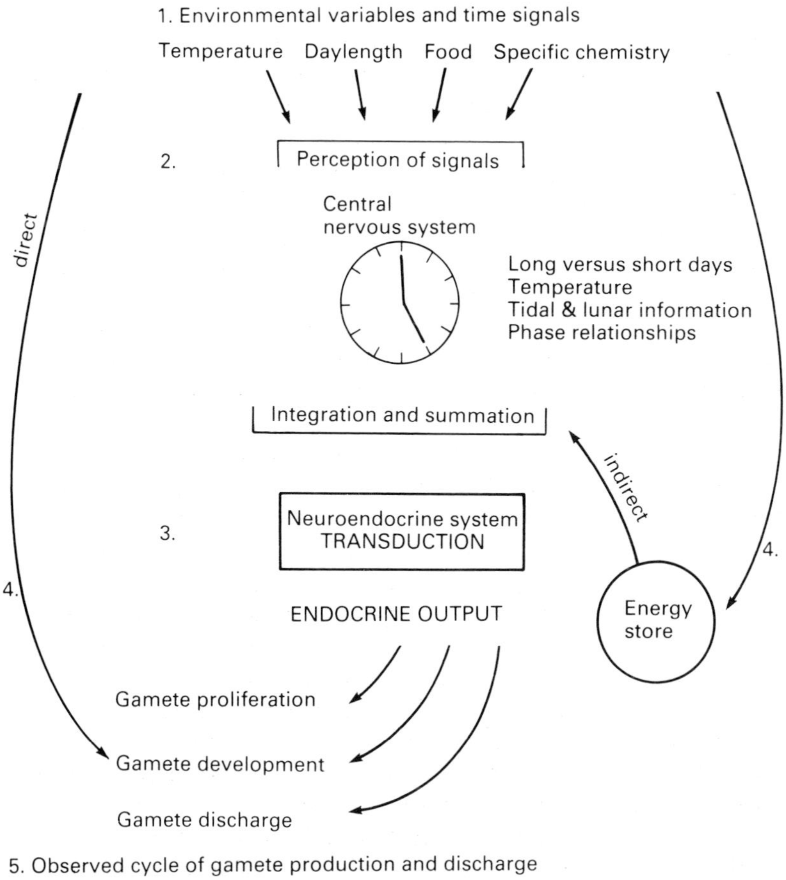

1. Environmental variation
2. Perception of change
3. Transduction of signal
4. Direct and indirect effects on limiting processes
5. The overt cycle of gametogenic activity and reproduction

Fig. 14.8. A diagrammatic representation of the chain of control elements involved in regulating an externally synchronized reproductive cycle (after Olive, 1985).

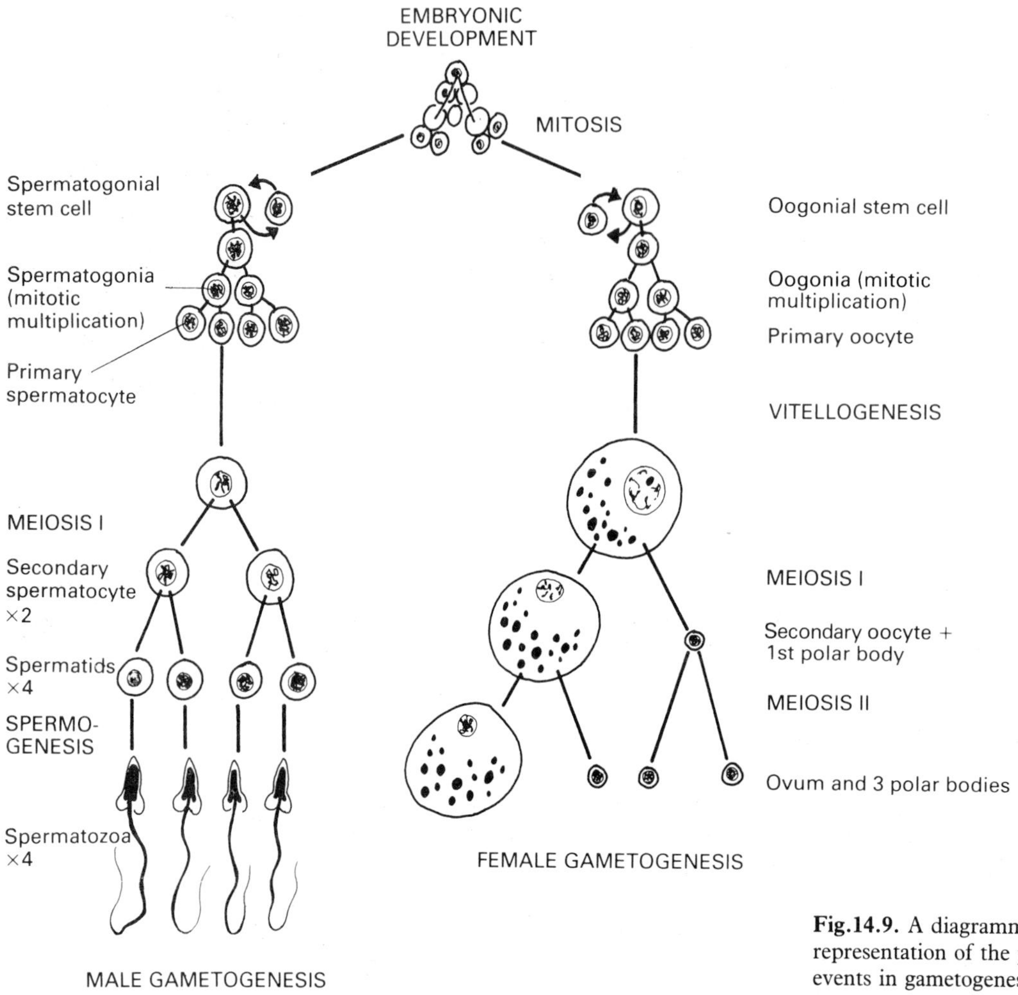

Fig.14.9. A diagrammatic representation of the principal cellular events in gametogenesis.

14.4.2.2 Oogenesis

Oogenesis is a protracted process during which food reserves are deposited in the developing egg, usually during an extended prophase, and meiosis is not completed until shortly before or after fertilization.

In some marine invertebrates, e.g. Sipuncula, Echiura and some Polychaeta, the developing oocytes are solitary cells suspended freely in the coelomic fluid. Such a pattern is described as *solitary oogenesis* (Fig. 14.11a); in these cases the metabolites to be stored in the oocyte cytoplasm are often incorporated as low molecular weight precursors (amino acids, simple sugars and monoglycerides) and built up into complex storage products, collectively called 'yolk', by synthetic organelles in the oocyte cytoplasm (as illustrated in Fig. 14.12). This pattern is described as *autosynthetic* and is particularly associated with solitary oogenesis, but the association is not absolute (see below).

Most invertebrates, however, exhibit *follicular* or *nutrimentary oogenesis* (Fig. 14.11b,c) in which the developing oocytes are intimately associated with an epithelium of somatic cells—called the follicle cells—

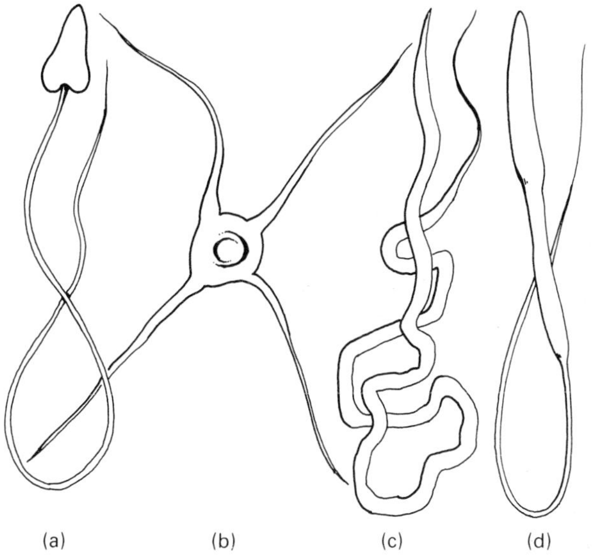

Fig. 14.10. (a) Spermatozoon of the primitive round-headed type. (b)–(d) Spermatozoa of a more advanced filiform type. Spermatozoa with a similar filiform shape may have very different internal structure and organization.

and/or sibling cells of the oocyte derived by incomplete cytokinesis during the mitotic divisions of the mother cell or oogonium, prior to the onset of meiotic prophase. The development of the complex ovary of this type, as in the dipteran *Drosophila*, is illustrated in greater detail in Fig. 14.13. There is increasing evidence that in the more complex types of oogenesis most of the high molecular weight materials deposited in the oocyte cytoplasm are synthesized not by the oocyte itself but by somatic cells elsewhere in the body, and the materials are then transported to the oocytes by the body fluids or blood vascular system as suggested in Fig. 14.14. This pattern of oogenesis is described as *heterosynthetic*. It was first discovered in insects but has subsequently been shown to be widespread, being characteristic of the Crustacea and the Mollusca, for example.

In the Polychaeta in which autosynthesis and solitary oogenesis were thought to be typical a remarkably wide range of patterns of oogenesis can be found, rivalling those of the insects in structural complexity. Some are now known to exhibit follicular

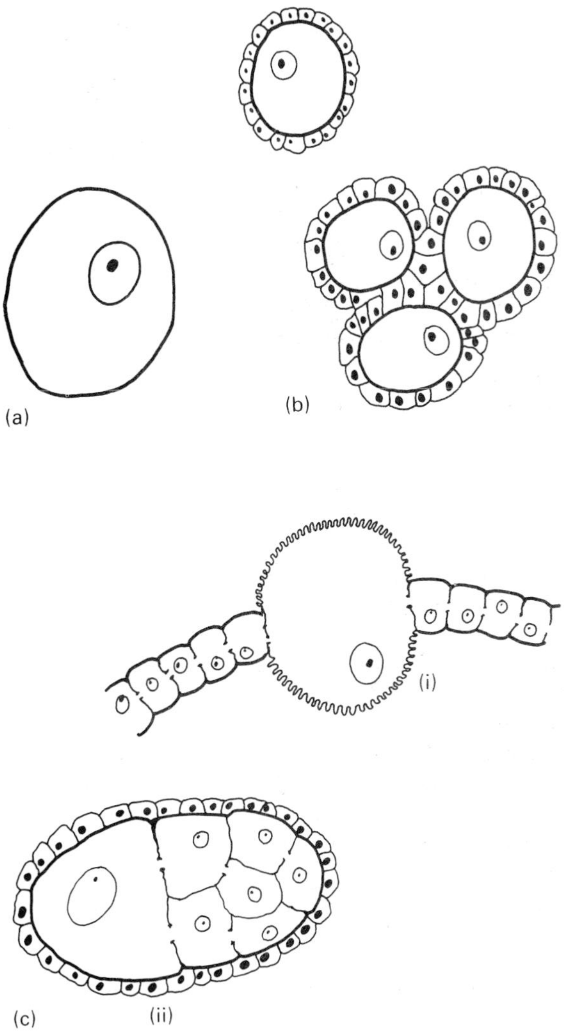

Fig. 14.11. Solitary, follicular and nutrimentary oogenesis. (a) *Solitary*: oocytes float freely in the body fluids and are not in association with any accessory cells. Characteristic of echiurans and sipunculans and some, but probably the minority, of polychaete annelids. (b) *Follicular*: oocytes intimately associated with an investment of somatic cells. (c) *Nutrimentary*: oocytes associated with sibling cells derived from the same primordium as the oocyte itself. Incomplete cytokinesis often results in the development of a syncitial complex in which oocytes and nurse cells are connected by cytoplasmic bridges or ring canals. (i) An example without associated follicle cells in the polychaete *Diopatra*. (ii) An example with associated follicle cells as seen in the dipteran insects, e.g. *Drosophila* (see also Fig. 14.13).

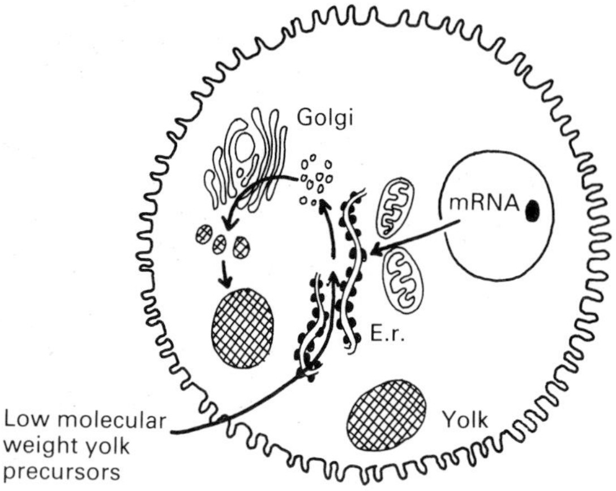

Fig. 14.12. Patterns of yolk synthesis in oocytes. The autosynthetic condition usually associated with solitary oocyte development. Low molecular weight precursors are absorbed by the microvillus surface of the oocyte, and the synthesis of high molecular weight storage products takes place in the oocyte cytoplasm using informational molecules derived from the oocyte nucleus.

and nutrimentary oogenesis and in the Nereidae where oogenesis is solitary, there is evidence of heterosynthesis.

In turbellarians the 'egg' is often a complex structure composed of a relatively yolk-free oocyte, combined with nurse cells which are packed with a yolk-like cytoplasm produced by the vitellarium (Fig. 14.15), all packaged together within a tanned protein

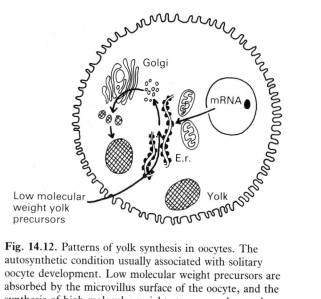

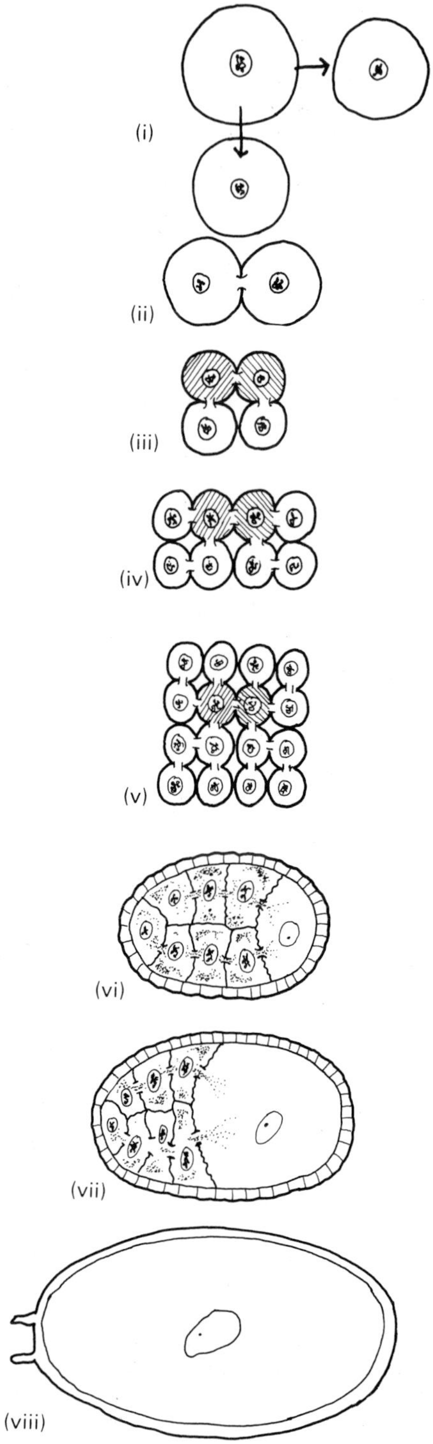

Fig. 14.13. Cellular aspects of development of the egg of a dipteran insect (e.g. *Drosophila*). (i) The stem cell divides to give a further stem cell and a primary germ cell; (ii)–(v) the primary germ cell divides four times but with incomplete cytokinesis to give a complex of 16 cells joined by cytoplasmic bridges. Note that the two shaded cells at each stage have more cell interconnections than any other cells in the complex. The oocyte will develop from one of these two cells. The other cells become nurse cells; (vi) the nurse cells act as an amplification system. Their nuclei release information molecules (mRNA) which are transferred to the oocyte cytoplasm. The whole complex becomes invested in a follicular epithelium formed by somatic cells; (vii) the oocyte grows at the expense of the nurse cells; (viii) the mature egg completely fills the follicle and is ready for further development into the embryo after fertilization and oviposition.

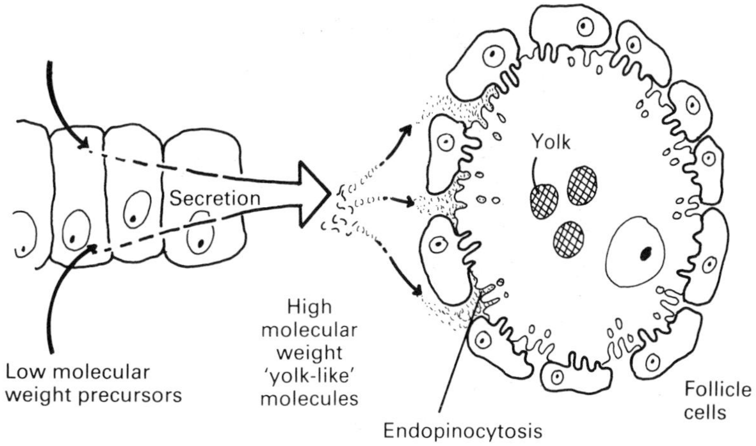

Fig. 14.14. The heterosynthetic condition, usually associated with a follicular mode of oocyte development but not necessarily so. Complex yolk precursors of high molecular weight are transported to the oocyte in the body fluids (blood system, haemolymph, coelomic fluid, etc.). These high molecular weight substances are manufactured from low molecular weight precursors by non-germ cells using informational molecules of the accessory cells. The diagram represents the situation in some insects.

coat. This type of egg formation is termed ectolecithal and is thought to be an advanced trait in platyhelminths; some free-living turbellarians have a more primitive endolecithal mode of egg development in which yolk is stored in the oocyte cytoplasm. Freshwater and terrestrial annelids and molluscs show similar reproductive adaptations from this point of view; in both, the eggs are laid within cocoons containing a nutritive albumen which supplements the storage materials deposited in the egg cytoplasm.

14.4.3 Synchronous reproduction of marine invertebrates

The dominant pattern of reproduction among marine invertebrates (Section 14.3.2) requires a high degree of synchronization of reproductive events within individuals and between different members of a population. The degree of synchronization can be very dramatic indeed. Table 14.6 records some of the dates and times when the breeding of the Pacific palolo worms (Polychaeta, Eunicida) has occurred during the last 100 years: spawning has a precise and fixed relationship to the time of the 3rd lunar quarter which first occurs after a date in early October. The timing is accurate to within 1 day and spawning also occurs at precisely the same time each day. A similarly precise timing of reproduction has been observed in the Japanese crinoid *Comanthus japonicus* (Echinodermata, Crinoidea) and Fig. 14.16 shows that the

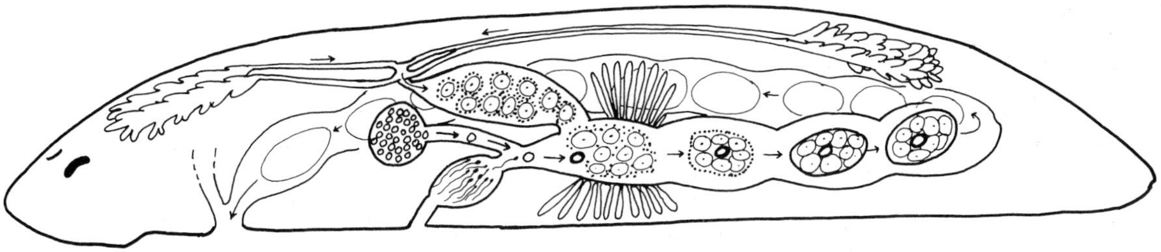

Fig. 14.15. Diagrammatic representation of a turbellarian showing the complex arrangement of glands associated with the production of the complex 'egg' which is made up of a fertilized ovum, extra-ovarian yolk cells and a protein coat. The arrows show the path of the egg as it passes from the ovary and is fertilized, and incorporated into the complex with protein and nutrient cells before it is finally released to the exterior (see also Fig. 3.46).

Table 14.6 A sample of data from a compilation showing the synchrony of emergence of sexually mature palolo worms (*Eunice viridis*) in the Samoan Islands

	Year	Third quarter of the moon		Dates of emergence	
		Oct.	Nov.	Oct.	Nov.
19 years	1843	16	14	15/16	
	1862	15	14	15/16	14/15
19 years	1874	31	30	31	1
	1893	31	29	31	1
19 years	1926	27	26	28	
	1927	17	15	17	
	1928	5	4		4
	1929	24	23	25	
	1930	14	13	14/15	
	1943	20	19	20	
	1944	8	7		7/9
	1945	27	26	28	

The data show that: (a) spawning never occurs before 8th October; (b) the worms spawn in October if the third lunar quarter falls after 18th October; (c) each 19th year spawning occurs on the same date. In addition, the time of day of emergence is also determined with precision.

whole cycle of gametogenesis spanning almost the whole year is constrained into this pattern. These are extreme examples perhaps, but the reproduction of most marine invertebrates in all phyla to some degree involves this kind of synchronization. The question arises as to how such cycles are maintained. It was formerly thought that variations in environmental temperature might provide the most important timing signals but this is clearly not sufficient to account for the two extreme examples cited above, and in general the temperature cycle seems to be subject to too much random variation (i.e. is too noisy a signal) to be solely responsible for the observed patterns.

The sea is in fact a complex rhythmic environment (Table 14.7) and reproductive cycles of marine invertebrates can have fixed phase relationships to all of these different cycles. It has been shown that such diverse animals as the Polychaeta, Echinodermata and Crustacea can all show clear-cut responses to

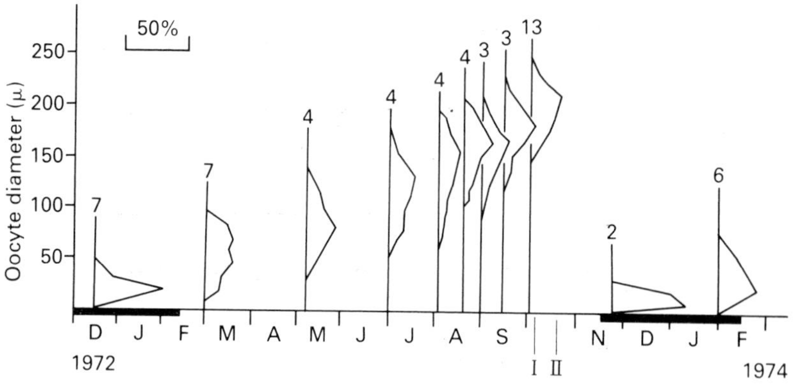

Fig. 14.16. The pattern of egg development in the crinoid *Comanthus japonicus*. The sequence of gametogenesis occupies 12 months and culminates in highly synchronized breeding on two days of each year with a precise correlation to the phases of the moon. The horizontal bar represents the time of gametocyte proliferation. (After Holland *et al.*, 1975.)

Table 14.7 Geophysical cycles in the marine environment

Name		Periodicity
Metonic	Recurrence of phases of the sun and moon	19 years
Annual	Cycle of earth about sun	1 year
Lunar	Cycle of moon about earth	29.5 days
Semi-lunar	Recurrence of phases of tidal and solar cycles (neap/spring tide cycle)	15 days
Tidal	1 lunar day	24.8 hours
Daily	1 solar day	24 hours
Semi-lunar	Recurrence of high or low tides with semi-diurnal tides	12.4 hours

Observations show that the reproductive activities of at least some marine invertebrates are correlated with each one of these cycles. This does not establish what is the causal factor (see text for discussion).

relative day length, which are at least as complex as the photoperiodic responses of the terrestrial insects to be described below. They can also show direct responses to cycles of moonlight, and can exhibit endogenous rhythms of circa-lunar and circa-tidal periodicity which can be entrained by exposure to appropriate external time-setting programmes (called zeitgeber) to run at exactly tidal or lunar rates. It is also becoming evident that underlying the overt annual reproductive cycles of some species are endogenous rhythms of circa-annual periodicity (Fig. 14.17).

The key elements in the reproductive cycles of marine invertebrates are the repeated cycles of gamete production and release. Such a cycle can be controlled through the cyclic production of hormones, which can be described by reference to their inferred functions as being of two basic types: those that have a gonadotrophic role and those that induce spawning or gametocyte maturation and/or activation, as illustrated in Fig. 14.18. This pattern of endocrine control has evolved independently many times during the course of invertebrate evolution and the substances involved in controlling these basic functions may be of quite different molecular structure. In the echinoderms the gonadotrophic function is associated with changes in the relative levels of the 'vertebrate' hormones progesterone and oestrone, while spawning is initiated by a cascade reaction starting with the

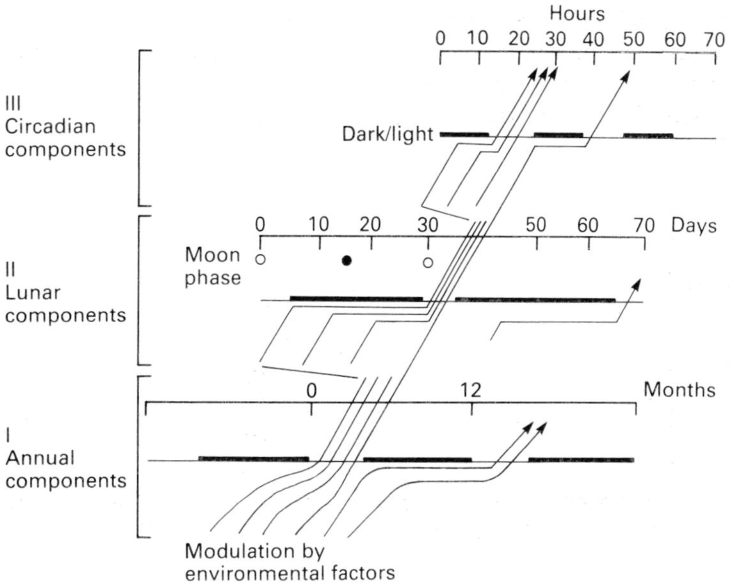

Fig. 14.17. Diagrammatic model for the extremely synchronized breeding of animals such as *Comanthus* (Fig. 14.16) or *Eunice* (Table 14.6). It is supposed that breeding is triggered by the interaction of endogenous rhythms, each of which may have external zeitgeber of annual, lunar or daily periodicity. The end result is breeding in 1 or 2 months of the year on days which have a precise relationship to the lunar cycle and at a precise time on each of those days. (After Olive, 1985.)

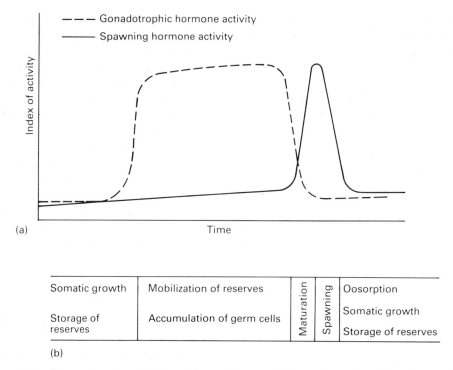

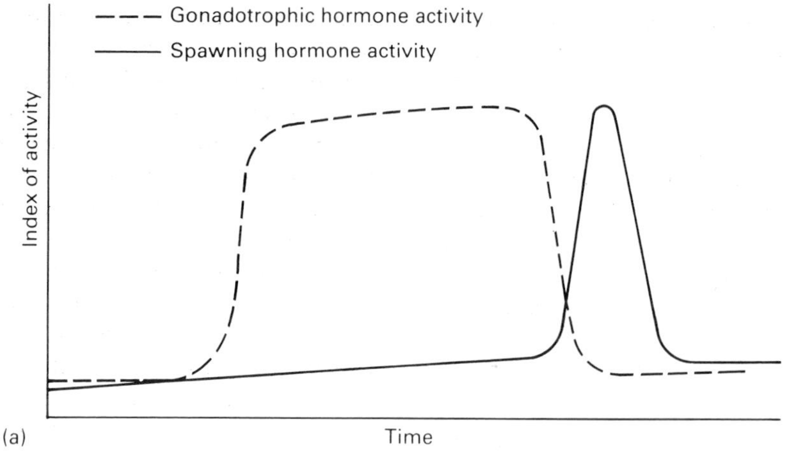

Somatic growth Storage of reserves	Mobilization of reserves Accumulation of germ cells	Maturation	Spawning	Oosorption Somatic growth Storage of reserves

(b)

Fig. 14.18. Control of annual reproductive cycles through the production and release of gonadotrophic and spawning-inducing hormones: (a) a generalized scheme; (b) cellular activities associated with different phases of the hormonal cycle.

release of a neurosecretory peptide from the radial nerves, which in turn induces production of the simple substance 1-methyl adenine by the ovary.

This substance causes contraction of the ovarian muscles and thus initiates spawning, but it also induces the production of a third messenger from the inner surface of the oocyte membrane which ultimately leads to germinal vesicle breakdown and renders the oocytes fertilizable. Similar functional schemes have been discovered in the Polychaeta and the Mollusca although the chemical nature of the substances involved are different and the hormonal mechanisms have presumably evolved independently.

Some marine invertebrates do not show repeated cycles of reproduction but breed only once per lifetime (see above). The endocrine systems controlling these reproductive cycles seem to be rather different in operation and are found in effect to control a normally irreversible switch from somatic growth to reproduction. The nereid polychaetes reproduce in this way and their maturation, and the somatic metamorphosis which often accompanies it, are controlled by the progressive withdrawal of a single hormone (see Chapter 16) essential for regenerative growth and for segment production; consequently, as the animals become mature, resources can no longer be deployed in these regenerative processes. This is adaptive since the animals inevitably die when they have reproduced; new segments could not contribute to their future fecundity and would thus have no value.

This is not true for worms in which reproduction occurs several times during a lifetime; in such species regenerative growth is not linked to sexual reproduction in this way. There seems to be a close relationship between the pattern of endocrine control of reproductive processes and the overall reproductive strategy adopted, but it is not clear which is the causal factor.

A similar situation is to be found in the Mollusca; most species exhibit iteroparous or repeated cycles of reproduction and are thought to have endocrine systems which fall into the general scheme outlined in Fig. 14.18. In most Cephalopoda, however, sexual reproduction involves a massive and rapid redeployment of reserves; stored metabolites are transferred to the developing germ cells and the animals do not survive breeding. The transition is stimulated by secretions of the optic glands, which could thus be said to have a gonadotrophic role, but the optic glands themselves are inhibited by the activity of the optic nerves. Isolation of the glands or cutting the optic nerves provokes sexual maturation and ultimately death, and the transition is normally an irreversible one.

14.4.4 Reproductive cycles and diapause in terrestrial and freshwater environments

The reproductive biology of terrestrial and freshwater invertebrates is strongly influenced by an ability to fertilize their ova internally and to lay fertilized and well-protected eggs. Linked to this is the ability to store spermatozoa and so fertilize eggs successively. In temperate, boreal and polar latitudes most invertebrates show seasonal reproductive patterns with prolonged periods of egg laying, interspersed with periods when sexual reproduction does not take place. Such a reproductive cycle is characterized not so much by the extreme synchrony of reproductive events as by the controlled transition from a state of reproductive activity to one of reproductive inactivity. Three states of reproductive inactivity can be recognized: *quiescence*, a direct and temporary response to adverse conditions which is reversed as soon as favourable conditions return; *facultative diapause*, a direct response to unfavourable conditions which once initiated will not reverse until some fixed period has elapsed; and *obligate diapause*, a phase of reproductive inactivity which recurs at specified times each year irrespective of the adverse conditions which it may be adapted to avoid.

Quiescence and facultative diapause are associated with the adult stages of pulmonate molluscs, earthworms and some insects; obligate diapause on the other hand is more frequently associated with earlier stages of development—eggs, larvae or pupae. Many populations of terrestrial and freshwater invertebrates in temperate and boreal regions, for instance, exist only as eggs during the winter months. The phenomenon has been particularly well studied in insects, and some of the factors which control the transition are now understood. Insects which are active in the summer and which enter diapause in the autumn are called 'long day insects', whereas others such as the silk worm *Bombyx mori* which are winter active are described as 'short day insects'. The terms are descriptive, but they are also apt, for it is indeed the day length which controls the transition from one physiological state to the other.

The changing cycle of day lengths associated with the progression of the seasons is a precise source of seasonal information in non-tropical regions of the world. The relative day length throughout the year at different latitudes is illustrated in Fig. 14.19; note that twice a year on 21st March and 21st September the day length is exactly 12 light:12 dark at every point on the earth's surface. Animals can often respond to the changing cycle of day lengths and this information can be used to control a seasonal, i.e. yearly, progression. The phenomenon is known as photoperiodism. The responses of many animals to the relative day length (or the relative night length) are usually non-linear; in effect light periods less than some critical length are interpreted as being short and those greater than the critical length are interpreted as long. This can be demonstrated by exposing colonies of insects to different light/dark regimes and recording the frequency with which diapause is induced (Fig. 14.20). The accuracy of the time measurement can be within 30 minutes or less and the phenomenon has attracted a great deal of interest from physiologists interested in the nature of biological clocks.

Studies with different insects suggest that rather

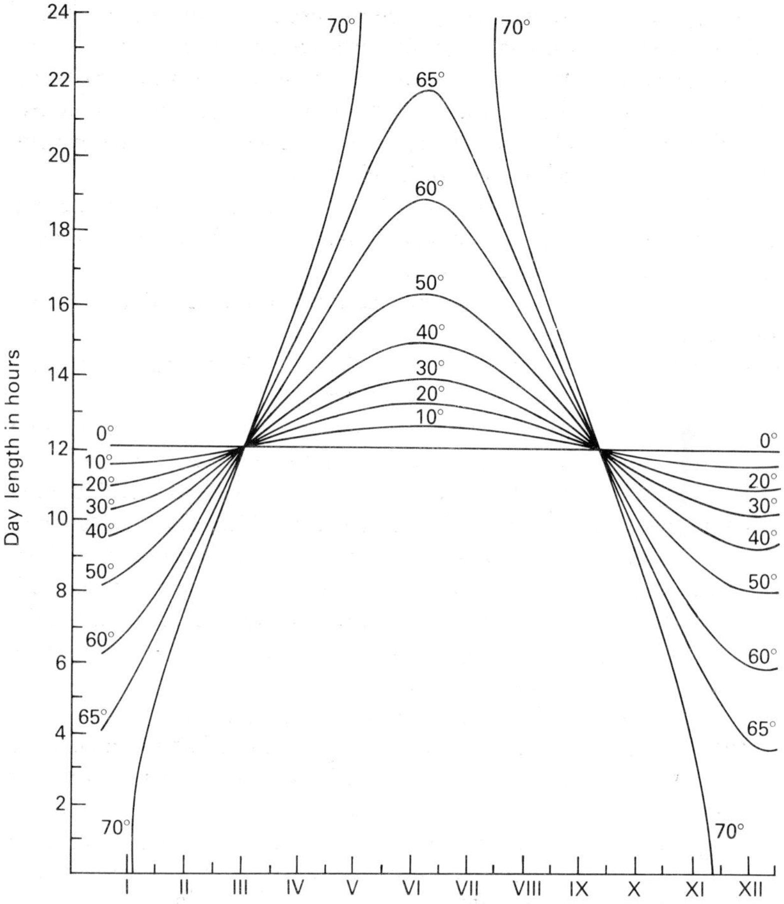

Fig. 14.19. Seasonal variation in day length as a function of latitude. Note that all points on the surface of the earth have exactly 12 hours of light and 12 hours of darkness twice in each year, and that the wavelength of the light/dark cycle is constant. Its amplitude, however, varies with latitude. (After Saunders, 1977.)

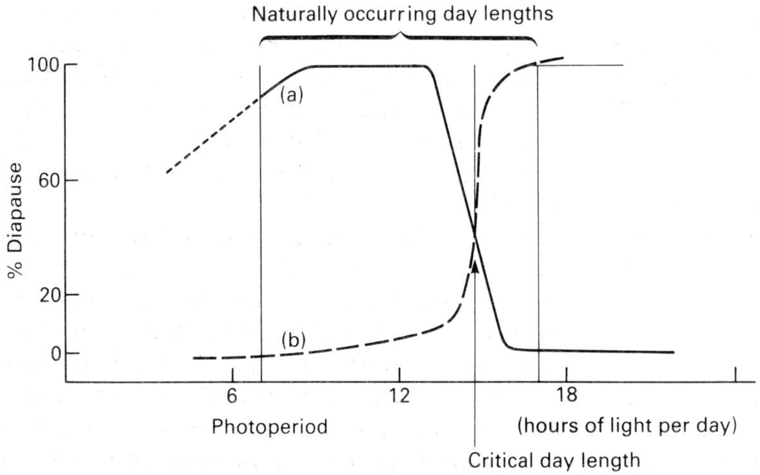

Fig. 14.20. Schematic representation of the photoperiodic response of (a) long-day and (b) short-day insects, where the response is measured as the percentage of individuals entering diapause when exposed to fixed photoperiods as shown.

different processes may be involved. In some, a circadian oscillator seems to be involved in the measurement of effective day length. This is most clearly demonstrated by experiments in which the insects are exposed to a sequence of experimental days in which the period of daylight is fixed but the night length has values from 12 hours or less up to 72 hours or more. Some of the experimental days are interpreted by the insects as being short, others as being long and this pattern suggests that there is a free running circadian (i.e. nearly 24 hours) oscillator in which there is a photosensitive phase which has a fixed phase relationship to the beginning of the cycle set either by the dawn or the dusk transition between light and dark. Figure 14.21 shows that in the resonance experiment such a photosensitive phase can come to fall in the light period in some photoperiodic regimes even though the duration of the light period is constant providing that the circadian cycle free runs through several cycles. There are other animals however where no resonance effects can be detected.

The aphid *Megoura viciae* for example shows no sign of resonance in an experiment such as that illustrated in Fig. 14.21. It shows evidence of an hourglass-like mechanism for time measurement as illustrated in Fig. 14.22. The key transition from parthenogenetic asexual reproduction to sexual reproduction, which is such a striking feature of the life cycle of this group of insects, is normally induced by the short days of the autumn. This is clearly of adaptive significance since the rich food resources of the spring and summer are about to disappear at that time. The transition is in effect controlled by the relative night length which is measured with a clock with the properties of an hourglass, i.e. one which measures elapsed time. If the period of darkness to which the aphids are exposed exceeds 9.75 hours, the short day length response is initiated and sexual morphs appear in the population which give rise to the overwintering diapause eggs.

14.5 Reproduction and resource allocation

The life histories and reproductive cycles of invertebrates are very diverse, yet the reproductive traits of individual species are usually firmly fixed. One of the challenges for invertebrate biology is the development of an evolutionary theory that will account for the diverse life histories that are observed in nature. No such theory has emerged, but some general principles can be identified.

The most widely accepted theory could be termed the 'demographic theory of life cycle'. Its basic assumptions are:

1 That natural selection acts on individual life-cycle traits.

2 That natural selection tends to optimize adaptive strategies (tends to maximize the fitness of individuals, where fitness is a measure of the contribution of genetic material to subsequent generations).

3 That individual life-cycle traits can evolve independently. Furthermore, it is likely that the resources available to an organism are limited and that where an organism has components of fitness associated with survivorship and fecundity there is a potential conflict and the possibility that trade-offs can be made between the two.

The demographic theory suggests that:

1 An increase in reproductive effort results in both an increase in reproductive output and a reduction in somatic investment.

2 An increase in reproductive effort will result in an increase in fecundity and/or increased survivorship of offspring.

3 A reduction in somatic investment will result in either reduced adult survivorship or reduced growth and hence reduced future reproductive output.

4 There is the possibility of a trade-off between *current reproductive output* and *residual reproductive value*, where residual reproductive value is the remaining lifetime product of survivorship and fecundity.

Where trade-offs are possible, different solutions may be optimal under different conditions, but the adaptive potentialities of organisms can be limited by their particular evolutionary history. It has been suggested that suppression of the veliger larval stage of gastropod molluscs, for example, involves the disruption of so many gene patterns that this step is irreversible. If this were so, one would expect a strong

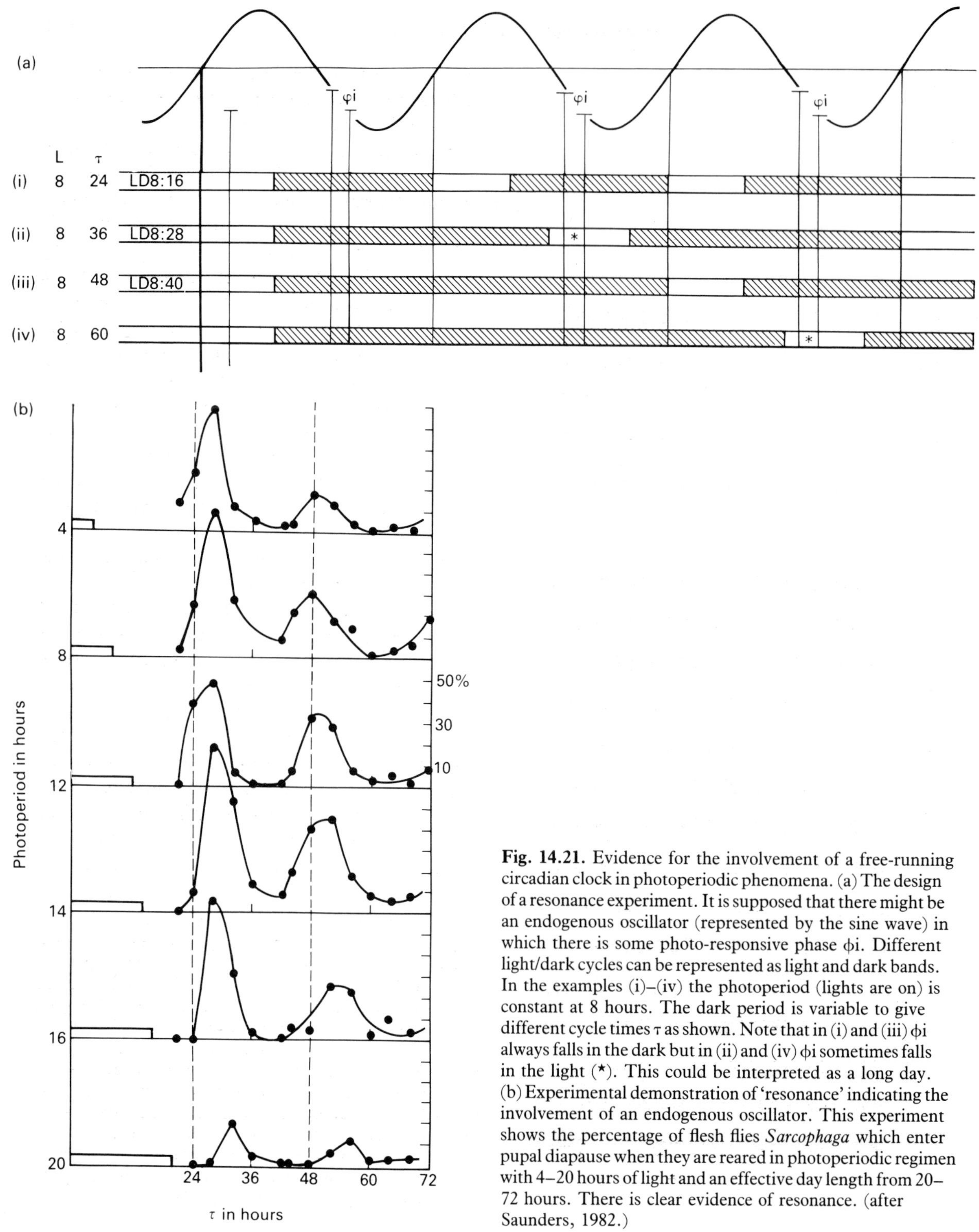

Fig. 14.21. Evidence for the involvement of a free-running circadian clock in photoperiodic phenomena. (a) The design of a resonance experiment. It is supposed that there might be an endogenous oscillator (represented by the sine wave) in which there is some photo-responsive phase ϕi. Different light/dark cycles can be represented as light and dark bands. In the examples (i)–(iv) the photoperiod (lights are on) is constant at 8 hours. The dark period is variable to give different cycle times τ as shown. Note that in (i) and (iii) ϕi always falls in the dark but in (ii) and (iv) ϕi sometimes falls in the light (*). This could be interpreted as a long day. (b) Experimental demonstration of 'resonance' indicating the involvement of an endogenous oscillator. This experiment shows the percentage of flesh flies *Sarcophaga* which enter pupal diapause when they are reared in photoperiodic regimen with 4–20 hours of light and an effective day length from 20–72 hours. There is clear evidence of resonance. (after Saunders, 1982.)

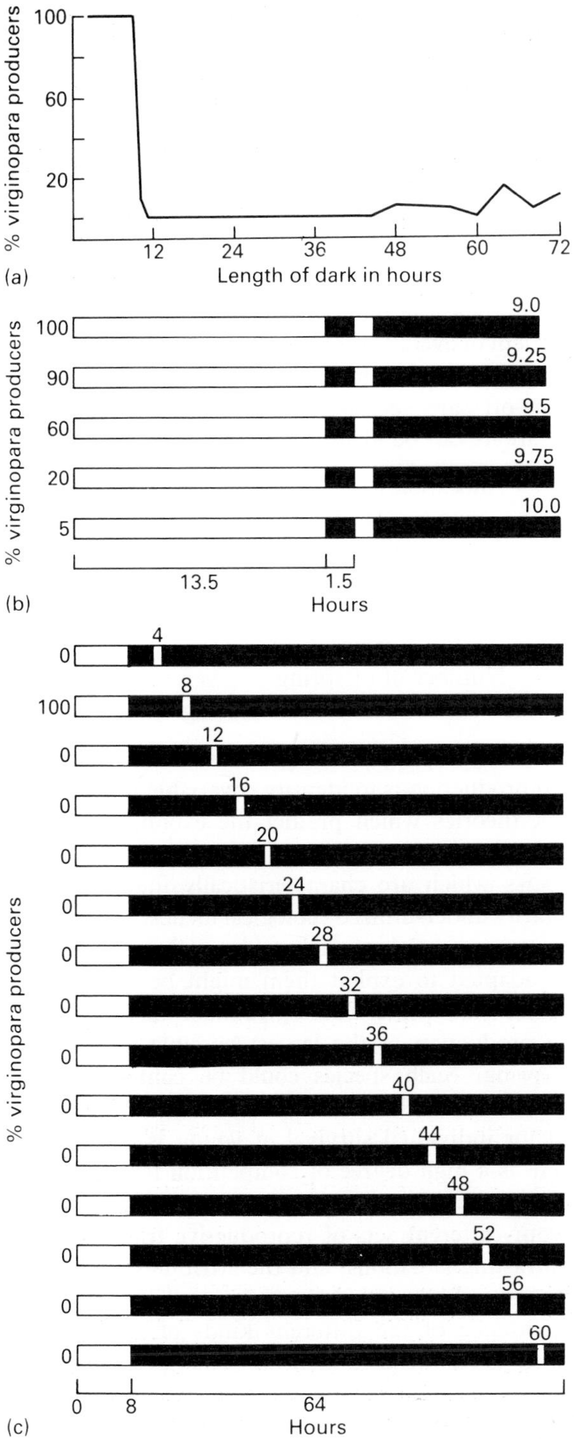

systematic bias in the distribution of gastropods which have retained and which have lost this larval stage, and this is in fact observed.

It has been recognized that in invertebrates there is an important trade-off between offspring survivorship and offspring number. This can be expressed as an apparent dichotomy between planktotrophic and lecithotrophic development or between brooding and non-brooding of embryos. It seems that resources may be allocated either to a relatively large number of offspring that are cheap to produce or to a smaller number of offspring that individually receive a higher allocation of resources and survive better. This choice is rarely made by individuals but seems in most cases to have been made during evolutionary history, the investment per propagule being a species-specific character. Often quite closely related species are found to differ from each other primarily in this characteristic.

A second important area of variability is in the relative amount of the available resources that are allocated to reproduction at any one time. We must suppose that any investment in reproduction reduces somatic investment and that the observed pattern of reproduction is an individual's solution to this conflict. Most invertebrates exhibit a pattern of reproduction that is far from random and there is usually a delay between the acquisition of resources and the allocation of those resources to reproduction. There are inevitably costs associated with that delay, due to the finite probability that death will occur before

Fig. 14.22. Evidence for an 'hour glass'-like system of night length measurement in the aphid *Megoura viciae*.
(a) Percentage of wingless parthenogenetic females (characteristic of the summer months) produced in colonies growing in light/dark cycles where photoperiod is held at 8 hours. Note the absence of resonance. (b) The percentage of wingless parthenogenetic females produced when dusk is followed by a 1.5-hour light break then a variable period of darkness. (c) the effect of a 1-hour light pulse at various times in a very long night. Long-day effects are produced only when the light break falls 8 hours after dusk. Such a break of the dark period consititutes a dawn before a critical night period has elapsed and has profound physiological effects. An earlier light break is interpreted as a new dusk; its effect is overridden by the following long night. (After Saunders, 1982.)

the allocation is made, and to the cost of maintaining the resource. To offset those costs there must be some benefits; these will arise when:

1 Conditions for offspring survival are not constant. This will often be the case when offspring exploit a temporary food resource or if the juveniles have a lower resistance to certain environmental conditions.

2 Fertilization is external. In these circumstances the probability that fertilization will occur is enhanced if gametes are released at the same time as those of other members of the population.

3 Reproduction is risky. In order to take part in reproductive activity individuals may have to incur higher risks of predation; under these circumstances it may be advantageous to delay reproductive activity until a large part or all of the resources to be allocated to reproduction have been acquired.

4 Adult experience confers enhanced survivorship on offspring. In some circumstances adults gain experience and abilities which confer a greater chance of survival on their offspring. This may be sufficient to offset the costs of delayed reproduction.

The biology of populations and the interactions between different members of a community are to a very large extent determined by the reproductive biology of the individuals. The life cycle is a pattern of allocation of resources between reproductive and somatic functions. Fecundity and survivorship and therefore the age structure of any population will, to a large extent, depend on this pattern. It will determine the frequency with which offspring are produced, and the relative scale of the allocation will be one of the factors which determines their survivorship. Unfortunately it is not yet possible to understand the functional significance of all the variety of life cycles and reproductive patterns exhibited by invertebrates nor the pressures which are responsible for their selection, but this is an area where new ideas are emerging.

It is generally accepted that life-history traits are subject to selection but it is equally recognized that there may be historical and functional constraints which prevent the selection of optimal solutions. Difficulties arise because of the apparent existence of multiple rather than single solutions. It is possible, for instance, to find closely related organisms in similar ecological niches which exhibit quite different patterns of resource allocation. Among marine invertebrates, for instance, it is not uncommon to find two similar species, in virtually the same habitat, one of which produces a relatively small number of large, yolky eggs and has direct development while the other produces a much larger number of small, less yolky eggs. It can be said that there is a trade-off between egg size and egg number, that this trade-off is independent of overall allocation of resources to reproduction, and that a number of different solutions to the trade-off are possible. Several other trade-off situations can be identified:

Allocation to individual offspring versus number of offspring;
Allocation to individual offspring versus survivorship of offspring;
Number of offspring of given size versus adult survivorship;
Number of offspring of given size versus number of future offspring.

One way of approaching the diversity of patterns of reproduction is to identify covariable traits and to seek theories which predict the evolution of such sets. As an example of this approach we can consider species which are characteristically found in temporarily under-utilized ecological niches. Such niches are transient and unpredictable, and species which are adapted to exploit them might be expected to exhibit traits which would give them the capacity for the production of the largest possible numbers of offspring. Such species could be contrasted with those found in climax communities where the potential for individual survival of young is limited and there is a high degree of competition for resources. Thus it could be argued that natural selection would favour different sets of reproductive traits in these contrasting situations, and the attributes of any one species could be regarded as a response to the relative importance of the different kinds of selection by which it has been influenced. This idea has had a very important influence on the development of the theory of reproduction and although it is now recognized that this does not provide a complete framework for an understanding of the evolution of reproductive traits, a slightly more detailed discussion will be useful.

The term '*r*-selection' has been used to describe the situation in which selection is primarily for the production of large numbers of descendants. The term is derived from the symbol '*r*' which is the constant defining the rate of population increase in the equations:

$$\text{a) } \frac{dN}{dt} = rN \quad \text{or b) } N_t = N_o e^{rt}$$

Selection for maximum values of *r* would favour early reproduction, high fecundity and short lifespan, and it might be supposed that '*r*-selected species' could be recognized by a combination of traits which would favour maximum production of potential descendants. It would also be expected that populations of such species would be unstable and characterized by large fluctuations in numbers. In contrast in conditions of high competitive interaction where there is little scope for growth in numbers, it would be expected that maximum fitness would be associated with delayed reproduction, low fecundity and the production of a relatively small number of well-provisioned offspring. Selection of this type was referred to as *K*-selection by reference to the constant *K* in the 'logistic' equation:

$$\frac{dN}{dt} = rN \frac{(K-N)}{K}$$

This equation defines the growth of a population to an asymptote which is called the carrying capacity of the environment. One of the difficulties with this analysis arises from the use of the terms *r* and *K* which are not strictly comparable. The Malthusian parameter *r* can be measured or estimated from a detailed population study; the carrying capacity of the environment, however, cannot as it is abstract and dimensionless.

The growth characteristics of populations can be determined from studies of survivorship and fecundity. One approach is through the construction of a life table which sets out the average survivorship and the average fecundity of each female in the population of a given age. Such a table can be constructed from observations on a population in steady state in which the ages of individuals can be determined, or it can be constructed by following the survivorship and fecundity of a given age class in the population. There are particular difficulties with the construction of a life table for invertebrates because it is not often easy to determine the age of individuals. Some techniques can, when used with care, obtain such information and sessile organisms such as the barnacles of rocky shores can be marked and observed over several years. An example of a life table obtained for a North American barnacle is given in Table 14.8. This table shows the average survivorship $l_{(x)}$ and the average fecundity $m_{(x)}$ at each age (x).

The sum of the products $l_{(x)}.m_{(x)}$, $\Sigma l_x m_x$, is the average lifetime reproductive output of a female born into that population. It is often referred to as R_o. The generation time can be estimated as the average age of an individual at death and can be calculated from the life table by the relationship:

$$T = \frac{\Sigma x l_{(x)} m_{(x)}}{\Sigma l_{(x)} m_{(x)}}$$

The Malthusian parameter is often estimated approximately from the relationship

$$r \simeq \frac{\ln R_o}{T} \text{ where } R_o = \Sigma l_{(x)} m_{(x)}$$

If '*r*' is positive as it is in Table 14.8, then the population is increasing in size, if '*r*' is negative the population is declining, and if it is zero the population is stable. Note that although '*r*' can be estimated from the life table this is not so for *K*.

All environments fluctuate to some extent but there are some that are more stable than others; these will exert mortality primarily on the small, weak offspring rather than the well-adapted adults. One would expect, therefore, that in stable conditions selection would favour extended lifespan and the progressive production of well-provisioned offspring. In fluctuating environments, on the other hand, mortality will be directed to the adults as well as the young. These arguments also suggest, as does the *r*-/*K*- selection

Table 14.8 Life table of *Balanus glandula* (from Hines, 1979)

	Age (x) months	Survivorship to age l_x	Average eggs at age (x) m_x	$l_x m_x$	$xl_{(x)} m_{(x)}$
Estimated mortality	0	1.000	0	0	0
prior to settlement	3	1.17×10^{-4}	0	0	0
	12	2.04×10^{-5}	20 504	0.418	5.016
	24	3.84×10^{-6}	66 814	0.256	6.153
	36	2.05×10^{-6}	113 125	0.231	8.335
	48	1.28×10^{-6}	140 742	0.180	8.641
	60	7.33×10^{-7}	159 435	0.117	7.008
	72	4.18×10^{-7}	170 892	0.075	5.151
	84	2.44×10^{-7}	176 922	0.043	3.629
	96	1.42×10^{-7}	180 540	0.026	2.469

$$R_o = \Sigma l_{(x)} m_{(x)} = 1.346$$

$$\text{Generation Time } T = \Sigma x\, l_{(x)} m_{(x)} = 46.402 \text{ months}$$

Intrinsic rate of population increase for time in months $r = \frac{\ln R_o}{T} = 0.006$ i.e. population would be growing slowly if this were to continue.

In this case, survivorship was determined by observing the number of survivors for each 1000 eggs from the ratio between egg production and numbers of spat settling. This is an estimate. Subsequently survivorship was observed directly. These data give column l_x.

Fecundity was estimated by measuring the mean basal diameter of the barnacles at each age and measuring the relationship between basal diameter and number of eggs per brood.

Finally the number of broods per year was estimated. The product gives an estimate of the mean fecundity at each age. These data give column m_x.

argument, that there will be coevolution of longer lifespan, low fecundity and late reproduction in stable, less fluctuating environments and short lifespan, higher fecundity and early reproduction in less stable environments. In other words we expect the trade-off relationships identified earlier in this section to be influenced by environmental stability in a predictable way. This is not entirely borne out by observation; earlier in this chapter the life cycles of marine invertebrates were described and there is a clear trade-off between offspring number and egg size. It is generally found, however, that those organisms which have pelagic planktotrophic development, with very high fecundity and massive juvenile mortality, are those which as adults have relatively stable populations, long lifespan and large body size. Species in the marine environment which do seem to be adapted to exploit temporary ecological niches often have relatively low fecundity, lecithotrophic development and direct rather than pelagic development.

In order to understand why individual animals reproduce as they do it will also be necessary to take into account the constraints which are imposed by the environments in which they live, the functional morphology of adult and developmental stages, and perhaps most important of all the evolutionary history of the organism which may limit the reproductive options open to it.

14.6 Conclusions

Life-history theory is a rapidly expanding field which

seeks a functional explanation of the diverse sets of reproductive traits which are exhibited by organisms such as invertebrates.

In this chapter, we have briefly reviewed some of the material which forms the database for the development of that theory. It is clear that sexual reproduction is the dominant mode of reproduction among all living things, although it may be combined with episodes of asexual reproduction in complex life histories such as those of many cnidarians, flatworms, annelids, small crustaceans and insects.

The widespread occurrence of sexual reproduction is associated with a bewildering variety of reproductive states and patterns of allocation to reproductive processes. This variety has been examined from a taxonomic, environmental and functional viewpoint. It is clear that all possible combinations of reproductive traits are not found equally in all environments, nor in all taxonomic groups. Reproduction in the sea is often associated with the release of free-living pelagic larvae and this trait is associated with many others, including external fertilization, mass epidemic spawning and the production of simple spermatozoa and energy-poor eggs.

Under special circumstances in the sea, and in all terrestrial and freshwater environments, pelagic larval development is suppressed. This is usually associated with internal fertilization, sperm storage and the production of energy-rich eggs.

All patterns of reproduction which differ from random are the product of the controlled allocation of resources to specific reproductive functions, and if there is a strong element of synchronization among members of the population there must be the input and neuroendocrine transduction of environmental information. This subject has been briefly touched on in this chapter, and will be discussed more fully in Chapter 16.

The analysis and description of reproductive processes in invertebrates as discussed here will provide the background to more detailed studies of ecology and community biology, as well as providing an introduction to the study of invertebrate development which follows (Chapter 15).

14.7 Further reading

Detailed background to the reproduction of invertebrates can be obtained by reference to two multi-volume treatises and a continuing series of review volumes:

Adiyodi, K.G. & Adiyodi, R.G. (Eds) 1983. *Reproductive Biology of Invertebrates.* Wiley, New York.
Vol. I. Oogenesis, oviposition and oosorption.
Vol. 2. Spermatogenesis and sperm function.
Further volumes pending.

Giese, A.G. & Pearse, J.S. (Eds) 1974–77. *Reproduction in Marine Invertebrates.* Academic Press, New York.
Vols I–IV and continuing.

Advances in Invertebrate Reproduction. Elsevier Science, Amsterdam.
Vol. 2. Clark, W. and Adams, T.S. (Eds) 1981.
Vol. 2. Engels, W. (Ed.) 1984.
Vol. 3. Porchet, M. (Ed.) 1986.

The following are monographs which address different aspects of invertebrate reproduction:

Begon, M., Harper, J.L. & Townsend, C.R. 1986. *Ecology: Individuals, Populations and Communities.* Blackwell Scientific Publications, Oxford.

Bell, G. 1982. *The Masterpiece of Nature: The Evolution and Genetics of Sexuality.* University of California Press, Berkeley.

Brady, J. 1979. *Biological Clocks* (Studies in Biology, 104). Edward Arnold, London.

Calow, P. 1978. *Life Cycles.* Chapman & Hall, London.

Charnov, E. 1982. *The Theory of Sex Allocation.* Princeton University Press, Princeton, New Jersey.

Cohen, J. 1977. *Reproduction.* Butterworth, London.

Greenwood, P.J. & Adams, J. 1987. *The Ecology of Sex.* Edward Arnold, London.

Maynard-Smith, J. 1978. *The Evolution of Sex.* Cambridge University Press, Cambridge.

Saunders, D.S. 1977. *An Introduction to Biological Rhythms.* Blackie, Glasgow.

15
DEVELOPMENT

Most multicellular animals begin life as a fertilized egg produced by the fusion of two very different haploid gametes. Fusion creates a diploid zygote and it is from this rather special cell that the adult multi-cellular organism is derived. The sequence of early cell divisions is referred to as cleavage; during this process a precisely organized cell division not associated with cell growth takes place. This subdivides the cytoplasm of the zygote in a predetermined way into a larger number of smaller cells which retain the spatial organization of the fertilized egg. This organization can have profound influence on the developmental fate of the cells in the early embryo. A small number of rather different patterns of cleavage are found among the invertebrates. The spiral pattern of many protostome phyla and the radial pattern of the deuterostome phyla are well-known examples.

The subsequent development of the embryo involves spatial reorganization to give the adult body plan, and, at a later stage, differentiation of cells to those found in the functional larva or adult. The developmental sequence thus involves:

1 Fertilization.
2 Activation of zygote metabolism and translation of maternal messenger molecules (mRNA).
3 Cleavage.
4 Activation of zygote nucleus and transcription of new zygote specific information molecules (mRNA).
5 Organogenesis.
6 Differentiation.

Development also frequently involves a metamorphosis when a differentiated larva adapted to one set of environmental conditions changes suddenly into a different adult which may be adapted to quite different conditions and which may have very different functions (see Chapter 14). Many of the invertebrates have provided particularly valuable models for the analysis of the biochemical mechanisms involved in cell differentiation and regional organization, and developmental studies are also important for the light they shed on the relationships between animals.

This chapter will examine the development of invertebrates with particular reference to experimental investigations of the underlying processes involved in the determination of cell fate. The review includes an analysis of what happens at fertilization and of the response of the egg. The descriptive work on cleavage patterns provides the basis for an experimental study of the onset of cell determination, or the fixing of the cells in their prospective developmental fate. A wealth of descriptive work on larval forms and metamorphosis provides the basis for an experimental study which challenges the boundaries of our understanding of development and differentiation. The chapter also includes a discussion of regeneration among the invertebrates. During regeneration, a complete pattern is reconstituted from a fragment of a pattern. The processes involved therefore mimic those of normal development.

15.1 Fertilization and the initiation of development

Fertilization is a complex process. Its principal components are:

1 Physical juxtaposition of the gametes.
2 Surface membrane interaction leading to the union of the two cells.

3 A physiological reaction at the surface of the egg leading to a block to further sperm entry, usually called the block to polyspermy.
4 Activation of the oocyte metabolism.
5 Fusion of the pronuclei to form the new diploid genome of the zygote.
6 The initiation of early cleavage.
Marine invertebrates with external fertilization have been especially important in establishing the general principles of this process.

The precise sequence of events depends on the state of maturation of the 'egg' when gamete fusion occurs. In some invertebrates the eggs and sperm interact prior to the completion of meiosis and the production of the polar bodies. In other cases meiosis is completed prior to fertilization. For example, the eggs of sea-urchins complete meiosis prior to fertilization but those of the starfish enter meiosis metaphase I immediately before their discharge (see Chapter 16), but they do not proceed with meiosis unless they are fertilized; the polar bodies therefore appear after, not prior to, the interaction between eggs and spermatozoon.

The eggs and sperm of many marine invertebrates are shed into sea water and it is in this medium that fertilization takes place. It might be expected that some attraction between eggs and sperm would be observed but this is, in fact, rather rare. Spermatozoa activated in sea water exhibit random movements which will, if the concentration of eggs and spermatozoa is sufficiently high, bring the gametes into physical contact, but only in a few cases has a chemokinesis been demonstrated, e.g. in urochordates, some molluscs, and the colonial hydroid *Campanularia*. In this hydroid, the eggs are encased in a flask-shaped structure, the gonangium, and sperm must reach the egg through its relatively narrow aperture. Cine film showed that the tracks of individual sperm were not random but were directed to the mouth of the gonangium (Fig. 15.1). A substance extracted from the mouth would attract sperm when dissolved in agar in microtubes.

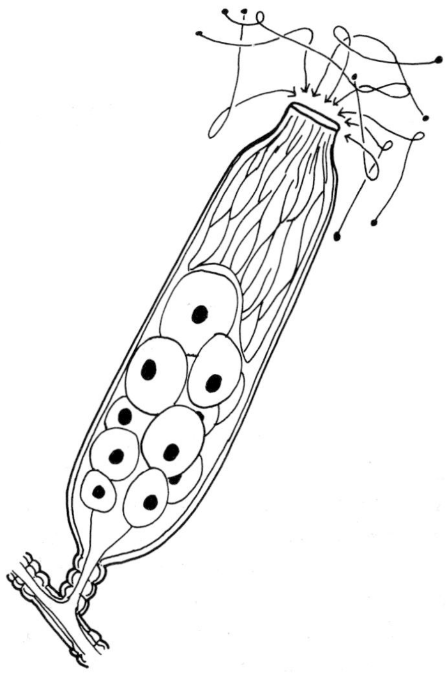

Fig. 15.1. A gonangium of the hydroid *Campanularia* and tracings of the tracks of spermatozoa as recorded by cinematography. This was one of the first cases where chemotaxis between eggs and spermatozoa could be demonstrated. It is certainly not unique; other examples include urochordates and molluscs but it is not known how widespread the phenomenon is. (After Miller, 1966.)

It is much more usual for randomly moving spermatozoa to fertilize the eggs by a mechanism which is still not fully understood. At the beginning of the century it was observed that the spermatozoa of sea-urchins, introduced to sea water in which eggs had been standing, became temporarily immobilized in an agglutination-like response. This was caused by the diffusion of a high molecular weight substance (in fact from the egg jelly which surrounds echinoid eggs) into the water. This substance was termed *fertilizin* and the corresponding detector protein on the spermatozoon *anti-fertilizin*. Most eggs have a species-specific sperm coagulant at the surface, but only in some cases will it diffuse into the surrounding medium, as it does in the sea-urchins.

The sperm of some invertebrates can be seen under the light microscope to undergo a physical change on contact with the egg (Fig. 15.2a) in which a needle-

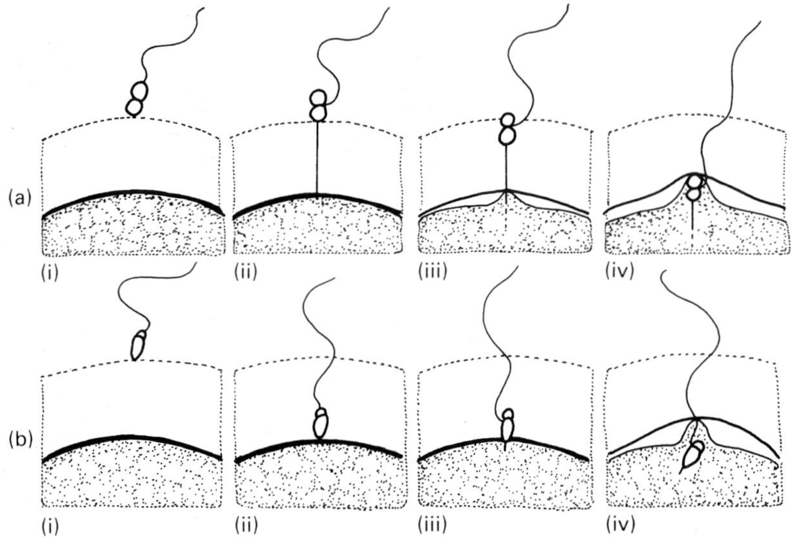

Fig. 15.2. Observations on the changes visible with the light microscope during the fertilization process of (a) a starfish (*Asterias*) egg and sperm and (b) a sea-urchin (*Arbacia*) egg and sperm.
(a) *Asterias*. (i) Approach of the spermatozoon to the egg which is surrounded by a jelly coat. (ii) Acrosome reaction. A fine filament seems to grow out of the small blister-like acrosome at the front of the sperm. (iii) The acrosome filament contacts the membrane, and the fertilization response occurs spreading away from the initial point of contact. (iv) The spermatazoon is engulfed by a cone-like eruption of oocyte cytoplasm.
(b) *Arbacia*. (i) Approach of the spermatozoon to the jelly coat. (ii) Passage of the spermatazoon through the jelly coat. (iii) Contact of the sperm head with the vitelline membrane. A very small acrosome filament may be visible as shown. (iv) Fertilization response and incorporation of sperm nucleus. (After Austin, 1965.)

like filament initially makes contact with the egg cytoplasm. This filament, the acrosome filament, develops from the most anterior part of the sperm—the acrosome. In others no such filament can be seen (Fig. 15.2b).

The electron microscope shows that some aspects of the sperm–egg interaction are essentially similar in all species, since they all involve membrane fusion. Figure 15.3 illustrates the changes which have been seen in a sea-urchin *Arbacia* and a polychaete worm *Hydroides* as the sperm approaches the egg. In each case there is first a fusion between the membranes of the acrosome vesicle and sperm plasma membrane. This releases the enzymic contents of the acrosome vesicle and the subsequent lysis permits the expanding acrosome tubules or filament to penetrate the egg coats and contact the egg plasma membrane. Further membrane fusion establishes a single membrane around both egg and sperm.

The initial reactions to this event are sometimes referred to as the 'block to polyspermy'. There are two components; the first is not visible but in some species involves a change in electrical potential at the surface of the egg. A few seconds later a visible cortical reaction spreads radially away from the point of sperm–egg contact. This reaction involves fusion of the membranes of the cortical granules, formed during late oogenesis, with the egg membrane and the release of their contents into the perivitelline space. The result is the raising of the fertilization membrane, as illustrated in Fig. 15.4.

A reaction of this type is almost universal; it can lead to the emission of a jelly coat, and an egg which has reached this stage of development cannot fuse with other sperm cells.

The egg now shows profound metabolic changes. The unfertilized egg is in some ways in a state of suspended animation. The binding of a spermatozoon to a receptor site initiates a sequence of events (summarized in Fig. 15.5) which release the egg from inhibition. There is a major increase in oxygen consumption which peaks during the first minute

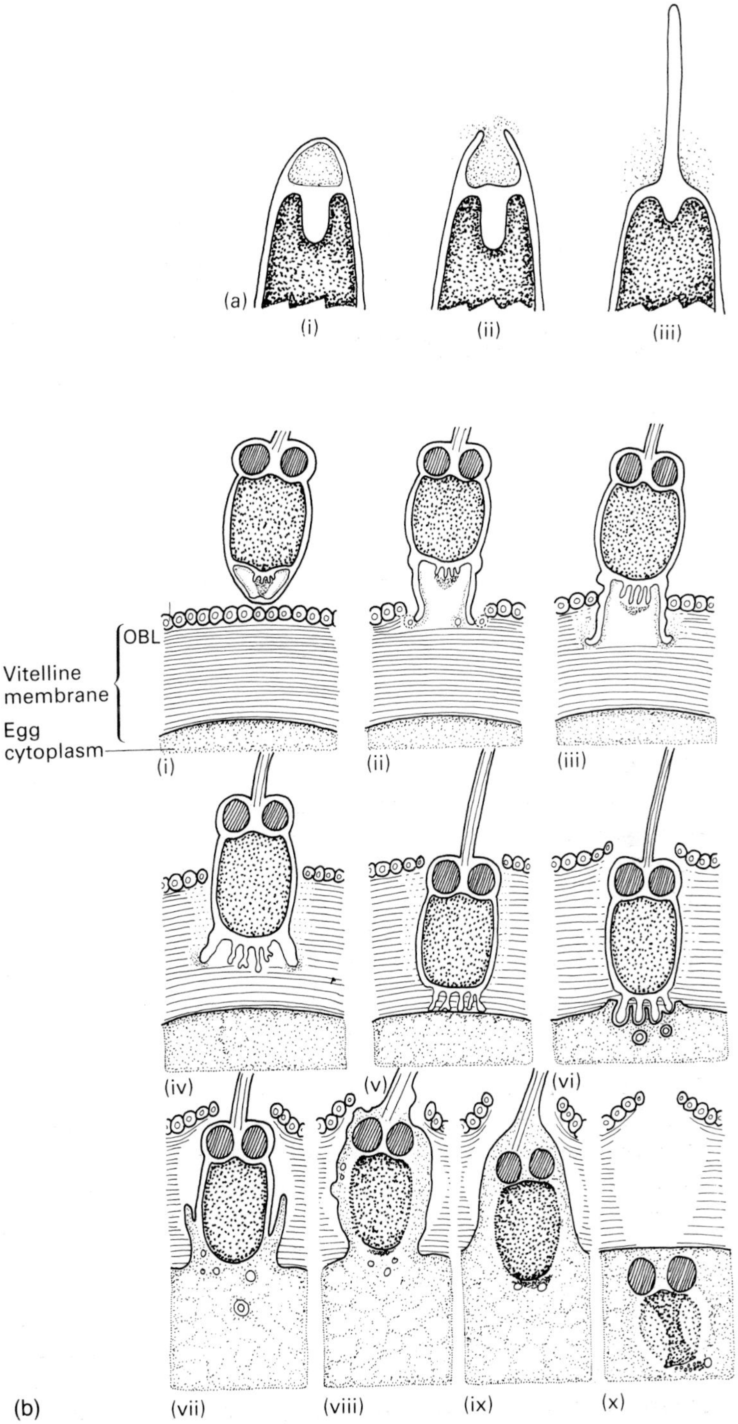

Fig. 15.3. Fertilization as seen with the electron microscope. (a) The acrosome reaction of the sea-urchin *Arbacia*. The acrosome filament of this species is too small to be seen easily with the light microscope. (i) The 'anterior' part of the sperm head as seen with electron microscope. (ii) Fusion between acrosome membrane and sperm membrane to release acrosome contents. (iii) Elongation of the acrosome filament. (b) Detailed analysis of the fertilization process of the polychaete *Hydroides*, as revealed by the electron microscope. (i) Approach of the spermatozoon to the outer border layer (OBL) of the vitelline membrane. The OBL is formed from the tips of the microvilli of the oocyte surface. (ii), (iii) Fusion of acrosome vesicle membrane causing release of the acrosome contents. Beginning of vitelline membrane penetration. (iv) Continuing penetration of sperm head and elongation of multiple acrosome tubules. (v), (vi) Contact and eventual fusion of acrosome tubules and oocyte plasma membrane. (vii), (viii) Incorporation of the sperm as oocyte cytoplasm moves into the joint oocyte/sperm membrane (caused by membrane fusion). (ix) Further incorporation. (x) Sperm nucleus and mitochondria incorporated into the egg cytoplasm. Note the hole left in the vitelline membrane. (Redrawn from electron micrographs of Colwin & Colwin, 1961.)

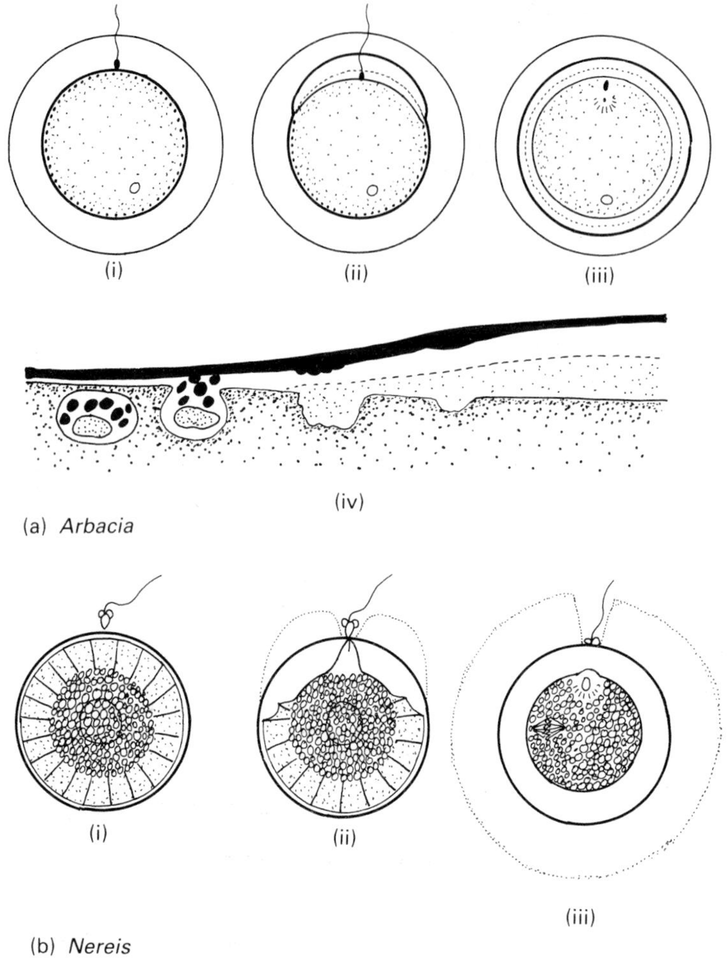

Fig. 15.4. The fertilization reaction: elevation of the fertilization membrane. (a) The sea-urchin *Arbacia*. (i) The sperm contacts the egg plasma membrane, cortical granules intact. (ii) Cortical granules break down and fertilization membrane is raised; the reaction moving progressively away from the point of contact. (iii) The fertilized egg. Note that the jelly coat is present prior to fertilization. (iv) Diagrammatic representation of the chain of events as seen in the electron microscope. (b) The polychaete *Nereis*. (i) The sperm approaches the egg. There is no jelly layer at this stage. (ii) Progressive spread of fertilization reaction from the point of contact. The outer part of the egg is composed of large, cortical alveoli which break down in this process. (iii) The fertilized eggs. In most species the contents of the cortical alveoli extrude through the fertilization membrane (ex-vitelline membrane) and form a new jelly coat. (After Austin, 1965.)

when the fertilization membrane is formed, but remains much higher than it was in the unfertilized egg. Subsequently there is an increase in the rate of protein synthesis, as masked messenger-RNA molecules, which were stored in the egg cytoplasm during oogenesis, are made available. The first cleavage will follow some predetermined time after fertilization, but important changes in the distribition of the cytoplasmic constituents may be observed prior to the first division.

In bilaterally symmetrical animals these cytoplasmic movements establish the primary axes of the future embryo by defining anterior, posterior, dorsal and ventral quadrants in the undivided egg. Such movements were first described in an ascidian and are illustrated in Fig. 15.6.

As we shall see in Section 15.3, the importance of the spatial distribution of different types of cytoplasmic constituent can hardly be emphasized too strongly.

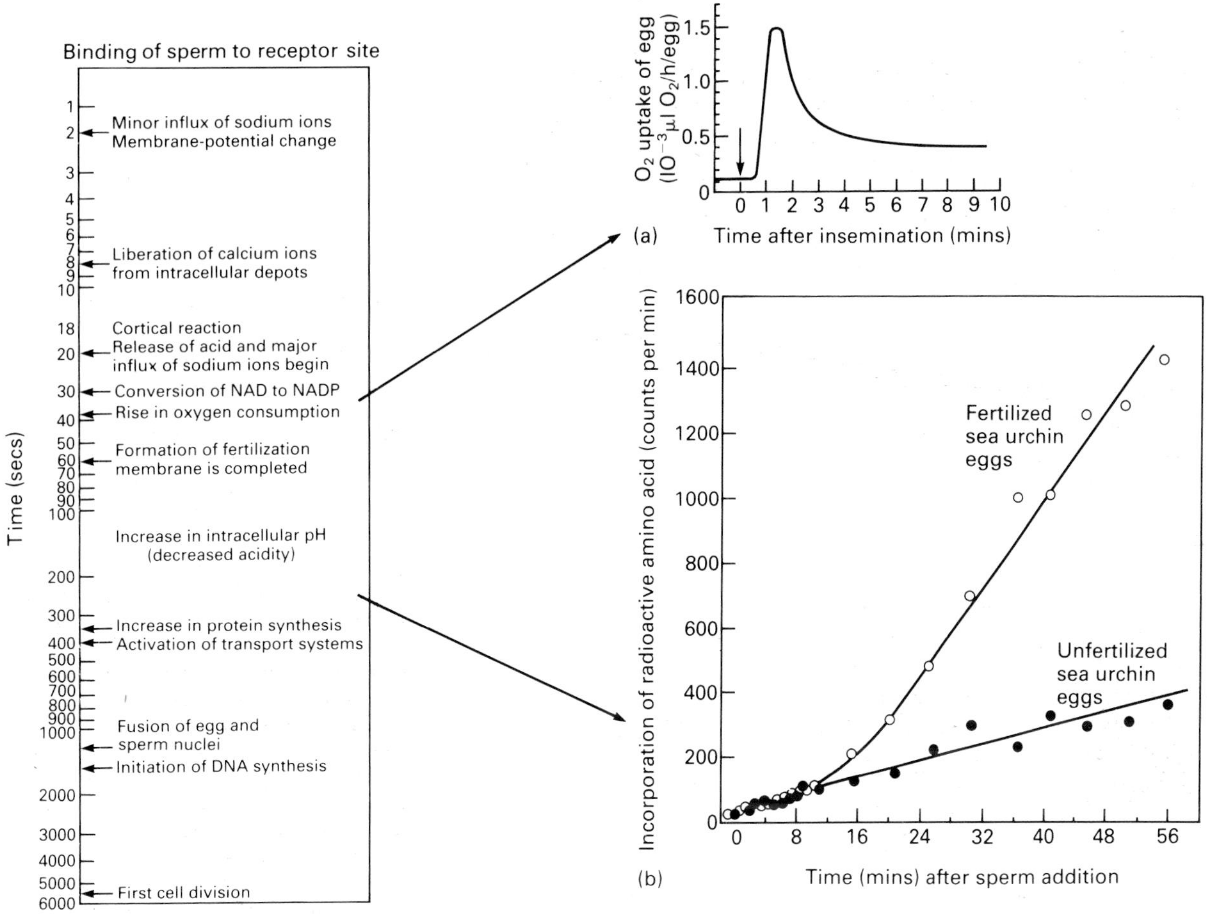

Fig. 15.5. The sequence of events following contact between sperm and egg membranes during sea-urchin fertilization (after Epel, 1977). Inset are examples of experimental data in support of this scenario. (a) Oxygen consumption (after Ohnishi & Sugiyama, 1963); (b) protein synthesis (after Epel, 1967).

15.2 Patterns of early development

15.2.1 Cleavage

Once mitosis has been initiated the embryos of most animals follow a precisely determined pattern of cleavage during which the cytoplasm of the relatively massive egg is subdivided into smaller cellular units. These cleavage patterns are such that any spatial organization in the fertilized eggs is retained into the multicellular embryo, and most give rise to a hollow ball of cells called a blastula.

Many of the protostome phyla (e.g. Nemertea, Annelida, Sipuncula, Echiura, Mollusca and perhaps Pogonophora) are said to exhibit spiral cleavage. The pattern is particularly well known from studies of annelids and molluscs, and this type of cleavage is described in some detail in Box 15.1, since it has proved to be of great importance in the experimental

Box 15.1 Spiral cleavage

This pattern of cleavage is characteristic of several protostome phyla including the Nemertea, Annelida, Sipuncula, Echiura and Mollusca. It is most easily seen in embryos with relatively little yolk such as those of the marine polychaetes and molluscs.

1 The unfertilized egg is radially symmetrical about an axis from the animal (less yolky) pole to the vegetal pole (Fig. i). All segments of such an egg cut along the animal-vegetal axis are equivalent.

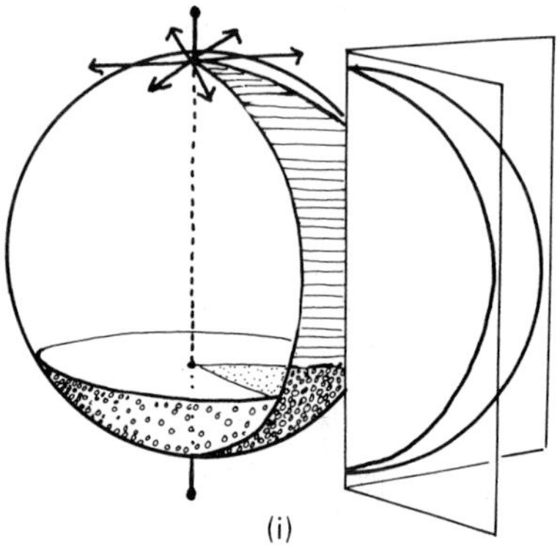

(i)

After fertilization the egg is bilaterally symmetrical. The principal planes of the embryo can then be recognized (ii–iv).

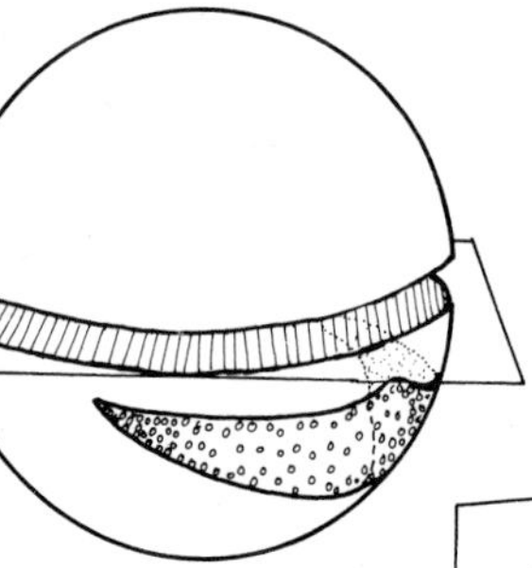

(ii)

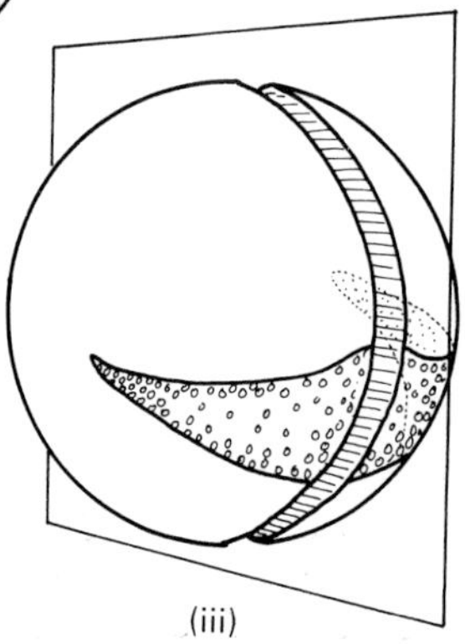

(iii)

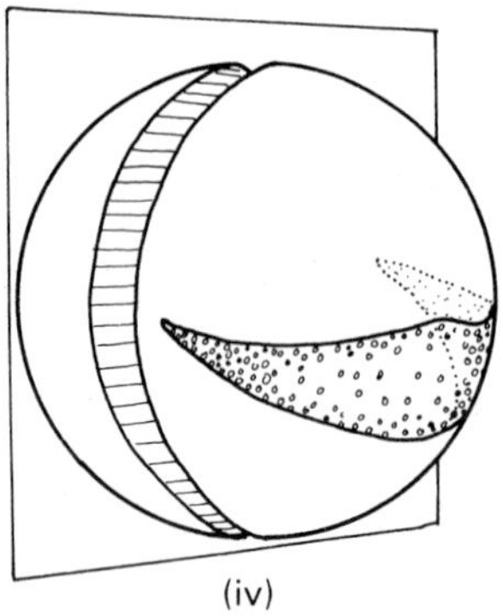

(iv)

(ii) Transverse—separates anterior from posterior.
(iii) Sagittal—separates left from right.
(iv) Frontal—separates dorsal from ventral.

2 The first two cleavage planes are longitudinal and bisect the angles between frontal and sagittal.

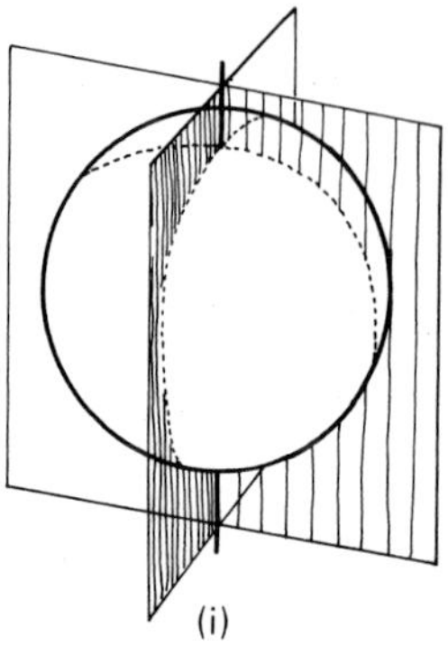

(i)

This separates the first two cells, AB and CD (ii); then the first four cells called A, B, C and D (iii).

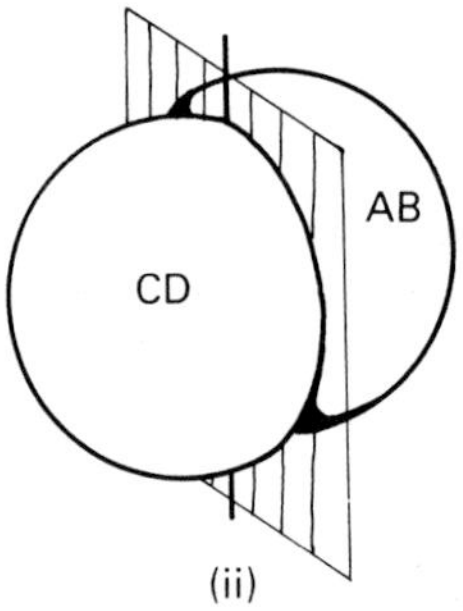

(ii)

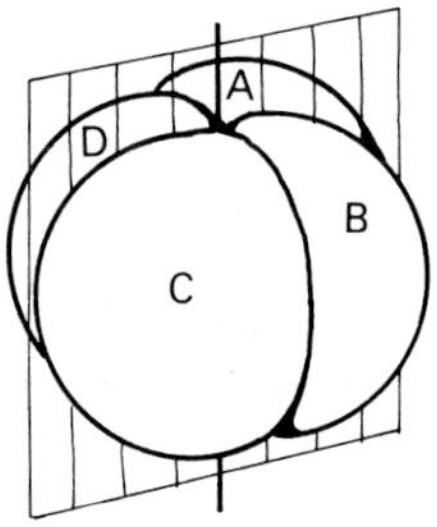

(iii)

The D cell is usually larger than the other three and by reference to this cell all subsequent cleavage cells can be identified.

3 The third cleavage plane is transverse and passes above

the equator. It separates four smaller cells at the animal pole from four larger ones at the vegetal (i).

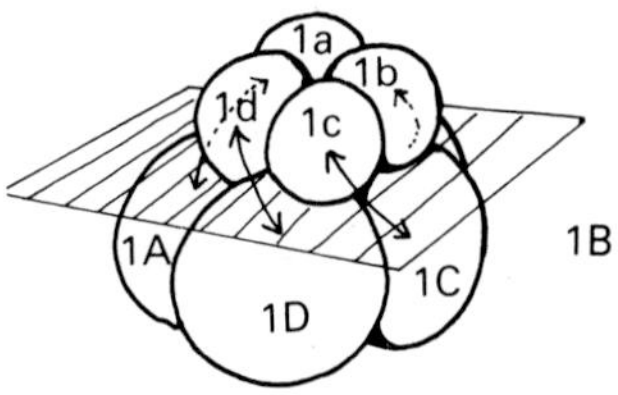

(i)

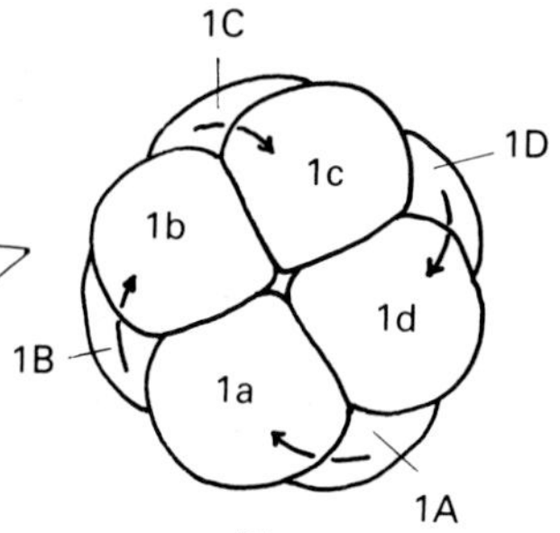

(ii)

The four smaller cells are referred to as the first quartet of micromeres. Because the cleavage plane is inclined to the long axis of the embryo, the micromeres lie over the inter-cell boundaries between the macromeres. When seen from the animal pole, the cells appear to be rotated in a clockwise direction (ii).

The eight cells of the embryo can now be individually identified as:

1a 1b 1c 1d
1A 1B 1C 1D

All subsequent divisions are transverse and result in subdivision of existing micromeres and the generation of new quartets of micromeres by unequal cleavage of the macromeres.

Fig. (iii) shows the 4th cleavage producing an embryo with 16 cells. At this stage they can be individually identified as:

Subdivision of 1st quartet	$1a^1$,	$1b^1$,	$1c^1$,	$1d^1$
	↕	↕	↕	↕
	$1a^2$,	$1b^2$,	$1c^2$	$1d^2$
2nd quartet	2a,	2b,	2c,	2d
	↕	↕	↕	↕
Macromeres	2A,	2B,	2C,	2D

Individual cleavage planes are inclined to the long axis such that there is alternately clockwise and anti-clockwise rotation of the micromeres.

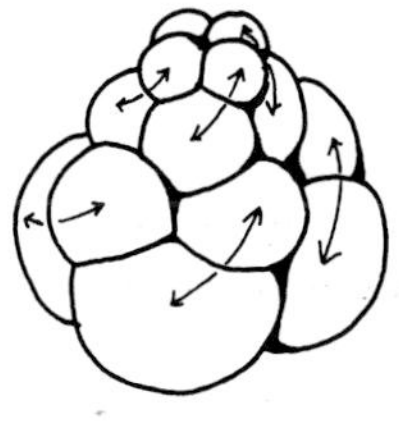
(iii)

4 A conventional scheme of notation has been developed which allows each cell to be identified up to the 64 cell stage. The formal scheme (completed only for the important D-cell lineage) is set out below:

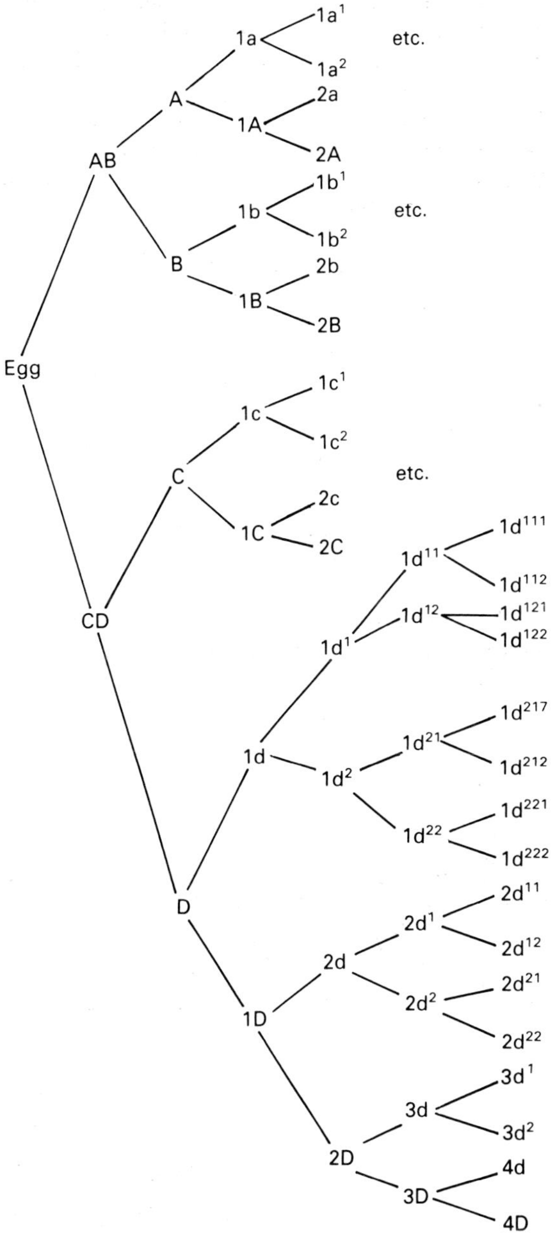

No. of cells	1	2	4	8	16	32	64
Cleavage:		1st	2nd	3rd	4th	5th	6th

As explained further in the text and in Boxes 15.2 and 15.3 the D-cell lineage is particularly important.

analysis of invertebrate development (see also Section 15.3).

The spiral cleavage exhibited by these phyla is also sometimes said to indicate a relatively close phylo-

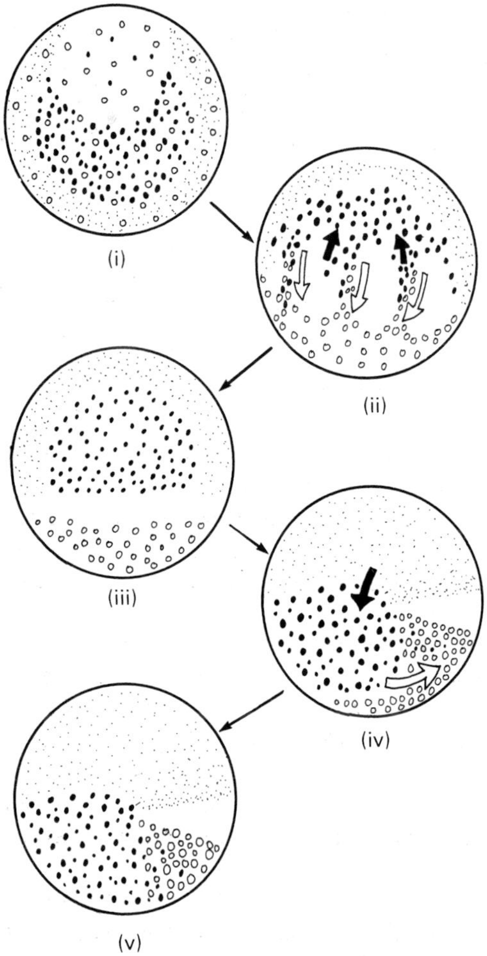

Fig. 15.6. Movements of the visibly different types of cytoplasm described in the early twentieth century in the ascidian *Styela partita*. These observations were important in demonstrating the potential importance of cytoplasmic factors in determining cell fate. (i) The radially symmetrical unfertilized egg. (ii), (iii) Migration of 'yellow' cytoplasm to the vegetal pole and segregation of 'grey' cytoplasm in the central mass of the egg. (iv) Segregation and migration of yellow cytoplasm and clear cytoplasm to one side of the egg. The egg becomes bilaterally symmetrical. (v) The bilaterally symmetrical egg with cytoplasm in positions occupied at the time of the first cleavage. (See also Box 15.3.)

genetic link between them. This thesis, however, has to be interpreted with care. These phyla are thought to be derived from flatworms, and some of the primitive flatworms also have a pattern of cleavage of this same fundamental spiral type. This is especially clear in the free-living polyclads and acoels, although in the latter there are only two large cells at the vegetal pole, the macromeres, and not, as is usually the case, four. Many other groups of phyla, including the pseudocoelomates or aschelminths, lophophorates and arthropods, as well as the deuterostomes, are also probably derived from flatworms (Chapter 2), but they will have quite different patterns of early cleavage, as will be described below. Indeed, in the platyhelminths spiral cleavage is unusual and is restricted to those few groups which have eggs in which the yolk is deposited in the oocyte cytoplasm. It is aberrant in those species in which yolk is provided by specialized yolk cells, as described in Section 14.4.2.2.

The observation that several phyla have this spiral pattern of cleavage in common need imply only that this has been a conservative feature in their evolution and it cannot be taken to indicate a relatively recent divergence from a common line of descent.

If the ancient flatworm pattern of cleavage was spiral then the observation of some other pattern in common between several phyla may suggest a more recent evolutionary divergence. This is shown, for instance, by the deuterostome phyla. Many of these exhibit a different pattern of cleavage, termed radial, in which the products of transverse divisions lie directly over each other. A relatively simple example, found in the holothurian *Synapta*, is illustrated in Fig. 15.7. All the cells are virtually identical in size and the embryo develops progressively to the hollow-ball blastula stage.

A more complex radial pattern is exhibited by the sea-urchins, which have been extensively used in experimental embryological research. Individual cells cannot be identified since the first two cleavages are equal, but different layers of cells can be identified in the 64 cell stage embryo, as explained in Box 15.2.

The yolky eggs of the arthropods, especially the insects, show yet another pattern of cleavage

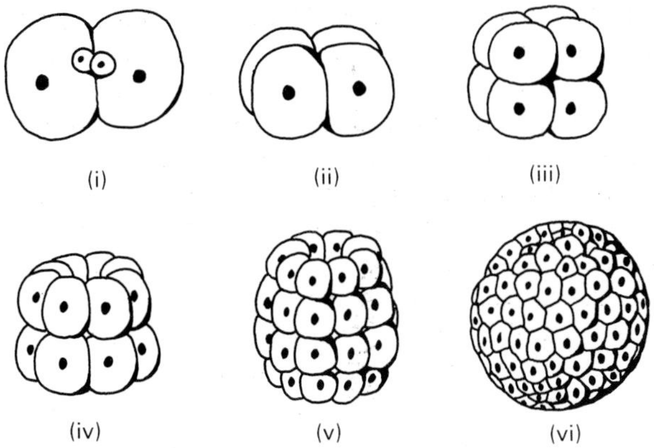

Fig. 15.7. Cleavage in the echinoderm *Synapta*: (i) two cell stage; (ii) four cell stage; (iii) eight cell stage; (iv) 16 cell stage; (v) late cleavage. (vi) Blastula stage.

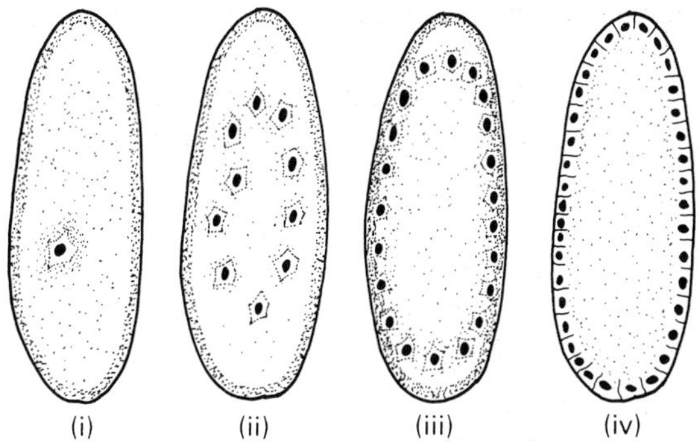

Fig. 15.8. Superficial cleavage of an arthropod (insect) egg. (i) The single nucleus lies in the central yolky part of the egg. (ii) Nuclear division results in the appearance of several nuclei but no cell boundaries. (iii) The nuclei move out to the peripheral ooplasm which is relatively yolk free. (iv) Radial cell walls appear between the nuclei.

described as superficial or endolecithal because of the distribution of the yolk as a central mass. The cytoplasm is restricted to a superficial layer and the initial nuclear divisions are not accompanied by cell division. The nuclei eventually move into the superficial layers and cell boundaries are formed around them (Fig. 15.8).

15.2.2 Gastrulation

With the major exceptions of the Placozoa, Mesozoa, Porifera, and the two coelenterate phyla, the invertebrates are triploblastic animals, that is to say, their bodies are derived from three embryologically distinct cell layers, the ectoderm, mesoderm and endoderm (some, for example the Pogonophora, now have no trace of endoderm in the adult). The embryo which results from early cleavage, the blastula, however, is only composed of a single layer of cells. If the egg is not very yolky the blastula is a hollow ball and the process of *gastrulation* which creates an essentially 3-layered embryo involves invagination of cells. This process can be seen very easily in the transparent eggs of the echinoderms (Fig. 15.9).

Primary mesenchyme cells, derived from the micromeres of the 64 cell embryo (see Box 15.2), invade the primary body cavity, the blastocoel. Then, by a process which involves changes in their relative adhesiveness, cells at the vegetal pole begin to roll inwards, forming a depression. In sea-urchin embryos, this inward movement may be aided by the direct contacts of filopodia formed between the secondary mesenchyme cells (see Fig. 15.9, iv) and the inner surface of the blastocoel. Such filopodia are not essential for gastrulation, however, and no such contacts are made in starfish embryos. It cannot therefore be concluded that pulling forces generated by the filopodia are essential for invagination. The movements are probably the result of forces generated by differential adhesiveness between the cells.

At the end of gastrulation, the embryo has an outer layer of ectoderm, an inner tube opening posteriorly and a number of mesenchymal cells. Such an embryo is called a *gastrula*. At this stage the definitive adult mesoderm has not yet been delaminated from the archenteron. The mesenchyme cells do not form the adult mesoderm but may contribute to skeletal and muscular elements in the larva which are lost at metamorphosis. The formation of the mesoderm is described in Section 15.4.

In deuterostomes the blastopore becomes the anus of the functional larva and often of the adult. The mouth will form as an ectodermal invagination.

Box 15.2 Radial cleavage

A pattern of cleavage which contrasts with that of the annelids and molluscs is seen, for example, in the echinoderms and in modified form in the chordates. In the sea-urchins (Echinoidea) a very precise pattern of cleavage is observed.

1 *The cleavage pattern*. The fertilized egg is bilaterally symmetrical (i). The first two cleavage planes are along the animal/vegetal axis, and they separate two, then four, equal cells (ii and iii).

The third cleavage is transverse and unequal, passing just above the equator (iv). It separates four '*mesomeres*' from slightly larger '*macromeres*'.

The fourth cleavage is different in the upper and lower layers of cells. In the upper half it is longitudinal to give a ring of three cells. In the lower tier it is transverse and very unequal to give four *macromeres* and four *micromeres*. In this example the micromeres are at the vegetal pole (v).

The next two divisions are transverse in the upper half, and alternately longitudinal and transverse in the lower half, to give a 64 cell stage embryo as seen in (vi).

2 *Cell identification*. Because the first four cells are structurally identical, individual cells cannot be identified. Cell layers, however, can be identified. These are conventionally identified as in the following scheme at the 64 cell stage:
Animal 1 layer = an1 = upper 16 cells.
Animal 2 layer = an2 = lower 16 cells of the animal hemisphere.
Vegetal 1 layer = veg1 = ring of eight macromere cells beneath the equator.
Vegetal 2 layer = veg2 = lower ring of eight macromere cells in the vegetal hemisphere.
Micromeres = mic = group of much smaller cells at the vegetal pole.

3 *Blastula*. The cleavage sequence produces a blastula as illustrated in (vii): a hollow ball of ciliated cells.

The initiation of gastrulation is marked by the invagination of the micromeres (viii). Gastrulation is further described in Section 15.2.2 and in Fig. 15.9.

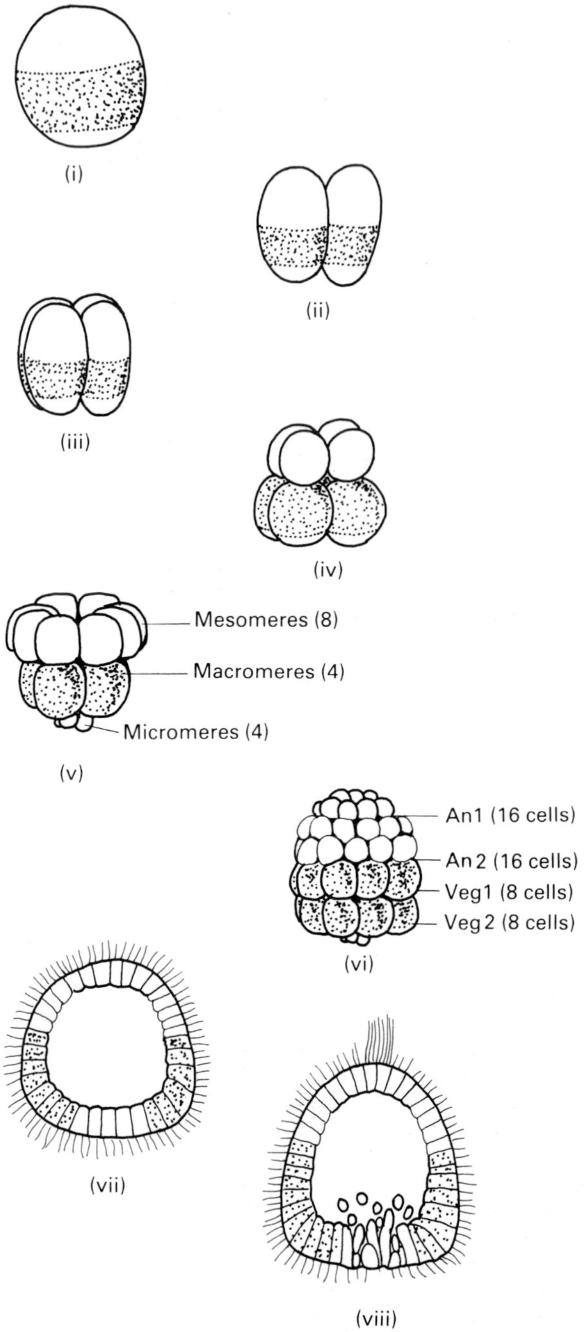

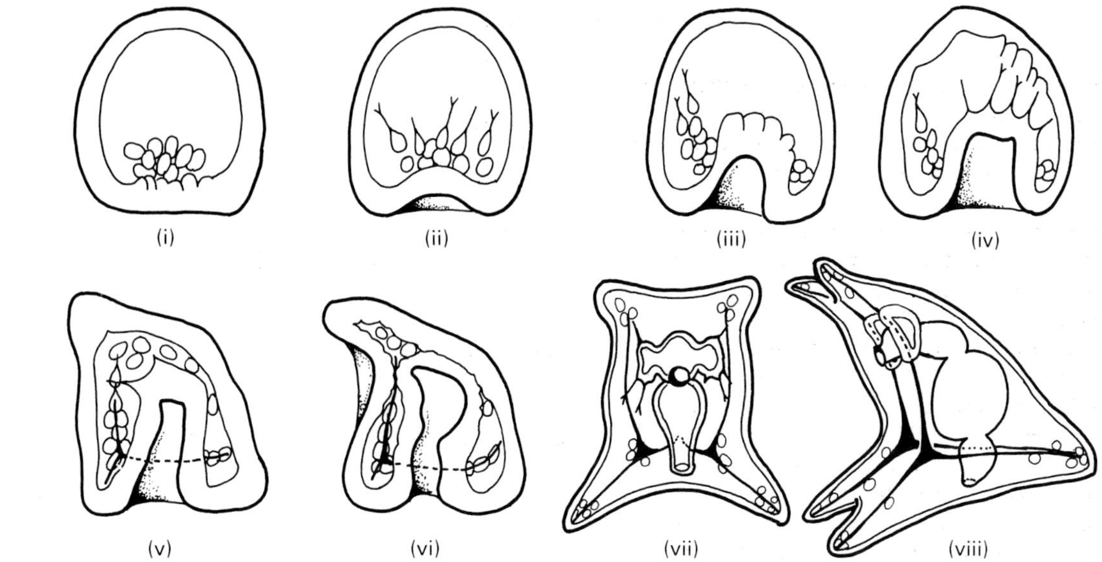

Fig. 15.9. Gastrulation in a sea-urchin. (i) Early gastrulation, invagination of primary mesenchyme. (ii), (iii) Migration of mesenchyme and the initiation of invagination. (iv) Filopodial contacts by secondary mesenchyme. (v) Late gastrula, appearance of skeletal spicules secreted by primary mesenchyme. (vi) Complete gastrula (prism stage embryo). Initiation of the oral field. (vii) Early pluteus seen from the oral surface. (viii) Early pluteus seen from the side. (After Trinkans, 1969.)

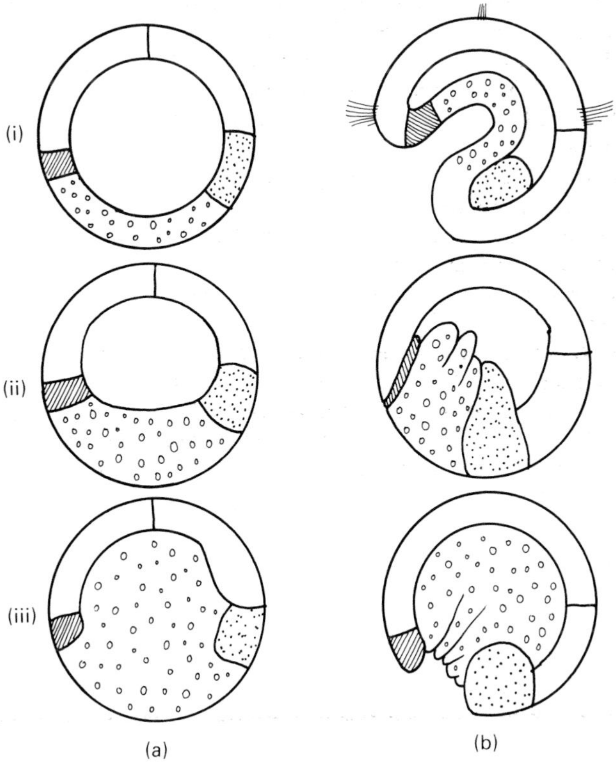

Fig. 15.10. Gastrulation in protostomes (these examples are based on studies of polychaetes with different amounts of yolk). (a) The structure of the blastula; (b) mid-gastrula. The morphologically important cells derived from 4d which form the adult mesoderm are shown with light stipple, the mouth-forming cells are cross-hatched and the gut cells are shown with open stipple. (i) Relatively yolk free form; (ii) moderately yolky form; (iii) yolky form in which gastrulation is by epiboly.

Gastrulation in the protostomes is rather different. It also involves the internalization of the presumptive endoderm and of the small number of cells from which the adult mesoderm will be derived. In forms with relatively yolk-free eggs, gastrulation involves invagination as in Fig. 15.10a, but in more yolky eggs the gastrulation movements involve an amoeboid-like migration of cells over the relatively inert yolk-filled endoderm (Fig. 15.10b,c).

The anterior lip of the blastopore in relatively yolk-free forms includes the cells which form the mouth of the larva, and it is this feature which is considered to be such a fundamental one of early development that several phyla can be grouped together as the protostomes.

The underlying cleavage pattern is frequently obscured in yolky eggs and gastrulation is also less easily observed. The blastocoel is frequently obliterated and the presumptive endoderm cells at the vegetal pole are internalized by the *epiboly* or overgrowth of the animal pole cells, as seen in a leech in Fig. 15.11. In some polychaetes, mouth formation is independent of the blastopore even though such animals would still be classified as protostomes.

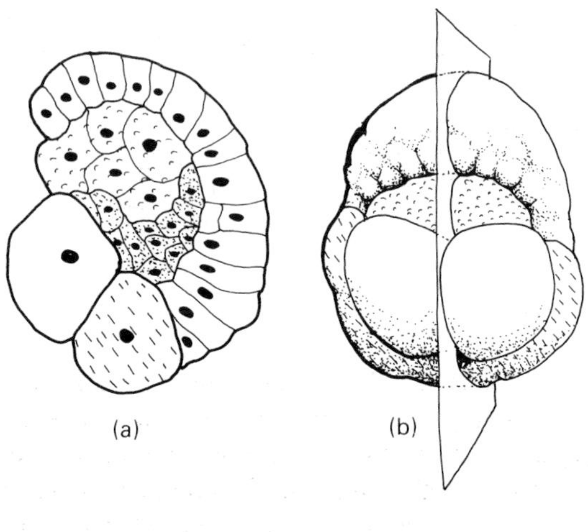

Fig. 15.11. Gastrulation in the yolky egg of a leech: (a) As seen in transverse section from the left-hand side; (b) in stereoscopic view from the vegetal pole showing the plane of the section.

Gastrulation is therefore a process which involves cell movement and internalization of cells to give the typically tribloblastic condition. It is possible to construct a fate map of the blastula in which the position of the future internal tissues is identified on its surface.

Figure 15.12 shows the fate map of a polychaete blastula and traces the cell movements which create the trochophore embryo. Note that cells derived from the cleavage products 2d and 4d have a very important role in the embryology of protostomes. Cells derived from 2d give rise to the adult ectoderm and from 4d give rise to the paired mesentoblast cells from which the future mesoderm is derived. Figure 15.12 shows the relative position of these cells prior to gastrulation and indicates the cell movements which cause invagination of the presumptive endoderm, and bring the mesentoblasts and ectoteloblast cells to their definitive position in the trochophore larva.

In the annelid larva illustrated in Fig. 15.12, all the segmented structures will derive from the products of the paired mesoderm bands and the ectoteloblast ring. Structures derived from other parts of the trochophore larva are the prostomium and the pygidium (see Section 4.14), and these can be considered asegmental.

A rather different mode of gastrulation is seen in the arthropods. In insects, the 'blastopore' is a furrow through which a ventral strip of prospective mesoderm cells destined to form body musculature, gut wall, gonads, etc. is invaginated. The sequence of events in a generalized insect is illustrated in Fig. 15.13, as seen in transverse section and in ventral view.

The interior of this embryo is not a hollow space but an enclosed mass of acellular yolk. The ganglionic cells of the segmented nervous system also invaginate at this time.

The insect embryo is invested by 'extra-embryonic membranes' and interesting comparisons can be made between the development patterns exhibited by the 'higher' and 'lower' invertebrates and the 'higher' and 'lower' vertebrates.

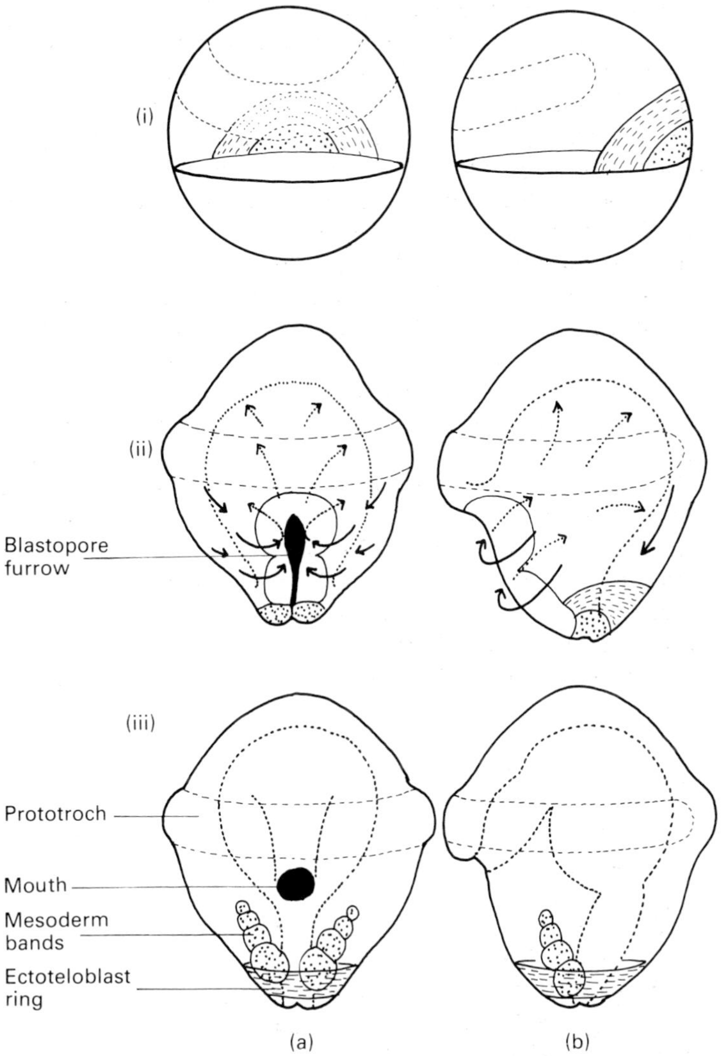

Fig. 15.12. Gastrulation in the protostomes, e.g. Polychaeta. (a) Ventral view. (b) Left-hand view. (i) Diagrammatic representation of a simple fate map. (ii) Gastrulation movements: heavy lines show the path of cells to the blastopore, broken lines the migration of invaginated cells. (Note that the stomodaeum or mouth is created by invagination of cells through the anterior blastopore. The mesoderm bands are derived from paired mesentoblast cells migrating through the posterior margin of the blastopores.) (iii) The definitive trochophore larva with paired mesoderm bands and ectoteloblast ring. These two elements form a segment blastema which remains at the anterior face of the pygidium through the adult life. The prostomium, peristomium and pygidium are therefore asegmental structures. (Partly after Anderson, 1964.)

15.3 Experimental embryology of invertebrates: the determination of cell fate

15.3.1 Introduction

A complex multicellular animal is derived from a single undifferentiated totipotent structure—the fertilized egg. During its development it proceeds to a state where it is composed of a larger number of cells, groups of which have different morphologies, chemical composition and function. Experimental embryology is the study of how populations of differentiated cells with specific functions arise in developing embryos, and some of the most favourable model systems for the analysis of the processes involved have been provided by the embryos and larvae of invertebrates.

During early embryonic development, maternal

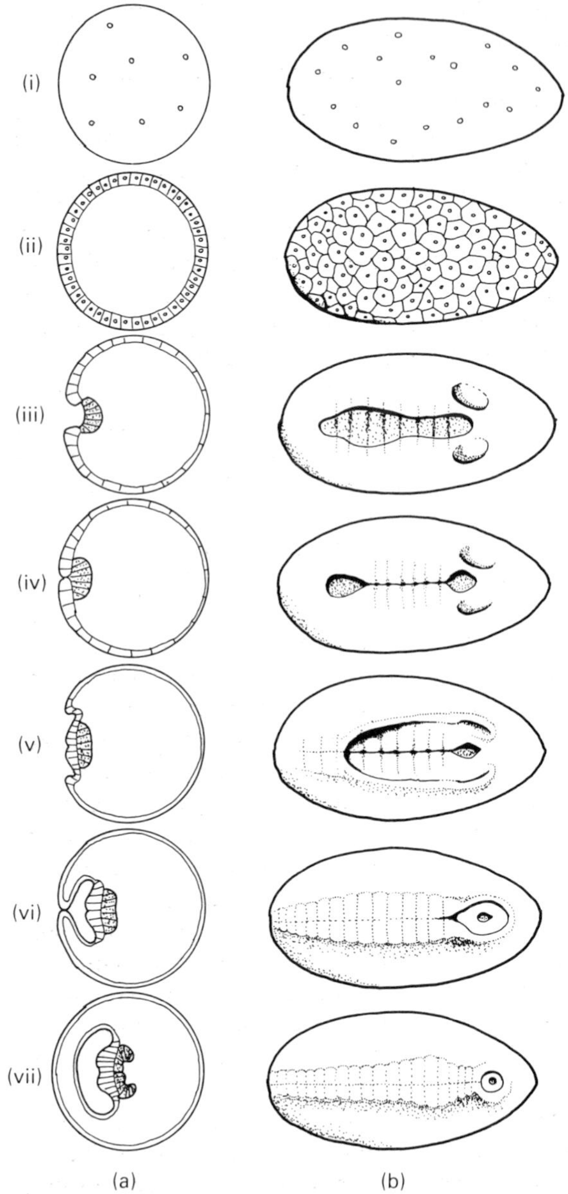

Fig. 15.13. Gastrulation in insects. (a) In cross-section. (b) Ventral view. (i) Early nucleus proliferation; (ii) superficial cleavage; (ii), (iv) invagination of mesoderm; (v), (vi) segmentation of the neuroectoderm, formation of extra-embryonic membranes; (vii) the completed gastrula. (After Slack, 1983.)

genes which have been stored in the oocyte cytoplasm are used to support the activities of the developing embryo.

Eventually, however, the zygote's own novel genome is called into play, and since many different types of cell eventually develop, there must be control over the information contained in the zygote nucleus. The potential control points are many and include each or several of the steps in the following sequence:

1 Genetic information in the zygote nucleus.
2 Allocation of information to daughter cells.
3 Transcription of genetic information.
4 Population of mRNA molecules.
5 Export of mRNA from nuclei to cytoplasm.
6 Population of informational molecules in the daughter cell cytoplasm.
7 Translation of information in the mRNA molecules.
8 Formation of cell-specific proteins.
9 Function of specific substances.

It is now generally accepted that in the majority of organisms the nuclei of differentiated cells contain the same genetic information as the nucleus of the fertilized egg. The evidence is based on the results of:

1 Experimental transplantation of nuclei from differentiated cells into enucleated egg cytoplasm;
2 Studies of the cytology of chromosomes during early cleavage;
3 Comparison of the banding patterns of the giant chromosomes found in several Diptera and other insects;
4 Studies of dedifferentiation and redifferentiation during regeneration (see also Section 15.5);
5 Analysis of DNA 'fingerprints' from different tissues.

There are, however, exceptions to this general rule. The primordial germ cells of the nematode *Ascaris* are, for instance, the only cells of the embryo to receive a full complement of chromosomes. In cells of other lineages chromosomes are lost during early cleavage as explained in Fig. 15.14.

If all the nuclei of the developing embryo have the same genetic constitution then differentiation must be an expression of an interaction between cytoplasmic information in the cell and its nucleus.

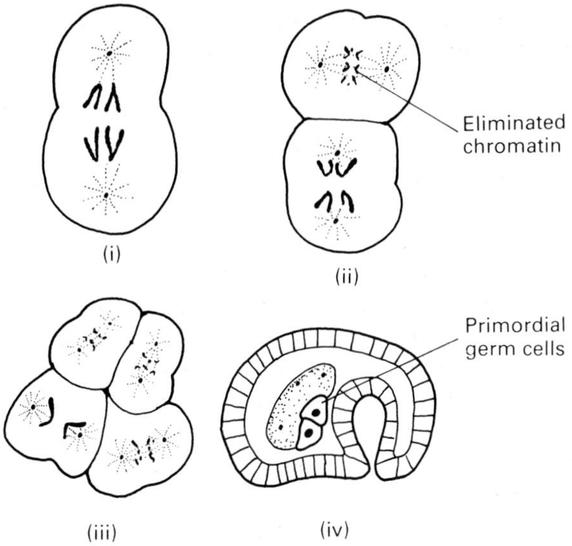

Fig. 15.14. An example of chromosome diminution during early embryonic development (*Ascaris*): (i) first cleavage, no chromosome loss; (ii) elimination of chromosome material during cleavage of the anterior cell; (iii) elimination of chromosome material and rearrangement of cells at the four cell stage. Only one cell retains the complete set of chromosomes and is therefore totipotent. (iv) The germ cell cytoplasm and cells with the full chromosomal complement can be traced to the two primordial germ cells of the gastrula.

15.3.2 Mosaic versus regulative development

By the early part of this century experimental embryologists had investigated the ability of isolated cells of invertebrate embryos to develop normally. In some cases, as in the sea-urchins, each one of the blastomeres at the four cell stage would, in isolation, give rise to a complete and perfectly normal pluteus larva (see Box 15.3, **1**). Such an embryo was said to be capable of regulation. In other cases the embryos developing from isolated blastomeres at the four cell stage were very deficient (see Box 15.3, **2** and 3). They were said to be mosaic embryos. It appeared that some cells inherited a particular type of cytoplasm and that this cytoplasm determined their developmental fate. Further analysis of these results requires knowledge of the normal developmental fate of individual cells, as expressed in a fate map of the uncleaved cytoplasm of the egg or of the blastula. This can be painstakingly built up by marking specific regions of cytoplasm and tracing the fate of the marked cells. A natural fate map of the ascidian *Styela partita*, however, was established by careful observation of the egg after fertilization (see Fig. 15.6). After fertilization the egg becomes visibly bilaterally symmetrical due to the movements of the cytoplasm. Prominent among its regions is a crescent of yellowish cytoplasm. This can be traced through the early cleavage stages to the muscle blocks of the tadpole larva. All the presumptive muscle cells inherit some of the yellow cytoplasm and none of the cells which do not become muscle cells receive any of it. The factors determining cell fate in invertebrates are discussed in more detail in Box 15.4.

These experiments with ascidians provide compelling evidence for the importance of localized cytoplasmic substances in the determination of cell fate. Experiments with insect eggs also show that the developmental fate of the nuclei can be determined by the cytoplasm into which they move. The cleavage pattern is described in Fig. 15.8. There is often a distinct cytoplasm at the posterior pole of the egg—the pole plasm. Only nuclei which become incorporated into cells with pole plasm have the capacity to form germ cells (i.e. remain totipotent) and embryos from which the pole plasm has been removed are sterile. Pole plasm of a genetically identified stock of *Drosophila*, when injected into the anterior part of an egg of a different strain of fly, causes the cells which form in this anterior region also to become germ cells.

Development however, is not to be understood as being solely the result of the pattern of distribution of cytoplasmic determinants. It also involves interactions between cells. Detailed experiments with the mud snail *Nassarius obsoletus* (often referred to by the generic name *Ilyanassa*) reveal how cytoplasmic determinants of cell fate and intercellular influences interact during development. They are explained briefly below and in more detail in Box 15.5.

Box 15.3 Regulative and mosaic development I

In the late nineteenth and early twentieth centuries, it became possible to study the developmental capacities of individual cells isolated from early cleavage stages of the embryos of marine invertebrates at the two or four cell stage. The results could be strikingly different.

1 *Echinoderm embryos*. Isolated cells of the four cell stage can each give rise to a normal (but small) pluteus. The small embryos were said to be regulative.

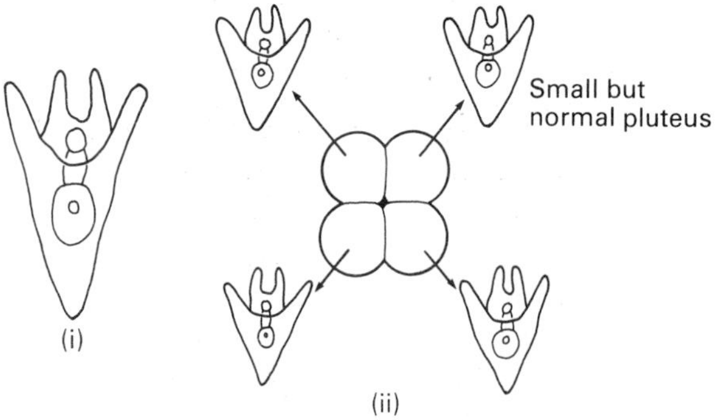

(i) The normal pluteus.
(ii) The results of cell isolation.
One isolated blastomere of the four cell stage divided to give two mesomeres, one macromere and one micromere, i.e. it follows the cleavage pattern of a typical embryo (see Box 15.2).

2 *Mollusc embryos*. Isolated cells at the two and four cell stage give rise to unbalanced or deficient embryos. Embryos developing from isolated D cells usually give rise to more normal embryos than those derived from A, B or C cells.

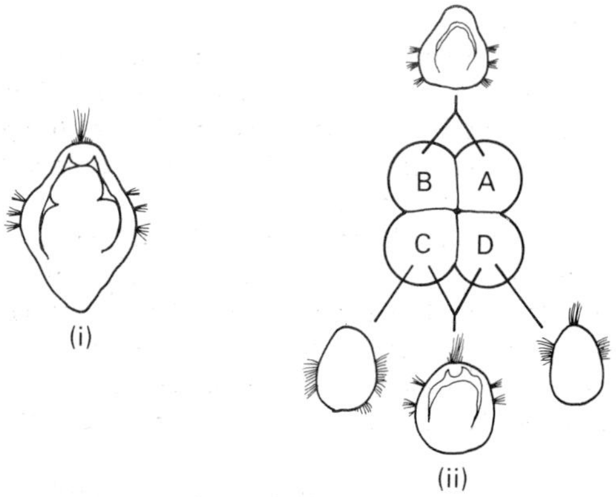

(i) A normal trochophore of *Patella*.
(ii) Abnormal trochophores from isolated cells as indicated.

3 *Ascidian embryos*. The eggs of ascidians often have visibly different regions of cytoplasm (see Fig. 15.6); these are segregated to different cells.

Embryos developing from isolated blastomeres are very different and their development reflects the structures they would have given rise to in an intact embryo.

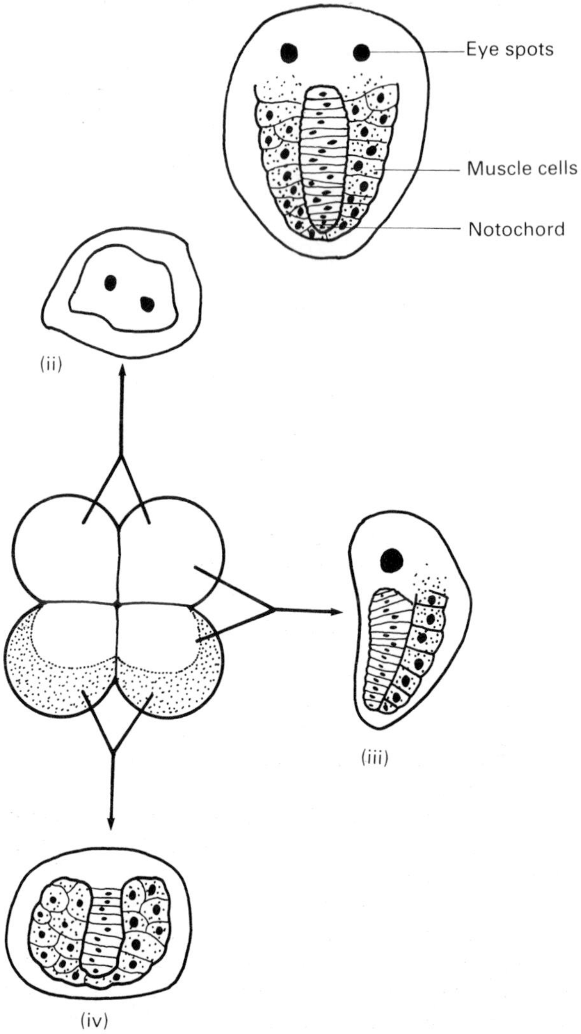

(i) The normal 'tadpole' embryo with paired eyes, notochord and paired muscle blocks;
(ii) Isolated anterior cells;
(iii) Isolated right-hand cells;
(iv) Isolated posterior cells.

Box 15.4 Cell fate, fate maps and cytoplasmic localization

1 *The principle of a fate map.* Marks can be placed on the surface of the uncleaved egg. If there is no mixing of the cells formed by cleavage, the genetic marks will subsequently appear in specific structures of the embryo.

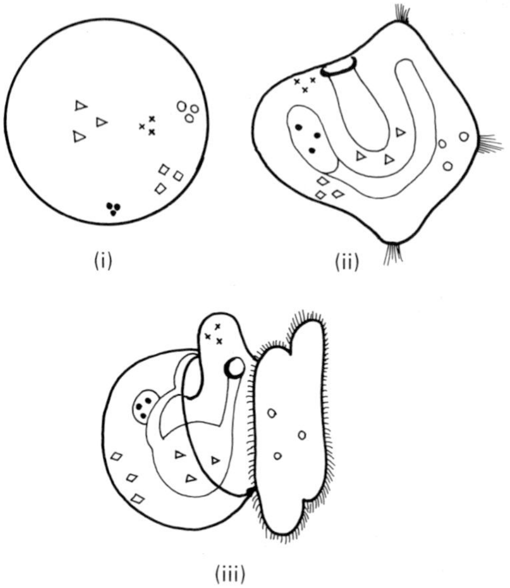

△△ presumptive gut
□□ presumptive mantle
•• presumptive operculum
xXx presumptive foot

A hypothetical example for a mollusc embryo.
(i) Marks on the surface of the egg;
(ii) The position of such marks during gastrulation;
(iii) The segregation of the marked cells to specific regions of the embryo.

2 *Natural markers.* After fertilization some embryos have clearly different regions of cytoplasm and bilateral symmetry. The figure shows how the different cytoplasmic regions of an ascidian egg (i) are segregated into the cells of the vegetal pole, (ii); during gastrulation these cells are invaginated, (iii) and (iv). After gastrulation, cells containing the visibility different cytoplasm will have given rise to different embryonic structures, (v) and (vi). The question arises in this case, 'do the natural cytoplasmic markers determine the fate of the cells?'

3 *In the ascidian, cytoplasmic factors do determine cell fate.* A region of yellowish cytoplasm acts as a natural marker for the muscle cells of the tadpole larva (Box 15.3, **3**).

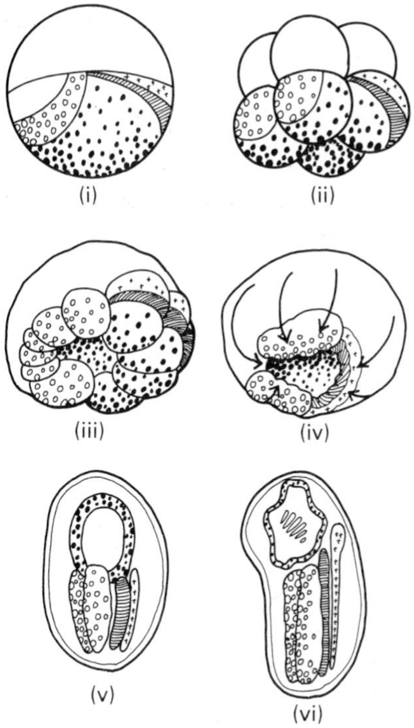

Gastrulation and cytoplasmic localization in ascidians.

4 The enzyme *acetylcholinesterase* is specific to the developing muscle cells and can be stained in the cytoplasm.

The following experimental results have been observed:
(a) The enzyme normally appears after 8 hours of development;
(b) The enzyme appears at the appropriate time even if cell division has been inhibited by the application of drugs;
(c) The appearance of the enzyme is inhibited by the application of Actinomycin D up to 5 hours or the application of Puromycin up to 7 hours;
(d) Compression of the egg during cleavage can lead to yellow cytoplasmic allocation to cells which *do not normally receive it*. The cells then show acetylcholinesterase activity.
These experiments suggest that the genes which code for the enzyme acetylcholinesterase are transcribed only in cells which inherit some factors normally associated with the yellow cytoplasm.

The timing of the gene transcription and translation is under the control of a cellular clock which operates independently of cell division.

Box 15.5 Regulative and mosaic development II

Experimental studies of the marine snail, *Nassarius obsoletus*.

1 Normal development of *Nassarius*. This snail exhibits spiral cleavage with a prominent polar lobe at the 1st and 2nd divisions. This ensures that the D cell (see Box 15.1) inherits a special pole plasm. The cleavage sequence is illustrated below.

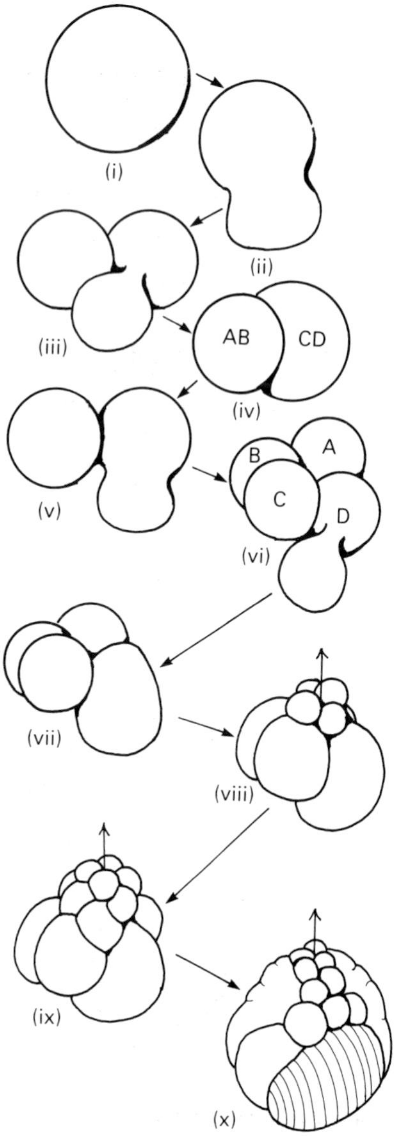

(i) Uncleaved egg;
(ii) Appearance of 1st polar lobe prior to 1st division;
(iii) 'Trefoil' appearance at 1st cleavage;
(iv) Two cell stage;
(v), (vi) Appearance of 2nd polar lobe during 2nd cleavage;
(vii) Four cell stage (note large D cell).
(viii) Eight cell stage shortly after the appearance of the 1st quartet of micromeres.
(ix), (x) Later cleavage stage.

The eight cell embryo is markedly asymmetrical and the D-cell lineage has its own rhythm of cell division which is different to that of the other cell lineages.

2 The special role of the D cells in *Nassarius* development.

Cell lineage studies show that specific larval structures develop from the products of individual cells which can be identified and named during early cleavage.

The cell lineage diagram shows tentative assignments for larval structures at the 29 cell stage using the nomenclature presented in Box 15.1.

Note that cells 2d and 4d have particularly important roles in the formation of the veliger. Cell 4d is called the mesentoblast cell.

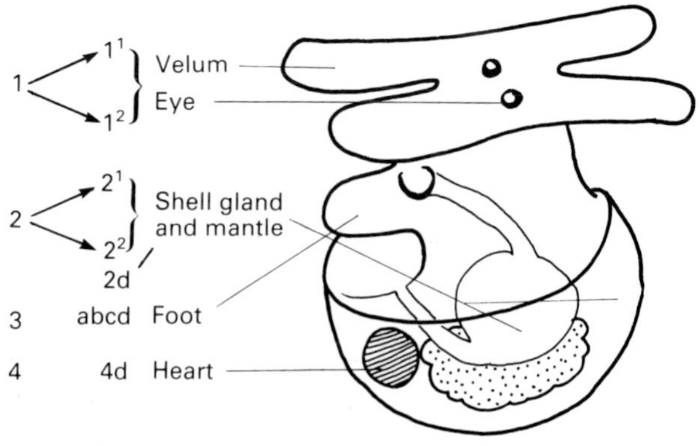

The organization of the veliger larva.

Box 15.5 continued

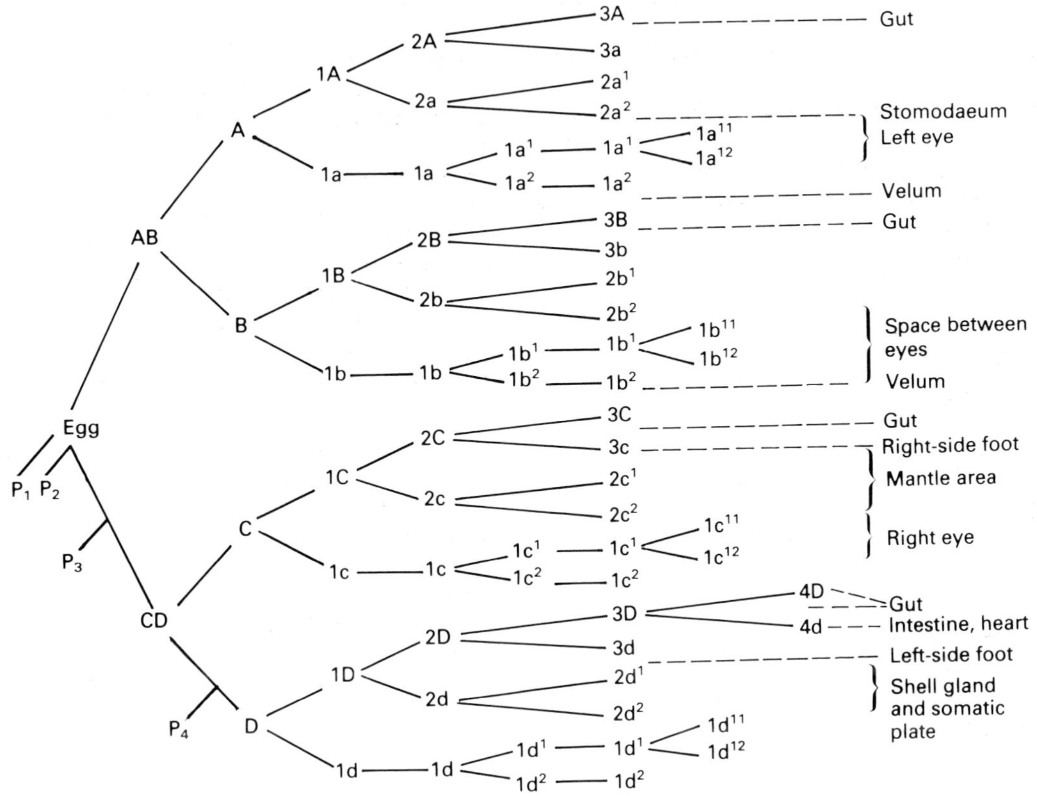

Diagram to show cell lineage.

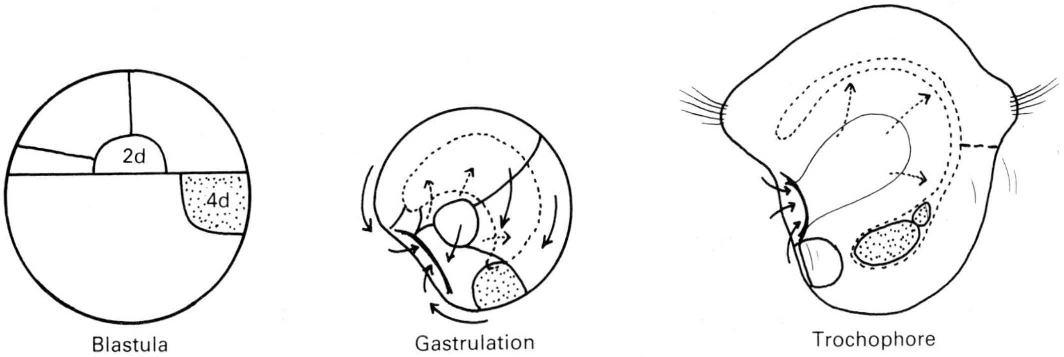

Figure showing the movement of cells 2d and 4d to their special positions during gastrulation.

Box 15.5 continued

3 The organizing role of the pole plasm. The polar lobe material can be removed during early cleavage by microsurgery. The results are dramatic.
The lobeless larvae lack all structural organization but show some signs of cell differentiation. Two typical examples of lobeless larvae are illustrated below.

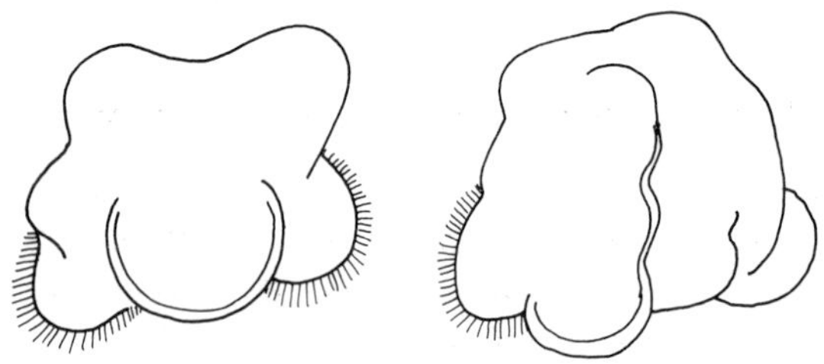

In a classic series of experiments, A.C. Clement tested the effects of removal of the D cell progressively during development. The results are summarized in the table below.

4 Interpretation: evidence of inductive interrelationships. Refer to the cell lineage diagram and the table below; note that: eyes form from derivatives of cell 1a and 1c, and shell gland from derivatives of cell 2d. Cells 1a and 1c are present in all embryos of the above series, but eyes only form in experiments (iv) and (v). Why? Answer: eyes only form in the presence of D-lineage cells at least to the formation of 3d.

The shell gland is only normal if 2d is in the presence of D-lineage cells at least until the formation of 3d.

Evidence of mosaic development: eyes normally form from 1a and 1c, and destruction of 1a or 1c results in the failure to develop eyes.

Conclusion: only derivatives of 1a and 1c are competent to form eyes, but these cells require the inductive influence of the D-cell lineage in order to become determined in that fate.

Heart normally forms from cell 4d, and only cell 4d is ever competent to produce the differentiated heart (see text for further discussion).

Summary of the effects of progressive ablation of the D-cell lineage in *Nassarius* (from data of A.C. Clement, 1962).

	Operation	Embryo	Defects found
(1)	D cell destroyed (see Box 15.1)	ABC	as lobeless embryos lacks: intestine, heart, shell, foot, statocysts, eyes
(2)	1D cell destroyed	ABC+1d	as for ABC
(3)	2D cell destroyed	ABC+1d+2d	as for ABC
(4)	3D cell destroyed	ABC+1d+2d+3d	shell variable, lacks intestine, lacks heart
(5)	4D cell destroyed	ABC+1d+2d+3d+4d	None

Nassarius obsoletus exhibits spiral cleavage (see Box 15.1) in which a prominent polar lobe appears at the 1st and 2nd cell divisions. The special pole plasm is allocated specifically to the D cell. Embryos from which the polar lobe is removed are symmetrical, the special timing of cleavage in the D-cell lineage is disturbed and the embryos develop into very deficient 'lobeless' larvae (see Box 15.5). 'Lobeless' larvae lack eyes, statocysts, a foot, velum, shell, heart and organized intestine.

It is possible, however, to recognize some products of cell differentiation. The cell lineage chart in Box 15.1 reveals the significance of these observations. Many of the lobe-dependent structures are not derived from the cells which inherit the polar lobe material. It is clear that the developmental fate of the cells derived from la and lc which form the eyes, or 2d which forms the shell, is determined not only by their cytoplasmic inheritance but also by their interactions with cells of the D-cell lineage during the critical period from the appearance of cell 2d to the formation of 4d. Similarly the only cells which are *competent* to give rise to the shell gland are derived from the single cell 2d, but they will only do so if cell 2d has been influenced by contacts with other cells in the D lineage.

In sea-urchin embryos the developmental fate of the different layers of cells is to an even greater extent determined by their relative position in the embryo. Their developmental fate is not fixed by their cytoplasmic inheritance. The principal components of the sea-urchin larva can be traced to the cell layers of the late cleavage stage embryo according to the following scheme:

an1 cells	apical ciliary tuft
an2 cells	oral field and stomodaeum
veg1 cells	ectoderm
veg2 cells	archenteron and coelomic pouches
micromeres	primary and secondary mesenchyme

This relationship is explained further in Box 15.6. The transverse divisions clearly separate material of different developmental potential, but the embryo does not develop as a simple mosaic of parts.

It seems that each cell layer develops according to its position relative to two gradients in the embryo, one declining from the animal towards the vegetal pole and another declining from the vegetal to the animal pole. Some of the critical experiments which lead to this conceptual framework are summarized in Box 15.6.

These differences between mosaic and regulative embryos seem to relate to the time when cells with specifically different cytoplasmic determinants are first segregated.

15.4 Larval development and metamorphosis

The life history of many animals is characterized by divergence of niches between the juvenile stage and the adults. This is especially so in many groups of free-living marine invertebrates (see Chapter 14), many parasitic species and in some insects. This imposes a requirement for a more or less sudden transition from the mode of life to which the juveniles are adapted to that of the adult. In many cases this is accompanied by a pronounced change in morphology which is described as metamorphosis.

15.4.1 Metamorphosis of marine larval forms

The ciliated larvae of marine invertebrates are adapted for a pelagic life and the ciliated bands and girdles which provide their locomotory power are not adequate for the larger adults (see Chapter 10). Metamorphosis in these animals therefore frequently involves the loss of these ciliated girdles and a transition to a mode of life in which muscle cells provide the locomotory forces.

In gastropods, metamorphosis of the veliger larva is accomplished progressively with a gradual reduction of the velum, which eventually becomes unable to support the developing snail and is replaced by the foot as the chief locomotory organ (see Fig. 15.15), but metamorphosis of the bivalve molluscs is often more rapid, with a sudden loss of the velum.

The conflicting demands of larval locomotion and the requirement for development towards the adult state are also illustrated by the polychaete annelids, in which segments are added progressively during em-

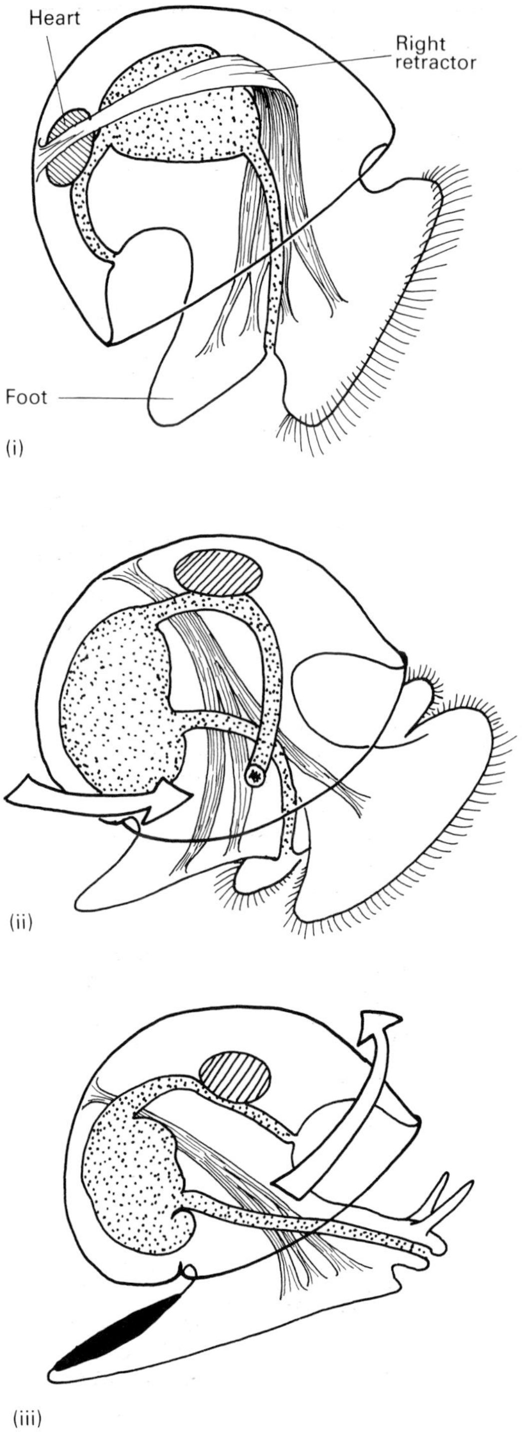

bryonic or larval life (see Section 15.2.2). The segments are derived from paired segment blastemas: ventral bands in the posterior region of the larva. The mesoderm cells in the blastema are derived from paired mesentoblasts formed from the cell 4d during typical spiral cleavage (see Section 15.2.1 and Box 15.1).

The segment blastema has mesodermal and ectodermal elements. The two bands of mesoderm-producing cells proliferate a series of blocks of tissue in which the coelom is formed; in these animals, unlike the echinoderms described below, the mesoderm is not derived from the cavity of the archenteron. This type of coelom is termed a schizocoel. The segments are produced by coordinated organogenesis of the mesodermal and ectodermal tissues in which the ventral nerve cord plays a major inductive and organizational role. As each newly proliferated segment develops in front of the pygidium, it increases the mass of the larva. In most cases each segment is provided with locomotory or flotation devices during the pelagic phase.

The development of the Echinodermata involves one of the most dramatic forms of metamorphosis in the animal kingdom. A fully developed echinoderm larva is bilaterally symmetrical; the dominant symmetry of the adult echinoderm, however, is pentaradial, although there is sometimes a secondary bilateral symmetry imposed on this (see Section 7.3.2). The coelomic pouches of echinoderms are derived from lateral out-pushings of the tip of the archenteron some time after the completion of gastrulation (see Fig. 15.9). Their formation is therefore

Fig. 15.15. Torsion during the development of a mollusc larva (e.g. veliger of a gastropod). At the start of torsion, the mantle cavity is posterior. When torsion has been completed it has an anterior position, and the topology of the adult snail has been achieved. (i) The early veliger with the asymmetrical right retractor muscle. (ii) The first part of torsion which is a fast 90° twist due to operation of the retractor muscle. This causes a partial rotation of the visceral mass about the foot. (iii) The second part of torsion, a further 90° twist of the visceral mass about the foot due to differential growth.

Box 15.6 Experimental analysis of sea-urchin development

1 *Cell layers*. Cleavage of the sea-urchin egg is described in Box 15.2. The principal layers of the 64 cell embryos are:
An1—16 cells
An2—16 cells
Veg1—eight cells
Veg2—eight cells
micromeres.

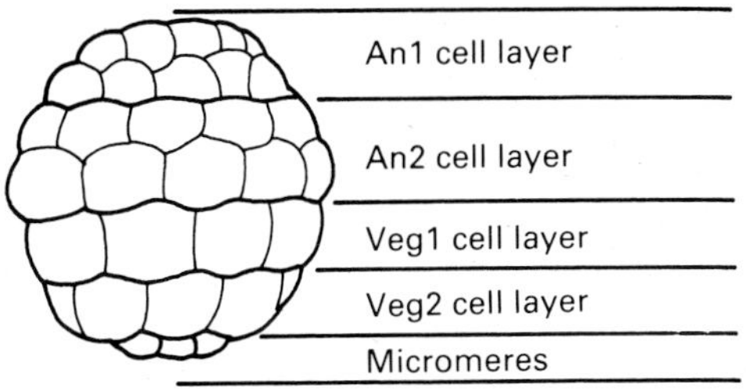

Relationship of the cell layers to the gastrula and early pluteus.

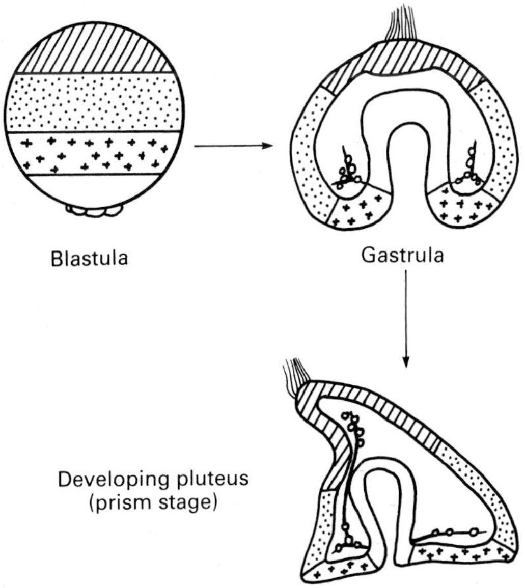

2 *The development of isolated half-embryos*. Isolated animal half-embryos give rise to permanent blastulas with an over-developed apical tuft. Isolated vegetal half-embryos give rise to more or less normal embryos with over-development of the gut. They are said to be vegetalized.

These observations could be compatible with self-differentiation of the primary cell layers, a 'mosaic' theory of development, but further experiments show this to be incorrect.

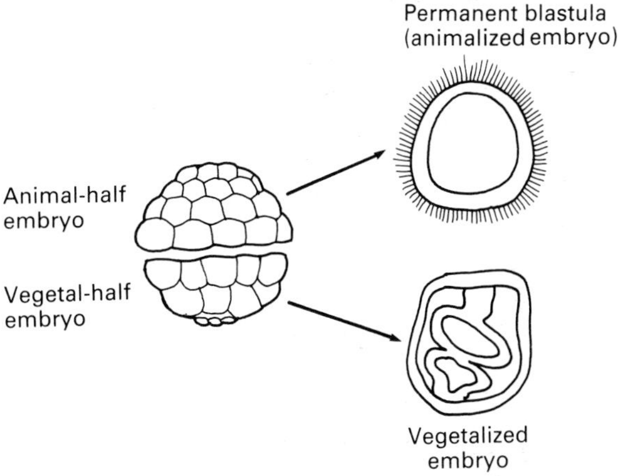

3 *Evidence for the gradient theory of development*.
An isolated animal half-embryo will develop into a normal pluteus if combined with four micromeres.
(i) Subdivision of the embryo.
(ii) Combination of animal hemisphere with 4 micromeres.
(iii) Gastrulation.
(iv) A normal pluteus.

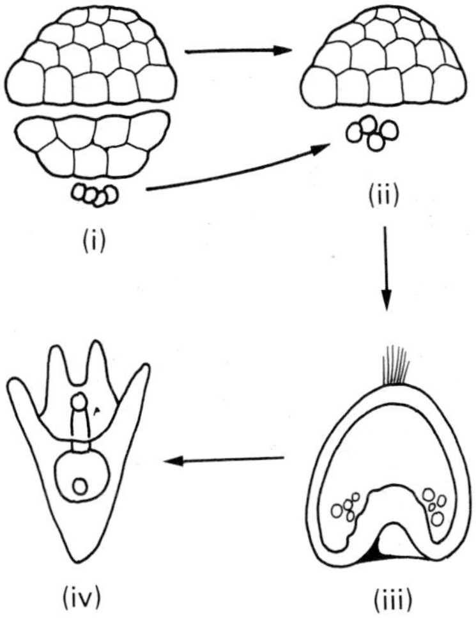

Box 15.6 continued

In the experiment described in **3** the four micromeres could be said to 'induce' archenteron-forming capability in the cells of the *an–2* layer. Cells of the *an–2* layer are competent to differentiate as archenteron, although they would not normally do so.

4 A large number of experiments have been performed on the development of isolated layers of the embryo in isolation and in combination.

The results of one such series is summarized below: Normal embryos can be produced by 'balancing' animal and vegetal tendencies.

an1 + 4 micromeres give a normal pluteus
an2 + 2 micromeres give a normal pluteus
veg1 + 1 micromere give a vegetalized embryo

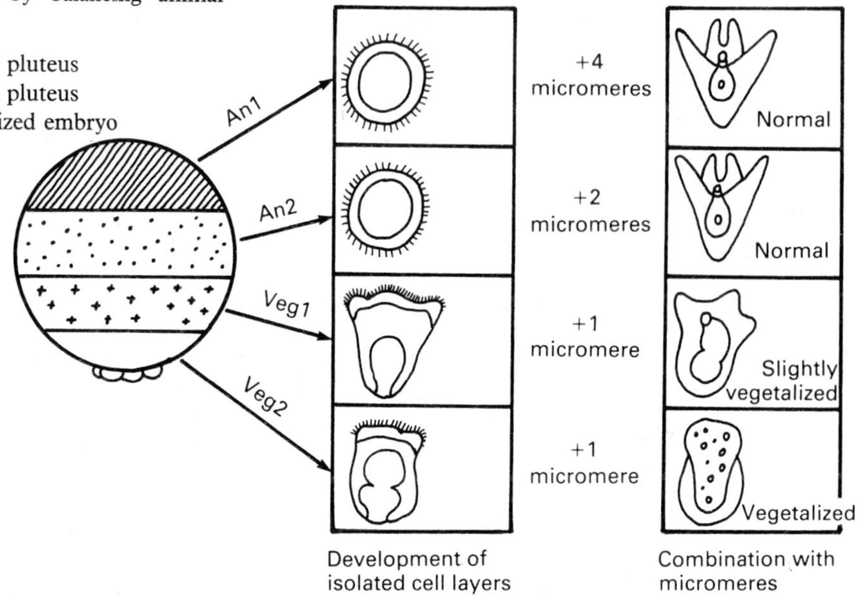

5 As shown, these results can be interpreted as the expression of a two-gradient system. It is supposed that in the normal embryo the relative position of a cell layer can be identified in relation to its position in the two gradients.

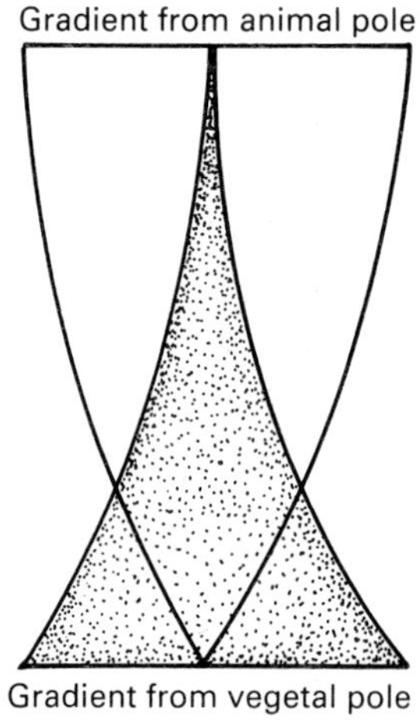

The gradient theory of development. Normal development requires a 'balance' between the two gradients

Box 15.6 continued

6 *Visualization of the gradient system*. Animalizing and vegetalizing agents. Some substances cause the development of abnormal animalized or vegetalized embryos like those developing from isolated animal or vegetal hemispheres.

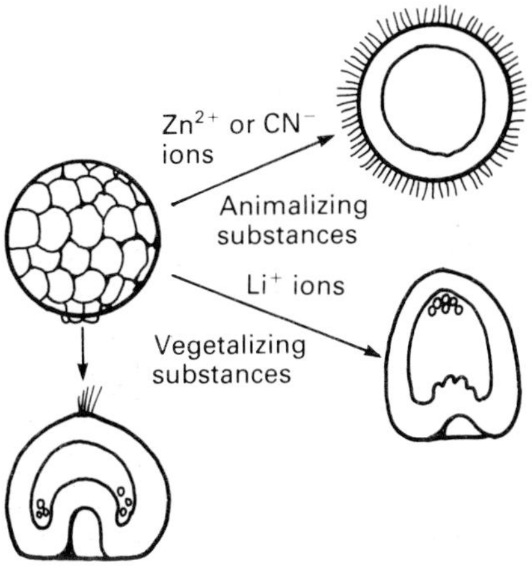

Li^+ ions have a powerful vegetalizing effect.
Zn^{2+} and CN^- ions cause an animalizing effect.

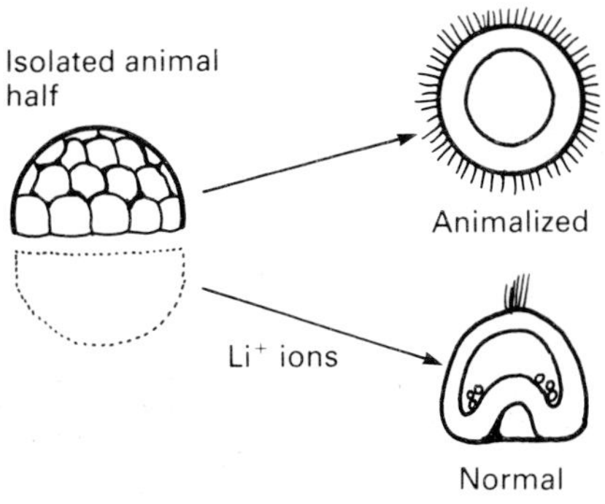

Li^+ ions can cause an isolated animal hemisphere to gastrulate normally, however.

7 Visualization of gradients of metabolic activity. Direct evidence for gradients of metabolic activity at the poles of the early gastrula is obtained by staining with the vital dye Janus green. This dye becomes pink, then colourless.

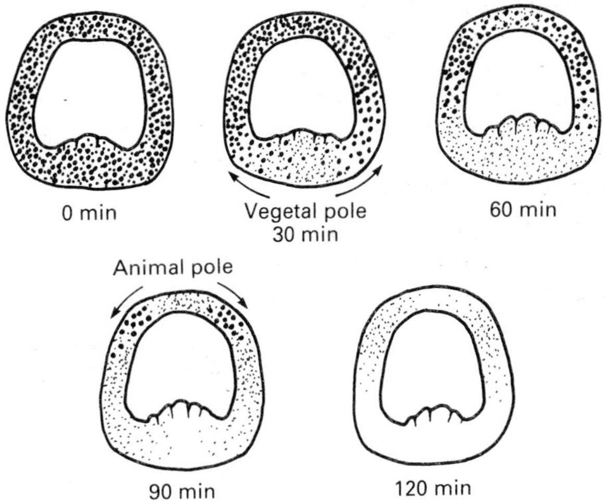

The course of reduction of the dye in normal embryos. Note the centres of reduction radiating from animal and vegetal poles.
Concurrence of evidence. Li^+ ions cause animal hemispheres to develop normally.

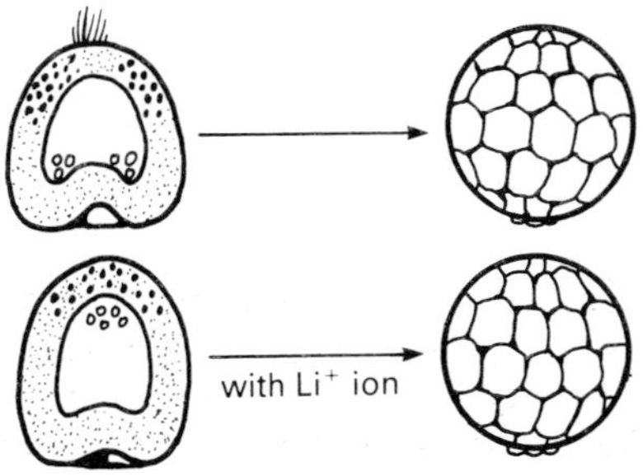

Li^+ ions suppress the animal pole centre of Janus green acidification.
8 The gradient system appears to have a physicochemical reality but the molecular basis of this system is not yet understood.

enterocoelic. A study of living echinoderms suggests that there were primitively three paired coelomic pouches: the axocoel, hydrocoel and somatocoel. The development of these coelomic spaces and the subsequent metamorphosis is illustrated in Fig. 15.16. In most living echinoderms the right-hand axocoel and hydrocoel are reduced or totally suppressed. The left hydrocoel becomes subdivided into the hydrocoel and an outgrowth forming the stone canal and the hydropore (Fig. 15.16, i–iv).

These primitive coelom anlagen are represented in a nine-day pluteus larva in Fig. 15.16,v. The left- and right-hand somatocoels spread over the stomach and form the body cavities of the adult. The left hydrocoel creates a pentaradial fluid-filled anlagen from which the water vascular system of the adult develops. The mouth of the future adult will form in the centre of the hydrocoel and this establishes the oral surface of the urchin. The oral/aboral axis of the adult is therefore approximately along the left/right axis of the pluteus larva (Fig. 15.16,vi). The developing sea-urchin or starfish appears as an imaginal disc which is unfurled at metamorphosis when the ectodermal epithelium of the larva shrinks and is discarded (Fig. 15.16,vii, viii).

Animals with an exoskeleton cannot grow and develop in a progressive manner but must proceed through a series of moults. At each moult the old skeleton is discarded and the body expands prior to hardening of the new external covering. This is well illustrated by the Crustacea. Their development frequently involves a series of morphologically distinct larval stages which finally moult to give the adult form (see Box 14.5 and Chapter 8). In the Crustacea growth will often continue through a sequence of moults after the adult morphology has been reached. In the winged insects, however, there is no further growth once the definitive adult stage has been reached.

In marine invertebrates the larvae are responsible for establishing the young adult in a suitable environment for its subsequent development. This is an important stage in the life history of the animal and for organisms which, as adults, are sedentary or sessile, the choice of a suitable substratum may be critical.

Marine biologists sometimes observe clouds of newly metamorphosed larvae settling apparently at random into substrata in which they will not survive; but this is not always the case. There are many studies which have revealed the precision with which pelagic larvae are able to 'choose' a suitable substratum. The processes involved include:

1 Behavioural sequences which bring the larva into contact with a suitable substratum;

2 Delayed metamorphosis in the absence of suitable substrata;

3 Discrimination and selection of a preferred substratum; and

4 Gregarious behaviour and the chemosensory detection of surfaces previously inhabited by adults or larvae of their own kind.

The larvae of the marine mollusc *Mytilus edulis* for instance exhibit a complex sequence of behavioural changes during their development.

The larvae of many species have been shown to be highly selective of the substrata on to which they will settle. Box 15.7 shows how multiple choice experiments reveal differences in the choice of settlement surface by larvae of closely related species of the tube-living *Spirorbis* worms, the adults of which are characteristically found on different substrata.

Fig. 15.16. (*Facing page.*) Development of the coelomic system of a sea-urchin and metamorphosis to the adult form. (i) Formation of the coelomic pouches from outpushings at the tip of the archenteron after the completion of gastrulation. (ii) The paired coelomic pouches. (iii), (iv) Subdivision of the pouches. (v) Differentiation of the 2nd left coelom as hydrocoel and stone canal/hydropore complex, and expansion of the 3rd left and right coelomic pouches. This stage is reached in *Psammechinus* after about 9 days of planktonic development. (vi) Organization of the pentaradial water vascular ring defining the oral surface of the developing echinoderm and establishing the oral/aboral axis. (vii) Progressive development of oral and aboral imaginal discs supported by expansion of the ciliated locomotory bands of the pluteus. (viii) *Psammechinus* embryo shortly before the completion of metamorphosis. Larval tissues will be discarded and the adult anlagen united to form the diminutive sea-urchin.

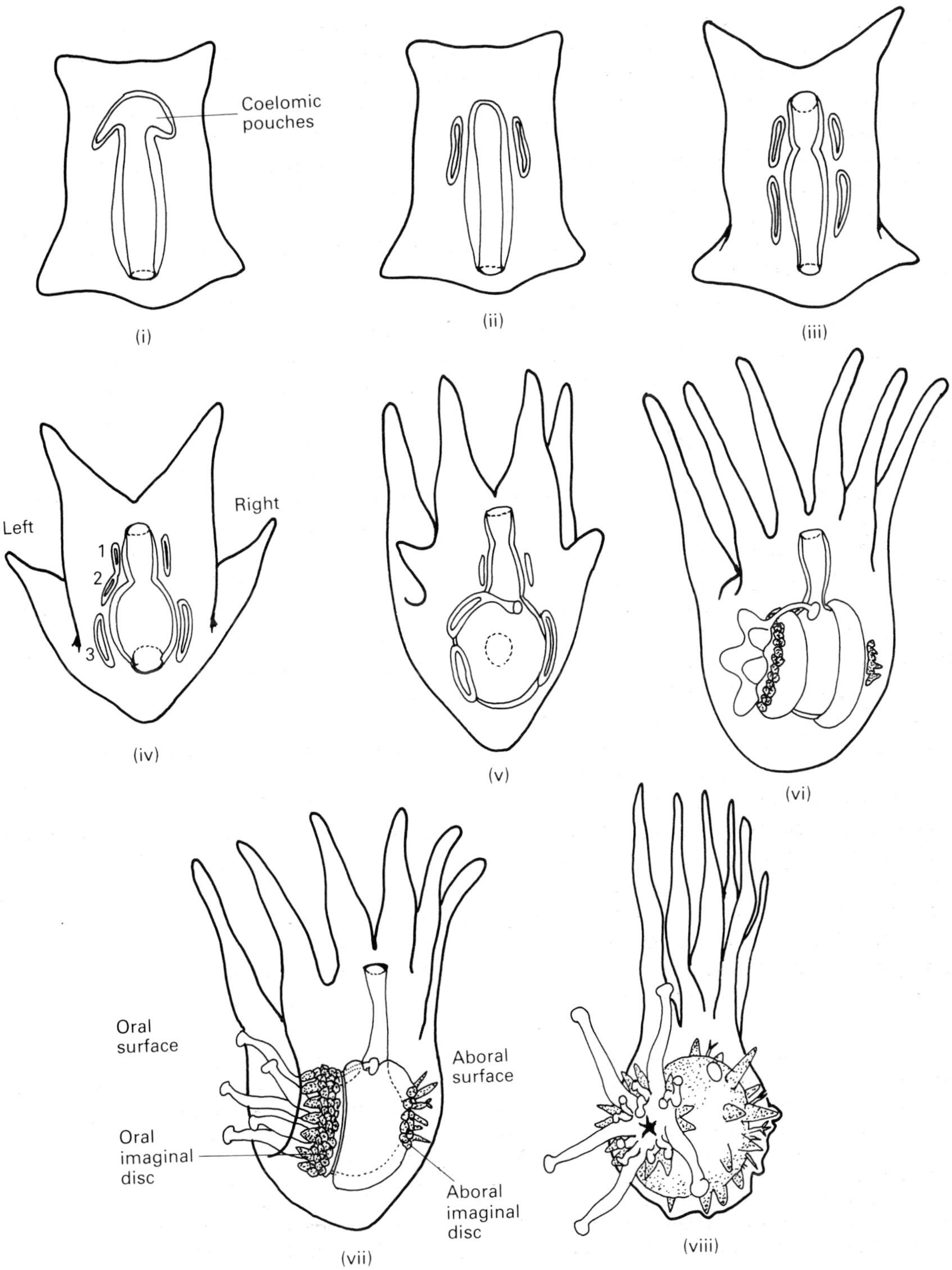
Coelomic pouches
(i)
(ii)
(iii)
Left
Right
1
2
3
(iv)
(v)
(vi)
Oral surface
Aboral surface
Oral imaginal disc
Aboral imaginal disc
(vii)
(viii)

The cypris larvae of barnacles also show a remarkable and complex behaviour prior to settlement, which enables them to choose good sites for metamorphosis. They respond to the texture of the surface (they prefer rough or pitted surfaces), but above all they respond to the presence of other barnacles, barnacle larvae, or the remnants of older barnacles of their own species.

15.4.2 The development and metamorphosis of insects

The phylum Uniramia includes the large group of animals commonly referred to as insects (the subphylum Hexapoda). There are, within this assemblage, animals with three different developmental patterns. The myriapod-like classes do not have wings, e.g. collembolans and thysanurans, and these develop gradually through a series of moults. Their morphology never undergoes any striking change and they cannot be said to undergo metamorphosis (Fig. 15.17a): they are ametabolous. A primitive feature of their development is the total cleavage of the eggs of many species, although some show an interesting characteristic in which cleavage eventually resembles that of the winged insects, but is at first total (Fig. 15.18).

The other insects show a more or less dramatic metamorphosis during their development and have a fixed number of larval instars prior to the adult form. In several orders the pre-adults, called 'nymphs' or, if aquatic, 'naiads', have external wing buds and metamorphosis is not extreme. These insects, which include the dragonflies and grasshoppers, are sometimes referred to as the exopterygotes and their development is said to be hemimetabolous. In the case of the grasshoppers and locusts illustrated in Fig. 15.17b, the nymphs occupy a similar niche to the adults and there is no major reorganization at the metamorphosis from the last larval instar to the winged adult. When the juveniles are aquatic there may be a rather marked change in morphology associated with the niche divergence between adults and juveniles. This metamorphosis, however, is not as dramatic as it is in those insects which have internal wing buds (the endopterygotes) where development is said to be holometabolous. In these a transitional pupal stage occurs between the final larval instar and the adult phase, as shown in Figs. 15.17c and 8.33. The pupa can be best interpreted as a much modified terminal larval instar. The larvae of the holometabolous insects belong to a number of different types, as illustrated in Fig. 15.19; they are often referred to by the general names caterpillar, grub, maggot, etc. As explained in Chapter 14 and Fig. 15.17c the larvae are specialized non-dispersive feeding stages, whereas the adults are specialized for dispersal and reproduction. The pupa is a stage during which locomotion and feeding are suspended while major reorganization of the body structure occurs.

In the holometabolous insects, groups of cells are set apart during early embryogenesis and they differentiate only during metamorphosis. At this time the groups of imaginal cells, which are organized either into rather diffuse nests or more commonly into well-organized 'imaginal discs', give rise to the entire adult integument, with its complex regional organization, to the salivary glands, and to other internal organs. A plan of the imaginal disc system of a dipteran larva and its relationship to the adult structure is shown in Fig. 15.20. The imaginal disc system of the fruit fly *Drosophila melanogaster* has come to have a special significance in contemporary biology. This is in very large part due to the unequalled wealth of genetic information which exists for this organism but also stems from the advantages of the imaginal disc system as a convenient model for the analysis of cell fate, and the processes of determination, differentiation and pattern formation which were discussed in Section 15.3 above.

15.4.3 The experimental analysis of the development of imaginal discs of *Drosophila melanogaster*

The imaginal discs of holometabolous insects are established during early embryonic life, and an early decision must be taken by the cells as to whether they will become functional cells in the larva or whether they will remain undifferentiated in the

Box 15.7 Metamorphosis and substratum choice by marine larvae

1 Many benthic organisms are found on a specific substratum which is characteristic for that species. Sometimes several closely related species occur sympatrically but on different substrata. The observed distribution of the adults is due to discrimination by the larvae. This has been demonstrated experimentally for several small, tubiculous polychaetes of the genus *Spirorbis*.

(i) *Spirorbis borealis* on *Fucus serratus*;
(ii) *Spirorbis tridentatus* on bare rock;
(iii) *Spirorbis corallinae* on *Corallina officinalis*.

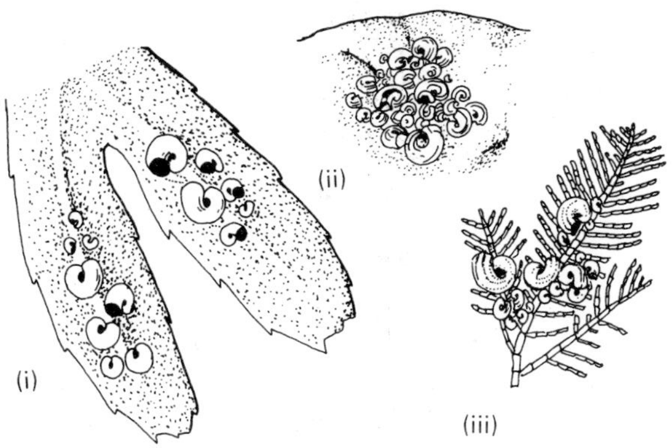

When presented with different substrata in a two-way choice experiment' the larvae of these different species show clear differences in their preferences (see table below).

Experiments on the choice of settlement substratum by larvae of different species of the polychaete *Spirorbis* (from data of De Silva, 1962).

Species	Substratum	Total number of larvae
Spirorbis borealis	*Fucus serratus*	1297
	Corallina officinalis	18
	Fucus serratus	457
	Filmed stone	295
Spirorbis tridentatus	*Fucus serratus*	0
	Filmed stone	52
	Corallina officinalis	0
	Filmed stone	55
Spirorbis corallinae	*Corallina officinalis*	63
	Fucus serratus	2

2 The cypris larva of an acorn barnacle (see below) is very choosy about a site for settlement.

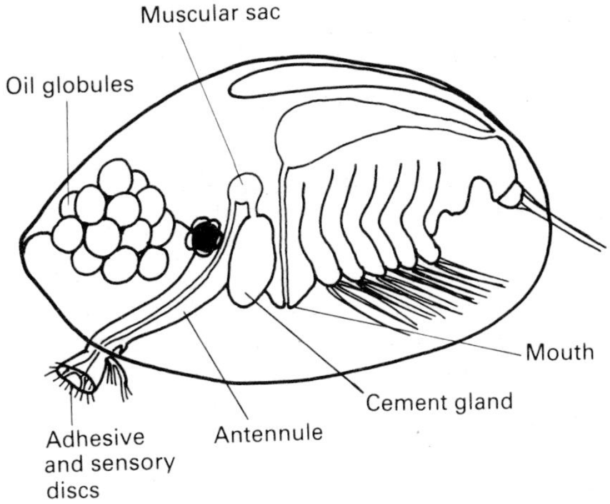

Before finally choosing to settle down the larva moves around a potential surface, showing exploratory behaviour and it may swim away and choose another site.

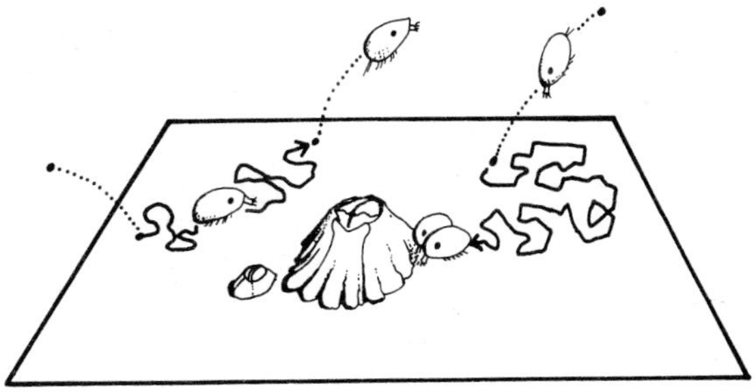

3 There are many factors which may increase the probability that a barnacle larva will settle on a particular surface.

Some factors stimulating settlement and metamorphosis:

1 A rough surface.
2 A pitted or grooved surface.
3 Remains of old barnacle tests.
4 The presence of newly settled cyprids.

Of these the most important are those due to the presence of other barnacles.

An unattractive surface can be rendered attractive by soaking the surface in extract of barnacle tissue. The attractive substance is a protein which can be detected by the cypris larva as a single molecular layer.

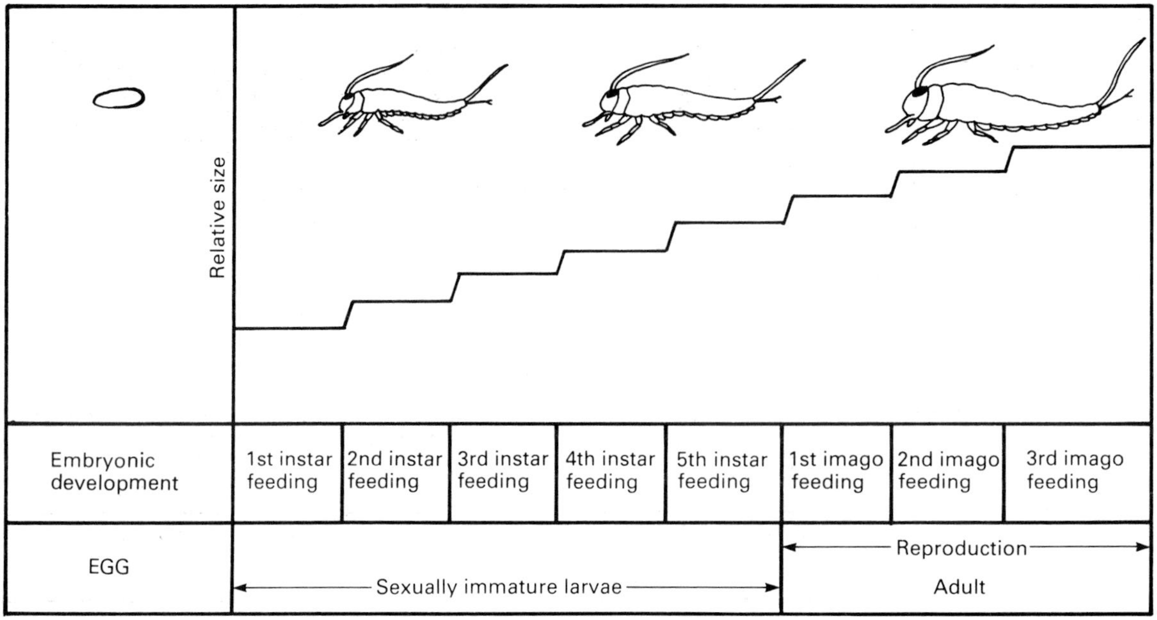
Relative size
Embryonic development
1st instar feeding
2nd instar feeding
3rd instar feeding
4th instar feeding
5th instar feeding
1st imago feeding
2nd imago feeding
3rd imago feeding
EGG
Sexually immature larvae
Reproduction
Adult

(a)

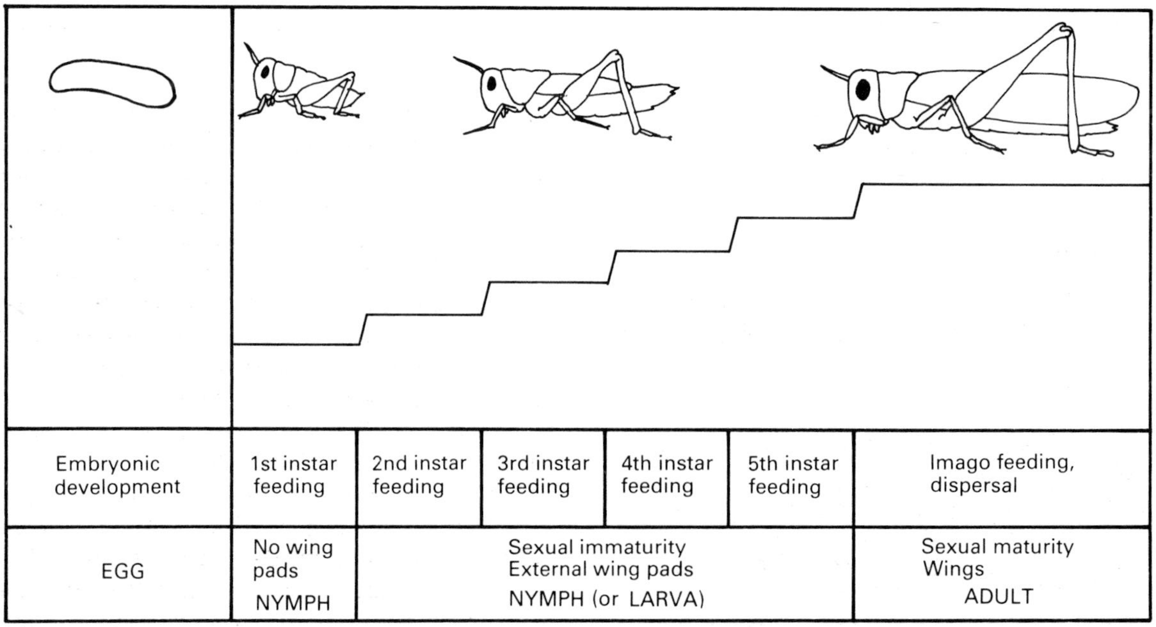
Embryonic development
1st instar feeding
2nd instar feeding
3rd instar feeding
4th instar feeding
5th instar feeding
Imago feeding, dispersal
EGG
No wing pads
NYMPH
Sexual immaturity
External wing pads
NYMPH (or LARVA)
Sexual maturity
Wings
ADULT

(b)

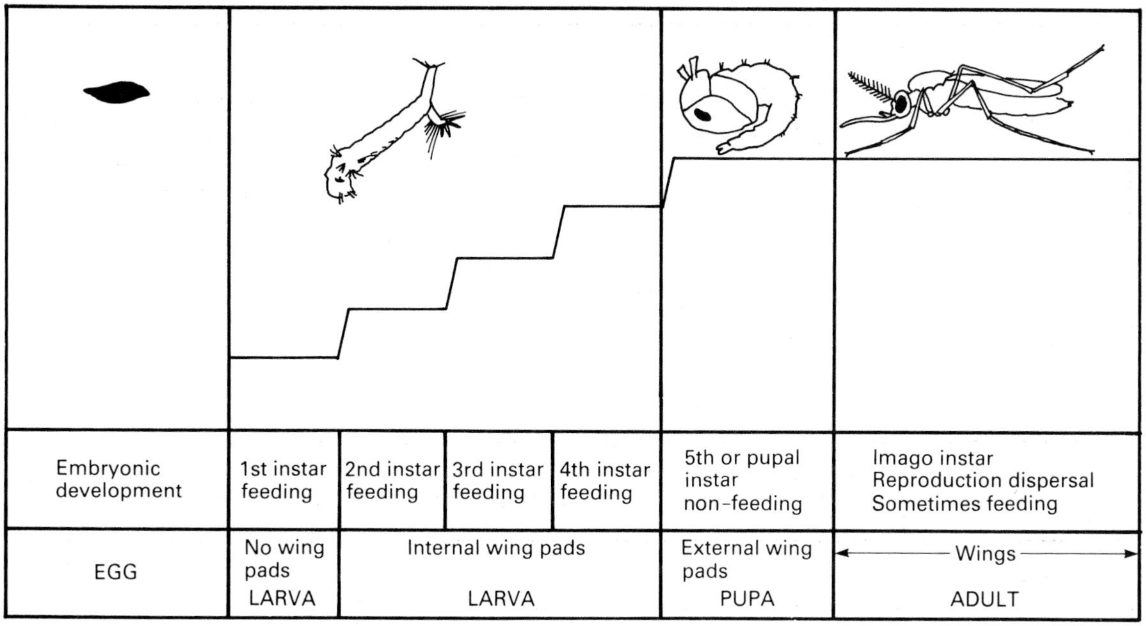

(c)

Fig. 15.17. Diagrammatic representation of growth and development in insects. (a) (*Facing page.*) *Ametabola*: no metamorphosis. In this example growth continues after breeding condition has been reached. (b) (*Facing page.*) *Hemimetabola*: incomplete or partial metamorphosis. In this example the wings appear as external wing pads from the 2nd instar. They are fully formed after the fifth and final moult. An increase in size occurs shortly after each moult. (c) *Holometabola*: complete metamorphosis. In this example there are four larval instars, a modified larval or pupal instar and an adult phase. Note growth is restricted to the period after each larval moult.

imaginal discs. The differentiation of these cells will be initiated by changes in hormonal milieu at pupation.

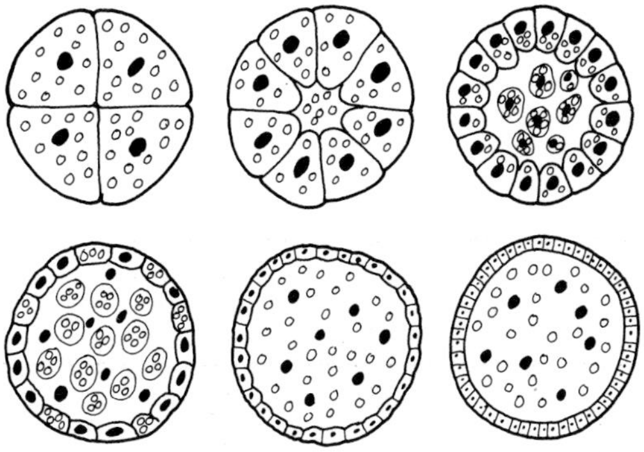

Fig. 15.18. Early development of a apterygote insect. Note that cleavage is at first total; cf. Fig. 15.8.

Although undifferentiated, the imaginal disc is a highly determined structure rather more like a mosaic embryo than a regulative one in its properties. There is, however, evidence for a gradient system underlying the determination of the disc as explained below. A fate map of a disc can be constructed by selective ablation of parts of it followed by implantation into a late larva. At metamorphosis an incomplete imaginal structure will develop according to which parts of the disc were removed. The fate map of the leg disc, for instance, has a concentric pattern (Fig. 15.21a). The outer zones give rise to proximal leg structures while the inner zones give rise to distal ones (Fig. 15.21b,c). The leg disc can be dissected and implanted into a young larva. When this is done compensatory growth takes place and the fragment will replace the missing parts and reconstitute an entire disc. However, the morphology of

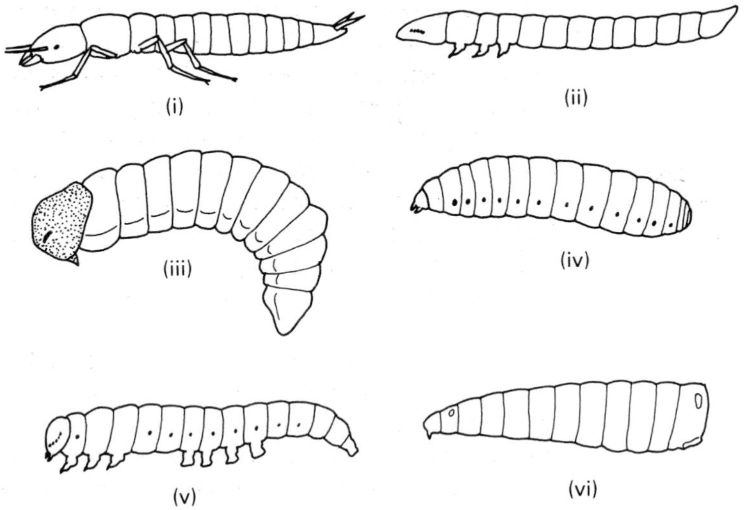

Fig. 15.19. Examples of holometabolous larval types. (i) Carabid larva—Coleoptera (beetles); (ii) elaterid larva or wireworm—Coleoptera; (iii) curculonid or weevil—Coleoptera; (iv) bee-grub—Hymenoptera (honey bee); (v) caterpillar larva—Lepidoptera (butterfly); (vi) maggot larva—Diptera (blowfly). In each case the morphology of the larva is very different from that of the adult.

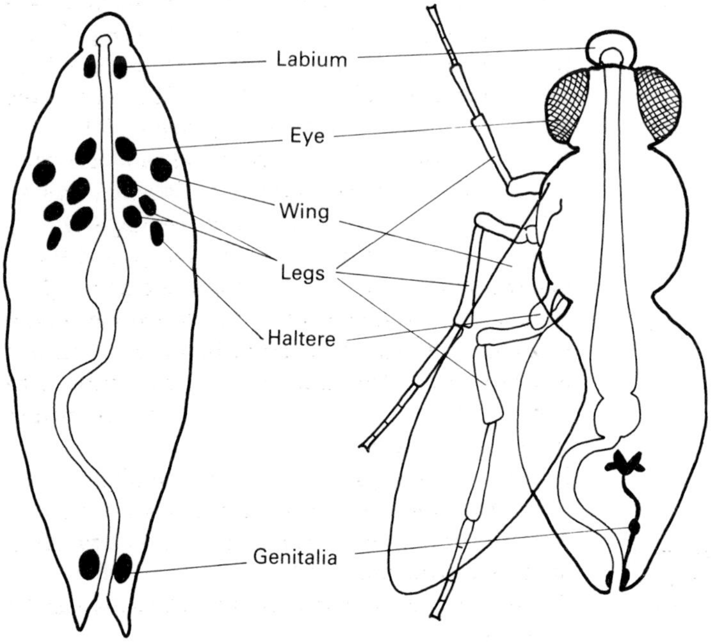

Fig. 15.20. The imaginal disc system of a dipteran fly (e.g. *Drosophila*). The position of the discs in the larva and the structures to which they give rise in the adult are indicated. Note that the alimentary system is not replaced at metamorphosis.

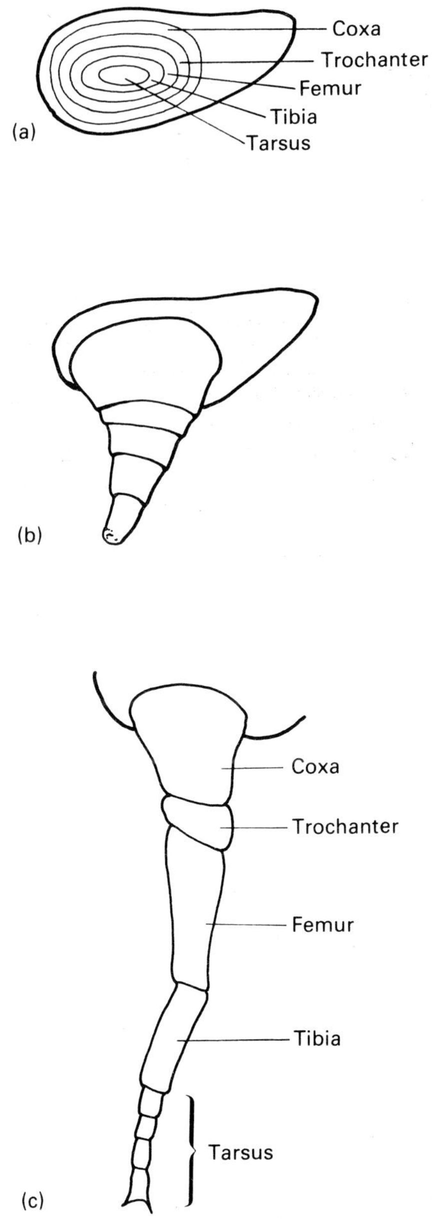

Fig. 15.21. The development of an insect limb from its imaginal disc. (a) Fate map of the imaginal disc as determined by selective ablation; (b) progressive development of the limb by outgrowth of the imaginal disc; (c) the completed limb with its component parts.

the resulting limb is influenced by the position within the limb of the original fragments.

If, for instance, the limb disc is subdivided into quartiles as in Fig. 15.22a, the upper medial quartile (A) will regenerate to give a complete, normal disc, but the lower medial quartile (C) will give a mirror-image duplication. The results are compatible with a gradient theory of organization. It may be supposed that there is a hierarchy of organization and that fragments high in the hierarchy can regenerate regions lower in the hierarchy, but fragments low in the hierarchy cannot regenerate regions of higher rank.

This model is represented diagrammatically in Fig. 15.22b. Models like this are valuable, but notice that although the model mimics observation it does not provide any clues about the biological basis of the mechanism involved. The imaginal disc system, however, is one where some of the most rapid progress is being made towards an understanding of the molecular and genetical basis for the determination of cell fate and cell identity in higher organisms.

An undifferentiated cell is often said to be *determined* in its developmental fate when experiments show that the pattern of its future differentiation is fixed and it is not influenced by the cellular environment in which it is placed. The cells of the imaginal disc of insects have this property. That cells may be determined in their developmental fate prior to the onset of biochemical differentiation was an important finding of developmental biologists.

A disc can be isolated from an embryo and transplanted to a different region of a host embryo (Fig. 15.23). It will continue (usually) to develop according to its original regional identity. In other words, if it is a prospective leg disc, the cells will continue to develop into a leg. Two elements of the determination can be recognized:

1 The overall regional identity of the disc.

2 The identity of individual cells within the disc which permits the differentiation of an organized integrated structure.

We shall see later that these two elements can be separated.

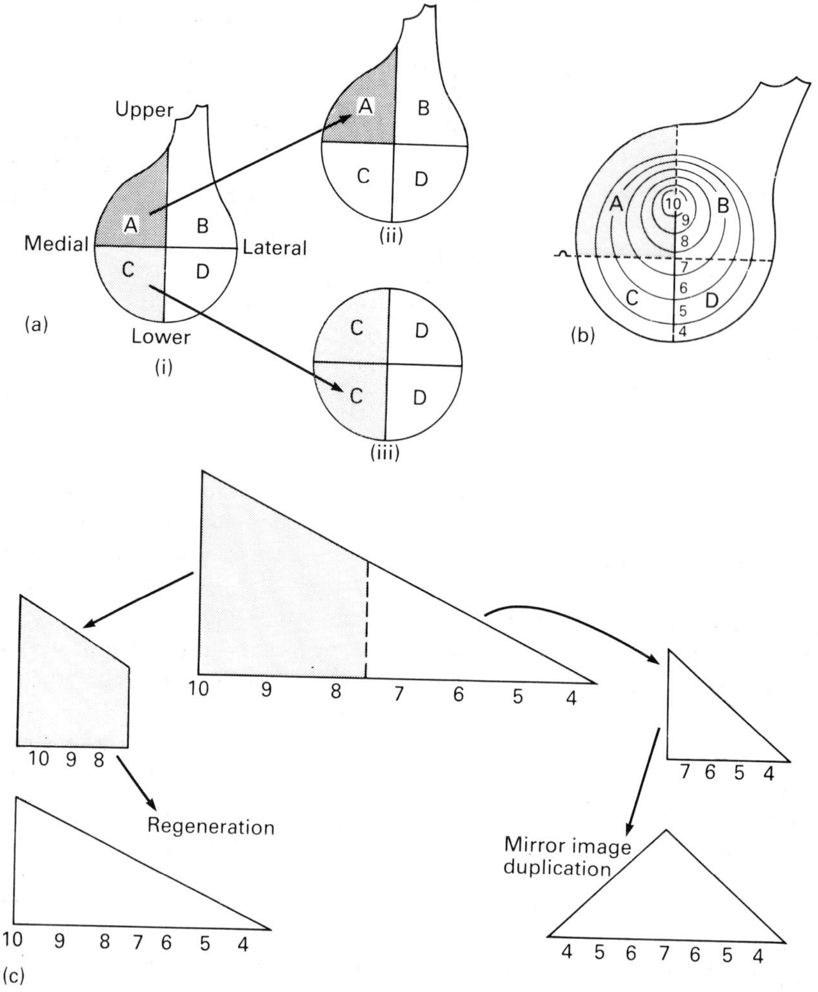

Fig. 15.22. Evidence of a gradient or hierarchical system of regional organization in the leg disc of *Drosophila*. (a) (i) Definition of quartiles in the intact disc; (ii) regeneration of the upper medial quartile gives a complete disc; (iii) regeneration of the lower medial quartile gives a mirror-image duplication. (b) Disc with supposed contour lines of developmental capacity. (c) A model for regional organization: regions of high rank can replace missing parts but regions of low rank replace with regions of lower rank.

The cells of the imaginal disc are capable of proliferation (mitotic division) and disc growth is a normal part of development. The disc is also capable of compensatory growth and (as suggested in Fig. 15.22) a disc fragment can reconstitute an entire disc.

The hormone milieu of an adult insect allows proliferation of imaginal disc cells but does not permit or cause differentiation of the disc. This will only take place if the disc is placed in the hormonal milieu of a late-stage larva and pupa.

Using these facts, experimental biologists have conducted many intriguing experiments, the results of which were often quite unexpected.

A disc can be maintained in its undifferentiated state long after it would normally have metamorphosed by serial culture in adult flies. The

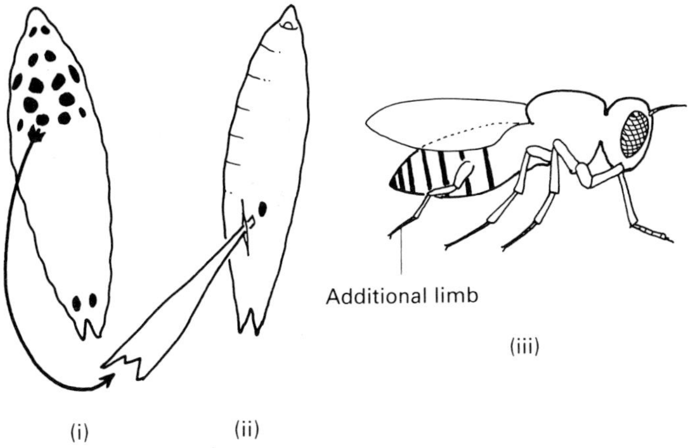

Fig. 15.23 The regional organization of an imaginal disc is fixed (but see also Box 15.8). The leg disc of a dipteran larva is (i) isolated by dissection and (ii) implanted into the abdominal region of a host embryo. Eventually the larva will complete metamorphosis; (iii) the resulting fly has an additional limb in the abdomen.

low levels of moulting hormone in the adult haemolymph do not allow the disc cells to undergo metamorphosis, although they can proliferate.

Disc fragments can be serially transplanted through many generations and their state of determination checked at any time by transplanting test implants into old larvae. The technique is illustrated in Fig. 15.24.

The cells of imaginal discs tested in this way normally retain their original regional identity and will continue to undergo metamorphosis into that structure, be it an eye, an antenna or a wing, etc., according to the developmental fate of the original disc. This rule of fixed regional determination, however, is not absolute; sometimes a disc fragment gives rise to a structure which is appropriate to a different region of the embryo (Box 15.8). This phenomenon is called transdetermination. What is particularly striking is that the various cells of the disc still behave in a co-ordinated determined manner: the cells behaving as if they were now part of a complete but different fate map.

Many observations have been made and certain transdeterminations are more frequently observed, whereas others are rarely or never seen (see Box 15.8 for more detail). When all the cells of the disc suddenly change their state of determination in this way it is as if there were some master control system and the genetical basis for this can be investigated through the study of certain mutants of *Drosophila* (see Box 15.9). The pattern of co-ordinated disc development appropriate to each specific segmental region is under the control of groups of genes which are known as the homeotic (sometimes homoeotic) genes.

The genetic mutant *Antennapedia*, for instance, causes a change in regional organization such that a partial leg develops where an antenna would be the appropriate structure. The normal wild type of the gene *Antennapedia* must be present to prevent an antennal disc developing into a leg (see Box 15.9).

Homeotic genes also control the differentiation of other structures which reflect regional organization. A list of some of these genes and their effects are summarized in Box 15.9. All the homeotic genes have the property of converting one segment into another or one part of a segment into another part of a segment. These genes offer some of the most exciting material for the investigation of regional and segmental organization in animals. It is known, for instance, that homeotic genes of *Drosophila* have base sequences in common with other genes in a wide variety of other organisms—these sequences are referred to as the homeobox—and this implies an ancient evolutionary origin for these genes of regional organization.

There are, however, fundamental questions which still have to be understood. The regional organization of the early insect embryo is maternally programmed; it is a product of the process of gametogenesis. Later

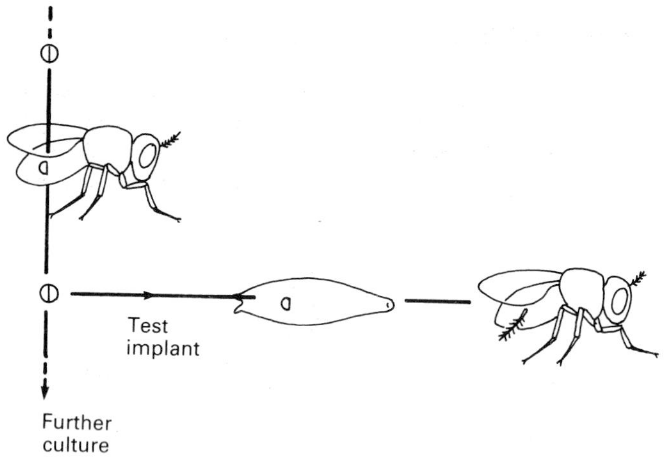

Fig. 15.24. The technique of serial transplantation of imaginal discs. A disc is removed from a donor larva, dissected and placed into an adult fly where it will proliferate and regenerate. The regenerated disc can be recovered from the host fly, dissected again and one component tested for developmental capacity in an old larva, the other fragment being transferred to a host adult for further culture. See also Box 15.8.

in development it seems that different sets of genes are transcribed in the imaginal discs according to position in the early embryo.

The homeotic genes are involved in this, but the molecular processes underlying such regional differentiation and the precise role of the genes has still to be elucidated.

15.5 Regeneration

15.5.1 Introduction

Regeneration can be defined as the capacity to replace, by compensatory growth and differentiation, parts of the body which are accidentally lost or which are autotomized. The ability to regenerate missing parts of the body in this way is a prominent feature of the biology of many of the soft-bodied invertebrates such as for example sponges, cnidarians, flatworms, nemertines, annelids and some echinoderms. Such animals also exhibit asexual reproduction by fission (see Chapter 14) and the two processes are obviously related. Invertebrates with hardened external coverings or very specialized external coverings such as the arthropod groups, the aschelminth phyla and the molluscs have poor powers of regeneration and do not usually reproduce asexually by fission. Regeneration in the arthropods is usually restricted to limb regeneration, which takes place when the animals moult.

Regeneration involves a number of processes which are similar to those taking place during normal development. These include:

1 Proliferation of undifferentiated cells, as in a blastula, and the construction of a blastema.

2 Pattern formation and the organization of cells in a spatial hierarchy.

3 Differentiation and the expression of pattern.

Regeneration therefore provides a convenient model for the investigation of developmental events. In some animals regeneration also involves dedifferentiation, which is not a feature of normal development.

In order for regeneration to occur it is essential that the organism responds to the loss of components of the body, and the response must involve both the proliferation of a segment blastema and the development of an appropriate pattern in the cells proliferated by that blastema.

Box 15.8 Serial transplantation of imaginal discs and the discovery of transdetermination

1 Using the technique of serial transplantation, the state of determination of a disc fragment derived from a single donor disc can be tested many times.

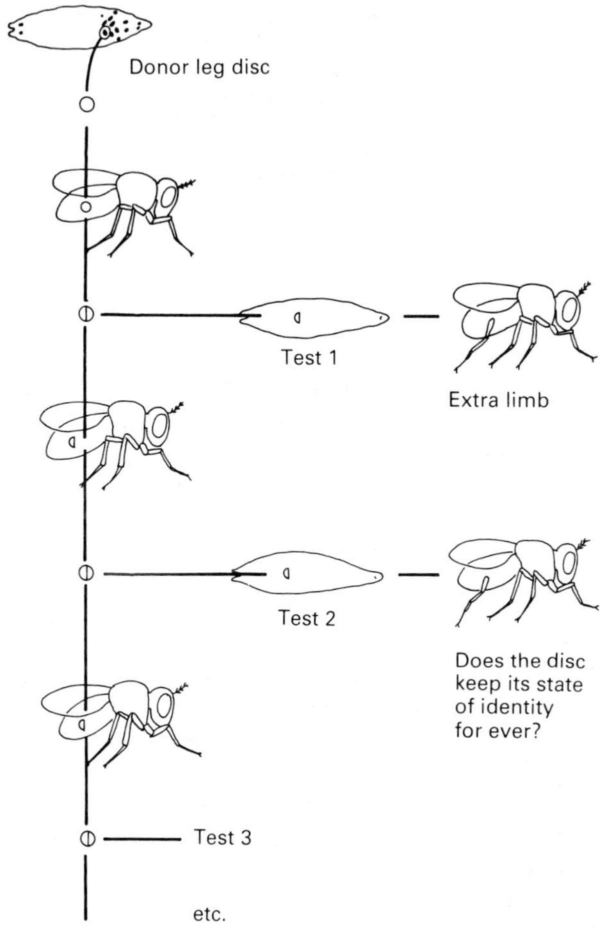

The technique of serial transplantation.

If the state of determination of the original disc is maintained, each test implant will reflect the character of the original disc. This is usually the case and the state of determination is stable.

2 If a large number of test implants are made, however, spontaneous changes in the state of determination may be observed. Not all spontaneous changes are possible. Some of those that have been observed in *Drosophila* are shown below.

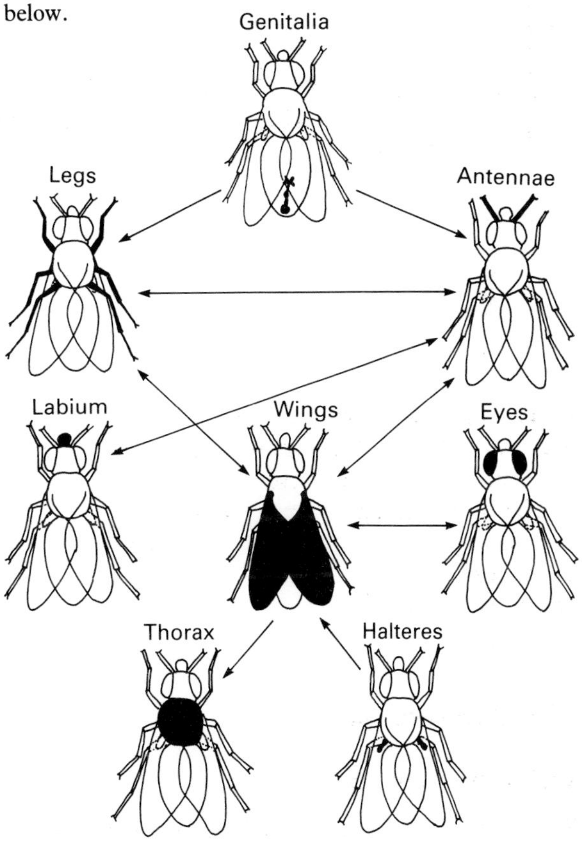

Arrows indicate the direction of spontaneous changes in the state of determination.

For instance, leg discs have been observed to change to give wings, antennae or mouthparts but not eyes, halteres or genitalia.

Note that all the different cells of the disc must undergo a spontaneous change and assume a new developmental fate, but they must retain local identity since they continue to give rise to structures with regional organization. Such a spontaneous change is referred to as *transdetermination*.

Box 15.9 Genetic control and regional organization: the homeotic genes

1 In cultures of the fly *Drosophila*, abnormal or monstrous flies are occasionally found in which the structures developing from an imaginal disc are of an inappropriate or incorrect type. For instance, a limb may develop from what would be expected to be the imaginal disc of an antenna.

Genetic analysis shows that these abnormalities are the consequence of mutation of single genetic loci. In other words, there are genes which are responsible for maintaining or ordering in some way the state of determination of individual imaginal discs. These master genes are called homeotic genes:

Mutant name	Symbol	Genetic locus	Effect
Antennapedia	$A_b$tp	3–48	Causes the antenna to develop as a leg (i).
Ophthalmoptera	Opt.	2–68	Causes the eye disc to develop as a wing (ii).
Hexaptera	Hx	2	Causes the dorsal pro-thorax to develop as a wing (and sometimes leg) (iii).

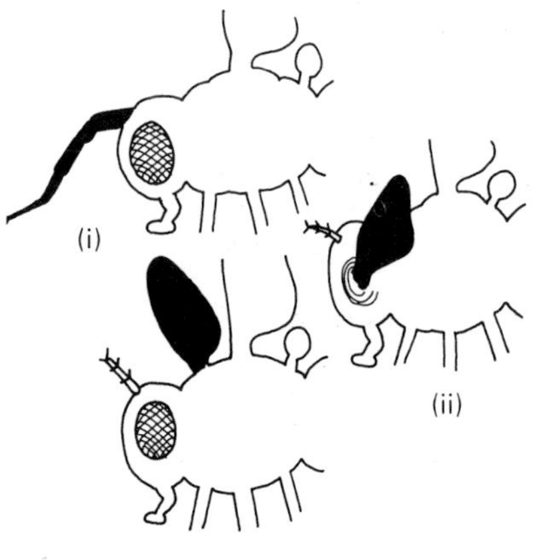

These are dominant mutations—others are recessive. In some ways the effects of the homeotic mutants are similar to the spontaneous changes observed in *transdetermination*.

2 The homeotic genes seem to be involved in the process which controls regional identity. This is especially clear in the *bithorax* complex of genes.

The *bithorax* complex is a series of closely linked genes on the 3rd chromosome of *Drosophila*, affecting some aspect of regional organization in the thorax or anterior abdomen. In fact, the genetic mutants reveal to us morphological boundaries not previously recognized.

The figure below shows a diagram of a fly broken down to its basic components. The homeotic mutants in the bithorax complex show that there is a regional boundary between anterior and posterior mesothorax (2nd thoracic segment), metathorax (3rd thoracic segment) and 1st abdominal segment.

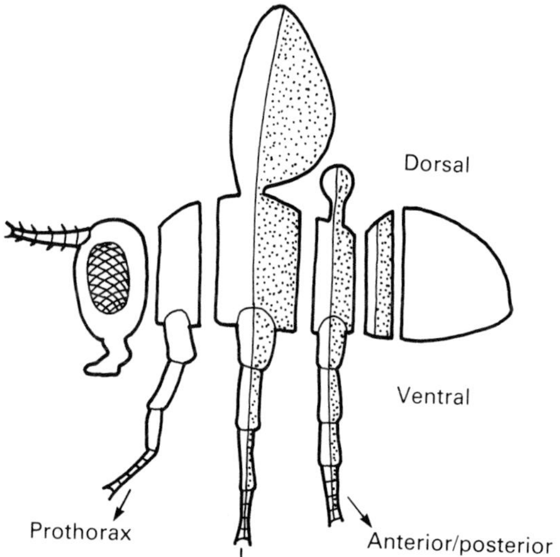

Diagram of *Drosophila* showing the normal configuration of thoracic structures.

The mutant *bithorax* (bx) is a recessive gene which causes the anterior metathorax (which normally produces a

Box 15.9 continued

haltere) to behave as an anterior mesothorax and produces an anterior part wing as shown in (i) below.

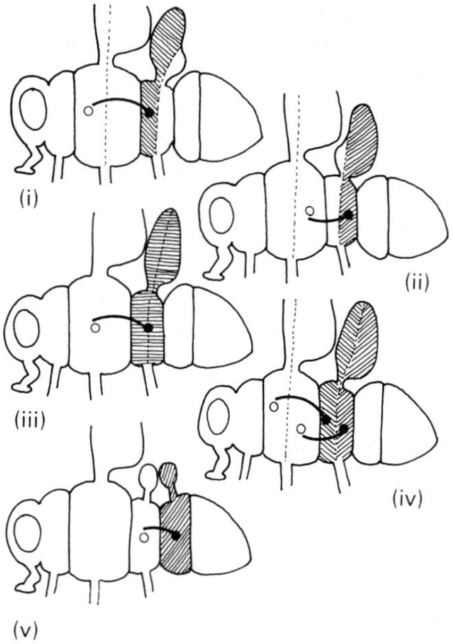

Gene	Gene symbol	Changes	To
(i) Bithorax	(bx)	Anterior metathorax (AMT) (anterior part of haltere)	Anterior meso-thorax (AMS) (anterior part of wing)
(ii) Postbithorax	(pbx)	Posterior meta-thorax (PMT) (posterior part of haltere)	Posterior meso-thorax (PMS) (posterior part of wing)
(iii) Ultrabithorax	(Ubx)	Metathorax (MT) (haltere)	Mesothorax (MS) (wing)
(v) Bithoraxoid	(bxd)	1st abdominal segment (AB1)	Metathorax (haltere)

Note that the double recessive condition $\frac{\text{bx} \quad \text{pbx}}{\text{bx} \quad \text{pbx}}$ (iv) resembles in its regional organization the dominant condition $\frac{\text{Ubx}}{\text{Ubx}^{+}}$ (iii).

15.5.2 The origin of the regeneration blastema

When regenerative growth occurs the new cells must be derived either from a reserve population of previously undifferentiated, totipotent cells or by dedifferentiation from previously differentiated cells. Considerable controversy has arisen over which of these two alternatives is involved. The Cnidaria have particularly good powers of regeneration and they have a pool of interstitial cells, from which the different classes of differentiated cells, such as the cnidoblasts, are normally derived and constantly replaced (Fig. 15.25).

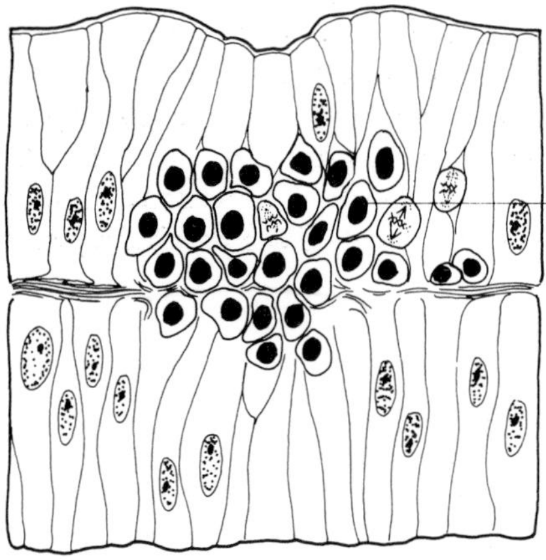

Fig. 15.25. Cross-section of a cnidarian showing interstitial cells and their proliferation prior to regeneration.

In *Hydra* the interstitial cells (or I cells) form a pool of mobile reserve cells which normally congregate in the ectoderm prior to asexual reproduction by budding. A wound provokes a similar response and constitutes a competing locus of attraction for the I cells, which form the regeneration blastema. In *Hydra*, though not in all cnidarians, I cells continue to proliferate throughout life and there is therefore a constant supply of cells for regeneration. The idea that the I cells are a self-maintaining population of reserve cells essential for regeneration is, however, an oversimplification. The I cells can be destroyed by chemical means without the loss of regenerative ability, and fragments of *Hydra* which do not normally contain I cells are capable of regeneration. Moreover, in suitable media, differentiated cells in explants of *Hydra* tissues can dedifferentiate into interstitial cells, multiply, and they are then capable of redifferentiation to cells of a different type. The normal route to differentiation may be via the population of reserve cells but this is not the only route.

Undifferentiated reserve cells, called neoblasts, are also implicated in the phenomenal regenerative prowess of the free-living flatworms. These animals have been utilized as favourable material for the study of regeneration for more than 100 years. Transverse section of a planarian leads to the reconstruction of two complete flatworms (Fig. 15.26a) and, similarly, small fragments including sagittally sectioned will reconstitute a complete worm (Fig. 15.26b).

The first stage of regeneration is the formation of a wound blastema and its subsequent invasion by neoblasts. The role of these reserve cells has been demonstrated by irradiation and transplantation experiments. Irradiation with X-rays at 3000 rad can prevent neoblast proliferation but this does not kill the organism. Such an irradiated animal will fail to regenerate; if, however, a fragment of a flatworm which has not been irradiated is implanted into the

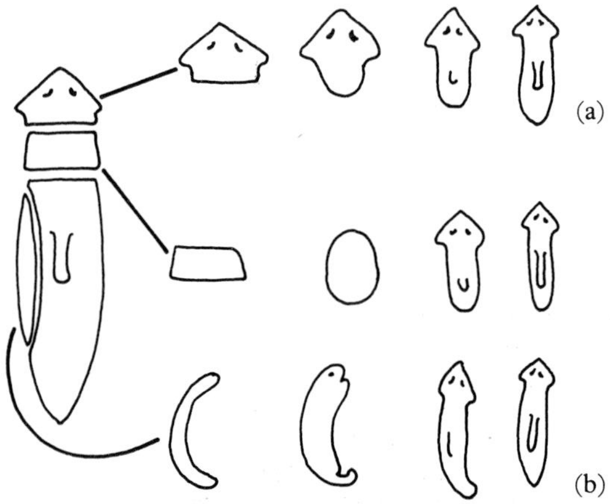

Fig. 15.26. Regeneration in flatworms. (a) Simple transection and subsequent regeneration of the missing posterior or anterior parts by the anterior and posterior fragments. (b) Even small fragments of fragments can reconstitute an entire flatworm, with the original tissue retaining its regional identity.

body of an irradiated host regeneration can occur. Moreover, if the cells of the implanted tissue can be identified, for instance if they have a different colour, the regenerated fragment has the characteristic colour of the implanted tissue (Fig. 15.27).

Totipotent reserve cells are not universally involved in regeneration and indeed a prominent role for such cells, such as occurs in the planarians, may be rather unusual.

In annelids the formation of the regeneration blastema does not involve a population of distinct totipotent reserve cells, but rather dedifferentiation and redeployment of differentiated cells derived from ectodermal and mesodermal layers. Loss of caudal segments, for example, is followed by wound healing which involves the migration of coelomocytes to the damaged surface and reconstitution of the growth zones which appear as bands of characteristic cells with spherical nuclei and a prominent nucleolus. In nereid polychaetes differentiation of the segment anlagen requires the presence of a cerebral growth hormone, and formation of the segmental ectodermal components (chaetal sacs and parapodia) requires the inductive influence of the ventral nerve (see below and Box 15.10).

15.5.3 Regeneration and regional organization

Many soft-bodied invertebrates are able to reconstitute a complete regionally organized structure from a very small fragment of the body (Fig. 15.26) or see Fig. 14.2 which shows the asexual reproduction of a polychaete from spontaneously autotomized fragments.

What all these examples have in common is the reconstitution of a complete pattern from part of a pattern. In each case the fragment retains its original polarity and compensatory growth re-establishes the fragment in its original position in the body plan. In Fig. 15.26 a head fragment of a planarian replaces

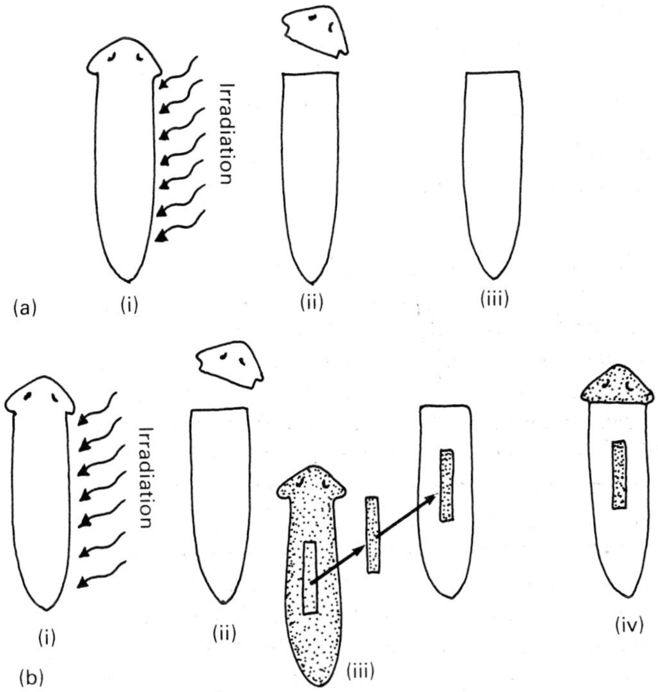

Fig. 15.27. Evidence for the special role of 'neoblast' cells in planarian regeneration. (a) (i) A specimen is irradiated with high energy X-rays; (ii) the anterior is removed; (iii) the posterior fragment fails to regenerate. (b) (i), (ii) As in (a); (iii) a small fragment from a genetically distinct specimen is implanted into the mid-region; (iv) the head can now regenerate; it has the genetic character of the implanted fragment.

Box 15.10 Positional information and caudal regeneration in annelids

1 *Segment number.* Annelids are composed of the following body regions (see also Fig. 4.51):

1 The prostomium.
2 A specific number of segments.
3 A postsegmental pygidium.

In some polychaetes, as in leeches, the number of segments is quite small, e.g.
Clymenella torquata—22 segments.
Ophryotrocha puerilis—25 segments.
In others, the number is much higher but nevertheless the segment number may be a species-specific character.

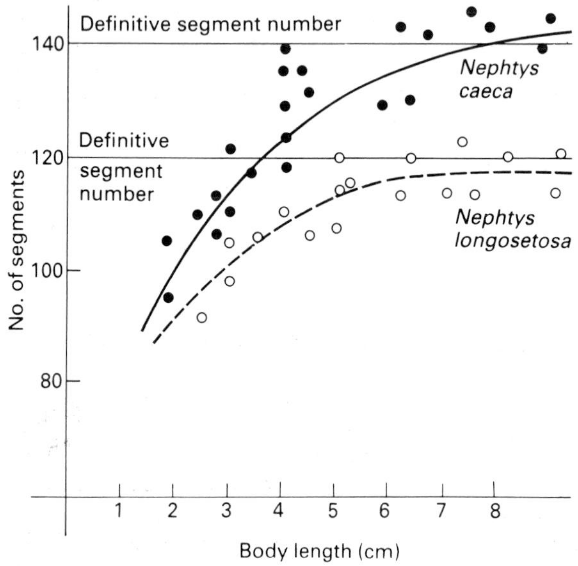

The relationship between segment number and body length in two closely related species of *Nephtys*.

In young animals the rate of segment proliferation is high, but in older animals growth is increasingly due to segment enlargement. The high rate of segment production characteristic of young worms is restored by amputation of tail segments. After segment amputation and wound healing a new segment proliferation zone is established. In *Nereis* the rate of segment proliferation is then directly proportional to the number of segments removed.

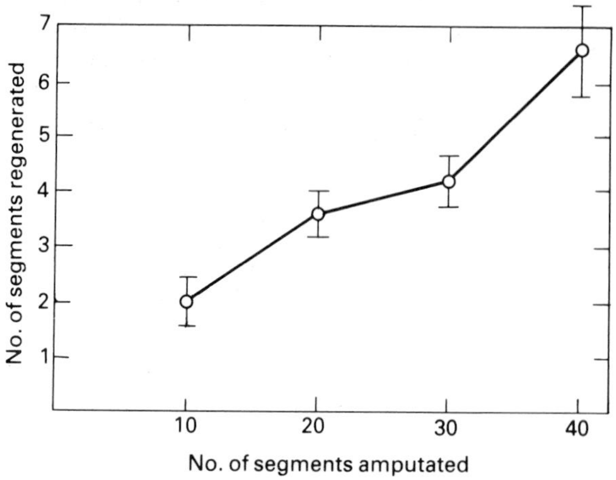

2 *Segment identity.* Each segment of the annelid body behaves as if it were part of the integrated whole. In all annelids each segment has its own particular structure and identity. This is particularly obvious in the tubiculous worm *Chaetopterus variopedatus* illustrated in Fig. 9.5. A single segment of this apparently complex worm is able to regenerate anterior and posterior segments to reform an entire worm!

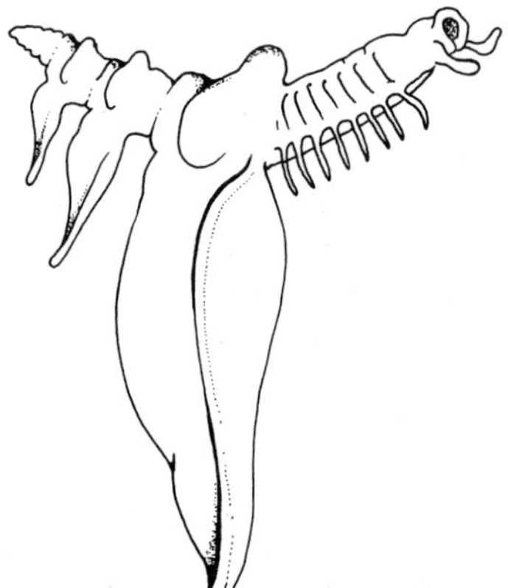

A stage in the regeneration of a complete worm from an isolated fan segment of *Chaetopterus variopedatus*.

Box 15.10 continued

The specific regional identity of each segment is also evident during regeneration of fragments of the worm *Clymenella torquata* which has, as an adult, exactly 22 segments.

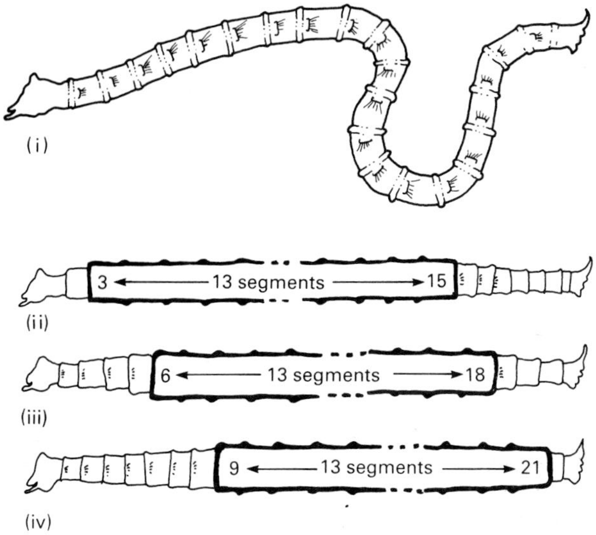

(i) The intact worm.
Compensatory regeneration of fragments of *Clymenella* each 13 segments long but taken from different regions of the body.
(ii) Segments 3–15.
(iii) Segments 6–18.
(iv) Segments 9–21.
Note that in each case the segments are re-established in their original position in the hierarchy.

3 *Morphallaxis.* In some polychaetes loss of cephalic segments causes a morphological rearrangement of the remaining segments—a process sometimes referred to as *morphallaxis.* It is as if the segments have redefined their position in an organized hierarchy.

This phenomenon is observed in the fan-worms, e.g. *Sabella*, is illustrated in the next column. Fan-worms are likely to lose the crown of feeding tentacles because of predation by fish, and these tentacles can be replaced.

Sabella has a prostomium with a complex crown of tentacles, a peristomium, a fixed number of thoracic segments with a distinctive arrangement of the parapodia, and a large number of similar abdominal segments (i). No more than three anterior segments are regenerated. If more than three are lost, posterior segments lose their chaetae and are converted into thoracic segments (ii–iv).

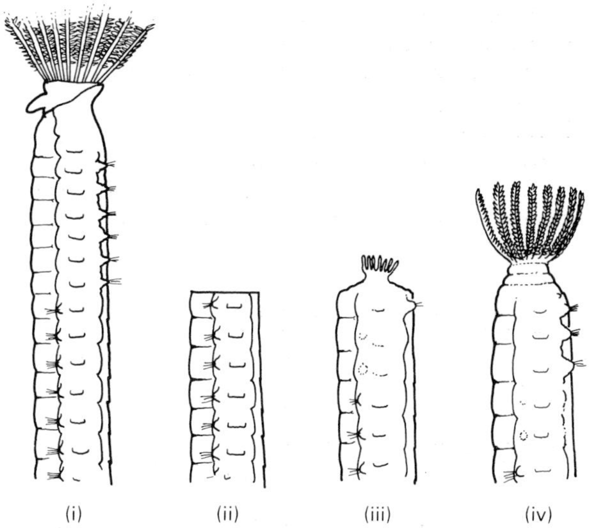

(i) The intact *Sabella*.
(ii) Loss of prostomium, peristomium and all thoracic segments.
(iii) Early stages of regeneration and morphallaxis.
(iv) Regeneration showing crown of tentacles, peristomium and one thoracic segment being formed. The appropriate number of abdominal segments are converted into thoracic segments.
In this way the feeding tentacles are replaced most quickly but it is also clear that each segment has its characteristic structure because of its position in an anterior/posterior gradient.

4 *A model of regeneration.* Many observations are compatible with the following simple model:
(A) The prostomium has positional value 0.
(B) The pygidium has a positional value equal to 1+ the species-specific segment number.
(C) The segment blastema exists at the anterior face of the pygidium.
(D) The rate of segment proliferation is a function of the difference between the value of the last segment and that of the pygidium and is zero when that value is unity.

Box 15.10 continued

(E) Segment proliferation continues until the difference between the positional value of the oldest segment and the pygidium is unity:

(F) Loss of caudal segments results in the reformation of the pygidium with its specifically high positional value.

This model is illustrated for the polychaete *Ophryotrocha*. It is applicable to normal embryonic growth (a) and to regenerative growth (b).

The experiment illustrated in (c) suggests that the nerve cord carries the positional information. Deflection of the nerve cord (in, say, segment 9) causes the formation of an additional pygidium. This supernumerary tail will now proliferate segments according to the above rules.

Similar two-tailed worms are occasionally encountered in the wild and can also be formed by grafting fragments of two worms together. In every case the rate of proliferation by each pygidium follows the rules for normal growth and is proportional to the difference between the positional value of the last segment and that of the blastema.

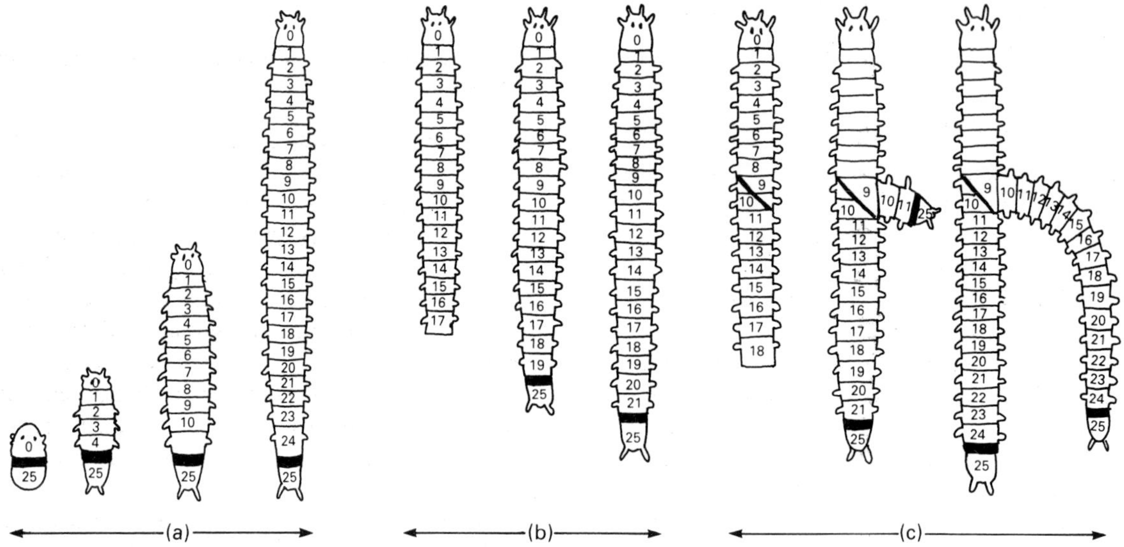

(a) Normal growth of *Ophryotrocha*. The prostomium has value 0. The pygidium has value 25. Growth establishes a sequence of segments by intercalation. (b) Regenerative growth in *Ophryotrocha*. After caudal ablation a number discontinuity exists between the last segment and the reconstructed pygidium. (c) Induction of a supernumerary pygidium by surgical interference with the ventral nerve cord.

the missing caudal region and a caudal fragment will replace the missing head. Each area of the body has a regional identity within the whole and this regional identity is retained during regeneration. In segmented animals the regional identity is more precisely defined and each segment has an identity in a linear hierarchy. In insect embryos there is a fixed number of segments and on each segment specific structures will develop. Genetic loci exist (see Box 15.9) which control or regulate the identity of those segments; note, however, that the bithorax genes show that there are morphological boundaries in addition to the more easily recognized segmental boundaries.

The arthropods are not able to replace lost segments but many of the annelids, which are also segmented animals, are able to do so. Each annelid segment is unique and forms part of a single, integrated whole. One way in which this is expressed is in the number of segments which, in most annelids, is fixed. Positional information seems to be given by the ventral nerve cord, as explained in Box 15.10. In polychaetes of the family Nereidae caudal regeneration requires the presence of a cerebral hormone but this seems to be a secondary adaptation related to the strictly semelparous reproduction of these animals (see Sections 14.3 and 14.4); nevertheless the rate of segment proliferation is still subject to positional control (Box 15.10).

15.5.4 Regeneration, growth and reproduction

Regeneration presents a demand on the resources of an organism and there is likely to be associated with it an enhanced resource allocation to somatic functions and a consequent reduction for processes of sexual reproduction. There is then a potential conflict between regeneration and sexual reproduction, and regulatory mechanisms which effectively control this antagonism between the two processes are to be expected. In particular, semelparous organisms engaged in the build-up of reproductive tissues should not divert resources to regenerative growth unless there is some compensatory increase in survivorship, fecundity or offspring survivorship.

A mechanism of this sort is to be found in the semelparous nereid worms which are, as shown above, capable of compensatory regenerative growth following the loss of posterior segments. As these worms approach maturity, resources are committed irrevocably to reproduction and they will not survive the single breeding episode. In these circumstances, segments regenerated during the final stages of reproduction would have little value. An endocrine mechanism ensures that regenerative growth does not take place at this stage of the life cycle. During sexual maturation secretion of the cerebral hormone is gradually reduced and reduction in the level of circulating hormone frees the final stages of gametocyte maturation (see Section 14.4) and initiates somatic changes associated with reproduction (see Fig. 16.35). The same hormone is essential for caudal regeneration and sexually mature nereid worms are not able to replace lost caudal segments. Polychaete worms which breed several years in succession do not have this endocrine mechanism. Regenerative growth proceeds even in mature specimens and indeed the reproductive value of regenerated segments remains high as they can contribute to reproductive output in subsequent years.

15.6 Conclusion: invertebrate development and the genetic programme

The experimental study of invertebrate development can be traced back to the late nineteenth century. It has therefore developed in parallel with the study of genetics, the discovery of the molecular basis of cell heredity and the rapidly developing field of molecular biology. An outstanding challenge for the future is the unification of these disciplines and the invertebrates provide a wealth of convenient model systems. Considerable progress has been made, and much of the material presented in this chapter can be reappraised from a molecular biological point of view.

The messenger molecules of the animal cell are mRNA sequences decoded from DNA sequences in the nucleus. In the early sections of this chapter we learned that the DNA sequences of the zygote nucleus are, in almost all animals, passed entire to the daughter cells during early cleavage.

These mRNA molecules can be investigated by the techniques of modern biochemistry. During oogenesis (see Chapter 14), maternal mRNA molecules are stored in the oocyte cytoplasm. These masked mRNA molecules are freed from inhibition at fertilization and they provide the material for early protein synthesis.

Ultimately the genome of the zygotic nucleus becomes the source of genetic information. However, the crucial role of the oocyte cytoplasm remains at the centre of differentiation and the creation of order. We have seen in a wide range of invertebrates that the function of a cell nucleus and the messages that are ultimately transcribed or translated are determined by its history in the embryo.

Sometimes specific cytoplasmic substances seem to call into play specific patterns of enzyme production (see the development of the tunicate *Styela*, for instance). In other examples, contacts between cells seem to call forth functional responses. Evidence for these interactions were presented for example in experiments with mollusc and sea-urchin embryos. Some of the most exciting developments are taking place in studies of insect embryos, especially that of the fruit fly *Drosophila melanogaster* where a wealth of genetic information can be combined with a convenient experimental model. It is likely that invertebrate embryos will continue to supply some of the best material with which to unravel the complexities of animal development.

15.7 Further reading

Berril, N.J. 1971. *Developmental Biology*. McGraw Hill, New York.

Brookbank, J.W. 1978. *Developmental Biology: Embryos, Plants and Regeneration*. Harper & Row, New York.

Browder, L.W. 1984. *Developmental Biology*, 2nd edn. Saunders, New York.

Epel, D. 1977. The program of fertilisation. *Scient.Am.* **237**, 129–38.

Gehring, W.J. 1985. The molecular basis of development. *Scientific American*, **253**, 137–46.

Oppenheimer, S.B. 1980. *Introduction to Embryonic Development*. Allyn & Bacon, New York.

Reverberi, G. 1971. *Experimental Embryology of Marine and Freshwater Invertebrates*. Amsterdam.

Slack, J.M.W. 1983. *From Egg to Embryo*. Cambridge University Press, Cambridge.

Stearns, L.W. 1974. *Sea Urchin Development: Cellular and Molecular Aspects*. Dowden, Hutchinson & Ross, Pennsylvania.

16
CONTROL SYSTEMS

Most of the earlier chapters in this section of the book have concentrated each on a single functional system—feeding, locomotion, respiration and the like—yet it is fundamental to our central thesis that selection acts not on individual attributes in isolation but on whole organisms. All the genes carried by individual animals succeed or fail together.

We have stressed that evolutionary advantage has two components: survival of the individual and its differential reproductive success. Both require the integration of information from the internal and external environments, and the successful carrying-out of appropriate responses. We have also emphasized that animals are continually subject to conflicting pressures. One activity will have repercussions on others and may, for example, expose the individual to increased risks of mortality: trade-offs have to be effected and bets hedged. Organisms therefore require control systems to regulate their development from zygote to breeding adult (as discussed in Chapter 15) and to obtain and process relevant data and then initiate, monitor and modify co-ordinated responses if they are to react in such a way as to maximize the probability of immediate survival. And these they must do at each and every stage of their lives. It is the neurobiological and endocrinological control systems of invertebrates that form the subject of this chapter.

16.1 Introduction

Research on invertebrate control systems is, in practice, undertaken for three main reasons. First, out of scientific curiosity; second, with a view to applications of economic and medical significance; and third, because they provide valuable 'model' systems for the study of learning, feed-back, etc.

A number of species (the 'laboratory rats' of the invertebrate world), mainly insects and molluscs, have been used extensively to investigate problems of fundamental significance, and their study has led to many of the most dramatic advances in neurobiology. Outstanding in this regard is the *giant axon* or nerve fibre of the squid *Loligo* (Box 16.1), but large neurones with cell bodies up to 1 mm in diameter are common in molluscs and it is almost routine for neurobiologists to insert as many as four electrodes into a single cell for purposes of recording, current injection, etc.

Many invertebrate nervous systems are characterized by a simplicity that has enabled the exhaustive description of their structure and of the functional relationships of their components to become a reality. Such knowledge remains only a distant dream with respect to vertebrates. Examples include the entire nervous system of the nematode, *Caenorhabditis elegans*, and the peripheral cardiac and stomatogastric ganglia of crustaceans. Furthermore, many *individual neurones* can be identified from specimen to specimen, and their structure, physiology and roles investigated. The tolerance of many invertebrates to experimental manipulation which would be impossible with vertebrates is also important.

The fruit fly, *Drosophila*, is uniquely placed to contribute to the study of the genetics of behaviour and neurobiology. Typically, the many mutants which have been produced have first been identified by their behavioural defects, but then provide material with which the morphological, biochemical and genetical bases of behaviour can be investigated. *Caenorhabditis* has also been extensively used for this purpose. It has six chromosomes (flies have four) and

Box 16.1 Giant fibre system of the squid, *Loligo*

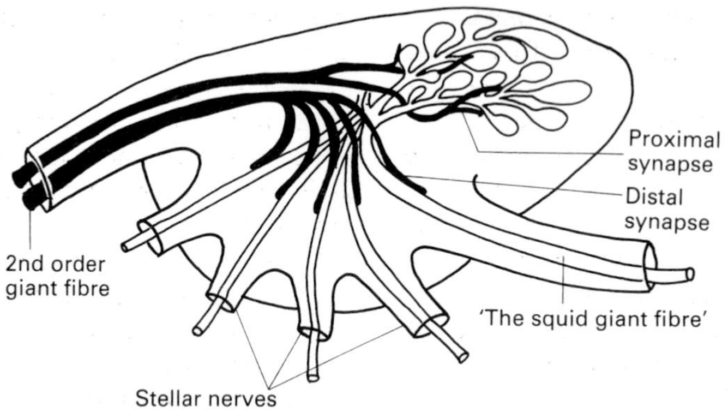

(a)

(a) Two first-order giant neurones send axons which fuse across the mid-line and innervate several second-order cells situated in the posterior part of the brain. Most of the latter are motoneurones and control muscles of the head and funnel directly, but two have axons which extend to the stellate ganglion, where they synapse with the third-order giant fibres (only one pair shown). (After Young, 1939.) (b) Stellate ganglion. Each second-order giant forms 'distal' synapses with the third-order fibres and another giant fibre forms 'proximal' junctions on these also. Each third-order fibre develops by fusion of the fibres of a number of cells and one is present in each stellar nerve (only five nerves shown). The most posterior and largest fibre is *the* giant fibre exploited with such good effect by neuroscientists. (c) Synaptic contacts formed by second-order and third-order giant fibres. Physiological coupling between the two elements is mediated by a multiplicity of junctions, at each of which the post-synaptic process is indented into the pre-synaptic. Arrows show direction of impulse flow.

(b)

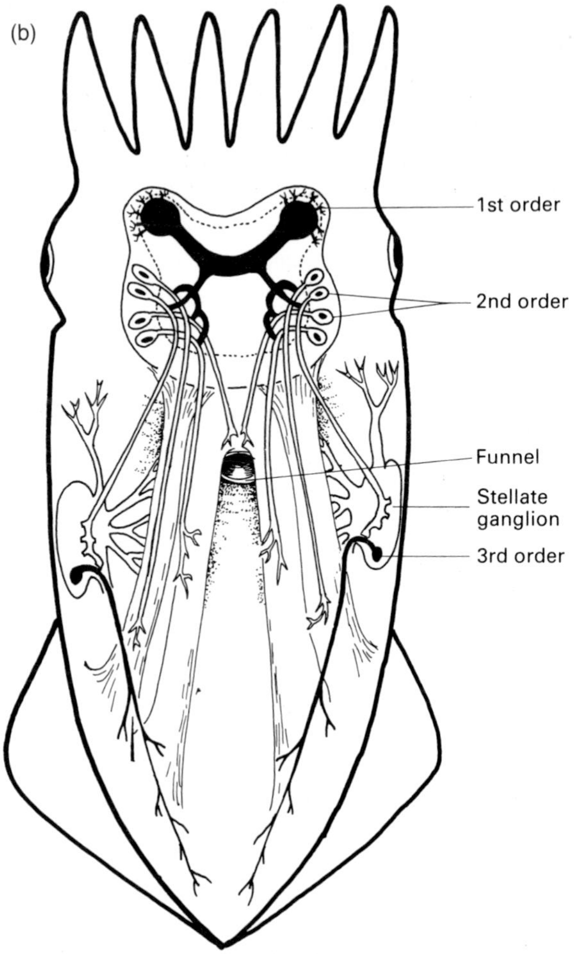

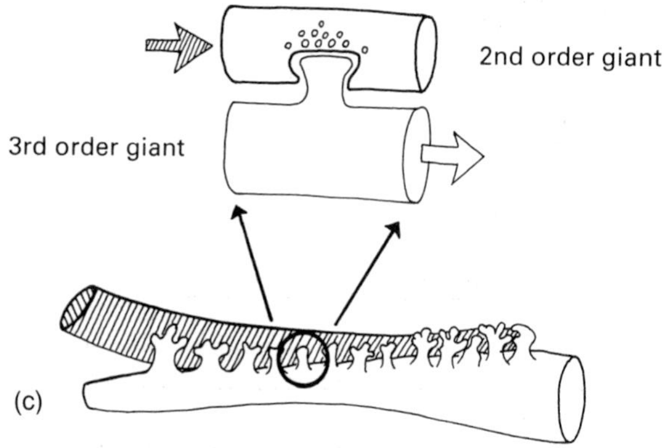

(c)

only 3000–5000 genes. Mutants can be maintained in nutritive media even when they are severely defective (e.g. paralysed) and reproduction is still possible because one of the sexes is a self-fertilizing hermaphrodite. According to a count in 1984, 228 mutations were known, affecting 14 genes, just among those rendering the animal insensitive to touch.

A range of techniques is now involved in the study of nervous systems. Foremost among them are those involving intracellular electrodes and oscilloscopes by means of which electrical potentials can be recorded (Fig. 16.1). A recent development is the 'patch-clamp' technique, by which a minute fragment of cell membrane can be attached to the orifice of a micro-electrode and the function of individual membrane channels studied. Light- and electron-microscopy are also widely used and are frequently combined with techniques that allow coloured dyes (e.g. Procion yellow) or dense materials to be injected into individual nerve cells, revealing the distribution of their processes and the contacts they make with other cells. In more recent years, there has been an explosion of information resulting from the application of the techniques of contemporary biochemistry and molecular biology. Neurotransmitters, hormones, receptors, etc. can be chemically identified and their genes sequenced, and antibodies to the molecules involved can be produced.

Neurobiology has a special relationship with *animal behaviour*. Observation on the behaviour of the whole animal can be compared with the study of a 'black box', such as a calculator or computer. Much can be learnt about the properties of the apparatus by observing it in operation and by comparing input and output, etc. Neurobiology involves 'opening the box' to study its internal structure, the properties of its various components, and the mechanisms by which it functions. As we shall see, neurobiology and endocrinology are most fruitful when carried out in the context of a thorough appreciation of the

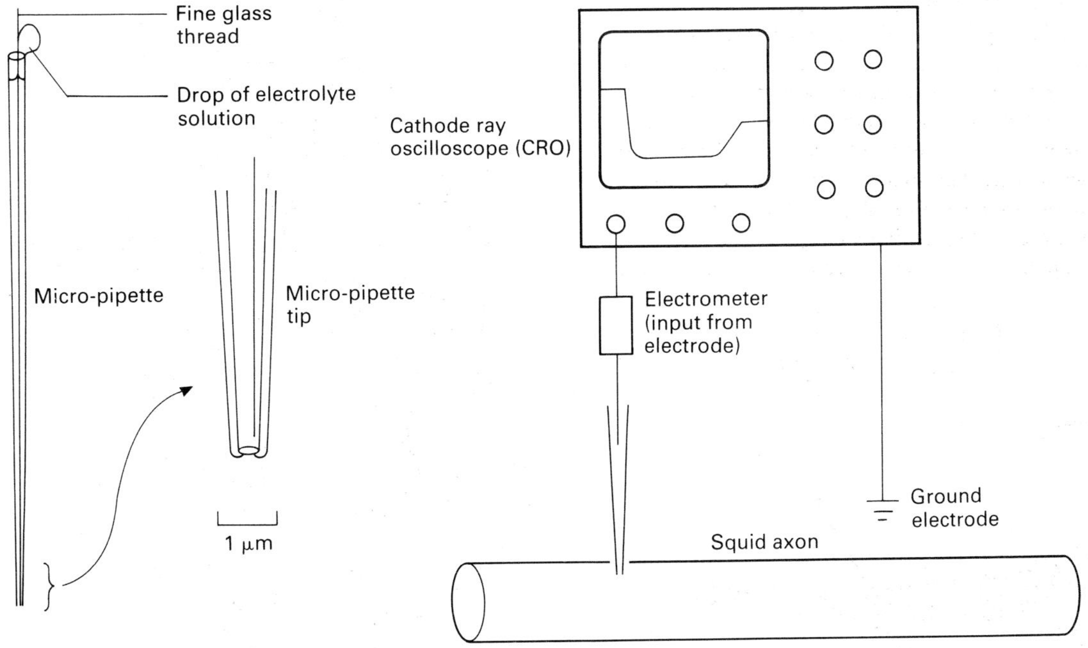

Fig. 16.1. Micro-electrode and the system used for recording intracellular electrical potentials (after Shepherd, 1983).

behaviour, physiology and wider biology of the organisms concerned.

16.2 Potentials

16.2.1 Membranes

According to the *fluid mosaic model*, the plasma membrane of the cell consists largely of a bilayer of lipid molecules whose hydrophilic poles extend outwards. It is usually separated from its neighbour, bounding an adjacent cell, by an electron-lucent, *intercellular space* about 20 nm across. The layer forms a major barrier to the diffusion of ions, etc. Embedded within the lipid bilayer are protein and glycoprotein molecules, many of which span it completely. They are thus well placed to act as 'channels' and 'pumps' by means of which ions can cross the membrane. Some components are anchored in position in specialized regions of the cell surface which are differentiated for reception of stimuli, nervous transmission, etc., but many float freely within the membrane.

The concentrations of ions within the cytoplasm and intercellular fluid, respectively, are affected by *passive movements* like diffusion, and the process of *active transport*. The membranes of nerve fibres have a 'sodium pump' responsible for a net, outward transfer (efflux) of Na^+ ions and the uptake of K^+. This movement occurs against the electrochemical gradient (see below) and is dependent on metabolic energy. The pump consists of an enzyme known as a Na^+–K^+-activated ATP-ase, so called because of its ability to catalyse the breakdown of ATP and thus tap its energy.

16.2.2 Potentials

The presence of *electrical potential differences* across the plasma membrane, a general characteristic of living cells, has its basis in the *differential distribution of ions*, and the *differential permeability of the membrane*. Usually, Na^+ and K^+ ions are most important. If the membrane is permeable mainly to K^+ ions, which are concentrated inside the cell, their diffusion across the membrane will lead to a build-up of negative charge inside the cell, since anions will be present in excess, and of positive charge outside (Fig. 16.2). This potential across the membrane will impede the further efflux of K^+, since opposite signs attract and similar ones repel, until the force due to the concentration difference is balanced by the force exerted by the potential difference. At this point the system is said to be in *electrochemical equilibrium*. In contrast, if Na^+, mainly present outside the cell, is the permeable ion, the inside of the membrane will be positively charged at the point of equilibrium. Such simple changes in permeability are the basis of much of the signalling that takes place within the nervous system.

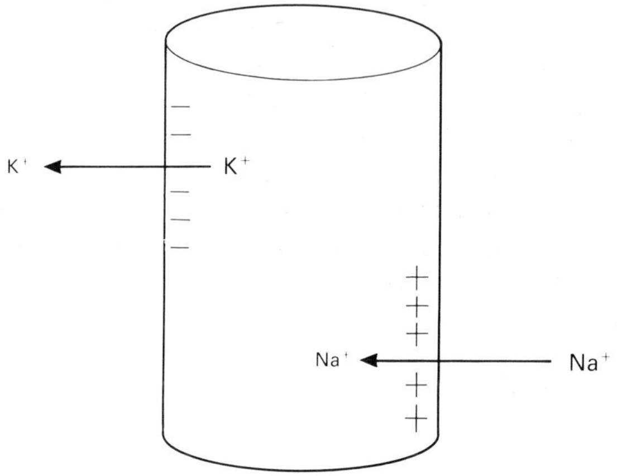

Fig. 16.2. Electrical potentials across the cell membrane result from differential permeability to ions, particularly K^+ (left) and Na^+ (right).

The potential normally associated with the inactive cell is called the *resting potential*. It is determined mainly by the relative concentrations of K^+ because the membrane is very permeable to this ion. This can be readily demonstrated by modifying the level of K^+ ions in the bathing medium and observing the effect. However, it has also proved possible to squeeze out the cytoplasm from the squid axon and replace it with experimental saline. The natural imbalance of K^+ ion concentrations can thus be reversed and the preparation dutifully performs, in accordance with the hypothesis, by reversing the polarity of the resting potential!

Changes in membrane potential can occur *spontaneously*, due to the operation of special channels through which Ca^{2+} ions leak into the cell, or in sensory cells on account of the influence of environmental stimuli. They are also generated at points of specialized contact between neurones called *synapses*, when they are usually caused by chemical substances (neurotransmitters) which are released by one partner and either excite or inhibit the other.

16.2.3 Conduction of graded potentials

The major function of nervous elements is the ability to carry information from one point in the body to another. This is achieved by the *conduction* of changes in membrane potential. Since a nerve fibre has the properties of an electric cable, local changes in potential will spread along it (Fig. 16.3). Such conduction is called *passive* or *electrotonic spread*, and the potential changes involved are both *graded* and *decremental*. Their magnitude depends on the level of the initial potential and the effect will gradually

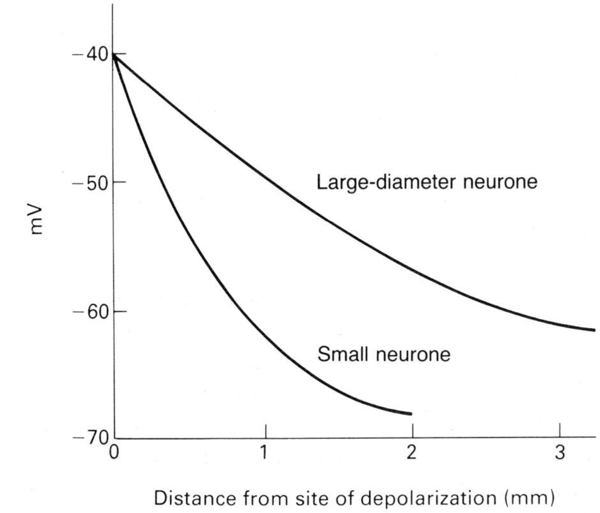

Fig. 16.3. Passive and decremental spread of depolarization down large and small axons, respectively. 'Leakage' is also affected by the efficiency of the glial insulation. (After Darnell *et al.*, 1986.)

fade as current leaks through the membrane—it is about halved during conduction down the axon of the photoreceptor cell in the fly eye. An analogy would be the effect of throwing a stone into a lake. The size of the ripples depends on the size of the stone and on the distance away from the source. Many nerve cells—'*non-spiking neurones*'—seem to function in this way entirely and all neurones have regions which work like this.

16.2.4 Action potentials

Partial depolarization of many axons triggers a rapid change involving loss and reversal of potential difference (Box 16.2). This *action potential* or, popularly, 'spike', shows a *threshold* effect since it is initiated only when the stimulus reaches a certain minimum level. It is an *all-or-nothing* response, since further increases in intensity of stimulation fail to enhance the response. In life, action potentials are triggered by spontaneous, receptor or synaptic potentials which spread passively from their sites of origin into an area of the membrane capable of responding actively. Current will then spread along the nerve fibre from this site. It depolarizes adjacent areas of membrane, stimulating them into action, causing further depolarization and so on. Consequently, an action potential is *actively propagated* along the membrane. It is not graded, but all-or-nothing; not decremental but self-perpetuating. An appropriate analogy is a trail of gunpowder ignited at one end with a match.

The velocity of conduction is dependent largely on the diameter of the fibre and consequently invertebrates typically have *giant fibres* to mediate escape reactions, etc. The marine annelid *Myxicola* holds the record with axons more than 1 mm in diameter, conducting at 20 m s^{-1}. In some cases, the speed of conduction is greatly increased by insulating lengths of the fibre with glial wrappings (Section 16.3.2). Such conduction is *saltatory*—it jumps from one region of uninsulated membrane to the next, being regenerated at each of these. Earthworm giant fibres have two such 'hot spots' on the dorsal surface in each segment; shrimp axons—the 'gold medallists' of the animal world for conduction velocity (200 m s^{-1}

Box 16.2 The action potential

The discovery of the basis of the action potential by Hodgkin and Huxley at Plymouth and Cambridge, and others, led to the award of the Nobel Prize in 1963. Using a technique known as 'voltage clamping', the current carried across the membrane, at different potentials, by individual types of ions was measured. This was done by replacing one of them with a substitute to which the membrane is impermeable; another method is to block individual types of ion channels with drugs: Na^+ channels by tetrodotoxin, extracted from puffer fish, K^+ channels by tetraethylammonium.

The development of action potentials has its basis in the presence in the membrane of *voltage-gated, ion-selective channels*. The Na^+ channels are presumed to change their form when depolarized, opening 'gates' through which Na^+ ions enter the axon, depolarizing the membrane still further. Na^+ channel opening is a transient phenomenon and 'Na^+ inactivation' rapidly ensues.

Voltage-gated K^+ channels are also present. They open after the Na^+ and the diffusion of K^+ ions through them (only one out of every 10^7 K^+ ions present is required per impulse) restores the resting potential and giving rise to the after-potential. A short recovery period follows during which the membrane is refractory and cannot be activated. Ca^{2+} channels make a major contribution to the action potential in many mollusc cells. The *shaker* mutant of *Drosophila* is so called on account of its palsied limb movements during ether anaesthesia. Its voltage-gated K^+ channels are absent or defective—the genes concerned have been identified—and in consequence, action potentials are broadened since repolarization is delayed.

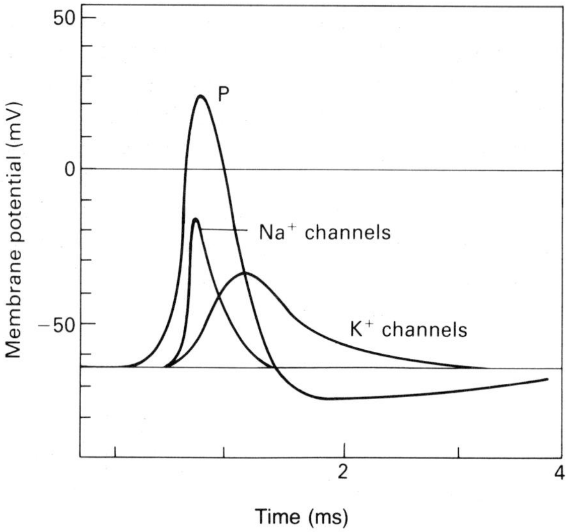

The graph shows the changes in membrane potential P (vertical axis) occurring at any given point along the axon, with time (horizontal axis), and the successive changes in the number of Na^+ and K^+ channels open which underlie the potential changes. (After Hodgkin & Huxley, 1952.)

at 20°C)—have them at the branch points of the axons. The myelin wrappings which insulate nerve fibres in vertebrates are not found in invertebrates.

16.2.5 Spiking and non-spiking neurones

Non-spiking neurones have been described mainly in arthropods and are commonly associated with receptors (e.g. the photoreceptors of insect eyes) and interneurones. The only motoneurones in this category are those innervating the muscles of the body wall in the nematode *Ascaris*. We simply do not know 'why' some neurones use graded potentials, whereas others employ action potentials. It cannot be explained simply on the basis of the length of the nerve fibres involved. Propagated potentials are doubtless indispensable for fibres extending to the extremities of limbs in large invertebrates. However, graded signals are employed by the barnacle photoreceptor with fibres up to 1 cm long, whereas many neurones with far shorter fibres, including some amacrine cells (Section 16.3.1), produce action potentials, as do tiny oocytes, gland cells, etc.

Many cells using graded potentials engage in almost continuous discharge of transmitter, and small changes in membrane potential (as little as 0.3 mV with respect to the photoreceptors of the fly) can modify this in either direction. Even when present in the small numbers typical of invertebrate systems, they can provide smoothly graded control, whereas larger numbers of spiking cells with summed outputs are thought to be required to produce the same effect. With action potentials, the message is apparently 'scarcely more complex than a succession of dots in the morse code' (Adrian, 1932). However, the combination of frequency, pattern and duration of the activity probably provides a highly sophisticated system of coding. Intermediates between the two extremes are common. Cells may have a small region of the membrane which generates a spike, but which is then conducted to the further terminals in a decremental fashion; voltage-gated Ca^{2+} channels along the axon may sustain a potential, which nevertheless remains graded; similar channels in the terminal membrane may amplify a graded signal on arrival.

16.3 Neurones and their connections

16.3.1 Neurones

Nerve cells are characterized by the possession of elongate processes and the ability to conduct electrical potentials. They are traditionally classified as: *sensory* (*afferent*) neurones which convey information into the central nervous system; *motoneurones* (efferent) which carry messages from the centre to the effectors (muscles, glands, etc.); and *interneurones* which link the previous two types (local interneurones are often distinguished from relay cells with longer processes). We can add *neurosecretory cells* that release hormones

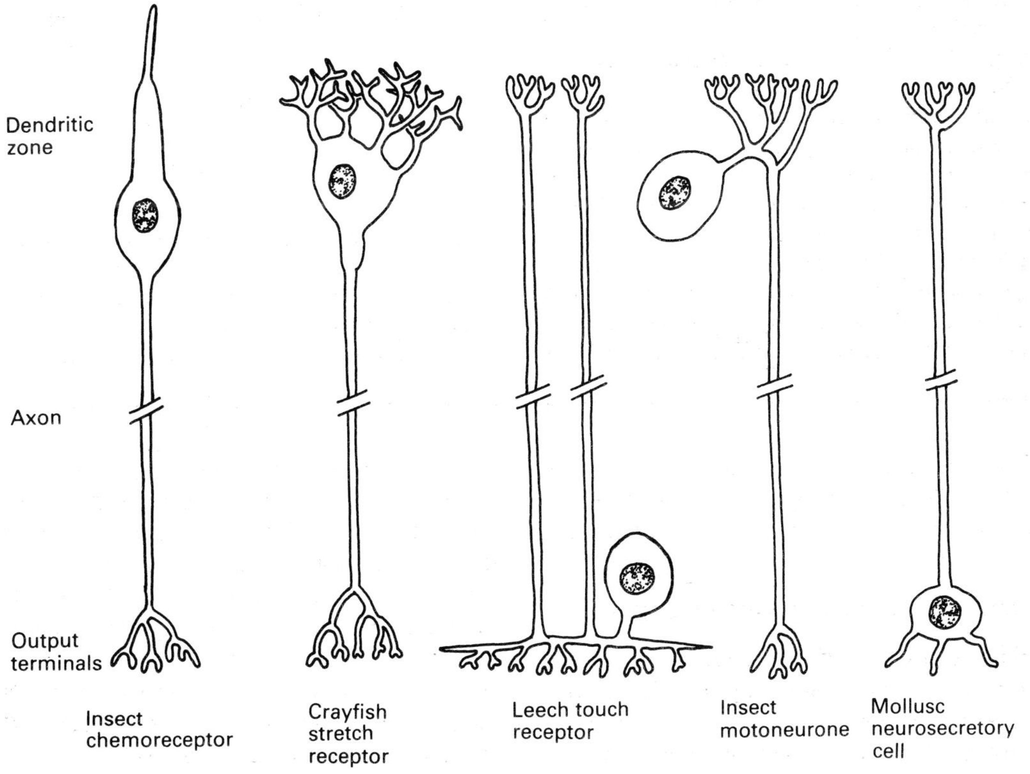

Fig. 16.4. Neurones of diverse morphologies and functions share the same pattern of functional organization, and show a dynamic polarity. Some neurones are monopolar, with a single process arising from the cell body (e.g. the insect motoneurone); others are bi- or multipolar.

into the blood stream. However, cells with the characteristics of both sensory and motoneurones, and other combinations, have also been described.

Nerve cells are often described in terms of a common pattern of functional morphology (Fig. 16.4). The *input* (or *dendritic) zone* is the sensitive receptor region of sensory neurones, and the synaptic region of other cells. Graded receptor or synaptic potentials arise here and spread across the membrane. Spiking neurones have an *impulse generation zone*, where graded potentials which reach the threshold trigger the generation of action potentials. The *axon* conducts potentials to the *nerve terminals* from which synaptic transmission occurs (i.e. communication with another cell). Note that the position of the cell body is irrelevant to these aspects of neurone physiology.

This concept of the *dynamic polarity* of neurones requires modification in that many dendrites are sites of information transmission as well as reception, and pre-synaptic terminals may receive information as well as transmit it (Fig. 16.5). Furthermore, some nerve cells have axons which probably conduct impulses in both directions and others (*amacrine* cells) lack a process recognizable as an axon and have branches that are probably capable of independent activity.

16.3.2 Glial cells

The second category of cells present in the nervous system are called *neuroglia*. The cells have a *mechanical* role, in supporting, separating and ensheathing nervous elements, an electrical role in insulating nerve fibres and enhancing the rate of nervous conduction, and the metabolic role of controlling the ionic environment within the nervous

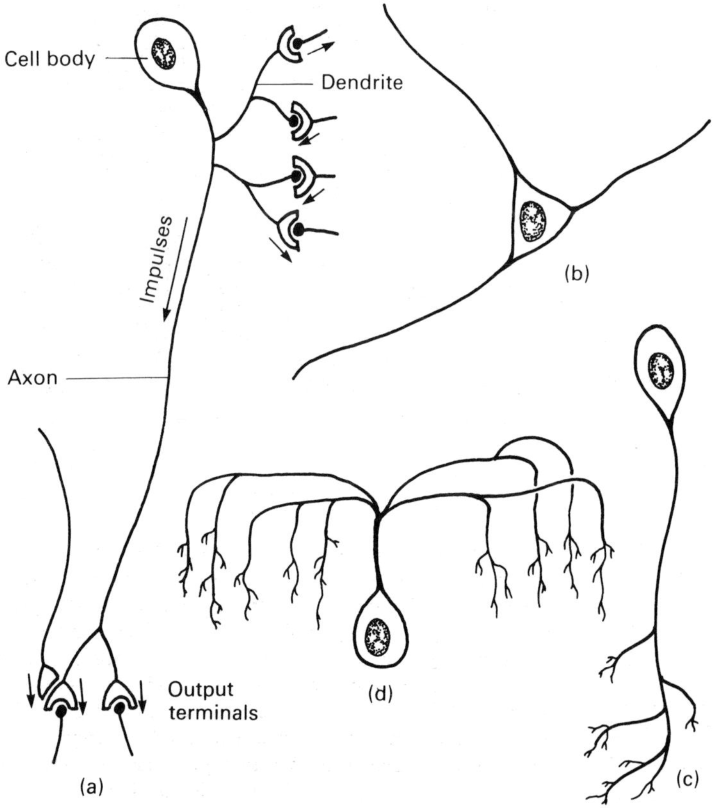

Fig. 16.5. (a) In motoneurones of the crustacean stomatogastric ganglion, and many others, the dendrites both receive and transmit information. Similarly, the output terminals may also be post-synaptic to other fibres. However, impulse traffic along the axon is, in nature, one way only. (b) Cnidarian isopolar neurone. (c) Neurone of polychaete corpora pedunculata. In neurones like these, dynamic polarity is apparently absent. (d) Amacrine cells, e.g. from the insect optic lobe, lack a major process identifiable as an axon.

system and breaking down neurotransmitters following their release from neurones.

16.3.3 Chemical synapses

Most neurones transmit information by means of the secretion of chemical substances called neurotransmitters or neuromodulators, but, whereas conventional gland cells disseminate their secretory products far and wide, neurones administer them in a highly localized and selective manner at specialized junctions with other cells, called synapses.

Chemical synapses are most frequently formed by the terminal regions of nerve fibres. The contacts may involve bulbous endings or swellings ('varicosities') formed along the length of the fibre. Many synapses have such a highly characteristic ultrastructure (Fig. 16.6) that the existence of a functional contact is usually presumed merely on the basis of observation with the electron microscope, although this correlation remains a subject for debate. It is also noteworthy that a 'synaptic contact' between two cells, detected electrophysiologically, may in fact be mediated by thousands of junctions (25 000 in the lobster stomatogastric ganglion) visible with the electron microscope.

Conventional neurotransmitters such as acetylcholine are stored within synaptic vesicles, and it is generally thought that release occurs by fusion of the vesicle membrane with the terminal membrane thickening with the subsequent discharge of the vesicle contents into the synaptic cleft—the process of *exocytosis*. Release apparently occurs at low levels even in resting terminals, but the arrival of an action potential (or a depolarization as small as 0.3 mV in a non-spiking neurone such as the fly photoreceptor) provokes exocytosis of large numbers of vesicles simultaneously. Secretory granules are storage sites for peptide neurotransmitters and modulators, and also show release by exocytosis (see Box 16.11). The fundamental mechanisms of synaptic function have been investigated using the synapses of the giant fibre system of the squid, since both pre- and post-synaptic terminals are large enough to be impaled with microelectrodes. Llinas and others have shown that the

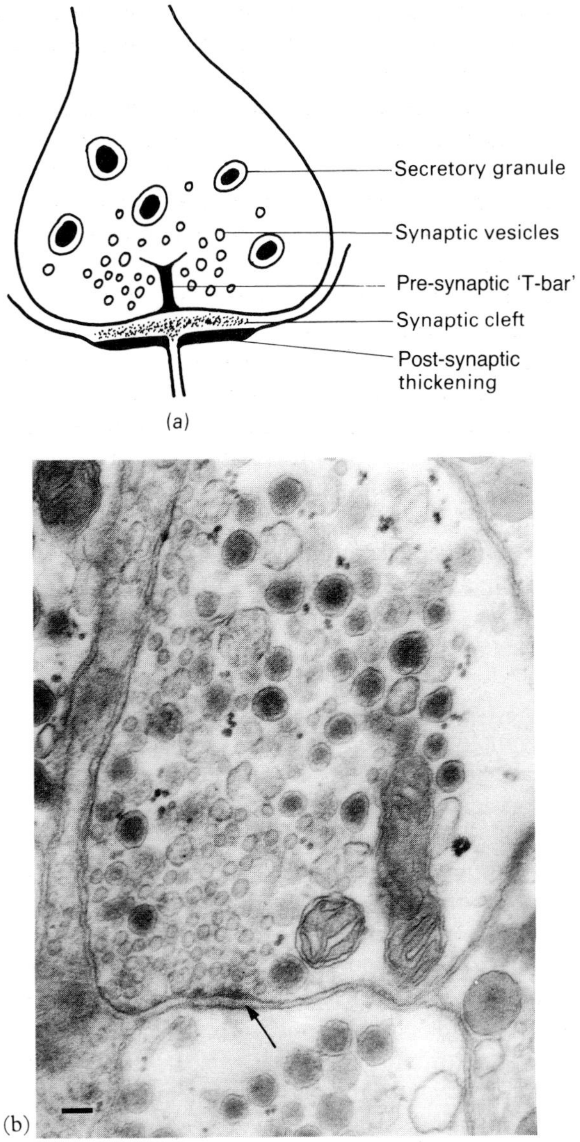

Fig. 16.6. (a) Pre-synaptic terminal of a chemically transmitting synapse in an arthropod. The synaptic vesicles are usually 30–50 nm in diameter and cluster around the pre-synaptic bar; secretory granules usually stand back and may dominate the body of the terminal. Many synapses lack one or more of the features shown—in echinoderms, neither specialized clefts, thickenings, nor projections are visible. (b) Electron micrograph of a synapse in the cerebral ganglion of the earthworm, *Lumbricus*. A diffuse, pre-synaptic thickening is present. Arrow, synaptic cleft. × 40 000. Bar, 100 nm.

arrival of an action potential leads to the opening of voltage-gated Ca^{2+} channels in the pre-synaptic membrane, and it is the elevated level of Ca^{2+} within the terminals which activates transmitter release rather than depolarization as such.

Following release, transmitter diffuses across the synaptic cleft and binds to protein receptor molecules embedded in the post-synaptic membrane. When activated, the receptor may function as a selective ion channel and such effects are typically extremely rapid, lasting for only a few milliseconds. Alternatively, a receptor may respond by activating the synthesis within the cell of a *second messenger*, such as cAMP (cyclic adenosine monophosphate) which has longer lasting metabolic effects on ion channels.

A single cell will usually receive a multiplicity of synaptic inputs, involving different transmitters, inducing the generation of excitatory (depolarizing) and inhibitory (usually hyperpolarizing) post-synaptic potentials, respectively. Cells are also influenced by neuromodulators—so-called because they have less clear-cut effects (e.g. Section 16.10.5). The actions of transmitters are rapidly terminated by enzymic degradation (acetylcholine, peptides) or re-uptake into the pre-synaptic terminals (amines, amino acids).

One advantage of chemical synapses is that they can amplify a signal. For example, transmitter (probably histamine) released from the terminals of the fly photoreceptor neurones in response to a given depolarization induces a change in membrane potential (hyperpolarization in this case) 7–14 times greater in the post-synaptic, second-order neurones. Another attribute of chemical synapses that is crucial to the function of the nervous system is their great *flexibility*—indeed there is evidence that this is the basis for various types of learning (Box 16.3).

16.3.4 Electrical synapses

One of the great controversies in the history of neurobiology concerned the mechanism of synaptic transmission. It is ironical that when the matter had apparently been finally resolved in favour of the chemical hypothesis, mainly as a result of work on peripheral juctions in vertebrates, electrical synapses were discovered in the crayfish (see Box 16.9) by Furshpan & Potter in 1959. Electrical synapses enable one cell to stimulate another directly, without intervention of transmitters and their receptors. Transmission depends on the presence of *gap junctions** at which protein molecules, forming hollow cylinders through which current can flow, span the intercellular space allowing depolarization in one partner to spread directly to the other. Transmission is rapid, involving a fraction of the synaptic delay of about 0.4 ms for chemical synapses. In many cases, transmission can proceed in either direction, but, in others, the synapses are *rectifying* and provide a pathway with low resistance in one direction only.

In some cases, electrical synapses are the means by which a number of cells are *integrated into a single functional unit*. The effect may be to synchronize the output of a group of cells (Fig. 16.7)—examples include the receptor cells within the units ('ommatidia') in the locust eye and neurosecretory cells which secrete spawning hormone in molluscs (as well as many gland and muscle cells). Neurones with giant fibres are usually linked with electrical synapses, giving the system a rapidity of function approaching that of a huge, single cell (see Fig. 16.25).

Embryonic neurones are extensively coupled by gap junctions, many of which are lost as ion channels and chemical synapses develop.

16.4 Organization of nervous systems

16.4.1 'Neuroid' system in sponges

No true nervous system can be formed in sponges since the 'tissues' are always in a state of flux. However, these organisms have contractile ability, both of the entire body and just the exhalent oscula. Some of them 'back-flush' to clear their canals and

*Gap junctions are not to be confused with *tight junctions* where fusion of the outer leaflets of the membranes occurs. Tight junctions prevent diffusion through the intercellular space and are the structural basis of *blood–brain barriers*, by means of which, in insects, cephalopods and vertebrates, the chemical environment of the nervous system is regulated.

Box 16.3 Cell biology of learning

The same basic types of learning can be observed throughout the animal kingdom, but only recently have their mechanisms been investigated, by Kandel and others, in terms of cell biology.

The mantle cavity in the mollusc *Aplysia* has a protective shelf and a siphon. Gentle touch activates a sensory pathway from each, and the neurones involved synapse directly on the motoneurones that control a defensive contraction of the ctenidium (a monosynaptic reflex arc—see Section 16.9.3). Sensory neurones innervating the tail stimulate interneurones which synapse on the sensory terminals.

The pre-synaptic terminals of the sensory neurones are the site of changes responsible for behavioural modification, i.e. for learning. Frequent stimulation of the respiratory structures, causing the repeated invasion of the terminals by action potentials, leads to a gradual inactivation of the Ca^{2+} channels (Section 16.3.3), a reduction in the quantity of transmitter released per stimulus, and a diminished behavioural response—the simplest form of learning, called *habituation*.

If an intense, noxious stimulus is applied to the tail, the response to a wide variety of other stimuli, including gentle touching of the respiratory structures, is enhanced. This form of non-associative learning, known as *sensitization*, is mediated by the release of the transmitter 5–HT by the interneurones, which initiates the following cascade of metabolic effects within the sensory terminals:

5–HT → Synthesis of cAMP → Activates protein kinase (an enzyme) → Phosphorylates special K^+ channel proteins in membrane
↓
K^+ channels inactivated → Action potentials longer lasting → Greater Ca^{2+} influx → Enhanced transmitter release

Conditioning is a form of *associative learning* (remember Pavlov's dogs?). Thus, if a series of mild stimuli applied to the gill shelf are, in each case, immediately followed by the noxious stimulation of the tail, the animal will start to respond far more vigourously to shelf stimulation alone, than to siphon stimulation for example. Such pairing of stimuli in the way described leads to the release of 5–HT from the interneurones, and to cAMP synthesis within *all* the sensory terminals, *at the same time* as the level of Ca^{2+} is elevated within the terminals of the stimulated, shelf sensory neurones *only*, by the arrival of action potentials. Ca^{2+} is thought to activate cAMP which thus has a potency in enhancing transmitter release beyond that seen during sensitization.

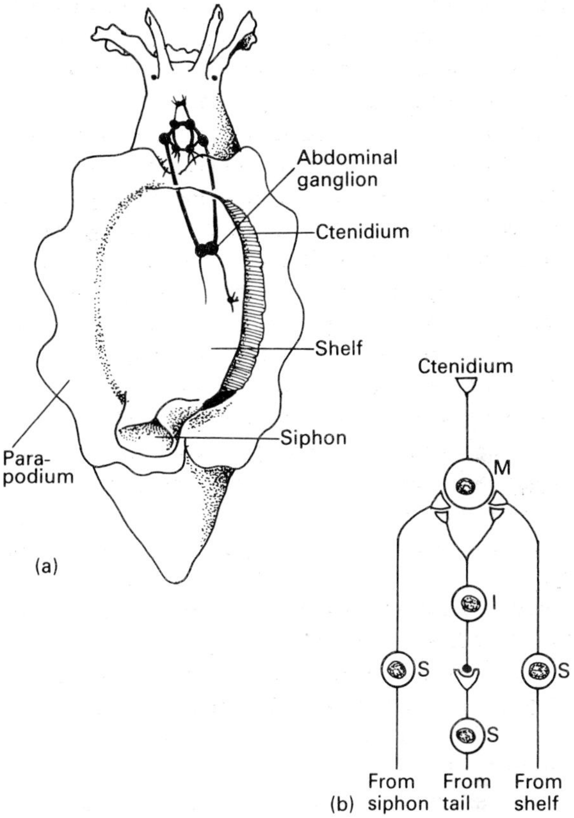

(a) View of nervous system and respiratory chamber of *Aplysia* from above. (b) Synaptic connections made by neurones within the abdominal ganglion used to investigate the cell biology of learning processes. S, sensory neurones; I, interneurone; M, motoneurone.

The forms of learning described above are *short-term*—their memory endures only for minutes or at most hours. However, training programmes can be undertaken which result in the development of *long-term memory* (more than 3 weeks in *Aplysia*). This depends not on transient metabolic effects but on *structural changes*. The number and size of synaptic sites, and the number of synaptic vesicles, are all significantly increased by long-term sensitization and reduced by habituation.

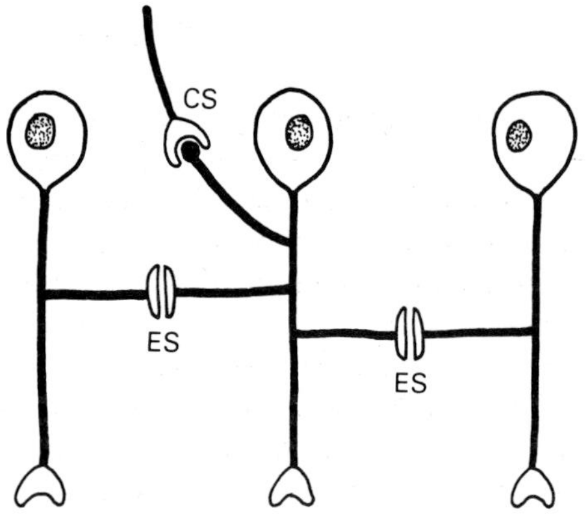

Fig. 16.7. Electrical synapses confer the properties of a single functional unit on a population of cells. CS, bulb indicates chemical synapse; ES, bars indicate electrical synapse.

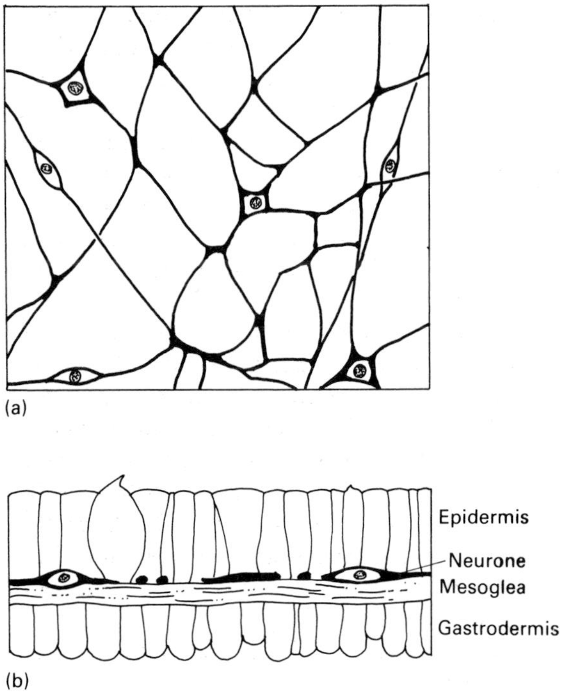

Fig. 16.8. (a) Surface view of a cnidarian nerve net composed mainly of isopolar and bipolar neurones (details of synapses not shown); (b) transverse section, showing nerve net within the epidermis. Where nerve nets are present in both epithelia, they may communicate across the mesoglea.

some eject their larvae by vigourous, apparently co-ordinated contractions. Many responses to stimulation are probably due only to the spread of mechanical effects—contraction of myocytes stretching adjacent cells, and so on. However, in *Rhabdocalyptus*, electrical conduction is probably involved since an electrical stimulus generates a response that is propagated over the whole body at about 0.25 cm s^{-1} and causes complete shut-down. Epithelial conduction (see below) represents another type of neuroid function.

16.4.2 Nerve nets

A nerve net is a diffuse, two-dimensional plexus of bi- or multipolar neurones (Fig. 16.8) and such systems constitute the characteristic grade of organization of cnidarian nervous systems. Their elements are located between the bases of the cells of the epithelium, above the mesoglea. They may be present in either or both of the two cell layers and show an conduction velocity in the order of 10, to that of 100 cm s^{-1}.

A distinctive feature is the way activity spreads out in every direction, from any point of stimulation. Early investigators cut tortuous patterns from sheets of cnidarian body wall and demonstrated that activity could spread through them—this presumably results from the presence of symmetrical (i.e. two-way) synapses, although both conventional, polarized junctions and reciprocal pairs of polarized synapses have also been described. In the stalk of the colonial *Namonia*, the cells are probably coupled by electrical synapses. Some nets are *through-conducting* whereas in others, conduction is *decremental*—it fades out with increasing distance from the point of initiation (Section 16.9.3).

A nerve net can show regional differentiation. For example, the mesenteries of sea-anemones have numbers of large, bipolar neurones, with a vertical orientation, which provide a rapid, through-conducting pathway. Furthermore, two distinct nerve nets may be present in the same epithelium. Scyphozoans (e.g. *Aurelia*) have a number of ganglia distributed around the margin of the bell. The ganglia receive sensory input from the 'diffuse nerve net', but provide motor output to the swimming muscles, via the more rapidly conducting 'giant fibre nerve net'—the first net to be described, by Schafer in 1879. The two nets cover much the same area of the under surface of the bell, but do not communicate directly.

The patterns of behaviour shown by cnidarians such as anemones include feeding, swimming, climbing on to the shells of commensal hermit crabs, 'walking' on the tentacles, moving the pedal disc, and burrowing. Their complexity and integration are very surprising in view of the apparently low level of organization of the nervous system.

In platyhelminths, a number of nerve nets are present and have the same functional characteristics as those described above. Two nerve nets combining sensory and motor functions may be present within and immediately beneath the ventral epidermis, and a third, purely motor net spread below the underlying muscle layer. Distinct nerve nets associated with the dorsal surface, the pharynx, and other parts of the gut have also been described. The nets have connections at a deeper level with nerve cords. Nerve nets are probably important in other groups, e.g. in echinoderms, the foot in molluscs, and the innervation of guts (even in vertebrates).

16.4.3 Central and peripheral nervous systems

An important distinction is that between the central nervous system, consisting of nerve cords and ganglia, and the peripheral nervous system consisting of nerves and receptors. The central nervous system acts as the destination of sensory input, the centre of synaptic integration, and the source of motor control, whereas peripheral nerves act mainly as conducting pathways (nerve nets combine both roles). Diminutive, *peripheral ganglia* are intermediate in that, in life, they are always connected to and influenced by the central nervous system, but are capable of a great measure of independent activity. Examples include the ganglia of the small, pincer-like organs of defence (pedicellariae) of some echinoderms (Fig. 7.18), the cardiac ganglia of crustaceans and the parapodial ganglia of polychaete annelids.

It used to be thought that virtually all sensory neurones had their cell bodies in the periphery in invertebrates, but it is now clear from many groups that many lie in the central nervous system. A curious feature in nematodes, and some platyhelminths and echinoderms (see Fig. 16.9e), is that instead of nerve fibres running from the motoneurones to the muscle blocks, the muscle cells have long, axon-like processes which extend to form post-synaptic terminals at the surface of the nerve cord.

Epithelial conduction with a velocity of 3–35 $cm\ s^{-1}$ is important in the medusae of hydrozoans and in the urochordates. In some cases, it enables the epithelium to be used like a single receptor organ with an enormous surface area, as in *Sarsia*, where mechanical stimulation anywhere on the outer surface of the bell initiates an action potential which spreads, via gap junctions between the cells, across the whole epithelium. Typically, epithelial conduction provides input for the nervous system (e.g. via electrical synapses with epidermal sensory neurones in the urochordate *Oikopleura*). It is also employed in motor control to spread the excitation originating in the nervous system to a sheet of muscle fibres or throughout a ciliated epithelium (e.g. of the urochordate pharynx).

In *hydrozoan medusae*, two circular nerve cords are situated close to the margin of the bell. The inner one is *medullary* in character, with nerve cell bodies (mainly bipolar) evenly scattered throughout, whose fibres extend parallel to each other. Medullary cords in *platyhelminths* may be arranged according to the *orthogonal pattern* (Fig. 16.9a), regarded as being the basic plan for the protostome invertebrates. The anterior regions of the nerve cords may be merely thickened, but in higher platyhelminths and in nemertines a well-differentiated cerebral ganglion (or 'brain') is formed.

In most *annelids*, pairs of ectodermal ganglionic rudiments in each segment become linked by connectives and typically form a single internal *ventral nerve cord* running the length of the body (Fig. 16.9b). Pairs of nerves extend to the periphery in

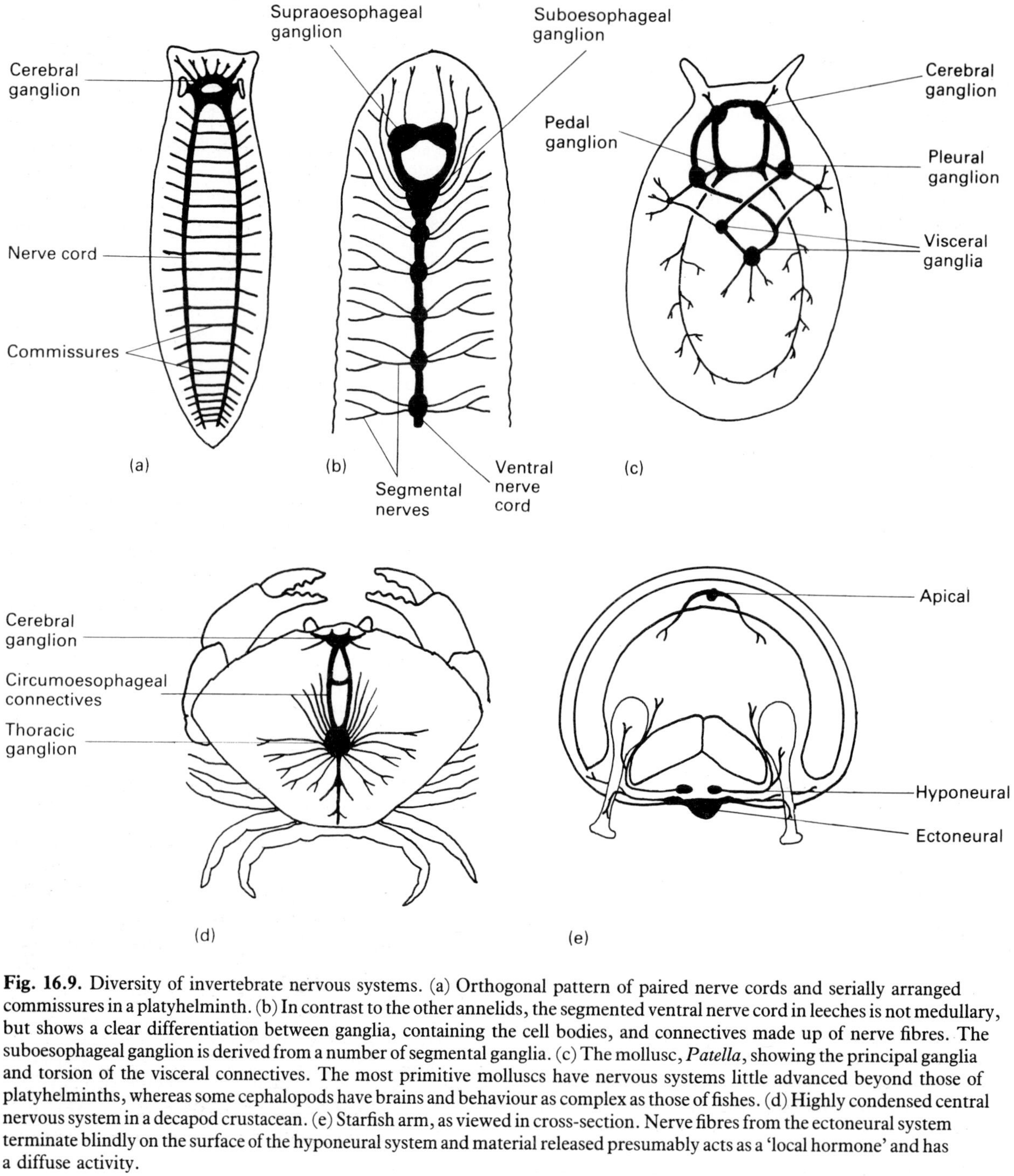

Fig. 16.9. Diversity of invertebrate nervous systems. (a) Orthogonal pattern of paired nerve cords and serially arranged commissures in a platyhelminth. (b) In contrast to the other annelids, the segmented ventral nerve cord in leeches is not medullary, but shows a clear differentiation between ganglia, containing the cell bodies, and connectives made up of nerve fibres. The suboesophageal ganglion is derived from a number of segmental ganglia. (c) The mollusc, *Patella*, showing the principal ganglia and torsion of the visceral connectives. The most primitive molluscs have nervous systems little advanced beyond those of platyhelminths, whereas some cephalopods have brains and behaviour as complex as those of fishes. (d) Highly condensed central nervous system in a decapod crustacean. (e) Starfish arm, as viewed in cross-section. Nerve fibres from the ectoneural system terminate blindly on the surface of the hyponeural system and material released presumably acts as a 'local hormone' and has a diffuse activity.

each segment. Anteriorly, a supra-oesophageal ganglion develops and many polychaetes have well-developed sensory tentacles, eyes and olfactory organs providing input to different parts of the brain (see Fig. 4.59). *Arthropods* also have segmented nerve cords, but in many groups the tendency for segmental rudiments to become incorporated into compound ganglia is taken far further (Fig. 16.9d). A well-developed brain is commonly formed, with regions comprising the proto-, deutero- and tritocerebrum, respectively. Nervous systems in *molluscs* consist of a number of ganglia, connected by commissures and connectives, from which nerves extend to the periphery. Features of interest include the figure-of-eight arrangement of the connectives (Fig. 16.9c)—a consequence of the process of torsion—in several gastropods. In many groups, ganglia tend to be concentrated to form a circum-oesophageal ring.

Nematodes have a number of longitudinal, medullary nerve cords (two main and four smaller in *Ascaris*) linked by a nerve ring round the oesophagus. In *echinoderms*, as exemplified by starfishes, the *ectoneural* system retains a primitive epidermal position (Fig. 16.9e) and combines sensory and motor roles. A radial nerve extends down the lower surface of each arm and there is a circum-oral

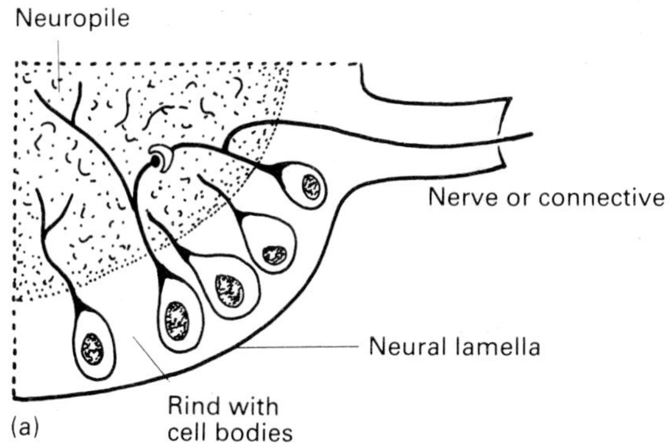

(a)

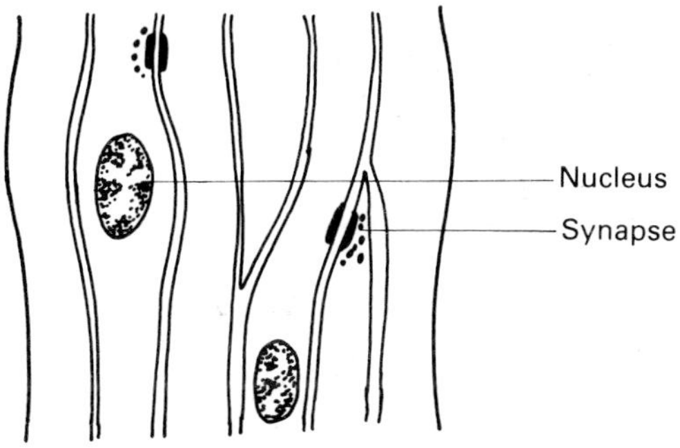

(b)

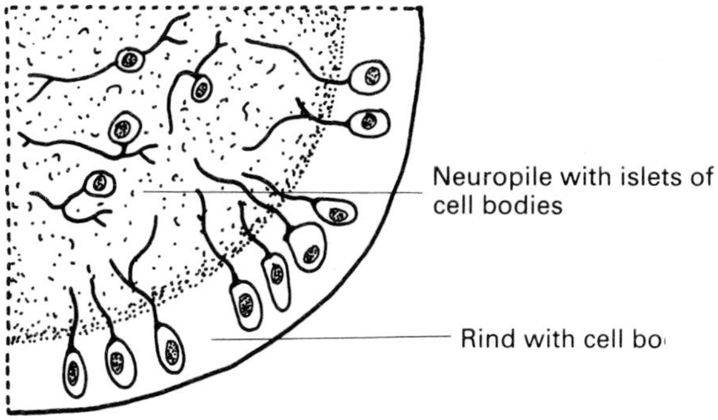

(c)

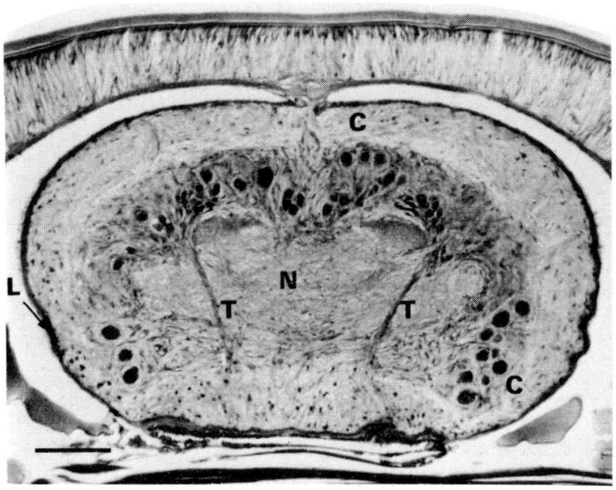

(d)

Fig. 16.10. Patterns of differentiation of neuropile. (a) differentiated neuropile; (b) parallel array; (c) advan islet; (d) transverse section of the cerebral ganglion of t polychaete annelid *Nereis*. C, neurone cell bodies; L, neural lamella or brain capsule; N, neuropile; T, tracts of neurosecretory axons extending to neurohaemal areas on the brain floor (cf. Section 16.11.3). × 125; bar 100 μm. (From Golding & Whittle, 1974.)

Box 16.4 Making the right connections

However great their final complexity, nervous systems develop from simple epithelia. Fibres extending out from the embryonic cells have tips forming *amoeboid growth cones* complete with pseudopodia, etc. The formation of the 'right' connections is crucial to the future function of the nervous system (Section 16.8.4), and the mechanisms involved have been investigated in insects such as grasshoppers, since their embryonic ganglia are fairly transparent and contain limited numbers of cells. Individual cells and even their growing axons can be reliably recognized.

Whereas some cells develop, others are *redundant* and are eliminated. Although the various ganglia in the nerve cord in the insect embryo possess the same number of cells, some cells in the abdominal ganglia are lost, whereas their counterparts in the thorax are developed to control the limbs (they survive even if the limbs are removed). Similarly, in the nematode *Caenorhabditis*, hermaphrodite worms have 302 neurones, whereas males retain an extra 79 in connection with reproductive functions. This process

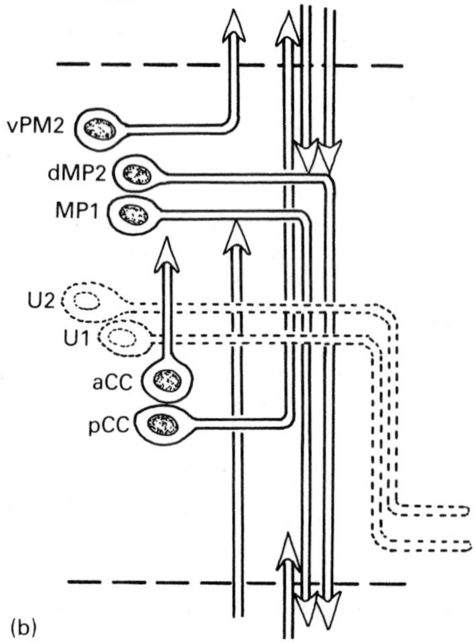

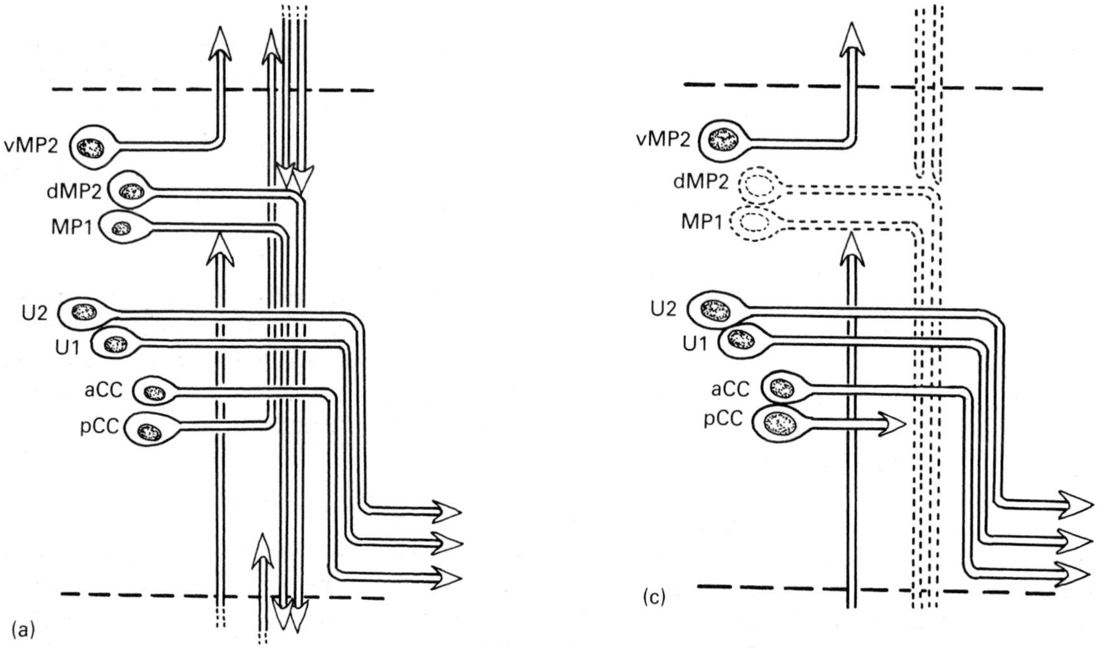

(a) Development of neuronal pathways in the grasshopper segmental ganglion on the seventh of 20 days of development; (b) Following removal of the pioneer neurones U1 and U2, aCC goes astray; (c) Removal of the pioneer neurones MP1 and dMP2 from this and adjacent ganglia—pCC goes astray. (After Bastiani *et al.*, 1985.)

Box 16.4 continued

of *programmed cell death* is genetically determined. The cells which die have the same developmental 'programme' within their genetic make-up as their surviving siblings, but, in their case, an inbuilt 'suicide programme' is switched on. One of the most interesting mutants in *Caenorhabditis* (of gene *ced–3*) is defective in this regard. Cells which normally die survive and form functional connections—apparently the animal doesn't even notice!

The outgrowth of fibres from different cells in ganglia of grasshopper embryos occurs in a distinct sequence, and the first to complete this process are called *pioneer fibres*. Whether growing from one ganglion to the next, or extending from epidermal sensory cells into the central nervous system, pioneer fibres follow stereotyped pathways. The growth cones appear to be guided by a variety of 'cues' provided by their surroundings, but particularly seem to follow basement membranes and glial cells.

Fibres growing out later often follow the pioneers since, in the absence of the latter, in experiments in which individual embryonic cells are removed by laser-beam microsurgery, they go badly astray. If the development of a cell is experimentally delayed, its axon will still follow the right route even though many additional features have appeared in the mean time.

Monoclonal antibodies have been produced from *Drosophila* central nervous system and used to show that the different fibres forming a bundle have a particular molecule in common. It is thought that this acts as a *recognition factor* and that a neurone is 'programmed' to follow first one of the chemical signposts and then another. Furthermore, although a given identified neurone may show significant differences in the completed patterns of its dendrites, etc., from one individual to the next, it forms the same synaptic connections in virtually every case.

---

ring. A deeper, exclusively motor *hyponeural* system is present, and an aboral, *apical* system is of major importance in crinoids. Ascidian tunicates have a single, cerebral ganglion situated midway between the two siphons, from which nerves run to the periphery. In the 'tadpole larva', a hollow neural tube thought to be homologous with that of the vertebrates extends back from the ganglion along the tail.

16.4.4 Differentiation of neuropile

At least three patterns of histological organization of nerve cords and ganglia may be distinguished (Fig. 16.10). Most commonly, regions of *differentiated neuropile* are present. Neurone cell bodies, often enveloped by glial cells, are segregated within the peripheral region or rind. Their branching processes contribute to the complex weft of nerve fibres forming the core or neuropile where the great majority of synaptic contacts are made, although some are formed on the cells (e.g. in molluscs).

A *parallel array* of neurones is seen in the nerve cords of nematodes and the inner ring of jellyfish. Cell bodies are not segregated histologically and synaptic contacts are made between adjacent fibres. In nematodes, neurones typically each have a single process (unbranched in *Caenorhabditis*; with two branch points at most in *Ascaris*) and the possible synaptic partners of a given fibre are restricted to those adjacent to it (these constitute the 'neighbourhood' of the fibre).

Lastly, an *advanced islet* pattern of organization is present in the brain of *Octopus*. Cell bodies are not only present as a rind, but are also scattered within the neuropile—a pattern comparable to that of vertebrate grey matter.

16.4.5 Segmental arrays

Many elongate animals whose locomotory patterns involve the production of waves of contraction travelling up or down the body (e.g. annelids; see Chapter 10) organize their muscles as a series of blocks or somites which can be activated in sequence. Within the nervous system, each segment is equipped with a 'standard set' of neural components, which

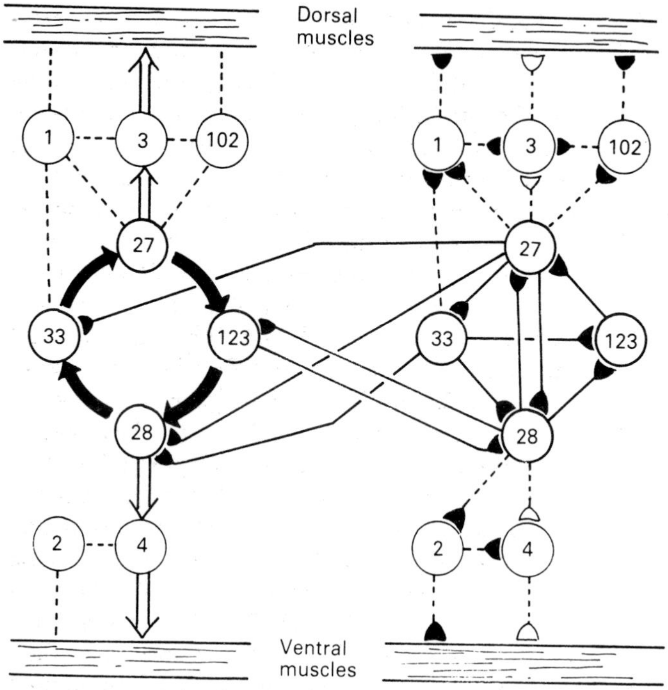

Fig. 16.11. Segmentally replicated set of neurones (identified by numbers) controlling swimming in the leech. The anterior ganglion (left) shows the oscillating pattern of nervous activity in the quartet of interneurones, which ensures that the excitatory motoneurones 3 and 4, inducing contraction of the dorsal and ventral muscles, respectively, are activated alternately. The posterior ganglion shows the synaptic connections within and between the ganglia. Clear bulbs, excitatory; filled bulbs, inhibitory chemical synapses; a rectifying electrical synapse is also formed between cells 33 and 28. (After Friesen *et al.*, 1976.)

thus show serial repetition along the length of the body. The relevant units and their interconnections have been identified in some invertebrates and a 'wiring diagram' of the system controlling swimming in the leech is given here (Fig. 16.11; also Box 16.4).

The set will usually include motoneurones with axons to the muscles, local interneurones to generate the appropriate pattern of activity (Section 16.10) and relay interneurones co-ordinating activity in adjacent segments—in the leech by inhibitory fibres. Not shown in the figure but also present are sensory cells to generate proprioceptive feed-back from the muscles and others whose activation initiates swimming.

Serial repetition is not restricted to segmented animals. The radial nerves of echinoderms and the axial cords in the arms of cephalopods show this effect, which involves groups of cells, areas of neuropile, and bilateral nerves extending to the periphery. In the nematode *Ascaris*, there are five sets of motoneurones, each of eleven cells, serially arranged along the body.

16.4.6 Hot lines

When an earthworm with its front end extended from the burrow encounters the proverbial 'early bird', intersegmental relays as described above, conducting at about 3 cm s^{-1}, are inadequate—a through-conducting pathway is required. Such pathways frequently consist of 'giant' axons (Section 16.2.4) originating, for example, in earthworms from a series of cells (one per ganglion), some of which have fused longitudinally to form multicellular syncytia, others being connected by electrical synapses. In hydrozoan medusae, a bundle of giant fibres, coupled with electrical junctions, permits the very rapid spread of excitation to all parts of the inner nerve ring and in consequence the symmetrical contraction of the bell required for swimming. In other animals, such pathways exert general stimulatory or inhibitory influences on segmental systems whose patterns of activity are controlled in detail by intersegmental relays (see Fig. 16.28). Of course, many fibres within nerve cords are intermediate between the two extremes described above and extend across a few segments.

16.4.7 Brain power

In the more complex invertebrates particularly, anteriorly situated ganglia are especially well developed. This tendency is part of the more general process of *cephalization*, or head development.

First, there can be no doubt that this phenomenon is largely related to the inevitable *concentration of sense organs* at the front end of mobile, bilaterally symmetrical animals. The role of the brain in the processing and integration of sensory input, and of responding to the latter, is well illustrated in insects. The protocerebrum with its large optic lobes receives information from the eyes, the deuterocerebrum from the antennae with their well-developed chemroceptors, and the tritocerebrum from the anterior region of the alimentary canal. A prominent feature of the protocerebrum in many arthropods, as of the cerebral ganglia of higher polychaetes and even platyhelminths, is the presence of *corpora pedunculata* or 'mushroom bodies'. They consist of large numbers of tiny neurones (less than 3 μm in diameter in insects) whose bundled axons form the 'stalks'. They are thought to be involved in the integration of sensory information provided by the different cephalic sense organs together with that from other parts of the body.

A second example of 'front-end business' concerns the presence of *neural machinery controlling feeding behaviour* which, for example, is abolished by removal of the supra- and/or suboesophageal ganglia in annelids. The physiology of feeding in the blowfly has been described in great detail by Dethier (1976) in his book *The Hungry Fly*. Detection of food substances by chemoreceptors on the tarsal (distal) segments of the legs, and then by those on the labellum (see Fig. 8.30e), results in the passage of nerve impulses to the suboesophageal ganglion, where interneurones and then motoneurones are activated inducing eversion of the proboscis and sucking movements by the pharyngeal pump.

Third, anterior ganglia act as *higher centres* in exercising *overriding control* of the reflex and spontaneous activities of 'lower' levels. During locomotion in the platyhelminth, *Notoplana*, food is passed directly to the mouth, whereas in the stationary animal it is passed first for inspection to the front end. A decerebrate specimen does not show the latter modification, nor is feeding inhibited when the gut is completely full. These experiments show that the brain receives information from a range of senses—in this case concerning the state of the gut and movement. It takes 'strategic decisions' and issues 'commands', blocking or switching reflex actions of lower centres, although it is the latter which possess the machinery for organizing the activity in detail. In a wide variety of invertebrates, decerebrate animals are generally characterized by hyperactivity.

Fourth, higher centres have a role in controlling the animals state of 'arousal' or 'mood', which has widespread effects within the nervous system. For example, neurones in the locust eye have their sensitivity to stimuli greatly enhanced by arousal mediated by nervous input from the brain; motor centres are stimulated simultaneously.

Finally, the brain plays a key part in higher nervous functions and advanced forms of behaviour such as those shown by *Octopus* (Fig. 16.12). Work by Young has shown that different parts of the brain are involved in the process of learning in relation to different types of sensory information. Visual pathways lead from the optic to the superior frontal, and thence to the vertical lobes, whereas tactile learning involves the inferior frontal, subfrontal and vertical lobes. Within each of these pathways, the information is relayed to a series of centres; for example, four centres for visual input have been identified within the vertical lobes. There are also fibres running back from the vertical to the optic lobes.

'Memory traces' are present at more than one level in any given system. Thus, removal of the vertical lobes substantially impairs memory of a visual process, but some impressions of the latter still remain in the other lobes. Furthermore, memory within one side of the brain, of a lesson learnt by

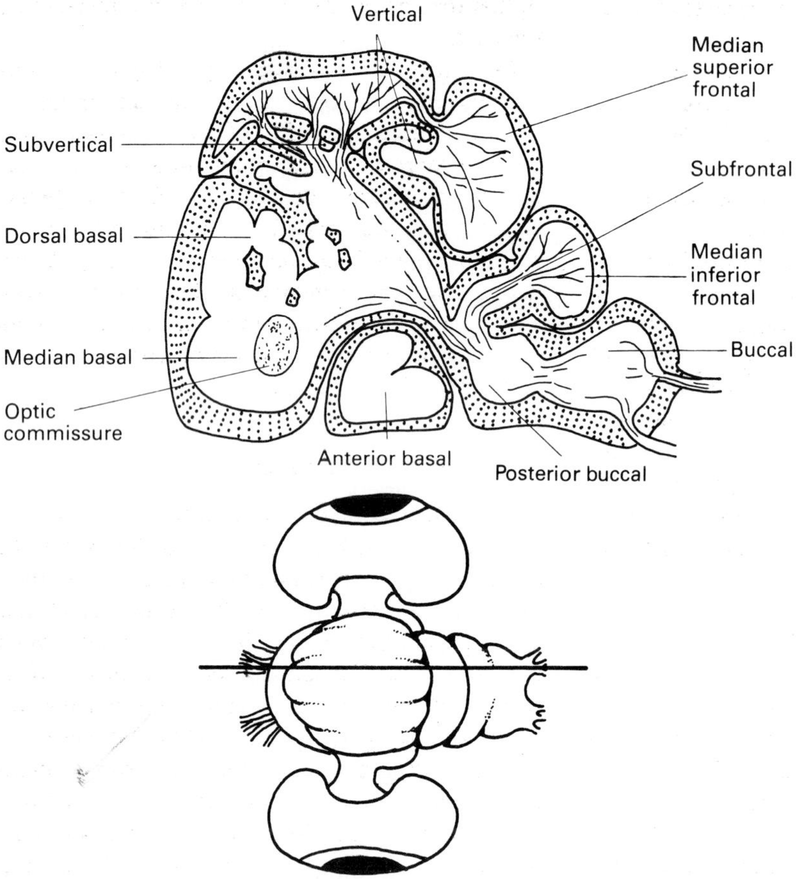

Fig. 16.12. Supra-oesophageal region of the brain of *Octopus*, as viewed in a longitudinal section (plan of section shown below), showing some of the many lobes (after Young, 1963).

unilateral presentation of stimuli, is shared over a period of hours with the other side.

Contrary to what may have been inferred from previous comments on learning (Box 16.3), higher nervous functions apparently require enormous numbers of neural units. For example, the 'highest' centres, the vertical lobes, contain 25 million tiny 'microneurones', many without axons. The lobes facilitate learning processes and this may be accomplished by the formation of a coded representation of a situation, and be the basis of the animals notable abilities to generalize from one subject to a similar one.

16.5 Spontaneity

16.5.1 Neural initiation

A regrettable side-effect of the major contribution to neurobiology made by studies of reflex arcs has been the tendency to regard nervous systems as being as dependent on external stimulation for their activity as a switched-off computer. Nothing could be further from the truth, as even the most superficial observation of the behaviour of organisms—from *Amoeba* to *Octopus*—will reveal. The capacity of the nervous system for spontaneous activity—its role in *initiating* events—is as crucial as its ability to *respond* to changes in the environment.

Endogenous initiation of activity can result as development of the nervous system is completed. However, much spontaneous activity is *rhythmical* and, when it occurs at short intervals of time, may be due to spontaneously generated action potentials, etc. (Section 16.2.2). The cells involved are called *pacemakers*. Such activity can also result from the mutual and alternating inhibition of two or more neurones. Synchronized, rhythmical bursts of activity ('brain waves') involving large numbers of neurones are well known in vertebrates including man. They are widespread in invertebrates also and generally show a high frequency (above 50 Hz) except in *Octopus*, in which they are, intriguingly, like vertebrates (below 25 Hz). Other rhythms have far longer periods and, as we shall see, have a different basis in cell biology.

16.5.2 Movement

In *Hydra*, series of spontaneous contractions of the epidermis occur, each series resulting in the withdrawal of the body to form a small ball. They are caused by series of 'contraction pulses'—electrical impulses which can be recorded from the surface of the body. Pulses are initiated in the epidermal nerve net in the subhypostomal region, since an animal treated twice with colchicine, which is selectively toxic to nerve cells, becomes entirely quiescent. Once initiated, the pulses are propagated in a through-conducting manner, probably in both the nerve net and the epidermal epithelium.

Spontaneous activity on the part of the marginal ganglia of schyphozoan jellyfish generates the rhythmical contractions of the bell involved in swimming. All the centres are spontaneously active, but one 'makes the running' for a time and then another will take over. The leader generates action potentials at intervals of, say, 2 seconds for an animal 2 cm across or 20 seconds for a 20-cm specimen. These spread widely across the under surface of the bell in the 'giant fibre nerve net'. They stimulate the swimming muscles and also reset the other ganglia so that they do not themselves fire and interfere with the rhythm.

In hydrozoan medusae (e.g. *Polyorchis*), spontaneous activity for the swimming rhythm is generated within the ring of giant neurones in the inner nerve ring—there appears to be no single focus for the pacemaker activity. In addition, tentacle contractions are caused by spontaneous activity within the ganglia of the outer nerve ring. As each new tentacle and ganglion is added during growth, it behaves independently until nervous connections are made; its activity is then synchronized with that of the rest of the system.

16.5.3 Autonomic functions

The annelid *Arenicola* lives in a burrow in the sand (Fig. 9.32), which it ventilates to aid respiration (Section 11.4.3). It shows periodic pumping movements, at intervals of about 40 minutes, which cause a current of water or bubbles of air to flow over its gills. A complex sequence of movements is involved: first locomotion to the tail-end of the burrow; then anteriorly directed, peristaltic waves of contraction of the body wall, associated with some anterior movement; lastly, the direction of ventilation is briefly reversed. This activity is not a reflex response to lack of oxygen, but is generated by multiple pacemakers, co-ordinated in life, present in the ventral nerve cord. Similarly, spontaneous feeding movements with a shorter time interval (about 7 minutes) arise from nervous elements in the oesophagus.

Together with respiratory movements, the heart beat is one of the most obvious rhythmical functions of animals. In many animals, spontaneous activity has its primary source in the muscle cells, and is only modulated by nervous influences. However, in crustaceans, the heart beat is *neurogenic* and has its source in the activity of the *cardiac ganglion* (Fig. 16.13). This small ensemble of neurones represents the nervous system in microcosm. All the cells show *spontaneous activity*, although in life they are co-ordinated by one which acts as a *pacemaker*. The ganglion is subject to the *influence of higher centres* (mediated by

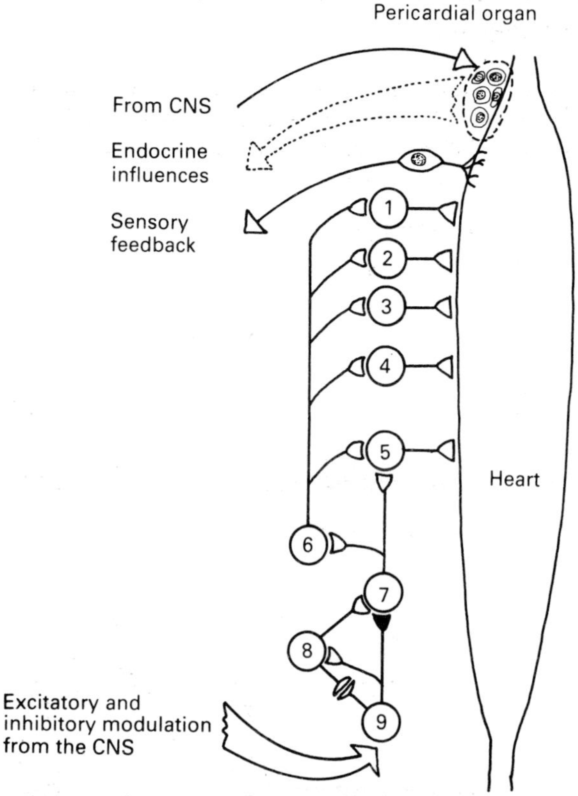

Fig. 16.13. The cardiac ganglion of the lobster shows spontaneous activity and generates the heart beat. 1–9, identified neurones. (After Shepherd, 1983.)

synaptic input from the central nervous system), to the *feed-back effects* of sensory neurones associated with the heart, and to *endocrine influences.*

16.5.4 Biological clocks

Some periodic phenomena imply that a mechanism is in operation for keeping track of the time of day or the passage of time, and such mechanisms are called *biological clocks* (Box 16.5). Clocks are indispensable to animals such as bees and ants which navigate by the sun. Honey bees can not only communicate the direction and distance of a food source to other workers by executing a figure-of-eight dance at the entrance to the hive (as shown by Von Frisch in 1950), but can also maintain an appropriate course over periods of time, making adjustments to compensate for the apparent movement of the sun.

Many organisms have the ability to respond to changing day-length, for example, by modifying their pattern of development or reproduction with the advent of the shorter days of autumn (see, for example, Section 14.3.4). This involves the capacity to 'measure' the length of day and/or night, and a notable feature of such mechanisms is that they are *temperature-compensated* (i.e. independent), unusual in ectothermic invertebrates, since major errors in the measurement of time would otherwise result.

16.6 Receptors

16.6.1 A fundamental characteristic

Sensitivity to environmental influences is a general characteristic of living cells and is shown even in the absence of obvious structural differentiation. For example, the protist *Amoeba* is photosensitive, and perception of day-length in insects apparently takes place within the brain and does not depend on the eyes. However, most animals develop a range of specialized *receptor cells*, which often form parts of multicellular *sense organs* (see Box 16.6). In the great majority of receptors, the sensitive part of the cell is structurally differentiated by the presence of cilia or microvilli (and in some cases, both) (see Fig. 16.16). In many cases, sensory cilia are modified in structure, although this seems to bear little relation to the sensory modality involved. The significance of these organelles is presumably related to the expanded surface area they provide. In some cases (e.g. the crayfish eye), intramembranous particles presumed to consist of visual pigment, which can be observed with the electron microscope, are concentrated within the microvillous membranes, implying that the latter are the locus of sensitivity. In others (e.g. *Drosophila*), the complement of particles within the microvilli resembles that of surrounding areas of the plasma membrane, suggesting a more diffuse sensitivity.

Box 16.5 Cell biology of a biological clock

Among the best known rhythms of biological activity are those associated with the regular alternation of day and night. A *circadian* (*circa dies*—about a day) rhythm of activity in the mollusc *Aplysia* is *entrained* (set in motion) by the influences of the daily cycle of illumination. However, if the animal is maintained under constant conditions, this rhythm persists for several days with only a small change in the length of each cycle. This is characteristic of circadian rhythms and enables organisms to initiate a response even before the arrival of the circumstances to which it is appropriate.

Rhythmical activity is widespread among neurones in the central nervous system, and it continues for several weeks even when ganglia are removed from the body and kept in organ culture. Different cells have their own rhythms, but in life these are synchronized by a 'master clock' located within the 'D cells' in the eye (not the photoreceptors, but cells which receive synaptic input directly from them). A circadian rhythm of activity can be detected within the isolated eye and the cells exert their influence not by hormone secretion as was once thought, but via axons which extend throughout almost the entire nervous system.

The fundamental mechanisms underlying the function of the clock have been investigated. Administration of anisomysin, which blocks protein synthesis within the cell by binding to a subunit of the ribosomes, eliminates the rhythm, and introducing a brief pulse of the substance stops the clock temporarily, so that it 'runs slow'. Jacklet (1981) has concluded that 'the daily synthesis of protein is a general requirement for circadian clocks'. The relevant proteins may be enzymes—protein kinases—which are known to influence ion channels and therefore impulse activity in other systems.

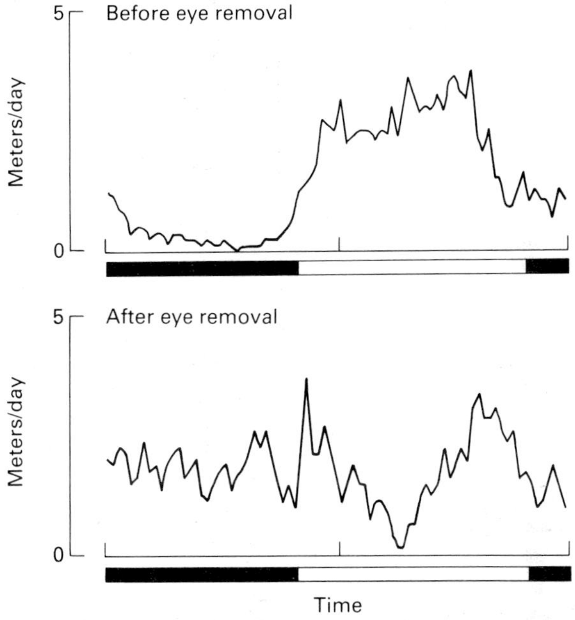

Locomotory activity of *Aplysia* throughout a night and a day before and 3 days after removal of both eyes. (After Strumwasser, 1974.)

16.6.2 Classification of receptors

Aristotle recognized what we call 'the five senses' and these have been joined by others, some quite exotic (e.g. sensitivity to magnetic fields). Types of senses are called *sensory modalities*. Contemporary classifications are based on the physical character of the stimulus concerned (e.g. light, mechanical, chemical, etc.). However, we may also distinguish between *exteroceptors* sensitive to external influences, *interoceptors* which respond to internal factors, and *proprioceptors* signalling movements or positions of muscles, joints etc., or between *phasic* receptors responding to changes in the environment, and *tonic* ones whose activity is related to the absolute level of stimulation—many receptors are a combination of the two.

Sensitivity to one modality may be exploited to provide information about another. For example, receptors sensitive to gravity called *statocysts* are specialized mechanoreceptors. They are widespread in invertebrates, for example being associated with the margin of the bells of jellyfish and present in *Octopus*. The organ consists of a *vesicle* containing

Box 16.6 Insect sensilla

Insects have combined receptor neurones with cuticular hairs, pegs, plates, pits etc. to produce an astonishing array of sensory organs called *sensilla*—the body just bristles with such tiny 'antennae'. All the cells composing a sensillum are produced by division from one mother cell. In nematodes, the chemoreceptor organs called *amphids* (Fig. 4.18) show a similar pattern of development.

Chemoreceptors, whether taste or olfactory, show close similarities with those of other animals. The dendrite (or 'inner segment') of each bipolar receptor neurone extends to the base of the organ, where it gives rise to one or more cilia (the 'outer segment'). One or many pores (up to 15 000 per hair) provide access by which chemical substances can reach the cilia via the special fluid, or 'lymph', in which the organelles are bathed.

Mechanoreceptors are well developed and are ciliated, as in the great majority of other animals. Movement of the bristle of a taste/touch organ distorts the cilium and stimulates the receptor cell. *Campaniform sensilla*, each consisting of a single neurone associated with a ridged, cuticular cap, detect stress within the cuticle. *Chordotonal sensillae* provide information regarding the positions or movements of joints in insects and crustaceans, and consist of receptors, each capped by a specialized *scolopale* cell, in association with strands of connective tissue. Such receptors also provide the sensory components for *tympanic organs*—auditory organs which may occur in several different parts of the insect body. Such organs consist of modified parts of the tracheal system, and have air sacs with tympanic membranes whose vibrations in response to sound stimulate the receptors associated with them.

Insects apparently lack specialized organs sensitive to gravity and acceleration, and depend on the many sense organs associated with joints, etc., to provide the relevant information. However, flies have *halteres*, dumb-bell-shaped, modified hind wings which oscillate rapidly during flight, acting rather like a gyroscope used in navigation, and signalling rotation.

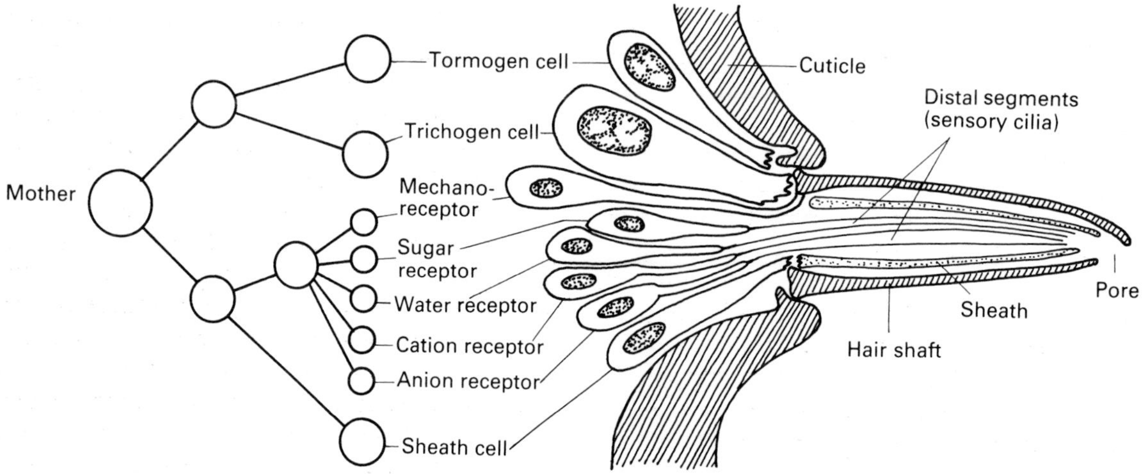

Development and structure of an insect sensillum. The tormogen cell secretes the socket, the trichogen cell the hair shaft (both contribute to the lymph); the sheath cell forms the sheath around the cilia. (After Dethier, 1976 and Hansen, 1978.)

one or more dense bodies, the *statoliths* (Fig. 16.14). As the statolith moves under the influence of gravity, it touches or distorts sensory cilia and activates the receptor cells. In the urochordate *Oikopleura*, the statolith is replaced by a melanin droplet which being lighter than water floats *upwards* and in this way indicates the direction of gravity.

The role of statocysts was shown in shrimp by Kreidl in 1893. These animals shed and replace the sand grains which form their statoliths when they moult. By providing iron filings instead of sand, Kreidl was able, with the use of a magnet, to make the shrimp swim upside-down, on their sides, etc., as the magnetic field simulated that of gravity. Statocysts can detect acceleration, but tubular systems in which movement of fluid activates receptors, comparable to the semicircular canals of vertebrates, are also present in some invertebrates (e.g. *Octopus*).

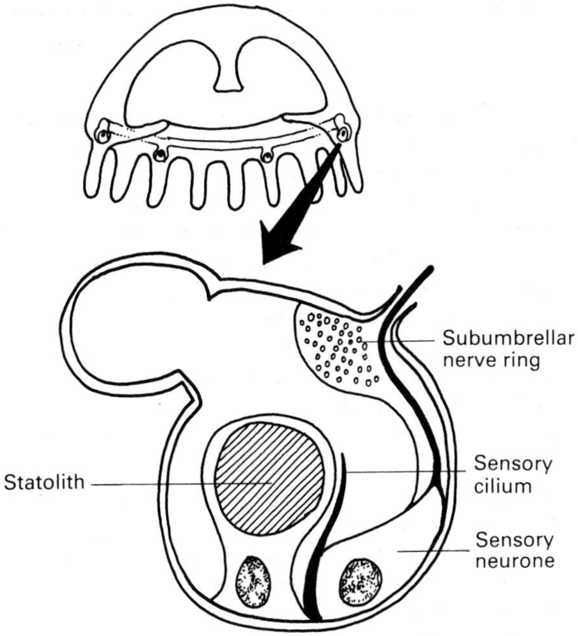

Fig. 16.14. Statocyst in a cnidarian (after Singla, 1975). (See also Figs 3.19 and 3.25.)

16.6.3 Specialists and generalists

Receptors for a given sensory modality have in common a *heightened sensitivity* to the form of energy involved, as shown by a lower threshold, and more intense response to stimulation. However, whereas some are *specialists*, others are *generalists*. Olfactory specialists each have a highly restricted spectrum of response to odours, with, for example, an acute sensitivity only to a single compound, a *pheromone* (Section 16.10.6), secreted as a chemical signal by another member of the species, or only to odours associated with food. Olfactory generalists respond to a wide variety of stimuli within the modality, but, since each has its own pattern of sensitivity, the animal is able to recognize any given substance by the unique *combination* of receptors activated.

16.6.4 Intensity coding

The information provided by receptors is not usually just 'on' or 'off', but also 'how much', and this is encoded in the frequency of the impulses generated in the nerve fibres, which is proportional to the log of the stimulus intensity, or to some comparable function. The range of stimulation intensity to which the organism is sensitive is extended by the control of receptor sensitivity exercised by *centrifugal fibres* (impulse traffic is directed towards the periphery) such as innervate the crustacean stretch receptors (see Fig. 16.24). This feature is particularly well developed in invertebrates. Lastly, different cells can operate across different parts of a wider range. In a classic study, Nicholls and others have shown that mechanoreceptors in the body wall of the leech, with identified cell bodies in each ganglion of the ventral cord, are of several different types. 'Touch' (T) cells are sensitive even to water currents in the medium. 'Pressure' (P) receptors start to respond when a pressure of about 7 g is applied by means of a probe. Finally, 'nociceptor' (N) units are activated only in response to severe damaging stimuli.

16.6.5 Sensory cells and sensory neurones

Many sense organs in vertebrates are made up of sensory cells lacking nerve fibres, which transmit

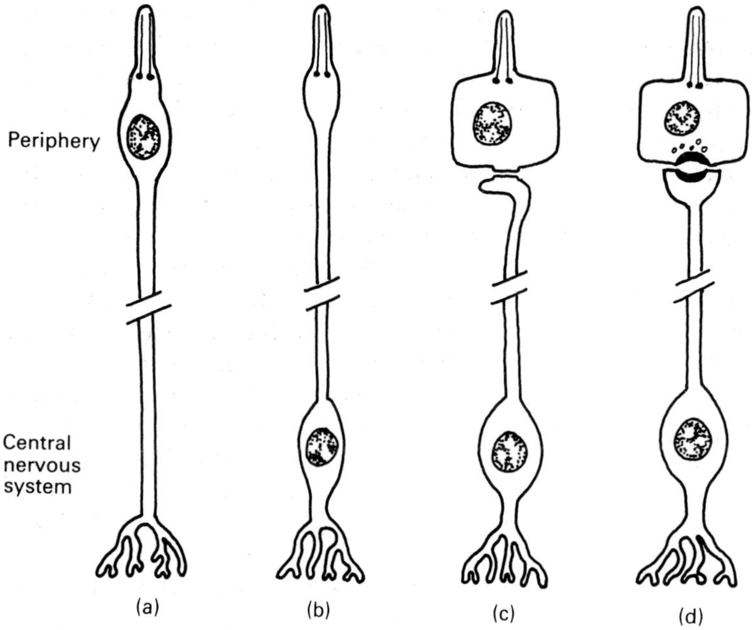

Fig. 16.15. Sensory neurones, as in the chemoreceptive nuchal organs of the polychaete annelids (a) *Nephtys* and (b) *Nereis*, respectively. Sensory cells in (c) *Oikopleura* and (d) *Tamoya* form electrical and chemical synapses, respectively, with sensory neurones.

information across chemical synapses to the terminals of sensory neurones whose axons extend to the central nervous system. (Despite the direction of information flow, the neurones were (confusingly) called primary, and the peripheral cells, secondary, in older literature.) In other cases, the receptor is itself a neurone and this is characteristic of the great majority of sensory systems in invertebrates. Exceptions include the Langerhans receptor in the urochordate *Oikopleura*, and photoreceptors in the cnidarian *Tamoya* (Fig. 16.15).

16.7 Vision

Vision is the sensory modality which provides us with most information about the world and the same applies to many of the most complex invertebrates, as shown by the refinement of their eyes.

16.7.1 Visual pigments

Sensitivity to light, a small band within the electromagnetic spectrum, is conferred by the possession of a *visual pigment*—a chemical substance which absorbs radiant energy, with the release of free energy. *Rhodopsin* is a compound of the carotenoid *retinal*, a vitamin A derivative, and the protein *opsin*. It is nearly universal, being found in organisms as diverse as the alga *Volvox* and man. Exposure to light converts the retinal from its 11-*cis* isomer to the 11-*trans* configuration, and then leads to bleaching of the pigment due to the dissociation of retinal and opsin. This change affects ion channels in the photoreceptor membrane, by promoting synthesis of a substance of the cAMP type (see Section 16.3.3), probably cGMP. Rhodopsin is regenerated by enzymic synthesis.

Rhodopsin exists in a variety of forms, depending on both the type of retinal and that of opsin. Fireflies active at dusk emit yellow light as signals to other members of the same species, whereas those active after dark produce green light; the maximal sensitivities of the retinal pigments in the different species are precisely tuned accordingly.

Colour vision is important in many crustaceans, insects and arachnids (surprisingly, *Octopus* is colour-blind) and is based on the possession of several types of receptors with pigments sensitive to different parts of the spectrum. For example, bees have receptors sensitive to yellow–green (540 nm), blue (440 nm) and ultraviolet light (340 nm), respectively. Butterflies are also sensitive to red light. In flies, the majority of receptors exhibit a curious *dual sensitivity* (blue–green *and* ultraviolet) due to the presence of both rhodopsin and another pigment. The other receptors show a range of spectral sensitivities, partly on account of the presence of *screening pigments* which cut out some wavelengths so that the receptors only respond to the others.

The orientation of *polarized light* can also be detected by some eyes. In flies, units at the margins of the compound eyes are specialized for this function. The pattern of polarization of the sky enables bees to navigate with reference to the position of the sun even when this is obscured by cloud. *Octopus* may use this ability to 'see through' the silvery camouflage of fish.

16.7.2 Ciliary and rhabdomeric eyes

The great majority of photoreceptors belong to one or the other of two types, as shown by Eakin (Fig. 16.16). In one class, the probable photoreceptive organelles are *cilia*, or derivatives of ciliary membranes, as in vertebrates. The cilia may have a normal or modified complement of tubules. In the second type, the receptor organelles are *microvilli*, whose ordered arrays, where formed, are called *rhabdomes*. It is noteworthy that dendrites bearing microvilli often possess a cilium or ciliary rudiment at the apex, suggesting that the ciliary complex may have an inductor role in rhabdome development. Furthermore, receptors with definitive ciliary and rhabdomeric organelles at different levels in the same cells have now been reported in serpulid annelids; in the cnidarian *Polyorchis*, microvilli from the cell membrane intermingle with those derived from ciliary membrane; and a few others too seem to be mixed.

Many photoreceptor membranes of either type show a massive, daily breakdown and renewal. In nocturnal dinopid spiders, the rhabdome occupies only 15% of the receptor cell volume during the day, but within one hour after sunset this has increased to 90%. The surplus is destroyed within 2 hours after sunrise. In a platyhelminth, the microvilli degenerate during the hours of darkness, but are reconstructed (presumably from membrane accumulated and stored as intracellular vesicles ready for use) between the fifth and tenth minute after exposure to light.

Most ciliary units are 'off' receptors. Darkness leads to the opening of Na^+ channels (usually), depolarization, and transmitter release from the terminals. In contrast, rhabdomeric receptors typically show 'on' responses—the above effects result from exposure to illumination. The distinction is illustrated in the scallop *Pecten*, whose eyes have an upper layer of ciliary ('off') receptors and a lower layer of rhabdomeric ('on') cells.

Ciliary receptors such as those of sabellid fanworms and those of *Pecten* seem well adapted to mediate 'shadow reflexes'—defensive reactions to sudden decreases in illumination. However, rhabdomeric 'on' receptors are involved in such a reflex in barnacles. Decrease in illumination hyperpolarizes the receptors, blocking the release of their *inhibitory* transmitter and thus allowing the activation of the innervated neurones in the brain. Furthermore, there are always exceptions to the rule, for example, microvillous receptors in the gastropod mollusc *Onchidium* apparently show an 'off' response. In salps (urochordates), both microvillous and ciliary types have 'off' responses. We can look forward to learning how the mixed ones work.

The two types of photoreceptors were long regarded as having a phylogenetic significance—ciliary receptors being representative of cnidarians and deuterostomes; microvillous receptors of the protostomes. However, many exceptions are now known. For example, starfish have microvillous receptors; the eyes of *Pecten* include both; the right larval ocellus in the platyhelminth *Pseudoceros* consists of several microvillous receptors, whereas the left has three microvillous, plus one ciliary cell.

Many flagellates have part of the flagellum specialized for receptor function and a stigma

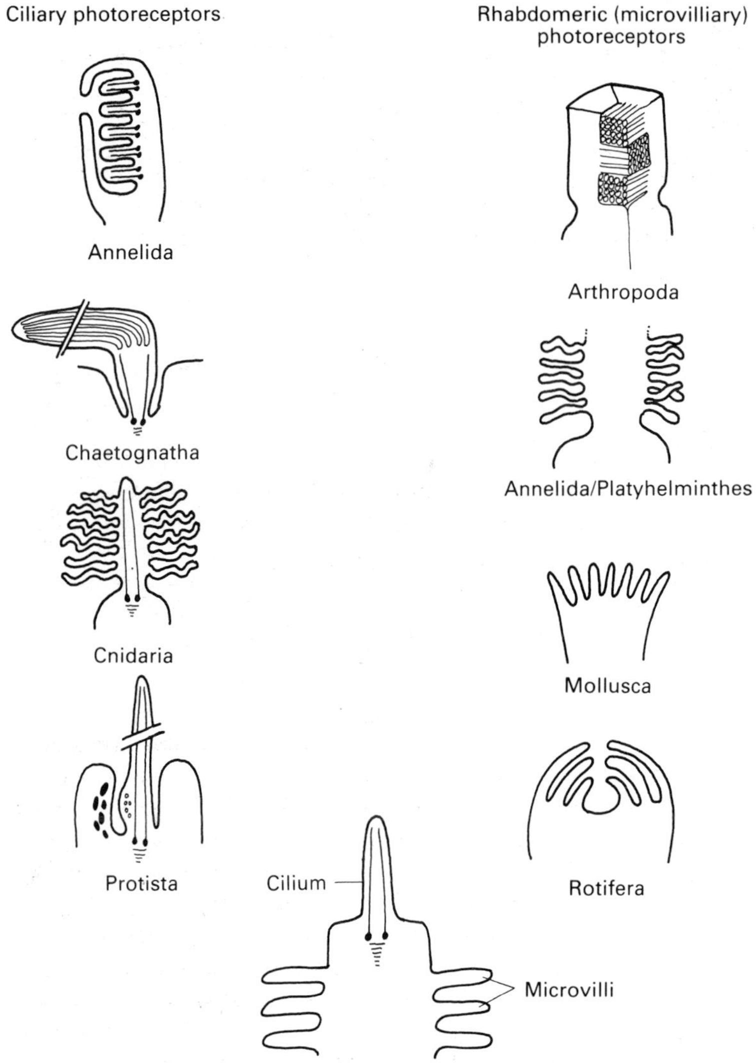

Fig. 16.16. Radiation of ciliary and rhabdomeric photoreceptors (largely after Eakin, 1968).

(pigment spot) confers directional sensitivity. Most extraordinary of all, some dinoflagellates have *ocelloids*, i.e. parts of the single cell resemble a simple eye with a cornea, lens and crystalline retina-like layer, backed by a pigment cup, and claimed to be able to form images!

16.7.3 Ocelli and eyes

Photoreceptor organs in invertebrates show a wide spectrum of grades of organization (Fig. 16.17). The simpler organs are called ocelli and only provide information regarding the intensity and direction of light. Caution should be exercised in attributing

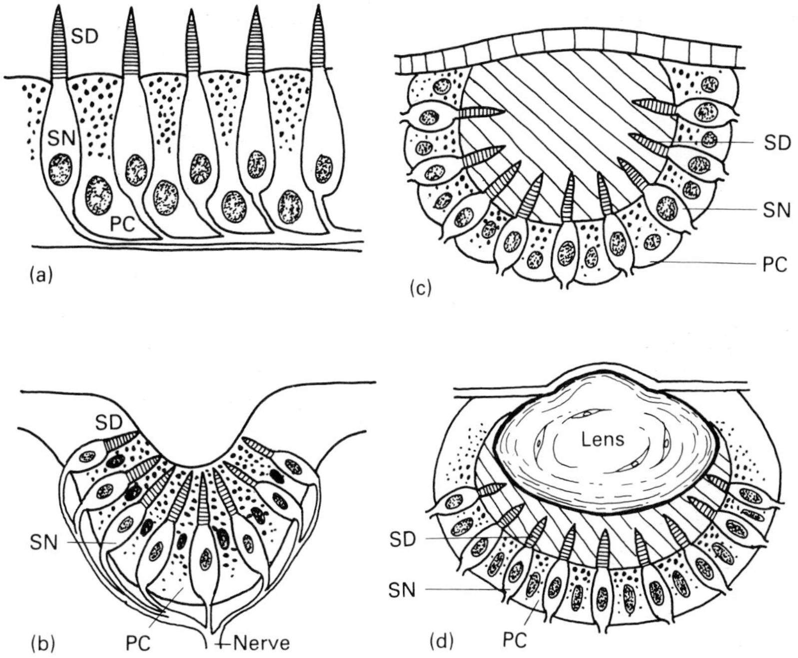

Fig. 16.17. Grades of organization of pigment-cup ocelli and eyes comparable to those shown by cnidarians: (a) photoreceptor epithelium in *Leukartiara*; (c) *Bougainvillia*; (d) *Tamoya*; and by prosobranch molluscs: (b) *Patella*; (c) *Nerita*; (d) *Valvata*. PC, pigment cell; SD, sensory dendrite; SN, sensory neurone. Dendrites bear cilia or microvilli.

evolutionary significance to such sequences since they are all drawn from present-day species. Furthermore, such sequences can be based on organs from each of many different phyla—as many as '40–65 phyletic lines of complexity' (Salvini-Plawen & Mayr, 1977) have been identified. For example, photoreceptor organs in cnidarians range from areas of pigmented epithelium to eyes with lenses (in *Tamoya*); in molluscs from the simple, open eye-cups of limpets, or the five-celled ocelli of some nudibranchs, to the eyes of cephalopods, with 20 million receptors, which bear comparison with our own.

Eyes which form well-focused images are called *camera eyes*. They depend, first, on the presence of an expanded sheet of receptors constituting a *retina*, which may either be *direct* (everted) or *inverted* depending on whether the receptive poles of the cells are directed towards or away from the source of the light, respectively (Fig. 16.18), and second, on the presence of a *device for focusing the light*, although when the lens lies adjacent to the surface of the retina and the retina is thick, image formation must be poor.

High-performance vision is of critical importance to predators such as jumping spiders, which need to be able to judge the speed of an object (all spider eyes are of the pigmented eye-cup plus lens type). The camera eye reaches its climax in *Octopus* and its relatives (Fig. 5.25b). It comes complete with eyelids, an adjustable pupil, a movable lens and extrinsic muscles by which, with the contributions of input from the statocyst, vision can be fixed on an object of interest, irrespective of body movements. Cephalopods hold the record for the largest eyes. Those of *Architeuthis*, the giant squid, can measure 40 cm in diameter and the retina could conceivably contain up to 10^{10} receptors (*cf.*, 10^8 in man). The eye of *Nautilus* is a puzzle, lacking both lens and cornea. It functions like a 'pin-hole' camera, but resolution and sensitivity are poor.

16.7.4 Lenses and mirrors

The convex cornea is responsible for much of the refraction of light in terrestrial species (e.g. spiders).

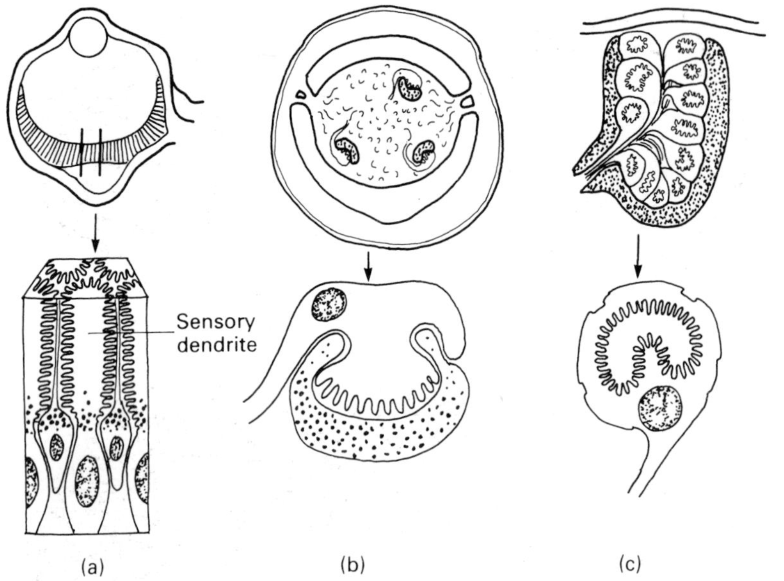

Fig. 16.18. Retinas of annelids. (a) Direct—eye-cup of the polychaete *Vanadis* (after Hermans & Eakin, 1974). (b) Inverted—eye-cups in the brain of the polychaete *Armandia* each with a single sensory neurone and one-celled pigment cup (after Hermans & Cloney, 1966). (c) Eye-cup with sensory neurones, each with vacuole-like 'phaosome', in the leech *Hirudo* corresponds to neither category. Phaosomes may be formed by both ciliary and microvillous photoreceptors (after Hess, 1897).

In aquatic animals, lenses must do all the work, but show little of the expected defect of spherical aberration. Matthiessen (in 1886) suggested that the lens has a high refractive index at its core, falling to lower values at the periphery, and indeed such lenses are found in cephalopods, some gastropod molluscs, some copepod crustaceans, and alciopid polychaetes—as well as fishes. Obviously, 'there really is only one way to make a decent lens, using biological material'! (Land, 1984).

Mirrors are important in focusing the light within many eyes, either by themselves or in combination with lenses. The mirror may redirect the light back through the receptors, virtually doubling the effective illumination (some spiders); or one layer of receptors receives light focused by the lens, and a second layer after it has been refocused by the mirror (some crustaceans).

16.7.5 Compound eyes

Compound eyes are organs consisting of a number of virtually identical units, or *ommatidia*, arranged in geometrical array. They are sometimes called *mosaic eyes* since the image is formed by the contributions of all the ommatidia, each of which provides only a small part. They are thought to be well adapted for detecting movement, but their resolving power (except in flies) is only about 1° at best, as compared with 1′ in vertebrates and *Octopus*. Whereas the retinas of pigmented eye-cups are concave and form inverted images, compound eyes and their retinas are always convex and the images erect (Fig. 16.19). Definitive compound eyes are found in many arthropods, but are not present in spiders, millipedes, etc. Eyes of this type are also present on the tentacles of sabellid fan-worms, in the bivalve mollusc *Arca*, and they form the 'optic cushions' situated at the tips of the arms in starfish.

The acuity of compound eyes depends on the angular separation of the ommatidia. Whereas the eye of *Daphnia* has 22 ommatidia, each with an angular separation from its neighbours of 38°, the crab *Leptograpsus* has several thousand, with a separation of 1.5°. The resolving power of the latter would be far greater, just as a mosaic with small components is superior to that with large ones, but this is correlated with a loss of sensitivity. Just as

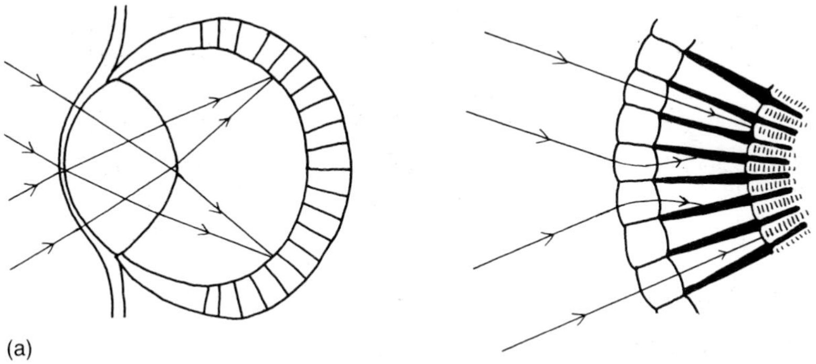

Fig. 16.19. (a) Contrasting patterns of organization of an eye-cup (left) and a compound eye (right). (b) Head of a robber fly. Anterior large facets of the compound eyes are approximately 60 μm in diameter. (Courtesy of Professor M. Land, University of Sussex.)

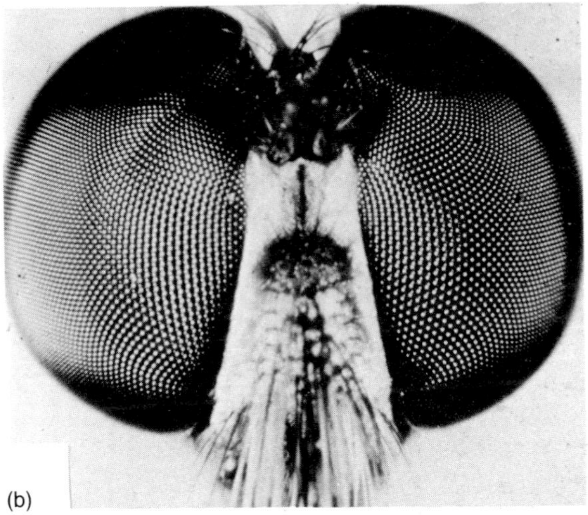

(b)

vertebrate eyes have a fovea, some crab eyes have fine-grained bands, made up of ommatidia with a small angular separation.

In arthropods, each ommatidium is an elongate unit consisting typically of a corneal lens, a crystalline cone, a number of receptor (*retinular*) cells (four in *Daphnia*, eight in crabs and flies, and ten to fifteen in *Limulus*), and associated pigment cells (Fig. 16.20). The retinular cells are arranged like the segment of an orange. Regularly arranged microvilli project from along the inner surfaces of each cell and constitute a *rhabdomere* and the rhabdomeres within an individual ommatidium together make up its *rhabdome*.

Some arthropod ommatidia depend on *diffraction* of light. Krill and moths have long, bullet-shaped crystalline cones which, according to Exner (1881), form 'lens cylinders' with a high refractive index along the axis and a progressively lower one towards the surface. However, it is now known that other eyes (e.g. in shrimps and lobsters) use *reflecting* optics, their pyramid-shaped 'cones' being 'silvered' with layers of crystals on their surfaces. The surfaces of such eyes have square instead of hexagonal facets.

Apposition eyes typically show some overlap between the fields of vision of adjacent ommatidia, but, in the crustacean *Phronima*, each ommatidium of the median eyes can accept light from a field 4° across, whereas the angle of separation of the ommaditia is only 0.5°. However, perhaps the most extraordinary compound eyes are those of stomatopod crustaceans. Whereas we form two images of an object, these animals form six, since each eye has three bands of ommatidia looking out in the same direction! These last two examples make it clear that no analysis of the capabilities of compound eyes is adequate if it fails to take account of the remarkable 'computing' capacity of the brain in analysing and interpreting the information provided by the receptors.

Types of compound eyes in arthropods differ in the degree and nature of the functional isolation of the ommatidia. Apposition eyes (Fig. 16.20a), found for example, in crabs and bees, are adapted for high light intensities. Pigment cells screen the rhabdome from all light apart from that entering via the lens of that particular ommatidium.

Superposition eyes (Fig. 16.20b) such as those of

Box 16.7 Neural superposition retina of the fly

Eyes with neural superposition optics provide resolving powers up to 100 times greater than that of other compound eyes. Within a given ommatidium, the rhabdome is 'open', the rhabdomeres remaining separate. Each rhabdomere receives light from a distinct angle and its axon extends to a cartridge shared not with others within the same ommatidium, but with those of surrounding ommatidia which share the same optical axis. Consequently, rays of light from a single source impinge on six, differently placed retinular cells in adjacent ommatidia. Nerve fibres from these cells follow a spiral path and all synapse on the same cartridge in the optic ganglion.

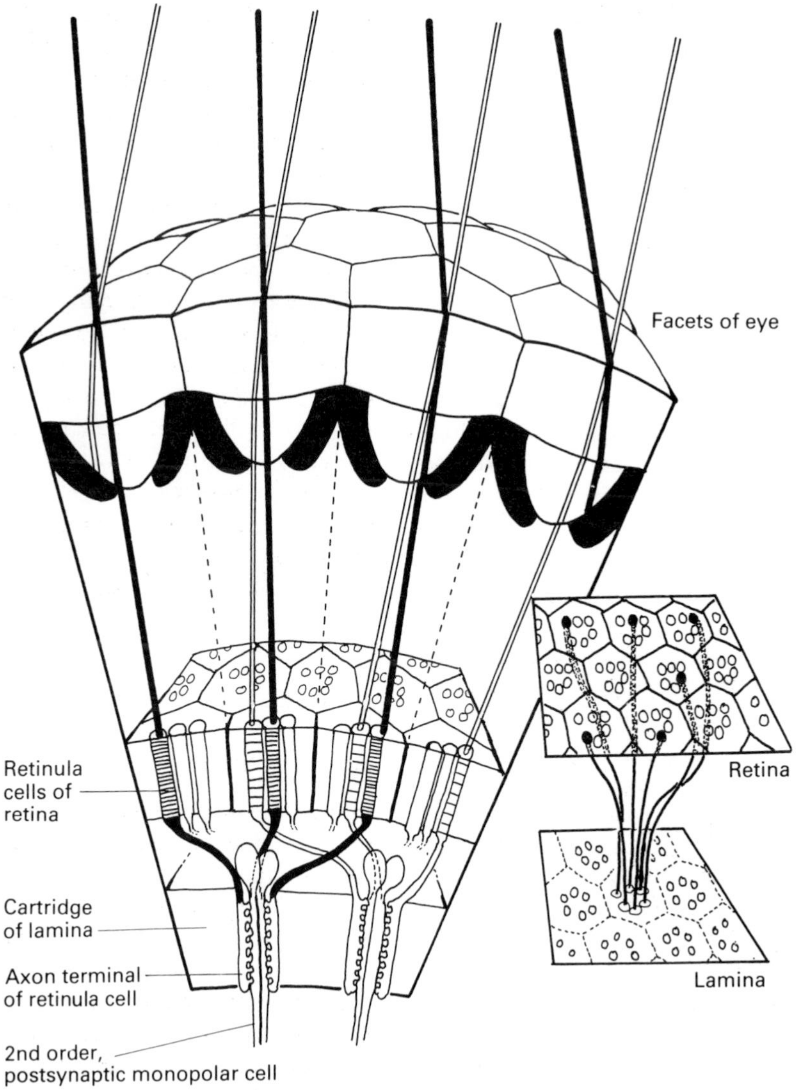

Paths of light rays from two points in the visual field (open and dense lines, respectively), their projection to retinular cells in adjacent ommatidia, and the convergence of the fibres of the receptors they activate upon optic cartridges in the lamina. Right, light rays from a single point impinge on different retinular cells in neighbouring ommatidia, but the axons of the latter converge upon a single cartridge. Only retinular cells 1–6 of the eight present in each ommatidium usually contribute to neural superposition. (After Strausfeld & Nassel, 1981.)

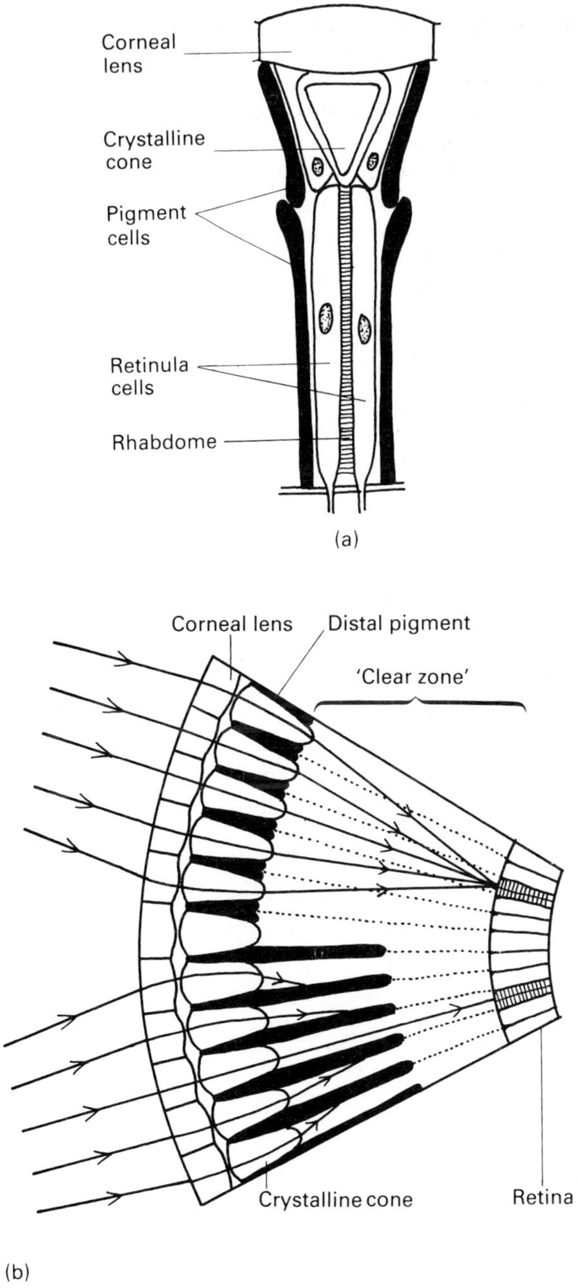

Fig. 16.20. (a) Longitudinal section of an ommatidium of an apposition, compound eye (after Wigglesworth, 1965). (b) Paths of light rays within a superposition eye when light adapted (lower half) and dark adapted (upper half).

shrimps, lobsters and moths, have a wide 'clear zone' interposed between the crystalline cone and the rhabdome, and the latter is shorter. They function like apposition eyes in conditions of high illumination, maximizing the resolution of the eye. However, in dim light, extension of pigment cells or migration of pigment within cells that remain fixed in position exposes the rhabdomes to light entering via surrounding ommatidia. The rays of light from a given source impinging on different ommatidia are redirected to amplify the input to a single rhabdome and this arrangement maximizes the sensitivity of the eye in dim light. However, resolution is impaired, presumably on account of the scattering of light during its passage across membranes.

In the eyes described above, there is no possibility of keeping separate the information contributed by different retinular cells within a single ommatidium, since the rhabdome is 'closed'—the microvilli interdigitate. Furthermore, each cell synapses on each of a small cluster of neurones (called a 'cartridge') situated immediately beneath the ommatidium. A different situation is encountered in the eyes of flies whose mechanism of *neural superposition* represents the summit of development of arthropod optics (Box. 16.7).

16.8 Sensory processing

16.8.1 Making sense of the world

Information regarding the external and internal environments must not only be gathered and *encoded* as changes in membrane potential in receptors, but *processed*—modified and transformed in ways that are adaptive to the animal concerned. For example, in the locust, the simple ocelli on the top of the head produce a poorly focused image and show massive convergence—information from about 1000 receptors in each is funnelled through a comparatively small number (25) of second-order neurones. During flight they provide a rapid, overall assessment of the position of the horizon, whereas, as we shall see, the lateral, compound eyes can distinguish finer points of detail.

The sophistication of the sensory and neural machinery of arthropods is apparent in the *recognition* of particular features or combinations of features, out of a multitude of similar alternatives. A bee learns to distinguish sites of importance 'as though it takes a panoramic snapshot of that position' (Collett, 1983). Some South American orchid bees visit daily a succession of the same flowers, each of whose positions they must remember, as they follow a stereotyped route over 20 km long.

A major aspect of sensory processing is that of *feature extraction*—the recognition, or selection for attention, of information of particular relevance to the animal. For example, arthropods show characteristic patterns of behaviour, such as the claw-waving, sexual displays of fiddler crabs, which are often very striking and stereotyped. They act as 'sign stimuli', 'releasing' particular behavioural responses, and mean nothing to other species. Apparently, the sensory and nervous systems are organized to detect, conduct and amplify such signals, whereas others are ignored.

Last, we can cite the behaviour of some male hoverflies. When a possible mate appears within the field of vision, the male sets a course of flight to intercept. In order to plot the right course, the distance, velocity and course of the object must be known. These features probably cannot be determined by observation, since a large, rapidly moving object at a distance will look much the same as a small, slower object close to the fly. Apparently, the fly 'assumes' that the object is the size of one of its own kind, and that it is travelling at the appropriate velocity. The distance at which the object can first be detected depends on its size, so this too can be inferred. The direction and speed at which the object moves across the visual field then indicates (to an appropriately programmed computer) its course, and the male sets off to intercept. Of course, if the object is a distant jumbo-jet, the would-be bridegroom will be disappointed, but there's always tomorrow!

16.8.2 Sensory systems

Information flows into the central nervous system through *sensory pathways*, being modified en route at a series of centres of synaptic interaction (Fig. 16.21). A series of *modules*—virtually identical synaptic centres—is present at a given level. They are often connected by local interneurones and are thus able to influence each other. The best known of such

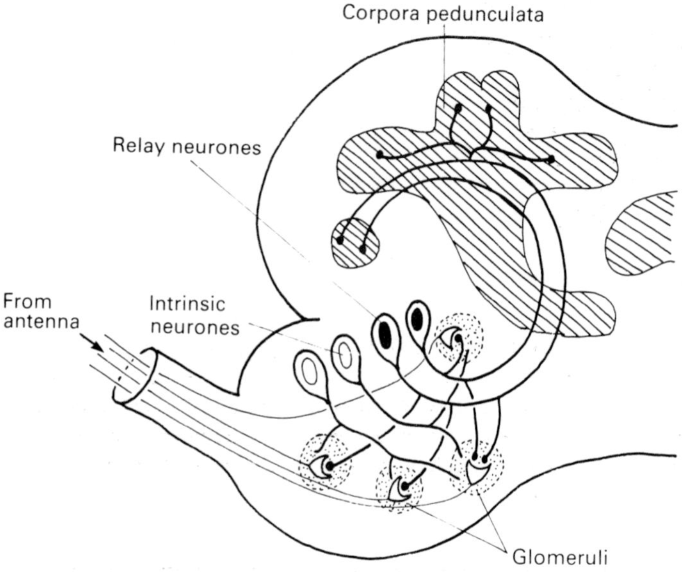

Fig. 16.21. Olfactory pathway in the cockroach. The 'glomeruli' are synaptic centres consisting of *triads*, involving input terminals, intrinsic neurones, and relay neurones conveying the processed information to higher centres. (After Boeckh *et al.*, 1975.)

Box 16.8 The optic lobe of the fly, *Musca domestica*

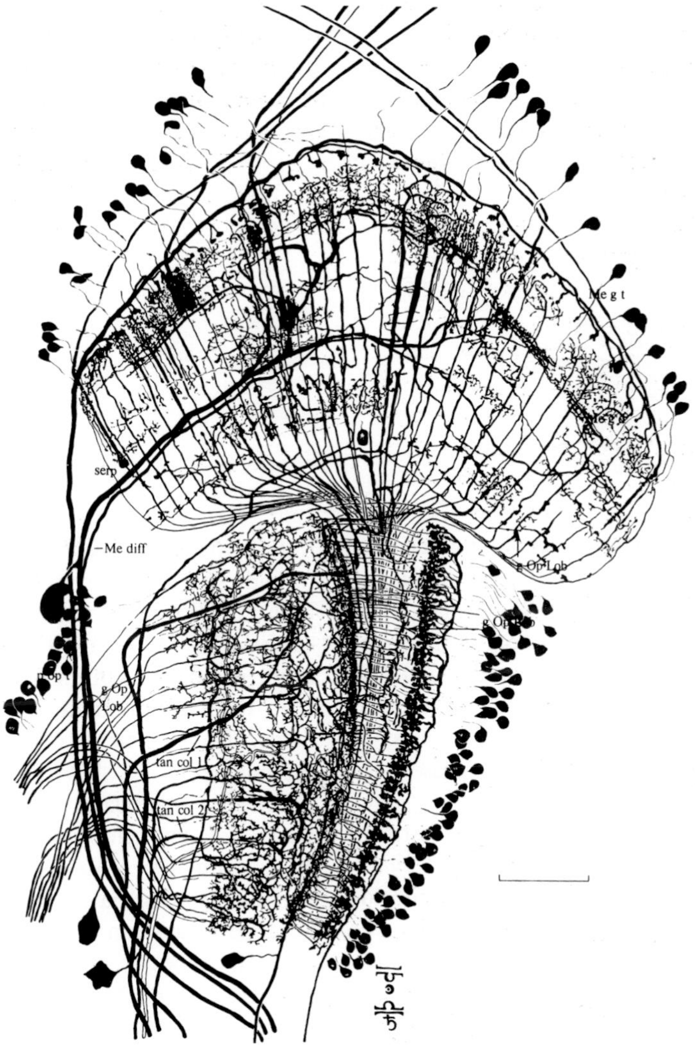

The complexity of the insect retina is something stupendous, disconcerting, and without precedents in other animals. When one considers the inextricable thicket of compound or faceted eyes; when one penetrates the labyrinth of neurones and integrating fibres of the three great retinal segments . . . one is completely overwhelmed.' (Ramon y Cajal, 1937).

In the lamina, all the cell types, and most of the connections they make, have now been identified and the majority studied electrophysiologically. Great progress has also been made in exploring the medulla and lobula. (Reproduced from Strausfeld & Nassel, 1981, with permission.)

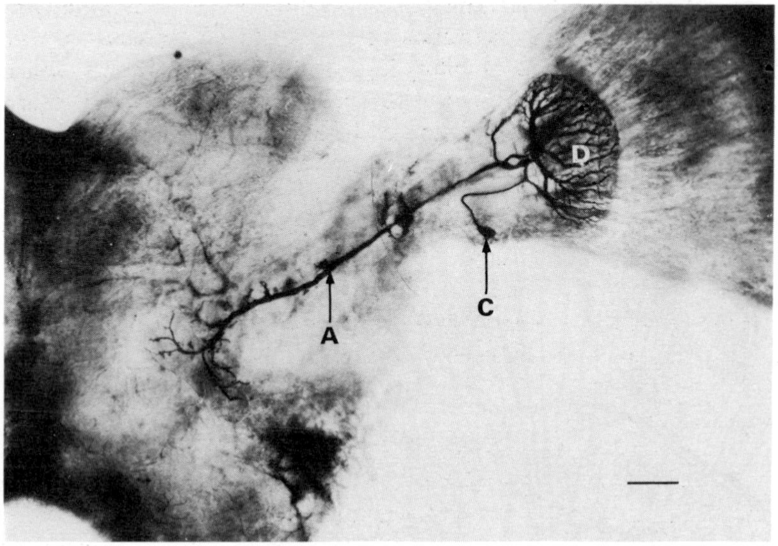

Fig. 16.22. LGMD neurone in the optic lobe of the locust injected with cobalt chloride solution. D, fan-like array of dendrites in the optic lobe; C, cell body; A, axon extending into the brain (left). × 80; bar, 100 μm. (Courtesy of Dr Claire Rind, University of Newcastle upon Tyne.)

interactions results in *lateral inhibition*, first described in 1949 in the classic work on the chelicerate *Limulus* by Hartline. The response of a single ommatidium when it alone is exposed to a given level of illumination is greater than when the whole area is illuminated, since the cell is then subject to the inhibitory influence of its neighbours. Sensory systems also show a *hierarchy* of organization—information passes to progressively higher and higher centres.

The optic pathway in insects involves three distinct regions of the brain—the *laminar*, *medulla* and *lobular* regions of the optic lobe (Box 16.8)—in addition to the protocerebrum. Each ommatidium within the eye has a module or *cartridge* of second-order neurones in the lamina, and then a *column* of third-order cells in the medulla. This principle of *retinotopy* means that a pattern of illumination falling on the eyes is reflected repeatedly in morphologically ordered patterns of activity at four different levels of the visual system.

16.8.3 Convergence and divergence

The *convergence* which characterizes insect ocelli is typical of many sensory systems. A related idea is that of the *sensory field*—the array of receptors that provide sensory input to a cell or centre in a nervous pathway. For example, the median giant fibre of an earthworm has an anterior sensory field—its main input is from receptors situated anteriorly, whereas the pair of lateral giant fibres have mainly posterior fields. The two fields overlap in the mid-region of the body.

Sensory pathways are also characterized by *divergence*, since information from a single receptor, or group of receptors, is conveyed into the central nervous system via multiple, or *parallel pathways*. Such pathways can be used to extract and segregate different types of information. Divergence is also characteristic of sensory systems in that although a system is usually responsive to only a single sensory modality, it provides output to a number of motor centres and thus influences several different types of behaviour.

16.8.4 Labelled lines

In contrast to endocrine signals which differ intrinsically by virtue of the chemical diversity of the hormones involved, many different items of information transmitted by the nervous system involve apparently identical patterns of electrical changes and the same chemical transmitter. The significance of the message depends on the neural elements involved, which are therefore said to be 'labelled'.

This principle can be seen in operation in the escape response of the cockroach or cricket. The lunging attack of a toad, a natural predator, creates a current of air which is detected by sensory hairs on the anal cerci of the insect (see, e.g., Fig. 8.31b) (removal of the cerci reduces the chance of escape). The hairs are arranged in a number of columns, which are sensitive to wind from different directions. The different columns form distinct combinations of connections with giant interneurones extending up to the thorax, and the various giants indirectly stimulate different motoneurones. For example, giant number 5 is activated only by puffs of air from the rear quadrant on the same side of the body, and excites the slow depressor motoneurones (and muscles) on that side to induce a turning motion away from the source of the air. When the cerci are rotated experimentally, the animal is 'fooled' and reacts as if the puff were coming from a different direction.

16.8.5 Feature extraction

A single photon is adequate to produce a measurable 'blip' in the membrane potential of a photoreceptor, and a single molecule is able in some cases to excite a chemoreceptor. Seeing that an individual may possess millions of receptors, it is clear that the animal must filter out from its sensory input those features of significance, or *salience*, and regard the rest as redundant.

As described above, a whole series of retinotopic projections extend into the optic lobe in the locust. Finally, they feed into the *lobular giant movement detector* (LGMD) neurone (Fig. 16.23), which has a large, fan-like array of dendrites so that it can receive stimuli from units covering any part of the visual field. The LGMD neurone is stimulated only if an object in the visual field is, first, *small*, so that the locust does not respond to movements of vegetation, etc. Information about large objects is filtered out partly by lateral inhibition (Section 16.8.2) and partly by activation of a parallel pathway which provides an inhibitory input to the neurone. Objects must also be *moving*. Units pre-synaptic to the LGMD neurone are phasic—each fires once and then becomes quiescent. Consequently, only a moving object, which excites a whole series of units, will provide a succession of stimuli on to the neurone and induce its activation. Last, the movement must be *novel*—repeated stimulation of the same synaptic inputs to the LGMD neurone causes neural depression and becomes ineffective.

16.8.6 Centrifugal control

Information flow in sensory pathways is not one-way—nerve fibres may be directed back from any level to more peripheral centres (i.e. centrifugally) and mediate, for example, negative feed-back effects. The movement detector neurones in the locust are influenced by both inhibitory and stimulatory centrifugal fibres from the brain. When the locust moves its head, the LGMD neurones are inhibited, since otherwise any small stationary object would move across the visual field and activate the system. Similarly, the system can be rendered more sensitive to visual stimulation when the organism enters a heightened state of arousal.

16.9 Neural bases of behaviour

16.9.1 Independent effectors

Some actions are undertaken by cells as adaptive responses to stimuli which they themselves receive—they are *independent effectors*. Some gland and muscle cells come into this category; chromatophores in echinoderms respond directly to light; cnidocytes (stinging cells in cnidarians) discharge upon contact with a foreign organism (although they are also influenced by the nutritional state of the animal). Similarly, co-ordination of the beat of cilia is brought about by mechanical interactions between adjacent organelles.

Protists are of necessity independent effectors, yet they show mechanisms of control identical to some of those operating in nervous function. Encounter with an obstacle anteriorly by *Paramecium* induces membrane depolarization and Ca^{2+} influx, causing reversal of ciliary beat. Stimulation posteriorly increases K^+ permeability, elevates the resting potential and accelerates the beat.

16.9.2 Units of behaviour

Most effectors are not independent. Behaviour typically owes its origin to spontaneous activity generated within the nervous system or represents a response to stimulation. Units of behaviour have generally been described as either *reflexes* or *fixed action patterns*. A reflex is a stereotyped, relatively simple motor action evoked by a specific stimulus; the action varies in strength and/or extent depending on the intensity of the stimulus.

The concept of fixed action patterns owes its origin to the development of *ethology*, the comparative study of species-characteristic behaviour, by Lorenz and others. Such patterns are stereotyped actions, often of considerable complexity, and do not vary with the strength of stimulation. They may be spontaneous, but if evoked by stimulation, this acts only as a trigger. In consequence, quite different stimuli can evoke the same pattern—for example, crayfish adopt a characteristic defensive posture in response to a wide variety of threatening stimuli. They are genetically determined and the expression of a precise 'programme' built into the 'wiring' of the nervous system. Feed-back influences have no role in the control of the pattern, as such.

We may conclude that the two concepts differ mainly in the role ascribed to stimuli. With respect to reflexes, stimuli have a role comparable to that of a person playing the piano, whereas, in fixed action patterns, they correspond at most to the action of switching on a recording.

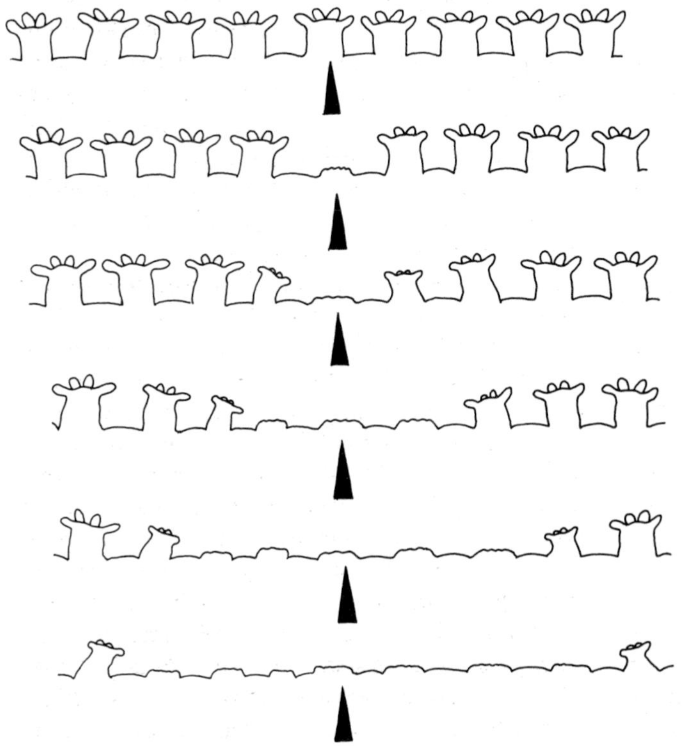

Fig. 16.23. Facilitated polyp withdrawal in the coral *Porites* in response to five successive, identical, electrical stimuli (after Shelton, 1975).

16.9.3 Reflex actions

Many cnidarians respond to light touches by a slow, local contraction, the extent of which depends on the intensity, number, and frequency of stimuli. We will call this a '*facilitated*' response ('decremental' and 'incremental' are also used), since one stimulus, though unable to evoke a given response itself, enables the following one to do so, and so on (Fig. 16.23). The classic explanation, developed by Pantin, depends on the process of pre-synaptic facilitation within the nerve net, by means of which a second or later impulse arriving at a synapse triggers a post-synaptic response which the first was unable to evoke.

Remarkably, facilitated reflexes are encountered in the hydroid *Cordylophora*, even though they are mediated in this case by electrical pulses within the epithelia (Section 16.4.3), indicating that many of the properties of chemical synapses can be mimicked by gap junctions. Last, in certain corals (Fig. 16.23), the reflexes are mediated by potentials in the colonial nerve net that are through-conducting—they are propagated throughout the colony with each and every stimulus! This paradox may be resolved by the progressive reduction in speed of conduction shown by a series of impulses. Inevitably, any given impulse will fall further and further behind the previous one, the time interval between the two will become progressively greater, and facilitation of neuromuscular junctions may be reduced.

Like the knee-jerk response in man, the reflex involving the *crayfish stretch receptor* is monosynaptic, the connection between the receptor and the motor neurone being a single chemical synapse. It contributes to the maintenance of posture, since if its associated muscle is stretched, activity in the receptor and then in the motoneurone which it drives will induce muscle contraction counteracting the effect. The circuit also has a role in implementing instructions from the central nervous system to the musculature (Fig. 16.24).

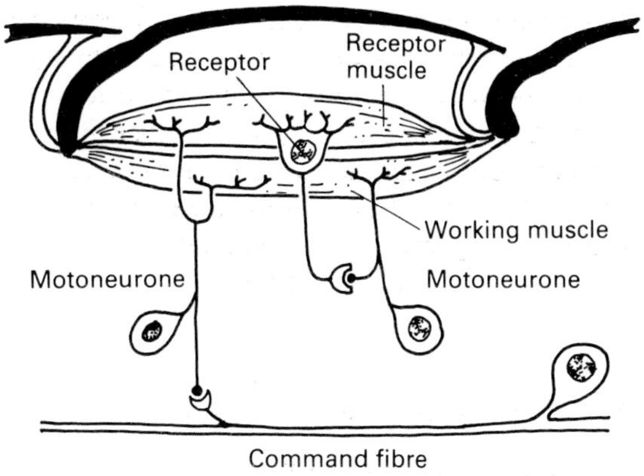

Fig. 16.24. Neuronal circuits involving the stretch receptor of the crayfish. If contraction of the 'working' and receptor muscles due to activation of the command fibre and motoneurone is impeded by external factors, the receptor is activated leading to further stimulation of the muscle. The receptor also mediates the monosynaptic stretch reflex. (Inhibitory innervation of the receptor cell by centrifugal fibres is not shown.) (After Kennedy, 1975.)

Many *escape* or *startle responses* involve giant axons (Section 16.2.4). Such elements are not present in leeches, but these do have within the nerve cord a fast, through-conducting pathway which mediates rapid contractions of the whole body. It consists of a series of 'S-cells', one per segment, whose longitudinally directed axons are coupled to those in adjacent segments by electrical synapses. The system illustrates the importance of electrical (and therefore fast) synapses in circuits mediating escape responses (Fig. 16.25). The various types of mechanoreceptor within the body wall also form monosynaptic reflex arcs with motoneurones in the same segment.

Perhaps the most curious reflex ever described is shown by the aggregate stage of some salps (urochordates), which form elongate chains of up to 20 'zooids'. Each zooid is connected to its neighbour by two epidermal plaques, transmitting in opposite directions (Fig. 16.26). The cells on the output side of any plaque have an ultrastructure resembling that of pre-synaptic terminals (remember that these are not neurones but epithelial cells). On the 'post-synaptic' side, separated by a thin layer of extracellular materal, are 6–12 ciliated receptor neurones. Stimulation of these cells results in the passage of impulses along their axons into the brain, where motoneurones are activated whose fibres extend to the epidermis, in which they evoke action

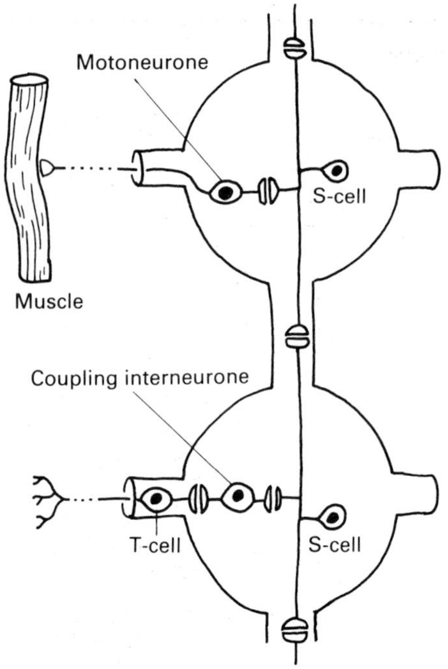

Fig. 16.25. Neuronal circuits involved in the startle response in the leech. Transmission from the sensory T-cells through the coupling interneurones, a series of S-cells, and into the motoneurones, is achieved exclusively by electrical synapses—only the neuromuscular junctions are chemically transmitting. Only one side of two adjacent ganglia is shown.

potentials. Excitation spreads by epithelial conduction to the output plaque of the zooid, and so on.

16.9.4 Fixed action patterns

The inherent capacity of the central nervous system to generate and co-ordinate integrated patterns of motor actions has been demonstrated in many groups of animals. 'The withdrawal of the mantle and closure of the valves in a clam, *Mya*, . . . copulatory movements . . . the flight rhythm, walking . . . respiratory movements . . . heart beat . . . swimming beat of jellyfish have all been shown with varying degrees of rigor (*sic*) to be centrally controlled' (Bullock, 1977).

The opisthobranch mollusc *Tritonia* reacts to touch by a predatory starfish with a bout of swimming lasting about 30 seconds. This involves alternate contractions of the dorsal and ventral muscles, and could conceivably involve either a centrally generated pattern, or a chain of reflexes—the initial stimulus evoking contraction of the dorsal muscles, which by sensory feed-back through reflex arcs stimulates the ventral muscles and so on. Investigations have shown that the normal, characteristic sequence of nervous activity in the motoneurones can be provoked, in brains kept *in vitro*, by an initial stimulus applied to the severed sensory nerves. Clearly, in this case, even if sensory feed-back is provided by the muscles, it is not essential for pattern generation.

16.9.5 Variations upon two themes

Many units of behaviour share at least some of the characteristics of both reflex actions and fixed action patterns (Box 16.9). In the locust, a single stretch receptor cell associated with the hinge of each wing is activated at the end of each upstroke. Here at least is a system which we might expect to work by means of a series of reflex arcs, but this is not the case, since isolated thoracic ganglia, deprived of sensory input, can show the pattern of activity associated with flight. Instead, impulses generated by the receptors in any given cycle exert a tonic stimulatory influence

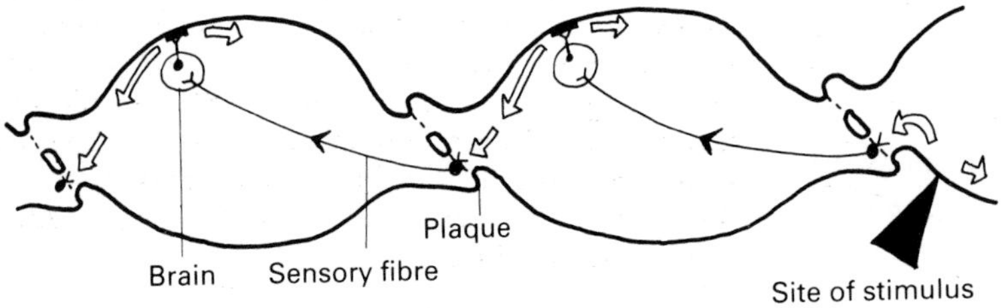

Fig. 16.26. Reflex arcs within each of a series of *Salpa* zooids. Each arc consists of a sensory pathway extending from an epidermal plaque to the brain, motor innervation of the epidermis, and epithelial conduction to the output plaque. Conduction of these pulses throughout the chain may induce its dissociation. (After Anderson & Bone, 1980.)

Box 16.9 Crayfish tail-flips

Crayfish tail-flips have been regarded as typical fixed action patterns. Sudden, threatening stimuli, intense enough to evoke a single impulse in the giant fibres, initiate a rapid, highly stereotyped, escape response—a 'tail-flip'—caused by the simultaneous contraction of the flexor muscles of the abdominal segments. The action is completed too rapidly to permit its modification by feed-back.

Activity in the giant fibres inhibits both the motoneurones to the extensor muscles (which are antagonistic to the flexors), and the muscle stretch receptors of the extensors, since they would otherwise interfere with the tail-flip, until maximum force is exerted by the flexors. The giant fibres are thus acting as *command neurones* (Section 16.10.2) whose output brings into play the various co-ordinated components of the action.

However, after the abdomen is flexed, there is a rapid re-extension of the body by contraction of the extensor muscles. This 'recovery stroke', an invariable part of the natural sequence, is *not centrally generated*, but is a *chain reflex effect*. The muscle receptors, once freed from their inhibition, are stimulated by being stretched and activate the extensors by the reflex arc described previously (Section 16.9.3). (Interestingly, the inhibition of the flexor muscles which accompanies the feed-back driven contraction of the extensors is an effect (albeit a delayed one) of giant fibre output.)

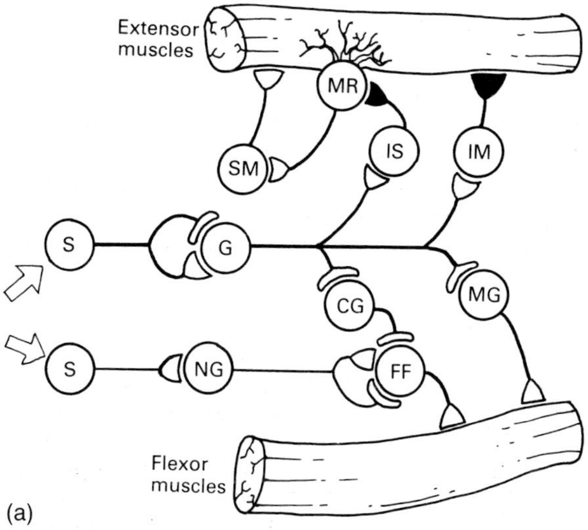

(a)

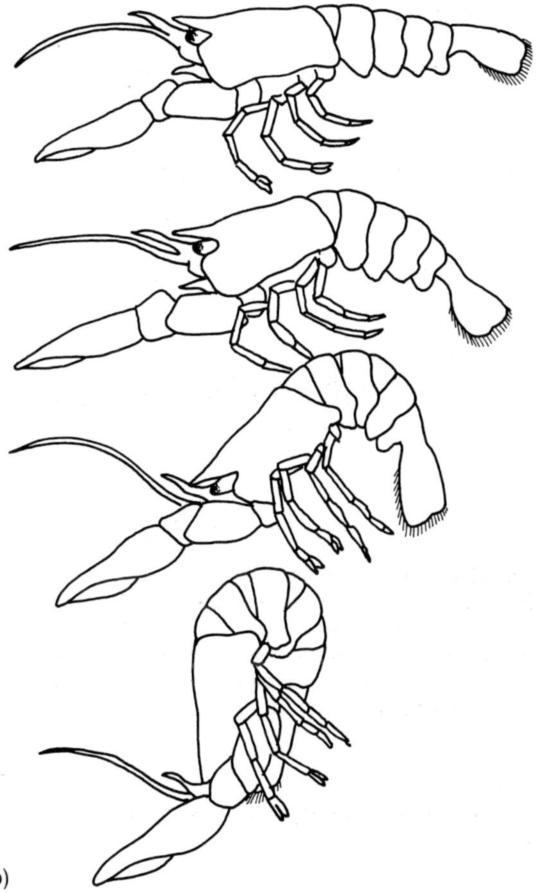

(b)

(a) Simplified neuronal circuits involving both giant neurones and non-giant pathways, present within each abdominal segment, controlling crayfish tail-flips. The systems are activated by environmental stimuli (arrows). CG, central giant neurones; FF, fast flexor motoneurones; G, giant command neurone; IM, inhibitory motoneurones; IS, inhibitory input to stretch receptor; MG, motor giant neurones; MR, muscle stretch receptor; NG, non-giant neural pathways; S, sensory neurones; SM, stimulatory motoneurone. The synapses between the giant fibre and the motor giants were the first electrical synapses to be discovered (see Section 16.3.4). (b) Crayfish tail-flip. (After Wine & Krasne, 1972.)

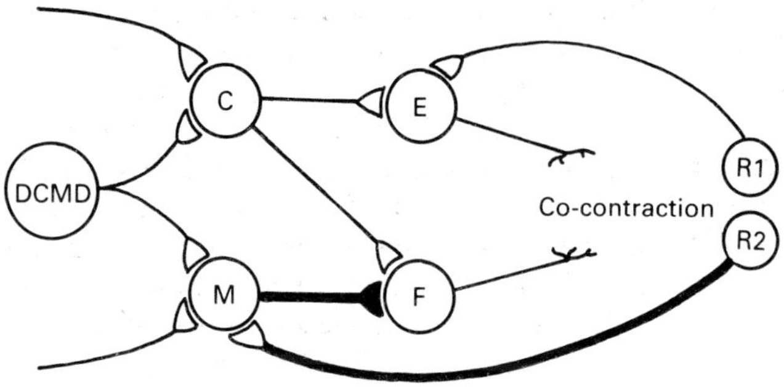

Fig. 16.27. Neuronal circuits controlling the locust jump. DCMD, movement detector neurone; C, C neurone; M, M neurone; E, F, motoneurones to the extensor and flexor muscles, respectively; R1, R2, sensory neurones mediating the feed-back effects of co-contraction. Heavy lines, nervous pathways leading to the inhibition of the flexors. (After Pearson, 1983.)

affecting the speed and strength of wing beat for several cycles.

The LGMD neurones of the locust (Section 16.8.5) provide reliable synaptic output to a pair of 'contralateral movement detector neurones' (DCMD) which descend to the third thoracic ganglion, where each innervates two further identified neurones ('C' and 'M') (Fig. 16.27). Each C neurone synapses on to the motoneurones of two antagonistic muscles found in the 'thighs' of the large legs—these are the extensor and flexor muscles for the tibia (the lower part of the leg).

First, the C cell induces both 'cocking' of the leg, locking the tibia under the insect, and 'co-contraction' of the two antagonistic muscles which distorts the elastic cuticle of the femur. Second, a *positive* feed-back influence from cuticular stress receptors acts on the motoneurones to reinforce extensor contraction. Last, feed-back activity in another set of receptors, evoked by co-contraction, stimulates the M cell. If this coincides with continuing input from the DCMD, etc., the M cell is activated, inhibiting the flexor motoneurones and triggering the jump as the energy stored in the distorted cuticle is released. Clearly, sensory feed-back is an integral part of the mechanism by which the jump is controlled and ensures that external stimuli cannot evoke the jump unless the mechanism is fully primed (see Section 10.6.4).

In conclusion, we must note the irreverence of nature with respect to our categories. The nervous system clearly does have the inherent capacity to generate integrated motor programmes. However, some fixed action patterns are 'more fixed than others'—for example, some rapid withdrawal responses in annelids mediated by the giant fibres are graded, whereas others are all-or-nothing—and it is interesting that the German term originally used

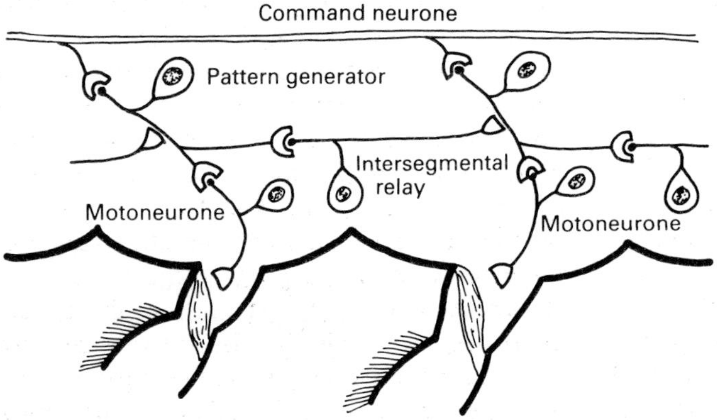

Fig. 16.28. Principal components of neuronal circuits controlling motor output of crayfish swimmerets (cells in two adjacent segmental ganglia are shown). More than one neurone of each type is usually present. (After Stein, 1971.)

by Lorenz stressed rather the inherited aspect. The exclusion of feed-back effects was regrettable. These are as much a consequence of the genetic make-up as the organization of the central nervous system, and constitute one of the main ways in which patterns of behaviour achieve the goals to which they appear to be directed.

16.10 Organization of motor output

16.10.1 Aspects of organization

Systems of neurones controlling motor output, like sensory systems, usually show a hierarchy of organization and this is well illustrated by the circuits controlling the abdominal appendages (pleopods or 'swimmerets') in crayfish (Fig. 16.28). Neurones at three levels can be distinguished, namely, *command cells*, *pattern-generating interneurones*, and *motoneurones*.

16.10.2 Command neurones

In the crayfish, both stimulatory and inhibitory cells are present, which extend to each of the segmental ganglia. Such cells were first discovered in crustaceans by Wiersma, who observed that activation of particular, single cells will evoke co-ordinated patterns of behaviour. A command neurone is a *decision-maker*. Just as sensory pathways show divergence and provide output to a multiplicity of motor centres (Section 16.8.3) so command cells within motor systems are the focus of convergence and integration of input from a diversity of sense organs. The resulting activity of the cell, or lack of it, determines whether or not the motor action is initiated. Such activity is therefore both *necessary* and *sufficient* to generate the action.

The giant fibres involved in crayfish tail-flips (Box 16.9) have been regarded as model command units. However, more detailed examination has revealed that the response to direct stimulation of the giant fibres is not quite the same as that evoked by natural stimulation—natural flips tend to be more vertical and to last longer. The giants thus determine the broad thrust of the action, but their activity is not sufficient for a perfect response, for which parallel, non-giant pathways are indispensable.

Command cells typically do not determine the 'details' of motor activity—they do not specify the timing or pattern of impulses within the interneurones and motoneurones they control. Exceptions include the giant fibres encircling the margin of the bell in medusae, and the giant fibres of annelids and other elongate animals, which are responsible for the virtually simultaneous contractions of widely separated parts of the body that are characteristic of escape reactions.

In many motor systems, no single cell, or group of cells, seems to correspond to a command unit (e.g. Fig. 16.27). Furthermore, although Mauthner's cells in teleosts and aquatic amphibia are command neurones, the concept has little general relevance to vertebrates.

16.10.3 Pattern generation

The generation of patterns of motor output is often a function of neither command cells, nor motoneurones, but of interneurones interposed between the two. One of the most common motor patterns involves two muscles acting as antagonists, and the neural machinery must ensure that only one of the two is stimulated at a time. One example, involving a closed loop of interneurones, controls swimming in the leech (Fig. 16.11).

In the crayfish, one non-spiking interneurone in each segment shows regular, spontaneous oscillations in membrane potential in phase with the swimmeret beat. Depolarization and transmitter discharge stimulate the motoneurones of the muscles responsible for the power stroke, but inhibit the motoneurones responsible for the recovery stroke. Although the command cells determine whether or not the interneurone is active, the latter is responsible for generating the *pattern* of activity. This can be shown by accelerating or retarding the oscillation experimentally. Following this the rhythm is found to be *reset*; it does not just resume a pattern in time with the previous one. Other interneurones mediate

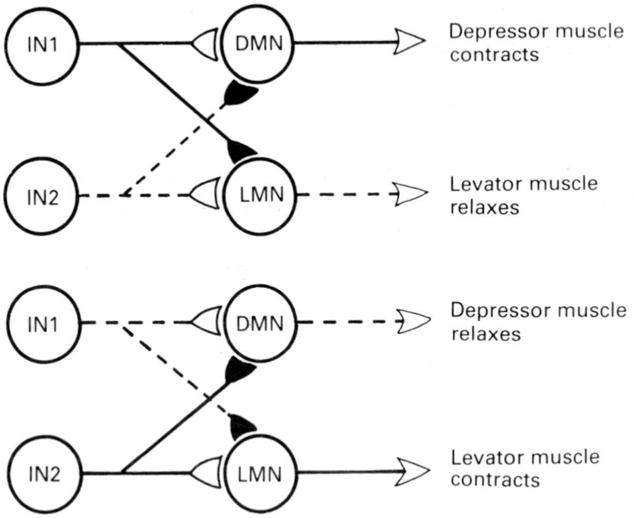

Fig. 16.29. Neuronal circuit controlling gill ventilation in the crab. Two non-spiking interneurones (IN1, IN2) are alternately active and drive the depressor muscle motoneurones (DMN) and levator muscle motoneurones (LMN), respectively. Dashed lines, inactive fibres; clear bulbs, excitatory synapses; filled bulbs, inhibitory synapses.

the intersegmental co-ordination responsible for the metachronal rhythm of swimmeret beating.

In a comparable system in the thoracic ganglion of the crab, which controls gill ventilation, one non-spiking neurone functions like that described above, but a second is present with oscillations that are out of phase with the first (Fig. 16.29). The two are presumably linked to maintain this functional relationship.

By having a number of interneurones, capable of generating a variety of patterns, individual motoneurones can be used in different ways and the unnecessary duplication of fibres providing neuromuscular innervation is avoided. Thus, Burrows has found that a single motoneurone to a leg muscle in the locust is influenced by at least twelve interneurones.

16.10.4 Motoneurones

Although generation of motor pattern frequently comes within the province of interneurones, many motoneurones show mutual interactions and are thus also involved in this process. In the nematode, *Ascaris*, five large interneurones extending the length of the body provide tonic stimulation for swimming. The excitatory motoneurones supplying the dorsal muscle in each 'segment' of the body both stimulate the inhibitory neurones for the ventral muscle and, through them, inhibit the ventral excitatory neurones (Fig. 16.30). Activation of ventral excitatory cells has similar effects on dorsal neurones. Activity oscillates between the dorsal and ventral neurones with a frequency and intensity dependent on the degree of stimulation by the interneurones, evoking contractions alternately in their muscles. Other cells form relays between adjacent regions of the body and co-ordinate their activity.

The prediction that pattern generation by motor units would necessitate duplication of the latter seems to be borne out in the case of *Ascaris*, since different types of motoneurones may control forward and backward swimming. It probably matters little to the animal because of its limited repertoire of actions, and this may apply to many motor systems in invertebrates.

The rule that the same set of motoneurones is used for different purposes, depending on the input they receive, is not generally followed by units involved in escape reactions. In squid, the giant fibres mediate the jet-propelled startle reaction, whereas smaller fibres control muscle contractions involved in respiratory movements. In the medusa *Aglantha*, the escape reaction consists of one to three violent

Box 16.10 Control of muscle contraction in crustaceans

Individual muscles in crabs are composed of fibres with different structural and functional characteristics. Some fibres readily generate action potentials and show a rapid development of tension, whereas others use mainly graded potentials and contract more slowly. The muscles have a complex *polyneuronal* innervation and, in contrast to vertebrate somatic muscle, 'decision-making' is often delegated to the periphery where conflicting influences collide within the innervated structure, as in many other invertebrates.

A single, fast-conducting axon sends branches mainly to the various fast muscle fibres; its synapses rapidly become fatigued upon repetitive stimulation. A fine-diameter fibre directs its terminals more towards slower muscle fibres and upon repetitive stimulation shows facilitation. Thus the degree of contraction is controlled, not by recruitment of extra motor units as in vertebrates, but by a graded control of the muscle fibres. In addition, inhibitory fibres discharge GABA both at synapses with the excitatory terminals (*pre*-synaptic inhibition), blocking release of the excitatory transmitter glutamate, and at synapses on the muscle fibres (*post*-synaptic inhibition) to damp down any excitation that gets through.

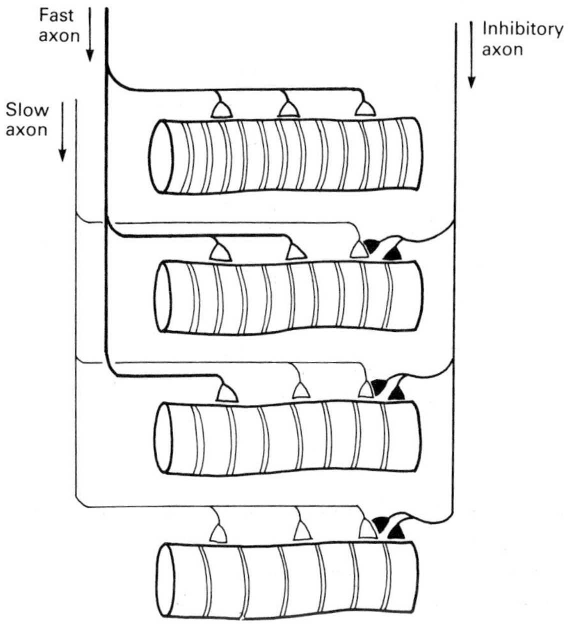

Innervation of muscles in the crab (after Atwood, 1973).

contractions of the bell, mediated by eight giant motor fibres running up beneath the subumbrellar surface and by smaller diameter, lateral motoneurones. The normal swimming contractions involve just the latter. Similarly, the lateral giant fibres of the crayfish excite not only (via a pair of central giants) the five to nine 'fast flexor' motoneurones used in flexible escape responses, but also a pair of giant motoneurones in each segment which are reserved exclusively for the tail-flips (Box 16.9).

16.10.5 Neuromuscular innervation

The control of muscular activity by motoneurones is mediated by the discharge of chemical neurotransmitters at synapses (neuromuscular junctions) formed between nerve terminals and muscle fibres (Section 16.3.3; see also Box 16.10). The presence or absence of propagated action potentials within individual muscle fibres, and/or transmission via gap junctions between fibres, is also of great importance. There is a wide variety of patterns of innervation of muscles and glands. Thus, the heart of ascidian urochordates (as of some insects) is unusual in *lacking innervation* and contractions are thus myogenic, possibly with endocrine modulation. In contrast, the heart beat in crustaceans is neurogenic, being controlled by *excitatory motoneurones only*. Somatic muscle in *Ascaris* is under *dual* (excitatory and inhibitory) *control*.

The extensor tibiae muscle in the locust is controlled by *four nerve fibres*. Excitatory fibres and an inhibitory fibre release the transmitters glutamate and GABA, respectively. In addition, terminals from a DUM (dorsal unpaired median) neurone secrete the neuromodulator octopamine (Table 16.1). On its own it seems to have little effect, and its primary role

Table 16.1 Chemical mediators in invertebrates

Neurotransmitters

(i) Acetylcholine $H_3C{-}\overset{\displaystyle O}{\overset{\|}{C}}{-}O{-}(CH_2)_2{-}N^+{-}(CH_3)_3$

(ii) Glutamate (amino acid) $H_3N^+{-}CH(C{=}O{-}O^-){-}CH_2{-}CH_2{-}\overset{\displaystyle O}{\overset{\|}{C}}{-}O^-$

(iii) Octopamine (amine)

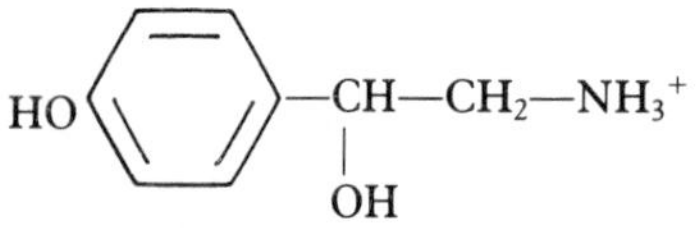

Hormones

(i) Hydra head activator (peptide)
pGlu—Pro—Pro—Gly—Gly—Ser—Lys—Val—Ile—Leu—Phe

(ii) 1-Methyl Adenine

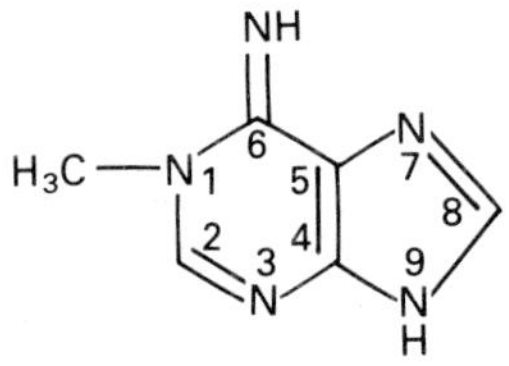

(iii) Ecdysone (steroid)

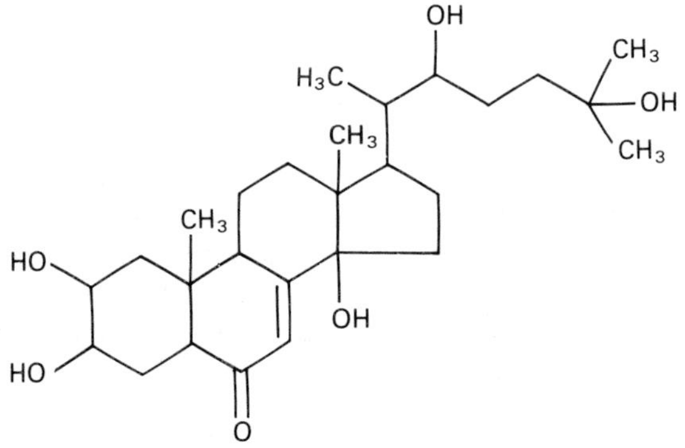

(iv) Juvenile hormone I

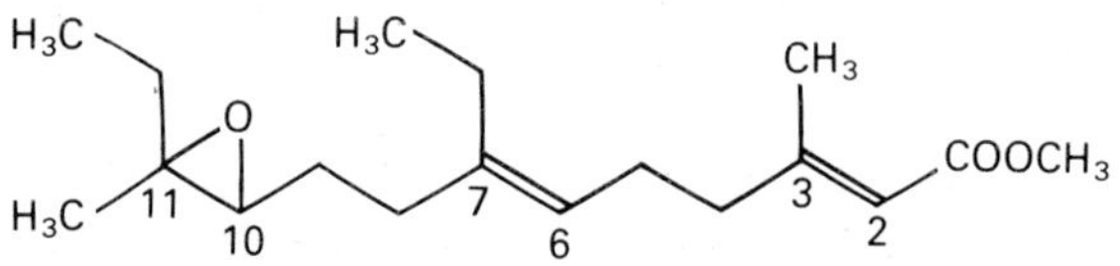

Pheromone

Bombykol

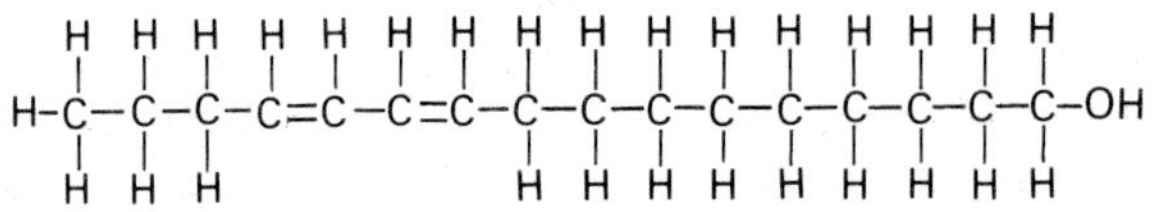

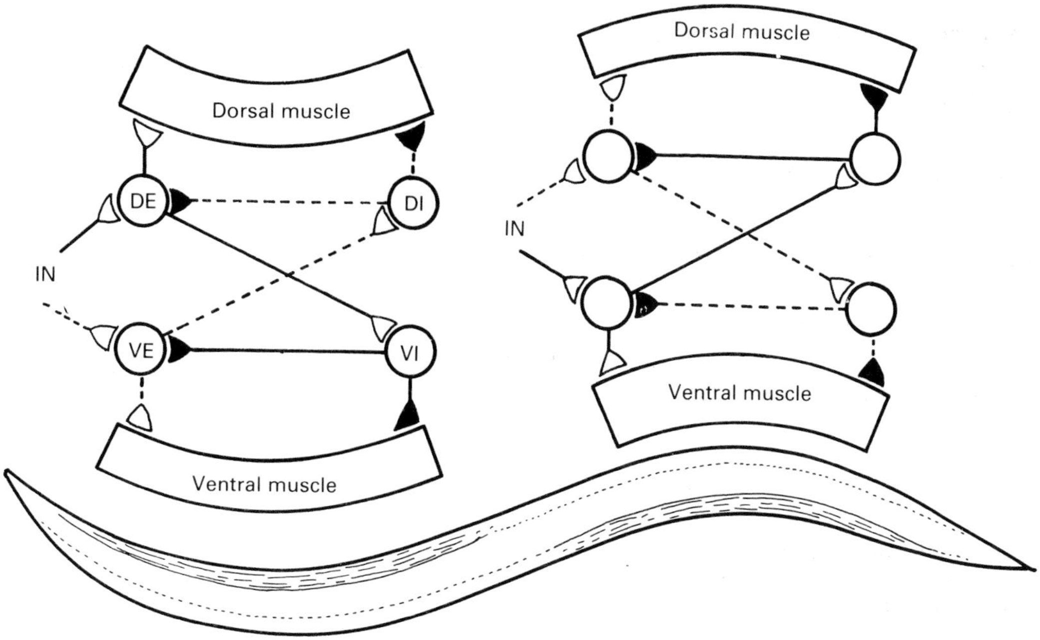

Fig. 16.30. Neuronal circuit controlling swimming in *Ascaris* which involves the alternate contraction of dorsal (left) and ventral muscles (right). IN, interneurones; DE, VE, DI and VI, dorsal and ventral excitatory and inhibitory motoneurones, respectively. Clear bulbs, excitatory synapses; filled bulbs, inhibitory synapses. (After Stretton *et al.*, 1984.)

is to confer on the muscle, which otherwise generates the more prolonged tension required for the maintenance of posture, the ability to respond to motor excitation by more intense contraction and rapid relaxation appropriate for locomotion.

The adductor muscles of bivalve molluscs show great tenacity and can hold the valves closed for hours or even days—a valuable defence against predation. Contraction is induced by activation of excitatory motoneurones which release acetylcholine, but, once tension has developed, the muscle enters a state of 'catch', during which oxygen consumption is low, and whose maintenance is independent of further stimulation. Relaxation is brought about by the activation of inhibitory fibres whose transmitter is apparently 5–HT.

In insects with direct flight muscles, impulses in the nerve fibres 'beat time' for the contractions. In contrast, the specialized, indirect flight muscles of, for example, flies have a frequency as high as a thousand beats per second, whereas the number of action potentials in nerve and muscle fibres is only five to ten. The neural influence has just a *tonic effect* and puts the muscle into a 'active state'. Contraction and relaxation are internal 'reflex actions' of the muscle fibres, triggered (see Section 10.6.3) by exposure to stretch, etc., which accompanies the wing beat.

16.11 Chemical communication

16.11.1 Spectrum of chemical communication

Neurotransmitters are the secretory products of neurones which are typically discharged in close proximity to other cells, whose activity they influence (Table 16.1). In consequence, neural action is highly specific in both space and time. In contrast, *hormones* are chemical messengers elaborated by cells which are, in most cases, aggregated to form *glands*, and they influence other, distant cells to which they are carried in the *blood stream* or other body fluids. Glands secreting hormones are called *endo*crine because they secrete their products *internally*, into the circulation, whereas *exo*crine glands such as mucus or enzyme-secreting glands discharge their products *externally*, usually into ducts.

Whereas the function of a typical neurone may be compared with the use of the telephone, endocrine function resembles the use of a megaphone. Hormones often have effects that are widespread within the body, but this does not mean that they cannot show great specificity of action. They are dependent for their effects on the possession by the target cells of *receptors*. For example, different categories of chromatophores, even those having the same coloured pigment, may be sensitive to different hormones.

Other substances, called *pheromones*, are secreted into the environment and have effects of biological significance on other individuals. The phenomenon is encountered at the protist level, mediating aggregation of individuals of amoeboid slime moulds. At the other extreme, the social insects are 'walking batteries of exocrine glands' (E. O. Wilson, 1975).

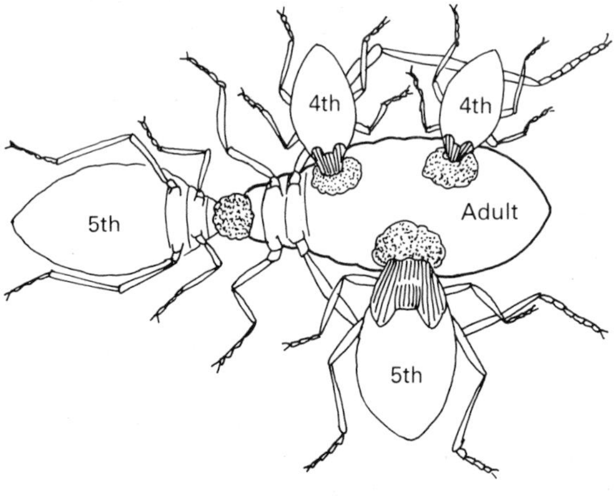

Fig. 16.31. Two fifth- and two fourth-stage larvae joined, with wax seals, in parabiosis and sharing the circulation with an adult *Rhodnius*. The larvae provide moulting and juvenile hormones and induces the adult to moult again, when it shows regression to a more juvenile form. (After Wigglesworth, 1940.)

16.11.2 Criteria for endocrine status

A number of criteria must be met to establish the endocrine status of an organ. In summary, *surgical removal* of the organ, but not of other organs (the latter is the *control* for the experiment) must abolish the effect its hormone is thought to evoke. *Replacement therapy* must be successful—i.e. reimplantation of the organ (but not of other organs), or injection of extracts, should restore the effect. This requirement can also be met by the *parabiosis* or grafting technique (Fig. 16.31). Alternatively, cells or organs can be maintained *in vitro*. Introduction of the gland or its extracts into the culture can then be used to assess its action.

16.11.3 Neurosecretion

In 1928 Ernst Scharrer postulated that some nerve cells—*neurosecretory cells*—combine the properties of neurones and endocrine gland cells. This idea was pursued in invertebrates by Berta Scharrer and others who showed that such cells are widespread within these animals. Studies on insects, particularly, led the way in providing conclusive evidence for the theory.

Neurosecretory cell bodies are typically clustered within the central nervous system to form ganglionic nuclei (Fig. 16.32). Axons extend from the cell bodies and form swollen terminals in close association with blood spaces. The terminals may be aggregated to form a discrete body, or *neurohaemal organ*, or spread across a *neurohaemal area* at the surface of a ganglion or a nerve (Fig. 16.33). Neurosecretory material is produced in the cell bodies, transported down the axons by axoplasmic flow, and stored in the terminals. Viewed with the electron microscope, the material consists of minute secretory granules, usually 100–200 nm in diameter, which are budded off by the Golgi apparatus.

Activity within neurosecretory cells can arise either spontaneously or via synaptic excitation. It leads to

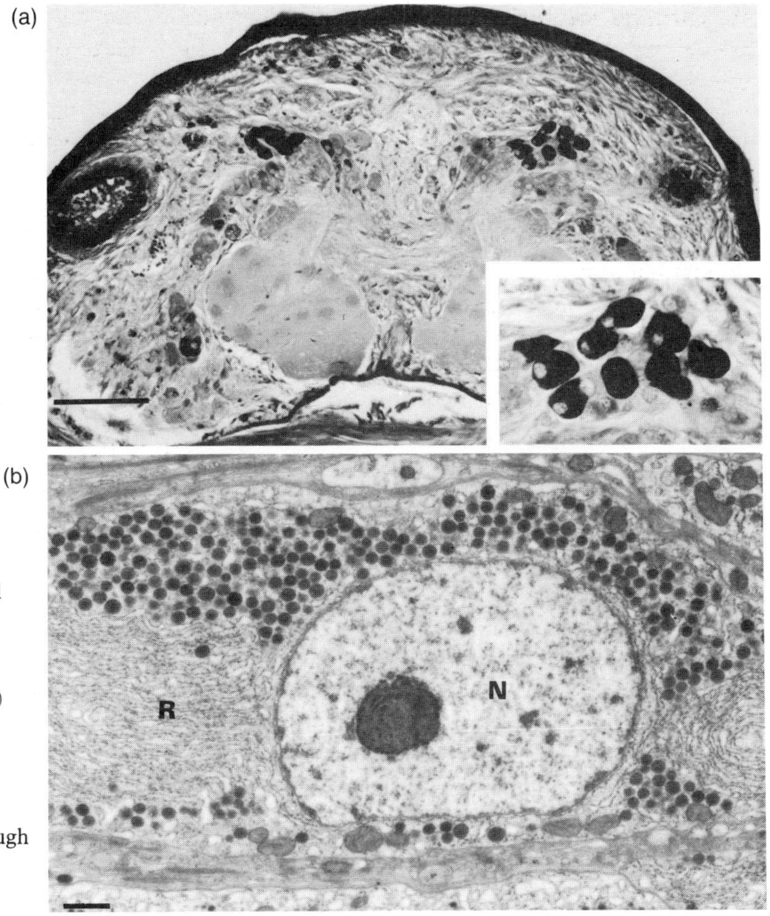

Fig. 16.32. (a) Neurosecretory cells (inset) in the cerebral ganglion of the polychaete annelid *Harmothoe*. The cytoplasm is charged with secretory material stained deeply with paraldehyde fuchsin, one of the 'Gomori' stains; the nuclei are unstained. × 125; bar, 100 μm. (From Golding and Whittle, 1977.) (b) Electron micrograph of neurosecretory cell in the earthworm *Lumbricus*. The secretory material consists of large numbers of minute granules about 200 nm in diameter. N, nucleus with nucleolus; R, rough endoplasmic reticulum. × 6500; bar, 1 μm.

the passage of action potentials down the axons, influx of Ca^{2+} ions into the terminals and the discharge of hormone into the blood stream. Release is accomplished by the process of *exocytosis* (Box 16.11).

16.11.4 A distinction blurred

More recent findings have clouded the distinction between neurosecretory cells and 'ordinary' neurones. Neurones resembling neurosecretory cells have been described in cnidarians and platyhelminths which lack vascular systems. In *Hydra*, a *head activator* (Table 16.1) is present and it was purified by Schaller and her colleagues in 1981. It promotes tentacle regeneration by body fragments maintained *in vitro*, at a concentration of 1 '*Hydra*-equivalent' per ml, or less than 10^{-10} mol/l. Interestingly, using antibodies to this substance, a system of neurones containing the same peptide has been found in the brains of mammals.

Furthermore, some neurones have now been described which share the *function* of classical neurosecretory cells (i.e. they secrete hormones), but lack the distinctive *cytology* of the latter. Conversely, other neurones have the *cytology*, but not the *function* of typical neurosecretory cells: they do not release their products into the circulation, but extend to make direct contact with the target cells.

16.11.5 Neuropeptides

The essential similarity of neurosecretory and other neurones is shown by the fact that the secretion of

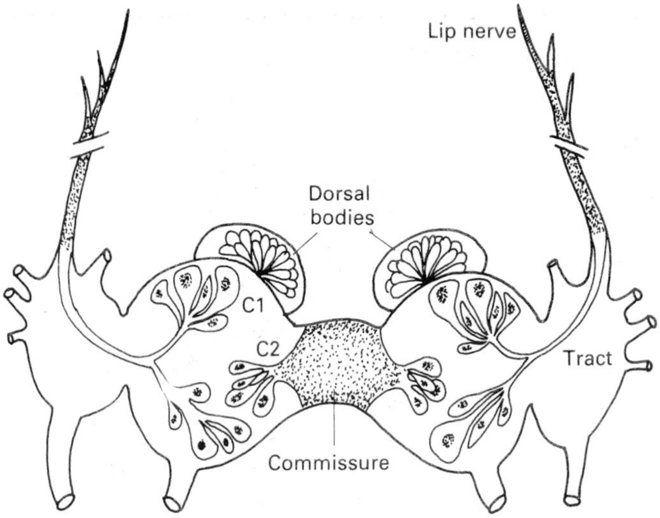

Fig. 16.33. Cerebral ganglia of the mollusc *Lymnaea*, showing fibre tracts leading from two types of neurosecretory cells to neurohaemal areas at the surface of the commissure and lip nerves. The dorsal bodies are non-nervous endocrine organs. (After Joosse & Geraerts, 1983.)

peptides, originally thought to be restricted within the nervous system to neurosecretory cells, is now known to be a characteristic of many and perhaps all neurones, including those producing conventional transmitters. Furthermore, typical neurosecretory cells can exert a co-ordinated influence on both the physiology and the behaviour of the animal. Within the 'bag cells' of the mollusc *Aplysia*, neurosecretory material is synthesized as a high molecular weight protein, or *prohormone*. This is divided into shorter fragments inside the secretory granules by the action of enzymes, so that secretory release involves the discharge not of a single hormone, but of a 'cocktail' of peptides. One, the 'egg-laying hormone', reaches the gonad via the blood stream and induces spawning. Another stimulates the nerve terminals within the neurohaemal complex ensuring their maximum participation. A number of them, including the egg-laying hormone, diffuse throughout the abdominal ganglion and stimulate particular neurones whose activity drives the highly characteristic pattern of behaviour that accompanies spawning.

16.11.6 Neuroendocrine systems

Nervous, neurosecretory, and endocrine systems do not exist in the body in functional isolation from each other, but form integrated, and in advanced invertebrates, complex *neuroendocrine systems*. A reflex pathway consisting of sensory reception, processing and integration by interneurones, and control of effectors may have an endocrine step which mediates the last part. The endocrine system is thus usually controlled by the nervous system, but, conversely, its products often have profound influences on the latter (Box 16.12).

16.11.7 Pheromones

The first pheromone to be identified was *bombykol*, (see Table 16.1), a substance secreted by females of the silk moth, *Bombyx mori*, the culmination in 1959 of work by Butenandt and his colleagues involving the extraction of half a million abdominal glands. Amazingly, a single molecule is thought to have a detectable effect on an olfactory receptor on the antenna of the male, and 200 molecules eliciting 200 impulses are sufficient to evoke a behavioural response. The male moth sets off upwind with a characteristic zigzag flight pattern.

More recent research has established the importance of *blends* of compounds. A combination of chemicals may be required to elicit the initial effect—for example, as a sex attractant. Alterna-

Box 16.11 Neurosecretory release by exocytosis

Material is discharged from neurosecretory neurones, as from many conventional gland cells, by *exocytosis*. During this process the granule membrane fuses with the plasma membrane of the axon terminal, so that the contents of the granule can diffuse out into the blood. This was first established in invertebrate studies, particularly that by Normann (1965) on the blow-fly, *Calliphora*.

More recent studies, first on annelids and then on a wide variety of invertebrates, have indicated that 'ordinary' nerve terminals—those making synaptic contacts with other cells—also show exocytosis of secretory granules. This means that synaptic terminals possess two distinct secretory mechanisms.

Synaptic vesicles may be formed locally, within nerve terminals, and take up conventional transmitters such as acetylcholine from the cell sap where such substances are synthesized. The vesicles typically discharge their contents at synaptic thickenings (Fig. 16.6). In contrast, neuropeptides are synthesized, and packaged within granules, in the cell body. After transport to the nerve terminal, the granules discharge their contents by exocytosis across apparently unspecialized areas of membrane.

Following exocytosis, the membranes of vesicles and granules are detached from the plasma membrane and internalized within the terminal (the process of *endocytosis*). However, whereas vesicles are *recycled* (i.e. refilled and re-used), membranes of granules are probably *degraded*.

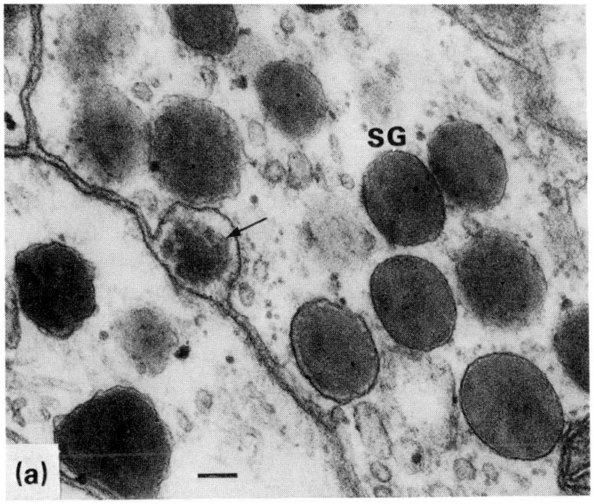

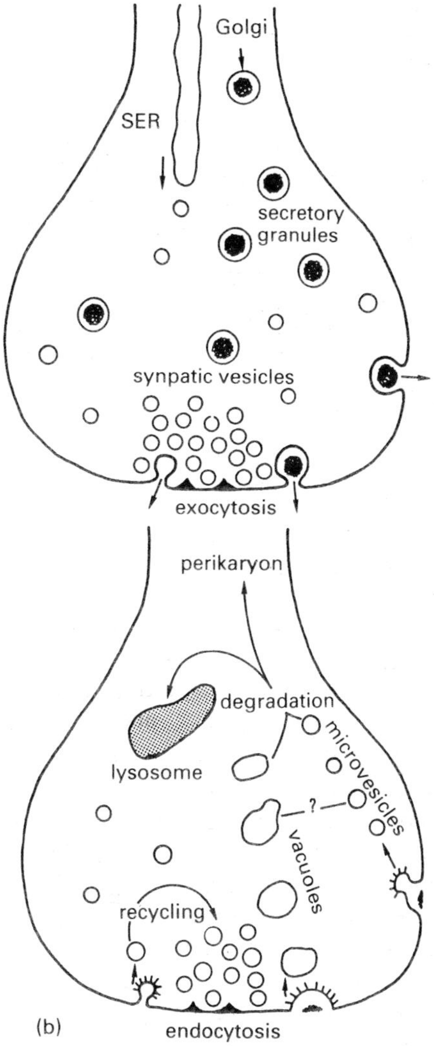

(a) Electron micrograph showing exocytosis of a secretory granule from within a synaptic terminal in the cerebral ganglion of the earthworm *Lumbricus*. Arrow, extruded granule core; SG, secretory granules. × 60 000; bar, 100 nm. (b) Synaptic terminals have two distinct secretory mechanisms, involving vesicles and granules, respectively. Neurosecretory endings have an identical ultrastructure. (From Golding & Pow, 1987.)

Box 16.12 Neuroendocrine systems in the locust

The diversity and complexity of neuroendocrine systems in some invertebrates are well illustrated in the locust.

System 1. A *diuretic hormone* is synthesized by neurosecretory cells in the brain, and transported along the axons to the storage lobe of the corpus cardiacum (a neurohaemal organ) from where it is released into the blood. The hormone stimulates urine production by the Malpighian tubules. This is a first-order endocrine system—only one hormonal step is involved.

System 2. Nerve fibres extend to the glandular lobes of the corpus cardiacum, where they form typical synaptic contacts with gland cells which they control by releasing the neurotransmitter, octopamine. When stimulated the cells release the peptide *adipokinetic hormone*, identified by Mordue and his colleagues in 1976, which mobilizes lipid from the fat body to fuel flight.

System 3. Another hormone secreted by neurosecretory cells within the brain/corpus cardiacum complex is called *prothoracotrophic hormone* (PTTH). It stimulates the non-nervous prothoracic glands to secret the steroid hormone *ecdysone*, which after modification by the tissues, triggers moulting. This is a second-order endocrine system, with two endocrine steps.

System 4. The secretion of JH by the corpus allatum is apparently controlled by both direct innervation and blood-borne factors.

Feed-back. Endocrine systems typically have feed-back loops. For example, high levels of circulating 20-hydroxyecdysone affect the cerebral neurosecretory cells and inhibit ecdysone synthesis by the prothoracic glands.

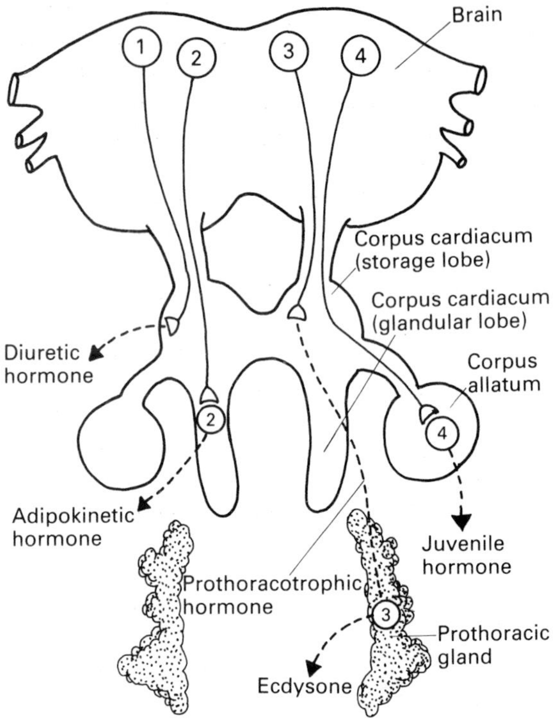

Brain and endocrine organs in the locust (the relative sizes of the corpus cardiacum and corpus allatum are greatly enlarged).

tively, one compound may have a primary, long-range action (e.g. as an attractant) and another, a secondary, short-range effect (e.g. as a copulatory releaser). Even bombykol is secreted in combination with another, related compound—*bombykal*—which has inhibitory effects on flight. At close range, males of some moths are themselves stimulated to secrete sex pheromones evoking responses by the female. The study of such substances is obviously of immense potential significance for the control of insect pests.

Pheromones are of great importance in the regulation of caste structure in social insects. The best known example is the secretion of the *queen substance* by the mandibular glands of honey bee queens. Its two components—9-oxydecanoic acid and 9-hydroxydecanoic acid—act synergistically to inhibit both gonadal development within the workers, and their construction of special cells in which larvae would develop as queens. When the queen is removed from the hive, these inhibitions are lifted and sexually active females are produced to replace her. The same substance is also important as a sex attractant, and during swarming and settling.

Chemicals secreted by members of one species may

have important effects on those of another and are called *allomones*. The relationship of many symbiotic organisms is established in this way. In some insects, a substance acting as a pheromone to members of the same species may act as an allomone to those of another, for example, attracting parasites. In the annelid *Nereis*, the declining concentration of the brain hormone induces breeding (following which the animal dies), and sporozoan parasites 'tune in' to this signal and co-ordinate their own reproduction with that of the host.

An interesting sequence of chemical signals leads up to reproduction in the moth *Antherea polyphemus*, whose larvae feed on the leaves of the red oak. An *allomone* produced by the oak leaves stimulates the moth to secrete a neurosecretory *hormone*, which induces exposure of the genitalia and release of a *pheromone*. Males are attracted to the female and reproduction takes place at a site providing food for the next generation.

16.12 Roles of endocrine systems

Hormones are involved in the control of a wide variety of processes in invertebrates—a small and diverse selection are described below.

16.12.1 Endocrine trigger

Some hormones are discharged in response to a particular stimulus and act as a 'trigger', having rapid, all-or-nothing effects. In starfish, spawning can be induced in both sexes by injecting extracts of radial nerves into the body cavity of a ripe animal. Strangely, the peptide hormone involved ('gonad stimulating substance' or GSS) is present throughout the radial nerves and nerve ring and is apparently produced by the subcuticular supporting cell layer (i.e. not by true neurones). How the hormone is transported across the nerve cords, and whether the coelomic fluid is used to transport it to the tissues, is debatable. Clearly there are still problems to be resolved concerning this extraordinary system.

Gonad stimulating substance induces gamete maturation in ovaries or ovary fragments *in vitro* and this seemed to indicate that its effect must be direct. In fact, GSS stimulates the synthesis and release of a second substance by the follicle cells of the gonad. This 'meiosis inducing substance' (MIS) was shown by Kanatani and his colleagues in 1969 to consist of 1-methyl adenine (Table 16.1).

MIS has a curious combination of endocrine and para-endocrine roles. First, it affects the follicle cells that produce it, so that they strip away from the gametes and allow the latter to be expelled through the gonoducts. Second, MIS diffuses across the intercellular spaces and induces meiosis within the oocytes. Last, MIS diffuses into the coelomic fluid, stimulating muscle contractions which aid spawning, and in some species evoking brooding behaviour.

16.12.2 Endocrine balance

Some functions are controlled not by the secretion of a single hormone, but by a balance of antagonistic factors. Sugar metabolism, and salt and water balance, are often regulated in this way, which provides for the rapid reversal of an endocrine effect.

In crustaceans, colour change is physiological in character, since it involves no change in the *quantity* of pigment, but is mediated by the *movement* (or 'migration') of pigment granules within cells called *chromatophores* (Fig. 16.34). The cells remain fixed in position and shape.

Hormones controlling chromatophores are typically released in response to environmental stimuli (background illumination, etc.), but some show circadian or tidal rhythms. Two antagonistic hormones are usually involved, one inducing pigment dispersion and the other pigment concentration. However, in some cases a multiplicity of factors affect chromatophores of a single colour, and different patterns of colouration of the body can be produced. Hormones also influence migration of retinal pigments which mediate changes in the physiology of vision in the compound eye (Fig. 16.20b).

16.12.3 Endocrine pattern—control of development in *Nereis*

The pattern of endocrine control shown in nereids (ragworms) is comparatively simple and similar

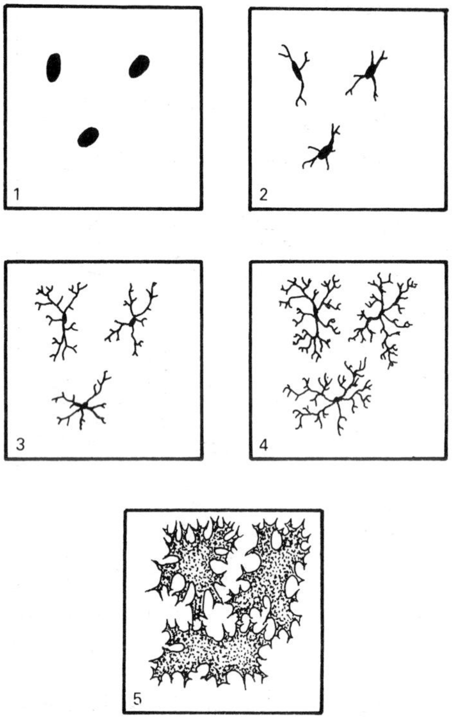

Fig. 16.34. Chromatophores showing the arbitrary scale by which the extent of pigment dispersion is assessed.

patterns are shown by nemertines and some other invertebrates (but not by most other annelids). As first shown by Durchon in 1948, decerebration results in *precocious maturation*. The final stage of the life cycle, in most species, sees the transformation of the body into the *heteronereis* form (Fig. 4.60) adapted for swarming and spawning in the surface waters of the sea. This process too is inhibited by the brain hormone.

Decerebration at earlier stages of life leads to the development of abnormal oocytes, or oocyte degeneration, and incomplete metamorphosis. This suggests that the hormone has a complex, dual action, incorporating a *trophic* effect by means of which the early development of gametes and of the body is sustained, and an *inhibitory* effect on processes appropriate to the final stages of development. In females, the trophic effect on the gametes may be indirect, since the synthesis of special yolk proteins (*vitellogenins*), which are essential for oocyte development, takes place within coelomocytes. The hormone is also indispensable for the proliferation of body segments (see Section 14.4.3).

During the early stages of life, a high level of hormone is secreted, but later the rate of secretion is reduced. Environmental influences may have a role in this regard. However, a substance produced by the maturing gametes inhibits the endocrine activity of the brain, which will accelerate gamete development and lead to the production of yet more gamete substance, and so on (a positive feed-back effect). It is debatable which of the two, brain or gametes, has the *action* and which the *feed-back action*.

A host of processes are involved in segment proliferation, gametogenesis and metamorphosis, and all have been thought to be controlled by the simple pattern of secretion of the 'brain hormone'. In contrast, electron microscopy suggests that the brain is the source of a multiplicity of hormones. Furthermore, the presence of *ecdysteroids* (Section 16.12.4) has recently been reported in these and other non-arthropod invertebrates, and in leeches they probably control moulting. Doubtless, the endocrine pattern will turn out to be more complex than envisaged.

16.12.4 Endocrine pattern—control of development of insects

Because of the dramatic character of insect development, with its series of moults, and metamorphosis in some groups from larva to pupa to adult, (Chapter 15), this subject has attracted a great deal of attention from endocrinologists over the last 50 years. The complexity of developmental processes in arthropods is reflected in that of the neuroendocrine system by which these processes are controlled (Box 16.12).

The importance of the brain was first demonstrated in the classic experiments of Kopec, from 1917 to 1923 (Fig. 16.35). The larva of the gypsy moth *Lymantra* will only pupate if exposed to the influence of the brain for a certain minimum, *critical period*. The hormone involved, *prothoracotrophic hormone*

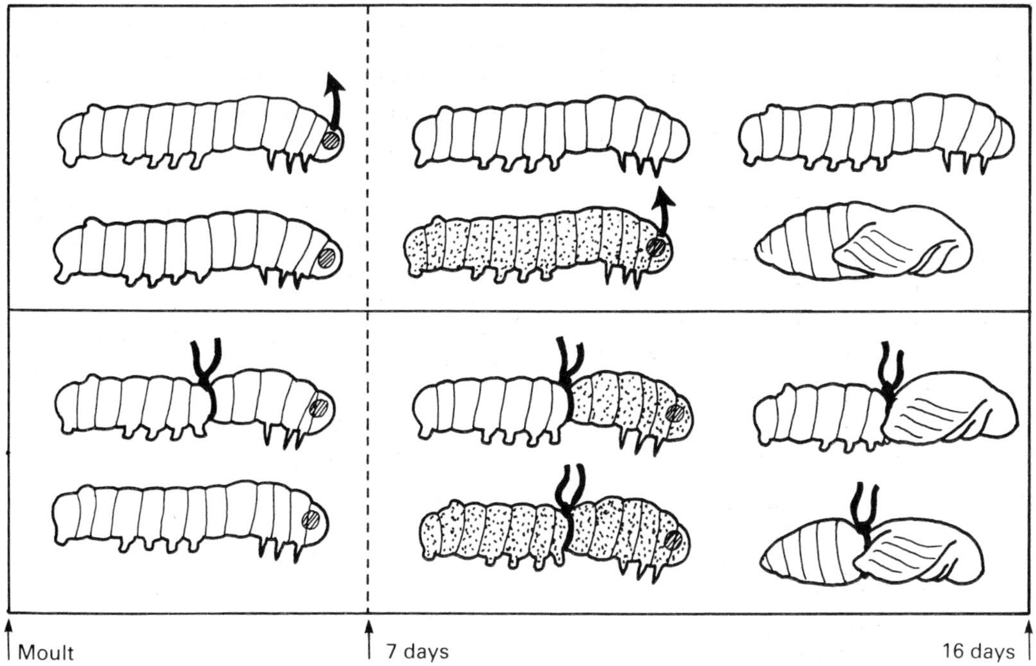

Fig. 16.35. Experiments showing the role of the brain in the control of pupation in *Lymantra*. After a critical period, hormone has been released from the brain in sufficient quantity, so that decerebration or ligation subsequently fails to block development.

(PTTH), is indispensable for moulting at each of the stages in the life cycle. In the moth *Manduca*, a single neurosecretory cell on each side of the brain seems to be the source of the hormone and nerve terminals within the corpus cardiacum and corpus allatum (Box 16.12) (mainly the latter) are the sites of storage and release. The stimulus responsible for triggering hormone release varies in different insects. In the bug *Rhodnius*, stretching of the abdomen by ingestion of a blood meal provides the trigger (the animal can be 'fooled' by blowing it up with air) whereas, in *Manduca*, a critical weight must be attained.

Recent work by Ishizaki and Gilbert and their colleagues has been directed towards the purification of PTTH. Two peptides capable of stimulating ecdysone secretion by the prothoracic glands have been isolated from the brain of *Bombyx* by the former, with molecular weights of *ca* 5000 and 22 000, respectively. The smaller one consists of two chains of amino acids linked by disulphide bridges, and this has led to speculation concerning the phylogenetic significance of the similarity of this compound with insulin and the insulin-like growth factors of vertebrates. The respective physiological roles of the two compounds have yet to be established.

The role of the prothoracic glands was demonstrated in experiments using pupae of the cecropia silk moth. Isolated abdomens fail to moult when active brains are implanted within them, whereas, if both active brains and inactive prothoracic glands are used, development proceeds. This indicates that the role of the brain hormone is to simulate the prothoracic glands or their homologues to secrete a second principle, the steroid *ecdysone* (Table 16.1). Karlson and his colleagues used 1000 kg of silkworm pupae to obtain 200 mg of the pure substance, which in 1963 became the first invertebrate hormone to be chemically identified. However, ecdysone is modified by the tissues to form the related 20-hydroxyecdysone, and this is the definitive moulting hormone.

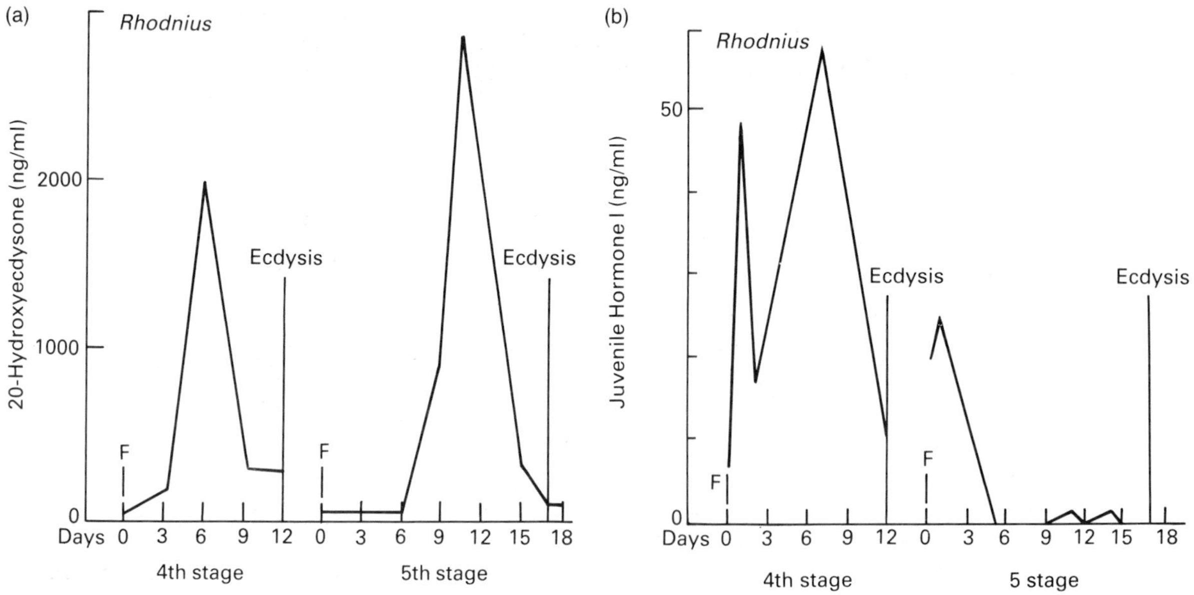

Fig. 16.36. Levels of circulating 20-hydroxyecdysone (a) and juvenile hormone (b) during the last stages in *Rhodnius*. (After Baehr *et al.*, 1978.)

A number of sensitive *assays* (tests for measuring the quantity of hormone) are available for such compounds (collectively known as *ecdysteroids*). The *Calliphora* test used by Karlson depends on the ability of ecdysone to induce pupation in ligated fly larvae. *Radioimmune assay* has been extensively used recently. Antibodies to an ecdysone–protein complex will bind radioactive ecdysone. The extent to which they do so depends on the amount of unlabelled ecdysone present (since this will compete for the binding sites) and thus provides a measure of the latter. In *Rhodnius*, a single peak is seen prior to each moult (Fig. 16.36).

Whereas PTTH acts like a typical peptide hormone in that its effects are mediated by receptors within the plasma membranes of the prothoracic glands, and the intracellular synthesis of cAMP, steroids penetrate within the cell and, after binding to cytoplasmic proteins (possibly), stimulate the genetic machinery in the nucleus. Flies provide valuable material with which to investigate this problem, because their salivary glands contain 'giant' chromosomes ten times longer and one hundred times thicker than normal. During development, various bands within the chromosomes become greatly thickened and such 'puffs' are sites of gene activity involving intense RNA synthesis. It has been shown experimentally that ecdysteroids induce the appearance of a specific sequence of early and late chromosome puffs identical to that normally observed prior to moulting.

A third hormone, *juvenile hormone* (JH) (Table 16.1), is secreted by the corpus allatum (Box 16.12). There are three main forms of JH, all closely related molecules, and any given insect will secrete any one or a combination of them. The role of JH is most clearly indicated by experiments in which the corpus allatum is removed from, or transplanted between, individuals at different stages of development (Fig. 16.37). High levels of JH are produced during the early stages of life. They ensure that, when moulting is triggered by ecdysone secretion, it results in the production of another larval stage. During the fifth and last larval stage in *Rhodnius*, greatly reduced levels are secreted (Fig. 16.36) and now ecdysone-induced moulting results in the development of the adult stage (or first in the pupa and then in the larva

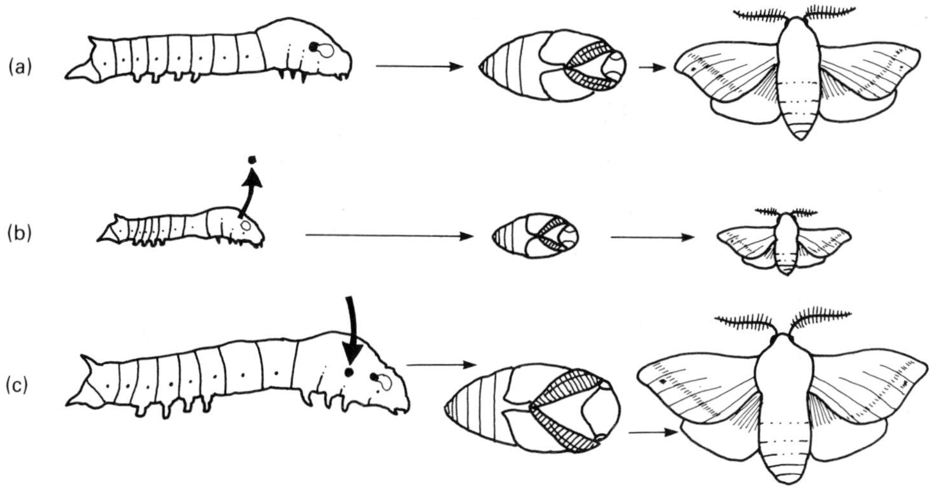

Fig. 16.37. Experiments to show the role of the corpus allatum in *Bombyx*. (a) Normal development, with fifth-stage larva, pupa and adult; (b) accelerated development following corpus allatum removal from a third-stage larva; (c) giant, sixth-stage larva, produced by implantation of an active corpus allatum into a fifth stage, develops into a giant pupa and adult. (After Turner, 1966.)

as, for example, in Lepidoptera).

Juvenile hormone is protected from degradation in the body during early stages of development by a specific blood protein to which it binds. Three mechanisms ensure that the titre is low prior to the final moult: the rate of synthesis is reduced; a specific enzyme is produced that breaks down JH (the binding protein notwithstanding); and secretion of PTTH and hence of ecdysone is only possible once the JH titre has been reduced.

Juvenile hormone is also involved in the control of reproduction in adult insects. It is being utilized, together with related chemical compounds, in pest control, and has the obvious advantages of being non-toxic and biodegradable. Hopefully, insects will find it difficult to become resistant to a substance upon which they depend for normal development.

Eclosion hormone is secreted by the brain and triggers the complex sequence of actions involved in the emergence of an insect from the old cuticle (this occurs some time after the production of a new cuticle by the epidermis under the influence of ecdysone). In the cricket, *Teleogryllus*, four major phases have been distinguished. First, the cuticle is split apart, loosened and attached to the substratum; then the insect climbs out of the old skin; the adult wings, legs, etc. are inflated and given their proper form; lastly, the old cuticle is eaten. The phases are composed of a stereotyped and co-ordinated sequence of, in all, at least 48 specific movements, which occur in repeated bouts, and last for 4 hours. Clearly, the neurosecretory cells which secrete eclosion hormone qualify as 'command neurones' (Section 16.10.2).

The new cuticle soon becomes hard and dark due to 'tanning' induced by the neurosecretory hormone *bursicon* that is produced, after emergence, by the brain or the abdominal ganglia.

It is interesting to compare the endocrine pattern of insects with those of other arthropods, particularly crustaceans. Crustaceans have the same steroid moulting hormone, 20-hydroxyecdysone, secreted by non-nervous endocrine glands, the *Y-organs*. Production of the steroid is controlled by a neurosecretory system in the eye-stalk. However, whereas PTTH in insects has a *stimulatory* action, the hormone involved in crustaceans *inhibits* the Y-organs so that eye-stalk removal accelerates moulting or even results in a series of rapid moults. Lastly, new research has shown that JH is present in Crustacea. It is secreted by the *antennary glands* and has a role similar to that in insects.

A major difference is the presence of *sex hormones*

in crustaceans. As shown by Charniaux-Cotton in 1954, a male hormone is secreted by the androgenic glands, which are associated with the vas deferens, and one or more female hormones are produced by the ovary. These factors promote gametogenesis and the differentiation of secondary sexual characteristics. In most insects, the genetic constitution apparently influences sexual differentiation directly, without the mediation of the endocrine system.

16.13 Conclusions

We will conclude our survey of invertebrate control systems with two examples which exemplify recent progress made towards explaining behaviour and physiology in terms of neural mechanisms.

The soil nematode, *Caenorhabditis elegans*, has been studied in great detail by Brenner and his colleagues at Cambridge. The worm is almost transparent, only 1 mm long, and when cultured in the laboratory has a life cycle of 3.5 days at 20°C. The structure of the nervous system has been reconstructed in its entirety from photographs of serial sections, taken with the electron microscope.

The hermaphrodite has exactly 302 neurones, each with a predictable position and shape. Furthermore, the 'wiring diagram' of the system is now complete—all of the synaptic contacts made by each of the cells are known. About 600 gap junctions and around 7000 chemical synapses are formed, although their number and position may vary.

The embryological cell lineage of every neurone has been followed, and the stages at which particular neurones differentiate during development are known, as are the changes made in synaptic connections.

A study of the marine opisthobranch mollusc, *Pleurobranchia californica*, by Davis and others in California, has been most successful in uncovering the neural mechanisms underlying principles of behaviour which may be observed in almost any animal.

Different aspects of behaviour are organized as a *hierarchy*. Thus the escape response takes precedence over feeding—and stimuli for the former activate inhibitory synapses on the single 'command' neurone which controls feeding. *Motivation* for feeding shows the usual dependence on recent consumption, and sensory pathways activated by the latter inhibit the feeding command neurone. Competing stimuli evoke variable responses depending on their relative strengths and the animal must have a capacity for *decision-making*, since it usually does only one thing at a time. This has its basis in the integration of sensory input within the nervous system. For example, if a stimulus for feeding is accompanied by one for withdrawal of the oral veil, the former prevails, since two neurones become active which block the motor output responsible for withdrawal. In contrast, if the food stimulus is weak, or, if the animal has fed, different circuits are activated and withdrawal takes precedence. The *endocrine system* is also involved, since the spawning hormone exerts an influence on neurones in the buccal gangion, inhibiting feeding.

Behaviour is both *orderly* and *adaptive* and in this it contrasts with the variety of competing influences impinging on the animal. The results of invertebrate neuroscience have uncovered some of the patterns of neural organization which underlie these characteristics. Nevertheless, complete wiring diagrams notwithstanding, we are still no further to knowing what these animals 'feel', or even whether they feel at all, and this should caution us with respect to our expectations for the future analysis of more advanced nervous systems.

16.14 Further reading

Bullock, T.H. & Horridge, G.A. 1965. *Structure and Function in the Nervous System of Invertebrates*. Freeman, San Francisco.

Eaton, R.C. (Ed.) 1984. *Neural Mechanisms of Startle Behaviour*. Plenum Press, New York.

Halliday, T.R. & Slater, P.J.B. (Eds) 1983. *Animal Behaviour. Vol.1, Causes and Effects*. Blackwell Scientific Publications, Oxford.

Mill, P.J. 1982. *Comparative Neurobiology*. Ed. Arnold, London.

Nighnam, K.L. & Hill, L. 1977. *The Comparative Endocrinology of the Invertebrates*, 2nd edn. Ed. Arnold, London.

Roberts, A. & Bush B.M.H. (Eds) 1981. *Neurones Without Impulses: Their Significance for Vertebrate and Invertebrate Systems*. Cambridge University Press, Cambridge.

Shelton, G.A.B. (Ed) 1982. *Electrical Conduction and Behaviour in "Simple" Invertebrates*. Clarendon Press, Oxford.

17

BASIC PRINCIPLES REVISITED

17.1 Basic physiological features of phenotypes

Animals require resources as building blocks to make new tissue (somatic and reproductive) and replace spent somatic tissues, and also as fuel to power these processes. Unlike autotrophic organisms, that can elaborate organic requirements from inorganic constituents plus an energy source such as sunlight, animals require to *feed* from other organisms (considered in Chapter 9). With the advent of this *heterotrophy*, there evolved a need to be able to move to find and capture resources and to avoid being eaten by other animals. Once locomotion evolved (see Chapter 10) there would have been coevolutionary pressure on predators and their prey to move more effectively. Locomotion also requires power (Chapter 11). As well, there is a need to invest resources in various processes and structures that provide a defence against exploitation by other organisms, and this was addressed in Chapter 13.

Macromolecules acquired in the food are simplified to their subunits by an enzymically mediated process known as *digestion*, prior to their absorption into the tissues for either (a) utilization in the *resynthesis* of macromolecules (*anabolic processes*) or (b) breakdown to release energy to power these processes (*catabolic processes*). Resources in excess of requirements, particularly amino acids and 'spent' tissue proteins, are also catabolized and *excreted* (see Chapter 12).

The major fuel for the metabolism (= anabolism + catabolism) of organisms is carbohydrate, usually monosaccharide, but it can be stored prior to use as polysaccharide (often glycogen) and/or as fat. Energy is released from this fuel by catabolic oxidation but since there was a time, after the origin of life but prior to the origin of photosynthesis, when the earth's atmosphere was devoid of oxygen (Chapters 1 and 2), then primitively this occurred *without* oxygen by the transfer of electrons to organic components, which in reduced form accumulated as end products. With the advent of an oxygenated atmosphere, more complete oxidation became possible and more efficient processes evolved; this required the supply of O_2 from the outside world and the removal of CO_2 which forms as an end product in the process. *Aerobic respiration* is now most widespread in the animal kingdom, but *anaerobic* respiration is still found as the major process of catabolism in some species. A universal compound, important in the short-term storage and transfer of the energy released in these catabolic processes, is the phosphorylated form of adenine (a nucleotide). Adenine triphosphate (ATP) is generated from the diphosphate form, ADP, by the reactions associated with respiration. It returns to this state after releasing its energy and is recycled. Respiratory processes were considered in Chapter 11.

These fundamental features of physiology are summarized in Fig. 17.1. Note that the resource input to the organism is limited by the resource-acquiring processes and structures of feeding. Hence, even when the availability of food in the environment is unlimited, the amount that can be made available to metabolism is finite and limiting. The more that is invested in one metabolic demand, the less that is available for others. The rates and routes of use are controlled by enzymes and are therefore ultimately specified by genes. The way the limited resources are allocated crucially influences the biology of organisms at a number of levels: allocations between anabolism

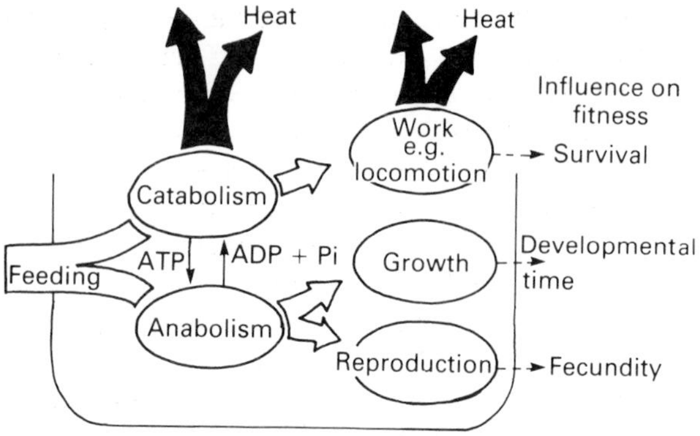

Fig. 17.1. Allocation of resources (derived from the food) between anabolic and catabolic processes is the basis of animal physiology (after Calow, 1986).

and catabolism influence the *physiology* of the organism; allocations between different activity demands influence the *behaviour* of the organism; allocations of the limited products of anabolism between different structures influence the *form* of the animal; and between somatic and reproductive structures the *reproductive* and *life-cycle* biology of the organism. Moreover, the extent to which these different demands are supported will importantly influence survivorship and fecundity and hence fitness. Therefore, according to Darwinian principles, those gene-determined patterns of resource allocation and utilization (often called *strategies*) that maximize the transmission of the genes that code for them will be selectively favoured. Hence, though the physiologies of all organisms are based upon a common organization, they will have been 'tuned' by natural selection according to the ecological circumstances in which they operate—and this is what is meant by *adaptation*.

It will be clear from this description of organisms that, functionally, they operate as integrated wholes. There are trade-offs in metabolic investments that might lead to trade-offs between the components of fitness; an increased investment in locomotion might enhance survival but mean that less resource is available for reproduction. The principle of organismic integration is also important for morphological structures since the development of one structure must be compatible with others. There are two consequences of this. First, there have to be proximate (immediate) *controls* over the development of form, the expression of behaviour and of physiological function. The control systems (see Chapter 16) involve chemical and electrical signals. Second, there is ultimate selection for integration; a gene-determined trait has to be compatible with the organismic environment in which it is expressed as well as bringing survival and reproductive gains by interaction with the external environment. Hence, genes and the traits they specify are selected *within the context of the organism*. This *organismic* or *holistic* orientation, which differs to some extent from the 'gene pool philosophy' that views organisms as collections of dissociable, 'selfish' genes, is the one we used in the preceding chapters.

17.2 The primacy of replication and reproduction

A very important, perhaps *the* most important, investment of resources by organisms is into reproduction. These are used to form the propagules that are the vehicles of genetic transmission.

Sometimes, as noted earlier (Section 14.1.1), a cell or group of cells separates from the parent with a perfect replica of its genome. At the heart of this

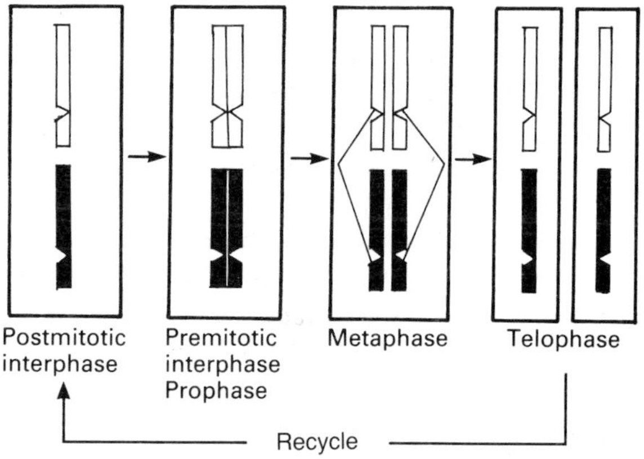

Fig. 17.2. The behaviour of a pair of chromosomes in mitosis (after Paul, 1967).

process, asexual reproduction, is cell division by mitosis (Fig. 17.2). The cell or cells in the propagule is/are formed by mitosis and these, in turn, reproduce a replica organism by mitosis.

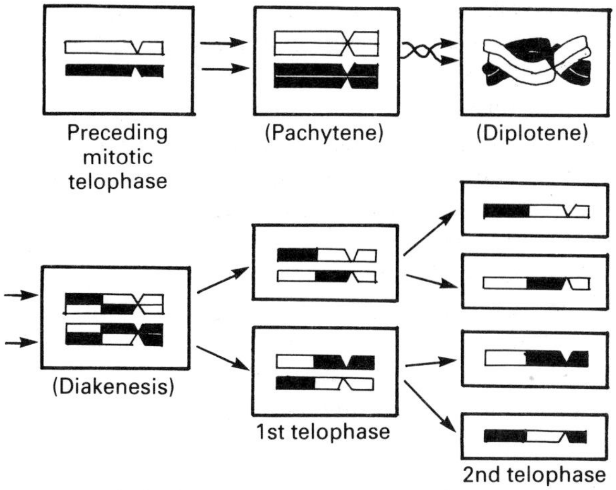

Fig. 17.3. The behaviour of a pair of chromosomes in meiosis. Notice that the crossing-over processes mean that the chromosomes of the progeny do not contain genes in the same arrangement as the parent (after Paul, 1967).

Alternatively, single germ cells (gametes) are produced, usually in specialized organs (the gonads), within the organisms. These must usually fuse with other germ cells before a new organism is reproduced. The germ cells are formed from a process of cell division which results in cells that have only half the DNA and number of chromosomes of the ordinary somatic cells. This is meiosis (Fig. 17.3). The full chromosome complement is reinstated by fusion with other germ cells (syngamy, fertilization) and since these are often derived from different parents, the progeny contain a mix of two genetic programmes. This is known as *sexual reproduction*. Processes of reproduction were considered in detail in Chapter 14.

17.3 Ontogeny

By definition the products of reproduction are smaller and usually simpler than the parent. The products of sexual reproduction are unicellular whereas the parents have many, functionally differentiated cells. Hence *development*, or *ontogeny*, must involve cell division to achieve a size increase (*growth*) and cell specialization (*differentiation*). Since specific cells occur in specific places in an organism there has to be *patterning*, and since adults have complex shapes, whereas the products of reproduction are usually more or less spherical, there has to be some shaping (*morphogenesis*).

Cell division (*cleavage* of the fertilized egg) occurs by the process of mitosis. This faithfully replicates the original genome and has been described above. Some of the early embryologists (particularly August Weismann, 1834–1914) thought that *differentiation* involved the progressive jettisoning of unwanted parts of the hereditary material in specific cell lines. But an understanding of mitosis, and the realization that some invertebrates can be fully regenerated from small pieces of somatic tissue suggest that every somatic cell contains a more-or-less complete copy of the original genome. Hence, differentiation must involve switching, whether on or off, of specific parts of the programme in different cells. In this way, though containing the same genetic programme, different cells come to produce different proteins and hence function in different ways.

A well-known model for this switching comes from work on the bacterium, *Escherichia coli*. Take one bacterial cell and allow it to divide into two; then culture one of the products in a medium containing glucose as energy source and the other containing another sugar, lactose, rather than glucose. The cells with lactose cannot use it directly, but produce an enzyme, β-galactosidase, that converts it to glucose. The cells in the other medium do not produce this enzyme. However, cells in both media have the same genome, and yet they produce different enzymes. They can therefore be said to be differentiated with respect to each other.

Cells in the glucose medium contain genes that code for β-galactosidase, yet these are normally repressed (switched off) by a *repressor* protein. This turns off specific gene transcription by binding to a specific operator. When lactose is present in the cell it is converted to allolactose and this interacts with the repressor in such a way that it no longer inhibits the operator and there is read-off of the appropriate RNA that acts as template for the enzyme. The enzyme is said to be *induced* and the metabolic machinery needed to handle the lactose is switched on. The model is illustrated in Fig. 17.4.

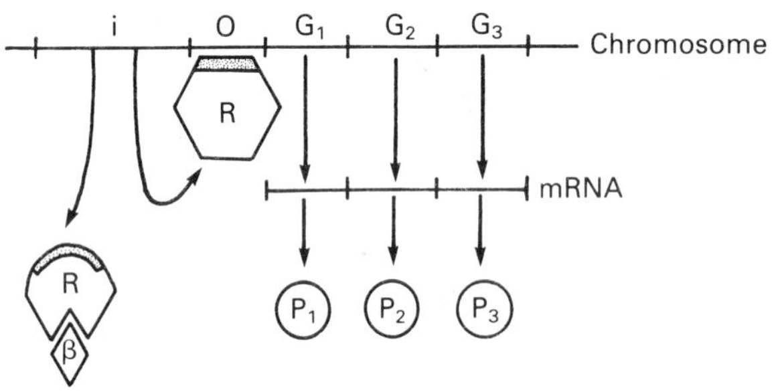

Fig. 17.4. Model of gene control. Gene i codes for a repressor, R, that binds to a specific part of the chromosome, O, and inhibits transcription of G_1, G_2 and G_3. If the substrate (β) for the enzymes P_1, P_2 and P_3 is present it binds to R changing its shape and not allowing it to bind to O. Under these circumstances G_1, G_2 and G, are transcribed, P_1, P_2 and P_3 are produced and β is metabolized. (After Maynard Smith, 1986.)

As well as this regulation by switching on, there could, at least in principle, be regulation by switching off, but this is clearly less economic. Little is known about the molecular basis of gene regulation in the differentiation of eukaryotic cells, yet it is likely to follow similar principles to the ones discussed above.

Pattern formation requires a pre-pattern of switching factors. The pre-pattern could either be a detailed set of instructions of how cells should respond in particular positions or be less detailed information from which cells can 'determine' their positions within the embryo and respond appropriately. Because the genetic instructions occur *within* cells and it is very unlikely that detailed information could flow from outside the cell or even the cell cytoplasm itself to the genetic programme inside the nucleus (a prohibition known as the 'central dogma of molecular biology'), this suggests that control through positional information is most likely. But still there is a puzzle of how the pre-pattern of positional information is established in the first place given that all cells in an embryo contain the same genetic information. And the answer, in gametic forms, might be that it comes from the cytoplasm of the egg cell. This is often visibly non-homogeneous (e.g. pigment and yolk being non-uniformly distributed through it) and even the products of the first cleavage may contain obviously different cytoplasms and hence conditions for expression of the two genomes.

Note that this process is similar to that proposed by Weismann, except that it is cytoplasmic substances rather than the nuclear material that are being partitioned and separated during the process of cell division. It also allows the possibility of a maternal influence, through the egg cytoplasm, on development. Finally, these controls, whatever they are, can either act rigidly, such that the fates of cells are specified at an early stage in development (a *determinate* process) or more plastically, such that experimental interference can change the fate of particular cells even up to a relatively late stage in development (an *indeterminate* process).

Morphogenesis involves a combination of cell movement and differential growth. The first process has been well described for the early stages of development in sea-urchins (Fig. 17.5a and see Chapter 15)—principally because the embryo is transparent, enabling cells inside to be observed and even recorded with time-lapse cinematography. At an early stage the embryo consists of a hollow ball of cells—the blastula. After this stage, some cells have to move inside to form internal organs, like the gut. This process is known as gastrulation and occurs in two stages. First there is inward movement of individual cells and then wholesale invagination. Migration begins when cells lose contact with each other, move into the blastocoel (inner cavity of blastula) and then migrate along the inner surface by extensions, filopodia, that pull them along. The invagination occurs in two stages: a slow phase of inward bending followed, after a lag, by rapid invagination. The first phase probably occurs by changes in adhesiveness of cells; they adhere less to each other but remain attached to a basement membrane. This

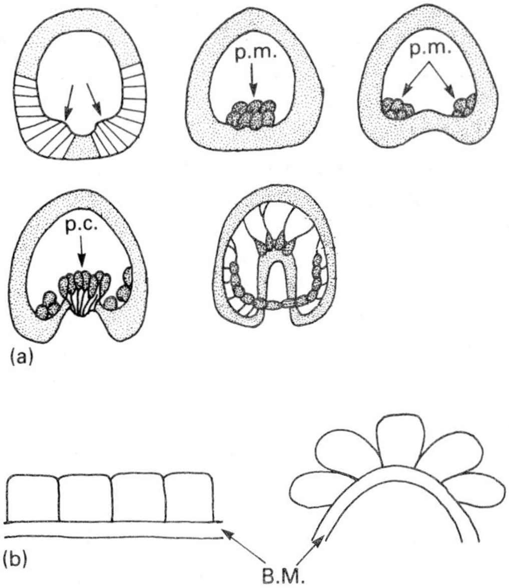

Fig. 17.5.(a) Early development in the sea urchin. p.m. = primary mesenchyme cells—these migrate to the 'corners' of the embryo using filopodia. p.c. = pear-shaped cells. (After Gustafson & Wolpert, 1967.) (b) Change in shape of cells at point of invagination. B.M. = basement membrane.

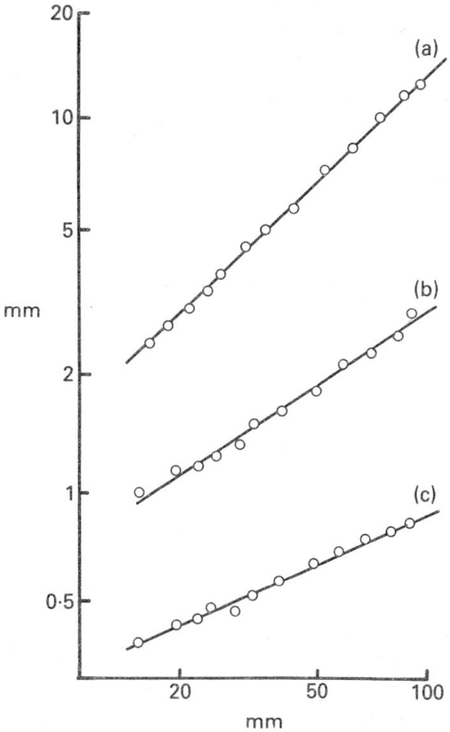

Fig. 17.6. Example of allometric growth in the insect *Carausius morosus*. Length of posterior part of prothorax (a), width of head (b), and diameter of eye (c) plotted against total length of body. Co-ordinates are logarithmic. (After Wigglesworth, 1965.)

causes them to become pear shaped and could cause inward bending as shown in Fig. 17.5b. The rapid phase probably occurs by the cells on this in-tucked region forming filopodia that contract after making contact with the roof of the blastocoel. These processes, involving cell–cell membrane interactions and active migration by filopodia, are probably fairly typical of morphogenetic changes in general. A major question, which remains unresolved, however, is how such processes are controlled to result in a completely organized individual.

Finally, differential growth of organs, external and internal, can cause shaping. Julian Huxley (in his *Problems of Relative Growth*, 1932) discovered that if you plot the logarithm of the size of one part of an organism against the logarithm of the size of another (or of the size of the whole organism) the points often fall on straight lines (Fig. 17.6). Equal intervals on a logarithmic scale represent equal multiplication factors so that these straight lines suggest that one organ multiplies itself (grows) at some specific rate relative to another. Clearly allometric relationships will have a profound effect on the shaping of an organism.

Here, then, we have summarized the basic processes of development—differentiation, pattern and morphogenesis—and made some comments about how they might be controlled. In Chapter 15 we discussed the details of these processes as they apply to invertebrates.

17.4 Ontogeny and phylogeny

Since both ontogeny and phylogeny are *apparently* progressive it is tempting to assume that the latter occurs by additions to the former. Then ontogeny *recapitulates* phylogeny—a view expressed by Ernst Haeckel (late nineteenth century). Ontogeny of the 'higher' form is, according to this view, supposed to repeat the adult forms lower down the scale of organization. Von Baer, a contemporary of Haeckel, argued instead that no higher animal repeats any adult stage of lower animals but, because development always proceeds from undifferentiated to differentiated state, the initial phases of development must be conserved in different phyla. Hence, it is these embryonic forms, rather than adults, that are repeated in different phyletic lines, and this is certainly consistent with the principles of development enunciated in the last section.

Terminal addition fits neatly into the Lamarckian theory of acquired characters, since the latter are usually acquired later in life and added on to existing structures. Mendelian genetics, on the other hand, undermines this view, for a mutation can bring a change at any stage of development. Indeed it is now known that genes control rates of development and changes in these can cause profound alterations by either arresting development (removing the 'old' adult) or causing it to go on beyond the normal end point. Walter Garstang (1868–1949) was one of the first zoologists to recognize the importance of genetics for development and the possibilities for evolutionary change allowed by these kinds of alteration to the rate of development.

Figure 17.7 classifies the complete spectrum of evolutionary shifts that might arise in this way. Each square contains a *developmental trajectory*, i.e. index of shape against size or age. The solid lines show the ancestral trajectory and the broken lines the descendant. *Start* = initiation of development; *stop* = cessation of development. In the top line development is decelerated or truncated, and the changes are referred to as paedomorphosis. Either by reduced rates of development (neoteny) or truncation (progenesis) embryonic, larval or juvenile features turn up in adults. The bottom line shows the reverse and is referred to as peramorphosis.

Walter Garstang was particularly interested in larval forms and thought that paedomorphosis had had a dominant effect in evolution. And indeed paedomorphosis could have been important in a number of instances (Chapter 2). For example, among living species sexual maturity is attained at a small size in some male echiurans, crustaceans (and fish) which attach 'parasitically' to the much larger (sometimes by several orders of magnitude) females. The evolution of six-legged insects from many-legged ancestors might well have occurred by paedomorphosis (Fig. 17.8). Finally, a widely accepted theory of vertebrate evolution suggests that they are derived

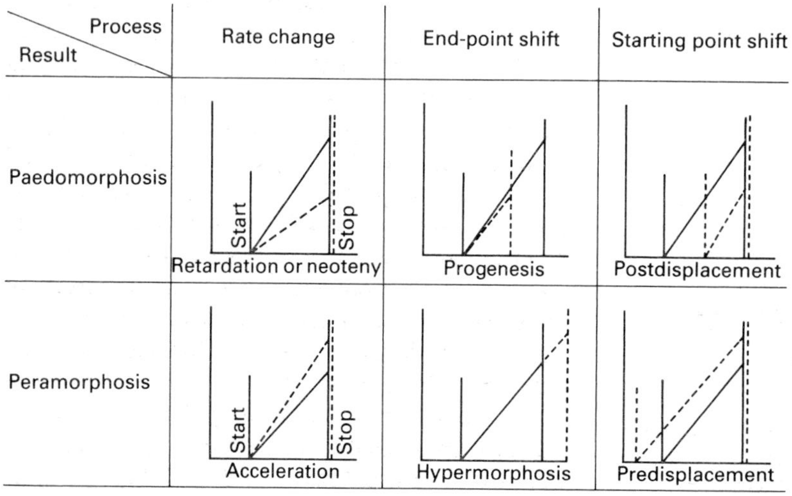

Fig. 17.7. Classification of possible evolutionary shifts in development. Abscissae = developmental time or size; ordinates = morphological changes (after Calow, 1983).

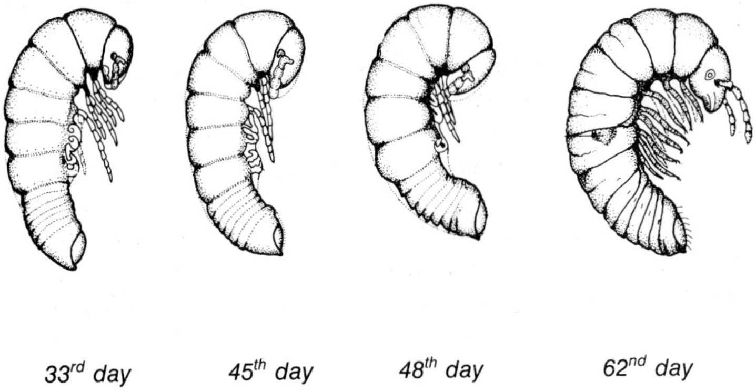

Fig. 17.8. Stages in myriapod development. The six-legged stage, 33rd day, is sometimes used as evidence for a paedomorphic origin of insects (see also Gould, 1977).

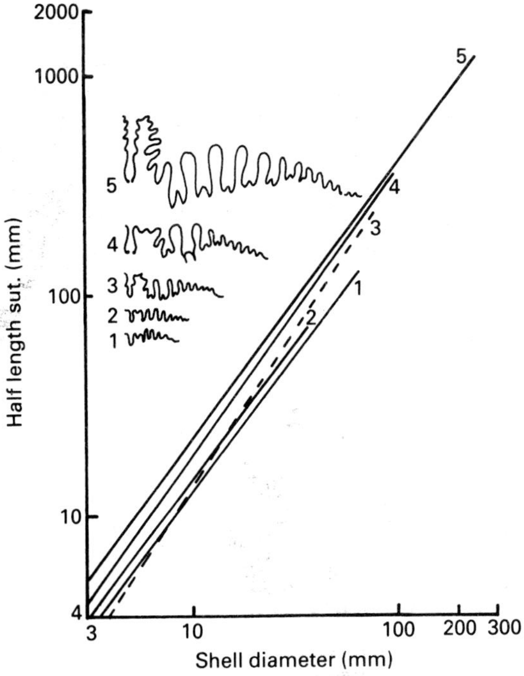

Fig. 17.9. Mixture of developmental changes in ammonites. 1 = ancestor; 2–5 = descendants. Older parts of the stage are to the right. (After Newell, 1949; see also Calow, 1983.)

from some larval stock resembling the free-swimming 'tadpole' larva of present-day tunicates (see Fig. 7.30).

Yet accelerative shifts are also a possibility. Consider the evolutionary trends in the patterns of sutures found on the shells of ammonites that are depicted in Fig. 17.9. Here the descendants' slopes are steeper than those of the ancestors so there has been *acceleration*. Also, sutural growth of descendants goes beyond that of ancestors, suggesting *hypermorphosis* (Fig. 17.7). Thirdly, the descendants' trajectories are 'higher' than those of the ancestor so the descendants begin with something of a start and this suggests some *predisplacement* (Fig. 17.7).

To summarize then: ontogeny does not recapitulate phylogeny, but because development has to do certain common things irrespective of taxa, effect increases in differentiation, size, pattern and morphogenesis (see last section), and has to do them on the basis of processes common to all animal life—mitosis, differential expression of a genome, cell movement, differential growth—there has to be, as von Baer suspected, some commonality in early development. Nevertheless, small modifications in the *rates* of these processes, which in principle can be effected by small genetic changes, are likely to have had profound effects on phylogenesis. Such changes might indirectly have been instrumental in causing some of the quantal jumps between levels of organization alluded to above (Chapters 1 and 2), a view that was certainly advocated by Walter Garstang.

17.5 Further reading

Calow, P. 1983. *Evolutionary Principles*. Blackie, Glasgow & London.

Gould, S.J. 1977. *Ontogeny and Phylogeny*. Harvard University Press, Cambridge, Massachusets.

Huxley, J.S. 1932. *Problems of Relative Growth*. Methuen, London.

Maynard Smith, J. 1986. *The Problems of Biology*. Oxford University Press, Oxford.

Paul, J. 1967. *Cell Biology*. Heineman, London.

GLOSSARY

abdomen term applied to the posterior zone of any body divided into three distinct regions of which the anterior region is the head (cf. **thorax**)

aboral surface of body opposite to that bearing the mouth, i.e. in animals with the mouth in the centre of the upper or lower surface

acclimation change in physiology as a result of the exposure of an organism to a changed environment

acoelomate without a body cavity, other than those of the gut lumen and within organ systems

acron anterior, non-segmental region of an arthropod

acrosome filamentous tube(s) at anterior end of spermatozoon (q.v.) that contacts and fuses with the cell membrane of the ovum (q.v.) during fertilization (q.v.)

active transport movement of solutes against a concentration gradient by a process that uses energy

adenosine triphosphate (ATP) the main energy-carrying molecule in living organisms

adductor muscle muscle closing, or holding closed, the valves of a shell

ageing irreversible, deteriorative changes in individuals with time causing increased vulnerability and reduced vitality [= senescence]

amoebocyte cell capable of amoeboid movement

anabiotic see **cryptobiotic**

anaerobic without air; often used to describe those energy-yielding metabolic (q.v.) processes that proceed without oxygen

annulate descriptive of cylindrical organism, organ, etc. of which the external surface is divided into a chain of rings or 'annuli' by furrows, giving the appearance of segments

anoxic without free oxygen

antennae filiform and often elongate chemosensory appendages on the head of some arthropods and polychaetes, of onychophorans, etc.

apodeme internally directed process of the exoskeleton of an arthropod

apomixis see **asexual reproduction**

apterygote wingless

arrhenotoky form of parthenogenesis (q.v.) in which unfertilized eggs develop into (haploid) males whilst fertilized eggs develop into (diploid) females [= haplodiploidy]

article one of the units comprising a jointed appendage; i.e. the rigid section between two adjacent joints

asexual reproduction form of multiplication not involving meiotic reduction division and fusion of gametes (q.v.) (cf. **sexual reproduction**) [= apomixis]

autotomy self-induced amputation of an appendage or of a body region, as, for example, a means of escape from a predator or during budding (q.v.) or fission (q.v.)

basement membrane amorphous sheet underlying an epithelium (q.v.), composed of a type of collagen (q.v.) and carbohydrate

benthic appertaining to the bottom or substratum of aquatic systems (cf. **pelagic**)

bilateral symmetry symmetry in which the body can be divided into two, and only two, mirror-image halves

bioturbation disturbance of benthic (q.v.) sediments by animal activity

bipectinate the state of a mollusc ctenidium (q.v.) that retains the primitive condition of having gill filaments on either side of the central axis (cf. **monopectinate**)

biradial state in which each of the four quadrants of a spherical organism or embryological stage is the same as that of the opposite quadrant but differs from the adjacent ones

biramous having two branches (cf. **uniramous**); used of an arthropod limb

blastema group of undifferentiated dividing cells from which differentiated cells may be derived (see **differentiation**)

blastocoel cavity within the blastula (q.v.)

blastopore aperture through which the embryonic gut in a gastrula (q.v.) communicates with the external environment

blastula hollow ball of cells formed from a zygote (q.v.) by cleavage (q.v.) during embryology

budding form of asexual multiplication (q.v.) in which a new individual begins life as an outgrowth from the body of the parent; it may then separate to lead an independent existence, or remain connected or otherwise associated to form a colonial (q.v.) organism

bursa copulatrix sac receiving spermatozoa (q.v.) during the act of copulation (q.v.)

carapace protective exoskeletal shield covering all or part of the dorsal and lateral surfaces of an arthropod

catabolism see **metabolism**

cephalic appertaining to the head

cephalization development of a head during phylogeny (q.v.) or ontogeny (q.v.)

cephalothorax region of the body of some crustaceans formed by the fusion of head and thorax (q.v.)

cerci pair of variously shaped appendages on the last abdominal segment of many insects

chaeta small, stiff, projecting, chitinous (q.v.) bristle of annelid, pogonophoran and echiuran worms (cf. **seta**)

chelate state of an arthropod limb that terminates in a pair of pincers or forceps

chemoautotrophy mode of bacterial nutrition in which the organism can synthesize all its nutritional requirements by chemosynthesis (q.v.)

chemosynthesis synthesis of organic compounds according to the general equation

$$CO_2 + 2H_2X \rightarrow (CH_2O) + H_2O + 2X$$

using the energy (and often the reducing power) released by the oxidation of inorganic materials such as Fe^{2+}, CH_4, NH_3, NO_2^-, H^+, S, etc. [= chemolithotrophy] or of pre-existing organic compounds (acetate, formate, etc.); carried out only in bacteria under anaerobic conditions (cf. **photosynthesis**)

chitin a nitrogen-containing polysaccharide

chloroplast eukaryote organelle, occurring in several protists and most plants, in which photosynthesis takes place

choanocytes characteristic flagellated cells of poriferans

chromatophore a pigment-containing cell

chromosome thread-like structure in the cell nucleus that contains genetic information

cilia projecting propulsive organelles that beat in a single plane by means of a stiff propulsive and a lax recovery stroke; usually relatively short and occurring as several to many per cell; non-motile forms are specialized for sensory reception; found only in eukaryotes (q.v.) (cf. **flagella**)

cirrus eversible (q.v.) copulatory organ (cf. **penis** and **gonopod**)

clade group of organisms all of which share the same ancestral form (cf. **grade**)

cleavage mitotic cell divisions subdividing a zygote (q.v.) into a multicellular but undifferentiated embryo (q.v.); cell growth does not occur during this process

cleidoic descriptive of an egg (q.v.) enclosed within a protective coat

clitellum secretory epithelium forming a cocoon, especially in clitellate annelids

cloaca dilated terminal region of gut receiving the discharge of some other organ system(s)

cnidocyte cnidarian cell containing nematocysts (q.v.)

coelom fluid-filled cavity within tissues of mesodermal (q.v.) origin and enclosed by a mesodermal membrane (cf. **pseudocoel** and **haemocoel**); range from large hydrostatic body cavities, bounded by a peritoneum (q.v.) (see

schizocoel and **enterocoel**), to, by definition, the epithelium (q.v.) lined spaces within mesodermal organs

coelomocyte cell suspended in coelomic (q.v.) fluid

coelomoduct (a) a blind-ending osmoregulatory/excretory gland and duct of mesodermal (q.v.) origin (cf. **nephridium**), or (b) open mesodermal ducts from coelomic (q.v.) body cavities to the exterior which serve for the discharge of gametes or of coelomic fluid

cohort group of individuals in a population all born at approximately the same time

collagen a fibrous protein usually associated with connective tissue or cuticular lattices

colonial descriptive of organisms produced asexually which remain associated with each other; in many animals, retaining tissue contact with other polyps (q.v.) or zooids (q.v.) as a result of incomplete budding (see also **modular**); also used to describe those sexually-produced individuals that form semi-permanent aggregations in space

commensal descriptive of an organism that lives in close proximity to another of a different type, e.g. in its burrow, without so far as is known affecting it

commissure cross connection between nerve cords or ganglia (q.v.) [= connective]

compound eye single eye formed from few to many individual optical units, ommatidia, each with its own lens, field of view, receptor cells, etc. (cf. **ocellus**)

contractile vacuole intracellular, membrane-bound vacuole, concerned with osmoregulation, that fills with liquid and suddenly contracts, expelling the liquid to the exterior

copulation act of sperm transfer by an organ of one individual into the body or a duct/sac of another; a widespread, but not the only, precursor to internal fertilization

corona locomotor apparatus of rotiferans consisting of rings of cilia (q.v.)

coxal sacs eversible (q.v.), thin-walled vesicles associated with the base of the legs of some uniramians and serving to take up environmental water; similar vesicles also occur in onychophorans

cryptic 'hidden', e.g. by virtue of resemblance to the surrounding area

cryptobiotic capable of entering a resistant state of suspended animation during periods of environmental adversity (usually lack of water) and therefore able to inhabit temporary aquatic environments; also termed anabiotic

ctenidia gills of a type confined to molluscs comprising (primitively) banks of filaments issuing from each side of a central axis

cuticle non-cellular and resistant body covering secreted by the epidermis (q.v.) and moulted periodically; often ornamented and/or thickened locally into plates

cyst encapsulated, desiccation resistant stage in life history

cytochromes respiratory enzymes, located in mitochondria, of similar structure to haemoglobin

definitive host host in which a parasite (q.v.) reproduces sexually (q.v.)

degrowth shrinkage of a starved animal

depolarize to decrease the electrical potential difference, (usually) across the cell membrane (cf. **hyperpolarize**)

deposit feeding consumption of detrital (q.v.) materials and/or associated organisms on and/or in the substratum

dermis non-muscular layer of body wall of mesodermal (q.v.) origin located beneath the epidermis (q.v.)

determinate development see **mosaic development**

detritus particulate decomposed or decomposing organic matter associated with water or the substratum, together with those microbial organisms colonizing it

deuterostome state in which the blastopore (q.v.) does not form the mouth, although it may form the anus (cf. **protostome**); also used to describe the animals displaying this state

diapause resting phase in life history in which metabolic activity is low and adverse conditions can be tolerated

differentiation process whereby totipotent embryonic cells become specialized to carry out different functions

dimorphic occurring in two distinct (usually morphological) forms (cf. **monomorphic** and **polymorphic**)

diploid having two of each chromosome in each somatic (q.v.) cell (cf. **haploid** and **polyploid**)

direct development development without a larval (q.v.) stage (cf. **indirect development**)

eclosion emergence of an insect, particularly that of the adult from a pupa (q.v.)

ectoderm outer germ layer, i.e. that covering the gastrula (q.v.) (cf. **endoderm** and **mesoderm**)

ectotherm an organism that derives its heat from the environment rather than from its own metabolism

egg general term for the initial stage of animal development—ovum (q.v.), zygote (q.v.), or a complex of cells or developing embryo, food reserves, etc. contained within a common shell or capsule

embryo early stage of development hatching from, or retained within, the egg (q.v.) and not capable of independent life

endoderm inner germ layer, i.e. that forming the embryonic gut of the gastrula (q.v.) (cf. **ectoderm** and **mesoderm**)

endopod the inner branch of a biramous (q.v.) arthropod limb (cf. **epipod** and **exopod**)

enterocoel coelom (q.v.) formed by the evagination of pouches from the embryonic gut (cf. **schizocoel**)

epiboly spreading of mobile yolk-free cells over yolky ones in gastrulation (q.v.)

epidermis outermost cellular covering of body, formed from ectoderm (q.v.)

epifaunal descriptive of benthic (q.v.) animals associated with the substratum surface (cf. **infaunal**)

epipod process arising from the basal article(s) (q.v.) of an arthropod limb (styli (q.v.) are probably epipods) (cf. **endopod** and **exopod**)

epithelium sheet or tube of tissue (q.v.) covering a free surface, e.g. forming the lining of a cavity

eukaryote having internal membrane-bound nucleus and organelles within the cell (cf. **prokaryote**); all organisms except bacteria are eukaryote

eutely state in which cells do not divide in the adult; the latter therefore has a fixed (and usually small) number of cells and growth occurs only by cell enlargement

eversible capable of being protruded by being turned inside out; extension is usually achieved hydraulically (q.v.) and retraction by muscular action

evolution origin and subsequent change over time

exhalent descriptive of the outward respiratory or feeding current (cf. **inhalent**)

exopod outer branch of a biramous (q.v.) arthropod limb (cf. **endopod** and **epipod**)

fate map formal description of the spatial distribution of cells or regions in a zygote (q.v.), embryo (q.v.) or imaginal disc (q.v.) that would normally give rise to the different parts of an organism

fermentation enzymatically-mediated breakdown of organic compounds, and the generation of ATP, under anoxic (q.v.) conditions

fertilization process of fusion of gametes (q.v.) to create a zygote (q.v.)

filopodia filament-like projections of cell cytoplasm

fission form of asexual multiplication (q.v.) involving division of the body into two or more parts, each or all of which can grow into new individuals

flagella projecting propulsive organelles that beat in a rotary or corkscrew manner; usually relatively long and occurring as one or two per cell; found only in eukaryotes (q.v.) [bacterial flagella are of a fundamentally different form] (cf. **cilia**)

flame cells ciliated cells at proximal end of a protonephridium (q.v.)

flosculae minute projecting sense organs, terminating in a number of papillae, found on the trunk of some priapulans, loriciferans, kinorhynchs and rotiferans

fore-gut anterior section of gut of ectodermal (q.v.) origin and lining (cf. **mid-gut** and **hind-gut**)

fumarole region of earth's crust out of which issues heated water and reduced substances

furca paired processes, of variable form, issuing from the telson (q.v.) of crustaceans

gamete haploid germ cell capable of fusing with a germ cell of the opposite sexuality to form a zygote (q.v.) (see **spermatozoon** and **ovum**)

ganglion discrete body of nervous tissue containing neurones (q.v.)

gastrula embryological stage succeeding the blastula (q.v.) in which the single layer of cells is converted into a two-layered state by cell migration, ingrowth, etc., and in which mesoderm (q.v.) cells are proliferated

gastrulation process whereby a gastrula (q.v.) is derived from a blastula (q.v.)

gemmule multicellular asexual propagule (q.v.) of some poriferans contained within a protective coat

gizzard muscular region of gut in which food may be ground

glia accessory cells associated with neurones (q.v.) [= neuroglia]

gonochoristic having separate sexes [= dioecious] (cf. **hermaphroditic**)

gonopod modified leg serving as a copulatory organ (cf. **cirrus** and **penis**)

gonopore orifice through which gametes are discharged

grade group of animals sharing the same type of bodily organization but without having inherited it from a common ancestral form (cf. **clade**)

haemocoel body cavity formed by blood sinuses, often deriving from the blastocoel (q.v.) (cf. **coelom** and **pseudocoel**)

haplodiploidy see **arrhenotoky**

haploid having only one of each chromosome in each somatic (q.v.) or sex cell (cf. **diploid** and **polyploid**)

hermaphroditic capable of producing both ova (q.v.) and spermatozoa (q.v.), either at the same time or sequentially (cf. **gonochoristic**)

heterotrophic mode of nutrition that requires the intake of preformed organic compounds

hind-gut posterior section of gut of ectodermal (q.v.) origin and lining (cf. **fore-gut** and **mid-gut**)

histolysis breakdown of tissues

homiotherm an organism whose body temperature is maintained at a more or less constant level (cf. **poikilotherm**) [= homoiotherm and homeotherm]

hydraulic operated by water pressure

hydrostatic descriptive of skeletal systems in which muscular forces are transmitted by water within body cavities or tissues

hyperpolarize to increase the electrical potential difference, (usually) across the cell membrane (cf. **depolarize**)

hypostome mound of tissue bearing the mouth in cnidarians

hypoxia conditions of low oxygen availability

imaginal disc group of undifferentiated cells in a larva (q.v.) from which a particular organ system will develop

indeterminate development see **regulative development**

indirect development development via a larval (q.v.) stage (cf. **direct development**)

infaunal descriptive of benthic (q.v.) animals that live buried or in burrows within the substratum (cf. **epifaunal**)

inhalent descriptive of the inward respiratory or feeding current (cf. **exhalent**)

instar one of several larval (q.v.) stages separated from the other such stages by a moult

integument non-muscular layers of body wall

intermediate host see **secondary host**

interstitial (a) appertaining to the spaces (interstices) within sediments, or, when used of cnidarian cells, (b) totipotent epidermal cells

introvert eversible (q.v.) and retractable anterior region of body

larva a juvenile phase differing markedly in morphology and ecology from the adult

lecithotrophic development at the expense of internal resources (i.e. yolk) provided by the female parent; used especially of marine larvae (cf. **planktotrophic**)

lemnisci tubular sacs associated with the proboscis of acanthocephalans and at least some bdelloid rotiferans; they are essentially infoldings of epidermis that probably function as hydraulic reservoirs

littoral intertidal

lophophore a system of hydraulically-operated, ciliated, feeding tentacles, formed as outgrowths of the body wall, that surround the mouth but not anus

lorica vase-shaped protective case formed by thickened cuticle (q.v.)

macroevolution genesis of taxonomic variety; i.e. evolutionary changes at or above the species level (cf. **microevolution**)

macromeres large, yolk-filled cells in the early embryo (q.v.) (cf. **micromeres**)

macrophagous feeding on relatively large food particles (cf. **microphagous**)

Malpighian tubule blind-ending, tubular, excretory diverticulum of gut

mandible jaw

mantle cavity region of environment enclosed within the confines of the shell of molluscs and brachiopods, within which are located the respiratory and feeding organs respectively

maxilla primary mouthpart of an arthropod additional to and located posterior to the mandible (q.v.)

medusa one of the two body forms of cnidarians: pulsatile, usually pelagic (q.v.), disc-, bell- or umbrella-shaped and often gelatinous (cf. **polyp**)

mesenchyme diffuse connective-tissue cells set in a jelly-like matrix

mesoderm germ layer elaborated between the ectoderm (q.v.) and endoderm (q.v.)

mesoglea (or **mesogloea**) thick or thin, cellular or acellular layer of jelly-like material between outer and inner cell layers of coelenterates

mesosome second body region of tripartite oligomeric (q.v.) animals; its body cavity, the mesocoel, may support a lophophore (q.v.) (cf. **prosome** and **metasome**)

metabolism chemical processes occurring in organisms to break down structures and substances (catabolism) and to build them up (anabolism)

metachronal rhythm pattern of synchronized movement of cilia (q.v.) or of multiple limbs in which the movement of each element has a fixed phase relationship to the others

metameric with a body largely comprising a linear series of from several to many segments (q.v.) (cf. **monomeric** and **oligomeric**)

metamorphosis drastic change in body form required to convert a larva (q.v.) into the adult

metanephridium open nephridium (q.v.) with an extracellular duct (cf. **protonephridium**)

metasome third body region of tripartite oligomeric (q.v.) animals; its coelom (q.v.), the metacoel, forms the main body cavity (cf. **prosome** and **mesosome**)

microevolution changes in gene frequencies observed within a single population over time (cf. **macroevolution**)

micromeres small cells, without yolk, in the early embryo (q.v.) (cf. **macromeres**)

microphagous feeding on small or minute food particles (cf. **macrophagous**)

microtriches see **microvilli**

microvilli numerous small finger-like projections of the free surface of cells responsible for absorption and, in specialized form, for sensory reception; in cestodes, they are termed microtriches

mid-gut region of gut of endodermal (q.v.) origin and lining (cf. **fore-gut** and **hind-gut**)

mimicry resemblance to an object or to another organism potentially resulting in concealment by virtue of 'mistaken identity'

mixonephridium see **nephromixium**

modular descriptive of a colonial animal that consists of repeated and connected, asexually produced units (or 'individuals') (see **polyp, zooid,** and **colonial**)

monomeric with a body not partitioned or divided internally into segments (q.v.) (cf. **oligomeric** and **metameric**)

monomorphic with only a single body form (cf. **dimorphic** and **polymorphic**)

monopectinate descriptive of an advanced mollusc ctenidium (q.v.) with gill filaments on only one side of the central axis (cf. **bipectinate**)

mosaic (= **determinate**) **development** development in which the cells of the embryo (q.v.) have their developmental fate fixed at an early embryological stage (by inheritance of maternal cytoplasm) so that the early embryo comprises a fixed pattern in which there is little capacity for the replacement of missing elements (cf. **regulative development**)

mucus mixture of mucoprotein (mucopolysaccharide bound to protein) secreted by mucous cells

naiad aquatic nymph (q.v.) of certain insects differing rather more from the adult form as a result of specific adaptations for aquatic life; e.g. for the capture of aquatic prey and/or for the uptake of dissolved respiratory gases

nanoplankton plankton (q.v.) of 2–20 μm size in largest dimension

natural selection evolutionary mechanism proposed by C.R. Darwin, based on differential survival and reproductive success in resource-limited environments

nekton pelagic (q.v.) animals capable of making progress against natural water flow (cf. **plankton**)

nematocyst intracellular organelle of cnidarians with an eversible (q.v.) coiled thread, used for prey capture, defence, etc.; contained within cnidocytes (q.v.)

neoteny see **paedomorphosis**

nephridium osmoregulatory/excretory organ of ectodermal (q.v.) origin (cf. **coelomoduct**)

nephromixium metanephridium (q.v.) like organ with regions of both ectodermal (q.v.) and mesodermal (q.v.) origin [= mixonephridium]

neurone cell specialized for the conduction of electrical signals and the transmission of information [= nerve cell]

neuropile region of nervous system in which nerve fibres and their terminals form synapses (q.v.)

neurosecretory cell neurone (q.v.) with a glandular function, usually producing hormones

notochord elastic dorsal skeletal rod, derived from highly vacuolated cells bound within a common sheath, characterizing the chordates

nymph juvenile insect differing little from the adult, except in size and in the development of organ systems found only in the adult (e.g. wings and gonads); characteristically their wing buds develop externally (as distinct from insect **larvae**)

ocellus a simple light-sensitive organ (cf. **compound eye**)

oligomeric with a body comprising a few (two or three) segments (q.v.) (cf. **monomeric** and **metameric**)

ontogeny the course of development of an individual organism from zygote (q.v.) to adult

opisthosoma posterior region of the body of those chelicerates in which the body is visibly divided into two distinct sections (cf. **prosoma**); sometimes used in a comparable fashion in other types of animals with two body regions

oral appertaining to the mouth

organ one or more tissues (q.v.) comprising a structural and functional unit

osphradium chemoreceptory tissue or organ in the mantle cavity (q.v.) of molluscs

ostia pores; for example through which water enters the body (in poriferans) or through which blood enters the heart (in animals with an open blood system)

oviparous egg laying

ovipositor tubular organ of some insects used to place eggs in specific microhabitats

ovum female gamete (q.v.)

paedomorphosis juvenilization process whereby either the adult retains juvenile features ('neoteny') or the organism becomes reproductively mature whilst still a juvenile in form and age ('progenesis')

parasite an organism that lives within or attached (permanently or temporarily) to another and causes it harm

parenchyma diffuse tissue (q.v.) of vacuolated cells that often fills the space between epidermis and gut in acoelomate (q.v.) animals

parthenogenesis form of asexual multiplication (q.v.) in which the ovum (q.v.) develops into a new individual without fertilization (q.v.)

pectines comb-like opisthosomal (q.v.) sense organs of scorpions

pedicellariae compound, articulating spines that function as forceps or pincers in certain echinoderms

pelagic appertaining to the water mass of an aquatic system (cf. **benthic**)

penis erectile (not eversible—q.v.) copulatory organ (cf. **cirrus** and **gonopod**)

periostracum proteinaceous covering of a mollusc or brachiopod shell

pericardial cavity cavity within which a heart is situated

peristalsis waves of contraction of circular and longitudinal muscles passing along a tubular organ or organism and having a propulsive effect

peritoneum the mesodermal bounding membrane of a coelomic (q.v.) body cavity

phagocytosis process whereby pseudopodia (q.v.) of an amoeboid cell flow around a particle to engulf it within a vacuole

pharyngeal cleft hole in the wall of the pharynx (q.v.) that extends right through the body to open at the surface; serves to permit the discharge of water taken in through the mouth

pharynx region of fore-gut (q.v.), located posterior to the buccal cavity and anterior to the oesophagus

photoperiodism ability to exhibit physiological responses consequent on changes in relative day-length

photoreceptor a cell sensitive to light

photosynthesis synthesis of organic compounds using the energy in sunlight, through the chlorophyll molecule, according to the general equation

$$CO_2+2H_2X \rightarrow (CH_2O)+H_2O+2X$$

In the oxyphotosynthesis of some bacteria and all photosynthetic eukaryotes, X = oxygen; in the anoxyphotosynthesis of some other bacteria, X is never oxygen—it is often, but not always, sulphur, H_2X then equalling H_2S

phylogeny the course of evolutionary descent and relationship

pinnule a small side branch of a tentaculate organ

pinocytosis ingestion of small liquid droplets by a cell

plankton pelagic (q.v.) organisms that effectively are suspended in the water and cannot make progress against its movement (cf. **nekton**)

planktotrophic feeding, at least in part, on materials captured from the plankton (q.v.); used especially of marine larvae (cf. **lecithotrophic**)

plasmodium multinucleate amoeboid mass bounded by a single cell membrane

pleopods the abdominal (q.v.) appendages of many crustaceans, often used in swimming

plexus network

poikilotherm an organism whose body temperature varies with that of its environment (cf. **homiotherm**)

polymorphic occurring in more than two distinct body forms (cf. **monomorphic** and **dimorphic**)

polyp one of the two body forms of cnidarians: a sedentary (q.v.) or sessile (q.v.) cylinder attached aborally (q.v.) and with a ring of tentacles around the mouth; often form colonial (q.v.) systems; sometimes used interchangeably with **zooid** (q.v.)

polyphyletic a group of organisms derived from more than one ancestral form

polyploid having more than two copies of each chromosome in each somatic (q.v.) cell (cf. **haploid** and **diploid**)

preoral cavity space in front of the mouth in which the mouthparts function or external predigestion takes place

primary host see **definitive host**

proboscis general term for any trunk-like process on the head or anterior body associated with feeding

proglottid serially repeated body unit of a cestode

prokaryote lacking internal membrane-bound organelles and a nuclear membrane within the cell (cf. **eukaryote**); bacteria are prokaryote

propagule reproductive body that separates from the parent; it may be multicellular (vegetative) or cellular (gametic); if cellular, it may be produced by meiosis (sexual) or by mitosis or aberrant forms of meiosis not leading to genetic reduction (asexual)

prosoma anterior region of the body (which includes the head) of those chelicerates in which the body is visibly divided into two distinct sections (cf. **opisthosoma**); sometimes used in a comparable fashion in other types of animals with two body regions

prosome first body region of tripartite oligomeric (q.v.) animals—its body cavity is the protocoel (cf. **mesosome** and **metasome**); and also the term used for the prosoma (q.v.) of copepods

prostomium anterior, non-segmental region of an annelid

protoeukaryote the hypothetical, probably phagocytic (q.v.) host cell which, together with various endosymbiotic prokaryotes (q.v.), formed the first eukaryote (q.v.) cell

protonephridium blind-ending nephridium (q.v.) with an intracellular duct (cf. **metanephridium**)

protostome the state in which the blastopore (q.v.) forms the mouth (cf. **deuterostome**); also used to describe animals that display this state

pseudocoel any body cavity that is not a coelom (q.v.)

pseudocopulation close association during gamete (q.v.) discharge of mating pairs of animals with external fertilization; the ova (q.v.) are thus fertilized (q.v.) immediately on leaving the female gonopore(s) (q.v.)

pseudofaeces faecal-like pellets of material taken out of suspension in the water by filter feeders but subsequently rejected (i.e. particles collected but not ingested)

pseudopodium temporary lobular protrusion of protoplasm formed during the movement, phagocytosis (q.v.), etc. of amoeboid cells

pupa non-motile transitional stage in the development of some insects occurring between the larval stages and the adult

pygidium posterior, non-segmental region of an annelid

radial cleavage type of cell division in which the cleavage plain is parallel or perpendicular to the polar axis of the blastula (q.v.) (cf. **spiral cleavage**)

radial symmetry symmetry about any plane passed perpendicular to the oral/aboral axis

ramus a branch (e.g. of a limb)

regeneration replacement by compensatory growth and differentiation (q.v.) of lost parts of an organism

regulative (= **indeterminate**) **development** development in which the embryo (q.v.) is able to compensate for missing cells and still produce a normal larva or adult, because the developmental fate of its cells is fixed only at a late stage (cf. **mosaic development**)

rejuvenate make young again

respiratory pigment molecule that combines reversibly with oxygen and so functions as a carrier or store

rhabdites see **rhabdoids**

rhabdocoel general term for groups of turbellarian flatworms that have a simple gut without lateral branches or diverticula

rhabdoids rod-like structures, of uncertain function, in epidermis (q.v.) of flatworms and flatworm-like animals; some arise from gland cells and are termed 'rhabdites'

rhabdome ordered array of photoreceptive microvilli (q.v.); e.g. within a compound eye (q.v.)

rostrum loose term for any median anterior projection of the body

scalids cuticular and epidermal projections, of a variety of forms (including spiniform, club-shaped, feathered and scale-like), disposed in whorls around the introvert (q.v.) of kinorhynchs, loriciferans, priapulans and larval nematomorphs; with sensory, locomotory, food capture or penetrant function

schizocoel coelom (q.v.) formed within blocks of mesodermal (q.v.) tissue by cavitation (cf. **enterocoel**)

sclerite a plate comprising part of an exoskeleton

sclerotized chemical hardening (and darkening) of areas of cuticle (q.v.); results from a tanning process.

secondary host host in which a parasite (q.v.) reproduces not at all or only asexually (q.v.)

sedentary tending not to move far

segment a semi-independent, serially repeated unit of the body; segmentation may affect only the body wall and associated structures or almost the whole body

semelparous breeding only once and then dying

septum membrane separating one region of the body from another

sessile permanently attached to a substratum; not capable of locomotion

seta bristle-like projection of cuticle (q.v.), with or without cellular material (cf. **chaeta**)

sexual reproduction form of multiplication in which there is exchange of chromosome material during meiosis, and in which gametes (q.v.) combine in the process of fertilization (q.v.) (cf. **asexual reproduction**)

skeleton a system for the transmission of muscular forces and/or for providing support for the body

somatic appertaining to the body as distinct from the sex cells

spermatheca sac in which a recipient animal stores spermatozoa (q.v.) prior to discharge of its ova and their subsequent fertilization (q.v.)

spermatophore a packet of spermatozoa (q.v.) enclosed within some protective covering

spermatozoon male gamete (q.v.), usually capable of active locomotion [= sperm]

spinneret external nozzle through which silk-producing glands discharge

spiracle opening at the body surface of part of a tracheal (q.v.) system

spiral cleavage type of cell division in which the cleavage plane is oblique to the polar axis of the blastula, there being alternate clockwise and anti-clockwise rotation about the polar axis during the sequence of transverse divisions following the 4-cell stage (cf. **radial cleavage**)

spontaneous generation notion that living organisms could arise directly and spontaneously from non-living materials (e.g. mud)

squamous epithelium epithelium (q.v.) composed of flattened cells

statoblast multicellular asexual propagule (q.v.) of some bryozoans contained within a protective coat

statocyst organ sensitive to gravity and/or acceleration

sternite ventral element of the segmental exoskeleton of arthropods

stolons stalk- or root-like structures by which animals may be connected to each other or to the substratum or from which asexual buds may be liberated

stylet hard, pointed, dart-like structure used for penetration of cells or tissues

styli minute, paired, unjointed appendage-like processes associated with the bases of the legs in some myriapods, and present in an equivalent position on some of the abdominal segments, and rarely also on those of the thorax, of most apterygote insects

subchelate descriptive of an arthropod limb in which the terminal article is reflexed back over the penultimate article to form a distally-hinged grasping organ

subumbrella the lower, usually concave surface of a medusa (q.v.)

superficial cleavage pattern of cleavage (q.v.) in which the zygote (q.v.) gives rise to a syncitium (q.v.); the nuclei of the syncitium move towards the surface; and cell boundaries are then organized around the nuclei

suspension feeding capture and consumption of materials suspended in water; capture is usually effected by some form of filter

symbiosis an intimate association between two dissimilar organisms which interact with each other; one is usually dependent on the other

synapse junction, across which information is transferred, between two cells at least one of which is a neurone (q.v.); the transmitting cell is 'pre-synaptic', the receiving one 'postsynaptic'

syncitium multicellular structure in which cell boundaries are partially or completely absent, ranging from cytoplasmic masses containing many nuclei without any apparent separation into the component cells through to networks of almost complete cells which are in cytoplasmic continuity through intercellular bridges

tegument syncitial (q.v.) external epithelium of parasitic platyhelminths

teleological purposeful or goal-directed

telson posterior, non-segmental region of an arthropod

tentacle any slender, flexible, projecting structure; often sensory, sometimes used for food capture

tergite dorsal element of the segmental exoskeleton of arthropods

test external, or almost external, protective body covering, usually composed of a number of elements

thorax term applied to the middle zone of any body divided into three distinct regions of which the anterior region is the head (cf. **abdomen**)

tissue associated cells of the same (or of a few) type(s) performing the same function; usually bound together by intercellular material (cf. **organ**)

tonic sustained

totipotent cells of a multicellular organism that are capable of differentiating (q.v.) into any specialist cell

trachea tube conveying air from the external environment directly to the tissues

tracheoles terminal capillary-like distributaries of a trachea (q.v.)

triploblastic embryonic condition in which three tissue layers—ectoderm (q.v.), mesoderm (q.v.) and endoderm (q.v.)—can be recognized

trochophore early larval stage of many marine animals characterized by a complete, double pre-oral band of cilia

tubicolous tube-dwelling

ultrafiltration passage of fluid under pressure through a semipermeable membrane

umbrella the upper, usually convex surface of a medusa (q.v.)

uniramous having a single branch (cf. **biramous**); used of an arthropod limb

uropods the last pair of abdominal (q.v.) appendages of decapod crustaceans, which together with the telson (q.v.) form a tail fan

urosome term applied to the opisthosoma (q.v.) of copepods

vital force mysterious, non-physical force once thought to give 'life' to organisms and to direct development and evolution

viviparity development of an embryo (q.v.) within the body of the parent using, in part, resources passing directly from parent to embryo

warning coloration distinctive, bright, contrasting scheme (e.g. black and yellow; black and red) often associated with noxiousness or toxicity in potential prey species

zoochlorellae general name given to symbiotic chlorophyte algae found within the tissues of various, mainly freshwater, invertebrates

zooid a modular individual in a colonial (q.v.) system produced by repeated incomplete budding (q.v.); applied to all such animals other than cnidarians

zooxanthellae general name given to symbiotic dino-flagellate algae found within the tissues of various, mainly marine, invertebrates

zygote single cell produced by union of a sperm (q.v.) and an ovum (q.v.) at fertilization

ILLUSTRATION SOURCES

Alexander, R. McN. 1979. *The Invertebrates*. Cambridge University Press, Cambridge.

Alldredge, A. 1976. *Sci. Am.*, **235**(1), 94–102.

Anderson, D.T. 1964. *Embryology and Phylogeny in Annelids and Arthropods*. Pergamon Press, New York.

Anderson, P.A.V. & Bone, Q. 1980. *Proc. R. Soc. Lond(B)*, **210,** 559–74.

Atkins, D. 1933. *J. Mar. Biol. Ass., UK,* **19,** 233–52.

Atwood, H.L. 1973. *Am. J. Zool.*, **13,** 357–78.

Austin, C.R. 1965. *Fertilisation*. Prentice Hall Inc., New Jersey.

Baehr, J.C., Porcheron, P. & Dray, F. 1978. *C.R. Acad. Sci. (Paris),* **287D,** 523–5.

Barnes, R.D. 1980. *Invertebrate Zoology*, 4th edn. Saunders, Philadelphia.

Barnes, R.S.K. & Hughes, R.N. 1982. *An Introduction to Marine Ecology*. Blackwell Scientific Publications, Oxford.

Bastiani, M.J., Doe, C.Q., Helfand, S.L. & Goodman, C.S. 1985. *Trends in Neurosciences,* **8,** 257–66.

Bayne, B.L., Thompson, R.J. & Widdows, J. 1976. In: B.L. Bayne (Ed.) *Marine Mussels: Their Physiology and Ecology*. Cambridge University Press, Cambridge.

Becker, G. 1937. *Z. Morph. Ökol. Tiere,* **33,** 72–127.

Belk, D. 1982. In: S.P. Parker (Ed.) *Synopsis and Classification of Living Organisms,* **2,** 174–80. McGraw-Hill, New York.

Bergquist, P.R. 1978. *Sponges*. Hutchinson, London.

Berrill, N.J. 1950. *The Tunicata*. Ray Society, London.

Biscardi, H.M. & Webster, G.C. 1977. *Exp. Geront.*, **12,** 201–205.

Blower, J.G. 1985. *Millipedes*. Brill, Leiden.

Boeckh, J., Ernst, K-D., Sass, H. & Waldow, U. 1975. In: D. Denton (Ed.) *Olfaction and Taste,* **V,** 239–45. Academic Press, New York.

Boss, K.J. 1982: In: S.P. Parker, (Ed.) *Synopsis and Classification of Living Organisms,* **1,** 945–1166. McGraw-Hill, New York.

Brill, B. 1973. *Z. Zellforsch.*, **144,** 231–45.

Buchsbaum, R. 1951. *Animals Without Backbones,* Vol. 1. Pelican, Harmondsworth.

Bullough, W.S. 1958. *Practical Invertebrate Anatomy,* 2nd edn. Macmillan, London.

Cain, A.J. & Sheppard, P.M. 1954. *Genetics,* **39,** 89–116.

Calkins, G.N. 1926. *The Biology of the Protozoa*. Baillière Tindall & Cox, London.

Calow, P. 1985. Causes de la mort i costos d'autoproteccio. In: *Biologia Avui*. Fundacio Caixa de Pensions, Barcelona.

Calow, P. 1986. In: R. Peberdy & P. Gardner (Eds) *The Collins Encyclopedia of Animal Evolution,* pp. 90–1. Equinox, Oxford.

Calow, P. & Read, D.A. 1986. In: S. Tyler (Ed.) *Advances in the Biology of Turbellarians and Related Platyhelminthes,* pp. 263–72. D.W. Junk Publishers, Dordrecht.

Carpenter, W.B. 1866. *Phil. Trans. Roy. Soc. Lond.*, **156,** 671–756.

Caullery, M. & Mesnil, F. 1901. *Arch. Anat. Micros.*, **4,** 381–470.

Clark, A.H. 1915. *US Natn. Mus. Bull.*, **82,** Vol. 1(1), 1–406.

Clark, R.B. 1964. *Dynamics in Metazoan Evolution* Clarendon Press, Oxford.

Clarke, K.U. 1973. *The Biology of the Arthropoda*. Arnold, London.

Clarkson, E.N.K. 1986. *Invertebrate Palaeontology and Evolution*, 2nd edn. Allen & Unwin, London.

Clement, A.C. 1962. *J. Exp. Zool.*, **149,** 193–215.

Cloudsley-Thompson, J. 1958. *Spiders, Scorpions, Centipedes and Mites*. Pergamon Press, London.

Cohen, A.C. 1982. In: S.P. Parker (Ed.) *Synopsis and Classification of living Organisms,* **2,** 181–202. McGraw-Hill, New York.

Colwin, L.H. & Colwin, A.L., 1961. *J. Biophys. Biochem. Cytol.*, **10:** 231–254.

Conway Morris, S. 1979. *Ann. Rev. Ecol. Syst.*, **10,** 327–49.

Conway Morris, S. 1985. *Phil. Trans. Roy. Soc. Lond. (B),* **307,** 507–82.

Corliss, J.O. 1979. *The Ciliated Protozoa,* 2nd edn. Pergamon Press, Oxford.

Cuénot, L. 1949. In: P-P. Grassé (Ed.) *Traité de Zoologie,* **VI,** 3–75. Masson, Paris.

Danielsson, D. 1892. *Norw. N-Atlantic Exped. (1876–1878) Rep. Zool.*, **21**, 1–28.
Darnell, J.E., Lodish, H.F. & Baltimore, D. 1986. *Molecular Cell Biology*. Scientific American Books Inc., New York.
Davies, I. 1983. *Ageing*. Edward Arnold, London.
Dehorne, A, 1933. *Bull. Biol. Fr. Belgique*, **67**, 298–326.
Dethier, V.E. 1976. *The Hungry Fly*. Harvard University Press, Cambridge, Mass.
Dixon, A.F.G. 1973. *Biology of Aphids*. Studies in Biology No. 44. Edward Arnold, London.
Durchon, M. 1967. *L'endocrinologie chez le Vers et les Molluscs*. Masson, Paris.
Eakin, R.M. 1968. *Evol. Biol.*, **2**, 194–242.
Elner, R.W. & Hughes, R.N. 1978. *J. Anim. Ecol.*, **47**, 103– 16.
Epel D. 1977. *Sci. Am.*, **237**(5), 129–38.
Fewkes, J. 1883. *Bull. Mus. Comp. Zool., Harvard*, **11**, 167–208.
Fingerman, M. 1976. *Animal Diversity*, 2nd edn. Holt, Rinehart & Winston, New York.
Fox, H.M., Wingfield, C.A. & Simmonds, B.G. 1937. *J. Exp. Biol.*, **14**, 210–18.
Fraser, J.H. 1982. *British Pelagic Tunicates*. Cambridge University Press, Cambridge.
Fretter, V. & Graham, A. 1976. *A Functional Anatomy of Invertebrates*. Academic Press, London and New York.
Friesen, W.Q., Poon, W. & Stent, G.S. 1976. *Proc. Nat. Acad. Sci. (USA)*, **73**, 3734–8.
George, J.D. & Southward, E.C. 1973. *J. Mar. Biol. Ass. UK*, **53**, 403–24.
Gibson, P.H. & Clark, R.B. 1976. *J. Mar. Biol. Ass. UK*, **56**, 649–74.
Gibson, R. 1982. In: S.P. Parker (Ed.) *Synopsis and Classification of Living Organisms*, 823–46. McGraw-Hill, New York.
Gilbert, L.E. 1982. *Sci. Am.*, **247**(2), 102–7B.
Glaessner, M.F. & Wade, M. 1966. *Palaeontology*, **9**, 599–628.
Gnaiger, E. 1983. *J. Exp. Zool.*, **228**, 471–90.
Golding, D.W. & Pow, D.V. 1987. In: M.C. Thorndyke, & G.J. Goldsworthy (Eds) *Neurohormones in Invertebrates*. Cambridge University Press, Cambridge.
Golding, D.W. & White, A.C. 1974. *Tissue & Cell*, **6**, 599–611.
Golding, D.W. & White, A.C. 1977. *Int. Rev. Cytol. Suppl.*, **5**, 189–302.
Goodrich, E.S. 1945. *Quart. J. Micros. Sci.*, **86**, 113–393.
Gordon, D.P. 1975. *Cah. Biol. Mar.*, **16**, 367–82.
Grassé, P.-P. (Ed.). 1948. *Traité de Zoologie*, **XI**. Masson, Paris.
Grassé, P.-P. (Ed.). 1965. *Traité de Zoologie*, **IV**. Masson, Paris.
Green, J. 1961. *A Biology of Crustacea*. Witherby, London.
Gupta, B.L. & Berridge, M.J. 1966. *J. Morphol.*, **120**, 23–82.
Gustafson, T. & Wolpert, L. 1967. *Biol. Rev.*, **42**, 442–98.
Hackman, R.H. 1971. In: M. Florkin, & B.T. Scheer (Eds) *Chemical Zoology*, **6**, 1–62. Academic Press, New York.
Hansen, K. 1978. In: G.I. Hazelbauer (Ed.) *Taxis and Behaviour Receptors and Recognition*, **5B**, 231–92. Chapman & Hall, London.
Hardy, A.C. 1956. *The Open Sea. The World of Plankton*. Collins, London.
Hedgpeth, J.W. 1982. In: S.P. Parker (Ed.) *Synopsis and Classification of Living Organisms*, **2**, 169–73. McGraw-Hill, New York.
Hermans, C.O. & Cloney, R.A. 1966. *Z. Zellforsch.* **72**, 583–96.
Hermans, C.O. & Eakin, R.M. 1974. *Z. Morph. Tiere*, **79**, 245–67.
Hescheler, K. 1900. In: A. Lang (Ed.) *Lehrbuch der Vergleichenden Anatomie der Wirbellosen Thiere*, 3rd edn. Fischer, Jena.
Hess, R. 1987. *Z. Wiss. Zool.*, **62**, 247–83.
Higgins, R.P. 1983. *Smithsonian Contrib. Mar. Sci.*, **18**, 1–131.
Hodgkin, A.L. & Huxley, A.F. 1952. *J. Physiol.*, **117**, 500–44.
Holland, N.D., Grimmer, T.C. & Kubota, H. 1975. *Biol. Bull.*, **148**, 219–42.
Holt, C.S. & Waters, T.F. 1967. *Ecology*, **48**, 225–34.
Hughes, T.E. 1959. *Mites or the Acari*. Athlone, London.
Hummon, W.D. 1982. In: S.P. Parker (Ed.) *Synopsis and Classification of Living Organisms*, **1**, 857–63. McGraw-Hill, New York.
Hyman, L.H. 1940. *The Invertebrates. Vol. I. Protozoa through Ctenophora*. McGraw-Hill, New York.
Hyman, L.H. 1951. *The Invertebrates. Vol. II. Platyhelminthes & Rhynchocoela*. McGraw-Hill, New York.
Imms, A.D. 1964. *A General Textbook of Entomology*, 9th edn., revised reprint. Methuen, London.
Ito, Y. 1980. *Comparative Ecology*. Cambridge University Press, Cambridge.
Jägersten, G. 1973. *The Evolution of the Metazoan Life Cycle*. Academic Press, New York.
Jeannel, R. 1960. *Introduction to Entomology*. Hutchinson, London.
Jones, A.M. & Baxter, J.M. 1987. *Molluscs: Caudofoveata, Solenogastres, Polyplacophora and Scaphopoda*. Brill, Leiden.
Jones, J.D. 1955. *J. Exp. Biol.*, **32**, 110–25.
Jones, M.L. 1985. In: S. Conway Morris *et al.* (Ed.) *The Origin and Relationships of Lower Invertebrates*, 327–42. Clarendon Press, Oxford.
Joosse, J. & Geraerts, W.P.M. 1983. In: A.S.M. Saleudin & K.M. Wilbur (Eds) *The Mollusca*, Vol. 5. Academic Press, New York.

Kandel, E.R. & Schwartz, J.H. 1982. *Science*, **218,** 433–43.

Kennedy, D. 1976. In: J.C. Fentress (Ed.) *Simpler Networks and Behaviour*. Sinauer, Sunderland, Massachusetts.

Kershaw, D.R. 1983. *Animal Diversity*. University Tutorial Press, Slough.

Krebs, J.R., Erichsen, J.T., Webber, M.I. & Charnov, E.L. 1977. *Anim. Behav.*, **25,** 30–8.

Kudo, R.R. 1946. *Protozoology*, 3rd edn. Thomas, Springfield.

Lacaze-Duthiers, F.J.H. de. 1861. *Ann. Sci. Nat. (Zool.)*, **15,** 259–330.

Lamb, M.J. 1977. *Biology of Ageing*. Blackie, Glasgow.

Lemche, H. & Wingfield, K.G. 1959. *Galathea Rep.*, **3,** 9–71.

Lester, S.M. 1985. *Mar. Biol.*, **85,** 263–8.

Lewis, J.G.E. 1981. *The Biology of Centipedes*. Cambridge University Press, Cambridge.

McFarland, W.N., Pough, F.N., Cade, T.J. & Heiser, J.B. 1979. *Vertebrate Life*. MacMillan, New York.

MacKinnon, D.L. & Haws, R.S.J. 1961. *An Introduction to the Study of Protozoa*. Clarendon Press, Oxford.

McLaughlin, P.A. 1980. *Comparative Morphology of Recent Crustacea*. Freeman, San Francisco.

Manton, S.M. 1952. *J. Linn. Soc. (Zool.)*, **42,** 93–117.

Manton, S.M. 1965. *J. Linn. Soc. (Zool.)*, **45,** 251–483.

Marcus, E. 1929. *Klassen und Ordnungen des Tierreichs*, **5,** 1–608.

Margulis, L. & Schwartz, K.V. 1982. *Five Kingdoms*. Freeman, San Francisco.

Marion, M.A-F. 1886. *Arch. Zool. Exp. Gén. (2)*, **4,** 304–26.

Meglitsch, P.A. 1972. *Invertebrate Zoology*, 2nd edn. Oxford University Press, Oxford.

Millar, R.H. 1970. *British Ascidians*. Academic Press, London.

Miller, R.L. 1966. *J. Exp. Zool.*, **162,** 23–44.

Miyan, J.A. & Ewing, A.W. 1986. *J. Exp. Biol.*, **116,** 313–22.

Moore, R.C. (Ed.). 1957. *Treatise on Invertebrate Paleontology, Part L. Mollusca,* **4.** University of Kansas Press, Lawrence.

Moore, R.C. (Ed.). 1965. *Treatise on Invertebrate Paleontology, Part H. Brachiopoda*. University of Kansas Press, Lawrence.

Morgan, C.I. 1982. In: Parker, S.P. (Ed.) *Synopsis and Classification of Living Organisms*, **2,** 731–9. McGraw-Hill, New York.

Morgan, C.I. & King, P.E. 1976. *British Tardigrades*. Academic Press, London.

Mortensen, T. 1928–51. *A Monograph of the Echinoidea*. 5 vols. Reitsel, Copenhagen.

Muscatine, L. *et al.*, 1975. *Symp. Soc. Exp. Biol.*, **29,** 175–203.

Newell, N.D. 1949. *Evolution*, **3,** 103–240.

Nichols, D. 1962. *Echinoderms*, 3rd edn. Hutchinson, London.

Nichols, D. 1969. *Echinoderms*, 4th edn. Hutchinson, London.

Noble, E.R. & Noble, G.A. 1976. *Parasitology*. Lea and Febiger, Philadelphia.

Ohnishi, T. & Sugiyama, M. 1963. *Embryologia*, **8,** 79–88.

Olive, P.J.W. 1980. In: D.C. Rhoads & R.A. Lutz (Eds), *Skeletal Growth in Aquatic Organisms*. Plenum Press, New York.

Olive, P.J.W. 1985. *Symp. Soc. Exp. Biol.*, **39,** 267–300.

Olive, P.J.W. 1985b. In: *Syst. Association*, series 28, 42–59. Oxford University Press, Oxford.

Oschman, J.L. & Berridge, M.J. 1971. *Federation Proceedings, Federation of American Societies for Experimental Biology*, **30,** 49–56.

Pashley, H.E. 1985. *The Foraging Behaviour of* Nereis diversicolor *(Polychaeta)*. PhD Thesis, University of Cambridge.

Pearson, K.G. 1983. *J. Physiol. (Paris)*, **78,** 765–71.

Pennak, R.W. 1978. *Fresh-water Invertebrates of the United States*, 2nd edn. Wiley, New York.

Phillipson, J. 1981. In: C.R. Townsend & P. Calow (Eds) *Physiological Ecology*, 20–45. Blackwell Scientific Publications, Oxford.

Pierrot-Bults, A.C. & Chidgey, K.C. 1987. *Chaetognatha*. Brill, Leiden.

Pringle, J.W.S. 1975. *Insect Flight*. Oxford University Press, Oxford.

Rice, M. 1985. In: S. Conway Morris, J.D. George, R. Gibson & H.M. Platt (Eds) *Origins and Relationships of Lower Invertebrates*. Clarendon Press, Oxford.

Ritter-Zahony, R. von. 1911. *Das Tierreich*, **29,** 1–35.

Robbins, T.E. & Shick, J.M. 1980. In: *Nutrition in the Lower Metazoa*, Pergamon Press, Oxford.

Russell-Hunter, W.D. 1979. *A Life of Invertebrates*. MacMillan, New York.

Sanders, D.S. 1982. *Insect Clocks*, 2nd edn. Pergamon Press, Oxford.

Sanders, H.L. 1957. *Syst. Zool.*, **6,** 112–28.

Savory, T.H. 1935. *The Arachnida*. Edward Arnold, London.

Schepotieff, A. 1909. *Zool. Jb. Syst.*, **28,** 429–48.

Schmidt-Nielsen K. 1984. *Scaling*. Cambridge University Press, Cambridge.

Sebens, K.P. & De Riemer, K. 1977. *Mar. Biol.*, **43,** 247–56.

Sedgwick, A. 1888. *Quart. J. Micros. Sci.*, **28,** 431–93.

Shelton, G.A.B. 1975. *Proc. R. Soc. Lond. B*, **190,** 239–56.

Shepherd, G.M. 1983. *Neurobiology*. Oxford University Press, Oxford.

Sheppard, P.M. 1958. *Natural Selection and Heredity*. Hutchinson, London.

Shick, P.M. & Dykens, J.A. 1985. *Oecologia*, **66**, 33–41.
Silva, P.H.D.H. de. 1962. *J. Exp. Biol.*, **39**, 483–90.
Singla, C.L. 1975. *Cell Tiss. Res.*, **158**, 391–407.
Sleigh, M.A., Dodge, J.D. & Patterson, D.J. 1984. In: R.S.K. Barnes (Ed.) *A Synoptic Classification of Living Organisms*, 25–88. Blackwell Scientific Publications, Oxford.
Smart, P. 1976. *The Illustrated Encyclopedia of the Butterfly World*. Hamlyn, London.
Smyth, J.D. 1962. *Introduction to Animal Parasitology*. The English Universities Press, London.
Smyth, J.D. & Halton, D.W. 1983. *The Physiology of Trematodes*. Cambridge University Press, Cambridge.
Snodgrass, R.E. 1935. *Principles of Insect Morphology*. McGraw-Hill, New York.
Snow, K.R. 1970. *The Arachnids: An Introduction*. Routledge & Kegan Paul, London.
Southward, E.C. 1980. *Zool. Jb. Anat. Ontog.*, **103**, 264–75.
Southward, E.C. 1982. *J. Mar. Biol. Ass., UK*, **62**, 889–906.
Spengel, J.W. 1932. *Sci. Res. Michael Sars N. Atlantic Deep Sea Exped.*, 5(5), 1–27.
Stein, P.S.G. 1971. *J. Neurophysiol.*, **34**, 310–18.
Sterrer, W.E. 1982. In: S.P. Parker (Ed.) *Synopsis and Classification of Living Organisms*, **1**, 847–51. McGraw-Hill, New York.
Stiasny, G. 1914. *Z. Wiss. Zool.*, **110**, 36–75.
Strausfeld, N.J. & Nassel, D.R. 1981. In: H. Autrum (Ed.) *Handbook of Sensory Physiology*, Vol. VII/6B. Springer-Verlag, Berlin.
Stretton, A.O.W., Davis, R.E., Angstadt, J.D., Donmoyer, J.E. & Johnson, C.D. 1985. *Trends in Neurosciences*, **8**, 294–9.
Strumwasser, F. 1974. *Neurosciences Third Study Program*, 459–78.
Treherne, J.E. & Foster, W.A. 1980. *Anim. Behav.*, **28**, 1119–22.
Trench, R.K. 1975. *Symp. Soc. Exp. Biol.*, **29**, 229–65.
Trinkans, J.P. 1969. *Cells into Organs*. Prentice Hall, New Jersey.
Trueman, E.R. & Foster-Smith, R. 1976. *J. Zool., Lond.*, **179**, 373–86.
Turner, C.D. 1966. *General Endocrinology*. Saunders, Philadelphia.
Valentine, J.W. & Moores, E.M. 1974. *Sci. Am.*, **230**(4), 80–9.
Wallace, M.M.H. & Mackerras, I.M. 1970. In: C.S.I.R.O., *The Insects of Australia*, 205–16. Melbourne University Press, Melbourne.
Warner, G.F. 1977. *The Biology of Crabs*. Elek, London.
Waterman, T.H. 1960. *The Physiology of Crustacea*. Academic Press, New York.
Welsch, U. & Storch, V. 1976. *Comparative Animal Cytology*. Sidgwick & Jackson, London.
Wenyon, C.M. 1926. *Protozoology*. Baillière, Tindall & Cox, London.
Whittington, H.B. 1979. In: M.R. House (Ed.) *The Origin of Major Invertebrate Groups*, 253–68. Academic Press, London.
Widdows, J. & Bayne, B.L. 1971. *J. Mar. Biol. Ass., UK*, **51**, 827–43.
Wigglesworth, V.B. 1940. *J. Exp. Biol.*, **17**, 201–22.
Wigglesworth, V.B. 1972. *Principles of Insect Physiology*, 7th edn. Chapman and Hall, London.
Wine, J.J. & Krasne, F.B. 1982. In: D.C. Sandeman & H.L. Atwood (Eds) *The Biology of Crustacea*, Vol. 4, 241–92. Academic Press, New York.
Wright, A.D. 1979. In: M.R. House (Ed.) *The Origin of Major Invertebrate Groups*, 235–52. Academic Press, London.
Wrona, F.J. & Davies, R.W. 1984. *Can. J. Fish. Aquatic Sci.*, **41**, 380–85.
Yager, J. & Schram, F.R. 1986. *Proc. Biol. Soc. Wash.*, **99**(1), 65–70.
Young, J.Z. 1939. *Phil. Trans. Roy. Soc. Lond. B.*, **229**, 465–503.
Young, J.Z. 1962. *The Life of Vertebrates*, 2nd edn. Clarendon Press, Oxford.
Young, J.Z. 1963. *Nature (Lond)*, **198**, 636–30.
Zullo, V.Z. 1982. In: S.P. Parker (Ed.) *Synopsis and Classification of Living Organisms*, **2**, 220–8. McGraw-Hill, New York.

INDEX